Construction Law

William Allensworth, Ross J. Altman,
Allen Overcash, and Carol J. Patterson
Editors

FORUM ON THE CONSTRUCTION INDUSTRY

Defending Liberty
Pursuing Justice

Cover design by ABA Publishing.

Printed in the United States of America.
14 5

Cataloging-in-Publication Data is on file with the Library of Congress.

Construction Law / William Allensworth, Ross J. Altman, Allen Overcash, & Carol J. Patterson, eds.

ISBN-10: 1-60442-323-4
ISBN-13: 978-1-60442-323-5

Discounts are available for books ordered in bulk. Special consideration is given to state bars, CLE programs, and other bar-related organizations. Inquire at Book Publishing, ABA Publishing, American Bar Association, 321 North Clark Street, Chicago, Illinois 60654-7598.

www.ababooks.org

Contents

About the Editors xxi
About the Authors xxiii
Preface xxvii
Acknowledgments xxix

1 Construction Law: The Historical Perspective **1**

1.01 Introduction 1
A. Construction in the Ancient World 1
B. Construction's 19th-Century Transformational Events 3
1.02 Construction Dispute Resolution 11
1.03 Construction Law Scholarship 14

2 Participants in the Design and Construction Process **17**

2.01 Overview 17
2.02 The Owner 18
2.03 The Design Team 20
A. The Architect 20
B. The Engineers and Consultants 21
2.04 The Construction Team 22
A. The Contractor 22
B. The Subcontractors 24
2.05 The Supporting Cast 25
A. Construction Lenders 25
B. Insurance Companies 27
C. Sureties 28
D. Title Insurance 28
E. Public Authorities 29
2.06 Assembling the Project Team 29

3 The Owner's Role 35

3.01 Overview 35
3.02 Providing the Building Site 35
A. Access to the Building Site 36
B. The Right to Use the Project Site for Its Intended Purpose 37
C. Title 37
D. Zoning 38
E. Availability of Utilities 39
F. Archaeological and Historic Sites 40
G. Environmental Concerns 41
3.03 Providing Information to the Architect and Contractor 42
A. Surveys 42
B. Geotechnical Information 43
3.04 The Owner's Implied Warranty to Supply Accurate Information 43
A. The *Spearin* Doctrine 43
B. Limitations on the Application of *Spearin* 46
1. Limitations on the *Spearin* Doctrine 47
2. The *Spearin* Warranty Can Be Disclaimed 49
3.05 The Owner's Program 50
3.06 Paying for Construction 52
A. Financing the Project 52
B. Evidence of Financing 53

4 Project Delivery Systems 57

4.01 Overview 57
4.02 Evolution of Different Approaches 58
4.03 Importance of Project Delivery Systems 61
4.04 Design-Bid-Construct 64
A. Selection Factors 65
B. Risk Factors 67
C. Common Variations 69
4.05 Construction Management 71
A. Agency Construction Management 71
1. Selection Factors 72
2. Risk Factors 74
B. At-Risk Construction Management 74
1. Selection Factors 75
2. Risk Factors 76

4.06 Design-Build 78
A. Selection Factors 80
B. Risk Factors 81
C. Common Variations 83
4.07 Innovative Approaches 85
4.08 Methods of Project Delivery System Selection 90
4.09 Selection Constraints 91
A. Method of Construction Procurement 91
B. Compensation Mechanisms 91
C. Owner Capability 92
D. Industry Sector Custom 93
4.10 Conclusion 94

5 Contracting for Construction Projects 97

5.01 Contract Formation 97
A. For Purpose of Design Agreements and Construction Contracts 97
B. Contract Formation Generally 98
C. Commencement of Services or Work Prior to Contract 99
D. Creation of Contracts—Implied Contracts and Quasi-Contracts 100
5.02 How to Select a Contract Form 102
A. Fundamental Issues 102
B. Standard Industry Forms 102
C. Advantages and Disadvantages of Using a Manuscript Document 103
5.03 Overview of Standard Industry Forms 104
A. AIA Documents 104
B. AGC Forms 108
C. EJCDC Forms 108
D. Other Standard Industry Forms 109
5.04 What Is the Contract? 109
A. The Design Agreement (Owner-Architect) 109
B. The Contract Documents (Owner-Contractor) 110
C. The Integration Clause 111
D. Incorporation by Reference versus Enumeration of Documents 112
5.05 Internal Conflicts 112

5.06 Contract Interpretation 113
A. Rules of Contract Construction 113
B. Parol Evidence Rule 114
C. Ambiguity and Intent of Parties 116
D. Reasonable and Logical Interpretation of Entire Contract 118
E. Interpretation against Drafter 119
F. Course of Performance and Prior Dealings 120
G. Custom and Usage 121
H. Terms: Specific over General; Written over Printed 122
5.07 Parties to the Construction Contract 123
A. Owner and Contractor 123
B. Third-Party Beneficiaries 123
C. Collateral Assignment to Lender 124
D. References to Other Parties 124
5.08 Scope of Contract 125
A. Description of Services of Design Professionals 125
B. Description of the Work under a Construction Contract 125
C. Owner-Furnished Items 128

6 The Design Undertaking 131
6.01 Introduction 132
A. Brief Overview of Contractual Relationship among Parties and General Allocation of Design Function 132
1. Owner 132
2. Design Professionals 132
3. Contractor 133
B. The Role of the Owner and Owner's Consultants with Respect to Design 133
C. The Design Professional Team 134
1. Architectural Consultants 134
2. Engineering Consultants 135
3. The Civil Construction Project (Engineer as Lead Design Professional) 135
6.02 Statutory Regulation and Licensure of Design Professionals 136
6.03 Building Codes and the Design Professional 136
A. Traditional Building Codes 137

B. Socioeconomic "Codes" 138
1. Accessibility under ADA 138
2. "Green" Codes and Growing Mandatory Compliance 139
3. Historic Preservation 139
6.04 Standard of Care Applicable to Design Services 140
A. Nature of Standard of Care and Inapplicability of Strict Liability 140
B. Common Law Standard of Care 140
1. General Statement of Standard of Care 140
2. Proof of the Standard of Care and Violation Thereof 143
3. Variations from Code and Consequences for Design Professionals 146
C. Contractual Standard of Care 147
D. Implied Warranties and Design Professionals 147
1. Implied Warranties Generally in Construction 147
2. Different Treatment of Design Professionals and Implied Warranties 148
6.05 Implementation of the Design through Contractors 151
A. Value Engineering 151
B. Performance Specifications and Design-Build of Discrete Portions of Construction 152

7 Design Agreements 155
7.01 Importance of an Agreement between the Owner and Architect 155
7.02 The Scope of the Architect's Services 156
A. Phases of the Design Process 157
1. Schematic Design Phase 157
2. Design Development Phase 157
3. Construction Documents Phase 158
4. Bidding or Negotiation Phase Services 158
5. Construction Phase Services 158
B. Basic Services versus Additional Services 159
C. The Owner's Responsibilities 160
7.03 Designing to a Budget 161
A. Contractual Provisions Regarding Estimates 165

7.04 Ownership of the Design Documents 170
A. Statutory Rights 170
B. Contractual Provisions 174
7.05 Payment 175
7.06 Insurance 176
7.07 Contractual Limitations of Liability 176
A. Disclaimers and Exculpatory Provisions 176
B. Waivers of Consequential Damages 179
C. Contractual Limitations of Liability 180
7.08 Termination 185

8 Contractor Selection 189

8.01 Introduction 189
8.02 Private Procurement Generally 190
8.03 Public Procurement 191
A. General Considerations 191
B. Selection of Design Professionals 192
C. Selection of Contractors 194
1. Traditional Methodology for Selection of Contractors 194
2. Contractor Selection Criteria 199

9 Pricing Construction Contracts 219

9.01 Introduction 219
9.02 General Pricing Mechanisms for Construction Contracts 220
A. Stipulated-Sum Contracts 220
1. Scope Limitations within Contract Price 221
2. Percentage-of-Completion Payments and Potential Abuse 223
B. Cost-Plus Contracts 224
1. Cost of Work and Limitations Thereon 225
2. Reasonableness or Necessity of Costs Incurred 228
3. Contractor's Overhead and Profit 229
4. Estimates and Cost-Plus Contracts 230
C. Cost-Plus with Guaranteed Maximum Price 231
D. Unit-Price Contracts 232
1. VEQ Clauses 232
2. Unbalanced Bidding 234
E. Mixed Pricing Contracts and Change Orders 235

9.03 Specific Price-Related Terms Utilized In Construction Contracts 235
A. Pricing Mechanisms 235
1. Index Pricing 236
2. Time and Material Pricing 236
3. Default Pricing Through Third Party 237
4. Default Pricing Through Dispute Resolution 237
9.04 Contract Terms Related to Pricing 238
A. Material Price Escalation Clauses 238
B. Interim Payment Clauses 238
9.05 Comparison of Pricing Terms Based on Pricing Mechanism 239
A. Scope of Work 239
B. Changes 240
C. Sharing of Savings 240
D. Definition of "Cost of Work" 241
E. Treatment of Self-Performed Work 241
F. Allowances 241
G. Contingencies 241
H. Builder Rebates/Credits 243
I. Purpose of Budget 243
J. Audit Rights 243

10 Subcontractors and Suppliers 247

10.01 Subcontractors and Suppliers 247
A. Subcontractors versus Suppliers: What Is the Difference and Why Does It Matter? 247
B. Subcontractors and Suppliers: Telling Them Apart 248
10.02 Selection of Subcontractors and Suppliers 249
A. Owner Input into Selection Process 249
B. Formal Restrictions on Choosing Subcontractors and Suppliers 250
C. Socioeconomic Considerations 251
10.03 Subcontractor Bidding: Who Is Bound by the Bid? 252
A. The Bidding Process 252
B. Can the Prime Contractor Enforce the Subcontractor's Bid? 253
C. Can the Subcontractor Enforce Its Own Bid? 255
10.04 Flow-Down Obligations and Rights 256
A. Introduction 256
B. The Nature of Flow-Down and Flow-Up Provisions 257

C. The Rights and Liabilities of Parties under Flow-Down Provisions 258
D. Conditional Assignment of Subcontract to the Owner 259
10.05 Coordination of Subcontract Work 259
A. The Duty of Coordination 259
B. Duty to Coordinate Subcontract Work 260
C. Avoiding Coordination Obligations 261
10.06 Subcontractor Indemnity Obligations 262
A. Judicial Indemnity Authority 262
B. Subcontractor Indemnity Agreements 263
C. Anti-Indemnity Statutes 264
10.07 Payment to Subcontractors and Suppliers 265
A. The Payment Process 265
B. Bearing the Risk of Owner's Failure to Pay 266
C. Risk of Payment Provisions 267
D. Other Methods to Secure Subcontractor Payments 269
10.08 Subcontractor Claims 269
A. The Problem of Subcontractor Claims 269
B. Barriers to the Subcontractor Making Its Claim 270
C. The Pass-Through System and Its Problems 272
D. Liquidating Agreements 275

11 Contract Time and Completion 279
11.01 Significance of Time for Performance 279
11.02 Construction Scheduling—Critical Path Method Schedules 280
11.03 Legal Significance of Construction Schedules 282
11.04 Time of Commencement and Time for Completion 283
11.05 Waiver of Time for Completion 284
11.06 Grounds for Extending the Time for Completion—Compensable and Excusable Delays 290
11.07 Substantial Completion 297
11.08 Final Completion 302
11.09 Damages for Late Completion 304
11.10 Liquidated Delay Damages 305
11.11 No-Damage-for-Delay Clauses 307

12 Construction Scheduling 311
12.01 Introduction 311

12.02 Types of Schedules 312
A. Bar Charts 312
1. Advantages and Disadvantages of Bar Charts 313
B. Critical Path Method Schedules 315
1. Basic Principles 315
2. Types of CPM 317
3. Advantages and Disadvantages of CPM 319
12.03 Standard Schedule Specifications 323
A. Type of Diagram 323
B. Number of Activities 324
C. Resource Loading 324
D. Approval 325
E. Updating 325
F. Subcontractor Involvement 325
G. Float Use and Reporting 326
H. Major Revisions and Time Extensions 326
I. Prohibitions on Schedule Manipulations 327
12.04 Baseline Schedule 327
12.05 Schedule Updates 330
12.06 CPM Proof of Delay Claims 332
A. The "Classic Technique," Which Underlies CPM Analysis of Time-Related Claims 332
B. Methods for Proof of Delay Claims—An Overview 332
1. Bar Chart 333
2. "But For" Analysis/Collapsed As-Built 334
3. Total Time Analysis 335
4. Impacted As-Planned CPM 336
5. Chronological and Cumulative Approach/Time Impact Analysis 336
6. Window Analysis 337
C. The CPM Expert and the Need for Impartiality 338
12.07 Conclusion 341

13 Architect's Contract Administration 343
13.01 Introduction 343
13.02 Nature and Scope of Architect's Role in Contract Administration 344
A. Overview of Contract Administration 344

B. Overview of Agency Law Principles Relevant to Contract Administration 346
1. Architect's Authority to Bind the Owner 346
2. Owner's Liability for the Architect's Acts 347
3. Architect's Liability for Exceeding Authority or for the Improper Exercise of Authority 347

13.03 Requests for Information, Interpretations, and Clarifications 348

13.04 Quality Control 349
A. Review and Approval Procedures 349
B. Approving or Objecting to Subcontractors, Manufacturers, and Suppliers 349
C. Substitutions, Shop Drawings, and Other Submittals 350
D. Decisions on Aesthetic Effect 355
E. Site Visits and Inspections 355

13.05 Coordination 359

13.06 Approval of Progress Payments 360

13.07 The Right to Reject Work or to Stop Work 361

13.08 Issuing Change Orders and Responding to Proposals and Change Order Requests 362

13.09 Certification of Substantial Completion, Final Completion, and Other Performance 362

13.10 Architect's Role in Claims Process 363
A. Resolving Claims and Disputes 363
B. Architect's Role in Contractor Termination 364

13.11 Architect's Immunity for Performing Quasi-Judicial Functions 366

14 Payment **369**

14.01 Introduction 369

14.02 Basis for Payment 370

14.03 Payment Process 373
A. Applications for Payment 373
1. The Contractor's Certification 374
2. The Architect's Certification 375
B. Retainage 376
C. Owner's Payment Obligation 379
1. Timing 379
2. Mechanic's Lien 381
3. Stop Notice/Lien on Funds 382

Half Price Books
1835 Forms Drive
Carrollton, TX 75006
OFS OrderID 39492580

Visit our stores to sell your books, music, movies games for cash.

SKU	ISBN/UPC	Title & Author/Artist	Shelf ID	Qty	OrderSKU
S472220318	9781604423235	Construction Law	03--13--6	1	

SHIPPED STANDARD TO:
Main Address
800 Avondale Ave
1033837-1-0133
Grandview Heights OH 43212
j1gg094j97htty5@marketplace.amazon.com

ORDER# **113-0323979-1665061**
AmazonMarketplaceUS

4. Lien Waivers 382
5. Joint Checks 383
6. Overpayment 385
D. Changes 387
E. Interest 387
14.04 Substantial Completion/Substantial Performance 387
14.05 Final Completion/Final Payment 389
A. Owner's Acceptance of Work/Waiver of Claims 390
B. Contractor's Acceptance of Payment/Waiver of Claims 391
C. Liquidated Damages 392
14.06 Specific Payment Issues 393
A. "Pay If Paid" versus "Pay When Paid" 393
B. Prompt Payment Acts 394
C. False Claims 395
D. Lender Liability 396
14.07 Conclusion 398

15 Construction Safety 401

15.01 Introduction 401
15.02 Accident Prevention 402
A. State Legislation 402
B. Occupational Safety and Health Act 402
C. OSHA in Civil Litigation 404
D. Contractual Responsibility for Safety Programs 409
15.03 Recovery of Compensation for Construction Accidents 410
A. Workers' Compensation Laws 410
B. Tort Liability 411
1. Active Negligence 411
2. Liability Based Upon Control of the Workplace 412
C. Owners 412
D. Prime Contractors 415
E. Architects and Engineers 419
15.04 Risk Allocation and Transfer 423
A. Risk Allocation and Transfer, Generally 423
B. Insurance 424
C. Indemnity 424
D. Assumption of the Risk 428
E. Workers' Compensation's Effect on Indemnity and Contribution in Third-Party Actions 428
15.05 Conclusion 429

16 Changes 431

16.01 The Inevitability of Changes 431
16.02 The Purpose of a Changes Clause 432
16.03 The Change Order Process 434
16.04 Authority to Issue Changes 437
16.05 Notice Requirements for Claims for Changes 440
16.06 Adjustments to Contract Price and Completion Date 443
16.07 Constructive Changes 446
16.08 Duty to Perform the Changed Work 449
16.09 Cardinal Changes 450
16.10 Administration of Changes 452
16.11 Releases—Accord and Satisfaction 454

17 Differing Site Conditions 459

17.01 Introduction 459
17.02 Common Law 461
A. Contracts without Site Indications or Risk Allocation Clause 461
B. Alternate Legal Theories in the Presence of Representations or Indications 463
1. Warranty 463
2. Misrepresentation 464
3. Superior Knowledge/Concealment 465
4. Mutual Mistake 466
17.03 Regulatory and Other Standard Frameworks for Addressing Differing Site Conditions 467
A. Differing Site Conditions—FAR 52.236-2 468
B. Standard Form Provisions in Commercial Agreements 469
17.04 Elements of a Federal Differing Site Conditions Claim 471
A. Type I—Materially Different than Indicated 471
1. Indications 472
2. Reasonableness 473
3. Differ Materially 474
4. Reasonably Unforeseeable 475
5. Causation of Increased Costs 475
B. Type II—Conditions of an Unusual Nature 476
17.05 Limitations on Claims for Differing Site Conditions 477
A. Duty to Investigate 477
B. Disclaimers/Exculpatory Language 479
C. Notice 479

D. Relationship to the Variation in Estimated Quantities Clauses 481
17.06 Representative Treatment of Differing Site Conditions in State Courts 483
A. New Jersey 483
B. Virginia 485
C. Illinois 487
D. California 488
17.07 Conclusion 490

18 Termination and Default 493

18.01 Introduction 493
18.02 Contractual Termination Provisions 494
A. Termination by the Contractor for Cause 494
B. Termination by the Owner for Cause 496
C. Termination for Convenience 499
18.03 Material Breach 500
18.04 Termination by Agreement 505
18.05 Notice and Opportunity to Cure 505
18.06 Wrongful Termination 510
18.07 Termination and the Role of the Surety 511

19 Mechanic's Liens 513

19.01 Origin of Mechanic's Lien 513
19.02 Statutory Framework 515
19.03 Lien Categories 516
19.04 Requirements for a Lien 517
A. Contract 517
B. Improvement 518
C. Amount of Lien 520
D. Lien Entitlement 521
19.05 What Is Lienable? 522
19.06 Interests Subject to a Lien 524
19.07 Perfecting the Lien 526
19.08 Priorities 526
19.09 Public Projects 527
19.10 Waiver and Release 528
19.11 Bankruptcy 529

20 Insurance and Bonds **531**

20.01 Introduction 531
20.02 The Role of Insurance 531
20.03 Contractual Insurance Requirements 532
20.04 Workers' Compensation Insurance 535
20.05 General Liability Insurance Policies 536
20.06 Builders' Risk Insurance Policies 541
20.07 Waivers of Subrogation 541
20.08 Professional Liability Insurance 542
20.09 Surety Bonds—The Tripartite Relationship 542
20.10 Obligations of the Performance Bond Surety 544
20.11 Payment Bonds 551
20.12 Rights and Remedies of Sureties 552
20.13 Subguard Policies 557

21 Dispute Resolution Processes **559**

21.01 Importance to the Construction Industry of Processes for Resolution of Disputes 559
A. Unique Dispute Resolution Needs of Construction Projects 559
B. Historical Development of Dispute Resolution in Construction 560
21.02 The Spectrum of Construction Industry Dispute Prevention and Resolution Techniques 561
A. The Dispute Resolution Step Chart 561
B. The Four Stages of Dispute Resolution 562
1. The Dispute Prevention and Cooperation Stage 562
2. The Dispute Control Stage 562
3. The Facilitated Nonbinding Resolution Stage 563
4. The Binding Resolution Stage 563
21.03 Individual Construction Dispute Prevention, Control, and Early Resolution Techniques 563
A. Prevention and Cooperation Techniques 563
1. Realistic Allocation of Project Risks 563
2. Incentives to Encourage Cooperation 564
3. Partnering 564
4. Open Communications and Notice Provisions 565

B. Dispute Control and "Real-Time" Early Resolution Techniques 566
1. Step Negotiations 566
2. Geotechnical Baseline Summary Report 566
3. Escrow of Bid Documents 567
4. Dispute Resolution Boards and Standing Neutrals 567
C. Facilitated Nonbinding Resolution of Disputes 569
1. Mediation 569
2. Minitrial 570
3. Advisory Opinions 570
4. Nonbinding Summary Advisory Arbitration 571
5. Binding Arbitration 571
21.04 Design of Dispute Prevention and Resolution Systems 571
21.05 Principles of Arbitration 572
A. Basis in Statute 572
B. Contractual Nature of the Remedy 573
C. Attacking Arbitration 575
D. Modification of Award by Arbitrators 576
E. Confirmation or Vacation of Award by Court 576

22 Defective Construction 581

22.01 Overview of the Legal Analysis of Defects 581
A. What Constitutes "Defective" or "Deficient" Construction? 582
1. Types of Defects or Deficiencies 582
2. Manifestations versus Defects 583
B. Responsibility 584
22.02 Owner Claims against Contractors 585
A. Contract Claims 585
1. Failure to Comply with Drawings and Specifications 586
2. Breach of Express Warranty 587
3. Breach of Implied Warranty 587
B. Tort Claims 588
C. Statutory Claims 589
D. Notice Requirements 589
22.03 Claims against Design Professionals 590
A. Tort and Contract Claims Generally 591

B. Defective Drawings and Specifications 591
C. Construction Observation 595
D. Claims against the Design Professional by Contractors 598
22.04 The *Spearin* Doctrine and Conflicting Warranties 599
A. The *Spearin* Doctrine 599
1. Owner's Implied Warranty of Drawings and Specifications 600
2. Design Specifications 603
3. Performance Specifications 604
4. Conflicts between the Contractor's and the Owner's Warranties 605
22.05 Time-Based Defenses to Liability 607
A. Statutes of Limitation 607
B. Statutes of Repose 611
C. Laches 612

23 The Economic Loss Rule in Construction Law 615
23.01 Introduction 615
23.02 Economic Loss Rule 620
23.03 The Scope and Purpose of the Economic Loss Rule 622
23.04 The Economic Loss Rule's Application Where Parties Are in Privity of Contract 625
A. Claims of Defective Construction Products/Components (Privity) 625
B. Claims of Defective Construction Services (Including Poor Workmanship and Improper Furnishing of Materials) (Privity) 630
C. Claims of Defective Design Professional/Testing Services (Including Insufficient Design and Improper Contract Administration Services) (Privity) 635
23.05 The Economic Loss Rule's Application Where Parties Are Not in Privity of Contract 638
A. Claims of Defective Construction Products/Components (No Privity) 639
B. Claims of Defective Construction Services (Including Poor Workmanship and Improper Furnishing of Materials) (No Privity) 649

C. Claims of Defective Design Professional Services (Including Insufficient Design and Improper Contract Administration Services) (No Privity) 655

23.06 The Economic Loss Rule as Applied to the Tort of Negligent Misrepresentation 660

24 Damages **667**

24.01 Introduction 667

A. Overview of Chapter 667

B. "Measures of Damages" versus "Cause of Action" 668

24.02 General Measures of Damages 669

A. Breach of Contract and Related Claims 669

1. Expectation Interest (Breach of Contract) 669
2. Reliance Interest—Promissory Estoppel 674
3. Restitution Interest—Unjust Enrichment 675
4. Quantum Meruit 676

B. Tort Claims 677

24.03 Calculation and Proof of Compensatory Damages 678

A. Direct Methods of Proving Actual Damages 678

1. Proof of Actual Costs 678
2. Methods of Estimating Actual Damages 683

B. Alternate Methods of Proving Damages 686

1. Total Cost Method 686
2. Modified Total Cost Method 690
3. Jury Verdict Method 691

C. Consequential Damages in Construction Contracts 692

D. Punitive Damages 692

E. The Claims Industry and Recordkeeping 693

24.04 Mitigation of Damages 693

A. Duty to Mitigate Damages 693

B. Economic Waste 694

C. Betterment 695

24.05 Contractual and Other Limitations on Liability 695

A. Indemnification Clauses and Limitations on Liability 695

B. Waivers of Consequential/Incidental Damages 700

C. No Damages for Delay 700

Table of Cases 703

Index 733

About the Editors

William Allensworth is the managing partner of Allensworth and Porter, LLP, in Austin, Texas, and has taught courses in Construction Law and in Construction Litigation as an adjunct professor at the University of Texas School of Law since 1999. He is a Fellow of the American College of Construction Lawyers. He was a founding member of both the Construction Law Section of the State Bar of Texas and the Austin Bar Association.

Ross J. Altman is a partner in the Chicago office of DLA Piper, LLP (U.S.), and co-chair of the firm's Construction Law Practice Group. He represents clients in connection with projects, located in the United States as well as overseas, ranging from commercial high-rises to industrial plants and infrastructure. Mr. Altman serves on the American Arbitration Association panel of construction industry arbitrators and is an adjunct law professor and Fellow in the American College of Construction Lawyers.

Allen Overcash practices law with Woods & Aitken, LLP, in Lincoln, Nebraska. He is a member of the American Arbitration Association's Construction Arbitrator Master Panel, its Panel of Mediators, and its ICDR Panel of Arbitrators. He has served as a member of the Board of Governors of the American College of Construction Lawyers and the Governing Committee of the ABA Forum on the Construction Industry, and in 2002 he received the Forum's Cornerstone Award. He is currently an adjunct professor of construction law at the University of Nebraska Law College.

Carol J. Patterson is a partner in the New York City office of Zetlin & De Chiara, LLP, where her practice is focused on construction law. She has represented clients in connection with a wide range of commercial, residential, arts, and infrastructure projects. She writes and speaks frequently on construction law issues. She has served as president of the Association of Real Estate Women in New York City. She is a member of the Governing Committee of the ABA Forum on the Construction Industry.

About the Authors

Philip L. Bruner, Esq., is a mediator, arbitrator, and the Director of JAMS Global Engineering and Construction Group. Prior to joining JAMS in 2008, he practiced law for 43 years in the construction law field, the last 17 of which as a senior partner and founding head of the Construction Law Group in the international law firm of Faegre & Benson, LLP. Mr. Bruner is the co-author of BRUNER & O'CONNOR ON CONSTRUCTION LAW.

Carl J. Circo teaches construction law and other subjects at the University of Arkansas School of Law. He previously served on the faculty at the Benjamin N. Cardozo School of Law and also taught law as a visitor at the University of Nebraska and as an adjunct at the University of Missouri–Kansas City. For 20 years, he practiced commercial real estate and construction law in Kansas City with Stinson Morrison Hecker, LLP, where he continues to consult as the firm's Professional Development Director.

W. Stephen Dale is a member of Smith Patcher McWhorter, PLC, in Vienna, Virginia, and focuses his practice on construction law. His experience includes counseling construction and engineering firms on construction issues from contract administration to claims and litigation.

Deborah Griffin practices construction law in the Boston office of Holland & Knight, LLP. She is a past chair of the ABA Forum on the Construction Industry and a Fellow of the American College of Construction Lawyers. She has spoken and written extensively on construction, surety, insurance, and bankruptcy topics.

Mark J. Groff is a senior counsel in the Washington, D.C., offices of Akin Gump Strauss Hauer & Feld, LLP. He focuses his practice on construction and government contract issues. He is an adjunct professor at the University of Maryland, Department of Civil and Environmental Engineering Project Management Program, and at Virginia Tech University, Department of Building Construction, where he teaches graduate courses on the legal aspects of design, engineering, and construction.

James P. Groton is a retired partner of the Atlanta and Washington, D.C., law firm of Sutherland, Asbill & Brennan. He is a Fellow in the Chartered Institute of Arbitrators and the College of Commercial Arbitrators, and a Fellow and former President of the American College of Construction Lawyers; he has also served on the executive committee of the National Academy of Construction. He has received numerous dispute resolution and construction industry awards.

A. Holt Gwyn is a principal in the law firm of Conner Gwyn Schenck, PLLC, in Greensboro and Raleigh, North Carolina. Mr. Gwyn is a former Chair of the ABA Forum on the Construction Industry and a Fellow in the American College of Construction Lawyers. He is the Editor-in-Chief of the JOURNAL OF THE AMERICAN COLLEGE OF CONSTRUCTION LAWYERS. He is a frequent author and speaker at the bar and industry groups on construction and environmental law and on alternative dispute resolution issues.

R. Harper Heckman, Esq., practices in the Greensboro, North Carolina, office of Nexsen Pruet, PLLC. Mr. Heckman is past Chair of the Contract Documents Steering Committee of the ABA Forum on the Construction Industry. He speaks and writes frequently on construction law issues and is an adjunct professor of construction law at the University of North Carolina School of Law.

Stephen A. Hess developed and teaches a Construction Law Seminar at the University of Denver's Sturm College of Law and practices construction law and litigation as a member of the Colorado Springs office of Sherman & Howard, LLC. His writings have appeared in numerous national construction law treatises and construction law journals. Mr. Hess is the author of several books related to litigation. He serves as Associate Editor of THE CONSTRUCTION LAWYER, which is published by the ABA Forum on the Construction Industry.

Val S. McWhorter is a member of Smith Pachter McWhorter, PLC, and has focused his practice in the area of construction law for 30 years. His practice has included representing contractors in large, complex claims arising under federal, state, and local government contracts, as well as contracts between private parties. He is a frequent speaker on construction-related issues.

Lawrence C. Melton has been practicing construction law for over 30 years and is currently special counsel to Nexsen Pruet, LLC, in Columbia, South Carolina. He is on the adjunct faculty of the University of South Carolina School of Law where he has taught Construction Law and Litigation. He is the author of numerous publications on construction law. He is on the American Arbitration Association Board of Neutrals and panel of mediators.

Bruce Merwin is a real estate partner in the Houston, Texas office of Haynes and Boone, LLP, and heads their construction practice. He is the author of numerous articles on construction law and speaks regularly to lawyer groups and construction-related trade associations throughout the State of Texas.

Andrew D. Ness is the managing partner in the Washington, D.C., office of Thelen Reid Brown Raysman & Steiner, LLP. Mr. Ness is the author of numerous articles on construction law topics and speaks frequently on a wide range of construction law issues. He is a member of the Governing Committee of the ABA Forum on the Construction Industry and a Fellow of the American College of Construction Lawyers.

Dennis J. Powers is a partner in the Chicago office of the Construction Law Practice Group of DLA Piper, LLP (U.S.). He concentrates his practice in the representation of owners, developers, and contractors in construction disputes of all kinds, including defect and delay claims and extras and payment claims in courts and arbitration across the country.

Bruce H. Schoumacher is a shareholder and Group Co-Chair at Querrey & Harrow, Ltd., in Chicago, Illinois, where he practices litigation and construction law, antitrust law, and commercial and professional liability. He is the author of numerous publications related to construction law, including ENGINEERS AND THE LAW: AN OVERVIEW.

David A. Senter is a member in the Greensboro, North Carolina, office of Nexsen Pruet, PLLC, and Practice Group Leader for the Construction Practice Group. Mr. Senter concentrates his practice in the area of construction. He serves on the panel of Construction Arbitrators for the American Arbitration Association and has authored numerous construction-related articles in bar and trade publications. He has served on the Governing Committee of the ABA Forum on the Construction Industry.

Steven M. Siegfried is a partner in the firm of Siegfried, Rivera, Lerner, De La Torre & Sobel, PA, in Miami. Mr. Siegfried is the author of numerous articles and books on construction law. He teaches construction law as an adjunct professor at the University of Miami School of Law. He is a founding Fellow of the American College of Construction Lawyers.

Stanley P. Sklar is a Partner in Bell, Boyd & Lloyd, LLP, in Chicago, Illinois. He is a past President of the American College of Construction Lawyers and the Society of Illinois Construction Attorneys, and a member of the American

College of Real Estate Lawyers as well as the College of Commercial Arbitrators, where he serves as Treasurer. He also serves as a member of the American Arbitration Association Large, Complex Commercial Case panel.

Lorence H. Slutzky is a partner in the Chicago office of Robbins, Schwartz, Nicholas, Lifton & Taylor, Ltd. He represents a wide range of clients in complex construction transactions and litigation. Mr. Slutzky is a Fellow in the American College of Construction Lawyers, adjunct law professor of Real Estate Law at The John Marshall Law School in Chicago, and has served as President of the Society of Illinois Construction Attorneys.

Richard F. Smith is Senior Counsel to Smith Pachter McWhorter, PLC, located in Vienna, Virginia. Mr. Smith serves on the mediation and arbitration panels of the American Arbitration Association. He has been a member of the Governing Committee of the ABA Forum on the Construction Industry and a member of the Board of Governors of the American College of Construction Lawyers. He is a frequent speaker and author on construction and alternative dispute resolution issues.

Richard J. Tyler is a partner in the New Orleans office of Jones, Walker, Waechter, Poitevent, Carrère & Denègre, LLP, where he is a member of the Business and Commercial Litigation Practice Group, and chair of the firm's construction practice. His construction practice involves dispute resolution and transactional work on a wide variety of projects. He serves on the American Arbitration Association roster of neutrals. He frequently writes and speaks on construction law issues.

Jon M. Wickwire is a founding partner in the law firm of Wickwire Gavin, PC, a founding Fellow of the American College of Construction Lawyers, and the founding Chairman of the Project Management Institute College of Scheduling.

Alan Winkler is an attorney in the New Jersey office of Peckar & Abramson, PC. For the past 20 years, his practice has been focused on construction law, including construction contracts, counseling clients on issues arising during construction projects, and representing clients in arbitrations and litigations of various disputes. He has also lectured and written articles on construction law.

Preface

This text is unique in several respects, and we believe a word about its origins is appropriate. Construction law is now a relatively well defined practice area. The field supports several journals, many scholarly treatises, and countless continuing legal education programs. At the same time, construction law as an academic subject is only now starting to enjoy the growth that the practice area has experienced over recent decades. The American Bar Association Forum on the Construction Industry believes that the publication of this book will facilitate the expansion of construction law education into law schools, as well as provide a useful resource for construction law courses in other professional programs.

Several attributes of this book distinguish it from traditional law school textbooks. Assimilation of construction law principles requires some working knowledge of the construction contracting process, and the first few chapters of this textbook therefore focus on describing the business relationships among parties to the construction process. The subject matters of these chapters do not lend themselves to being taught through the standard casebook method, and indeed the study of case law itself is not always a fruitful means of conveying information about how the parties to a construction project go about doing business with each other. As a result, the text will be instructive to new construction lawyers as well.

The book covers the many facets of the construction process—from initial design through construction and concluding with the dispute resolution process. Once the student has acquired some knowledge of the parties' roles in the more common project delivery systems, he or she can begin to study the contracts that define the parties' respective rights and obligations, with special emphasis on form contracts used in the industry. We have done this largely through the prism of a typical small commercial construction project and the American Institute of Architects contract forms that reflect it.

We then explore the design undertaking and contractor selection, and the essential terms of scope, time, and price. The related issues of scheduling, payment, subcontracting, and the role of the architect in contract administration are then discussed in separate chapters, as are changes, differing site conditions, construction safety, and contractual provisions for the termination of the construction contract. The remainder of the text deals with issues arising out of construction disputes, including mechanic's liens, insurance and bonding, defective construction, damages, the Economic Loss Rule, and, of course, alternative dispute resolution systems.

This text is published by the Forum on the Construction Industry and is written by industry-leading practitioners, most of whom teach construction law. The selection of topics and the scope of treatment have been dictated by practical needs in the industry, and the student can assume with confidence that the subject areas presented are those that are most commonly faced in the field. We believe the text benefits from its identification of a core group of subjects with which a well-rounded construction attorney should be familiar on a practical level.

William Allensworth
Ross J. Altman
Allen Overcash
Carol J. Patterson

Acknowledgments

The editors wish to thank the many people who made this book possible. First and foremost, we thank the ABA Forum on the Construction Industry for its enthusiastic and unwavering support. The Forum was founded to provide opportunities for construction lawyers to contribute to the profession by offering a wide range of educational opportunities to the construction bar, including programs, journals, and books. This book is consistent with this mission. It was a long time in the making. The original idea for this project came from William Allensworth and other members of the "Professors Committee" of the American College of Construction Lawyers, who discussed the need for a comprehensive construction law textbook that was based on the development cycle of a construction project and was readily adaptable for use in the classroom. Responding to these discussions, Fred Wilshusen, the Chair of the Forum's Publications Committee, assembled a committee, consisting of himself, Adrian Bastianelli, and Robert Beaumont to explore the idea. They decided to put together a team to respond to this demand. William Allensworth was tapped to lead this effort as editor of the book, and he was soon joined by Ross Altman and Allen Overcash, and eventually by me as well.

The leadership of the Forum embraced this idea as an opportunity to advance the educational opportunities available for law students and other aspiring construction professionals. We especially appreciate the support we have received from Adrian Bastianelli, who initially championed the project as a member of the original subcommittee, then as Chair of the Forum's Publications Committee and Chair-Elect. Mark Heley, Fred's successor as Chair of the Publications Committee, continued to promote this project. We also appreciate the support we have received from Andrew Ness, the 2008–2009 Chair of the Publications Committee, and from a series of Forum chairs including John Heisse, Doug Oles, Ty Laurie, Mike Tarullo and Robert MacPherson.

We extend our thanks to all of our contributing authors. Stephen Hess not only wrote two chapters of this book, but also generously provided editorial assistance on other chapters. The editors offer many thanks to Ridgely Jackson, Penny Cobey, Mark Heley, and Richard Tyler for their incisive comments and suggestions. Many other Forum members put time and effort into this book, and provided editorial and proofreading assistance. Harper Heckman, David Scotti, Richard Tyler, and Joe Jones mobilized members of Divisions 2 and 3 to assist with the final review process. Marilyn Soltau and Dorothea Brandemour's care, attention, boundless patience, and good humor in connection with the preparation of the manuscript contributed immeasurably to the efficient production of the book. Christine Barnathan also deserves thanks for her contributions to the production process.

Finally, we thank ABA Publications and acknowledge our editors, Richard Paszkiet and Sarah Forbes Orwig, for their consistent encouragement and assistance in bringing this project to completion.

I was pleased to have the opportunity to join this superbly qualified editorial team. Like most projects of this type, it took much more time and effort than anyone expected at the outset. Our lively discussions provided an opportunity to stand at a distance from the legal issues we deal with on a daily basis and decide how to organize and present the material to students who are not familiar with the construction industry. The enthusiasm and professionalism that William, Allen, and Ross brought to this endeavor was inspiring. We hope that this book will contribute to the education of construction professionals in the years to come.

Carol J. Patterson

CHAPTER

1

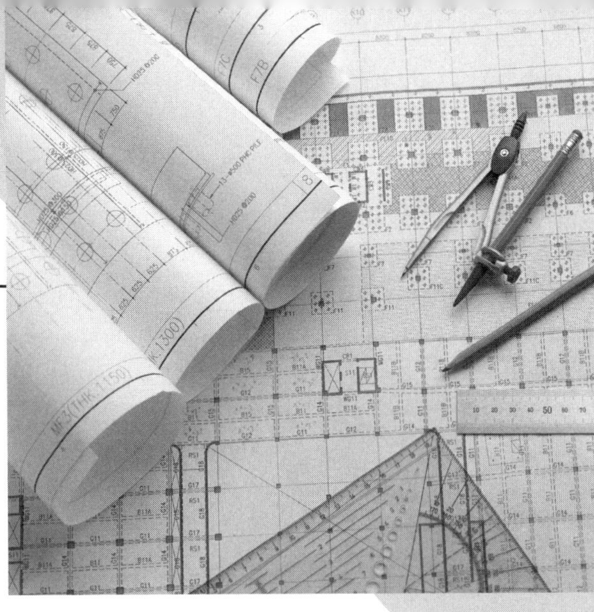

Construction Law: The Historical Perspective

PHILIP L. BRUNER

1.01 INTRODUCTION

A. Construction in the Ancient World

For more than 4,500 years, from primitive Mesopotamian fire-brick and early Egyptian cut-stone construction to the extraordinary structures of the modern-built environment, construction has been a hallmark of the advancement of human civilization.[1] Ever since mankind first promulgated rudimentary principles of law to regulate human rights and obligations arising out of societal

Portions of this chapter are adapted from articles by the author published in The Journal of the American College of Construction Lawyers and William Mitchell Law Review.

1. *See* Plutarch, Pericles (75 AD):

> That which gave most pleasure and ornament to the city of Athens, and the greatest admiration and even astonishment to all strangers, and that which now is Greece's only evidence that the power she boasts of and of her ancient wealth are no romance or idle story, was [Pericles'] construction of the public and sacred buildings.

See also 1 Philip L. Bruner & Patrick J. O'Connor, Bruner & O'Connor on Construction Law § 1:1 (West Group 2002) [hereinafter Bruner & O'Connor].

interaction, there have existed principles of law governing the built environment and the construction process. The earliest known principles of construction law were primitive and punitive. Under its "eye for an eye" system of justice, Hammurabi's Code dictated that builders be punished for injuries to others caused by collapse of their buildings. The code provisions pertinent to construction state:

> 229 If a builder build a house for someone, and does not construct it properly, and the house which he built fall in and kill its owner, then the builder shall be put to death.
>
> 230 If it kill the son of the owner, the son of that builder shall be put to death.
>
> 231 If it kill a slave of the owner, then he shall pay slave for slave to the owner of the house.
>
> 232 If it ruin goods, he shall make compensation for all that is ruined, and in as much as he did not construct properly this house which he build and it fell, he shall re-erect the house from his own means.
>
> 233 If a builder build a house for someone, even though he has not yet completed it, if then the walls seem toppling, the builder must make the walls solid from his own means.[2]

As classical antiquity gradually civilized the built environment, so too did it refine the governing law. By the reign of Rome's Caesar Augustus (27 BC to 14 AD),[3] construction risks inherent in building upon unsuitable soils[4] and building without managerial competence and cost control[5] were widely recognized.

---

2. THE CODE OF HAMMURABI §§ 229–33 (ca. 1750 BC) (R. W. King trans. 1910).

3. When he became Rome's first emperor in 27 BC, Gaius Julius Caesar Octavius (63 BC to 14 AD), great nephew of Julius Caesar, was given the name Augustus by the Roman Senate.

4. Jesus of Nazareth, who is said to have practiced carpentry as a boy, employed widely understood metaphors in his sermons, and concluded His Sermon on the Mount with this admonition:

> Everyone then who hears these words of mine and acts on them will be like a *wise man who built his house on a rock*. The rain fell, the floods came, and the winds blew and beat on that house, but it did not fall, because it had been founded on rock. And everyone who hears these words of mine and does not act on them will be like a *foolish man who built his house on sand*. The rain fell, and the floods came, and the winds blew and beat against that house, and it fell—and great was its fall! (Emphasis added.) *Matthew* 7:24–27 (New Revised Standard Edition).

5. *See* MARCUS VITRUVIUS POLLIO, DE ARCHITECTURA, Book X, Introduction (ca. 20 BC) (Morris Hickey Morgan trans., 1914). Known to history as "Vitruvius," he was chief engineer to Caesars Julius and Augustus and thus in his time could be called the "chief engineer of the civilized world." Vitruvius wrote a 10-volume treatise for Augustus on Roman construction practices, which survived the ravages of time to influence the architecture of the European Renaissance. Among other things, Vitruvius proposed to Augustus that Rome resurrect an ancient ancestral law of the Greek

Good construction practice under Roman law favored careful contractual articulation of the scope of work and allocation of construction risks.[6]

B. Construction's 19th-Century Transformational Events

For 1,900 years following the advent of Augustus's Imperial Rome—through Europe's Dark Ages, Renaissance, and Industrial Revolution—construction law was subsumed by broader and more generalized fields of law and by perceptions

---

City of Ephesus (the same place to which Saint Paul almost a hundred years later wrote his Letter to the Ephesians):

> In the famous and important Greek City of Ephesus, there is said to be an ancient ancestral law, the terms of which are severe, but its justice is not inequitable. When an architect accepts the charge of a public work, he has to promise what the cost of it will be. His estimate is handed to the magistrate, and his property is pledged as security until the work is done. When it is finished, if the outlay agrees with his statement, he is complimented by decrees and marks of honor. If no more than a fourth has been added to his estimate, it is furnished by the treasury, and no penalty is inflicted. But when more than one-fourth has been spent in addition on the work, the money required to furnish it is taken from his property.

Roughly two generations after Vitruvius wrote his treatise, Jesus of Nazareth used the same common problem as a metaphor:

> For which of you, intending to build a tower, does not first sit down and estimate the cost, to see whether he has enough to complete it? Otherwise, when he had laid a foundation and is not able to finish, all who see it will begin to ridicule him, saying, "this fellow began to build and was not able to finish." *Luke* 14:28–30 (New Revised Standard Edition).

6. *See* Vitruvius, *supra* note 5, Book I, chapter 1. Regarding "construction law," Vitruvius advised the architect—the "master builder" of those days—as follows:

> [A]s for principles of law, [an Architect] should know those which are necessary in the case of buildings having party walls, with regard to water dripping from the eaves, and also the laws about drains, windows, and water supply. And other things of this sort should be known to architects, so that, before they begin upon buildings, they may be careful not to leave disputed items for the householders to settle after the works are finished, and so that in drawing up contracts, the interests of both the employer and the contractor may be wisely safeguarded. For if a contract is skillfully drawn, each may obtain a release from the other without a disadvantage.

Roman builders had good reason to exercise care in contracting because the Roman legal doctrine of *pacta sunt servanda* ("contracts must be honored") imposed strict contractual liability unless nonperformance was excused under the doctrine of *rebus sic stantibus* ("provided the circumstances remain unchanged"). *See also* Restatement (Second) Contracts, Introductory Note to Chapter 11 ("Contract liability is strict liability. It is an accepted maxim that *pacta sunt servanda*, contracts are to be kept."). These ancient principles undergird the modern law of contract and its legal doctrines of sanctity of contract, *force majeure*, and impracticability. *See* 5 Bruner & O'Connor, *supra* note 1, §15:22.

of construction as local and parochial and as invoking primarily the "law of the shop," rather than the "law of the courts."[7] Then, beginning in the mid-1800s, American law governing construction was transformed by a series of revolutionary events:

- In 1857, the founding of the American Institute of Architects, which championed the practice of architecture as a specialized profession distinct from construction contracting, heralded the eclipse of the architect's historic role as "master builder"—the single person in charge of design and construction. Engineering associations thereafter were formed to promote engineering as a profession, separate from both architectural design and construction contracting, and in turn spawned a number of professional engineering subspecialties—electrical, mechanical, structural, civil, and geotechnical—to address emerging technical disciplines. Professional specialization accelerated after legislative enactment of state design professional registration laws in the United States, beginning with the State of Illinois in 1897. By the mid-20th century, the architectural profession was perceived as having abandoned its age-old role as "master builder."[8]

---

7. The earliest treatises on the English common law, which span a five hundred year period, make no mention of legal principles of construction law. *See* HENRY OF BRACTON, ON THE LAWS AND CUSTOMS OF ENGLAND (ca. 1230), *available at* http://hlsl.law.harvard.edu/bracton; WILLIAM BLACKSTONE, COMMENTARIES ON THE LAWS OF ENGLAND (1765–1769), *available at* http://www.yale.edu/lawweb/avalon/blackstone and http://www.lonang.com/exlibris/blackstone. In Blackstone's case, the omission is particularly telling because he was trained as an architect prior to going into law and frequently used architectural metaphors in his legal writings. *See* Wilfrid Prest, *Blackstone as Architect: Constructing the Commentaries*, 15 YALE J. L. & HUMAN. 103, 123 (2003).

8. *See* Carl M. Sapers, *Ruminations on Architectural Practice*, 25 CONSTR. CONTR. L. REP. ¶ 106 (April 20, 2001) (citations omitted):

> [T]he increasing complexity of construction projects . . . challenged the architect's historic role as the most knowledgeable player at the job site. As Professor Salvadore of Columbia University observed, architects came in the 1970s to know less and less about more and more until the architect is "sometimes said to know nothing about everything." Even if we stop short of Salvadore's caricature, it is clear that the architect was no longer venerated for his or her comprehensive grasp of all aspects of building. . . .
>
> During this same period, whenever the economy tightened, opposing forces claimed greater pieces of the architect's historic domain. Civil engineers claimed the right to design hospitals, office buildings, and court houses, interior designers claimed the right to design 60,000 square foot office build-outs. Mechanical engineers made arguments that, in the end, suggested that the shapely Hancock Tower in Boston was merely a chase for the mechanical system.
>
> Professionals became increasingly targets of the plaintiff's bar; in the 60s and 70s architects were conventionally sued if anything went wrong at the project.

- By 1888, in response to the rigid express contractual risk allocation imposed by the legal Doctrine of Sanctity of Contract,[9] which allocated almost all construction and completion risks to the contractor, unless the contract expressly stipulated otherwise, the American Institute of Architects and the National Association of Builders (predecessor to the modern Associated General Contractors of America) negotiated and cosponsored the so-called Uniform Contract—the first national attempt to create a standard construction contract form. Building upon that cooperative mutual relationship, the American Institute of Architects, from 1911 to the present, has published 13 editions of its standard construction documents with the endorsement of the Associated General Contractors of America.[10]
- By the mid-1800s, states had begun to enact mechanic's lien statutes to protect unpaid subcontractors, laborers, and materialmen who had performed work on private property, by granting such persons defeasible equitable interests in the improved real estate up to the value of their respective contributions. Such statutes, however, were construed to grant no lien rights in public property.[11]
- In 1894, troubled by contractor defaults on federal contracts during the financial panic of 1893 and by the absence of mechanic's lien protection on public projects, Congress enacted the Heard Act to require federal

---

> The fall of the house of privity made the architect a direct target of unhappy subcontractors and contractors. The rising tide of civil litigation elevated the role of the insurance industry. The insurance industry not only affected practice by describing conduct that would result in the loss of coverage, it insisted on a place at the table in the AIA Construction Industry documents being drafted. The effect of listening too closely to the cautions of a prudential insurance industry was that the architect further retreated from the dominant role he had once played. . . .

9. *See* Dermott v. Jones, 69 U.S. 1, 2, 7; 17 L. Ed. 762 (1864):

> It is a well settled rule of law that if a party by his contract charges himself with an obligation possible to be performed, he must make it good, unless its performance is rendered impossible by Act of God, the law or the other party. Unforeseen difficulties, however great, will not excuse him. . . . [The rule] rests upon a solid foundation of reason and justice. It regards the sanctity of contracts. It requires parties to do what they have agreed to do. If unexpected impediments lie in the way and a loss must ensue, it leaves the loss where the contract places it. If the parties have made no provision for a dispensation, the rule of law gives none. It does not allow a contract fairly made to be annulled, and it does not permit to be interpolated what the parties themselves have not stipulated.

10. *See* 2 Bruner & O'Connor, *supra* note 1, § 5:2; Justin Sweet, *The American Institute of Architects: Dominant Actor in the Construction Documents Market,* 1991 Wis. L. Rev. 317 (1991).

11. *See, e.g.,* Jordan v. Bd. of Educ. of Taylor's Falls, 39 Minn. 298, 39 N.W. 801 (1888) (mechanic's lien could not be foreclosed against public property).

contractors, as a condition of contract award, to post surety bonds to protect subcontractors, laborers, and materialmen against the credit risk of nonpayment and to protect the government against the performance risk of default. In 1935, Congress replaced the Heard Act with the more comprehensive Miller Act. All states followed suit by adopting their own "Little Heard" or "Little Miller" acts. These acts fostered the formation of the modern surety industry.[12]

- The early 1900s witnessed the emergence of a primary judicial vehicle for development of construction law principles—the modern theory of "contextual contract,"[13] which elastically allowed the judiciary to add contractual terms, conditions, and warranties implied by the transaction's surrounding circumstances[14] and complexity,[15] and to interpret express contractual language in conformance with industry usage, custom, and practice.[16] Contextual contract principles led courts to recog-

---

12. *See* 4 Bruner & O'Connor, *supra* note 1, §§ 2:8–2:9; Willis D. Morgan, *The History and Economics of Suretyship*, 12 Cornell L. Q. 153 (1926) and 13 Cornell L. Q. 487 (1927).

13. *See* Richard Speidel, *An Essay on the Reported Death and Continued Vitality of Contract*, 27 Stanford L. Rev. 1149, 1173–1174 (1975); 1 Bruner & O'Connor, *supra* note 1, §§ 2:8–2:9.

14. *See* Oliver Wendell Holmes, Jr., *The Path of the Law*, 10 Harv. L. Rev. 457, 466 (1897).

> You always can imply a condition in a contract. But why do you imply it? It is because of some belief as to the practice of the community or of a class, or because of some opinion as to policy, or, in short, because of some attitude of yours upon a matter not capable of exact quantitative measurement, and therefore not capable of founding exact logical conclusions. Such matters really are battle grounds . . . where the decision can do no more than embody the preference of a given body in a given time and place. We do not realize how large a part of our law is open to reconsideration upon a slight change in the habit of the public mind.

See also Todd D. Rakoff, *Social Structure, Legal Structure, and Default Rules: A Comment*, 3 S. Cal. Interdisc. L.J. 19, 20 (1999) ("When we look at the world of contracts as a whole, most of the contextualizing comes from having different norms—whether formulated as rules or as standards—for different types of transactions.").

15. *See* Karen Eggleston, Eric A. Posner, & Richard Zeckhauser, *The Design and Interpretation of Contracts: Why Complexity Matters*, 95 Nw. U. L. Rev. 91, 92 (2000) ("We argue that . . . the current tendency of scholars to focus on completeness and neglect complexity has resulted in an inadequate understanding of contracts and contract law.").

16. *See* Oliver Wendell Holmes, Jr., *The Theory of Legal Interpretation*, 12 Harv. L. Rev. 417 (1899); 1 Bruner & O'Connor, *supra* note 1, § 3:1, et seq. Rapid evolution and specialization of language continue to cause misunderstandings. Two hundred fifty years ago, Samuel Johnson, in the preface of his Dictionary of the English Language (1755), the first of the great English dictionaries, wrote:

> It must be remembered, that while our language is yet living, invariably by the caprice of everyone that speaks it . . . words are hourly shifting their relations, and can no more be ascertained in a dictionary, than a grove, in the agitation of a storm, can be accurately delineated from its picture in the water.

nize numerous implied conditions in construction contracts as a matter of law: The owner's implied duty of full disclosure; the owner's implied warranty of the adequacy of detailed design; the contractor's implied duty of good workmanship; the contractor's duty of inquiry and clarification; the mutual implied duty of cooperation; and, the mutual implied duty of good faith.[17] In addition, the judiciary fashioned contextual contractual principles of unconscionableness, disproportionality, and misrepresentation; and restitutionary principles of promissory estoppel. Moreover, Congress and state legislatures added their own contextual concepts by using the legislative-administrative process to preempt areas of law traditionally reserved for private bargainers and the courts.

- By the late 19th century, competitive bidding laws had been enacted in most jurisdictions to prevent chicanery and fraud in the award of public construction and other contracts by requiring public contracts to be awarded to those responsible bidders who submitted the lowest responsive bids. Known today as the "Design-Bid-Build" or "sealed bid" project delivery method, competitive bidding remains the most widely used procurement approach.[18]
- The 19th century's host of new specialized construction trades—electricians, plumbers, iron workers, steam fitters, and others organized to fabricate or install newly invented technologies—necessitated utilization by supervising "general" contractors of improved construction scheduling techniques, and led to the introduction and widespread use of bar charts in the early 1900s and of sophisticated critical path method networks beginning in the 1950s.[19]
- Following the Great San Francisco Earthquake in 1906, municipalities began to take more seriously and to enact comprehensive building and fire codes formulated by regional code organizations. Hundreds of local and regional building codes developed in the 20th century have been replaced by adoption of the new International Building Code introduced in 2003.[20]
- In the 1920s, the science of "soil mechanics" was founded by a young engineer, Carl Terzaghi.[21] This science, in the 1940s, led to the development

17. *See* 3 Bruner & O'Connor, *supra* note 1, §§ 9:78–9:103.

18. *See* 1 Bruner & O'Connor, *supra* note 1, §§ 2:22–2:153; Ralph Nash, Jr. & Michael Love, Jr., *Innovations in Federal Construction Contracting*, 45 Geo. Wash. L. Rev. 309, 310–13 (1977).

19. *See* 5 Bruner & O'Connor, *supra* note 1, §§ 15:3–15:6.

20. *See* M. Tobriner, The History of Building Codes to the 1920s (1984); 5 Bruner & O'Connor, *supra* note 1, §§ 16:1–16:2.

21. Interestingly, it was the law that drove Terzaghi to his new calling. After receiving a mechanical engineering degree in Austria in 1904, Terzaghi worked for a design-build firm.

of the Unified Soil Classification System,[22] which created a scientific framework and terminology for precise classification of soils for engineering purposes by geotechnical professionals worldwide.

- By the early 20th century, the increasing complexity of the construction process led the industry to promote specialized industry dispute resolution procedures invoking professional decision making and arbitration. By 1905, before any state had authorized enforcement of arbitration agreements or awards and at a time when the judiciary was hostile to arbitration under a perception that the forum was intended to divest courts of judicial business, the "Uniform Contract" of the American Institute of Architects and the National Association of Builders called for resolution of disputes by arbitration.[23] Thereafter, the American Institute of Architects' Standard General Conditions of Contract continued to provide for the resolution of disputes by arbitration. By 1925, Congress had enacted the Federal Arbitration Act, followed after 1955 by most states' adoption of the Uniform Arbitration Act. Thereafter, the

---

He designed a factory building with footings sized in accordance with the empirical formulae of the day, and had the site load tested with a typical two-foot-by-two-foot platform loaded to 150 percent of design load. No settlement occurred within 24 hours, and Terzaghi allowed the construction to proceed. As soon as the building was completed, it began to settle and crack. Terzaghi was sued and lost quite a bit of money, and as a result:

> He began to question the reasons for this failure. He was soon impressed with the high standards of engineering design related to concrete construction compared with the guesswork and ignorance associated with the bearing values of soils that support those structures. He decided to devote himself to this most backward, unscientific aspect of civil engineering practice—the study of soils.

Hyman Cunin, *Soils Part I: Engineering Aspects and Physical Properties*, THE CONSTRUCTION SPECIFIER 80 (May 1968).

22. *See* ASTM Standard D2487-98, STANDARD CLASSIFICATION OF SOILS FOR ENGINEERING PURPOSES (UNIFIED SOIL CLASSIFICATION SYSTEM).

23. *See* the UNIFORM CONTRACT art. XII (1905 ed.):

> In case the Owner and Contractor fail to agree in relation to matters of payment, allowance or loss referred to in Arts. III or VIII of this contract, or should either of them dissent from the decision of the Architects referred to in Art. VII of this contract, which dissent shall have been filed in writing with the Architects within ten days of the announcement of such decision, then the matter shall be referred to a Board of Arbitration to consist of one person selected by the Owner, and one person selected by the Contractor, these two to select a third. The decision of any two shall be final and binding on both parties hereto. Each party shall pay one-half of the expense of such reference.

judiciary openly embraced arbitration as a favored method of alternate dispute resolution.[24]

- In the 21st century, the construction industry remains the largest single segment of the production sector of the American economy[25] and probably of the world economy. The industry also remains one of the most technologically complex. The development of modern engineering principles, of sophisticated construction practices, and of new building and materials technologies produced a host of specialized design disciplines and construction trades to oversee the design and installation of highly specialized modern systems, equipment, and materials unknown prior to the 20th century—electricity, plumbing, heating and ventilating, lighting, telephones, fiber optic cables, elevators and escalators, fire suppression, curtain wall, roofing and insulation, sealants, reinforced concrete, paints and coatings, and high-strength steel and glass. Those disciplines, systems, equipment, and materials fostered an exponential increase in the complexity, size, and scope of the built environment—"skyscraper" office towers with deep foundations, large-bore tunnels, massive dams and power plants, subways and interstate highways, wastewater treatment plants, airports, and harbors.

---

24. *See* Moses H. Cone Memorial Hosp. v. Mercury Constr. Corp., 460 U.S. 1, 24–25, 103 S. Ct. 927, 941–42, 74 L. Ed. 2d 765 (1983) (enunciating pro-arbitration policy). *See also* 6 Bruner & O'Connor, *supra* note 1, §§ 20:1–20:4. *See also* U.S. Supreme Court Chief Justice Warren E. Burger, *Using Arbitration to Achieve Justice*, 40 Arb. J. 3, 6 (Dec. 1985):

> I cannot emphasize too strongly to those in business and industry—and especially to lawyers—that every private contract of real consequence to the parties ought to be treated as a "candidate" for binding private arbitration. In the drafting of such contracts, lawyers will serve their clients and the public by resorting to tested clauses the American Arbitration Association has developed to fit particular needs.
>
> . . .
>
> We must now use the inventiveness, the ingenuity and the resourcefulness of American businessmen and lawyers—the "Yankee Trader" innovativeness—to shape new tools to meet new needs. In the area of arbitration, the tools and the techniques are ready and waiting for imaginative lawyers to make use of them.
>
> If the courts are to retain public confidence, they cannot let disputes wait two, three and five years or more to be disposed of. The use of private arbitration is one solution, and lawyers should be at the forefront in moving in this direction.

25. *See* U.S. Department of Labor Bureau of Labor Statistics at http://www.BLS.gov/IAG/TGS/IAG23.htm; Thomas Stipanowich, *Restructuring Construction Law: Reality and Reform in a Transactional System*, 1998 Wis. L. Rev. 463, 465 (1998) ("The largest production sector in the United States economy is the construction industry. It accounts for annual expenditures of a half a trillion dollars per year and directly employs one of every 20 workers, represents as much as 13 percent of the gross national product, and touches the lives of every citizen.").

Construction today has acquired a legendary reputation for extraordinary factual and legal complexity,[26] which is not unjustly earned:

> Construction is an inherently complex business. Even casual observers of the construction process are struck by the enormous amount of information required to construct a project. Hundreds, even thousands, of detailed drawings are required. Hundreds of thousands of technical specifications, requests for information, and other documents are needed. Complex calculations are used to produce the design. For years, this complexity dictated a labor-intensive, highly redundant methodology for doing the work. Projects were fragmented and broken into many parts. Different entities undertook different parts of a project, both for design and construction. Therefore, the construction industry became exceptionally fragmented. On a project of even average complexity, there may have been from 5 to 15 firms involved in design. From 40 to 100 companies may have been engaged in construction. Many more companies supplied materials, professional services, and other elements necessary for completion of the project. It was effectively impossible to convey the sum of knowledge necessary to construct a facility in a set of plans and specifications. Stated another way, the information technology traditionally used for construction is inadequate.[27]

Construction's complexity has created recognized public safety risks, which in turn have led to increased governmental regulation of the construction process through legislative imposition of licensing laws, safety regulations, and building codes.[28] The recognized financial credit risks inherent in the multiparty construction process have led to legislative enactment of an assortment

---

26. The judiciary itself has embellished this legendary reputation. *See, e.g.*, Erlich v. Menezes, 21 Cal. 4th 543, 87 Cal. Rptr. 2d 886, 896, 981 P.2d 978 (1999), in which the Supreme Court of California observed:

> [T]he [owners] may have hoped to build their dream home and live happily ever after, but there is a reason that tagline belongs in fairy tales. Building a house may turn out to be a stress-free project; it is much more likely to be the stuff of urban legends—the cause of bankruptcy, marital dissolution, hypertension, and fleeting fantasies ranging from homicide to suicide. Justice Yegan noted below, "No reasonable homeowner can embark on a building project with certainty that the project will be completed to perfection. Indeed, errors are so likely to occur that few if any homeowners would be justified in resting their peace of mind on [its] timely or correct completion. . . ." The connection between the service sought and the aggravation and distress resulting from incompetence may be somewhat less tenuous than in a malpractice case, but the emotional suffering still derives from an inherently economic concern.

27. *See* John W. Hinchey, *Visions for the Next Millennium*, in 1 Constr. L. Handbook § 2.01[A] (1999).

28. *See* 5 Bruner & O'Connor, *supra* note 1, §§ 16:1, et seq.

of laws to protect owners and unpaid construction trades against the risks of contract default.

1.02 CONSTRUCTION DISPUTE RESOLUTION

Like other highly complex fields of human endeavor, the construction process has spawned its own unique customs, practices, and technical vocabulary, which in turn led courts and legislatures to develop legal principles consistent with industry realities. Construction law has derived much of its uniqueness from industry experience, customs, and perceived foreseeable risks, which shaped evolving principles of common law and statutory law applicable to the built environment. Oliver Wendell Holmes, Jr. reminds us that:

> The life of the law has not been logic; it has been experience. The felt necessities of the time, the prevalent moral and political theories, intuitions of public policy, avowed or unconscious, even the prejudices which judges share with their fellow-men, have had a good deal more to do than the syllogism in determining the rules by which men should be governed. The law embodies the story of a nation's development through many centuries, and it cannot be dealt with as if it contained only the axioms and corollaries of a book of mathematics. In order to know what it is, we must know what it has been, and what it tends to become.[29]

Under the weight of a century of contextual experience, construction law indeed is evolving into a "separate breed of animal."[30] Construction law today is a primordial "soup" in the "melting pot" of the law—a thick broth consisting of centuries-old legal theories fortified by statutory law and seasoned by contextual legal innovations reflecting the broad factual "realities" of the modern construction process. Construction law has been viewed by some academicians

29. *See* Oliver Wendell Holmes, Jr., The Common Law 1 (1881).

30. Paul Hardeman, Inc. v. Arkansas Power & Light Co., 380 F. Supp. 298, 317 (E.D. Ark. 1974) ("[C]onstruction contracts are a separate breed of animal; and, even if not completely sui generis, still [the] law must be stated in principles reflecting underlying and industry realities. Therefore, it is not safe to broadly generalize. True, general principles of contract law are applied to construction contracts, but they are applied under different operative conditions. Care must be taken, then, not to rely too uncritically on such cases as those arising from the sale of real or personal property. And even within the larger rubric of 'construction contracts,' it is manifest that the law, if sensitive to the underlying realities, will carefully discriminate between, say, a contract to construct a home and a contract to construct a 50-story office building; between a contract to build a private driveway and a contract to construct an interchange on an interstate highway. This is what one would expect a priori; this is, generally, what one finds when he reviews the actual development of the law.").

incorrectly as mere "advanced contract law"—a misunderstanding that arises from viewing construction law through the prism of a historically narrow academic discipline rather than through the kaleidoscope of complex legal and factual issues inherent in the construction process itself. Construction law is a "capstone" subject that includes the nuances of the "contextual" contract and the towering legal edifice constructed to govern legal relationships among the multitude of parties—design professionals, contractors, subcontractors, material suppliers, sureties, insurers, or tradesmen—typically engaged in varying degrees in construction projects.[31] These legal relationships include: (1) multiple express and implied contractual relationships; (2) tort relationships, rights, and obligations where contractual privity does not exist; (3) suretyship relationships invoking equitable principles governing construction bonds; (4) insurance relationships invoking principles applicable to products insuring construction and design risks; (5) agency principles applicable to construction industry participants and their representatives; (6) design professional rights and liabilities created by common law and statutory duties; (7) construction lender relationships and liabilities pertaining to project financing; (8) statutory rights and obligations created by statutes governing mechanic's liens, public contractor bonds, and public contract bidding; (9) statutory rights and obligations arising under the Uniform Commercial Code governing relationships for the purchase of construction materials and equipment; (10) public duties created by building codes, licensing laws, and health and safety laws; (11) problems of proof of causation of loss; (12) damage measurement and computation principles that recognize construction's imperfect world under doctrines such as "substantial performance" and "economic waste;" and (13) issues unique to construction dispute resolution, partnering, and alliancing.

Like other highly complex fields of law, the litigation of construction disputes relies heavily for proof of causation upon opinion testimony of experts—

---

31. The American and foreign jurisdictions that certify construction law as a specialized area of legal practice recognize the field to have significant breadth. *See* AMENDMENTS TO THE RULES REGULATING THE FLORIDA BAR, 875 So.2d 448 (2004), which establish standards by which a Florida lawyer may become a "board certified construction lawyer" and which define "construction law" as follows:

> "Construction law" is the practice of law dealing with matters relating to the design and construction of improvements on private and public projects including, but not limited to, construction dispute resolution, contract negotiation, preparation, award and administration, lobbying and governmental hearings, oversight and document review, construction lending and insurance, construction licensing, and analysis and litigation of problems arising out of the Florida Construction Lien Law, § 255.05, FLORIDA STATUTES, and the federal Miller Act, 40 U.S.C. § 2470.

a fact of life that can be frustrating to courts[32] and mesmerizing to juries[33]—and all too frequently results in detailed factual records of proceedings that appear "formidable" to finders of fact and reviewing appellate judges.[34] Some judges, overburdened by their judicial workloads, have little time for complex construction cases, and contend that construction cases invoke the "law of the shop" rather than a "law of the court" and should be settled by arbitrators or by other alternate dispute resolution devices.[35] Such views over the years

---

32. *See* E. C. Ernst, Inc. v. Manhattan Constr. Co., 387 F. Supp. 1001, 1005–1006 (S.D. Ala. 1974), in which a federal judge, during a pretrial conference, advised the parties:

> Being trained in this field, you are in a far better position to adjust your differences than those untrained in these related fields. As an illustration, I, who have no training whatsoever in engineering, have to determine whether or not the emergency generator system proposed to be furnished . . . met the specifications, when experts couldn't agree. This is a strange bit of logic. . . . The object of litigation is to do substantial justice between the parties' litigant, but the parties' litigant should realize that, in most situations, they are by their particular training better able to accomplish this among themselves. . . .

33. The common "lore" in construction litigation is that the more complex cases should not be tried to a jury and should be reserved for trial to the court or to experienced construction arbitrators.

34. *See* Blake Constr. Co., Inc. v. C. G. Coakley Co., Inc., 431 A.2d 569, 575 (D.C. 1981):

> [E]xcept in the middle of a battlefield, nowhere must men coordinate the movement of other men and all materials in the midst of such chaos and with such limited certainty of present facts and future occurrences as in a huge construction project such as the building of this $100 million hospital. Even the most painstaking planning frequently turns out to be mere conjecture and accommodation to changes must necessarily be of the rough, quick, and *ad hoc* sort, analogous to ever-changing commands on the battlefield. Further, it is a difficult task for a court to be able to examine testimony and evidence in the quiet of a courtroom several years later concerning such confusion and then extract from them a determination of precisely when the disorder and constant readjustment, which is to be expected by any subcontractor on the job site [or by any other party for that matter], became so extreme, so debilitating and so unreasonable as to constitute a breach of contract between the contractor and a subcontractor. This was the formidable undertaking faced by the trial judge in the instant case. . . .

35. For example, former U.S. Supreme Court Chief Justice Warren E. Burger expressed this view (believed to be shared by many in the judiciary) as follows:

> The obligation of the legal profession is, or has long been thought to be, to serve as healers of human conflicts. To fulfill that traditional obligation means that there should be mechanisms that can produce an acceptable result in the shortest possible time, with the least possible expense and with a minimum of stress on the participants. That is what justice is all about.
>
> . . .

have led to wide industry use of alternative dispute resolution procedures such as arbitration,[36] and innovative dispute review board, mediation, and stepped claims settlement practices.[37]

1.03 CONSTRUCTION LAW SCHOLARSHIP

Although construction is the largest segment of the production sector of the United States gross domestic product, and quite likely of the world's gross domestic product, American legal and economic scholars have paid little attention to the industry.[38] The articulated reasons for such academic oversight, although anecdotal, suggest that legal academicians have been unwilling to

---

> My overview of the work of the courts from a dozen years on the Court of Appeals and now 16 in my present position, added to 20 years of private practice, has given me some new perspectives on the problems of arbitration.
>
> One thing an appellate judge learns very quickly is that a large part of all the litigation in the courts is an exercise in futility and frustration. A large proportion of civil disputes in the courts could be disposed of more satisfactorily in some other way.
>
> . . .
>
> My own experience persuades me that in terms of cost, time, and human wear and tear, arbitration is vastly better than conventional litigation for many kinds of cases.
>
> In mentioning these factors, I intend no disparagement of the skills and broad experience of judges. I emphasize this because to find precisely the judge whose talents and experience fit a particular case of great complexity is a fortuitous circumstance. This can be made more likely if two intelligent litigants agree to pick their own private triers of the issues. This is not at all to bypass the lawyers; they are key factors in this process.
>
> The acceptance of this concept has been far too slow in the United States.

Burger, *supra* note 24.

36. Arbitration was mandated by The Uniform Contract as early as 1905 and by subsequent editions of the Conditions of Contract promulgated by the American Institute of Architects. For background, *see* 6 BRUNER & O'CONNOR, *supra* note 1, §§ 20:1–20:3.

37. *See* James P. Groton, Robert A. Rubin, & Bettina Quintas, *A Comparison of Dispute Review Boards and Adjudication*, 18 INT'L CONSTR. L. REV. 275 (2001); James P. Groton, *The Progressive or "Stepped" Approach to ADR: Designing Systems to Prevent, Control, and Resolve Disputes*, in CONSTRUCTION DISPUTE RESOLUTION FORMBOOK 1–33 (Robert F. Cushman et al. eds. 1997). See Chapter 21 of this book for more detailed discussion of alternative dispute resolution in the construction arena.

38. *See* William A. Klein & Mitu Gulati, *Economic Organization in the Construction Industry: A Case Study of Collaborative Production Under High Uncertainty,* 1 BERKELEY BUS. L. J. 138, 139 (2004); *see also* Diane H. Kay, *The Education of the Construction Work Force in the Post-Industrial Era,* 27 THE AMERICAN PROFESSIONAL CONSTRUCTOR 25 (April 2003).

acquire practical understanding of the complexities of the construction process and hence have been unable to develop significant capability to contribute to the development of law undergirding the construction industry.[39] Those few academicians who have mastered construction law suggest that their academic colleagues have more interest in "public law,"[40] a subject less dependent on custom and practice and less factually intensive. Although there have been occasional cries within academia over the years for more scholarly study of construction law issues,[41] American academicians continue to contribute little to the teaching and development of construction law.[42]

The American Bar Association Forum on the Construction Industry and the American College of Construction Lawyers are seeking, through the writing and publication of this textbook, to introduce a generation of fledgling lawyers to the practice of construction law. While it necessarily will be taught primarily by adjunct faculty, it is this author's hope that the book will inspire an interest in the study of construction law, and that some of the students will in turn be inspired to teach others.

---

39. *See* Stipanowich, *supra* note 25, at 496.

40. Professor Emeritus Justin Sweet of the Boalt Hall School of Law at the University of California, Berkeley, was one of the few 20th-century legal academicians to devote a career to construction law. In his article, *Construction Law: The Need for Empirical Research*, 23 CONSTR. LITIG. RPTR. 3 (Jan. 2002), he offers the following reason for academia's lack of empirical scholarship in the construction law field:

> One is the lack of full-time law teachers with interest in Construction Law. The best and often only empirical work comes out of the law schools. Law teachers can involve statisticians and sociologists in their studies. Money can be found, though I admit not easily. Yet you can count on one hand the number of full-time teachers of Construction Law, maybe not even that many. Law teachers come out of certain schools, clerk for important judges, and are interested mainly in Public Law. This pool does not produce many teachers who want to spend their time in Construction Law.

41. Professor Edward Patterson of Columbia, in his article, *Builder's Measure of Recovery of Breach of Contract*, 31 COL. L. REV. 1286 (1931) observed:

> The economic importance of the building industry, the frequency of litigation involving this type of contract, and the inadequacy of judicial analyses of the complex problems of [construction] damages warrants academia's attention.

42. In contrast, European law schools have perceived the importance of construction law as a scholarly endeavor. For example, the University of Strathclyde offers an LLM Degree in Construction Law.

CHAPTER

2

Participants in the Design and Construction Process

ROSS J. ALTMAN

2.01 OVERVIEW

A building or other structure is not an object simply purchased by a consumer or produced by a single person or entity. The design and construction of a project is a collaborative process requiring the talent, execution, and coordination of many different people and organizations. As the size, cost, complexity, or unusual features of a project increase, the number of participants needed to design and construct that project likely will increase as well. The financial, technical, business, and regulatory challenges involved in even a small commercial project demand the participation of many diverse participants. Although each participant makes a separate contribution to the project, the activities of one inextricably bear upon the performance of the others.

It should come as no surprise that assembling a construction transaction is a complicated task. The interdependence of each member of the design and construction process on the performance of the other members may cause difficulties. Like any complex process, there is ample opportunity for failure, breakdown, and difficulty along the way. Many project participants will not have worked together previously and may not have worked on a project similar to the one for which they are retained. The participants may have different policies and manners in which they conduct their business, and the policies and manners of one may conflict with those of the other participants. On international projects, there will be language barriers, multiple cultures, and the practical difficulties created by long distances and different time zones. Despite such obstacles, those involved in the production of a project must be well organized and coordinated to deliver a project successfully. This book explores the multitude of legal issues that arise out of a process that has collaboration as its goal but that in practice often fosters disputes and controversy. To consider such legal issues, however, it is first necessary to understand the roles and objectives of each of the typical participants in the design and construction process.

2.02 THE OWNER

The owner initiates the project. The owner furnishes the project site, is responsible for satisfaction of certain regulatory requirements necessary to develop and use the project site, and must pay for the development of the project. As the person or entity producing the project, the owner is responsible for ensuring that all activities involved in the delivery of the project are performed. Many of those activities are well beyond the capability of the typical owner, however, and are delegated to third parties who have the skill, experience, and manpower that the owner may not possess.

There are many different types of owners. Some owners are skilled and possess prior experience with the design and construction process. Other owners may undertake a project only once. Issues that are of particular importance to one owner may not be particularly meaningful to another; cost concerns are paramount to many owners, while other owners are more flexible with respect to cost when certain time requirements are satisfied. The manner in which a project is delivered and the allocation of responsibilities among the participants to the project can be influenced significantly by the various characteristics of the owner.

Whether the owner is a public or private entity will affect greatly many fundamental decisions about the delivery of the project. Public owners, such as federal, state, and local governments, are obligated to procure and manage construction projects pursuant to a statutory and regulatory framework.[1] Public entities are not free to engage in innovative project delivery methodologies,

1. Public entities must comply with a complex set of procurement statutes when awarding a contract, and then only enter into contracts that conform to applicable regulations.

absent enabling legislation.[2] Private owners, on the other hand, have significant flexibility in how they conduct the design and construction of a project. Although a private owner may have certain business policies and practices that impose limitations, such restrictions are likely to be less burdensome on the private owner than those imposed by law on the public owner.

The nature of the owner's interest in the project also can affect the manner in which the project is delivered. If the project is simply an investment vehicle, the owner may focus more on achieving an anticipated rate of return from that investment, rather than on satisfying the design and construction requirements of a third party who is purchasing or leasing a portion of the project. An owner who intends to occupy and use the project may place greater emphasis on technical and performance aspects of that project. The owner likely will view such a project as a capital investment and from a longer-term perspective.

The manner in which the owner pays for a project also can influence the design and construction process. Owners who undertake projects with their own funds are free from the restrictions commonly imposed by a construction lender, and often are able to make decisions and take action on an expedited basis. Because late or defective completion of a project adversely affects the ability of a construction lender to recover the loaned amount and anticipated interest, construction lenders commonly require various contractual provisions and project controls aimed at ensuring (1) that the contractor completes the project in conformance with the drawings and specifications, (2) that the payments are made only for work properly completed, and (3) that the loan remains in balance.[3]

The level of an owner's knowledge and experience in matters pertaining to the design and construction of the project, however, may have the greatest effect on the activities of the various participants to the design and construction process. An owner who has developed other projects often has the requisite skill and ability to play an active role in the design and construction process and to assume risk commensurate with the control such an owner exerts over the project. An inexperienced or novice owner is far more likely to transfer risk and delegate responsibility to third parties, thereby yielding control over the delivery of the project to those third parties.

---

2. For example, most public entities enact special legislation prior to proceeding with a public-private partnership for a construction project or other form of procurement that varies from a competitively bid process.

3. In other words, that the undisbursed balance of the loan is greater than the remaining cost to complete the project.

In addition to the owner's breadth of experience, the manner in which the owner conducts itself often is derived from the type of business entity in which the owner is organized. For example, parties with whom the owner contracts need to know who is authorized to act on behalf of the owner. That determination is relatively easy when dealing with an individual or a partnership. The scope of authority of various persons associated with the owner, however, is less clear when the owner is a limited liability company or corporation. Moreover, larger entities with complex business structures often are unable to act quickly and must build consensus within their own organization prior to making a decision. In many circumstances, contracts attempt to solve this issue by specifically providing for designation of a representative who is authorized to act on behalf of the owner.[4]

2.03 THE DESIGN TEAM

A. The Architect

Although some owners have in-house design capability, the vast majority of owners retain third-party design professionals to prepare the drawings and specifications from which the project is built.[5] For a commercial building project, an architect usually leads the design effort. Architects perform professional design services, which either begin with the preparation of a program for a project or are based upon a program supplied by the owner.[6] The architect's services generally are performed in various phases. During each successive phase of design services, the drawings and specifications become more detailed and finalized.[7]

As a practical matter, the design is never really complete. Even if possible, the time and cost to describe in the drawings and specifications every detail of a project would be prohibitive. The construction team must fill in the blanks

4. *See* American Institute of Architects, AIA Document A101–2007, *Standard Form of Agreement Between Owner and Contractor* § 8.3 (2007) [hereinafter AIA Document A101–2007].

5. For a description of the design process, see Chapter 6.

6. Schematic design, the first phase of the architect's design services under AIA contract forms, is based on the owner's program. American Institute of Architects, AIA Document B101–2007, *Standard Form of Agreement Between Owner and Architect* § 3.2.1 (2007) [hereinafter AIA Document B101–2007]. The owner is obligated to provide the architect with certain information, including "a written program which shall set forth the Owner's objectives, schedule, constraints and criteria including space requirements and relationship, flexibility, expandability, special equipment, systems and site requirements." *Id.* § 5.1.

7. *See id.* §§ 3.2–3.4.

by constructing what is "reasonably inferable" from the contract documents.[8] Also, certain aspects of the project design may pertain to products or systems that are proprietary or require particular knowledge unique to such products or systems. It would be inefficient for the architect to design such aspects of the project, and these tasks typically are left for the construction team to accomplish through the shop drawing process.[9]

The architect often assists the owner in procuring the construction contract and administering the construction process. Services provided by an architect during the construction phase, however, typically are limited to undertaking specified activities that give the owner some assurance the project is being constructed in accordance with the approved drawings and specifications, and providing certain administrative functions.[10] Notwithstanding the fact that an architect participates in the construction phase of the project, the activities of the architect generally are described as providing services, and not as work.[11] Architects very rarely accept any responsibility for construction means, methods, techniques, sequences, and procedures, or for safety measures and precautions.[12]

B. The Engineers and Consultants

The professional design services required of the architect typically include structural, mechanical, and electrical engineering services.[13] Depending on the nature of the project, other specialized engineering or consulting services might be expected. Although some larger architectural firms possess engineering capability, such an in-house staff customarily does not perform overly sophisticated engineering tasks.

Consequently, an architect commonly enters into subconsulting agreements with engineers and other specialty design consultants.[14] The architect

---

8. *See* American Institute of Architects, AIA Document A201–2007, *General Conditions of the Contract for Construction* § 1.2.1 (2007) [hereinafter AIA Document A201–2007].

9. The shop drawing process is described in *id.* § 3.12.

10. *See* AIA Document B101–2007, *supra* note 6, § 3.6.

11. Generally speaking, the parlance of the industry is that design professionals provide "Services" and construction contractors perform "Work."

12. Such responsibility is disclaimed routinely. *See* AIA Document B101–2007, *supra* note 6, § 3.6.1.2; AIA Document A201–2007, *supra* note 8, § 4.2.2.

13. *See* AIA Document B101–2007, *supra* note 6, § 3.1.

14. Note that for a larger project, the architect may identify certain subconsultants at the time the owner and architect enter into an agreement. American Institute of Architects, AIA B103–2007, *Standard Form of Agreement Between Owner and Architect for a Large or Complex Project* § 1.1.12 (2007).

creates the overall design concept for the project and makes many fundamental design decisions. Nonetheless, much of the actual design is carried out by others under contract with the architect; the architect manages and coordinates the efforts of those subconsultants. This can get complicated. Even third-party engineers retained by an architect may need to retain specialists to deal with particular matters. Notwithstanding the delegation of duties, the architect usually remains responsible to the owner for the services of all of the architect's direct and indirect subconsultants.[15]

Finally, some aspects of the project may require highly specialized design skills or knowledge that an architect typically does not possess and might not consider itself capable of managing. An architect may be reluctant simply to serve as a conduit between the owner and the provider of such specialty services, especially if such services are considered high risk, such as geotechnical or environmental engineering. In those circumstances, the architect typically insists that the owner contract directly with specialty design professionals.

2.04 THE CONSTRUCTION TEAM

A. The Contractor

Construction companies enter into contracts with the owner to build a project and, therefore, are known as "contractors." When contracting with an owner directly, a contractor is referred to as a prime contractor. Under some circumstances, there may be multiple prime contractors on a project. Many owners, not surprisingly, prefer to retain one prime contractor to assume principal responsibility for construction of the project. That prime contractor commonly is referred to as a general contractor.[16]

The contractual obligations of the general contractor are referred to as the "Work," which term is defined in construction contracts.[17] Although the principal objective of the general contractor is to complete construction of the proj-

15. Although such responsibility is not explicitly stated in AIA documents, the architect delegates the performance of its services to subconsultants. Therefore, the architect generally remains responsible to the owner for such services. *See* 5 PHILIP L. BRUNER & PATRICK J. O'CONNOR, BRUNER & O'CONNOR ON CONSTRUCTION LAW § 17:71 (West Group 2002).

16. A prime contractor responsible for performing work only in a single trade (e.g., carpentry, concrete) usually would be referred to as a "trade contractor." The term "general contractor" implies responsibility for multiple disciplines or trades.

17. *See* AIA Document A201–2007, *supra* note 8, § 1.1.3, which provides that the "Work means the construction and services required by the Contract Documents . . . and includes all other labor materials, equipment or services provided or to be provided by the Contractor to fulfill the Contractor's obligations."

ect, the Work involves many other related activities, such as reviewing field conditions and construction documents,[18] maintaining labor harmony at the project site,[19] supplying certain materials and other items necessary for the general requirements of construction,[20] obtaining construction permits,[21] preparing shop drawings,[22] maintaining the project site's security,[23] cleaning up the project site,[24] testing and start-up of a project,[25] and maintaining safety precautions and programs.[26]

It would be unusual for the drawings and specifications prepared by the design team to instruct the general contractor how to build the project. The general contractor decides on its own which construction means, methods, techniques, sequences, and procedures to employ in order to complete the project.[27] Such decisions are important because the manner in which the project is constructed affects the cost and time of the general contractor's performance.

General contractors have different approaches to constructing any given project. For example, a general contractor may decide to use certain equipment or a methodology that it believes will cost less or save time compared to other approaches favored by its competitors. Owners and architects want to avoid any allegation that they may have interfered in the general contractor's decisions on how to construct the project, thereby adversely affecting the general contractor's ability to satisfy its contractual obligations. Therefore, owners and architects commonly disclaim any responsibility for the general contractor's construction means, methods, techniques, sequences, or procedures.[28]

The general contractor's approach to constructing a project also is important because it bears directly on the safety of construction workers and others

---

18. *See id.* § 3.2.

19. *See id.* § 3.4.3.

20. *See id.* § 3.4.1.

21. *See id.* § 3.7.1.

22. *See id.* § 3.12.5.

23. *See id.* § 10.2.3.

24. *See id.* § 3.15.

25. Such responsibilities are common to contracts for industrial projects where an engineering process is involved, such as a power plant or manufacturing facility, and for specialized building systems such as elevators. Because such activities are often crucial to confirm the adequacy of the completed project, separate contracts sometimes are required to describe such services and applicable standards, as well as to allocate responsibility. *See* American Institute of Architects, AIA Document B211–2007, *Standard Form of Architect's Services: Commissioning* (2007), regarding the retention of an architect to provide commissioning services.

26. *See* AIA Document A201–2007, *supra* note 8, § 10.2.1.

27. *See id.* § 3.3.1.

28. *See* AIA Document B101–2007, *supra* note 6, § 3.6.1.2; AIA Document A201–2007, *supra* note 8, § 4.2.2.

who may come into contact with the work. The owner and architect want to limit the potential for claims against them arising out of bodily injuries, which can involve significant damages. This provides another reason why most construction contracts provide that the general contractor, solely, is responsible for construction means, methods, techniques, sequences and procedures.[29]

B. The Subcontractors

Performance of construction work requires a wide range of skill and expertise. Most projects are technically complex and depend on knowledge of particular products or systems that may be proprietary. Although some general contractors perform construction work with their own forces (i.e., their own employees), such self-performed work is usually limited to a particular trade with which the general contractor is familiar, such as carpentry or concrete work. Most general contractors do not find it efficient to have a large staff of employees to perform work in each trade. It is unlikely that a general contractor could keep such a large number of employees busy on a full-time basis or adequately maintain the level of skill and expertise that specialization can bring.

Therefore, most general contractors function as managers of the construction process. The general contractor retains separate contractors with specialized skill and knowledge for each trade on a project-by-project basis as needed. Such specialized trade contractors, when retained by a general contractor directly, are referred to as subcontractors. On a large project, a general contractor might retain 50 or more subcontractors to perform various portions of the work.[30] Although the work is performed by subcontractors, from a legal standpoint, the general contractor simply is delegating its obligations to another party. The general contractor, therefore, remains responsible to the owner for the performance of all work of the general contractor's subcontractors.[31]

The scope of work delegated to a subcontractor also may include items that the subcontractor customarily does not perform with its own forces; for example, work that involves highly specialized experience or the supply of materials that the subcontractor does not either keep in stock or manufacture. When the subcontractor retains a separate contractor to perform such specialized work at the site, that separate contractor customarily is referred to as a sub-subcontractor.[32] In such an arrangement, the subcontractor might be referred to as a first-tier

29. *See* AIA Document A201–2007, *supra* note 8, § 3.3.1.

30. *See* Karl Sabbagh, Skyscraper: The Making of a Building (Penguin Books 1989) for a description of the development of a large project.

31. *See* AIA Document A201–2007, *supra* note 8, § 3.3.2.

32. *See id.* § 5.1.2.

subcontractor, whereas the sub-subcontractor is sometimes referred to as a second-tier subcontractor. A subcontractor might also retain suppliers. Unlike subcontractors of any tier, suppliers do not perform work at the site, but rather only supply materials or equipment for the project.[33]

Even though the subcontractor may delegate a portion of its work to third parties, the subcontractor remains contractually liable to the general contractor for all work within the scope of the subcontract between the general contractor and the subcontractor, regardless of whether such work is performed by the subcontractor's own forces or through lower-tier subcontractors or suppliers. The subcontractor must supervise and coordinate all such activities.[34]

2.05 THE SUPPORTING CAST

Although the principal players in a project are the owner, the members of the design team, and the members of the construction team, it would not be possible to bring a project to fruition without the supporting services of many other business entities, as discussed next.

A. Construction Lenders

Few owners have sufficient cash on hand to pay for the design and construction of a project. Even those owners that have the assets and liquidity required to pay for a project frequently prefer to use such funds to pay for expenses pertaining to their core business or for other investments. Therefore, owners typically pay for a construction project through a combination of debt and equity. The debt component may take various forms, such as a traditional construction loan or bond financing.

A construction lender generally does not make a loan unless the owner or other investors make a certain level of equity investment. Additionally, construction financing is not a long-term loan. Construction lenders expect the loan to be paid off within a short time following completion of the project. In most circumstances, this is accomplished by the owner obtaining a replacement or permanent loan, the proceeds of which pay off the construction lender after project completion and allow the owner to amortize the costs of construction over a longer period of time. Because of the risk involved in construction activities,

33. The American Institute of Architects' AIA series of form documents do not define the term "supplier." For a more detailed discussion of the distinction between subcontractors and suppliers, see Chapter 10.

34. *See* American Institute of Architects, AIA Document A401–2007, *Standard Form of Agreement Between Contractor and Subcontractor* §§ 4.1.1–1.2, 4.6.1 (2007).

the interest rate charged by a construction lender is higher than that charged by a lender providing a loan secured by a mortgage on a completed project.

Sometimes projects are financed through the sale of bonds, which are purchased by investors who expect to be repaid their purchase price, plus a stated amount of interest. The bonds might not be retired or paid off until long after the project is completed, perhaps 10, 20, or even 30 years after the bonds are issued. Bonds are used frequently by public agencies to finance capital improvement projects, although private entities, particularly corporations, may issue bonds to finance construction projects. The interest rate to be paid under a bond depends on a number of factors, such as credit enhancement techniques, the rating of the bond provided by an outside agency, and tax considerations.[35]

Regardless of the source of construction financing, any entity providing funds or credit enhancement for the project will want to make sure that certain measures are taken to ensure that such funds and the agreed-upon interest are repaid. Such measures might include the following:

- Without completion of the project, the lender's ability to recover the principal loan amount and anticipated interest is severely impaired. Therefore, to provide assurance that the project will be completed, lenders often mandate compliance with various insurance specifications, as well as require a contractor to furnish payment and performance bonds.
- Construction lenders are not interested in taking over a partially completed project in order to recover their investment; that is not the lender's core business and is often a losing proposition. Therefore, as a means of facilitating project completion by the borrower, a construction lender typically requires an allocation of risk in the construction contract that does not overburden its borrower (the owner).
- A construction lender also often mandates certain procedures to ensure successful completion of the project. For example, a lender may require peer review of certain aspects of the project design that are particularly important or innovative. Construction lenders also require procedures ensuring that progress payments are made only for work properly completed and that the balance of the amount remaining to be paid to the general contractor equals or exceeds the value of work to be per-

35. The interest rate earned on the principal amount of the bond relates directly to the risk of payment by the bond issuer to the bondholder. Higher risk translates into a higher rate of interest.

formed.[36] Such procedures often involve use of an architect retained by the construction lender to provide independent advice with respect to the general contractor's applications for payment and completion of the work.

B. Insurance Companies

The participants in the design and construction process all are engaged in business activities with substantial risk. Design errors, work injuries, and defective construction, just to name a few potential problems, can result in significant damage, loss, claims, and disputes. Quality control efforts and funding of loss reserves may help, but they are not always cost-effective methods to control or manage risk. Few business entities would remain viable for any length of time without implementing some method of risk transfer.

As a result, owners, design professionals, contractors, and others associated with the development or operation of a project, routinely procure and maintain insurance.[37] Various types and forms of policies have evolved over time to provide a mosaic-like coverage for the risks most commonly associated with a construction project. As a practical matter, no single insurance policy will provide complete coverage. Therefore, design agreements and construction contracts often specify requirements for a number of insurance policies, most commonly commercial general liability insurance, worker's compensation insurance, employer's liability insurance, motor vehicle liability insurance, professional liability insurance, and environmental liability insurance, as well as other terms and conditions related to insurance.

The insurance industry wields a great deal of influence over the design professions and the construction industry. Moreover, as a condition of coverage, insurers may require compliance with safety programs or mandate other measures that help ameliorate their exposure to claims and loss payments.[38] Insurers further manage their risk by entering into agreements, sometimes called treaties, with reinsurers, who provide insurance to the primary insurer (i.e., the insurance company that issues the underlying policy to the named insureds). The participants in a project often attempt to include language in contracts that limit their liability to amounts and risks that are covered by applicable insurance.

36. The lender may also want the unpaid claims to cover a contingency.

37. For a more detailed discussion of insurance and its importance to the construction industry, see Chapter 20.

38. This is particularly so with a so-called owner-controlled insurance program (an OCIP) and other similar programs commonly referred to as wrap-ups.

C. Sureties

Payment and performance bonds issued by sureties sometimes are required to provide security for the contractor's performance of its contractual obligations. In essence, the payment bond guarantees the flow of payment to certain subcontractors and suppliers, and the performance bond guarantees completion of the construction work and certain obligations that flow out of the work. Generally, only construction work is bondable, not design services. The performance of construction activities is measurable and, therefore, is underwritten routinely by surety companies. The performance of design services, however, is not easy to assess, and, therefore, sureties are reluctant to write bonds for such activities.[39]

Unlike insurance, suretyship is not a transfer of risk, but rather a form of credit enhancement. An insurer expects to pay losses and offsets such payments with premiums that it receives from its customers and from income generated by investing premiums. In other words, in the event of a valid and covered claim, subject to the terms of the policy, the insurer will bear the loss. A surety, on the other hand, charges a fee for conducting a very thorough underwriting process, which may result in the surety issuing a bond to a project owner (also known as an obligee). By issuance of the bond, the surety agrees to stand behind and extend its credit on behalf of a contractor (also referred to as the principal). A surety, however, does not intend to bear any loss. Rather, the surety requires the contractor (the principal) to indemnify the surety from all loss, damage, cost, and expense the surety may incur as a result of having issued a bond, and may even require collateral or other security by which to satisfy the indemnity obligation.

By law, many public entities must require payment and performance bonds on construction projects. Private owners have no such obligation. Many private owners prefer to avoid the expense of a bond or may choose to bond only major subcontractors. Letters of credit sometimes are used as an alternative to a surety bond, particularly on international projects. Insurance products, such as subcontractor default insurance, also provide an alternative to bonds. Such surety bond alternatives, however, are not always permitted on public projects.

D. Title Insurance

Title insurance provides protection for defects in title, liens, and other encumbrances on real property. There are both policies issued to owners and policies issued to lenders (both construction lenders and permanent lenders). If title

39. Design services are bonded sometimes in connection with a Design-Build project.

insurance is provided during construction, then at the time of each progress payment to the contractor, a "date-down endorsement" is issued to increase the amount of insurance by the amount of payment made.

In addition, title insurers often serve as escrow agents for disbursement of funds to contractors, even when not issuing title insurance. When acting as an escrow agent, the title insurer reviews lien waivers, sworn statements, and other documentation to confirm that title is free and clear of claims, liens, and other encumbrances prior to making disbursements of funds or issuing any insurance. If a lien or other encumbrance for which the contractor is responsible is disclosed, an adequate amount of funds may be withheld.

E. Public Authorities

For the health, safety, and general protection of the public, the design professions and construction industry are regulated. Such regulations take many forms and differ somewhat among jurisdictions. Design professionals and some trades in the construction industry must be licensed and registered. Government agencies issue permits to authorize commencement of construction, and issue various certificates to allow occupancy. Government agencies also perform inspections of work in progress and require various tests to demonstrate compliance with minimum safety requirements.

The design and construction process is governed by many codes. For example, a zoning ordinance will dictate permitted uses of real property as well as various requirements that affect the design of a structure. The design of a project also must comply with applicable building codes, which may be comprised of separate codes for structural design, plumbing, electrical, HVAC, and fire protection. Finally, building codes and statutes, such as the Occupational Safety and Health Act, affect the manner in which construction work is performed.

2.06 ASSEMBLING THE PROJECT TEAM

The number of participants on any given project extends well beyond the owner, architect, and general contractor, the parties most closely associated with the project. Behind the scenes, a wide range of subcontractors, suppliers, engineers, and consultants, as well as other entities needed to support the design and construction process, perform various services or work that contribute to the completion of a project. It simply would not be possible to design and construct a project without the collective efforts of those many persons and entities.

Collective efforts, however, do not translate automatically into collaboration. The sheer number of participants in a project gives rise to an almost endless array of interrelationships to consider and coordinate. Each project participant likely has a contractual relationship with one or more other project participants, and may have separate functional relationships beyond those with whom they have contracts. The full extent of interrelationships between the project participants is vast.

Consider a relatively straightforward contractual structure for a small commercial building project. In this scenario, the owner undertakes development of a building for its own use and, perhaps, intends to lease some space to third parties. To accomplish its objectives, the owner retains an architect and a general contractor, enters into lease agreements with a tenant or two, and also endeavors to complete the project as needed to facilitate its own business operations and relationships. Figure 1 offers a simplified depiction of some of the contractual relationships that are reasonably certain to occur with such a project.

In reality, Figure 1 could highlight many more relationships that pertain to a project, and the chart could continue to extend outward from the owner to reveal other parties further removed from the project, but who nonetheless affect its progress or the economics of the overall transaction. Moreover, Figure 1 does not show functional relationships, such as the role of public authorities or dealings between parties that may not have a direct contractual relationship. Those relationships would be an overlay on top of the contractual relationships shown in Figure 1.

No single project relationship, whether that relationship is contractual or functional, stands alone or is isolated from the other project relationships. Every project participant is dependent upon the other project participants to some extent, even when a direct contractual relationship does not exist. In fact, the circumstances that affect one project participant can bear directly on the performance and risk assumed by other project participants, many of whom are several "links" away from each other or the owner in the organizational structure of the project team.

The contractual relationships depicted in Figure 1 show how one party may be adversely affected by the performance of those to whom it may be only indirectly related. For example, assume that a subcontractor performs its work late or in a defective manner. Other subcontractors whose work is scheduled to commence upon completion of the work that is behind schedule or requires corrective action will be delayed and run the risk that their work will not perform properly if connected to the defective work of another. Separate contractors with whom the owner enters into agreements also may be delayed. Although those various construction entities may have a functional relation-

FIGURE 1
Sample Project Contractual Relationships

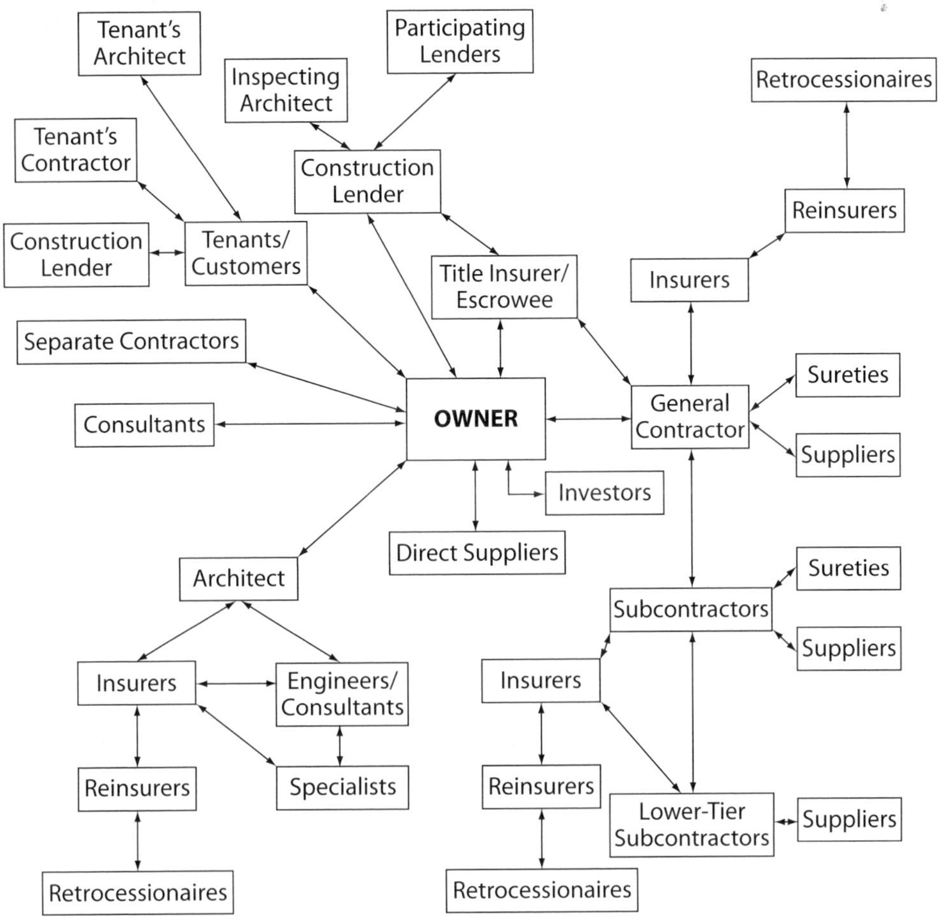

ship, such as a practical need to coordinate their work, there is no direct contractual relationship between them.

Issues also arise even when no apparent functional relationship exists between project participants. Assume that a project is a manufacturing facility intended to produce a particular product, and that a customer of the project owner demands a change in the design of that product or the supply of that product on an accelerated schedule. The stress on the relationship between the project owner and its customer will impact the project and will flow into the relationships more directly involved with the design and construction of the project. Changes in the project design may be needed, as may changes in the schedule and in the construction means and methods originally contemplated by the general contractor. The project cost may go up as well.

In sum, whenever stress is placed on any of the relationships that pertain to a project, that stress has the potential to resonate throughout all of the other project relationships. Those parties more closely linked to the relationship directly under stress are more likely to feel the adverse effects, but that is not always the case. Problems confronted by any one of the project participants easily can reverberate throughout the project team, resulting in defective or delayed performance, disputes, and claims.

Determining the entities that will participate in a project, and their respective roles, also is affected by the type of project. Construction is not limited to general commercial and residential building. There are many other types of projects and construction industry sectors, each of which has its own customs and practices.[40] The organizational chart in Figure 1 might look very different if the project were a highway or a manufacturing facility.

For example, it is unlikely that an architect would lead the design effort for a project that is of an industrial nature or whose principal purpose is not human occupancy. An architect may not be involved at all (e.g., an engineer might be the appropriate entity to lead the design team for an industrial project). Furthermore, given the importance of certain specialized pieces of equipment to the operation or performance of such projects, manufacturers and suppliers may have a more prominent role in the organization of the project team. Even the contracts used for such a project often do not bear much resemblance to those used in the general building sector.

Because every project presents different challenges, the team engaged to deliver a project must be assembled accordingly.

---

40. ENGINEERING NEWS-RECORD, a trade publication, periodically publishes a list of the largest contractors and design professionals, often identifying the percentage of business for each entity in nine different industry sectors, including general building, manufacturing, power, water supply, sewer/solid waste, industrial process/petroleum, transportation, hazardous waste, and telecommunications.

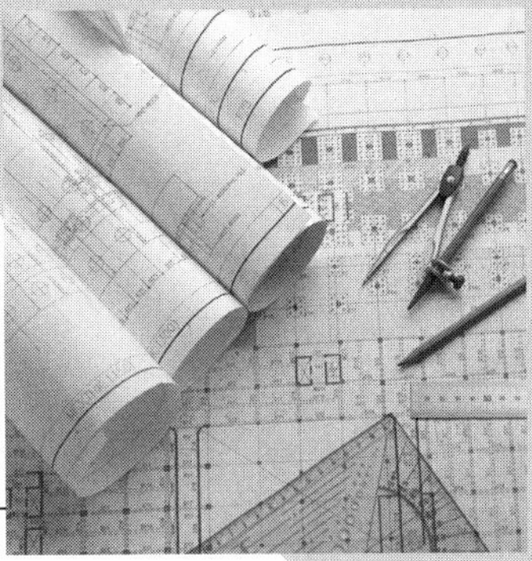

QUESTIONS

1. If you were counsel to an architect or contractor, what concerns might you have if the owner were a large institutional corporation as opposed to a small partnership?
2. What advantages might an owner have when funding construction out of its own funds as opposed to financing construction through an institutional lender?
3. What advantages and disadvantages might arise for an owner who retains a large architect with in-house engineering capability? What benefits might accrue to an owner who contracts separately with an architect and various engineers?
4. Describe how the failure or default of one project participant might adversely affect others involved in a project. Can any measure be taken to protect against such situations?

CHAPTER

3

The Owner's Role

LORENCE H. SLUTZKY AND DENNIS J. POWERS

3.01 OVERVIEW

To initiate a project, the owner must establish site and building requirements and identify potential project locations that fulfill those requirements. The owner must investigate the site's availability, access to the site for construction, any restrictions imposed by the title, environmental issues, and regulatory restrictions imposed by local and other governmental entities. This relevant site information must be provided to the architect and contractor. Commencement of construction is typically coordinated with the requirements of regulatory authorities and the issuance of the necessary permits. If the owner wishes to finance the project, the owner and the project must qualify for a construction loan. In conjunction with these activities, the owner also must consider, at least in the northern latitudes, the timing of the initial phases of construction that are affected by weather.

3.02 PROVIDING THE BUILDING SITE

Selection of a building site requires more than just an evaluation of the location for the building's ultimate use. Other factors must be reviewed and evaluated, such as access to the site for construction, soil conditions, zoning restrictions, the availability of utilities, environmental issues, the existence of easements over the property, and covenants that follow title that restrict what can be built. All of these factors must be considered in evaluating a building site and included in the design and construction equation.

A. Access to the Building Site

Prior to the owner issuing the notice to proceed to the contractor, access to the building site must be secured for the type of construction to be undertaken. Unless the responsibility to obtain access to the site is expressly delegated to the contractor, site access is the owner's responsibility. The owner owes to the contractor an implied obligation to provide unencumbered site access.[1] The obligation to provide site access remains in effect throughout the course of the contractor's performance, and can be breached even if the owner is not directly responsible for the contractor's inability to gain access to the project site.[2] If access to the work area is impeded by work of another contractor, the owner likely will be deemed to have breached its implied obligation—and in some jurisdictions, an implied warranty—to provide timely and adequate site access to the obstructed contractor.

Ensuring that a contractor has adequate site access may require a construction easement agreement with adjacent landowners that permit the use of the adjacent property for the limited purpose of constructing the building. A failure by an owner to procure the necessary rights of way and easements can lead to liability to the contractor for the delay if the absence of rights of way and easements prevent the contractor from accessing the site. For example, the construction may require an easement to allow the boom of a crane to operate over an adjacent property because title to property in most jurisdictions includes the air rights above the property. In other jurisdictions, a property owner's air rights only may extend to the space used by the property owner and, thus, such an easement will not be required.[3] Easements over adjoining property also may be needed for access to the construction site. Some jurisdictions provide a statutory mechanism for an owner to seek court approval to enter adjoining land for construction purposes.[4] Industry form contracts gen-

1. *See* Carter v. State, 259 Ga. App. 798, 578 S.E.2d 508 (2003).

2. Hoffman v. United States, 166 Ct. Cl. 39, 340 F.2d 645 (1964).

3. Geller v. Brownstone Condominium Ass'n, 82 Ill. App. 3d 334, 402 N.E.2d 807 (1985), adjoining property's air rights only extend to the space used by the adjoining property owner.

4. *See, e.g.*, Illinois' Entry on Adjoining Land to Accomplish Repair Act, 765 ILCS 125/0.01, and New York's Access to Adjoining Property to Make Improvements or Repairs Act, N.Y. Real Prop. Actions and Proceedings Law § 881.

erally require the owner to secure and pay for necessary approvals and easements required for construction.[5]

B. The Right to Use the Project Site for Its Intended Purpose

The owner must possess the unencumbered right to use the property for construction. Restrictions, such as rights of way, leases, easements, utilities, encroachments, or licenses that grant rights to others in the property, may interfere with construction operations. Before beginning construction, the owner must obtain a title report and survey to determine whether these restrictions exist and where they are located on the site. An owner will not be able to unilaterally relocate electric and phone lines or a sewer line that are the subject of enforceable easements. In most standard form agreements between the owner and architect, it is the owner's duty to provide a survey to the architect identifying the location of easements and utilities.[6]

The impact of the construction activities and the completed project on adjacent landowners also must be evaluated, as the project cannot adversely impact or damage adjacent property. If a court deems that an activity or condition that damages adjoining property is inherently dangerous, which construction work often is, the owner will be liable for that damage even if an independent contractor performing the work causes the damage.[7] Most states have adopted Section 839 of the Restatement (Second) of Torts, which will govern a claim by an adjacent or neighboring landowner for nuisances created by the construction process or the completed structure.

C. Title

Obviously, the owner must hold the necessary ownership interest in the project site to permit construction. The title to the property may contain restrictions on the use of the property and what may be constructed on the property. Unless the proposed use violates public policy, restrictions on the use of the property

---

5. *See* American Institute of Architects, AIA Document A201–2007, *General Conditions of the Contract for Construction* § 2.2.3 (2007) [hereinafter AIA Document A201–2007]. *But see* D. C. McClean, Inc. v. Arlington County, 249 Va. 131, 452 S.E.2d 659 (1995), contractor bore responsibility to obtain certain access easements.

6. *See, e.g.*, American Institute of Architects, AIA Document B101–2007, *Standard Form of Agreement Between Owner and Architect* § 5.4 (2007) [hereinafter AIA Document B101–2007] and American Institute of Architects, AIA Document B103–2007, *Standard Form of Agreement Between Owner and Architect for a Large or Complex Project* § 5.5 (2007) [hereinafter AIA Document B103–2007].

7. City of Columbus v. Barngrower, 250 Ga. App. 589, 552 S.E.2d 536 (2001).

and what may be constructed on a piece of property are generally enforceable.[8] These restrictions can include adherence to neighborhood or condominium association rules, which can be very extensive and indeed affect the manner, methods, and means of construction.

Construction may be undertaken by an entity that does not hold the fee simple interest in the property, but merely holds a long-term lease of the property, often referred to as a "ground lease." A "ground lease" grants the lessee use of the property for an extended period of time. Upon expiration of the term, the property, including any improvements constructed by the lessee, often reverts back to the owner of the property. Construction projects also are undertaken by lessees of property, such as tenants of office space that build out their own space. A tenant's construction must comport with the terms of the lease. Moreover, an owner will not be able to perform construction that will interfere with the quiet enjoyment of a valid leasehold interest in the property. A covenant of quiet enjoyment is implied in every lease agreement.

D. Zoning

Virtually all communities limit the type, size, and use of structures based on the location of the structure within the community. Zoning is an outgrowth of the tort of nuisance and is the general plan to control and direct the use and development of property in a municipality by dividing it into districts according to the present and potential use of the property.[9] Zoning may prevent high-rises from being built in single-family residential neighborhoods and the construction of heavy industrial buildings in retail areas.

A municipality has the power to divide the entire community into planned districts of such number, shape, and area, and of such different classes of use as deemed best suited for the growth, lifestyle, and existing uses in the community. Zoning districts generally consist of low-density residential, high-density residential, business or retail, light industry, and heavy industry. Zoning restrictions may be cumulative, where any use permitted in a more restrictive district may be permitted in a less restrictive district, or exclusionary, where all uses not explicitly permitted are excluded. Compliance with the applicable zoning law is of critical importance to the owner and the uninterrupted progress of the construction project.

---

8. Bowman v. Walnut Mountain Property Owners Ass'n, 251 Ga. App. 91, 553 S.E.2d 389 (2001).

9. Devaney v. New Haven Board of Zoning Appeals, 132 Conn. 537, 45 A.2d 828 (1946).

Most municipalities provide a mechanism for an owner to obtain approval to build a project that does not strictly comply with the existing zoning. Thus, an owner may seek a variance from the existing zoning to permit construction of a structure that otherwise would not comply with the current zoning. Variance request procedures often require public hearings to permit nearby property owners to weigh in on the variance request.

Absent an express contract provision governing who shall be responsible for ensuring compliance with zoning requirements, most jurisdictions impose a duty upon the architect to design a project to conform to zoning requirements. Industry form contracts require the owner to supply to the architect surveys that include zoning information.[10]

E. Availability of Utilities

The availability of utilities and the costs associated with bringing utilities to the site are important considerations in the project budget and schedule. A project in an undeveloped area is often referred to as a "greenfield" project, and may require substantial economic resources to bring electric, gas, water, and sewer lines, perhaps miles to the site. Similarly, although an industrial project may have utilities, they may be inadequate to service the planned building and use of the property and may require extensive upgrades. All of these factors must be considered in the economic evaluation and the scheduling of the project.

Unless the parties agree otherwise, the owner is responsible to provide utilities to the site for the completed building. The parties may contractually allocate responsibility for providing temporary power and heat for use during construction, and many industry form contracts allocate this responsibility to the contractor.[11] The AIA General Conditions of the Contract for Construction state that unless the contract documents provide otherwise, the contractor shall provide and pay for water, heat, and utilities.[12] If this responsibility is allocated to the owner, the owner's failure to timely provide utilities can create liability to the contractor if the absence of these utilities impacts construction. Industry form contracts require the owner to supply to the architect and contractor surveys identifying utility locations.[13]

---

10. AIA Document B101–2007, *supra* note 6, § 5.4 and AIA Document B103–2007, *supra* note 6, § 5.5.

11. AIA Document A201–2007, *supra* note 5, § 3.4.1.

12. *Id.*

13. AIA Document B101–2007, *supra* note 6, § 5.4; AIA Document B103–2007, *supra* note 6, § 5.5; AIA Document A201–2007, *supra* note 5, § 2.2.3.

F. Archaeological and Historic Sites

Discovery of Native American artifacts, human skeletal remains, earthworks, or other archaeological and paleontological artifacts on a site can have a significant impact on a project's progress and budget. Most jurisdictions require owners to notify the state archeologist or historic preservation agency of the discovery of archaeological significant objects. These jurisdictions also require the owner to cease construction activity to afford the appropriate state agency the opportunity to perform a survey of the site. Express agency approval is often required before resuming construction activities. For example, the State of Illinois reserves to itself the exclusive right to explore, excavate, or collect, through the Illinois Historic Preservation Agency, all archaeological and paleontological resources found upon or within any public lands and prohibits any person from knowingly disturbing any archeological or paleontological resource.[14] Alaska requires notification of the Department of Natural Resources if historic, prehistoric, or archeological sites, remains, or objects are discovered during construction. Approval by the Alaska Department of Natural Resources must be obtained before continuing construction.[15] Also, the Federal Antiquities Act provides for criminal penalties for any person who excavates or damages any archeological resources situated on lands owned or controlled by the United States.[16]

Form industry contracts require the contractor to immediately suspend operations if it encounters human remains or recognizes the existence of burial markers, archeological sites, or wetlands, and impose upon the owner the duty to obtain any necessary government authorization to resume construction operations. Form industry contracts further contemplate the contractor seeking adjustments in the project schedule and additional compensation due to the impact of such discoveries upon a project.[17]

The definition of a historic site in some jurisdictions can be quite broad. Historic sites can include any parcel or contiguous grouping of parcels of real estate under common or related ownership or control where any significant improvements are at least 40 years old, or any earthwork, village, burial ground, historic or prehistoric ruin, or other location that is or may be the source of important archeological data. A construction project that is deemed a historic site will impose impediments to the owner proceeding with development or alteration of the property.

---

14. 20 ILCS 3435/3.
15. AS 41.35.070.
16. 16 U.S.C. § 470ee.
17. AIA Document A201–2007, *supra* note 5, § 3.7.5.

The National Historic Preservation Act (NHPA) requires that the head of any federal agency having jurisdiction over a federally assisted undertaking take into account the effect of the undertaking on any site or object included in or eligible for inclusion in the National Register of Historic Places and provide the Advisory Counsel on Historic Preservation a reasonable opportunity to comment with regard to the undertaking.[18] Further, the National Environmental Policy Act contains provisions for evaluating the degree to which federal projects will adversely affect districts, sites, structures, or objects listed or eligible for listing in the National Register of Historic Places or that may cause loss or destruction of significant scientific, cultural, or historic resources.[19] Preconstruction archeological studies or environmental impact studies that assess the potential loss of historic resources—and, if necessary, mitigation plans to protect the historic resources—are common in areas that contain numerous historic sites.

G. Environmental Concerns

Construction projects can present a litany of environmental issues that will dramatically impact the progress of a project. Environmental issues arise in a variety of ways. They can range from contaminated soils discovered during new construction to the discovery of asbestos or lead-containing materials in a renovation project. There are a host of federal statutes and regulations dealing with environmental issues which, together with state regulation, create a great deal of regulation for an owner to negotiate.

Contaminated soils are governed by the Comprehensive Environmental Response Compensation and Liability Act of 1980 (CERCLA), as well as state regulation and, in some instances, municipal ordinances.[20] Industry form contracts permit a contractor to cease work if contaminated soils or other hazardous materials are encountered to permit the owner to test the materials and remove them or safely contain them.[21] These contracts also contemplate that the contract time will be extended and the contract sum increased for the contractor's reasonable costs of shutdown, delay, and start-up due to encountering hazardous materials.[22] Such site conditions, if known to the owner, must be disclosed to a contractor to avoid claims by the contractor for additional costs and additional time due to the contaminated nature of the site.

---

18. 16 U.S.C. § 470 and 36 C.F.R. § 800.4.
19. 40 C.F.R. § 1508.27(b)(3)(8).
20. Comprehensive Environmental Response, Compensation and Liability Act of 1980, 42 U.S.C. § 9601, et seq.
21. AIA Document A201–2007, *supra* note 5, § 10.3.1.
22. *Id.* § 10.3.2.

Also, an owner must be cognizant of the Clean Water Act and its wetlands protection provisions.[23] The Environmental Protection Agency (EPA) broadly defines a wetland, thus increasing the likelihood that an owner may be faced with wetlands issues. Special permits are required to work in areas that are deemed wetlands and special protections exist for wetlands. Further, the Clean Water Act imposes penalties and remedial requirements if construction debris makes its way into protected rivers and streams, and most states and municipalities require the control of sedimentation and its migration onto adjoining properties and into waterways.

Further, construction waste must be disposed of in compliance with federal and state regulations. If the waste is "hazardous," a series of federal regulations, including the Resource Conservation and Recovery Act of 1976 (RCRA), govern its transportation, handling, and disposal.[24] Even if the waste is not hazardous, there may be state statutes that govern the disposal of construction waste due to its unique nature.

If asbestos is encountered on a project, notification to the EPA may be required by federal emissions standards.[25] These standards also govern how the asbestos should be removed. Most states also regulate when and how asbestos is removed. Moreover, the federal Occupational Safety and Health Administration (OSHA), which sets forth safety standards and hazard communication standards for the construction industry, also governs how asbestos is to be handled. Failure to adhere to environmental and occupational safety laws governing asbestos can result in criminal prosecution.

3.03 PROVIDING INFORMATION TO THE ARCHITECT AND CONTRACTOR

A. Surveys

An owner's obligation to obtain and provide to the architect and contractor engineering information such as a survey of the site that shows the property lines, location of utilities, rights of way, easements, and location of other potential impediments to the construction process is generally dictated by the terms of the contract. The AIA form General Conditions of the Contract for Construction requires the owner to "furnish surveys describing physical characteristics, legal limitations and utility locations for the site of the Project, and a legal description of the site."[26]

23. Clean Water Act, 33 U.S.C. § 1311, et seq.

24. 42 U.S.C. § 6901, et seq.

25. 40 C.F.R. § 61.140.

26. AIA Document A201–2007, *supra* note 5, § 2.2.3.

Once the owner furnishes such information to the architect or contractor, the architect and contractor have the right to rely on that information. In *Cadral Corp. v. Solomon, Cordwell, Buenz & Assoc., Inc.*,[27] the owner provided the architect with a survey. The architect had, by letter to the owner, defined the requirements of the survey but had not expressly requested the inclusion of "restrictions." After excavation was complete and the caissons and some of the foundations were in place, a subsequent survey (required by the bank financing the construction loan) revealed that the west wall extended beyond the 10-foot building line. The owner filed a breach of contract and malpractice action against the architect, alleging that the architect ignored building-line restrictions. The court determined that sufficient evidence existed to support the jury conclusion that the survey provided by the owner, although depicting a line designating a "10 ft. building line," was insufficient to apprise the architects of critical information regarding the building line.

B. Geotechnical Information

The bearing capacity and nature of the site's soils are important considerations in the design of a building. A frequent source of owner-design professional disputes has been the discovery of unknown soil conditions and responsibility for the unanticipated additional costs for design and construction. Industry form design contracts transfer the responsibility to secure soil borings and analysis of the bearing capacity of the soils to the owner. The AIA Standard Form of Agreement Between Owner and Architect requires the owner to "furnish services of geotechnical engineers, which may include but are not limited to test borings, test pits, determinations of soil bearing values, percolation tests, evaluations of hazardous materials, seismic evaluation, . . . with written reports and appropriate recommendations."[28]

3.04 THE OWNER'S IMPLIED WARRANTY TO SUPPLY ACCURATE INFORMATION

A. The *Spearin* Doctrine

The owner impliedly warrants the accuracy of the information supplied to a contractor, including the information contained in the drawings and specifications, and sometimes site condition information. On a project in which the owner is supplying the drawings and specifications, the owner impliedly

27. 147 Ill. App. 3d 466, 497 N.E.2d 1285, 1287 (1st Dist. 1986).

28. AIA Document B103–2007, *supra* note 6, § 5.6 and AIA Document B101–2007, *supra* note 6, § 5.5.

warrants to the contractor the adequacy of the drawings and specifications for their intended purpose. This warranty is often referred to as the *Spearin* Doctrine because it arises from the U.S. Supreme Court opinion in *United States v. Spearin*.[29]

In *Spearin*, the contractor agreed to construct a dry dock for the United States. The work included the relocation of an existing sewer line. Flooding caused by a condition at the site damaged the relocated sewer line constructed by the contractor and the dry-dock worksite, and the United States requested the contractor to reconstruct the damaged sewer line at the contractor's cost. The Supreme Court ruled that the contract documents, which described the character, dimensions, and location of the sewer, but did not disclose the existence of the condition that caused the damage, created a warranty by the government that, if so constructed, the sewer would be adequate. Justice Brandeis's opinion states, in part:

> The general rules of law applicable to these facts are well settled. Where one agrees to do, for a fixed sum, a thing possible to be performed, he will not be excused or become entitled to additional compensation, because unforeseen difficulties are encountered. *Day v. United States*, 245 U.S. 159; *Phoenix Bridge Co. v. United States*, 211 U.S. 188. Thus one who undertakes to erect a structure upon a particular site, assumes ordinarily the risk of subsidence of the soil. *Simpson v. United States*, 172 U.S. 372; *Dermott v. Jones*, 2 Wall. 1. But if the contractor is bound to build according to plans and specifications prepared by the owner, the contractor will not be responsible for the consequences of defects in the plans and specifications. *MacKnight Flintic Stone Co. v. The Mayor*, 160 N.Y. 72; *Filbert v. Philadelphia*, 181 Pa. St. 530; *Bentley v. State*, 73 Wisconsin, 416. See *Sundstrom v. New York*, 213 N.Y. 68. This responsibility of the owner is not overcome by the usual clauses requiring builders to visit the site, to check the plans, and to inform themselves of the requirements of the work, as is shown by *Christie v. United States*, 237 U.S. 234; *Hollerbach v. United States*, 233 U.S. 165, and *United States v. Utah &c. Stage Co.*, 199 U.S. 414, 424, where it was held that the contractor should be relieved, if he was misled by erroneous statements in the specifications.
>
> . . .
>
> The contention of the Government that the present case is to be distinguished from the *Bentley Case, supra*, and other similar cases, on the ground that the contract with reference to the sewer is purely collateral, is clearly without merit. The risk of the existing system proving adequate might have rested upon Spearin, if the contract for the dry-dock had not contained the provision for

29. 248 U.S. 132, 39 S. Ct. 59 (1918).

> relocation of the 6-foot sewer. But the insertion of the articles prescribing the character, dimensions and location of the sewer imported a warranty that, if the specifications were complied with, the sewer would be adequate. This implied warranty is not overcome by the general clauses requiring the contractor, to examine the site, to check up the plans, and to assume responsibility for the work until completion and acceptance. The obligation to examine the site did not impose upon him the duty of making a diligent enquiry into the history of the locality with a view to determining, at his peril, whether the sewer specifically prescribed by the Government would prove adequate. The duty to check plans did not impose the obligation to pass upon their adequacy to accomplish the purpose in view. And the provision concerning contractor's responsibility, cannot be construed as abridging rights arising under specific provisions of the contract.[30]

The contractor was not required to repair the defective sewer at its expense because the specifications that dictated the character, dimensions, and location of the sewer constituted a warranty that the sewer was adequate for the existing conditions. The Court also ruled that the government's implied warranty superseded contract language that required the contractor to thoroughly check the drawings and examine the worksite, as those requirements do not require a contractor to verify that the improvements as designed by the government are adequate to accomplish their purpose.

Where the contractor performs the work according to specifications and the work is faulty due to defective specifications, the contractor may assert the *Spearin* Doctrine as a defense to a claim by the owner for the cost to remedy the defect. A contractor is entitled to rely on the drawings and specifications supplied by the owner and is not liable for defects in construction arising from defective drawings and specification.[31]

Contractors can use the *Spearin* Doctrine as a shield from claims by the owner and as a basis for a claim by the contractor against the owner if the contractor incurs loss and/or delay due to the deficiency of the drawings and specifications provided to the contractor by the owner.[32] These damages may include the costs incurred by the contractor in attempting to comply with the defective specifications of *Spearin* and the delay in the completion of the project.

---

30. *Spearin,* 248 U.S. at 136, 39 S. Ct. at 169.

31. *Spearin,* 248 U.S. at 136, 39 S. Ct. at 61; Martin K. Eby Constr. Co., Inc. v. Jacksonville Transp. Auth., 436 F. Supp. 2d 1276, 1308 (M.D. Fla. 2005). Pennsylvania Dep't. of Transp. v. Jones D. Morrisey, Inc., 682 A.2d 9 (Pa. Comm. Ct. 1996).

32. Big Chief Drilling Co. v. U.S., 26 Ct. Cl. 1276, 1992 U.S. Cl. Ct. Lexis 454 (1992).

B. Limitations on the Application of *Spearin*

In *Dugan & Meyer Constr. Co., Inc. v. Ohio Dep't of Admin. Services*,[33] litigation between a contractor and The Ohio State University ensued over delays arising from a $20 million contract for the construction of three buildings at the university. The contractor argued that under *Spearin*, the university breached its implied warranty of sufficiency of the drawings and specifications because the drawings and specifications were improperly prepared, causing the contractor delay damages. The university argued that the contract's no-damage-for-delay clause trumped the contractor's *Spearin* Doctrine claim for inaccurate and incomplete drawings. Although there was evidence of the inadequacy of the drawings and specifications, such as untimely owner responses to 700 requests for information, 250 field work orders, and 85 architectural supplemental instructions, the Ohio Supreme Court held that the contractor was not entitled to delay damages but was limited to the remedy specified in the contract—an extension of time. In discussing the application of the *Spearin* Doctrine to the contractor's claim, the court stated:

> . . . we decline the opportunity to extend the *Spearin* Doctrine from job-site-conditions cases to cases involving delay due to plan changes.
>
> Moreover, the contract in this case included terms that addressed the contractor's remedy when changes were made to the plans. This court has long recognized that "where a contract is plain and unambiguous, it does not become ambiguous by reason of the fact that in its operation it will work a hardship upon one of the parties thereto and a corresponding advantage to the other, [and] that it is not the province of courts to relieve parties of improvident contracts." *Ohio Crane Co. v. Hicks*, 110 Ohio St. 168, 143, N.E. 388 (1924). In addition, "unless there is fraud or other unlawfulness involved, courts are powerless to save a competent person from the effects of his own voluntary agreement." *Ullmann v. May*, 147 Ohio St. 468, 72 N.E. 63 (1947).
>
> We have previously affirmed these principles in a case involving a competitively bid public construction contract. In *S & M Constructors, Inc. v. Columbus*, 70 Ohio St. 2d 69, 434 N.E. 2d 1349 (1982), the contractor agreed that it would make no claim against the city even if the conditions of the subsurface as reported to contractors before the bidding differed materially from actual subsurface conditions encountered during the project. We held that the "no claim" provision was unambiguous and was enforceable in the absence of a showing of fraud or bad faith on the part of the city. We observed that the *Spearin* Doctrine does not invalidate an express contractual provision: "Where one agrees to do, for a fixed sum, a thing possible to be performed, he will not

33. 113 Ohio St. 3d 226, 864 N.E. 2d 68 (2007).

> be excused or become entitled to additional compensation because unforeseen difficulties are encountered." [Citations omitted]
>
> The contract in the case at bar contained several relevant provisions, which were valid under Ohio law when the contract was signed. The parties agreed, "Time is of the essence to the Contract Documents and all obligations thereunder," and the contract provided a specific procedure to be followed in the event of project delay. The contract also contained a no-damages-for-delay clause, which provided that "extension of time granted pursuant to paragraph GC 6.2 shall be the sole remedy which may be provided by the Department" and that the Contractor shall not "be entitled to additional compensation or mitigation of Liquidated Damages for any delay listed in paragraph GC 6.2." The contract provided that the contractor's failure to request, in writing, an extension of time within ten days after the occurrence of a condition necessitating an extension of time "shall constitute a waiver . . . of any claim for extension or for mitigation of liquidated Damages."[34]

The *Spearin* Doctrine has been adopted by most jurisdictions,[35] but some states, such as Pennsylvania, have refused to apply it. For example, in *Stabler Constr., Inc. v. Comm. of Pennsylvania,*[36] the court stated, "*Spearin* involved a contract with an agency of the federal government, and the *Spearin* court determined the rights and obligations of the parties to that contract by applying federal law. The instant case does not involve or require the application of federal law. Therefore, *Spearin* is not binding on this court."[37]

1. Limitations on the Spearin *Doctrine*

The contractor only will have the benefit of the *Spearin* Doctrine if it performs in accordance with the drawings and specifications furnished by the owner.[38] Thus, if the owner can establish that the contractor failed to follow the owner's drawings and specifications even if they are defective, the owner is insulated from any claims brought by the contractor with respect to the sufficiency of the drawings and specifications.[39]

---

34. *Dugan & Meyer Constr. Co., Inc.*, 113 Ohio St. 3d at 231, 864 N.E. 2d at 73–74.

35. Fidelity and Deposit Co. v. City of Sheboygan Falls, 713 F.2d 1261 (7th Cir. 1983); *Martin K. Eby Constr. Co., Inc.*, 436 F. Supp. 2d at 127.

36. 692 A.2d 1150, 1153 (Pa. 1997).

37. *Id.* at 1153; *see also* Rhone Poulenc Rorer Pharm., Inc. v. Newman Glass Works, 112 F.3d 695 (3d Cir. 1997).

38. Appeal of Conrad Bros., Inc., 1996 WL 675811 (B.C.A. 1996), *aff'd* 135 F.3d 778 (Fed. Cir. 1998).

39. *See* Tyger Constr. Co. v. United States, 31 Fed. Cl. 177 (1994); McConnell v. Corona City Water Co., 149 Cal. 60, 85 P. 929 (1906); Montrose Contracting Co. v. County of Westchester, 80 F.2d 841 (2d Cir. 1936).

Under the *Spearin* Doctrine, the owner may escape liability to the contractor for defective drawings and specifications if the design defects are "glaring or obvious" or reasonably could have been discovered by the contractor.[40] A contractor has a duty to discover defects in the drawings or specifications that are reasonably discoverable or patent and to warn the owner of the defects.[41] This doctrine applies even where an owner supplies comprehensive design specifications because a contractor is not justified in "blithely proceeding with its work in the face of obvious and recognized errors"; rather, the contractor has a duty either to take appropriate steps to address obvious design defects or demand that the owner take action.[42]

Further, a contractor has a duty to read and understand all of the project documents and to inform itself of the owner's representations concerning the work. Such an inquiry should cause the contractor to discover obvious or potential discrepancies in the contract documents. Discovery of a patent discrepancy triggers a duty on the contractor to inquire.[43] Indeed, a contractor who is on notice of an incipient problem but neglects to solve it or inquire of the owner as is required by most contracts, cannot rely on the principle that ambiguities in contracts are interpreted against the drafter. The contractor is under an affirmative duty to call attention to the inconsistency. If the contractor fails to call it to the attention of the owner, it may be construed against the contractor, which often results in an obligation to perform work that was reasonably necessary to complete the project.[44] It is also important to note that industry form contracts often require the contractor to advise the architect or owner of any known defects in the contract documents.[45]

For example, in *Caddell Constr. Co., Inc. v. United States*,[46] the contractor argued that the design specifications were so defective as to constitute a breach

---

40. PCL Constr. Services., Inc. v. United States, 47 Fed. Cl. 745, 785 (2000).

41. *Id.*; St. Paul Fire & Marine Ins. Co. v. Pearson Constr. Co., 547 N.E.2d 853, 858 (Ind. 4th Dist. 1989).

42. Allied Contractors, Inc. v. United States, 381 F.2d 995, 999 (Ct. Cl. 1967); *see also* Graham Constr. Co., Inc. v. Earl, 362 Ark. 220, 2005 WL 1041149, at *3 (2005), "A competent and experienced contractor cannot rely upon submitted specifications and plans where he is fully aware, or should have been aware, that the plans and specifications cannot produce the proposed results." (*citing* Housing Auth. of the City of Texarkana v. Johnson Constr. Co., 264 Ark. 523, 573 S.W.2d 316 (1978)); *PCL Constr. Services, Inc.*, 47 Fed. Cl. at 794; Mann v. Clowser, 59 S.E.2d 78, 85 (Va. 1950).

43. *See* Gardner-Zenkes Co. v. State, 109 N.E.2d 729, 790 P.2d 1010 (1990).

44. Beacon Constr. Co. v. United States, 314 F.2d 501 (1963).

45. AIA Document A201–2007, *supra* note 5, § 3.2.2.

46. 78 Fed. Cl. 406.

by the United States of the *Spearin* implied warranty. For evidence of the defective specifications, the contractor relied primarily on the almost 300 requests for information that it was required to send to the government. The Federal Court of Claims rejected this argument, reasoning that design specifications need not be paragons of perfection, but only reasonably accurate, and the fact that the drawings required repeated clarification was not an indication that the drawings were defective. Although the court agreed that the drawings contained errors and inconsistencies, it found that they were workable and, therefore, not defective. The court in *Caddell Construction* restricted application of the *Spearin* Doctrine to a fundamental design defect that dramatically impacted the project, as opposed to a number of smaller defects, even if cumulatively they dramatically impacted the project. Nor does the existence of a contractual completion date constitute a warranty by the owner that the project can be completed within that time period.[47]

2. *The* Spearin *Warranty Can Be Disclaimed*

The owner's implied *Spearin* warranty can be disclaimed by a properly drafted specific disclaimer of the adequacy of the drawings and specifications.[48] Courts also have held that a contractor's express warranty will take precedence over the owner's *Spearin* Doctrine implied warranty. In *Rhone Poulenc Rorer Pharmaceuticals Inc. v. Spectrum Glass Products, Inc.*,[49] the court held that the *Spearin* Doctrine's implied warranty from the owner must yield to the contractor's express warranty that all work will be free from faults and defects and the contractor's express duty to remove and replace any materials that the owner or its architect deem defective.[50] There, the court reasoned that the parties had contractually allocated to the contractor any risk that the glass specified by the owner would not be appropriate for its application.

---

47. American Ship Building Co. v. United States, 654, F.2d 75 (Ct. Cl. 1981).

48. A contract between the owner and contractor that provides "omissions from the drawings or specifications or the misdescription of details or work which are necessary to carry out the intent of the drawings and specifications shall not relieve the contractor from performing such omitted or misdescribed details of the work, but shall be performed as if fully and correctly set forth and described in the drawings and specifications" may insulate an owner from a contractor's *Spearin* Doctrine claim for defective plans and specifications. Connor Bros. Constr. Co. v. United States, 65 Fed. Cl. 657, 669 (Fed. Cir. 2005); R. B. Wright Constr. Co. Through Rembrant, Inc. v. United States, 919 F.2d 1569 (Fed. Cir. 1990).

49. 112 F.3d 695; see further discussion at Chapter 22.

50. However, a contract provision that merely provides that a contractor has inspected the project site is not specific enough to disclaim the owner's *Spearin* implied warranty. *See* Green Constr. Co. v. Kansas Power & Light Co., 1 F.3d 1005 (10th Cir. 1993).

3.05 THE OWNER'S PROGRAM

The owner is responsible for preparing what is commonly referred to as a program. The program is a written narrative and graphic document that summarizes major design objectives and constraints for the project, which might include matters such as price, aesthetic preferences, development schedules, budget constraints, physical and site limitations, space requirements and relationships, functional issues, maintenance and operating requirements, technological concerns (material, equipment and system needs), energy requirements, financing issues, safety issues, and other matters pertinent to the scope of the work that is to be designed.[51]

Architectural programming is performed prior to preparation of any drawings and specifications. In fact, the program is the basis for the design of the project and, therefore, is an extremely important document. The programming process warrants diligence, as the ability to influence the design of a project is greatest during architectural programming. Before ideas are committed to drawings and specifications, the design is nothing more than a collection of concepts and ideas at best. All aspects of the design are still flexible, and changes can be accomplished with relative ease and efficiency. As the design becomes more developed, however, making any change is often challenging to execute. One change may also require other changes, causing additional design fees and schedule delays.

Under most standard industry contract forms, the owner is responsible for furnishing the architect with a written program.[52] The architect, in turn, reviews the program and provides a preliminary evaluation of the program. Only after the architect has reached an understanding of the project requirements, as embodied in the program, does the architect proceed with the preparation of a preliminary design illustrating the scale and relationship of the project components.[53] The design continues to evolve in greater detail and level of completeness in each successive design phase, but always based on the program. The architect relies on the design objectives set forth in the program, and the architect is entitled to additional compensation if the owner requests a deviation from the direction given in the requirements set forth in the program.[54]

---

51. For a general description of programming, *see* HANDBOOK OF PROFESSIONAL PRACTICE 519–25 (The American Institute of Architects 2003).

52. AIA Document B101–2007, *supra* note 6, § 5.1, describes the items that are required to be set forth in a written program the owner must provide to the architect.

53. *Id.* §§ 3.2.3 and 3.2.4.

54. *Id.* § 4.3.1, which provides, in pertinent part, that changes in many items that constitute information provided in the program are considered "Additional Services."

Preparing a program can be a complicated process, depending on the nature of the project. Given the importance of the program in determining the obligations of the architect and the criteria from which the project design is derived, it is somewhat surprising that this task is typically delegated to the owner in agreements between owners and architects. Many owners do not possess the skill to effectively carry out this responsibility on their own. As a consequence, architects often provide programming services as part of so-called "additional services" for which additional compensation is received.[55] Because architectural programming is closely related to architectural design, it is considered a specialty; therefore, sometimes professionals other than architects are retained to provide such services.[56]

Two of the more important elements established in the owner's program are the project budget and schedule. Both must be periodically updated to reflect changes in the scope of the project. The AIA Owner-Architect Agreement requires the owner to "establish and periodically update the Owner's budget for the Project, including (1) the budget for the Cost of the Work . . . ; (2) the Owner's other costs; and, (3) reasonable contingencies related to all of these costs."[57] This argument also requires the owner to notify the architect if the owner significantly increases or decreases the owner's budget for the work, and provides that the owner and architect shall agree to a corresponding change in the project's scope and quality.[58]

The importance of price and time is relative. For the established manufacturer, getting the plant operational is of paramount importance to achieve or maintain market share and profit margins. In this situation, price may be secondary. For the economically challenged owner, the cost of the project may be of paramount importance. To the public entity that must issue publicly approved bonds, price is also the primary concern, but accurate estimation of the cost of construction is indispensable to adequately finance the project.

The project schedule can be an important and crucial tool for the management of the project. If the contract between the owner and contractor requires compliance with the project schedule, the owner will be able to enforce the schedule and its interim milestones and seek damages for the contractor's failure to adhere to the schedule.[59] Industry form contracts generally require the contractor to prepare and submit a schedule to the owner and architect

---

55. *Carter*, 259 Ga. App. 798, 578 S.E.2d at 519.

56. *Id.* at 520.

57. AIA Document B101–2007, *supra* note 6, § 5.2 and AIA Document B103–2007, *supra* note 6, § 5.3.

58. AIA Document B101–2007, *supra* note 6, § 5.2 and AIA Document B103–2007, *supra* note 6, § 5.3.

59. AIA Document A201–2007, *supra* note 5, § 3.10.1.

consistent with the time limits imposed by the contract. The contractor also is usually required to periodically revise the schedule as required by the project conditions.[60] Because construction rarely proceeds as originally planned, the contractor's duty to periodically update the schedule is important for the success of the project in order to keep all of the parties advised of the current timeline for certain activities. Periodic updates are necessary to coordinate all of the subcontractors' and suppliers' activities.

3.06 PAYING FOR CONSTRUCTION

The owner is responsible for paying for the project, which, in turn, is often governed by the financing that the owner can obtain.

A. Financing the Project

A significant percentage of major construction projects utilize financing, usually in the form of a construction loan. Indeed, the viability of a project often hinges on the owner's ability to secure financing. Upon completion of construction, the construction loan is repaid with funds from a permanent loan or other sources that may be available to the owner. The construction loan agreement will require a security interest in favor of the lender in the project that gives priority to the construction lender over other creditors of the owner.

Because many private construction projects are owned by an entity created for the sole purpose of constructing and owning the building, the building often will be the only asset of the borrower. However, such a lender's contractual priority security interest may not ensure priority over all potential claimants, due to statutory priorities that may exist. For example, contractors' mechanic's lien rights may in some instances have priority over a lender's rights, at least to the extent the contractor has enhanced the value of the property, and taxing bodies such as municipalities may have priority for taxes due. The lender will pay particular attention to attempting to obtain priority over contractors' lien rights.

Some lenders may require an owner to obtain agreements from contractors to subordinate their lien rights to the lender's mortgage. A request to a contractor to agree to subordinate its mechanic's lien rights to the lender's mortgage is almost always met with resistance from the contractor, as a contractor's mechanic's lien right is powerful protection. To be truly effective, such subordination must be obtained from all of the subcontractors.

A lender also will seek to ensure that the construction loan proceeds are used only for their intended purpose—to pay the contractors. Payment made

60. *Id.* § 3.10.3 requires the contractor to perform the work in general accordance with the most recent schedules submitted to the owner and architect.

by the owner to the contractor may be conditioned upon receipt of the contractor's payment application, certified by an independent architect or engineer engaged at the owner's expense to verify that the percentage complete represented by the contractor is accurate. The lender also may require that the certification attest that changes to the contract amount do not cause the balance due to the contractor to exceed the loan balance. If the balance of the loan is exceeded by the current balance owed to the contractor, the owner may be obligated to make up the difference with its funds. The objective is to assure that the construction loan is always "in balance," with the available loan proceeds equal to or more than the amount needed to complete construction.

The construction loan documents also may require that lien waivers for a particular payment be tendered by all contractors working on the project. To ensure that this procedure is followed, lenders may utilize a title company or other escrowee to collect lien waivers before disbursing the loan proceeds.

The lender also may require input into the terms of the construction contract and often will require modifications to the construction contract to protect its interests. As a term of the construction loan, lenders commonly require a collateral assignment of the construction contract from the owner to the lender. Under this assignment, if the lender declares a default of the construction loan and is required to foreclose, or obtains title to the project by agreement with the owner, under the collateral assignment, the lender will step into the shoes of the owner if the lender chooses to do so to permit it to complete construction with the existing contractor.

The construction lender also may require that the contractor provide performance and payment bonds. These bonds reduce the risk that the project will not be completed or that liens will impair title to the project. A performance bond gives the lender assurances that should the contractor default, file bankruptcy, or materially breach the contract, funds will be available to complete the project. A payment bond also provides assurances to the lender that a source of funds exists to defend and pay lien claimants if the general contractor fails to do so. For a more complete discussion of performance and payment bonds, see Chapter 20.

B. Evidence of Financing

Prior to commencement of the work, a contractor generally is entitled to evidence that the owner has made financial arrangements to pay for the project.[61] If such language exists in the contract, submission of evidence of adequate funds or financing is a condition precedent to the contractor commencing or continuing construction.

---

61. AIA Document A201–2007, *supra* note 5, § 2.2.1.

The ability to fund a project is of obvious importance. Contractors do not wish to expend resources and assume liabilities to third parties without a strong prospect of receiving payment from the owner. Although a contractor usually has lien rights as security for payment for work performed, such rights may be subordinate to those of others and inadequate alone to protect the contractor's interests. Therefore, the need to provide reasonable assurance of the ability to pay is not particularly controversial, at least prior to the commencement of construction. The parties can discuss the concerns of the contractor, and the contractor need not commence work unless it is satisfied. The requirement to provide evidence of financial arrangements after commencement of construction, however, is where most controversies occur. Issues such as under what circumstances the owner must provide evidence beyond that previously furnished, and what constitutes evidence adequate to discharge the owner's obligations can prove to be menacing.

The AIA documents attempt to clarify some of those issues by providing that the contractor may demand evidence of financial arrangements after commencement of the work when "the owner has failed to make payments to the Contractor as the Contract Documents require or if a change in the Work materially changes the Contract Sum."[62] Although such language is helpful, it does not resolve all potential disputes on the topic from the perspective of either party and may create additional interpretation issues with respect to good faith payment disputes and standards of materiality. The ability of the owner to understand and to satisfy its obligations regarding the supply of evidence is important, as the contractor often has the right to terminate the contract for cause if such evidence is not promptly forthcoming.[63]

What constitutes reasonable evidence of financial arrangements remains a question of fact that must be evaluated under the circumstances. Acceptable evidence has been suggested to include a construction loan commitment, a good credit report, a corporate resolution or government appropriation, or other similar documentation that the owner has the financial ability to pay for the work.[64] Neither party, of course, wishes to argue about the adequacy of evidence of financial arrangements after the project has commenced, but the prospect for such disputes exists unless agreement is reached upfront as to what evidence is acceptable. The owner may have additional concerns regarding the confidentiality of such evidence once it is furnished.

62. *Id.* §2.2.1.

63. *Id.* §14.1.1.4.

64. *See* 2 Philip L. Bruner & Patrick J. O'Connor, Bruner & O'Connor on Construction Law, §5.39 (2002).

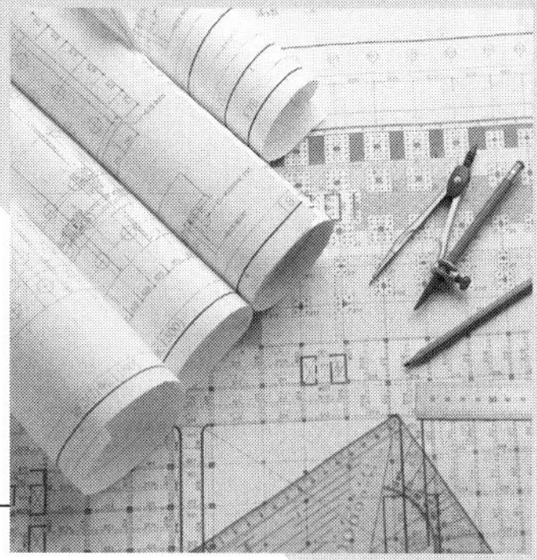

QUESTIONS

1. A contractor incurs additional costs and delays due to its inability to access the worksite as it had planned because another contractor performing work for the owner has not completed its work. Even if the owner's conduct is not the cause of the second contractor's delay in completing its work, who bears responsibility for the first contractor's additional costs due to its inability to perform the work as it originally planned? How does the analysis change if the contract between the owner and contractor provides that a contractor is not entitled to damages for project delays, but is only entitled to extensions of time?
2. Even though the contractor performs the work in compliance with the design drawings and specifications, the work proves to be faulty, and the owner seeks compensation from the contractor. What defenses does the contractor have to the owner's claim for defective work? What recourse does a contractor have if, due to a series of errors and inadequacies in the drawings and specifications that require numerous clarifications and corrections, a contractor incurs additional costs due to the extended duration of the project?
3. What factors and site conditions should be considered and evaluated in selecting a site for construction? Once construction begins, what site conditions must an owner and contractor be cognizant of due to federal, state, and municipal regulation?

CHAPTER

4

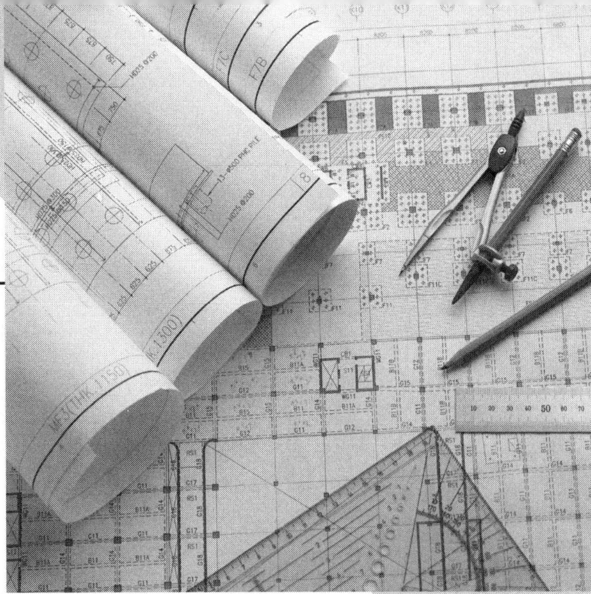

Project Delivery Systems

ROSS J. ALTMAN

4.01 OVERVIEW

A wide range of physical, human, financial, and other resources must be marshaled and coordinated to design, procure, and construct a project. Performance of the tasks required to deliver even a simple project often demand multiple design professionals, consultants, specialized contractors, and product and system vendors. The larger or more complex a project, the greater the number of persons and entities engaged in the delivery of that project. If performed as intended, the design, procurement, and construction of a project is a collaborative exercise. Predictably, however, many challenges arise when attempting to orchestrate efficient interaction among so many project participants.

The owner determines how to organize the process through which the project is designed, procured, and constructed. Although many issues must be considered, the structure of the method used to manage the overall process of design, procurement, and construction is driven principally by two factors. First, how are the responsibilities for design, procurement, and construction allocated among the project participants? Second, what is the sequence of major activities that comprise design, procurement, and construction? The answers to those questions largely define the organizational structure of the design, procurement, and construction process and describe the fundamental elements of what is referred to as the "project delivery system."

4.02 EVOLUTION OF DIFFERENT APPROACHES

Methods of building construction and technology available prior to the mid-1800s were rather limited by today's standards. A master mason or carpenter, sometimes described as a master builder, was commonly appointed in charge of the entire project. The cathedrals of Renaissance Europe and many other significant structures that fill the pages of architectural history books were delivered by such a methodology.[1] Subsequent advances in technology and an increase in project complexity, however, produced a need for design specialization. Architecture emerged as a profession, and the design function was severed from the performance of construction. The Design-Bid-Construct project delivery system (described in Section 4.04) was fashioned to capitalize on this bifurcation of design and construction skills.

Throughout most of the 20th century, Design-Bid-Construct remained the most popular method of delivering a project in the United States. The public sector continues to rely heavily upon Design-Bid-Construct, due in part to requirements dictated by procurement statutes. The private sector is not bound by such limitations, however, and is free to experiment with and develop project delivery systems that respond to the particular desires of an owner. Because the objectives of owners differ and are subject to market and other forces that change with time, new methods to deliver a project continue to evolve. No single project delivery system is suitable for every project, and each project delivery system has its own advantages and disadvantages.

Unfortunately, the terminology used to define project delivery systems and identify various approaches is neither precise nor applied consistently. Some terms are used interchangeably, while others are used to describe a variety of different project delivery systems.[2] Certain qualities or features are associated

---

1. *See* Frances Gies & Joseph Gies, Cathedral, Forge and Waterwheel (Harper Collins 1994); Richard A. Goldthwaite, The Building of Renaissance Florence (John Hopkins University Press 1980); and Ross King, Brunelleschi's Dome (Walker & Company 2000).

2. The terms "Design-Build," "Turnkey," and "Build-to-Suit" often are used interchangeably. Although differences probably are intended by each term, the distinctions identified by those in the design professions and construction industry are not consistent. At the other end of the spectrum, a wide range of project delivery systems falls under the umbrella of "Construction Management."

sometimes with a particular project delivery system, yet there is no uniform nomenclature. Even among those in the design professions and construction industry, different criteria are used to define project delivery systems. Indeed, the same project delivery system may go by different labels, depending on who makes the reference.[3]

Therefore, when referring to a project delivery system by name or descriptive phrase, the contracting parties cannot assume that each party ascribes the same meaning to such terminology. For contracting parties to reach a mutual understanding of what is contemplated by such references, they must distinguish one project delivery system from another. Because the terminology surrounding project delivery systems is not uniform, it is often necessary to describe the underlying and defining characteristics of a particular project delivery system to achieve that understanding.

Some in the design professions and construction industry distinguish further between project delivery systems and management techniques. According to those who espouse this distinction, a project delivery system is considered a method to allocate responsibilities with respect to design, procurement, and construction, while a management technique is viewed as a device or method used to coordinate the process of design, procurement, and construction.[4] The distinction is interesting, but the terminology commonly adopted to describe project delivery systems is more fluid and simply does not always recognize such differences. Many arrangements that might be defined more precisely as a management technique are often referred to as a project delivery system.[5]

Identifying the defining characteristics of the project delivery system is a bit discretionary, and many conclusions are possible. There are almost endless variations of project delivery systems, along with the opportunity to customize

---

3. At-Risk Construction Management is also identified as CMeR, CMc, and CM/GC.

4. For example, planning, staffing, budgeting, scheduling, and monitoring programs are all methods used to manage the process of design, procurement, and construction. Such methods might be used in connection with any type of project delivery system.

5. Program management (in many ways a surrogate for the owner's staff, used to analyze facilities, develop programmatic requirements, and monitor overall project development for an owner) and partnering (an effort to promote cooperation among all project participants through education, team-building exercises, and adoption of a "charter" agreement) are examples of management techniques that occasionally are referred to as types of project delivery systems. Agency Construction Management (see Section 4.05(A) in this chapter), which is described routinely as a project delivery system, perhaps more accurately is a management technique applied to the Multiple Prime project delivery system.

such methodology.[6] Nonetheless, when examining the defining characteristics, most project delivery systems can be organized into four general categories:

- *Design-Bid-Construct.* Defining characteristics include: (1) three principal participants (owner, architect, and contractor); (2) two separate prime contracts (owner-architect agreement and construction contract); and (3) generally linear sequence of design, procurement, and construction activities.
- *Construction Management.* Defining characteristics include: (1) three principal participants (owner, architect, and construction manager); (2) two separate prime contracts (owner-architect agreement, construction management agreement); and (3) potentially overlapping design, procurement, and construction activities.
- *Design-Build.* Defining characteristics include: (1) two principal participants (owner and design-build entity); (2) one prime contract between the owner and design-build entity; and (3) generally overlapping design, procurement, and construction activities.
- *Innovative Approaches.* Certain highly customized and other novel project delivery systems emerge from time to time. Defining characteristics are not possible to identify in advance or across the board with respect to all such approaches. Many of these methods are viewed by some as management techniques or procurement methods, as opposed to project delivery systems. As a general proposition, however, innovative approaches that have gained recent popularity usually emphasize forms of collaboration among the principal participants. Many such approaches are intended to align interests through a risk share/gain share arrangement that does not fall neatly into any of these other three categories.

Each of these four general categories of project delivery systems is described and examined in greater detail in Sections 4.04 through 4.07.

---

6. A research report by the Construction Industry Institute identifies at least 10 principal project delivery system categories, including "traditional Design-Bid-Build, traditional Design-Bid-Build with early procurement, Project Management (agent), Construction Management (agent) with early procurement, Construction Management (at risk), Design-Build or Engineer Procure Construct, multiple Design-Build or Engineer Procure Construct, multiple Prime Contractors, staged Design-Bid-Build, and Turnkey." *Project Delivery and Contract Strategies*, Construction Industry Institute, RR165-11, December 1, 2003. For an overview of different project delivery systems, see a paper by Allen Overcash titled *Project Delivery Methods: Modern Construction Problems*. Allen L. Overcash, Presentation of Paper, *Project Delivery Methods: Modern Construction Problems*, American Bar Association Forum on the Construction Industry, in Toronto, Ontario, Canada (September 29–30, 2005).

4.03 IMPORTANCE OF PROJECT DELIVERY SYSTEMS

The project delivery system dictates the overall legal and commercial framework of the transaction, including risk allocation, obligations of the parties, and sequence of design, procurement, and construction activities. The contractual relationships among the project participants, and form of contracts used, are also a direct consequence of the delivery system applicable to the project.[7]

Some forms of project delivery facilitate use of "fast track" construction, in which construction commences before completion of the entire design. When executed successfully, a fast track or "phased" construction project is completed earlier than if construction were to commence only after completion of the entire design. If completion of construction is accelerated, the project can generate revenue for the owner sooner, and construction financing debt service is reduced. Fast track construction, however, limits the owner's ability to make design changes, increases the likelihood of design problems, and requires a difficult coordination effort. Fast track construction also creates pricing uncertainty, with a so-called guaranteed maximum price proposed before the design is final. In theory, however, the financial benefits of early completion outweigh any additional cost to complete the project that may arise out of overlapping design, procurement, and construction.

The mere quantity of project delivery systems, and the permutations of each, is potentially confusing. However, so many options exist because every project presents its own unique set of challenges. Each project delivery system offers a transaction structure that is better suited for a particular assortment of challenges than for others. Although a list of issues that concern an owner in a typical project is almost endless, some of the more common items include the following:

- *Cost.* Most owners hope to complete the project for the lowest price possible. Some owners, however, are willing to pay a higher price if, in exchange for an incremental increase in cost, the owner will receive an appropriate increase in value (e.g., lower cost of operating and maintaining the completed project, longer useful life of materials, equipment, and systems). Cost predictability is also important. How soon will the owner know the likely final cost to complete the project?

7. *See* Shively v. Belleville Township High Sch. Dist. No. 201, 329 Ill. App. 3d 1156, 769 N.E.2d 1062 (2002). In *Shively*, the parties disagreed as to the type of project delivery system that had been employed on an already completed project. The *Shively* court analyzed the contract between the parties and concluded that many features of the contract were consistent with an Agency Construction Management project delivery system.

How reliable is that price? What is the need for price contingencies to cover the likelihood of unforeseen conditions or cost volatility in the marketplace? Some owners prefer to transfer risk to the other party and pay a premium, if that risk transfer reduces the possibility of future claims for cost adjustments. Other owners are willing to absorb some risk and the price uncertainty that comes with acceptance of risk, when the trade-off provides a reasonable opportunity to achieve a lower final contract price.

- *Time.* The ability to complete a project on time or ahead of a specified schedule is generally important to owners. Aside from the owner's desire to have beneficial use and occupancy of a project, an ongoing project absorbs an owner's resources. Perhaps more importantly, the adage "time is money" holds true for most owners. Earlier completion usually translates into less debt service. For owners who finance the cost of construction, this can mean a huge savings. The earlier a project is completed, the sooner the project is generating revenue or replacing less efficient facilities. Completion schedules also may represent far more than dates by which to realize desired goals. An owner may have firm obligations to third parties that are discharged only upon completion of the project, such as a commitment to furnish a prospective tenant with a completed space by a date specified in a lease. The owner may be in a seasonal business, where failure to open as of a certain date results in significant loss of sales. Business cycles also may drive a project. If a project will be used to produce a product with a limited market advantage or useful life, such as a pharmaceutical plant or chip factory, then any delay in completion has very damaging ramifications.
- *Design Quality.* Of course, the owner wants an error-free design, with drawings and specifications that are complete enough to avoid disputes and to produce reliable pricing. The owner also wants its aesthetic and functional objectives satisfied by the design. The technical requirements and performance criteria also should be met. Some owners want the ability to exercise an active role in the design. The owner may intend to develop a distinctive, "signature" quality project, in which the physical appearance of the project is identified with, or is an emblem for, the owner. In such circumstances, aesthetic issues are paramount.[8]

8. Although the Transamerica Pyramid in San Francisco, California, is no longer the headquarters of the Transamerica Corporation, the building is depicted in the logo of the company and its distinctive shape remains associated with Transamerica. A more novel example is the headquarters for basket maker Longaberger Company in Newark, Ohio, which looks like a seven-story replica of a woven wood basket.

- *Reduction in Claims*. The design and construction process can be contentious, adversarial, and fraught with claims. Owners are frustrated by such matters, which demand time and are a drain on resources, as well as affect the project adversely. The owner may want to take measures to diminish the possibility of disputes among the principal project participants. It is important to all owners that construction be completed in conformance with the construction documents and be of high quality. Owners also may be willing to pay for better construction quality when the trade-off is a reduced probability of warranty issues.
- *Other Issues*. Some typical concerns might include safety and minimization of job-site accidents and injuries. Other owners may be more concerned with simplifying any administrative burden that a project might generate, particularly if the owner does not have in-house construction expertise or if existing resources already are overstretched. Each owner will have other concerns that pertain to its specific circumstances.

So, how does an owner find a match between a project delivery system and its need to deal with the particular project challenges? Initially, the owner must determine which project objectives are most important to accomplish, given the constraints and market conditions under which the owner operates. Certain project delivery systems are more likely than others to achieve the project objectives with which the owner is most concerned. The probability of a project delivery system to achieve specific objectives of an owner might be described as the "selection factors" for that project delivery system. Not all project objectives, however, can be addressed by any one project delivery system, and prioritizing project objectives involves making some hard choices.

For example, if an owner desires price certainty prior to commencing construction, then a project delivery system in which the design, procurement, and construction activities overlap is usually not an option. Without a complete set of construction documents, pricing information is less reliable and often subject to various assumptions, clarifications, and exclusions from the contractor. In other words, certain project objectives compete with one another and often are mutually exclusive. Consequently, the owner must identify those objectives that are more important. Those choices also may involve a judgment by the owner that the risks created by failing to address objectives can be managed effectively, and, therefore, such objectives are deemed less important. Project objectives that are not the basis for the selection of a project delivery system might be described as "risk factors," since the transaction structure is not designed to minimize the adverse effects of failing to achieve those objectives. Although it is always possible to fine-tune and make adjustments to a project delivery system in order to lessen the potential adverse effects of a risk factor, it is not possible to eliminate them entirely.

By selecting a project delivery system, the owner attempts (1) to optimize its ability to achieve certain project objectives, while accepting responsibility to manage the risk of other factors, and (2) to set up a transaction structure that incentivizes conduct and performance by the parties to the transaction in a way that is consistent with project goals. The options are infinite. Every project delivery system has its advocates, each of whom will claim the superiority of the advocate's preferred approach.[9] No project delivery system is perfect, however; nor can any such system solve all of the challenges that an owner confronts. Ultimately, the owner must balance selection factors against risk factors and determine which approach is best for its project.

4.04 DESIGN-BID-CONSTRUCT

Historically, Design-Bid-Construct has been the most commonly used method of project delivery in the United States.[10] The method is known by many similar names, such as design-bid-build, design-award-construct, or the "traditional approach." The traditional system is a three-party arrangement and involves two principal contracts. The owner contracts with an architect for the design of the project and enters into a separate contract with a general contractor for the overall construction of the project. The architect and the general contractor do not have a direct contractual relationship. Construction usually commences only after the architect completes a final, detailed set of drawings and specifications. Figure 1 provides a diagram of the contractual arrangement in Design-Bid-Construct.

The Design-Bid-Construct delivery system is intended to take advantage of the separate expertise of design professionals and construction contractors, and, therefore, features two separate contracts held by the owner. Design and construction obligations are delegated to entities with training and expertise in those tasks. Under this system, the architect is responsible for the design of the project. To assist in that undertaking, the architect may retain a team of engineers and other design consultants.[11] A general contractor is responsible

9. Trades groups have formed around various project delivery systems, such as the Construction Management Association of America (the CMAA) and the Design-Build Institute of America (the DBIA).

10. Joint Committee of the American Institute of Architects and the Associated General Contractors of America, *Primer on Project Delivery* 3 (2004).

11. *See* American Institute of Architects, "Exhibit A—Initial Information," AIA Document B101–2007, *Standard Form of Agreement Between Owner and Architect* § A2.5 (2007) [hereinafter AIA Document B101–2007], in which the architect identifies its principal consultants.

FIGURE 1
Design-Bid-Construct

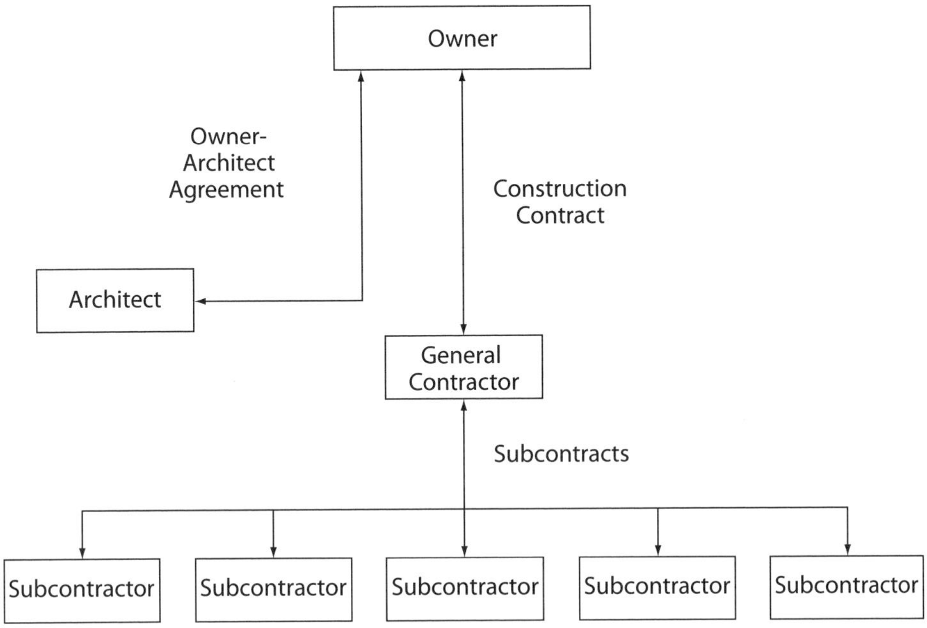

for the construction of the project as depicted in drawings, specifications, and other documents prepared by the architect and the design team. The general contractor generally retains and is in charge of a team of subcontractors and suppliers to complete the construction obligations.[12]

A. Selection Factors

The Design-Bid-Construct approach possesses certain qualities that make it attractive to owners. Many of the advantages derive from the fact that the design is largely complete before a construction contract is awarded and construction commences. Therefore, the drawings and specifications provided to the contractor for purposes of bidding and construction should be well developed, coordinated, and of high quality. In theory, a thorough and well-prepared

12. *See* American Institute of Architects, AIA Document A201–2007, *General Conditions of the Contract for Construction* § 3.3.2. (2007) [hereinafter AIA Document A201–2007], which provides that "[t]he Contractor shall be responsible to the Owner for acts and omissions of the Contractor's employees, Subcontractors and their agents and employees, and other persons or entities performing portions of the Work for or on behalf of the Contractor or any of its Subcontractors."

set of construction documents is less likely to be a source of difficulty or dispute during construction.

Under this approach, the owner selects the architect, who will work directly with the owner. Therefore, the owner has the opportunity to engage an architect that has a particular design style or other attributes that appeal to the owner. The owner has a direct contractual relationship with the architect and usually is involved in the development of the project design on a fairly regular basis.[13] During construction, the architect often acts as the agent of the owner and provides advice to, and advocacy on behalf of, the owner.

Because the drawings and specifications are complete when the construction contract is procured, cost of construction should be more predictable. The Design-Bid-Construct approach also lends itself to competitive bidding, which produces a number of residual benefits for the owner. First, the general contractor is not left to guess as to what is included in the work; therefore, it is possible to provide a fixed price to complete the work.[14] Second, subject to contract terms, cost overruns become the risk of the general contractor. Therefore, the general contractor assumes the risk of volatility in labor and material pricing. Finally, the pressure of competitive bidding provides the owner with some assurance that the contract sum reflects pricing available in the marketplace.

Because the architect and general contractor have separate contracts with the owner, the Design-Bid-Construct methodology has checks and balances built into it. The architect observes construction in progress and acts as the owner's agent to the extent empowered to do so under the agreement between the owner and architect.[15] The architect advocates in construction matters on behalf of the owner, who frequently does not possess meaningful construction knowledge or capability. The general contractor, on the other hand, is

---

13. Under the AIA document approach, the design process moves through schematic, design development, and construction documents phases (see Chapter 6 for a more detailed discussion). The architect is not permitted to commence services in a design phase without approval from the owner of the documents produced in the prior phase. *See* AIA Document B101–2007, *supra* note 11, §§ 3.2.5, 3.2.7, 3.3.1, 3.3.3, 3.4.1, 3.4.5 (2007).

14. *See* American Institute of Architects, AIA Document A101–2007, *Standard Form of Agreement Between Owner and Contractor* § 4.1 (2007), which specifies the "Contract Sum" for the Performance of the Contract.

15. *See* AIA Document B101–2007, *supra* note 11, § 3.6.1.2, which provides that "[t]he Architect shall have the authority to act on behalf of the Owner only to the extent provided in this Agreement." AIA Document A201–2007, *supra* note 12, § 4.2.1 provides that "[t]he Architect will have the authority to act on behalf of the Owner only to the extent provided in the Contract Documents."

motivated to identify deficiencies and errors in the documents prepared by the architect and can serve as a second set of eyes with respect to some aspects of the design.

Members of the design professions and construction industry are very familiar with the Design-Bid-Construct process. The roles of the parties are well understood and assumed with relative ease and comfort. The owner has a single point of responsibility for design and another single point of responsibility for construction. The administrative burdens of the owner are also easy to identify, and outside assistance is readily available if needed.

B. Risk Factors

The characteristics of the Design-Bid-Construct approach, however, are not well suited for all projects and pose certain risks. This approach relies heavily upon the architect for the entire design of the project. There is little opportunity to receive input during the design process from the construction team on matters for which contractors and suppliers often have more expertise than design professionals, such as value engineering, cost estimating, constructability, construction scheduling, labor and material market analysis, equipment availability, design of specialized equipment and proprietary systems, and other similar issues. As a result, the project design might not achieve the objectives of the owner to the fullest extent possible. Although such input may be received after the general contractor is retained, any changes to the design at that point in time will cost more and cause delays in project completion.

When using Design-Bid-Construct, construction does not start until the design is complete. The process is linear and sequential. As a result, project completion occurs at a date later than if the design, procurement, and construction phases were to overlap. A longer project schedule has disadvantages to the owner. First, delays in locking in price provide a greater opportunity for escalation in construction costs. Second, many projects are intended to generate revenue directly or indirectly, whether through rental income, through sales activities that occur in the completed project, by producing items that are sold, or by furnishing facilities that support revenue generation. But, if the ability to commence the beneficial use and occupancy is delayed, then so is the revenue stream generated by such use and occupancy. Delays in commencement of construction also create the risk of interest rate inflation, and, therefore, significant increases in the cost of debt service over the life of a construction loan.

When a construction contract is subject to a fixed price, the general contractor's entrepreneurial interests may be at odds with the owner's expectations regarding quality of construction and cost control. A general contractor is reluctant to provide any work that is not called for clearly in the drawings and specifications. Performing such work simply is adverse to the economic assumptions upon which the general contractor bid the project. If questions arise regarding scope or the type of work to be provided, the general contractor will make decisions consistent with its financial interests. The owner may not agree with positions taken by the general contractor. As no set of drawings and specifications is ever 100 percent complete, reasonable inferences must be made from those documents.[16] Expectations regarding what is reasonable to infer from construction documents is a fertile source of disputes.

The three-party arrangement that is the hallmark of Design-Bid-Construct may make claims difficult to sort through. All parties will have a natural tendency to defend their work and services. Architects are inclined to argue that problems are the result of contractor error, and contractors are predisposed to contend that problems are caused by the architect's design errors or omissions. In many instances, it is difficult to determine fault, and a construction-related dispute that is not resolved amicably is always expensive to adjudicate. Because neither the architect nor the contractor wishes to be held responsible for the obligations of the other, each will negotiate vigorously for certain exculpatory language in contracts to guard against such risk. Even if the owner is not at fault for problems encountered on a project and possesses legitimate claims through which to recover damages for its losses, that will be of little consolation when such claims cannot be resolved promptly due to "finger-pointing" between an architect and a general contractor.

Also, the owner may find itself the focus of claims between the design team and the construction team. In those states with a more orthodox application of the Economic Loss Rule, the probability of encountering this situation is amplified.[17] For example, if a general contractor believes that it has suffered economic losses for which the architect or another member of the design team is responsible, the Economic Loss Rule may prohibit the general contractor from bringing a cause of action against any of those parties directly. The general contractor may have no option but to pursue a claim against the party with whom it has a contract—the owner—and place the burden on the owner to bring a third-party action against the architect.

Claims also can be a problem with respect to subcontractors and other lower-tier entities that do not have a direct contract with the owner. The rela-

---

16. *See* AIA Document A201–2007, *supra* note 12, § 1.2.1, which provides that "performance by the Contractor shall be required only to the extent consistent with the Contract Documents and reasonably inferable from them as being necessary to produce the indicated results."

17. See Chapter 23.

tionship of lower-tier entities to the owner may be very remote, and payment to lower-tier entities for their work or services flows from the owner through other parties prior to receipt by the lower-tier entity. There may be substantial delays between the date the work or service is performed by a lower-tier entity and the date upon which payment for that work or service is received.

As a result, lower-tier entities must absorb a cost to finance their work for the owner. For many lower-tier entities to remain financially viable, that finance cost is included in the contract sum of the lower-tier entity and is passed upstream, eventually reaching the owner, and thereby ultimately increasing total project costs. From a practical standpoint, however, lower-tier entities are often thinly capitalized. The financial stress created by delays in payment to them is often too great. Disputes spring from delays in payment, and sometimes companies fail as a result.[18]

C. Common Variations

As with all project delivery systems, variations of the Design-Bid-Construct approach have been developed. Sometimes modified versions are motivated by a desire to optimize a selection factor or, more likely, in an attempt to diminish the potentially adverse affects of a risk factor in the use of Design-Bid-Construct. One of the most common variants of Design-Bid-Construct is referred to as Multiple Primes. In this method of project delivery, there is no general contractor. Rather, the owner contracts directly with each trade. In essence, the entities that are referred to as subcontractors under the Design-Build-Construct approach become prime contractors under the Multiple Prime approach. The owner still contracts with the architect separately, and the process of design, procurement, and construction remains linear. Figure 2 provides a diagram of the contractual arrangement in the Multiple Prime approach.

In furtherance of various public policy reasons, some states require the use of the Multiple Prime approach under their procurement statutes for public projects.[19] The Multiple Prime approach provides opportunity for participation by a greater number of trade contractors. Because each trade contract is

---

18. The Surety Information Office cites a study by BizMiner (a company that analyzes U.S. industries) and concludes that "of the 850,029 building (non-single-family), heavy/highway, industrial buildings/warehouses, hotel/motel and multifamily home construction, and specialty trade contractors operating in 2004, only 649,602 were still in business in 2006—a 23.6% failure rate." *See* Surety Information Office, *Why Do Contractors Fail*, *at* http://www.sio.org/html/whyfail.html (last accessed June 18, 2008).

19. *See* N.Y. State Fin. Law § 135, sometimes referred to as the Wicks Law. Other states have similar legislation, including New Jersey, Ohio, Pennsylvania, and Wisconsin.

FIGURE 2
Multiple Primes

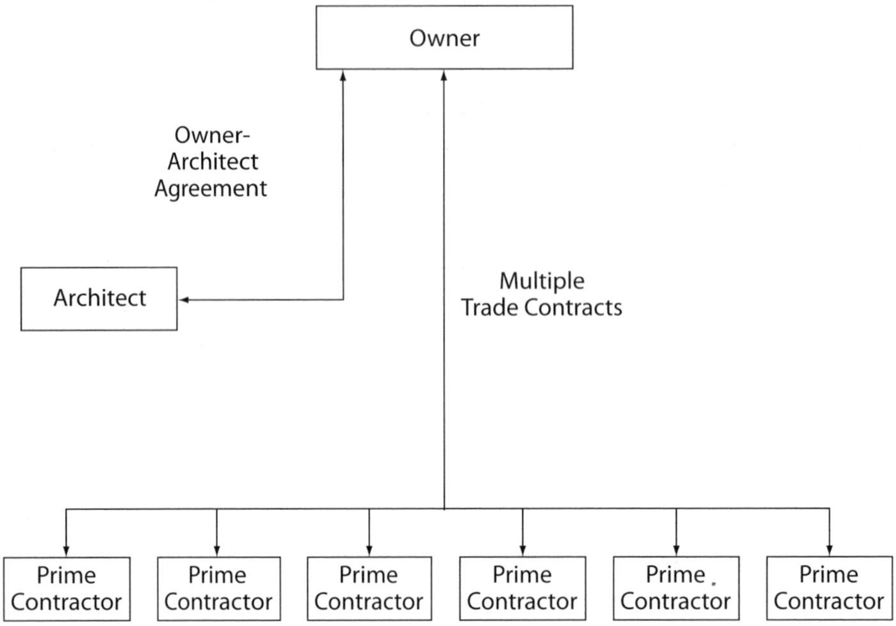

awarded through competitive bidding, the likelihood for exclusionary practices in the award of contracts to the trades is reduced. The use of competitive bidding also provides some assurance of a market price. In practice, however, the Multiple Prime approach may not achieve all of the benefits that it is designed to attain. Nonetheless, those potential benefits are important considerations to a government agency charged with shepherding public assets.

When the private sector considers the Multiple Prime approach, the reasons are similar, but not always identical, to those cited by the public sector. Removing a tier of fee or markup from the construction team (the cost of retaining a general contractor) is attractive on some projects. An owner also may like the idea of actively deciding which trade contractors to retain and administering each such contract. The Multiple Prime methodology provides the owner with significant control. Furthermore, eliminating a general contractor from the construction process removes a link in the payment chain, resulting in less financial stress on the trades and those with whom the trades contract. Accordingly, the need for lower-tier entities to build into their price a cost to finance delays in payment is diminished somewhat.

Notwithstanding the advantages of the Multiple Prime method, any change to a project delivery formula may cause new challenges to emerge. Although the owner in a Multiple Prime project may enjoy increased control over the trades, with that control comes a heavy coordination and administrative responsibility. Many owners do not have the ability to discharge those obligations successfully. An owner must possess substantial construction skill and experience to act as its own general contractor. Moreover, because a general contractor no longer stands between the owner and the lower-tier contractors, the owner in a Multiple Prime job must deal directly with claims from the numerous trades. Those claims might not be easy to resolve. In addition to customary disagreements between the design team and the construction team, finger-pointing also can be expected among and between the various trade contractors.

4.05 CONSTRUCTION MANAGEMENT

The term "construction management" applies to a variety of project delivery systems. There is no universal definition, and the references to "construction management" are vague, at best, without further clarification. The role of a construction manager may range from an additional consultant to the owner who functions in a purely advisory role, to that of a constructor with extensive obligations pertaining to the construction and completion of the entire project. Construction management concepts developed out of increasing dissatisfaction expressed by owners regarding the inability of the traditional delivery system to effectively furnish reliable cost estimating and control, construction scheduling, alternative designs, and constructability analysis. Construction managers generally possess expertise in such areas, which basically bridges the gap between professional design services and construction activities. To derive the full benefit of such skills, owners usually retain construction managers during, or prior to commencement of, the design process.

A. Agency Construction Management

One common form of construction management often is referred to as Agency or Pure Construction Management. The agent construction manager is paid a fee to provide professional management services consistent with the interests of the owner. Under the agency version of construction management, the construction manager is a consultant to the owner and may contract with trade contractors as the owner's agent. The agency construction manager coordinates the trades but does not guarantee cost or time of completion, nor does an agency construction manager assume many other traditional general contractor obligations.[20] Figure 3 provides a diagram of the contractual arrangement in the Agency Construction Management approach.

In many respects, the Agency Construction Management approach simply takes the Multiple Prime method and augments it with an advisor to the owner. The owner outsources to that advisor the coordination and other responsibilities that the owner retains in a Multiple Prime approach that it is not capable of handling itself, due either to lack of necessary skills or available resources. The advisor is retained in an effort to ameliorate some of the risk factors inherent in the Multiple Prime approach. Consequently, many in the design professions and construction industry view Agency Construction Management as a management technique applied to the Multiple Prime approach and not as a separate project delivery system. Regardless of that debate or how it might be resolved, Agency Construction Management has found its way into the parlance of project delivery systems. It is the basis of many contract forms and, therefore, the concepts and objectives behind the use of Agency Construction Management warrant understanding.[21]

1. Selection Factors

Essentially a permutation of the Multiple Prime approach, Agency Construction Management possesses many of the same advantages found in the Multiple Prime approach. Adding the agent construction manager to the Multiple Prime arrangement addresses some of its shortcomings. For example, the owner retains the agent construction manager during the design process. As a result, the owner obtains construction-related advice on a timely basis as the drawings and specifications evolve. Also, because of the early participation in the project by a construction advisor, the process lends itself to overlapping the design, procurement, and construction phases of a project. Implementing phased, or fast track, construction is a natural objective in engaging a construction manager during the design phase of a project.

Another notable characteristic of Agency Construction Management is that the interests of the owner are aligned with the party that is principally responsible for managing the construction process, thereby eliminating the inherent

20. *See generally* the Construction Management Association of America, *Agency Series Model Contract Documents, available at* http://cmaanet.org (follow CMAA Bookstore link; then follow CMAA Publications); and ConsensusDOCS 510, *Owner/Construction Manager Agreement/General Conditions (Cost of the Work, Preconstruction Opt)* (2007) (Note: The publisher, ConsensusDOCS, is a group comprised of approximately 20 construction associations).

21. *See supra* note 20.

FIGURE 3
Agency Construction Management

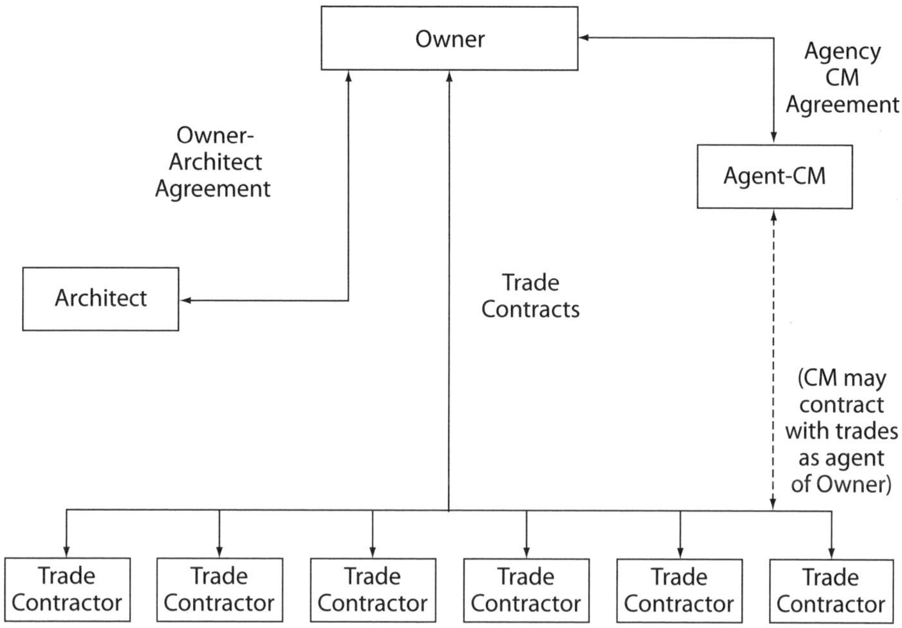

conflict between the interests of the owner and those of a general contractor. The agent construction manager does not assume the risks of a general contractor, such as responsibility to complete the project for a specific cost or within an agreed upon time. Therefore, the agent construction manager does not have the same entrepreneurial interest in the project that a general contractor, or prime trade contractor, would have. The owner pays the agent construction manager a fee for services. The agent construction manager does not have other pecuniary interests that motivate performance of its services.

Without having a stake in the cost or time of project completion, the agent construction manager is less influenced by self-interest and is more likely to provide impartial advice to the owner. That advice is a key reason for choosing a construction management approach as the project delivery system. Therefore, it makes sense to use a process that enhances the value and quality of that advice. In addition, the agency relationship between the construction manager and the owner allows the agent construction manager to advocate on the owner's behalf. Because the agent construction manager's interests are more aligned with the owner than with the interests of the trades, the agent construction manager can do an effective job of evaluating and defending claims asserted by the trade contractors during construction of the project.

2. *Risk Factors*

The weaknesses of the Multiple Prime approach that are not addressed through the retention of the agent construction manager obviously remain present in the Agency Construction Management method. Most notable among them, there is no single point of responsibility for performance and completion of construction. The agent construction manager only coordinates such matters. Although variations are common, the degree to which the agent construction manager retains liability for construction cost, delays in completion, quality of work, safety of the work, and other construction-related matters is usually limited by the contract between the owner and the agent construction manager. Such limitations, more than any other criticism, are enough to cause reluctance among many owners to consider Agency Construction Management for their projects.

Construction management in all its forms is a relatively new phenomenon, with a short track record compared to other project delivery systems. The obligations and risks assumed by the principal parties, in some respects, still are evolving and do not have a standard industry approach. This leads to some confusion. It also provides the potential for protracted contract negotiations, since each contract tends to be highly customized. Uncertainty also stems from the fact that many see the agent construction manager as providing services that are somewhat professional in nature. Therefore, questions arise as to whether a license is required to provide such services and what kind of insurance is appropriate for the agent construction manager to procure and maintain. Bonding also may be an issue. Finally, all of these variables in combination mean that the body of law available to define the nature of Agency Construction Management is relatively limited.

B. At-Risk Construction Management

The other project delivery approach most commonly referred to as construction management is a hybrid of Agency Construction Management and Design-Bid-Construct. At-Risk Construction Management, as it usually is described, involves retention of a construction manager during or prior to commencement of the design process. The at-risk construction manager provides the same type of pre-construction consulting services that are provided by an agent construction manager. Cost estimating, construction scheduling, value engineering, and constructability analysis, to name a few, are all of great benefit in optimizing the design of the project.

The at-risk construction manager, however, does not act as the agent of the owner. Trade contractors are retained by the at-risk construction manager directly. The trades, therefore, are subcontractors of the at-risk construction manager. The subcontractors have no contractual privity with the owner and look to the at-risk construction manager for payment. In fact, the at-risk construction manager is often referred to as a "constructor" and is the single point of responsibility for the performance of construction.[22] Figure 4 provides a diagram of the contractual arrangement in the At-Risk Construction Management approach.

1. *Selection Factors*

The At-Risk Construction Management method is an effort to combine the characteristics of Agency Construction Management and Design-Build-Construct that are perceived as more favorable to owners. Unlike the agent construction manager, the at-risk construction manager is responsible for the overall construction of the project. During the construction process, therefore, the at-risk construction manager is very much like a general contractor in the Design-Build-Construct approach. As a constructor, not an agent of the owner, the at-risk construction manager remains obligated for the quality, cost, and timely completion of construction, along with all other responsibilities customarily assumed by a general contractor.

Similar to Agency Construction Management, however, the owner in the At-Risk Construction Management approach also receives the benefit of pre-construction consulting services necessary to optimize the design process. With the benefit of such pre-construction advice, the drawings and specifications produced by the architect should describe a project that presents fewer constructability challenges and that is more likely to be completed in accordance with the budget and schedule requirements of the owner. At-Risk Construction Management also facilitates the opportunity to overlap the design, procurement, and construction phases of a project, shortening the duration of a project. Therefore, At-Risk Construction Management is generally associated with fast

22. American Institute of Architects, AIA Document A121–2003, *Standard Form of Agreement Between Owner and Construction Manager Where the Construction Manager Is Also the Constructor* (2003); *see also* AIA Document A201–2007, *supra* note 12, § 1.1.2 (no contractual relationship between the Owner and a Subcontractor); § 5.1.1 (definition of Subcontractor); §9.6.4 (Owner has no obligation to pay a Subcontractor except as may be required by law).

FIGURE 4
At-Risk Construction Management

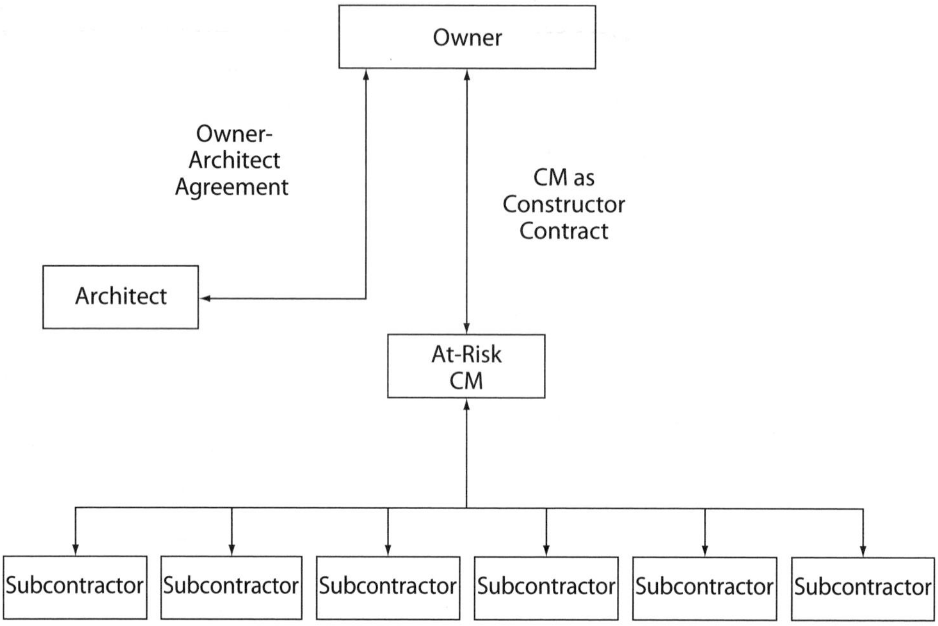

track construction or some other technique that allows aspects of procurement and construction to commence prior to the completion of the drawings and specifications by the architect.

2. *Risk Factors*

As a hybrid approach, At-Risk Construction Management has some of the disadvantages of the project delivery systems with which it shares certain characteristics (i.e., Agency Construction Management and Design-Build-Construct). As is the case with Agency Construction Management, there is no substantial body of law from which to gain guidance on legal issues that may arise in connection with At-Risk Construction Management. Further, although the at-risk construction manager has responsibilities very much analogous to those of a general contractor, the pre-construction role of the at-risk construction manager poses questions surrounding licensing, insurance, and bonding.

Many of the risk factors inherent with Design-Bid-Construct also remain unsolved by At-Risk Construction Management. Subcontractors and suppliers, particularly those in lower tiers, bear the financial burden of receiving payment long after they provide their work and services. The division of design and construction responsibility during the construction phase of an At-Risk Construction Management project can cause adversarial relationships between the architect and the at-risk construction manager, each assuming that the other is principally responsible for any project problem.

Perhaps the most troublesome risk factor is that the economic incentives of the at-risk construction manager are not entirely aligned with the interests of the owner. Like a general contractor in a Design-Bid-Construct project, the ability of the at-risk construction manager to earn a profit on the project is linked directly to its ability to satisfy certain contractually specified criteria. Consequently, the at-risk construction manager will resist vigorously any condition or circumstance that might cause it to incur unanticipated cost or require additional time to complete construction. Claims are the likely result.

The at-risk construction manager should not be expected to perform work beyond the minimum required by the construction contract. For example, providing items that are less desirable to the owner or sequencing work in a certain manner may save the at-risk construction manager time and money. Such measures may be entirely permissible under the terms of the construction contract, which typically provide a general contractor or an at-risk construction manager with some latitude in discharging its obligations. Even during preconstruction services, the at-risk construction manager might be inclined to recommend a particular course of action that it believes will facilitate its ability to satisfy contractual obligations, even when better alternatives from the standpoint of the owner might be available.

Issues with respect to ensuring accurate and reliable pricing are common when using an At-Risk Construction Management approach. First, the at-risk construction manager generally is retained before design services are complete. Because the drawings and specifications are most likely not well developed at that time, any price furnished by the at-risk construction manager early in the design process is only an estimate and probably contains a large contingency to protect the at-risk construction manager from costs that might become apparent when the drawings and specifications are better defined.

Usually, the at-risk construction manager does not propose a binding "guaranteed maximum price" until the drawings and specifications achieve a level of completion where the at-risk construction manager is reasonably certain of what is required to complete the work. That might occur when the drawings and specifications are anywhere from 50 to 90 percent complete, depending on the nature of the work.[23] Depending on the at-risk construction manager's level of comfort with the accuracy and completeness of the drawings and specifications, the guaranteed maximum price will be based on numerous assumptions, clarifications, and exclusions. If the final drawings and specifications are inconsistent with any such qualification, the at-risk construction manager will expect the guaranteed maximum price to be increased.

Finally, At-Risk Construction Management contracts are procured more often through negotiation than through competitive bidding. Because the at-risk construction manager is retained during the design phases, there is no project design to bid.[24] Therefore, the majority of owners retain an at-risk construction manager on the basis of particular skill and expertise, believing that (1) the ability of the at-risk construction manager is of primary importance to the owner, and (2) market pricing will occur through competitive bidding of subcontracts. Some owners will retain a construction manager initially for only pre-construction activities and then seek competitive bids from other at-risk construction managers when the design is largely complete, but this generally is not an effective means of achieving market pricing for the construction phase. The familiarity of the original at-risk construction manager with the project is usually an advantage that rarely can be overcome by other bidders.

4.06 DESIGN-BUILD

Under the Design-Build approach, an owner contracts with a single entity, who undertakes to both design and construct the project. Many owners prefer a single point of responsibility for the entire project and desire to unburden themselves from the day-to-day contract administration responsibilities that are endemic to running most construction projects. Because Design-Build tends to be a very integrated process among the design and construction teams, the design, procurement, and construction phases almost always overlap. A number of variations of Design-Build are possible, many of which are described in greater detail in Section 4.06(C). Figure 5 provides a diagram of the contractual arrangement in the basic approach to Design-Build.

Design-Build is increasingly popular. Some statistical analyses predict that Design-Build eventually will be used more often than any other project

23. *See* Skidmore, Owings & Merrill v. Intrawest I L.P., 87 Wash. App. 1054 (Div. 1 1997). The *Skidmore* court found that determining the degree to which drawings and specifications are complete is not an exact science. Various approaches abound. In *Skidmore*, the architect was found to have breached an obligation to furnish 90 percent complete drawings.

24. In such circumstances, some owners seek bids on fee and certain overhead expenses.

FIGURE 5
Design-Build

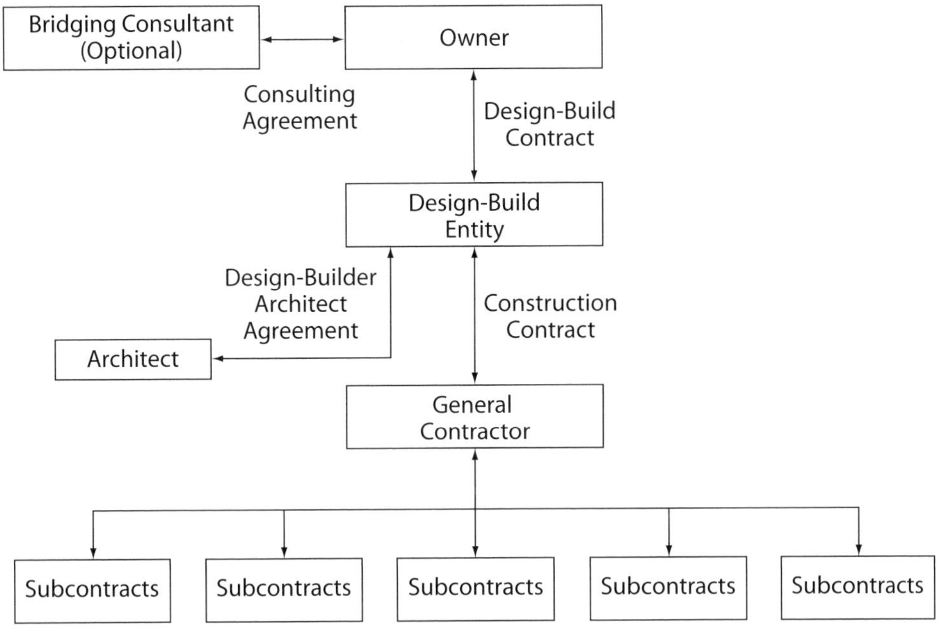

delivery system.[25] Although touted primarily by those who promote Design-Build, such projections are noteworthy. Indeed, Design-Build is gaining use in the public sector as well, where it had been virtually prohibited under procurement statutes until recently. Recent surveys reveal that many states have legislation allowing for use of Design-Build to deliver construction projects.[26]

25. *See* 2 Philip L. Bruner & Patrick J. O'Connor, Bruner & O'Connor on Construction Law § 6:2 (West Group 2002) [hereinafter Bruner & O'Connor]; G. William Quatman, *Design Build for the Design Professional*, Ch. 1, in Construction Law Library (Aspen Law and Business 2001). The Design-Build Institute of America further projects that by 2010 or shortly thereafter, the number of nonresidential construction projects in the United States delivered via the Design-Build approach will exceed the number of such projects delivered using the traditional Design-Bid-Build project delivery system. *See* graph of *Non-Residential Design and Construction in the United States*, http://forms.dbia.org/ind_info/mkt_chrt.html (last accessed June 18, 2008).

26. According to the Federal Highway Administration of the U.S. Department of Transportation, as of August 2006, 38 states had laws allowing the use of Design-Build for transportation procurement. Brandon Davis & Nancy Smith, *50-State Survey of Transportation Agency Design-Build Authority* (2006), *at* http://www.nossaman.com/articles.aspx (last accessed June 18, 2008). The Design-Build Institute of America cites similar statistics for transportation procurement, updated in March 2006. *See* map of *2006 Design-Build State Laws for Transportation Procurement*, *at* http://forms.dbia.org/ind_info/transportmap.htm (last accessed June 18, 2008).

The number of states and other government agencies that permit Design-Build is expected to increase.

Design-build entities often are construction contractors who either have in-house design expertise or are comfortable taking responsibility to subcontract with an architect for such services. Sometimes the design-builder is a joint venture or special-purpose vehicle comprised of entities that, collectively, possess the requisite skills to accomplish the project, such as a joint venture formed by a contractor and an architect.[27] The joint venture or special-purpose vehicle may self-perform the project, or perform design and construction through subcontracts with its constituent entities or other third parties. It is uncommon for an architect or other design professional to act alone as a design-builder.

A. Selection Factors

For many owners, the Design-Build approach is a convenient way to transfer risk. Overall responsibility for design and construction resides with the design-build entity. In some ways, the classic contractual arrangement in Design-Build is nothing more than the Design-Bid-Construct approach pushed down one tier. The risk factors that an owner confronts in the Design-Bid-Construct approach still exist, but they now belong to the design-build entity with whom the owner contracts. The design-builder retains the architect and general contractor. The owner is relieved of all the challenges that go along with holding those contracts. As a consequence, the owner may devote more of its resources to its core business or other pursuits, as opposed to the project.

Many entities, in addition to the architect, are responsible for the overall design of the project. It is common for architects to delegate aspects of project design to the construction team. For example, product manufacturers and specialty trade contractors generally are far more knowledgeable than architects with respect to building equipment, technology, and systems (e.g., elevators, security systems, and curtain wall systems). Incorporating the knowledge of specialty trade contractors and product manufacturers directly into the initial design process has many benefits. First, it helps speed the design process and facilitates use by phased or fast track construction. Second, such collaboration produces design documents that conform to the practicalities of current market conditions. Contractors and product manufacturers are more likely to know what building equipment and systems are readily available, how such items

27. A design professional–led team and other options are possible, but less common in the general building sector.

are efficiently utilized for a project, and how to best customize them for the particular project being designed. Design-Build, in part because of such collaboration, often leads to innovation and creative solutions to an owner's needs.

B. Risk Factors

When an owner transfers risk to the design-builder, it also must cede much control over the project. First, in Design-Build, the architect does not answer to the owner directly, but rather reports to the design-builder. Because the owner does not have a direct contractual relationship with the lead design professional, the owner has a limited ability to control the design process. Therefore, if aesthetics are a paramount concern of the owner, the Design-Build approach may not be a good match. Moreover, the owner will not have the independent advice of the architect during the construction phases and, therefore, no check or balance on the performance of the construction team.

The design-builder produces the project design in response to performance criteria or specifications furnished by the owner. The performance criteria specify various "outputs" to which the completed project must conform. For example, structural loads, heating and air conditioning requirements, and other performance criteria are described. It is up to the design-builder, however, to determine how to satisfy the performance criteria. As long as the completed project satisfies the required performance specifications, the design-builder has discharged its design obligations. That means much rides on the quality of the performance specifications.

Because the Design-Build approach relies on satisfaction of performance criteria, the owner must be certain to specify all appropriate "outputs" clearly and comprehensively. Those requirements set forth the design obligations of the design-builder. Performance criteria are more effective if they describe requirements that lend themselves to measurement, such as the technical characteristics of building systems and specialty equipment. In the case of an industrial plant, certain performance aspects of the facility and the quality and quantity of items produced are capable of measurement. The easier it is to measure the "output," the easier it will be to verify if the completed project complies with the applicable performance criteria.

The owner, however, may have expectations with respect to the aesthetic and functional qualities of a project. Those characteristics are not easily measured. Therefore, the design-builder has greater latitude with respect to such matters than the owner might like if the design-builder's contractual obligations are governed only by performance specifications. The most direct solution is for the owner to describe aesthetic and functional requirements as completely as possible through design specifications to assure that the finished product will meet those expectations.[28]

The procurement by an owner of a Design-Build contract also creates numerous challenges. A common approach is for the owner to request proposals from competing design-build entities. The potential design-builders base their proposals on the performance criteria furnished by the owner. Each proposal likely offers a solution that differs from the other proposals. How does the owner compare and choose among proposals that are "apples and oranges"? Moreover, pricing data provided with a Design-Build proposal is based necessarily on a very rudimentary design and, therefore, lacks reliability. Because many aspects of the design will not be determined by a design-builder until after the proposal is accepted, the design-builder simply may reverse engineer the design in order to meet the budget to which it has made a commitment. Thus, final project quality or features are somewhat variable in order for the design-builder to assure itself of an acceptable return on its investment.

Preparing a proposal is a difficult exercise for a design-builder, even if only basic design information is required as part of its proposal. The design-builder faces a dilemma. The design-builder arguably is best served by developing the design to a point where the design-builder is reasonably confident in its ability to predict all elements needed to produce a final design. If the design-builder is comfortable with the design, reasonable projections are possible for cost, time, performance, characteristics, and other matters. The design-builder can reduce contingencies accordingly and submit a more reliable proposal. Such an effort, however, consumes limited resources and is very costly. The design-builder will not recover those costs if another party is awarded the contract or if the project is abandoned. But, the alternative of putting forth only minimal effort in preparing a proposal is a risky venture. The resulting proposal may not be responsive to the owner's needs and, therefore, a wasted effort. Moreover, the less-developed proposal may put the design-builder at risk. The design-builder has made a wager that it will be able to complete the project for the cost and

---

28. Design specifications set forth in precise detail the materials to be employed and the manner in which the work is to be performed. The contractor may not deviate from the specifications and is required to follow them as one would a road map. A design specification, also known as a descriptive or prescriptive specification, prescribes the exact materials and procedures of assembly. Performance specifications set forth a standard to be achieved and, unlike a design specification, do not specify in specific materials or assembly direction. The contractor is expected to exercise its ingenuity in achieving that objective or standard of performance, selecting the means and assuming a corresponding responsibility for that selection. This permits the contractor to choose a system or construction procedure that is most attractive to the contractor that still fulfills the design criteria. This topic is discussed in greater detail in Chapter 6.

time proposed, although the design-builder may not have studied the project in detail and as appropriate to make such a judgment.

Another approach to procuring a Design-Build contract is for the owner to negotiate with a preferred provider. This method requires a certain degree of optimism from the owner. Negotiated Design-Build contracts generally come in two parts. During the first phase, the design is developed to a point where a reasonably reliable guaranteed maximum price and related data are submitted to the owner. If the price is too high for the owner or the proposal is otherwise unacceptable, the owner can try to work with the design-builder to arrive at a better proposal, but success is not guaranteed.

Starting over with another design-builder may not be a viable option, if the owner cannot invest additional time. The owner also must determine whether it has the right to use the proposal of the current design-builder if the owner proceeds with a different design-build entity. Design-builders sometimes price their design services very competitively, expecting to make their profit during the construction period. Under such circumstances, a design-builder may not be interested in simply turning over its design for exploitation by the owner and another design-builder.

In large part due to the professional services provided by the design-builder, some jurisdictions may require the design-builder to be licensed. Insurance companies also scrutinize Design-Build projects closely. The line between professional design services and construction work is not always clear in a Design-Build project. Special attention must be exercised to ensure that risks customarily associated with a project are properly insured. Also, sureties sometimes are less willing to issue bonds on such projects or may exclude certain activities from the bond. Sureties are accustomed to defining and underwriting construction risks, but design risks are less measurable.

C. Common Variations

Concentrating all design and construction obligations in the design-builder prevents the owner from selecting an architect with a particular style and from exerting control over the design process that an owner customarily desires. This situation is ameliorated sometimes when the owner uses a so-called bridging technique, in which the owner retains an architect to prepare a preliminary design that subsequently is turned over to a design-builder for completion. A bridging consultant typically prepares an initial design through approximately 35 percent complete design development documents. A lesser or greater level of completion can be performed, but the owner must balance its desire to dictate the project design against depriving the design-builder of design flexibility.[29] Limiting the options available to the design-builder during the design process diminishes the potential for innovation and efficiencies normally achieved through the Design-Build process.

By retaining an architect directly, the owner obtains greater assurance that the completed project will conform to the owner's aesthetic and functional goals. Often, the bridging consultant remains associated with the project, even after the initial design is turned over to the design-builder. The bridging consultant provides the owner with independent advice throughout the completion of the design and during the construction phase. This is helpful to the owner because the architect of record reports to the design-builder. Although the bridging consultant is an additional expense, it may be money well spent by the owner who desires greater influence over the design or who is unsophisticated in construction matters.

Design-Build also is used in connection with many civil and industrial projects. However, structures and facilities, such as bridges, roads, power plants, manufacturing facilities, and waste water treatment plants are products more of engineering than architecture. In fact, an architect may not be involved on such a project. Civil and industrial projects also frequently require and are dependent upon the supply and installation of very large, specialized equipment. Therefore, the Design-Build approach is adapted for use on civil, industrial, or similar projects. Due to the emphasis on engineering and equipment procurement, the type of Design-Build process used on such projects is generally referred to as Engineer-Procure-Construct, or EPC.

EPC projects are generally very large, complicated, and high-risk propositions. Contracts are heavily negotiated. Most EPC projects contain complex economic terms, which usually include a rigorous completion/acceptance protocol, liquidated damages for delays and failure to achieve numerous performance guarantees, as well as various limitations of liability. Because these projects often involve the installation and assembly of large pieces of specialized equipment, it is common for the owner to have paid a significant amount of money before anything is delivered to the site.[30] Project financing is typical for an EPC-delivered facility. The lender will rely on revenue generated by the project as security for the construction loan. Because of the limited recourse

29. 2 BRUNER & O'CONNOR, *supra* note 25, § 6.35.

30. In contrast with an EPC project, an owner in the general building sector typically makes progress payments to the general contractor in relative proportion to the percentage of work installed at the project site. Therefore, under this scenario, the owner will be in a position to take possession (and likely have title) to the work for which it has paid.

nature of project finance, an EPC project may involve use of independent engineers and an allocation of risk with a focus protecting the lender's interests.

Another common variation of Design-Build is referred to occasionally as Turnkey. Although there is no standard definition of Turnkey, it usually describes a project where an owner retains a design-builder to deliver a project, but without payment made by the owner to the design-builder until the project is completed. These projects are often real estate transactions, with a construction component. The design-builder may own the parcel of land on which the project is constructed. Upon completion, there will be a closing where a purchase price is paid to the design-builder and the project is turned over to the owner. The owner only "turns the key" in order to commence the use and occupancy of the project.

On occasion, the design-builder retains ownership of the project and provides the owner with a long-term lease to use and occupy the project. Those projects are sometimes referred to as Build-to-Suit.

4.07 INNOVATIVE APPROACHES

From time to time, various innovative forms of project delivery systems emerge. Some are truly customized, one-of-a-kind approaches that could not easily be duplicated or adapted for use on other projects. Others have a broader application. Most of the innovations focus on techniques that increase collaboration among the major project participants, and some attempt to create an arrangement for the principal project participants to share jointly in the success or failure of the project. Such approaches to project delivery occur most commonly in infrastructure projects and with other complex, high-profile undertakings that warrant creative efforts to facilitate delivery of a project. Nonetheless, some traditional, institutional owners with large portfolios of construction projects also examine ways to improve the project delivery.

A range of creative approaches to project delivery is used in connection with projects that fall within the rubric of public-private partnerships, or P3. The term "public-private partnerships" applies broadly to a number of different arrangements through which the private sector is called upon to provide services that customarily are delivered by the public sector. Sometimes, this is nothing more than outsourcing certain services, such as operation and maintenance of a public facility. But public-private partnerships also involve the construction of public works projects and often require the ongoing operation and maintenance of the completed project by the private sector.

A typical model for a public private partnership involving construction of public works is Build-Operate-Transfer (BOT). Used for many projects overseas, BOT and similar methods of project delivery are still an emerging market in the United States.[31] In the so-called BOT approach, a government agency retains a project company to design and construct a project, very much analogous to the Design-Build approach. In addition, however, the project company takes a concession to operate and maintain the project for a specified period of time following completion. Even more novel, the project company generally is responsible for financing its own activities and work. The public agency, at most, may make partial payments to help defray costs of construction, but often makes no payments until the project is complete and operational. A simple example of how a BOT project is organized contractually is depicted in Figure 6.

The project company, usually a special-purpose entity, makes profit (it hopes) from the operating concession. Most concessions last between 25 and 50 years, but some are even longer. Upon expiration of the concession, the project company must restore the project to acceptable condition and then turn over the project to the public agency that authorized the project. Either a portion of the revenue generated by the project (e.g., tolls) is paid to the project company, or payments are made in accordance with predetermined formulae based on performance criteria (e.g., availability of the improvement, quality of service). While the return on investment by the project company can be rewarding, such projects are risky.

The concession agreement is a lengthy and complicated document. Among many issues, it must address performance criteria for the completed project and standards of operation and maintenance, the economics of the transaction, permissible adjustments in fares (for projects such as mass transit systems and toll roads) or other fees and charges, transferability of the concession, and anything else that might come up during the life of the concession. It is difficult to predict what the future holds. From the standpoint of the project company, such projects are more financial transactions than construction projects. The project company must be confident that it will recover its costs and make an appropriate return. It is not enough for the project to be a construction success. The project must be a commercial success, as well.

The BOT model allows public agencies to proceed with projects that otherwise might not be viable due to funding limitations, inadequate resources, or

---

31. Many Web sites track trends in P3 procurement and P3 projects (potential and actual) such as P3/Americas, *available at* http://www.p3americas.com (last accessed June 18, 2008).

FIGURE 6
Build-Operate-Transfer

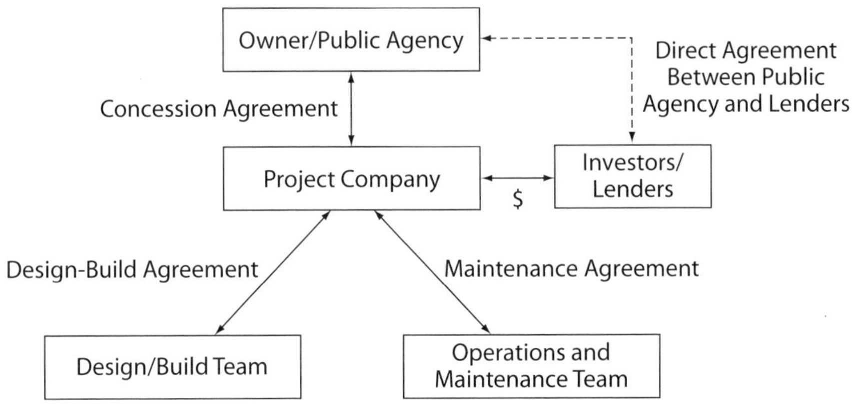

other priorities. As demands on public agencies increase to repair or replace outdated infrastructure and provide other public improvements, the BOT approach and other public-private partnerships may become more mainstream. From the standpoint of project delivery systems, the BOT method transfers many traditional owner responsibilities to the project company and aligns the interests of the public agency with the project company.[32]

In carrying out its construction obligations, the project company is free to employ a traditional delivery system, construction management, or some form of Design-Build in implementing the project through the team that it retains. Given the nature of these projects, the project company usually prefers an EPC-type approach, but some projects warrant different methods.[33]

32. For example, the project company will be keenly interested in a well-completed project if it is obligated to operate and maintain the project.

33. Delegating to a single point of responsibility makes it easier for the project company to manage its risk. Nonetheless, commercial conditions may be such that no single entity is willing to accept full responsibility for the entire design, procurement, and construction of a project nor accept joint and several liability exposure in a joint venture of some nature. Under such circumstances, the project company must retain separate design and construction teams. Similarly, the entities that perform design or construction activities may have no interest in the risks that flow from the operation and maintenance of the completed project during the concession period, and, therefore, a separate agreement will be entered into by the project company with an "operator."

Another source of innovation in project delivery systems is found in the concept of Integrated Project Delivery. The philosophy behind Integrated Project Delivery has been defined as:

> a project delivery approach that integrates people, systems, business structures and practices into a process that collaboratively harnesses the talents and insights of all participants to optimize project results, increase value to the owner, reduce waste, and maximize efficiency through all phases of design, fabrication and construction.[34]

Integrated Project Delivery is driven out of a desire to maximize the benefits of new information technologies, such as building information modeling, and dissatisfaction with the more traditional project delivery systems that are viewed as inhibiting innovation, impairing the coordination of design and other major activities, and producing overly adversarial relationships.

The implementation of Integrated Project Delivery is still a work in progress.[35] All project delivery systems that draw upon principles espoused by Integrated Project Delivery are predicated upon a fundamental premise that the primary participants in a project are willing to place the interests of the project ahead of their own self-interests. Therefore, an Integrated Project Delivery approach will not work well with some traditional forms of project delivery. Similarly, to maximize the benefit intended by Integrated Project Delivery, use of collaborative tools, such as building information technology, is recommended.

The core concepts of Integrated Project Delivery are visible in an approach to project delivery known as an Alliance, which has been pioneered to a great

---

34. *See* AIA Contracts Committee & AIA California Council, *Integrated Project Delivery: A Guide* (2007), *at* http://www.aia.org.ipdg (last accessed June 18, 2008).

35. In May 2008, the AIA published new contract forms for use with the Integrated Project Delivery methodology. The AIA offers two approaches to integrated project delivery, both of which mandate the use of building information modeling. The so-called transitional agreements describe an approach that utilizes a familiar AIA style contracting model, and include: (1) American Institute of Architects, AIA Document B195–2008, *Standard Form of Agreement Between Owner and Architect for Integrated Project Delivery* (2008) (an owner-architect agreement); (2) American Institute of Architects, AIA Document A195–2008, *Standard Form of Agreement Between Owner and Contractor for Integrated Project Delivery* (2008) (an owner-contractor agreement with a guaranteed maximum price amendment); and (3) American Institute of Architects, AIA Document A295–2008, *General Conditions of the Contract for Integrated Project Delivery* (2008) (a "shared" general conditions document). A second, more complete integration and sharing of risk and reward between the principal project participants is embodied in American Institute of Architects, AIA Document C195–2008, *Standard Form Single Purpose Entity Agreement for Integrated Project Delivery* (2008) (an agreement for the formation of a special-purpose entity for integrated project delivery). For another approach to integrated project delivery, see ConsensusDOCS 300, *Standard Form of Tri-Party Agreement for Collaborative Project Delivery* (2007).

extent in Australia.[36] In an Alliance approach, the primary participants are all parties to a single contract. Those signatories collectively form a team to manage the project. Overall project goals are established and memorialized in an appendix to the Alliance contract. Decisions of the management team, for the most part, must be unanimous. Staff may be comingled on the basis of the best person for the job. Every effort is made to align the interests of all the contracting parties.

Each party is reimbursed at its cost, which is verified independently. Everything is "open book." Other compensation terms vary among Alliance approaches. Nonetheless, there usually are mutually-agreed-upon formulae through which the parties share in both the profits and losses of the project, based on metrics tied to how well the completed project complies with predetermined project goals. To further enhance cooperation, the parties to the Alliance agree to resolve disputes amicably. Adjudication of disputes is not permitted for anything other than the most egregious of accusations, such as fraud.[37]

The selection of the right parties to participate in an Alliance is obviously crucial. If an Alliance is to succeed, each participant must buy into the basic premise that its coparticipants in the Alliance will conduct themselves appropriately and in the interests of the Alliance, even when operating in a "blame-free" environment. Once that assumption is deemed plausible, the parties essentially are making a risk-management decision that the benefits of reducing the time and resources normally devoted to resolving disputes and claims that plague the design and construction process will outweigh the inability to recover certain losses or damages that might be suffered through the participation in the Alliance.[38]

---

36. Project alliances were used to deliver major oil and gas projects in Western Australia, as well as infrastructure projects. *See* Department of Treasury and Finance, State of Victoria, Australia, *Project Alliancing Practitioners' Guide* (April 2006), *at* http://www.dtf.vic.gov.au/CA25713E0002EF43/pages/asset-management-project-support-project-alliancing. Use of alliancing appears more common with respect to projects that are time- or quality-driven. Such projects have been predominantly industrial or infrastructure projects, although a notable exception is the National Museum of Australia, which opened in Canberra in 2004.

37. *See* AIA Contracts Committee & AIA California Council, *Integrated Project Delivery: A Guide* (2007), *at* http://www.aia.org.ipdg (last accessed June 18, 2008).

38. Project alliancing principles are also found in Lean Project Delivery, which is largely derived from manufacturing objectives to produce value, as defined by the customer, without producing waste. By integrating the team that delivers the project, in theory, much of the back-and-forth of the normal project delivery approach can be eliminated. Production becomes more "real time." Lean Project Delivery also relies on an integrated agreement among the key members to a construction undertaking. Like an Alliance, the single, multiparty contract seeks congruence among the interests of the primary project parties, including a sharing of project successes and failures.

4.08 METHODS OF PROJECT DELIVERY SYSTEM SELECTION

With a wide variety of options available, how does an owner select a delivery system for a project? As discussed in Section 4.03, the selection process should endeavor to identify a delivery system that improves the potential for the owner to accomplish its project objectives, while reducing the potential for serious project risks.

Owners who build infrequently may not know which project delivery system to use and may not care. They rely on their advisors to make such selections, or the choice is a default to the approach that was used on a prior project. For other owners, choosing the project delivery system often is driven by a single, overriding concern. Either cost or time considerations, for example, may be so important that all other factors pale in comparison.

For owners who wish to evaluate a variety of project delivery systems, however, a number of different methods exist. All involve consideration of metrics that pertain to particular project goals, such as cost, schedule, quality, and claims avoidance but operate with different levels of formality and sophistication. After assembling data pertaining to such metrics, the more simple methods employ decision-trees or filters to eliminate options deemed less suitable for a project. Other methods are more diagnostic and, through the application of arithmetic formulae and analytical matrices, attempt to identify the optimal delivery system.[39]

Little quantitative analysis exists to determine preference for one project delivery system over another. Few studies have been performed. Some of the most striking data produced, however, reveal that all project delivery systems have a significant number of projects with poor results for each metric under review, even if statistically the overall or median results for a project delivery system are quite favorable.[40] Therefore, no project delivery system is a guarantee of good results. The delivery system must be properly suited for the project, and the project participants must implement the chosen system correctly and execute their responsibilities well, in order for the project to succeed.

---

39. For a more thorough discussion, *see* Ross J. Altman, Presentation of Paper, *Selection of Project Delivery Systems and Related Implications*, American Bar Association Forum on the Construction Industry, in Toronto, Ontario, Canada, April 29–30, 2005. Examples of analytic methods are described by: (1) Georgia State Financing & Investment Commission, *Selecting the Appropriate Project Delivery Option*, 2 Project Delivery Options, Recommended Guidelines (2003), *available at* http://gsfic.georgia.gov/; (2) *Owner's Tool for Project Delivery and Contract Strategy Selection User's Guide*, Construction Industry Institute, IR165-2 (2003).

40. *See Project Delivery Systems: CM at Risk, Design-Build, Design-Bid-Build*, Construction Industry Institute, RR133-11 (April 1, 1998).

4.09 SELECTION CONSTRAINTS

The owner must confirm that there are no major constraints that would impair or altogether prevent the implementation of a preferred project delivery system. The analysis might be as simple as acknowledging that the corporate culture of the owner, or its past experiences, do not allow the proper implementation of a particular project delivery system. In the case of a public entity, rules and regulations governing procurement might prohibit use of a certain project delivery system, no matter how attractive that system might be for a project. Issues commonly vetted in a post-selection analysis are described next.

A. Method of Construction Procurement

The method through which the owner procures construction services affects significantly which project delivery systems are best suited for the project. The most common methods for procurement of construction services include (1) lowest competitive bid, where selection is based solely on cost; (2) best value, where selection is based on a combination of cost and other factors; and (3) qualifications, where selection is based on qualifications and, perhaps, other noncost factors. There is little point in using a project delivery system that is not compatible with the required or desired method of procuring construction services. For example, when use of a lowest competitive bid is the mandated or desired method of procurement, then using a project delivery system that emphasizes overlapping design and construction phases probably is not advisable. If project scope is not well developed at the time of bid submission, the constructor would likely compensate for the lack of information by including a contingency that increases the price; otherwise, the constructor is forced to assume a disproportionate amount of financial risk.

B. Compensation Mechanisms

Several different mechanisms are available to compensate the prime constructor of a project. Compensation mechanisms used most frequently by owners include lump sum (stipulated sum), cost-plus-fee with a guaranteed maximum price, unit price, cost-plus-fee without a guaranteed maximum price, and time and materials.[41] The level of design completion at the time of contract award,

---

41. Different standard industry construction contract forms have been published for various different compensation mechanisms. For example, AIA Document B101–2007 is used when the contractor's compensation is a stipulated sum, but AIA Document A102–2007 is used when the contractor's compensation is based on the cost of work plus a fee (with a guaranteed maximum price). See also Chapter 9 for a discussion of compensation mechanics.

the willingness of the owner to accept a price risk, or some other allo-cation of risk and reward play into the approach ultimately selected. As with procurement methods, certain compensation mechanisms are more compatible with particular project delivery systems than others.

For example, a compensation mechanism that limits the owner's budget risk and transfers more financial exposure to the constructor (e.g., lump sum, guaranteed maximum price) requires the owner to yield more project control to the constructor. Therefore, such a compensation mechanism is less compatible with project delivery systems that involve overlapping of design, procurement, and construction phases, unless the owner is comfortable with a performance-based delivery approach (e.g., Design-Build) that permits a range of design solutions as opposed to one based on prescriptive or design specifications where the constructor has less design input.

Depending on factors unique to the project under consideration, different compensation mechanisms may be more appropriate than others. For example, changing from lump sum to a cost reimbursable compensation mechanism increases an owner's ability to change design, but reduces an owner's ability to control costs. If a suitable compensation mechanism cannot be identified for the project delivery system under consideration, then a different project delivery system might be needed.

C. Owner Capability

Each project delivery system requires a different skill set and level of involvement from the owner. A project delivery system that appears optimal based on the criteria of a selection approach is doomed to failure if the owner does not possess the ability to implement properly that project delivery system. To a certain extent, an owner may outsource its role in the project, but that alternative is not always desirable or without risk.

Issues pertaining to the owner's capabilities and expectations include, among others, the following:

- ❒ To what extent does the owner intend or desire to be involved in the production of the design of the project?
- ❒ Does the owner have skilled and available resources to administer the project delivery system?
- ❒ Does the owner have past experience, good or bad, with the project delivery system?
- ❒ Can the owner make timely decisions?
- ❒ Can the owner identify a qualified pool of designers and constructors?

- ❒ Do legal/corporate considerations restrict the owner?
- ❒ Can the owner facilitate project team cooperation and communications?
- ❒ Can the owner administer multiple contracts on the project?
- ❒ Does the owner have the ability or desire to take responsibility for design?
- ❒ What are the funding needs and availability?

Responses to such questions provide guidance as to whether the project delivery system under consideration is suitable.

D. Industry Sector Custom

Within the design professions and construction industry, there are industry sectors that over time develop their own customs and practices. Many industrial facilities are dependent on specialized equipment and, consequently, often rely on an EPC form of project delivery. A paper mill project illustrates this point well. The heart and soul of a mill is the paper-making machine. Everything else in the mill is ancillary and exists to support the paper-making machine. The project objectives of the mill owner pertain principally to the speed at which the paper-making machine will run, what periods of downtime are expected, and various characteristics of the product that is produced by the paper-making machine.

Over the years, manufacturers of paper-making machines have developed proprietary designs, as well as tremendous amounts of know-how regarding the fabrication and method of assembly of those machines. The ability of the original equipment manufacturer to utilize that know-how in the design of a machine generates many efficiencies. Although each paper-making machine involves a varying degree of custom design, many component parts and design features are more or less standard to the original equipment manufacturer. To further assure economical fabrication and assembly of paper-making machines, original equipment manufacturers make significant investments in machine tools and skilled labor forces.

It would be possible for a mill owner to completely engineer a paper-making machine to the mill owner's own unique specifications, either with its own in-house staff or through outside consultants and then retain a constructor or vendor to fabricate that particular machine. Procuring a paper-making machine in this manner, however, undoubtedly would require fabrication of component parts and assembly techniques vastly dissimilar from those that are readily available. The additional cost and time to deliver a paper-making machine in such a fashion would be difficult to justify. Therefore, acquiring the machine through use of an EPC delivery system that capitalizes on the experience and skill of the original equipment manufacturer is the preferred approach.

4.10 CONCLUSION

No single project delivery system is perfect, and more than one choice can be right for a project. Selecting a project delivery system, however, is not an easy task and involves a certain amount of conjecture. Nonetheless, the process of selection must be undertaken diligently. Industry experience and data, both anecdotal and quantitative, confirm that the choice of a project delivery system is extremely important. The allocation of obligations and their sequence greatly influence the ability of the owner to achieve its project goals. Various tools exist, with different levels of sophistication, to assist the owner in selecting a project delivery system and to remove some of the guesswork from the process.

QUESTIONS

1. What project delivery system selection factors might be important to an owner undertaking the development of a manufacturing facility?
2. If an owner wishes to maintain control over the aesthetic and functional design of the project, which project delivery systems would be preferable?
3. If an owner needs a stipulated sum contract, which project delivery systems are less suitable?
4. What features would you consider incorporating into an Alliance or Integrated Project Delivery approach?

CHAPTER

5

Contracting for Construction Projects

BRUCE MERWIN

5.01 CONTRACT FORMATION

A. For Purpose of Design Agreements and Construction Contracts

Construction transactions are complex. They involve multiple parties and are technically complicated. Such transactions require heavily documented financial arrangements, multistep payment procedures, and unique business terms and conditions. Finally, construction transactions also are complex from a legal standpoint. Therefore, it makes sense to prepare well-thought-out contracts to memorialize the agreements among the appropriate parties in connection with the design and construction of a project. The process of preparing and negotiating such contracts often helps the parties arrive at a better "meeting of the minds" and thereby diminishes the opportunity for disputes to arise out of frustrated expectations. Furthermore, a contract provides a written record of the deal between the parties to help unravel any disagreement that might occur.

The author thanks Jacob Fain, an associate at Haynes and Boone, LLP, for his valuable assistance in the preparation of this chapter. The author also thanks Joe F. Canterbury, Jr. for permission to use portions of his *Texas Construction Law Manual*, 3rd edition, with respect to Contract Formation.

The design and construction process involves reciprocal promises among the many project participants. Owners, architects, and contractors are primarily interested in five considerations affecting their respective contracts. First is the description of the scope of services or work for the project. Second is the timeframe within which the project will be completed, often described as the project schedule. Third are the parameters regarding completion of the project within the owner's budget, as amended by change orders. Fourth is the quality of the services and work, and various administrative matters that ensure such quality. Fifth is the interest each party to the contract has in its respective rights and remedies in the event that the other party breaches any of its promises. The design agreement and construction contract, in essence, should be a roadmap to successful completion of the project and the establishment of procedures for resolving disputes if problems occur during the design and construction process.

To summarize, design agreements and construction contracts should both recognize that the primary purpose of the contract is to do the following: (1) inform each party what it must do and to what it is entitled, (2) inform each party of its rights when the other party does not perform as promised, and (3) reflect the reality of contract administration and not require one party to agree to procedures that will not likely be followed or will significantly increase the cost of construction. Each party's attempt to protect its respective interests should not cause the design services agreement or construction contract to get out of balance to the point that the contract terms themselves become a more significant risk factor than the value and complexity of the services or work itself.

B. Contract Formation Generally

What constitutes a design agreement or a construction contract? As defined in BLACK'S LAW DICTIONARY, a contract is a promise or set of promises to which the law attaches legal obligations or an agreement between two or more parties creating legal obligations.[1] For the contract to exist, there must be an offer, acceptance, and consideration tendered by both parties.[2] If either of the elements of mutual assent or consideration is missing, then there is no contract.[3] Because of the complexity involved in a design agreement or construction contract, for the most part, such contracts are written agreements executed by both parties.

---

1. BLACK'S LAW DICTIONARY (8th ed. 2004).
2. Harco Energy, Inc. v. Re-Entry People, Inc., 23 S.W.3d 389, 392 (Tex. App.–Amarillo 2000).
3. Hubbard v. Shankle, 138 S.W.3d 474, 481 (Tex. App.–Fort Worth 2004).

It is possible to have an oral contract involving construction, and the parties are not required to explicitly express their intent in writing. A meeting of the minds can be implied from, and evidenced by, the parties' conduct and course of dealing.[4] The contract, to be enforceable, must be reasonably definite in its terms and must sufficiently describe the essential elements of the agreement. However, it may be difficult to establish oral agreements or contracts based solely on conversations between the parties.

C. Commencement of Services or Work Prior to Contract

Although not the preferred approach, owners often commence design and construction activities prior to execution of contracts for design services or construction work. Because of the number of issues that must be addressed in design agreements and construction contracts, there often is great time pressure on owners to commence design services while the balance of the contractual matters is being worked out or the economic viability of the project is being evaluated. When the advantages of accelerating completion of the design may outweigh the risk of proceeding without having a full contract in place, owners sometimes issue to an architect or other design professional an early start letter or limited authorization to proceed. Recognizing the risk in commencing design without a complete contract, an owner might build various controls into such an early start arrangement by, for example, authorizing performance of only a limited scope of design services with the hope that the deal can be finalized prior to completion of the services for which an early commencement is authorized.

Construction contracts are usually more complex than agreements with design professionals and, on large or particularly difficult transactions, can take months to draft and negotiate. The potentially adverse effects arising out of delays in commencement of construction, therefore, are amplified. For example, many projects require long lead-time items, such as materials, equipment, or systems that must be specially fabricated. There may be a limited window of opportunity to get onto a manufacturer's schedule in order for such items to be delivered to the site and available for incorporation into the work as necessary to conform to the owner's contemplated schedule. Any delay could be costly, and an owner might want to commence the ordering of such items while final terms of a construction contract are negotiated.

Given the volatility in labor and material costs, it is often critical to the owner to lock in prices for materials, reserve the use of equipment or production time of subcontractors, or otherwise start the project prior to finalizing

4. Harrison v. Williams Dental Group, P.C., 140 S.W.3d 912, 916 (Tex. App.–Dallas 2004).

the terms of a construction contract. Loan agreements and leases entered into by developers often have fixed deadlines with financial penalties or additional costs if construction is not started by a specific date that requires commencement prior to the negotiation of a construction contract.

Therefore, early start letters often outline a limited scope of work for a specific price or cost of the work, plus a fee, not to exceed a certain amount, and a specific timetable for performance of such work. If the owner and contractor are not able to reach agreement with respect to the construction contract, then such early start letters provide for the termination of the owner-contractor relationship, final payment of the contractor, and either assignment of subcontracts and purchase orders from the contractor to the owner or termination of such subcontracts and purchase orders. Early start letters also generally provide that they will be superseded by the construction contract entered into between the owner and contractor.

Letters of intent for construction contracts are similar to limited authorizations to proceed, except that most letters of intent are non-binding with respect to the material terms of the construction contract. The purpose of the letter of intent is to express a good faith intention by the owner and contractor to enter into a construction contract once drawings and specifications have been further refined, pricing is further refined, and the contractor has been able to obtain bids from subcontractors and suppliers. Letters of intent, although granting no legal rights, may serve the practical purpose of allowing a contractor to show such letter to subcontractors and suppliers for the purpose of obtaining bids from such subcontractors and suppliers.[5]

D. Creation of Contracts—Implied Contracts and Quasi-Contracts

While express contracts are stated by the parties, implied contracts arise from the actions and conduct of the parties.[6] The facts and circumstances indicate that a mutual intent exists between the parties to a contract.[7] The existence of such a contract becomes a question for the trier of fact.[8]

---

5. Letters of intent may be binding in a limited respect with respect to confidentiality provisions, the obligation to negotiate in good faith for a certain period of time, and also an obligation on the part of the contractor to perform limited work for a specific price. This is similar to early start letters.

6. Haws & Garrett General Contractors, Inc. v. Gorbett Bros. Welding Co., 480 S.W.2d 607, 609 (Tex. 1972).

7. Gillum v. Republic Health Corp., 778 S.W.2d 558, 569 (Tex. App.-Dallas 1989).

8. *Haws & Garrett*, 480 S.W.2d at 610.

In *Haws & Garrett General Contractors, Inc. v. Gorbett Bros. Welding Co.*,[9] a general contractor was engaged in bridge construction, and his superintendent made a telephone order for a crane and an operator. The crane subcontractor prepared a form contract containing a broad hold-harmless provision. However, the contract was not signed until the work was done. The structure was damaged during the work, ostensibly because of the subcontractor's negligence. In response to a lawsuit brought by the general contractor, the subcontractor argued that, since there was no express agreement as to the type of indemnification provision prior to performance of the work, such an agreement should have been an implied condition of the contract.[10]

The Texas Supreme Court, however, rejected this claim of implied agreement on the grounds that the subcontractor did not prove that prior contracts between the parties had been preceded by any negotiations regarding the indemnification clause or that any prior agreements with respect to the clause had been made by authorized personnel.[11] Presumably, if the parties' prior course of dealing had evidenced an agreement with respect to this clause, it could have been an implied term of the bridge contract and would have insulated the subcontractor from negligence liability.

Quasi-contract is generally considered no contract at all, but rather the creation of the court, where appropriate, to correct a substantial injustice and avoid unjust enrichment. For example, with respect to extras, courts seem to be relying on the theory of an implied contract, without indicating whether they are relying on a contract implied in fact (i.e., a true contract) or a contract implied in law (i.e., a quasi-contract). If the relief granted is based upon a theory of an implied in fact contract, the remedy is in *quantum meruit*.

Quantum meruit is a Latin phrase that means "as much as is deserved." The rationale underlying the *quantum meruit* doctrine is that where one knowingly accepts a benefit from another who is not a volunteer and who does not confer the benefit as a gift, the law may infer a promise by the recipient to pay reasonable compensation. However, if the benefited party is not aware of the work being done at the time it is being performed, there is no basis for a recovery in *quantum meruit*. The concept is one of equity, since it would be unjust to allow a party to knowingly receive benefits and/or services without paying for the value of work completed even in the absence of a binding contract.[12]

9. *Id.* at 607.
10. *Id.* at 608–09.
11. *Id.* at 610.
12. De Leon v. Saldana, 745 S.W.2d 55, 57 (Tex. App. San Antonio 1987), writ denied (1988).

5.02 HOW TO SELECT A CONTRACT FORM

A. Fundamental Issues

Although the terms of different agreements for design services may vary significantly, the basic framework of design contracts is similar. Construction contracts, however, come in a variety of shapes and forms. The basic framework of a construction contract is driven predominately by two issues. First, what is the applicable project delivery system? And, second, what is the mechanism used to determine the compensation of the contractor? The terms of a construction contract will be very different depending on the answer to those two questions. Consequently, numerous options exist among construction contract forms in order to provide a starting place document that conforms to the fundamental aspects of the transaction, thereby reducing the need for undue customization in connection with the applicable project.

Once the project delivery system and compensation mechanism are established, the owner and contractor need to determine whether to use a standard industry form or a manuscript (originally drafted) document that has been prepared by one of the parties to the construction contract for use on the project. There are a number of industry-prepared forms available for each type of project delivery system and compensation framework, although the American Institute of Architects (AIA) forms are the most commonly used for commercial building construction in the United States.

B. Standard Industry Forms

A number of advantages can be identified for using standard industry forms for most private construction projects, including, without limitation, the following:

- Many form documents enjoy a reputation of relative fairness to the parties and are accepted by the construction industry as a reasonable starting point for use in writing and administering construction agreements.
- The economic realities on any project benefit from the quick conclusion of negotiations on the part of many construction contracts, and standard forms are more likely to be understood and negotiated quickly than manuscript documents. If the parties have to analyze and review each of the risk allocation and other substantive provisions of a manuscript document, the contracting process will take much longer and cost much more than when using AIA and other standard industry forms.

- Most standard forms contemplate related forms with respect to the role of the other parties in the construction process, such as the architect, engineer, and lender.
- Many of the standard industry forms are linked together with forms published by the same trade association, and, therefore, they fit and accommodate the allocation of risk and responsibility provided in other contractual relationships on the same projects.

The disadvantages of using standard industry forms are as follows:

- Standard forms may be out of date or fail to reflect current law or customary practices. Standard forms, by their nature, cannot address jurisdictional issues.
- Most standard industry forms contain provisions that have a bias toward the industry group that publishes and promotes the form. To protect against such bias, the other parties to the agreement must know how to modify the key provisions of the standard industry form.
- On specialized projects, a standard industry form may not be appropriate because it does not deal directly with the important issues raised by the specific scope of work. The administrative costs and procedures included in the standard industry form also may be inappropriate. In other words, standard industry forms often describe only the "lowest common denominator" applicable to a range of projects.
- Forms generally work best for a particular sector of the construction industry. Most AIA forms, for example, contemplate general building projects and are not well suited for other sectors that involve heavy or complex construction or industrial projects.

C. Advantages and Disadvantages of Using a Manuscript Document

Custom form construction contracts generally are used by large institutional developers continuously involved in construction projects and by commercial developers and commercial contractors on smaller projects, including Design-Build projects. Custom form contracts also are used by developers with in-house construction divisions that may or may not be profit centers for the developer. Most owners and contractors developing a custom form start with one of the AIA or other standard industry forms, and modify the form substantially. In some cases, the organizational format of the standard form is retained, which facilitates review by the other party. In other cases, provisions from the AIA form or other standard form are included in the custom form with little or no modification.

The major advantage of custom forms is that they are drafted for a specific project. Since the design and construction process reflects a complex web of different but interrelated contractual relationships, the custom form can define the precise relationship of the following separate parties in the context of the subject project: owners and developers, architects, contractors, engineers, construction managers, trade subcontractors, suppliers, consulting and testing, design engineers, surety companies, construction lenders, and inspection services. For the owner drafting the custom form, another advantage is the ability to overcome the perceived bias of the standard industry forms toward the architect and contractor.

The disadvantages of using manuscript documents are primarily based on lack of uniformity among, and familiarity with, the terms of such documents. Use of the custom form undoubtedly will raise a host of concerns for the other contracting party, including trying to determine which of the rights and procedures found in the AIA or other standard industry forms have been omitted or substantially modified in the custom form. Almost inevitably, the use of a custom form will cause the other party to the agreement to incur additional expenses due to the time necessary to understand the custom form; and sometimes will cause the contractor to increase its bid on general principles to cover perceived risks or to cover the legal fees incurred in negotiating the form. Finally, the lack of experience with the particular custom form is generally a major stumbling block in reaching a quick conclusion of negotiations with respect to the custom form.

5.03 OVERVIEW OF STANDARD INDUSTRY FORMS

A. AIA Documents

AIA documents are divided into several letter-designated series, and each series pertains to a particular type of form or use. The various forms in each series are drafted carefully to complement documents in each other series. Therefore, an AIA form of agreement between owner and architect and an AIA form of agreement between owner and contractor should have consistent terms. The most common series of AIA documents are the A-Series, B-Series, C-Series and G-Series, discussed in greater detail next.

The A-Series applies to the owner-contractor relationship. There are two principal types of documents outlining agreements between the owner and contractor, which include:

- ❒ *Owner-Contractor Agreements (the 2007 editions of which are numbered A101, A102 (formerly the A111), A103 (formerly the A114), and A107.*

The owner-contractor agreement sets forth the basic economic deal between the parties. Key terms pertaining to time, cost, and other basic parameters of construction are added to the text of the contract form. Different AIA forms of owner-contractor agreements are available for various project delivery systems, with options for different compensation mechanisms that are commonly employed with each.

- *The General Conditions of the Contract for Construction (A201 and variations).* General conditions are an integral part of the contract for construction. They establish the rights, responsibilities, and relationships of the owner, contractor, and architect. Different versions of AIA Document A201 are available with respect to certain project types. For example, a modified version of the A201 is published for projects delivered through an At-Risk Construction Management approach. A different version of the A201 also is available for use on projects procured by the federal government. The A201 typically is adopted by reference in certain other AIA documents, such as owner-architect agreements, owner-contractor agreements, and contractor-subcontractor agreements. Thus, it often is called the "keystone" document.

Since conditions vary by locality and by project, supplementary conditions often are added to amend or supplement portions of the general conditions as required by the individual project. The A511–1999, *Guide for Supplementary Conditions*, provides a guide for creating supplementary conditions for the A201.

The B-Series of AIA documents pertain to the relationship between the owner and the architect. In addition to the principal forms of agreement between owner and architect, there also are forms that address the role of an architect for projects delivered on a construction management basis, such as the owner-construction manager (CM) agreements (B141/CMa, B801/CMa, A101/CMa, and A201/CMa–1992).

The C-Series documents govern the relationship between the architect and its consultants, such as engineers and other design professionals. They are drafted to "flow down" to such consultants the obligations that the architect has to the owner. Architects want to avoid situations where their obligations to the owner are greater than the obligations that the architect's consultants (i.e., the architect's consulting engineers) have to the architect. For example, the obligations of the architect's consulting engineers (mechanical and structural), as to their respective portion of the design work to the architect, with respect to basic services, additional services, insurance coverage, and indemnification should be consistent with the obligations of the architect to the owner with respect to such issues.

The G-Series documents include a number of forms commonly used in the day-to-day administration of the project, such as insurance specifications, change orders, certificates for payment, applications for payment, and schedules of values.

When using the AIA documents, it makes sense to see them as families of related documents—families that relate to the various delivery approaches that an owner and architect may employ to structure projects. With respect to the AIA documents:

- There are a number of different B-Series forms for the owner-architect agreements; selection of the appropriate form is based on the scope of the project and role of the architect.
- The type of contractor compensation employed (stipulated sum, cost-plus-fee) requires different A-Series owner-contractor agreement forms.
- The choice of general conditions form tracks the owner-contractor agreement form; there is a separate form for federal work (AIA Document A201 S/C).
- There is a separate family of Design-Build documents.
- A number of A-, B-, and G-Series documents are designated for use in projects utilizing a construction manager.
- There are specific forms for interior projects, as well as for projects of abbreviated or limited scope.

The AIA Documents Committee revises the AIA documents periodically, usually about every 10 years. There are new 2007 editions for most of the AIA construction contract forms. However, many contractors will prefer to continue using the 1997 edition of the construction contracts for two primary reasons. First, the specific 2007 revisions are slightly more favorable to owners than contractors. Second, contractors are familiar with the provisions of the 1997 forms and resistant to change. However, the AIA has stated it would discontinue providing electronic format access to the 1997 edition as of May 31, 2009. To determine which edition is applicable to a particular project, the drafter of the contract will need to review the Request for Proposal prepared by the owner or architect or the Project Manual prepared by the architect. The bid proposal section of the Project Manual should list the AIA documents that are applicable to the project.

In the private construction industry, the AIA contract forms have dominated the field for decades. The following reasons probably account for such continued dominance:

- Although the AIA forms are written by a trade association for architects, the AIA seeks input and endorsements from as many industry groups as possible. As a result, the AIA documents enjoy a wide reputation of relative fairness to parties and are accepted by the construction industry as a natural starting point for use in writing, revising, and administering construction agreements.
- Use of a custom form undoubtedly will raise a host of concerns for the other contracting party, including trying to determine which of the rights and procedures from the AIA forms have been omitted or substantially modified in the custom form. This may result in resentment and additional expense on non-construction matters.
- Most construction projects involve multiple parties, most of whom are familiar with the AIA documents.
- The economic realities of most projects require the quick conclusion of negotiations on the part of many construction participants.
- The AIA forms contain contractual provisions and language that have been frequently interpreted in the courts.
- The experience factor gives the AIA forms a strong edge over competing form contracts or custom contracts of an owner or contractor.
- The AIA provides a wide variety of forms to accommodate the differences of the rules of the road for different types of construction projects, and the AIA forms are linked to fit and accommodate the allocation of risk and responsibility provided in the other AIA forms for other contractual relationships on the same project.
- The history of the AIA forms, their widespread familiarity and use, and the scrutiny and commentary the AIA forms have received from industry groups operate as an inexpensive substitute for legal review and advice in individual transactions. However, it remains important for the parties to the AIA (or other standard task group) forms to retain counsel and make sure that such forms are consistent with the actual intent of the parties. A limited review by counsel would be better than blind reliance that the AIA forms adequately cover all of the important legal issues relevant to the specific project, without the necessity of revising the AIA forms in some manner. When parties must stop to analyze and allocate each of the risks that they take under a non-AIA form construction contract, the contracting process can take much longer and cost much more than it does when using AIA forms.
- AIA documents are available in electronic format on a computer disk, so that they can be modified in a more efficient manner than striking paragraphs by deletion, interlineations, and/or use of addenda pages

or an addendum to the entire form. The electronic system currently in use includes the "additions and deletions report." This alerts the other party to the transaction that the original AIA document has been edited and avoids misleading the party that it is an unaltered standard AIA document.

B. AGC Forms

The Associated General Contractors of America (AGC), the nation's oldest and largest construction trade association, is an organization of construction contractors and industry-related companies. The AGC provides a full range of services satisfying the needs and concerns of members, including drafting over 70 industry-tested contracts and forms. AGC jointly drafts these contracts with owners, contractors, subcontractors, and sureties. The most recently published forms are known as ConsensusDOCS and are also available in electronic format. The stated goal of ConsensusDOCS is to better the industry by providing standard contract documents that strive to reflect the consensus for best practices.

ConsensusDOCS are organized into five series:

- General Contracting
- Collaborative Agreements
- Design-Build
- Construction Management and Program Management
- Subcontracting

The AGC drafting process purports to craft documents that balance perspectives and allocate risks and responsibilities fairly. For example, with respect to the Design-Build series of AGC documents, the drafters contend that those forms reflect industry "best practices" and provide the framework for a cooperative relationship between the parties to the Design-Build project. Like the AIA series of documents, AGC documents have different forms for the different project delivery systems and compensation mechanisms. All of the AGC forms are interrelated.

C. EJCDC Forms

The Engineers Joint Contract Documents Committee (EJCDC) develops, publishes, and updates standard contract documents for the design and construction of engineered facilities. The EJCDC construction series (C-Series) is comprised (as of 2007) of 21 documents, including a narrative guide for use in establishing and administering the contractual relationship between a project owner and a contractor.

The EJCDC documents are drafted with the active participation of representatives of certain owner groups; risk managers; professional liability insurers; surety and insurance experts; construction lawyers; various professional societies, including but not limited to the supporting societies; and construction managers. The EJCDC documents are intended principally for use on projects designed by engineers, including public infrastructure; water and wastewater treatment and conveyance facilities; utility work; solid waste handling and disposal facilities; transportation projects; production and processing facilities; site development work; environmental remediation projects; street, curb, and gutter work; tunneling and excavation projects; and similar applications.

D. Other Standard Industry Forms

The Design-Build Institute of America (DBIA), founded in 1993, is an association of leaders in the design and construction industry utilizing Design-Build and integrated project delivery methods to achieve high-performance projects. The DBIA has published 10 contracts for Design-Build construction for use by the various participants in the Design-Build process, contractors, designers, architects, engineers, owners and managers, and suppliers. The DBIA model contracts cover all project forms of compensation.

The Construction Management Association of America (CMAA) is a national organization designed to promote the growth and acceptance of construction management as a professional discipline that can add significant value to the entire construction process, from conception to ongoing operation. The CMAA publishes seven principal forms designed to cover the two general approaches to construction management.

5.04 WHAT IS THE CONTRACT?

A. The Design Agreement (Owner-Architect)

The agreement between an owner and architect is comprised generally of a single document and exhibits. The exhibits might include typically a timetable for the performance of services, a list of required insurance coverages, an index of key personnel and consultants, a digital data protocol, and other terms or conditions that vary or complete the underlying form of agreement or that are specific to that particular transaction. Some industry forms utilize an exhibit through which the parties identify the business terms that are unique to their particular transaction, providing an opportunity to keep the form language as intact as possible.[13]

B. The Contract Documents (Owner-Contractor)

The term "contract documents" identifies the documents, in whatever form or medium, that describe the work that is to be performed by the contractor and the contractual relationship between the owner and contractor in connection with the performance of such work. The contract documents collectively form the construction contract and typically include the following instruments:

1. Owner-contractor agreement, which is an agreement between the owner and the contractor containing the essential business elements of the owner-contractor relationship, such as the identity of the parties, a description of the work, the price, the time for commencement and completion, procedures for payment, and a list of the contract documents;
2. General conditions of the contract for construction, which expand upon the information provided in the agreement between the owner and the contractor and provide more complete contract terms governing the parties, such as the owner's obligations to the contractor; the services that the contractor will provide; insurance and bonding requirements; changes in the work; procedures for requesting an extension in time; payment procedures, site safety, suspension and termination procedures; and, although the architect is not a party to this document, the construction administration services that the architect will provide;
3. Supplementary or special conditions of the contract for construction, modifying or supplementing the agreement and general conditions of the contracts;
4. Drawings prepared by the design professionals, which are the graphic and pictorial expressions of the work to be performed, and which may be in electronic as well as printed media;
5. Specifications prepared by the design professionals, which are the written technical requirements for the materials, equipment, systems, and standards for the work;
6. Addenda modifying or supplementing any of the foregoing instruments that are issued before the construction contract is executed;

13. *See* American Institute of Architects, AIA Document B101–2007, *Standard Form of Agreement Between Owner and Architect* § 3.2.1 (2007) [hereinafter AIA Document B101–2007], which requires the completion of an exhibit to specify fundamental terms of the agreement.

7. Other documents/exhibits that describe certain procedures applicable to the work, including governmental requirements, such as special safety requirements, union bidding requirements, affirmative action hiring requirements, minimum wage requirements, etc.; and
8. Modifications or changes to the contract in the form of construction change directives, field orders, change orders, or other modifications that become part of the contract documents after commencement of the work.

Most of the industry-prepared construction contracts include some form of general conditions of the contract for construction, similar to the A201, as a separate document. These general conditions documents, like the A201, are incorporated by reference into owner-architect agreements, owner-contractor agreements, contractor/subcontractor agreements, and the architect's subcontracts with consulting engineers. By separating these general conditions contracts from the standard form of agreement between the owner and contractor, the owner and/or contractor can deliver the instrument to a necessary party to the construction project without disclosing the business terms of the owner's agreement with the contractor. Another important benefit separating the standard form agreement from the general conditions is the ease of dealing with subcontractors and their ability to recognize and understand the relationship between their respective subcontracts and the construction contract.

C. The Integration Clause

Many construction contracts contain an "integration clause" in which owner and contractor agree that "the Contract represents the entire and integrated agreement between the parties hereto and supersedes prior negotiations, representations or agreements, either written or oral."[14] The provision then explains how the construction contract may be amended or modified. Practically every well-drafted agreement will have a provision similar to the integration clause contained in the A201 to eliminate oral and written negotiations that have not been included in the contract instrument. It is this type of clause that intends to prevent the invariably lengthy and fluid negotiation process from becoming part of the contract, except to the extent memorialized in the construction contract itself. The integration clause terms expressly provide that the construction contract supersedes all prior negotiations, proposals, bids, representations or agreements, whether written or oral.

---

14. American Institute of Architects, AIA Document A201–2007, *General Conditions of the Contract for Construction* § 1.1.2 (2007) [hereinafter AIA Document A201–2007].

D. Incorporation by Reference versus Enumeration of Documents

The agreement between an owner and contractor usually includes an enumeration of the contract documents to specify which documents comprise the construction contract. Some construction contracts may also incorporate various documents by reference. Regardless of which technique is used, problems arise if it is not clear which documents collectively form the construction contract.[15]

Also, to insure that the owner is not put at risk through the subcontracting process, almost all construction contracts contain a "flow down" clause. Such provision requires the contractor to provide in its subcontracts that each subcontractor, to the extent of its work, "be bound to the contractor by terms of the contract documents, and to assume toward the contractor all the obligations and responsibilities, including the responsibility for safety of the subcontractor's work, which the contractor, by these contract documents, assumes toward the owner and architect."[16] In other words, the work of the subcontractor must comply with the same requirements that apply to the contractor. The flow down of these obligations protects the owner by providing assurance that the subcontractor is required to comply with obligations consistent with those of the contractor, and also protects the contractor by ensuring that its obligations to the owner are no greater than the obligations that the subcontractor owes to the contractor.

5.05 INTERNAL CONFLICTS

With multiple contract documents collectively forming the construction contract, inconsistent provisions are a possibility. Some drafters attempt to address this issue through creation of a hierarchy of contract documents. A common approach is for modifications and change orders to have the highest degree of priority, followed by the agreement, the supplement to general conditions, the general conditions, and the drawings and specifications (which may be equal with differences to be resolved by the architect or the specifications should control) and any other ancillary document. The owner (with the architect's assistance) and the contractor should work together to make sure that the hierarchy is accurately stated. The contractor will want its bid or proposal to be a contract document and to govern and control other contract documents to protect the basis upon which the contractor has agreed to price, times, and other contractual matters.

---

15. *See* Atlantic Mutual Ins. Co. v. Metron Eng'g and Constr. Co., 83 F.3d 897 (7th Cir. 1996); Ritchies Food Distrib., Inc. v. Refrigerated Constr. Serv., 2004 WL 957659 (Ohio Ct. App. 2004).

16. AIA Document A201-2007, *supra* note 14, § 5.3.1.

The construction contract often will have more than one hierarchy to address particular types of conflicts. In addition to the hierarchy that addresses differences or conflicts between the contract documents, a second hierarchy may be included with instructions to resolve a specific type of conflict. For example, many construction contracts provide that in the event of inconsistencies within a particular contract document, contractor shall (1) provide the better quality or greater quantity of work or (2) comply with the more stringent requirement, either or both in accordance with the owner's interpretation. Other examples include instructions that large-scale drawings govern over small-scale drawings. Contract drafters have devised almost limitless rules through which to resolve internal conflicts in documents. Regrettably, hierarchies can have unintended results, and multiple hierarchies can themselves conflict, totally nullifying the intended benefit of including a hierarchy provision. Since each project is different, it is critical that the owner and architect work together to make sure that the hierarchy is accurately stated.

Many construction contracts do not use a hierarchy of documents. Instead, they may include a general instruction that permits the architect to resolve a conflict or may use a default mechanism obligating the contractor to comply with the more stringent requirement. Other construction contracts simply provide that the contract documents must be read as a whole. Indeed, the AIA forms provide that "[t]he intent of the Contract Documents is to include all items necessary for the proper execution and completion of the Work by the Contractor. The Contract Documents are complementary, and what is required by one shall be as binding as if required by all."[17] In other words, AIA contract documents are to be read as whole and reconciled accordingly, rather than setting forth a prioritization of documents that might lead to unpredictable, and undesirable, results. But conflicts can and do occur, so judicial rules of contract interpretation have developed over the years to assist in resolving conflicting requirements, including those that are not resolved by a document hierarchy.

5.06 CONTRACT INTERPRETATION

A. Rules of Contract Construction

Courts, not juries, determine whether a contract is *ambiguous*.[18] If the court finds that it is unambiguous, the court gives it the only legal meaning to which

17. *Id.* §1.2.1.

18. Mark V, Inc. v. Mellekas, 845 P.2d 1232, 1235 (N.M. 1993).

the contract is susceptible.[19] The *interpretation* of an ambiguous contract, however, must be determined by the trier of fact.[20]

There are three primary rules of contract construction:

1. Apply the reasonable and logical meaning;
2. Construe the contract in its entirety; and
3. Apply normal and ordinary meaning of a contract's terms.[21]

The language of a contract should be given its plain grammatical meaning, unless it definitely appears that the intention of the parties would be defeated thereby.[22] In *Reilly v. Rangers Mgmt., Inc.*, the Texas Supreme Court stated:

> A court should construe contracts from a utilitarian standpoint bearing in mind the particular business activity sought to be served and need not embrace strained rules of interpretation which would avoid ambiguity at all costs. Courts will avoid where possible and proper a construction which is unreasonable, inequitable, and oppressive.[23]

If the court finds the contract to be clear after application of these primary rules, there will be no resort to secondary rules of construction, such as evidence of custom and usage.[24] If, however, after applying these primary rules of interpretation, a written instrument remains reasonably susceptible to more than one meaning, extraneous evidence is admissible to determine the true meaning of the instrument.[25]

B. Parol Evidence Rule

Before parties sign an agreement that outlines their respective rights and obligations, there are often preliminary negotiations that are not intended to establish contracts in and of themselves. When the written contract is finally

---

19. R & P Enters. v. LaGuarta, Gavrel & Kirk, Inc., 596 S.W.2d 517, 519 (Tex. 1980); Weekley v. Weekley, 27 S.E.2d 591, 595 (W.Va. 1943).

20. *In re* Nelson Co., 959 F.2d 1260, 1264 (3d Cir. 1992); Reilly v. Rangers Mgmt., Inc., 727 S.W.2d 527, 529 (Tex. 1987).

21. *See, e.g.*, SW. Bell Tel. v. PUC, 467 F.3d 418, 422 (5th Cir. 2006).

22. Wash. Constr. Co. v. Spinella, 84 A.2d 617, 619 (N.J. 1951).

23. *Reilly*, 727 S.W.2d at 530 *(citing* Neece v. AAA. Realty Co., 322 S.W.2d 597, 602 (1959)); *see also* Hicks v. Smith, 330 S.W.2d 641, 646 (Tex. Civ. App.–Fort Worth 1959) writ ref'd n.r.e.

24. George v. El Paso County Water Control & Imp. Dist. No. 1, 332 S.W.2d 144, 148 (Tex. Civ. App. El Paso 1960) writ ref'd n.r.e.

25. *Enters*, 596 S.W.2d at 519; Brown v. Royal Maccabees Life Ins. Co., 137 F.3d 1236, 1247 (10th Cir. 1998).

signed, it may fail to deal with all issues discussed in preliminary negotiations, or it may deal with such issues differently than in the discussions. The application of the parol evidence rule works to bar these preliminary negotiations from being considered when a dispute arises over a written contract. The rule has been stated in CORBIN ON CONTRACTS as follows:

> When two parties have made a contract and have expressed it in a writing to which they have both assented as the complete and accurate integration of that contract, evidence, whether parol or otherwise, of antecedent understandings and negotiations will not be admitted for the purpose of varying or contradicting the writing.[26]

Although the word "parol" means oral, the parol evidence rule does not limit itself only to prior oral discussions.[27] Any prior writing that is not incorporated into the final agreement may be excluded as well.[28] Further, and again contrary to its very name, the parol evidence rule is not considered a rule of evidence at all; rather, many courts regard it as a rule of substantive law.[29]

The parol evidence rule only applies to prior or collateral communications that vary or contradict the terms of a written contract.[30] Parties always have the power to modify or supplement a written contract after its execution.[31] The added material, however, must satisfy all of the elements of a contract with respect to mutual assent and consideration.[32]

If it appears from the face of the written contract that the contract was intended to be a complete and final expression of the entire agreement (an integration), courts will not permit either party to testify as to prior understandings for the purpose of varying or contradicting the contract as written.[33] The rationale is that one purpose of creating a written agreement is to memorialize the applicable terms and to exclude all other understandings to the contrary.[34] To allow any prior or collateral agreement to vary or contradict the terms of the final written agreement would defeat its purpose.

---

26. CORBIN ON CONTRACTS § 573 (1993).

27. Altrutech, Inc. v. Hooper Holmes, Inc., 6 F. Supp. 2d 1269, 1275 (D. Kan. 1998).

28. *Id. See also* Segari v. Uchello, 44 So. 2d 722, 724 (La. Ct. App. 1950), stating some narrow exceptions to the general rule.

29. *See, e.g.*, Weber v. Pascarella Mason Street, LLC, 930 A.2d 779, 787 (Conn. App. Ct. 2007) (*quoting* Alstom Power, Inc. v. Balcke-Durr, Inc., 849 A.2d 804, 811 (Conn. 2004)).

30. *Id.*

31. A.W. Wendell & Sons, Inc. v. Qazi, 626 N.E.2d 280, 287 (Ill. App. Ct. 1993).

32. Hathaway v. Gen. Mills, Inc., 711 S.W.2d 227, 228 (Tex. 1986).

33. Cavě Hill Corp. v. Hiers, 570 S.E.2d 790, 794 (Va. 2002).

34. Filippi v. Filippi, 818 A.2d 608, 619 (R.I. 2003).

In determining whether a document is an integration, courts view the surrounding circumstances in which the agreement was reached as well as the actual written words within the contract.[35] As discussed in Section 5.04(C), many construction contracts contain a "merger" or "entire agreement" clause. An example of such a clause is found in *Advertising and Policy Committee of the Avis Rent A Car System v. Avis Rent A Car System*: "The agreement supersedes all prior agreements whether written or oral between the parties hereto [and] contains the entire agreement of the parties, [and] no representations, inducements, promises, or agreements, oral or otherwise, between the parties not embodied herein shall be of any force or effect."[36]

There are some cases, however, where parol evidence can apply. A party may introduce parol evidence to show that a contract was never formed or that the contract is unenforceable due to fraud, accident, or mutual mistake.[37] Also, if it is determined that the writing was not intended to encompass the parties' final agreement, the writing may be supplemented by proof of other terms and agreements outside of the document.[38] Further, testimony is allowable to place the written word in the context of the specific factual situation surrounding its execution.[39]

C. Ambiguity and Intent of Parties

In construing contracts, the fundamental objective is to ascertain the intent of the parties as expressed by the language of the instrument, regardless of the subjective intent of the parties.[40] "Where the parties reduce an agreement to a writing, which, in view of its completeness and specificity, reasonably appears to be a complete agreement, it is taken to be an integrated agreement unless it is established by other evidence that the writing did not constitute a final

---

35. *See* Money v. Fort Hayes State Univ. Endowment Ass'n., 64 P.3d 458, 462 (Kan. App. 2003) ("Kansas law permits the admission of parol evidence to show that even an unambiguous written agreement was not a complete integration. . . .").

36. Adver. & Policy Comm. of the Avis Rent-A-Car Sys. v. Avis Rent-A-Car Sys., 780 S.W.2d 391, 394 (Tex. App.–Houston [14th Dist.] 1989), writ granted, *judgment vacated*, 796 S.W.2d 707 (Tex. 1990), vacating in accordance with terms of settlement agreement).

37. Branstetter v. Cox, 496 P.2d 1345, 1347 (Kan. 1972); *Weber*, 930 A.2d at 779, 787 (*quoting Alstom Power*, 849 A.2d at 811).

38. *Weber*, 930 A.2d at 787 (*quoting Alstom Power*, 849 A.2d at 811).

39. Commercial Union Ins. Co. v. Martinez, 635 S.W.2d 611, 613 (Tex. App.–Dallas 1982), writ ref'd n.r.e.

40. Missouri ex rel Vincent v. Schneider, 194 S.W.3d 853, 859–60 (Mo. 2006).

expression."[41] Many courts will look to the four corners of the instrument to ascertain the intent of the parties.[42] The courts look not to what the parties meant to say, but to the meaning of what they did say.[43]

In the case of *Nat'l Union Fire Ins. Co. of Pittsburgh, PA v. CBI Indus., Inc.*, the Texas Supreme Court articulated the law on ambiguity in a contract:

> If a written contract is so worded that it can be given a definite or certain legal meaning, then it is not ambiguous. If however, the language is subject to two or more reasonable interpretations, it is ambiguous. Whether a contract is ambiguous is a question of law for the court to decide by looking at the contract as a whole in light of the circumstances present when the contract was entered. Only where a contract is first determined to be ambiguous may the courts consider the parties' interpretation and admit extraneous evidence to determine the true meaning of the instrument.[44]

Ambiguity does not arise merely because contracting parties disagree as to the correct interpretation of their contract.[45] Only if a court finds a contract to be ambiguous may the parties' differing interpretations be considered.[46] If the disputed provision is plain and unambiguous to a judge, the parties' construction is immaterial.[47]

In *Standard Constructors, Inc. v. Chevron Chem. Co.*, the Houston appellate court was presented with a dispute concerning claims for overtime charges for use of equipment.[48] In the contract documents, for certain pieces of equipment, "n/a" (agreed to mean "non-applicable") was listed instead of a rate.[49] The court rejected the argument that "n/a" should be construed to mean variable or dependent rates and held that the term "n/a" had one meaning in the

---

41. Restatement (Second) of Contracts § 209(3) (1981).

42. Missler v. Anne Arundel County, 314 A.2d 451, 457 (Md. 1974).

43. *Id.* (stating that "[t]he written language embodying the terms of the agreement alone will govern the rights of the parties irrespective of their subjective intent if that language is, on its face, clear and unambiguous").

44. Nat'l Union Fire Ins. Co. of Pittsburgh, PA v. CBI Indus., Inc., 907 S.W.2d 517, 520 (Tex. 1995) (citations omitted); *see also* Columbia Gas Transmission Corp. v. New Ulm Gas, Ltd., 940 S.W.2d 587, 591 (Tex. 1996), explaining that, in order to determine the parties' agreement, the court needed to examine the entire contract and circumstances surrounding its formulation.

45. *See* McJunkin Corp. v. City of Orangeburg, 238 F.2d 528, 532 (4th Cir. 1956).

46. *Id.*

47. *Id.*

48. Standard Constructors, Inc. v. Chevron Chem. Co., 101 S.W.3d 619, 625 (Tex. App.–Houston [1st Dist.] 2003), pet. denied.

49. *Id.*

contract—"no rate charge."[50] The court stated "[a] contract is not ambiguous merely because a party believes the contract does not mean what it actually says."[51]

D. Reasonable and Logical Interpretation of Entire Contract

A court should view a contract as would a reasonably intelligent third person acquainted with trade usage and surrounding circumstances—all interpretations must be reasonable.[52] Any interpretation that produces illogical results will be rejected. Words, terms, and phrases in the contract are to be given their ordinary and normal meaning unless the parties clearly have given them different interpretations.[53] According to the RESTATEMENT OF CONTRACTS:

> A. The ordinary meaning of language throughout the country is given to words unless circumstances show that a different meaning is applicable.
>
> B. Technical terms and words of art are given their technical meaning unless the context or a usage which is applicable indicates a different meaning.[54]

Courts continue to apply the reasonable person standard when interpreting contract documents while still considering the intent of the parties and the surrounding circumstances at the time of contract formation.[55] Courts presume that the parties to an instrument intend every word in it to have some meaning and purpose; therefore, they give effect to all parts of an instrument.[56] Where parties to a contract disagree as to the interpretation of a particular provision, the court will not look solely to the disputed provision but will consider all parts of the contract in an effort to determine the meaning of the provision in question.[57]

As the Texas Supreme Court stated:

> When provisions in an agreement appear to be in conflict, the court should examine and consider the entire writing, seeking to harmonize and reconcile the conflicting provisions to the greatest extent possible. In achieving this objective, the court should favor an interpretation that gives effect to all provisions

---

50. *Id.*
51. *Id.*
52. Triad Elec. & Controls, Inc. v. Power Sys. Eng'g., Inc., 117 F.3d 180, 191 (5th Cir. 1997).
53. RESTATEMENT OF CONTRACTS § 235 (1932).
54. *Id.*
55. Manzo v. Ford, 731 S.W.2d 673, 676 (Tex. App.–Houston [14th Dist.] 1987, no writ) (*citing* Skyland Developers, Inc. v. Sky Harbor Assocs., 586 S.W.2d 564, 570 (Tex. Civ. App. Corpus Christi 1979, no writ); Stitt v. Royal Park Fashions, Inc., 546 S.W.2d 924, 926 (Tex. Civ. App.–Dallas 1977, *writ ref'd n.r.e.*).
56. *See, e.g.*, *Standard Constructors*, 101 S.W.3d at 625.
57. *Manzo*, 731 S.W.2d at 675.

of the contract so that none will be rendered meaningless and the primary purpose of the agreement and intent of the parties will be effectuated.[58]

E. Interpretation against Drafter

An ambiguous agreement is construed most strictly against the party who drafted it.[59] Therefore, if an instrument is open to more than one reasonable construction, it will be construed most strictly against its author.[60] The applicable principle of law was stated by the Texas Supreme Court in *Temple-Eastex, Inc. v. Addison Bank* as follows:

> In Texas, a writing is generally construed most strictly against its author and in such a manner as to reach a reasonable result consistent with the apparent intent of the parties. If two constructions are possible, a construction rendering the contract possible of performance would be preferred to one that renders its performance impossible or meaningless.[61]

In *Austin Co. v. Vaughn Building Corp.*, when evaluating a breach of warranty claim made by the owner of a building against a general contractor, the Texas Supreme Court interpreted the following warranty clause in the contractor's form agreement: "Austin guarantees the work against defective workmanship and material for one year from the date of completion of the work as follows: upon written notice of any such defects, Austin will either make necessary repairs or at its option request owner to make such repairs, all at Austin's expense."[62] The contractor argued that, since written notice of roof leaks was not given until one year and 11 days after completion, it was not obligated under the warranty.[63] The court disagreed, holding that the clause should be construed against the contractor.[64] The court interpreted the provision as providing that the contractor had to repair defects for a period of one year after being notified of them, not one year after completion.[65]

---

58. *Id.* (*citing Enters*, 596 S.W.2d at 519).

59. Corbin on Contracts § 24.27 (1993).

60. Chevron Oil Co. v. E. D. Walton Constr. Co., 517 F.2d 1119, 1122 (5th Cir. 1975); Gulf Constr. Co., v. Self, 676 S.W.2d 624, 628 (Tex. App.–Corpus Christi 1984), writ ref'd n.r.e.; Netherland v. Winner, 662 S.W.2d 786, 788 (Tex. App.–Houston [14th Dist.] 1983), writ ref'd n.r.e.; Harris v. Phillips Pipe Line Co., 517 S.W.2d 361, 364 (Tex. Civ. App.–Austin 1974), writ ref'd n.r.e.

61. Temple-Eastex, Inc. v. Addison Bank, 672 S.W.2d 793, 798 (Tex. 1984); Borg-Warner Acceptance Corp. v. Tascosa Nat'l. Bank, 784 S.W.2d 129 (Tex. App.–Amarillo 1990), writ denied.

62. Austin Co. v. Vaughn Bldg. Corp., 643 S.W.2d 113, 115 (Tex. 1982).

63. *Id.*

64. *Id.*

65. *Id.*

In *Bd. of Regents of the Univ. of Texas v. S & G Constr. Co.*, the owner failed to supply drawings and instructions to be coordinated with existing contractual documents as required by the contract.[66] The owner sought to escape liability for its failure by claiming that oral, day-to-day field directions given by the architect on the project satisfied the contractual requirements.[67] The court accepted the contractor's arguments that the parties had intended that additional instructions be prepared and become part of the contractual documents and that day-to-day oral directives by the architect could not be considered as such.[68] In reaching this conclusion, the court followed the general rule that any doubt as to the meaning of the contract was to be resolved against the party who prepared the document.[69] This rule is particularly applicable where an agreement purports to exempt the drafter from liability.[70]

F. Course of Performance and Prior Dealings

Performance of the contract reflects the best evidence of the intentions of the parties to a contract. In the absence of an express term to the contrary, the court will consider evidence of the parties' course of performance first and foremost when a contract is deemed ambiguous and parol evidence is admitted.[71] The Uniform Commercial Code (UCC) defines course of dealing as "a sequence of previous conduct between the parties to a particular transaction which is fairly to be regarded as establishing a common basis of understanding for interpreting their expressions and other conduct."[72] In the context of transactions governed by the UCC, there is statutory language that recognizes that the parties have the best understanding of what they mean and that their actions are the best reflection of the meaning of their agreement. The UCC provides:

> (a) where the contract for sale involves repeated occasions for performance by either party with knowledge of the nature of the performance and

---

66. Bd. of Regents of the Univ. of Tex. v. S & G Constr. Co., 529 S.W.2d 90, 93 (Tex. Civ. App. Austin 1975), writ ref'd n.r.e., *overruled on other grounds by* Federal Sign v. Tex. So. Univ., 951 S.W.2d 401 (Tex. 1997).

67. *Id.* at 94.

68. *Id.*

69. *Id.*

70. *Manzo*, 731 S.W.2d at 676; Gen. Corrosion Servs. Corp. v. K Way Equip. Co., 631 S.W.2d 578, 580 (Tex. App.–Tyler 1982), no writ.

71. *See* U.C.C. § 2-202; Nolte v. Hudson Nav. Co., 16 F.2d 182, 184 (9th Cir. 1926) ("[W]hen parties choose such equivocal language, they must be content with the interpretation which they put upon it immediately thereafter, and to which they continuously adhered for nearly 25 years.").

72. U.C.C. § 1-205(1).

opportunity for objection to it by the other, any course of performance accepted or acquiesced in without objection shall be relevant to determine the meaning of the agreement.

(b) express terms shall control course of performance and course of performance shall control both course of dealing and usage of trade.[73] Actions taken by a contractor or any other party evidencing his or her understanding of the agreement can be considered in determining the intent of an otherwise ambiguous contract.[74]

In construing an ambiguous contract, the court also will look at other contracts performed by the parties in the past, or those currently being performed by the parties, which contain the same or similar provisions.[75] A contractor may be found to have impliedly agreed to a certain contractual interpretation by virtue of past performance on other projects.[76] Since course of dealing reflects the prior actions of the specific parties in dispute, it is given more weight by the court than evidence of the general trade custom and usage.[77]

G. Custom and Usage

Parties are presumed to have entered into a contract governed by the general customs and usages of the particular industry in which they conduct business.[78] When a construction contract is ambiguous and there is a usage or custom in the industry that is so generally known as to give rise to a presumption that the parties knew of it and contracted with reference to it, the rules of custom and usage can be applied to interpret the contract.[79] For example, the court in *Temple-Eastex, Inc., v. Addison Bank* found as follows:

> Where there is no clear evidence that the parties intended to use a standard or meaning different from customary practice in the industry, courts tend to give a term the meaning it usually has within the industry.[80] This rule is limited

---

73. *See* U.C.C. § 1-205.

74. Connelly v. Paul, 731 S.W.2d 657, 660–61 (Tex. App.–Houston [1st Dist.] 1987), writ ref'd n.r.e. ("The conduct of the parties, which indicates a construction that the parties have themselves placed on the contract, and it may be considered in determining the parties' true intent"); *Nolte*, 16 F.2d at 184.

75. U.C.C. § 1-205(1).

76. *See* Am. Mach. & Tool v. Strite–Anderson Mfg., 353 N.W.2d 592, 598–99 (Minn. App. 1934).

77. U.C.C. § 1-205(1).

78. *See* Restatement (Second) of Contracts § 220(1); Harrison v. Facade, Inc., 355 S.W.2d 543, 545 (Tex. Civ. App.–Dallas 1962), no writ.

79. *See Temple-Eastex, Inc.*, 672 S.W.2d at 798, explaining that custom and usage were used to interpret an ambiguous contract in the banking industry.

80. *See* U.C.C. § 1.205, cmt. 1.

by the possibility that, in the construction industry, the owner may require performance in excess of the standard normally accepted, and by the principle that custom and usage will not be used to vary or contradict plain contractual language.[81]

H. Terms: Specific over General; Written over Printed

Another basic rule of contract interpretation is that a specific or more particular clause must prevail over a general one.[82] For example, a construction contract generally requires that all work meet applicable building codes, but the drawings and specifications may set more stringent performance standards for portions of the work. The specific provisions of the drawings and specifications would be construed to supersede the general provision, and a contractor who simply performed to the standard required by building codes could be in default of the contract.

Parties to a contract should pay close attention to specific terms and be wary of their limiting effect on the whole of the contract. The court, in *Allied Chemical Corp. v. American Independent Oil Co.*, held that "the expression in a contract of one or more things of a class implies the exclusion of all not expressed, even though all could have been implied had none been expressed."[83] For example, if a painting schedule specifically called for some rooms in a school to receive enamel paint, the omission of a room from that list would eliminate the requirement for that specific room, unless the parties' intent to the contrary was affirmatively demonstrated.

Courts will hold that when the printed portions of a contract conflict with handwritten or typewritten portions, the handwritten or typewritten portions will control.[84] This principle has been applied in a number of contexts. For example, in *Montgomery Ward & Co., Inc. v. Dalton*, the court was presented with conflicting warranty provisions.[85] The homeowners relied on the content of the front page of the contract where there appeared a handwritten description of the materials to be used and the work to be performed, also including a handwritten guaranty stating: "20 yr. guarantee 9 yrs. all labor and materials,

81. *George*, 332 S.W.2d at 148.

82. W. Oil Fields, Inc. v. Pennzoil United, Inc., 421 F.2d 387, 389 (5th Cir. 1970).

83. Allied Chemical Corp. v. Am. Independent Oil Co., 623 S.W.2d 760, 763 (Tex. App.–Houston [1st Dist.] 1981), writ ref'd n.r.e.

84. *See* Corbin on Contracts §24.24 (1993).

85. Montgomery Ward & Co., Inc. v. Dalton, 665 S.W.2d 507, 510 (Tex. App.–El Paso 1983), no writ.

11 yrs. prorated 1/240 per mo. transferable."[86] Montgomery Ward relied on the printed contract terms on the back of the document that provided for a more limited warranty.[87] Referring to the handwritten guaranty on the front page of the agreement, the court stated, "[t]his written guaranty controls over the conflicting printed matter on the back of the instrument. The rule is well established that where the written and printed words and phrases in a contract are in conflict, the written words and phrases control.[88]

5.07 PARTIES TO THE CONSTRUCTION CONTRACT

A. Owner and Contractor

The owner and the contractor are the primary, if not sole, parties to the construction contract. With respect to Design-Build agreements and Construction Management agreements, there may be one additional party to the construction contract.

B. Third-Party Beneficiaries

Most industry contracts include prohibitions against third-party beneficiary rights, even though other parties to the construction process such as the architect, engineer, subcontractors, and suppliers are mentioned in various provisions of the construction contract.[89] Generally speaking, neither the owner nor the contractor wants the architect to be a third-party beneficiary and be entitled to enforce the provisions of the construction contract against the owner or contractor. Since most construction contracts provide the owner with rights with respect to the contractor's subcontractors and suppliers, the owner may want to be a third-party beneficiary of subcontracts, but will at least want a conditional assignment of such contracts.

The courts especially scrutinize the intention of the parties when a third party (someone not a party to the subject contract) claims that amounts are due to it as a third-party beneficiary of the contract. Materialmen or suppliers often advance this theory in an effort to obtain funds due from the general contractor to a subcontractor. Some courts consistently have refused to allow claims by

86. *Id.*
87. *Id.*
88. *Id.* at 512.
89. *See* AIA Document A201–2007, *supra* note 14, §1.1.2.

materialmen and laborers as third-party beneficiaries to funds retained under construction contracts with others.[90]

A party claiming third-party beneficiary status must establish either that the contracting parties intended to afford some benefit to that third party or that the contracting parties contracted principally for the benefit of that third party.[91] To establish that the contract was directly and primarily entered into for its benefit, the third party must show itself to be either a donee or creditor beneficiary of the contract, as opposed to an incidental beneficiary.[92] A donee beneficiary receives the performance promised under the contract as a pure donation, but a creditor beneficiary receives the performance promised in satisfaction of a legal duty owed to the creditor beneficiary by the promisee.[93] Courts focus on the intent of the contracting parties in assessing a third party's ability to enforce a contract.[94]

C. Collateral Assignment to Lender

AIA Document A201 and other industry contracts provide for the collateral assignment by the owner to the lender of the owner's rights under the construction contract as additional security for the lender's loan. In most cases, the lender requires a separate consent and assignment instrument to be executed by the owner and contractor that provides the lender with the ability to assume the owner's position under the construction contract or to terminate the construction contract in the event of the owner's default under its loan arrangement with the lender.

D. References to Other Parties

The references to the architect, engineers, subcontractors, suppliers, and lender are important because each of the other parties plays a significant role with

90. *See, e.g.*, Econ. Forms Corp. v. Williams Bros. Constr. Co., 754 S.W.2d 451, 458–59 (Tex. App.–Houston [14th Dist.] 1988), no writ, denying recovery for the lessor of certain materials used by a subcontractor and recognizing both that the intention of the parties to the contract controls the issue of whether a third party can enforce a contract provision and the presumption that parties generally contract for themselves and their own benefit; Corpus Christi Bank & Trust v. Smith, 525 S.W.2d 501, 504–05 (Tex. 1975).

91. Chi. R.I. & P. Ry. Co. v. Md. Cas. Co., 75 F.2d 596, 600 (8th Cir. 1935); McKeever v. Or. Mortgage Co., 198 P. 752, 753 (Mont. 1921).

92. *See* Corbin on Contracts § 774 (1993).

93. *Id.*

94. *See, e.g., Chi. R.I. & P. Ry. Co.*, 75 F.2d at 600. *But see* Lidia Hernandez v. Home Depot, U.S.A. et al., 2007 U.S. Dist. LEXIS 57395, which gives a somewhat broad interpretation to when a third party is intended to qualify as a beneficiary of a construction contract.

respect to the design, construction, and/or financing of the project, as applicable, and such role should be recognized in the construction contract and consistent with the rights of such parties under their separate agreements with the owner or contractor. For example, the architect will have certain rights with respect to construction administration and, in the event of a default by the contractor, may need to provide additional design services to correct defective construction. In such event, the contractor should be responsible for the additional design costs incurred by the owner.

5.08 SCOPE OF CONTRACT

A. Description of Services of Design Professionals

Design agreements are challenging documents to prepare. The subject matter of the agreement is unknown. There may be a program for the project to describe basic project parameters, but the project itself is just an idea. The agreement between the owner and architect, therefore, is really a description of the process through which that idea is developed and documents prepared to depict the project in a combination of two-dimensional drawings and written narratives. The scope of a design agreement comprises the phases of design services, the deliverables that are furnished as a result of the design services, and the construction administration activities that might be provided, all of which are described in greater detail in chapters that follow.

B. Description of the Work under a Construction Contract

The work to be performed by a contractor has been described as "the construction and services required by the contract documents, whether completed or partially completed, and includes all other labor, materials, equipment, and services provided or to be provided by the contractor to fulfill the contractor's obligations."[95] Owner-contractor agreements typically provide a large blank space to be completed as necessary to describe the work that the contractor is obligated to perform. Although this description is obviously the essence of any contract, it is often given only routine consideration when preparing a contract. The parties must not underestimate the difficulty of describing the work accurately and comprehensively, as any failure to ensure an adequate description of the work can frustrate reasonable expectations at a later date. If the owner desires the contractor to perform certain activities that fall outside the scope of the work, an adjustment in both cost and time most likely will be required.

95. AIA Document A201–2007, *supra* note 14, §1.1.3.

Since the work generally will be described by incorporating the drawings and specifications by reference, it is important for the owner's architect, or a third-party architect or engineer retained by the owner (if the project is a Design-Build project), to review the sufficiency of the drawings and specifications from the standpoint of producing the intended result in compliance with all applicable codes, laws, and regulations.

Additionally, owners generally include a provision in the construction contract requiring the contractor to execute the work described in the contract documents "and reasonably inferable by the contractor as necessary to produce the results 'intended' or 'indicated' by the contract documents."[96] The "reasonably inferable" language is important because no set of drawings and specifications is perfect.

With respect to the drawings and specifications, who should bear the risk of any inaccuracies or omissions in the drawings and specifications? Contractors undoubtedly will argue that they should be entitled to rely on the accuracy or completeness of the drawings and specifications prepared by the architect and engineers, but no set of drawings and specifications is perfect, as indicated previously. Therefore, if, in the course of constructing the improvements, the contractor discovers any error, inconsistency, or omission in the contract documents, the contractor should be obligated to report such item immediately to the architect and/or owner.

Owners may attempt to attribute to the contractor constructive knowledge of a defect if a reasonable and prudent contractor would have caught the mistake. The patent ambiguity rule (to wit: the rule requiring the contractor to notify the owner if the contractor is aware of conflicts or defects in the design documents apparent on the face of such documents) protects the owner, and the contractor may be charged with knowledge of patent ambiguities so as to preclude the contractor from recovering from such ambiguities. Compulsory disclosure of any conflict or ambiguity of which the contractor has actual knowledge guards against gamesmanship in the bidding process with respect to the contractor who uses the change process to increase the contract sum to cover the costs resulting from a conflict or ambiguity after being awarded the contract.

The owner wants to make sure that if the contractor performs any work in conformity with any contract document, knowing it to be inconsistent with any other contract document or not in compliance with applicable laws and codes, without first specifically reporting it and obtaining from the owner and/or architect written instructions on how to proceed, the contractor should be

96. *See* AIA Document, *supra* note 14, A201-2007.

obligated to correct such work, at owner's direction, without cost to the owner. The contractor also should be required to contact the owner and/or architect without delay upon discovery of problems in the drawings and specifications or in compliance with laws. Moreover, the failure to promptly notify the owner of any error or inconsistency, particularly if the contractor knowingly performs work that will need to be re-executed, should be the responsibility of the contractor. A contractor that agrees to this provision is not accepting design responsibility or liability, but only the liability to notify the owner and/or architect of a known potential problem for input from the owner and/or architect prior to execution of the work that is the subject of a potential conflict.

From the contractor's perspective, it is also important to provide a comprehensive description of the work to be performed by the contractor under the contract documents by reference to and incorporation of the drawings and specifications, as well as by referring to the contractor's qualifications or clarifications (which should have the highest degree of priority). Additionally, references to subjective terms like "good and workmanlike" construction should be avoided. With respect to subparagraph 1.2.1 of the A201, the contractor should attempt to delete (the owner should attempt to avoid deletion of) the language that "the intent of the contract documents is to include all items necessary for the proper execution and completion of the work by the contractor" and the contract documents are complementary and what is required by any one contract document shall be as binding as if required by all; and the contractor's performance should be required only to the extent consistent with the contract documents and reasonably inferable from them as being necessary to produce the "indicated" results, not "intended" results.

Finally, the reasonably "inferable" standard itself, from the contractor's viewpoint, is in need of repair to ensure that it is not abused. The simplest fix, adopted in some industry contracts, is to limit inferred work to work that is "incidental" to the work shown on the drawings and specifications. A more satisfactory approach is to define more carefully work that is not shown in the drawings and specifications as "reasonably inferred" or "incidental" to give the parties, and (if necessary) the court, a more thorough set of criteria on which omitted work can be attributed to the contractor.

There are several conditions that should serve to exclude an addition of work as being cast as part of what is "reasonably inferred" from the drawings and specifications. The most obvious is cost: When omitted work involves material additional cost and is not indicated in the drawings and specifications, it should not qualify as "reasonably inferred" from drawings and specifications. Second, it is the design professional's job to design, not the contractor's. If work requires any meaningful exercise of the design function, it should not qualify as "reasonably inferred" from the drawings and specifications.

Even when contractors are saddled with some elemental design function, they ordinarily are required to prepare submittals to have the design approved. If the omitted work usually would require anything like a submittal, but is not specified in the required submittals, that is a very good sign that the work is not "reasonably inferred" from the drawings and specifications. Finally, if the work requires material changes to sequencing or scheduling, it should not be construed as "reasonably inferred."

All these elements ultimately look to the materiality of the change. No doubt, there can be pitched battles as to what is material and what is not. But it is better to fight on those appropriate grounds and get to the heart of the responsibility in an equitable manner than argue over what is "reasonably inferred as necessary," a question that may bear little relationship to the fairness of attributing extra work to the contractor.

C. Owner-Furnished Items

In many commercial projects, the owner will furnish certain materials, equipment, or systems that are incorporated into the work. Sometimes, the owner takes advantage of an opportunity to purchase materials (lumber and flooring as an example), equipment, or fixtures at a discount in advance of the actual need for such items or wants the contractor to install specialty items obtained from older properties or at auctions. In such event, the owner expects the contractor to install such items and to remain fully responsible for furnishing the necessary services and other items necessary to install and incorporate such owner-purchased items into the work. The owner will want the contractor's installation of such owner-purchased items to be covered by the contractor's warranties.

In many instances, the contractor will attempt to accept no responsibility for owner-supplied items and require the owner to contract directly with separate contractors to install owner-supplied items. If, however, the contractor agrees to install owner-supplied items, limited liability, if any, should apply to the contractor for failing to reject nonconforming or defective owner-supplied items. The owner should be required to look directly to the supplier for any nonconforming or defective owner-supplied items, unless the owner-supplied items are damaged by the contractor during installation or during the contractor's performance of the work or the contractor had actual knowledge of the defects and nevertheless installed the owner-supplied items.

QUESTIONS

1. What are the primary considerations for entering into early start letters on construction projects?
2. Illustrate two situations where a contractor should be successful in pursuing a *quantum meruit* claim for work done for another party?
3. Why are industry construction contract forms, like the AIA documents, more frequently used than manuscript forms?
4. Which parties to the industry forms, architects' agreements, and construction contracts generally request more changes and why? What is the primary focus of the requested changes?
5. What circumstances could possibly support the introduction of parol evidence to defend the enforceability of a construction contract, as well as to permit parol evidence to supplement a construction contract?

CHAPTER

6

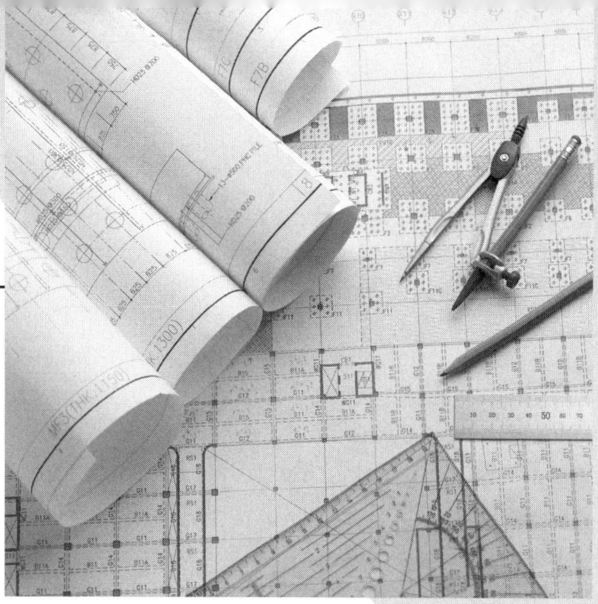

The Design Undertaking

STEPHEN A. HESS AND LAWRENCE C. MELTON

The function of this chapter is to explain the design process from a legal perspective with an examination of the sources of rights and responsibilities of each of the parties who participate in or benefit from the design process. Although there are obvious participants who bear most of the design responsibility (architects and engineers, in particular), the design process is not exclusively within the province of architects and engineers but instead is shared to some degree by many of the other participants as well.

Section 6.01 describes the different kinds of design professionals who contribute to the construction process, the various elements of the design process, and the professionals' traditional roles in a Design-Bid-Build project. Section 6.02 provides an overview of the regulation of design professionals, and Section 6.03 discusses the variety of building codes and related regulations that govern construction. Section 6.04 introduces the reader to standards of care in the design professional's practice. Section 6.05 concludes this chapter by describing the process by which a particular design is implemented through the construction process and some of the problems that arise during the course of construction.

6.01 INTRODUCTION

This section introduces the reader to the design function and its allocation among various participants in a traditional Design-Bid-Build project. It begins with a brief review of the typical Design-Bid-Build contract relationships (discussed in more detail in Chapter 4) and continues by discussing those consultants who ordinarily contract directly with the owner and their primary responsibilities. It then describes the structure of the architect's design team and the delegation of design responsibility through the architect to specialty consultants and engineers. The section concludes with a discussion of the contractor's limited role in design.

A. Brief Overview of Contractual Relationship among Parties and General Allocation of Design Function

Ultimately, the reader must appreciate the exact contractual relationships that govern construction of a particular project, as liability issues cannot be assessed meaningfully without a clear understanding of the various parties' interwoven contractual relationships. Although there are many different ways in which the design responsibilities on a project can be allocated, the following discussion is intended to characterize the chief design functions and to describe a typical allocation of design responsibility under the most common contract mechanism—Design-Bid-Build, also often referred to as the Traditional method of construction contracting. It is the project delivery system contemplated by the chief AIA documents.

1. Owner

The owner's primary responsibility is to provide a site for its project and then to make choices in communication with the lead design professional as to how the project will be implemented. The owner's choices with respect to design may be based on aesthetic reasons, financial considerations, functional requirements, or other factors; the independent consultants whom the owner may retain to assist in these areas are described later. In a Design-Bid-Build project, the owner contracts with a design professional and then enters into a separate contract with the contractor.

2. Design Professionals

The lead design professional's ultimate *design* goal is to draft drawings and specifications that govern the contractor's work and implement the owner's requirements. There are numerous constraints on how the design professionals accomplish this, and this chapter is devoted largely to identifying those standards.

In addition to the design function, the design professional may have many other *nondesign* responsibilities related to the construction that go well beyond drawings and specifications. These may include budgeting, procurement, and contract administration, among a variety of other roles. Chapter 7 of this book identifies these nondesign roles in more detail and discusses the legal parameters of a design professional's contracting to perform these nondesign functions. Generally, a design professional works under contract with the owner and has no contractual responsibility to or relationship with the contractor.

3. Contractor

The contractor's chief function is to construct the project in accordance with the drawings and specifications furnished on behalf of the owner by the design professional. At the same time, few construction projects are so simple that a contractor can take an initial set of drawings and specifications and complete the work without communicating any further with the design professional. Moreover, there are some elements of a project whose ultimate design is better left in the hands of the contractor or specialty subcontractor. The regular communication between the contractor and the design professional, together with the occasional design role undertaken by the contractor, give the contractor an important role in the design process as described later in the chapter.

This is true even though the contractor in a Design-Bid-Build project works under contract with the owner, and not with the design professional.

B. The Role of the Owner and Owner's Consultants with Respect to Design

Before the owner contracts with an architect or engineer for design of a structure itself, the owner often engages a number of design professionals whose responsibility is to advise the owner with respect to a number of general site development issues. These may include such specialists as transportation engineers, who study traffic flow and civil design issues to ensure that a proposed development is built consistent with transportation demands. They may also include environmental consultants, whose responsibility is to assess potential environmental impacts of a proposed development and to prepare reports and recommendations involving environmental concerns. Although it is conceivable that an owner could pass all such development issues on to a design professional who offers such "one-stop" contracting for services, the owner typically contracts for such services directly with a site development firm or a group of different site development professionals.

Perhaps the most commonly used professional retained directly by the owner is the geotechnical engineer. Geotechnical engineers perform various tests to evaluate the suitability of ground for construction purposes.[1] The geotechnical report and test data are then used to determine the feasibility, cost, and foundation design for the project. Geotechnical engineers are necessary and highly respected participants in the design process. Notwithstanding this fact, the AIA 201 family of documents treats the geotechnical engineer somewhat like a social outcast. The geotechnical engineer virtually always works directly for the owner and not as a subcontractor or joint venture partner with the architect. There are two reasons for the unique position of the geotechnical engineer: (1) errors of judgment in the geotechnical investigation and/or report can lead to extensive project cost overruns; and (2) the owner must ultimately decide how much money to invest in the geotechnical report, and that decision must be based on the advice of the geotechnical engineer to the owner and the owner's willingness to accept the risk of unsuitable ground. For these reasons, architects try to avoid even an administrative role in geotechnical engineering.

The fact that an owner usually hires the geotechnical engineer has important consequences for liability disputes. If a building settles after construction, for example, one cannot simply allow the owner to recover by allocating responsibility to the design architect or engineer. Instead, the owner may bear responsibility if the geotechnical engineer erred in recommending a particular foundation design.[2]

C. The Design Professional Team

The complexity of modern buildings requires the architect to use and to rely on the services of many other design professionals. At the outset, the architect will enter into a single, all-encompassing design contract with the owner. The architect then subcontracts out portions of the design responsibility to consultants. While it is also possible for the owner to hire design consultants directly, the architect is still likely responsible to lead the overall design effort.

1. Architectural Consultants

Architectural consultants, who themselves are generally aligned with the architectural side of design (rather than with the engineering side discussed later), include landscape architects, interior designers, lighting and acoustic special-

1. *See* 4 Philip L. Bruner & Patrick J. O'Connor, Bruner & O'Connor on Construction Law, § 14:2 et seq. (West Group 2002) [hereinafter Bruner & O'Connor].

2. Of course, the owner may then have a right to recover from the geotechnical engineer. The important lesson is that liability for design errors cannot be assessed without knowing who contracted with whom for particular engineering services.

ists, code consultants, telecommunications and security experts, and "green" building experts. The architect retains these specialists and may contractually accept responsibility to the owner for the suitability of the work of these subcontractors.

2. *Engineering Consultants*

As with architectural consultants, the architect usually farms out portions of the work to engineers. The architect may retain several different engineering specialists as its own subcontractors or retain one engineer whose responsibility includes obtaining the assistance of any other engineers necessary. Traditionally, these engineers include structural engineers, civil engineers, mechanical engineers (who deal with plumbing, heating, ventilation, air conditioning, and the like), and electrical engineers.

The chief difference between the retention of architectural consultants and engineering consultants is that the former is usually voluntary, while the latter is often necessary to the extent that a licensed engineer must stamp certain aspects of the drawings.

3. *The Civil Construction Project (Engineer as Lead Design Professional)*

The lead design professional is ordinarily an architect when the construction involves structures that are intended to be occupied by people—office buildings, hospitals, restaurants, theaters, hotels, and the like. However, a substantial amount of construction does not focus on occupied buildings (e.g., water treatment plants, bridges, oil refineries, tunnels, railways, roads, dams) even if there are related facilities that may be occupied such as a refinery's administration building. These latter projects belie accurate comprehensive labels, but for want of a better term, we will call the building of unoccupied structures "civil construction."

Civil construction projects ordinarily utilize engineers as the chief design professionals rather than architects. The engineer may retain an architect as a subcontractor for those limited portions of the project that require an architect's stamp on drawings, but the engineer will ultimately undertake responsibility for the final design of the project under contract with the owner. Are there important legal differences between engineers and architects as design professionals? As a general rule, no: The design professional's rights and liabilities are dictated much more by its role in the web of contracts among the parties than by whether the lead design professional is an architect or an engineer.

On a practical level, there may be substantial differences in the design approaches and contract administration styles of architects as opposed to engineers, although these are usually of more interest to the owner than to the lawyer. Perhaps the largest difference is in the initial contracting: Architects are likely to use standardized contract forms published by the American Institute of Architects, while engineers are more likely to use contract forms published by the Engineers Joint Contract Documents Committee. The choice between the two is determined by the owner, who may indeed prefer to use contract forms more to his liking.[3]

6.02 STATUTORY REGULATION AND LICENSURE OF DESIGN PROFESSIONALS

The responsibility for the safety of citizens is generally relegated to each state, and one aspect of safety involves the integrity of buildings and their compliance with codes. Just as states regulate the practice of medicine as part of their police power for the benefit of the safety of their citizens, so too they regulate the practice of architecture and engineering.[4]

In general, states exercise this authority by limiting the practice of architecture and engineering to those who qualify to obtain licenses from the state. If a person purports to practice architecture or engineering without a license, that person is subject to sanctions for the unauthorized practice of architecture or engineering, and ordinarily will lose any right to compensation for the work performed in violation of the licensing laws.[5]

6.03 BUILDING CODES AND THE DESIGN PROFESSIONAL

In addition to regulating design professionals so as to protect citizens, states (municipalities, counties, etc.) also provide for the safety of their citizens by regulating construction directly through building codes. This chapter describes the general scheme of building (and related) codes that impose specific requirements on construction.

The earliest regulation of construction may have focused on maintaining the quality of materials. Vitruvius, for example, reported that in order to assure that undried bricks were not used in construction, "at Utica in constructing

---

3. For a more detailed discussion of contract forms, see Chapter 5.

4. There has always been a "turf war" over what aspects of design constitute engineering and what aspects constitute architecture, as each discipline jealously guards the scope of its own practice. *See, e.g.*, Rosen v. Bureau of Prof'l and Occupational Affairs, State Architects Licensure Bd., 763 A.2d 962 (Pa. Commonwealth 2000) (rejecting complaint by architect that engineer's work on building impermissibly crossed over into providing services as architect).

5. *See generally* 4 Bruner and O'Connor, *supra* note 1, § 16:18 (and Supp. 2007).

walls they use brick only if it is dry and made five years previously, and approved as such by the authority of a magistrate."[6]

Building codes are an exercise of the State's police power to protect the public health, safety, and welfare.[7] Modern building codes are historically traceable to fire prevention. Following the Great Fire of 1666, for example, Parliament required that all buildings in London be constructed of brick and meet other strict requirements.[8] It is, in fact, difficult for people in this age who work in offices made of fireproof materials, equipped with automatic sprinkler systems, to imagine the fear of fire that obsessed the designers of wooden buildings lit by candles, heated by open fireplaces and lacking any source of running water. Nevertheless, ambitions and well-intended restrictions on construction have always run into the practical restrictions of cost and availability of resources. In colonial South Carolina, for example, a statute of 1713 required all buildings in Charleston, which had repeatedly been damaged by fires, to be made of brick, except upon petition that bricks "are no ways to be had but at such excessive rates, that wholly disenables them to build therewith, whereby their lands lie waste, to the sensible decay of Charlestown, and the disadvantage of the Province in general." That act was repealed in favor of a statute requiring only that chimneys, hearths, and fireplaces be built of brick or stone.[9]

A. Traditional Building Codes

Two preliminary issues confront anyone attempting to apply or understand a building code. The first is that there are many building codes, not one, and they can vary from state to state and from municipality to municipality within a state. There are four main model building codes: the International Building Code, the Uniform Building Code, the Standard Building Code, and the National Building Code. Most states have adopted one or the other. Texas has adopted portions of each. New York and Wisconsin have written their own codes.[10]

South Carolina, to take one example, requires municipalities and counties to adopt the "latest edition" of the building, residential, gas, plumbing, mechanical, fire, and energy codes as published by the Southern Building Code Congress International, Inc., and the National Electrical Code, published by the

6. Marcus Vitruvius Pollio, The Ten Books on Architecture 43 (ca. 25 bc), (Morris Hickey Morgan trans. Courier Dover Publications 1960).

7. *See, e.g.*, S.C. Code Ann. § 6-9-7

8. 18 Car. 2.c.I (1666).

9. S.C. Stat. at Large, No. 374 (1717).

10. The composition of the Wisconsin Code is instructive and can be reviewed at http://www.reedconstructiondata.com/building-codes/wisconsin/.

National Fire Protection Association. Municipalities are not required to adopt appendices to the standard codes. Municipalities may petition the Building Code Counsel for exceptions or for permission to adopt additional codes. Farm buildings are exempt from all building codes and are covered by the Federal Emergency Management Agency (FEMA) or the National Flood Insurance Program.[11]

B. Socioeconomic "Codes"

The codes identified so far generally restrict their attention to the physical integrity of structures and the safety of those who use them. On the other hand, a number of design constraints imposed by law (or by owners) take the force of codes even though they are not directed to structural integrity or the safety of citizens. These constraints may not rise to the level of safety concerns; nonetheless, they are sufficiently important to attract the support of legislatures. The following subsections identify some of the most important "socioeconomic" codes.

1. Accessibility under ADA

In addition to the safety requirements embodied in state and local building codes, design professionals are confronted by the Americans with Disabilities Act (ADA), a civil rights law rather than a building code. The Americans with Disabilities Act is devoted to eliminating discrimination against individuals with disabilities by ensuring full and equal enjoyment to public places, including physical access to public buildings.[12]

The applicability of the ADA to design professionals has proven to be controversial. Some courts have held that architects are responsible for following the mandates of the ADA,[13] while in a case of first impression, the U.S. District Court for the District of Columbia held that the statute does not apply to architects.[14] Subsequent litigation has not provided a definitive answer.[15]

---

11. Lawrence C. Melton, South Carolina Construction Law § 3.4 (2005).

12. The ADA comprises three parts: Title I, 42 U.S.C. § 12111, et seq., prohibits discrimination against the disabled with regard to employment; Title II, 42 U.S.C. § 12131, et seq., prohibits discrimination against the disabled by public entities; and, Title III, 42 U.S.C. § 12181, et seq., prohibits discrimination against the disabled with regard to access to commercial buildings. *See* PGA Tour, Inc. v. Martin, 531 U.S. 1049, 121 S.Ct. 1879, 1889, 149 L.Ed.2d 904 (2001). Actions under Title III of the ADA are limited only to injunctive relief. 42 U.S.C. § 12188(a).

13. United States v. Ellerbe Becket, Inc., 976 F. Supp. 1262 (D. Minn. 1997); Johanson v. Huizenga Holdings, Inc., 963 F. Supp 1175 (S.D. Fla. 1997).

14. Paralyzed Veterans of Am. v. Ellerbe Becket Architects & Eng'rs, 945 F. Supp. 1 (D.D.C. 1996).

15. *See, e.g.*, John Lonberg, et al. v. Saborn Theaters, Inc., et al., 259 F.3d 1029 (9th Cir. 2001).

2. "Green" Codes and Growing Mandatory Compliance

Governmental consciousness of environmental issues and the development of environmentally favorable building technologies have led to a growing movement toward "green" construction, or construction that minimizes the impact of construction with respect to both the structure itself and its use of natural resources in energy, water, and waste disposal. In its initial development, green building was primarily aspirational, and several competing sets of standards existed to guide the design professional. In recent years, however, more and more jurisdictions have adopted tax structures or other economic benefits that favor green building, or they go even further in *requiring* that new construction satisfy certain environmental standards, or green codes.[16] Of course, private owners are always at liberty to direct their design professionals to design in accordance with green codes that otherwise are voluntary. To date, there is no uniformly accepted green building code, although some—such as the program implemented by the U.S. Green Building Council—have developed sophisticated requirements, training programs, and even certification of green building professionals.[17]

3. Historic Preservation

Historic preservation codes require that construction on existing historic buildings be accomplished in a manner that preserves as much of a particular structure as possible or to model the surrounding neighborhood. Such codes may also prescribe the appearance of new construction so as to ensure that the historic atmosphere of a particular district is maintained. Because historic preservation codes effectively limit the uses (and some of the potential economic benefits) of private property, due process constraints help define the content of such codes.[18] Nonetheless, the U.S. Supreme Court has held that historic preservation codes do not, per se, constitute takings of private property requiring compensation under the Fifth Amendment.[19]

---

16. Typically, mandatory requirements are imposed by governmental entities for their own buildings. *See also* Ann E. Marimow, *Montgomery Aims to Make Green Homes Mandatory*, Washington Post, April 23, 2008 at A01, *available at* http://www.washingtonpost.com/wp-dyn/content/article/2008/04/22/AR2008042202839_pf.htm.

17. *See generally* http://www.usgbc.org.

18. *See, e.g.*, Donovan v. City of Deadwood, 538 N.W.2d 790 (S.D. 1995) (preservation commission could not constitutionally deny permit to demolish building even given viability of preservation code absent procedural due process safeguards).

19. Penn Cent. Transp. Co. v. City of New York, 438 U.S. 104, 98 S. Ct. 2646 (1978).

6.04 STANDARD OF CARE APPLICABLE TO DESIGN SERVICES

A. Nature of Standard of Care and Inapplicability of Strict Liability

When a tornado renders a house into 50,000 pieces, the owner does not recover the cost of a new house from the architect. On the other hand, when the skywalk at the Kansas City Hyatt Regency collapsed, the victims looked to the design professionals for compensation. What distinguishes the two disasters legally? At the very least, we do not hold design professionals responsible for all the adverse consequences that may befall a building, nor do we expect every design to be flawless in every respect.

Instead, we only expect architects and engineers to be able to protect us against certain dangers and flaws, and we understand that there may well be numerous inconsequential errors in any sophisticated set of design documents. In articulating the design professional's liability, in other terms, we do not hold design professionals strictly liable.[20] Instead, the law only imposes liability where there is fault. "Fault" in this sense simply means deviation from an identifiable standard of care. This section begins by discussing the most often utilized basis for liability—the standard of care traditionally imposed on the design professional at common law (ordinarily thought of as the "negligence" standard) (Subsection B). This is not the only theory, however, under which people may recover for a design professional's work. Accordingly, this section then turns to the viability and use of contractual standards of care and variations of the common law standard of care (Subsection C). Finally, this section discusses the existence *vel non* of implied warranties as governing the design professional's standard of care to a client (Subsection D).

B. Common Law Standard of Care

1. *General Statement of Standard of Care*

For purposes of liability for professional negligence, all construction professionals are held to a "standard of care" that can be generally defined as "the ordinary and reasonable degree of care required of a prudent professional

20. *See, e.g.*, Three Affiliated Tribes of the Fort Berthold Reservation v. Wold Eng'g, P.E., 419 N.W.2d 920 (N.D. 1988); Nelson v. Commonwealth, 368 S.E.2d 239 (N.E. 1988); Chaney Bldg. Co. v. City of Tucson, 716 P.2d 28 (Ariz. 1986); Carter v. Deitz, 556 So. 2d 842 (La. Ct. App. 4th Cir. 1990), writ denied, 566 So.2d 992 (La. 1990) and writ denied, 566 So. 2d 993 (La. 1990) and writ denied, 566 So.2d 960 (La. 1990) (mere errors of judgment insufficient to establish liability; plaintiff must establish negligence).

under the circumstances."[21] This standard of professional liability is both relatively old and relatively well established. It is, however, sufficiently broad to be almost meaningless outside the context of a particular case. The development of case law on the "standard of care" is enlightening.

The early case of *Coombs v. Beede*[22] is instructive in setting out the basic principles of professional liability that still apply today. In that case, an architect sued for a fee (the way many disputes over professional competence arise), and the owner defended by alleging that the services rendered were inadequate. In determining the duty owed by the architect to the owner, the court ruled broadly:

> The responsibility resting on an architect is essentially the same as that which rests upon the lawyer to his client, or upon the physician to his patient, or which rests upon any one to another where such person pretends to possess some skill and ability in some special employment, and offers his services to the public on account of his fitness to act in the line of business for which he may be employed. The undertaking of an architect implies that he possesses skill and ability, including taste, sufficient to enable him to perform the required services at least ordinarily and reasonably well; and that he will exercise and apply, in the given case, his skill and ability, his judgment and taste, reasonably and without neglect. But the undertaking does not imply or warrant a satisfactory result. It will be enough that any failure shall not be by the fault of the architect. There is no implied promise that miscalculations may not occur. An error of judgment is not necessarily evidence of a want of skill or care, for mistakes and miscalculations are incident to all the business of life.

The "standard of care" for professional liability is easy to state but often difficult to apply to the extent that it relies, in part, on standards developed within a trade itself rather than the traditional "reasonable man" approach. How does this impact application of the standard? Few cases have stated the rule and its implications better than *Hubert v. Aiken*,[23] in which a special referee ruled that an architect was not liable for a defective flue design in a steam heating system. The court rejected that position:

> I am unable to agree with the referee in his conclusion. The plaintiffs are architects of standing, who assume to be able to plan and superintend the construction of first-class apartment houses, to be heated by steam, and to be provided with every convenience demanded by the luxurious tastes of the day.

---

21. Joseph A. Demkin, The Architect's Handbook of Professional Practice, Glossary, 737 (13th ed. 2001).

22. Coombs v. Beede, 36 A. 104 (Me. 1896).

23. 15 Daly 237, 19 N.Y.St. Rep. 914, 2 N.Y.S. 711 (1888).

> They are not architects in a rural community, but in the first city in America. Steam-heating is, as we all know, common, if not a necessity, in all apartment houses of large size, and of a high class. It is true that houses of this description are of recent introduction; but they are now a very important part of our system of economics, for in some of the new streets they are more numerous than private residences, or tenements of the kind that formerly was in vogue. The architect who undertakes to construct a house that is to be heated by steam is groping in the dark unless he knows how large a chimney is required. It is as necessary that the architect should know what is needed to make the steam-heating apparatus serviceable, as it is that he should know how sewer gas is to be kept out of the house. No one would contend that at this day an architect could shelter himself behind the plumber, and excuse his ignorance of the ordinary appliances for sanitary ventilation by saying that he was not an expert in the trade of plumbing. He is an expert in carpentry, in cements, in mortar, in the strength of materials, in the art of constructing the walls, the floors, the stair-cases, the roofs, and is in duty bound to possess reasonable skill and knowledge as to all these things; and when, in the progress of civilization, new conveniences are introduced into our homes, and become, not curious novelties, but the customary means of securing the comfort of the unpretentious citizen, why should not the architect be expected to possess the technical learning respecting them that is exacted of him with respect to other and older branches of his professional studies? It is not asking too much of the man who assumes that he is competent to build a house at a cost of more than $100,000, and to arrange that it shall be heated by steam, to insist that he shall know how to proportion his chimney to the boiler. It is not enough for him to say, "I asked the steam-fitter," and then throw the consequences of any error that may be made upon the employer who engages him, relying upon his skill. Responsibility cannot be shifted in that way. In the case of *Moneypenny v. Hartland*, (twice reported, once in 1 Car. & P. 352, and then in 2 Car. & P. 378) it was held that if a surveyor be employed to erect a bridge and form the approaches to it, he is bound to ascertain for himself, by experiments, the nature of the soil, even although a person previously employed for that purpose by his employer has made such experiments, and has given him the result at his employer's request; and if the surveyor makes a low estimate, and thereby induces persons to subscribe for the execution of the work who would otherwise have declined it, and it turns out that, owing to his negligence and want of skill, such estimate is grossly incorrect, and that the work can be done, but at a much greater expense, he is not entitled to recover for his services.

Modern law is no different in principle from the law summarized in these cases, even though the technology referred to in the opinions would not be familiar to today's students.

2. *Proof of the Standard of Care and Violation Thereof*

As a general rule, professional negligence can only be proven by the testimony of an expert qualified in the same field as the person accused of negligence.[24] The expert must establish both that a standard of care is recognized and that the defendant failed to meet it. The following case is typical of courts' handling of such disputes.

McKee By and Through McKee v. City of Pleasanton
242 Kan. 649, 750 P.2d 1007 (Kan. 1988)

ALLEGRUCCI, Justice:

The plaintiff, Kelly McKee, is a minor, suing by and through his father and next friend, Jerry McKee. The plaintiff alleges that the defendant Reiss & Goodness Engineering, Inc., was negligent in not properly inspecting the construction of the Pleasanton, Kansas, swimming pool, and that the remaining defendants were negligent for failing to properly construct the swimming pool. The petition alleges that, as a result of the negligent actions of the defendants, Kelly McKee received serious injuries in an accident at the swimming pool.

. . .

The plaintiff's petition alleges that, on June 13, 1980, Kelly McKee received serious injuries after slipping on the coping tile used around the edge of the swimming pool. On December 31, 1986, the Linn County District Court granted summary judgment against the plaintiff on behalf of [design engineer] defendants Reiss & Goodness . . .

. . .

Although the district court's ruling touched briefly upon other matters, the essence of the court's decision was that the plaintiff had failed to present any expert evidence indicating that the defendants had negligently designed and constructed the swimming pool. During the course of the construction of the swimming pool, red quarry tile was substituted for the concrete coping originally called for by the plans and specifications for the pool. The main issue in the present case is whether the trial court properly found that the plaintiff had failed to provide satisfactory expert evidence that the use of quarry tile was a deviation from the defendants' obligation to use reasonable care in the design, inspection, and construction of the swimming pool.

24. Robert M. Schoenhaus, *Necessity of Expert Testimony to Show Malpractice of Architect*, 3 A.L.R. 4th 1023 (1981).

As the trial court's memorandum decision notes, the plaintiff retained two expert witnesses to testify regarding the use of the quarry tile coping. The first expert retained by the plaintiff was George W. Forman, a professor of mechanical engineering at the University of Kansas. Forman's testimony was limited to a comparison of the relative slipperiness of a piece of quarry tile to a piece of concrete coping. Mr. Forman conducted no quantitative measurements and did not attempt to estimate the coefficient of friction, but testified that the tile was not as slip resistant as the concrete coping. Forman expressly refused to testify as an architect or expert on the design and construction of swimming pools.

. . .

The second person retained as an expert witness by the plaintiff was Dr. Alan R. Caskey. The trial court found that, although he "has experience and educational background in the area of the administration of municipal parks and recreation," Caskey "is not experienced or trained in [swimming pool] design and construction."

Caskey holds a Ph.D. in parks and recreation administration. The only formal education Caskey has received in the area of swimming pool construction was an undergraduate class which had "elements" of pool construction in it, in addition to aquatic skills, programming, and safety. Caskey is neither an architect nor an engineer.

The only person to perform an examination to determine the wet coefficient of friction of the quarry tile used in the Pleasanton swimming pool was Robert Weiner, an engineer retained by one of the defendants. Weiner's tests indicated that the quarry tile used as coping on the swimming pool possesses a wet coefficient of friction of 0.60.

In *Seaman U.S.D. No. 345 v. Casson Constr. Co.,* 3 Kan.App.2d 289, 594 P.2d 241 (1979), the Kansas Court of Appeals stated:

> One purpose of the community medical standard and its requirement of expert testimony is to educate the fact finder as to otherwise alien terminology and technology and thus preclude his rendering judgment on something he knows nothing about. We have no qualms in applying a similar standard to the architectural profession. We agree that architectural procedures are sufficiently technical and outside the realm of ordinary knowledge to warrant application of an "architectural community standard" in most instances. 3 Kan.App.2d at 293, 594 P.2d 241.

In the present case, the plaintiff has failed to meet the obligation of providing expert evidence indicating that the quarry tile used in the Pleasanton swimming pool deviated from the reasonable standards used within the appropriate architectural community. As the trial court noted, the plaintiff's first witness,

George Forman, expressly refused to testify as an expert relating to the acceptability and safety of the quarry tile. The district court noted that the plaintiff's second witness, Dr. Alan Caskey, was neither an architect nor an engineer and had never personally designed and constructed a swimming pool. Moreover, in defining "slip resistant" coping as material possessing a wet coefficient of friction in excess of 0.50, Caskey expressly relied upon a reference book on swimming pool construction which expressly recognizes the use of quarry tile as coping material for swimming pools. Finally, the only quantitative examinations performed upon the quarry tile in the present case were those conducted by Robert Weiner. Weiner concluded, on the basis of his tests, that the quarry tile possessed a wet coefficient of friction of 0.60 and, thus, possessed a markedly better degree of slip resistance than that which Caskey indicated was the standard for swimming pool coping material. Thus, the trial court correctly concluded that the expert evidence in the present case indicated that the quarry tile used in the Pleasanton swimming pool exceeded the safety standards accepted by the local architectural community.

. . .

Although, in the present case, the material used as coping may have deviated from the type of material called for in the original plans and specifications for the swimming pool, this fact alone does not create liability on the part of the defendants as to third parties. The plaintiff must still demonstrate that the coping material actually used in the swimming pool was an unreasonable violation of the relevant architectural and engineering community standards. . . . In the present case, all of the expert witnesses testifying as qualified architects or engineers stated that the quarry tile used was an acceptable material for swimming pool coping. The only quantitative examinations performed upon the quarry tile indicate that it exceeds the coefficient of friction standard required for swimming pool coping material. The trial court, thus, did not err in concluding that the plaintiff had failed to present expert evidence to support the allegation that the coping material used in the Pleasanton swimming pool violated relevant architectural and engineering community standards.

The plaintiff cites no authority which either directly or indirectly supports the proposition that a deviation from the plans and specifications in a construction project automatically renders the contractor liable for negligence. While a violation of the plans and specifications may create contractual liability on the part of the contractor to persons in privity with him, something more must be shown before the contractor may be held liable for negligence.

In the present case, the use of quarry tile as a coping material in the Pleasanton swimming pool deviated from the original plans and specifications for the swimming pool. However, to recover, the plaintiff must demonstrate that the use of

> quarry tile constituted an unreasonable deviation from relevant architectural and engineering professional standards, or constituted a dangerous defect or condition that was known to the defendants or, by proper inspection of the pool, should have been known. In the present case, the plaintiff has clearly failed to make such a showing. All the expert testimony in the present case indicates that the use of quarry tile, while not called for in the original plans and specifications, was not a violation of the defendants' obligation to use reasonable care in the design, inspection, and construction of the swimming pool.
>
> . . .
>
> The judgment of the trial court is affirmed.

3. Variations from Code and Consequences for Design Professionals

Most states hold that a design professional's failure to comply with a building code requirement is negligence per se.[25] In a case that reached the Tenth Circuit Court of Appeals (applying New Mexico law), an engineer drafted plans and specifications for a building that failed to comply with the applicable building code. Although the engineer conceded this defect, he argued that he could not be held responsible for the ensuing damage because the code did not specifically call out that it was applicable to engineers. The Tenth Circuit rejected this argument:

> We think it would be anomalous to assume that the New Mexico legislature omitted to include design engineers in a comprehensive legislative scheme for the regulation of the construction industry in New Mexico with power in a board to promulgate and adopt minimum standards for the construction, alteration and repair of buildings. If the standards prescribed by the building code are to have any efficacy, they must be made applicable to every stage and phase of the regulated construction. The structural design is of course an integral part of the construction blueprint. And since the engineers do not contend

25. St. Joseph Hosp. v. Corbetta Constr. Co., Inc., 316 N.W.2d 51 (Ill. 1974) (architect liable for owner's costs in having non–code compliant wall covering replaced); Raymond v. Baehr, 163 N.W.2d 51 (Minn. 1968) (violation of code requiring fireproof materials in vent shafts held negligence per se, even though building code was not published as strictly required by state law); Kincaid v. Landing Dev. Corp., 344 S.E.2d 869 (S.C. App. 1986) (violation of a statute is negligence per se; technical error to so instruct jury when no ordinance adopting the code was offered into evidence, but jury verdict for homeowner affirmed when judge directed verdict as to liability on non-statutory grounds).

> that the code was not intended to benefit the architects as a class, we quite agree with the trial court that the statute is applicable to design engineers.[26]

C. Contractual Standard of Care

Notwithstanding the commonly recognized rules establishing the standard of care for architects and engineers, many design professional contracts state the standard of care explicitly. For example, the Standard Form of Agreement Between Owner and Engineer for Professional Services published by the Engineers Joint Contract Documents Committee provides:

> **[6.01 Standards of Performance]**
>
> A. The standard of care for all professional engineering and related services performed or furnished by Engineer under this Agreement will be the care and skill ordinarily used by members of the subject profession practicing under similar circumstances at the same time and in the same locality. Engineer makes no warranties, express or implied, under this Agreement or otherwise, in connection with Engineer's services.

AIA Document B201–2007 contains a similar provision.

What is the function of such a clause in the face of a common law rule? It is twofold. First, the use of a contractual standard in many jurisdictions allows invocation of the economic loss rule to ensure that the design professional will be held only to the precise contractual standard and not to some independent tort standard or tort cause of action.[27] Even when the substantive contractual standard is identical to the tort standard, the avoidance of tort liability may cut off punitive damages or provide other procedural relief. Second, the standard of performance section in this instance expressly limits the standard such that no express or implied warranties will expand the design professional's liability.

D. Implied Warranties and Design Professionals

1. *Implied Warranties Generally in Construction*

Under the *Spearin* Doctrine,[28] an owner impliedly warrants that the drawings and specifications are sufficient for construction of the work at hand. One might imagine that the owner, in turn, benefits from a similar implied warranty

---

26. Burran v. Dambold, 422 F.2d 133, 136 (10th Cir. 1970).

27. The economic loss rule is discussed in more detail in Chapter 23.

28. See Chapter 22.

to the effect that the design professional's drawings and specifications are suitable for construction. Somewhat surprisingly, the rule is not uniform.

2. *Different Treatment of Design Professionals and Implied Warranties*

a. Jurisdictions Imposing Implied Warranties

Some courts have found an implied warranty of fitness in the work of the design professional. In *Hill v. Polar Pantries*, 64 S.E.2d 885 (S.C. 1951), a refrigerator dealer provided drawings and specifications for construction of a cold storage warehouse that included a concrete floor that used vermiculite for insulation. The floor did not cure properly, buckled, and proved to be unusable. The court's ruling has been applied to a variety of people who hold themselves out as having specialized knowledge:

> It seems to be well settled that where "a person holds himself out as specially qualified to perform work of a particular character, there is an implied warranty that the work which he undertakes shall be of proper workmanship and reasonable fitness for its intended use, and, if a party furnishes specifications and plans for a contractor to follow in a construction job, he thereby impliedly warrants their sufficiency for the purpose in view." [citation omitted] These principles have been applied to building contracts. [citations omitted]

The latter statement—that a party impliedly warrants the sufficiency of drawings for construction—is the *Spearin* Doctrine itself, and *Hill*'s statement finds little difficulty in imposing the implied duty on the design professional. Some states follow suit.

b. Refusal to Imply Warranties for Design Professionals

On the other hand, a number of courts have held that the law will not imply any warranty on the part of the design professional's work. The following case illustrates the rejection of implied warranties in the context of an architect's general duty of care:

Ryan v. Morgan Spear Assoc., Inc.
546 S.W.2d 678 (Tex. Civ. App., 1977)

BISSETT, Justice.

This is a suit for damages for breach of contract, and, in the alternative, for negligence in the performance of professional services. Thomas Ryan (Dr. Ryan) and wife, Beverly Jean Mitchell Ryan, were plaintiffs in the trial court and Morgan Spear

Associates, Inc., an architectural corporation, was defendant therein. Following a jury trial, a take nothing judgment was rendered in favor of the defendant. The plaintiffs have appealed.

The plaintiffs, in their action for breach of contract, alleged that they entered into a written contract with the defendant for the preparation of plans and specifications for an animal hospital building which was to be built in Corpus Christi, Texas, and for the supervision of the construction of the building; that they were led to believe, and did believe, that the defendant had properly tested the soil upon which the building was located and had adequately and properly designed and specified a foundation that would support the building; that immediately after completion, the building began to deteriorate, which was caused by an inadequate and improper design of the foundation; that the defendant breached its contract with the plaintiffs by reason of the faulty design of the foundation; that the building was a total loss; and that their damages amounted to $77,129.57.

. . .

The plaintiffs, in points 1 and 2 contend that the trial court erred in [instructing the jury on negligence only] because the suit is "for breach of contract and an implied warranty that defendant had a duty to prepare plans and specifications for a building suitable for the purposes for which it was going to be used and the submission of 'negligence,' which was not an issue, placed an undue burden on plaintiffs."

. . .

It is a well settled rule that an architect must use the skill and care in the performance of his duties commensurate with the requirements of his profession, and he is liable in damages if he is negligent in performing such duties. It is stated in 6 C.J.S. Architects, § 27:

> The architect, by his contract, implies his possession of ordinary good taste, skill and ability, and a promise to exercise them reasonably, without neglect, and with a certain exactness of performance, in seeing that the work is properly done. The degree of skill required is such as would produce, if followed, a building of the kind called for, without marked defects in character, strength or appearance.
>
> In the absence of special agreement, an architect is not liable for faults in construction resulting from defects in the plans, as his undertaking does not imply or guarantee a perfect plan or a satisfactory result, it being considered enough that the architect himself is not the cause of any failure, and there is no implied promise that miscalculations may not occur. Thus, an architect is only liable for a failure to exercise reasonable care and skill.

. . .

Concerning the question of warranties, both express and implied, in a contract between an architect and the owner of a proposed building, the rule in Texas is clearly stated in 6 Tex.Jur.2d, Architects and Engineers, § 32, as follows:

> A warranty by an architect will not be implied unless there is the clearest reason for it, and the burden of showing such reason rests on the one seeking to establish the warranty. Thus, where there is no warranty, express or implied, that plans contain no defects that might cause the collapse of the building while it is in construction, no such warranty should be implied by the owner in favor of one who, with plans and specifications before him and ample time to consider them, obligates himself to carry the work to completion.

Although there is a contract between the parties in the present case, the cause of action for damages sustained by the plaintiffs is tortious in nature. Capitol Hotel Co. v. Rittenberry, supra; Pecos & N.T. Ry. Co. v. Amarillo St. Ry. Co., 171 S.W. 1103, 1105-06 (Tex.Civ.App.-Amarillo 1914, no writ). The contract between the plaintiffs and the defendant architectural corporation does not contain any special guarantees or warranties. There is no implied warranty on the part of the defendant to the plaintiffs that the plans and specifications for the foundation of the building were free from the alleged inherent defects. The contract is set out in general terms such that the only implication arising from the terms is that the architect will use reasonable care in preparation of the plans and supervision of the construction of the project.

Damages "due to breach of a written contract," as asserted by the plaintiffs, is not an issue in this case. Any liability on the part of the defendant for the damages sustained by the plaintiffs can only be sustained on proof that it was negligent and that such negligence proximately caused plaintiffs' damages.[29]

---

29. *See also* City of Mounds View v. Walijarvi, 263 N.W.2d 420 (Minn. 1978) ("We have reexamined our case law on the subject of professional services and are not persuaded that the time has yet arrived for the abrogation of the traditional rule [to the effect that "One who undertakes to render professional services is under a duty to the person for whom the service is to be performed to exercise such care, skill, and diligence as men in that profession ordinarily exercise under like circumstances."] Adoption of the city's implied warranty theory would in effect impose strict liability on architects for latent defects in the structures they design. That is, once a court or jury has made the threshold finding that a structure was somehow unfit for its intended purpose, liability would be imposed on the responsible architect in spite of his diligent application of state-of-the-art design techniques. If every facet of structural design consisted of little more than the mechanical application of immutable physical principles, we could accept the rule of strict liability which the city proposes. But even in the present state of relative technological enlightenment, the keenest engineering minds can err in their most searching assessment of the natural factors which determine whether structural components will adequately serve their intended purpose. Until the random element is eliminated in the application of architectural sciences, we think it fairer that the purchaser of the architect's services bear the risk of such unforeseeable difficulties.").

Where there is no implied warranty running from the design professional to the owner,[30] the owner faces the discomforting prospect that it might be liable to a contractor for drawings that are not suitable for construction yet be unable to prove that the drawings and specifications constitute a breach of the design professional's contract with the owner. This issue turns out to be more theoretical than practical, although the differential standards require attention when a project runs into trouble.

6.05 IMPLEMENTATION OF THE DESIGN THROUGH CONTRACTORS

Once the design professional has completed its design in a Design-Bid-Build project, the design documents are turned over to the contractor for construction. The design process is probably not complete at that point, however, as both planned and unplanned design issues arise during the course of construction. In addition, the contractor itself may have some limited role in design.

There are two common areas in which the contractor may become at least partially involved in design: (1) value engineering and (2) design-build of components of the construction.

A. Value Engineering

In "value engineering," a concept that construction lawyers encounter frequently in practice, the contractor uses its superior knowledge of construction materials, means, and methods to offer suggestions to the owner as to how equivalent construction can be completed cheaper than as originally designed (or as to how superior construction can be obtained within the same budget). Value engineering typically occurs in contracts under the terms of which the contractor can earn additional compensation by building less expensively than originally planned. The contractor rarely has the right to change a design unilaterally, but instead must propose design changes to the owner for approval; thus, the value engineering process is one of negotiation.

30. For additional cases rejecting implied warranties in the owner/architect relationship, *see* Gravely v. Providence Partnership, 549 F.2d 958 (4th Cir. 1977) (applying Virginia law); Ryan v. Morgan Spear Assoc., Inc., 546 S.W.2d 678 (Texas Civ. App. Corpus Christi 1977), writ refused n.r.e., (June 1, 1977) (implied warranty not recognized in Texas); Mississippi Meadows, Inc. v. Hodson, 299 N.E.2d 359 (3d Dist. 1973) (Illinois does not recognize implied warranty); Smith v. Goff, 1958 OK 100, 325 P.2d 1061 (Okla. 1958); Ressler v. Nielsen, 76 N.W.2d 157 (N.D. 1956); Paxton v. Alameda County, 259 P.2d 934 (Cal. App. 1st Dist. 1953).

B. Performance Specifications and Design-Build of Discrete Portions of Construction

The reader will encounter the difference between design specifications and performance specifications in Chapter 22 and is commended to those discussions for a more thorough review. At this juncture, it is enough to state that design specifications direct a contractor as to the specific manner in which construction is to be accomplished, and the contractor's responsibility is simply to ensure that it adheres to the design specifications. A performance specification, on the other hand, describes the overall *results* to be accomplished as measured by performance criteria, and the contractor's responsibility is to ensure that the construction satisfies those criteria. As a simple example, a design specification may call out four permissible models of pumps from which the contractor must choose, while a design specification might specify only the pump capacity for particular placement in a processing plant and leave the selection of the pump to the contractor.

With respect to performance specifications, the contractor in some sense bears responsibility for the "design" of the construction so as to ensure compliance with performance criteria. If the contractor's work fails to achieve the stated performance criteria, it is generally the contractor's liability for the insufficiency or inappropriateness of the design of that portion of the construction.

On a bigger scale, the contract between the owner and the contractor may overtly allocate to the contractor the design of a portion of the construction on a design-build basis, even when the contract itself is primarily a Design-Bid-Build contract. This typically occurs with respect to technical trades within the province of specialized subcontractors, such as the design of programmable logic controller (PLC) circuits for processing plant controls. Although it would be possible for the architect or engineer to hire such specialists, there are many cases in which the specialty subcontractor works for the general contractor on a design-build basis. Again, the importance of this relationship is that it puts the burden of proper design of the PLC system on the contractor rather than on the lead design professional.

QUESTIONS

1. If the architect would agree to include the services of a geotechnical engineer within his scope of work, would you recommend to the owner that he do so? If you were representing the architect, would you recommend against it? Why?
2. An architectural firm you represent has asked you to review language in one of its proposals, which states that the firm "designs to the state of the art" and that its work "is of the highest standard." How might this language affect the standard of care that would be expected of the firm?
3. What is the difference, if any, between warranting the suitability of plans and agreeing to prepare them in accordance with the standard of care to be expected of architects?
4. Would an architect's failure to prepare drawings that complied with the Americans With Disabilities Act be negligence per se?

CHAPTER

7

Design Agreements

LAWRENCE C. MELTON

7.01 IMPORTANCE OF AN AGREEMENT BETWEEN THE OWNER AND ARCHITECT[1]

Formation of the legal relationships among the many participants needed to perform a construction project begins, in many ways, with the owner-architect agreement. How construction is administered, the type of documentation that is to be furnished to the lender, how disputes are resolved, and many insurance requirements, for example, are described initially in the owner-architect agreement. Such provisions set the tone for other contractual arrangements in connection with the project that subsequently are established. Many provisions in a construction contract often are derived simply from certain terms and conditions found in the owner-architect agreement in order to maintain consistency between provisions in the two documents that are interrelated.

The author thanks Ridgely Jackson, an associate of the Chicago office of DLA Piper LLP (US), for her valuable assistance in the preparation of this chapter.

1. *See generally*, Ross J. Altman & Ridgely J. Jackson, *Owner-Architect Agreement: A Discussion of Key Provisions,* presented by Ross J. Altman, *Owner's Avoiding Claims*, in Washington, D.C. (February 13, 2008); updated from Ross J. Altman, *Generation Next*, presented at American Bar Association Forum on the Construction Industry, in Atlanta, Georgia, October 1997, and San Diego, California, November 1997.

The owner-architect agreement is a somewhat unusual document because the object of the agreement does not physically exist and is difficult to describe. The project may not be anything more than a vague idea to the owner or an operational need that must be addressed. An owner-architect agreement involves the development of ideas and the transformation of those ideas into drawings and specifications. Unfortunately, many agreements with architects describe the design process in jargon that is not well understood by those outside the design professions. Because many disputes in the design and construction industry are caused by nothing more than disappointed expectations, it is crucial that the owner-architect agreement accurately describes the design process in language that is understood well by all parties.

The design process was described in Chapter 6. This chapter provides an overview of how owners and architects contract for the performance of the design process. Like most agreements between owners and architects, Chapter 7 begins with a description of the professional design services that architects usually provide and the circumstances under which an owner may desire to expand the scope of services originally contemplated. Following that discussion is a summary of other key issues in an agreement between the owner and the architect, such as (1) when the architect may have responsibility to the owner if actual costs of construction exceed the owner's budget, (2) who owns the documentation and ideas created by the architect and the design team, (3) payment for professional design services, (4) architects' and engineers' professional liability insurance, (5) methods used by architects and engineers to limit their liability to an owner, and (6) termination of the agreement between the owner and the architect.[2]

7.02 THE SCOPE OF THE ARCHITECT'S SERVICES

The architect's activities usually are described as "services" and may be performed by the architect with its own forces or through subconsultants. Although architects are capable of performing a range of different services when requested by an owner, the typical scope of the architect's responsibilities includes architectural design and "usual and customary structural, mechanical, and electrical engineering services."[3] Most agreements between an owner and an architect describe a design process that is performed in phases.

2. A detailed discussion of the architect's services performed during the construction phase of a project is found in Chapter 13.

3. American Institute of Architects, AIA Document B101–2007, *Standard Form of Agreement Between Owner and Architect* § 3.1 (2007) [hereinafter AIA Document B101–2007]. Notwithstanding the AIA description of the architect's scope of services, from time to time certain sectors of

A. Phases of the Design Process

The agreement between the owner and the architect can take many forms. The design process, however, necessarily includes certain activities—beginning typically with schematic design and ending upon completion of construction.[4] The project design is not well developed in the early phases, but becomes more complete and detailed in each successive phase. Although the design phases may not be performed in a perfectly sequential manner (i.e., some aspects of the design may move along more quickly than others), the overall design gradually becomes more defined over time.

1. Schematic Design Phase

The first phase, the schematic design phase, generally requires preliminary evaluation by the architect of the owner's program, schedule, budget, site, and proposed delivery method; preparation of a preliminary design for the owner's approval; preparation of the schematic design documents (which may include a site plan, preliminary drawings or site plans, study models, and preliminary selections of major building systems and construction materials); and consideration of such issues as safety, alternative building methods, environmental, and budget constraints.[5] Schematic design documents generally illustrate only the scale and relationship of various project components.

The schematic design phase ends with delivery of the schematic design documents to the owner for the owner's review and approval. Once such approval is obtained, along with authorization for any adjustments in the schematic design or other project requirements, the architect proceeds with the design development phase.

2. Design Development Phase

The documents prepared in the schematic design phase are further developed in the design development phase and include drawings and other documents, such as plans, sections, elevations, typical construction details, and diagrammatic layouts of building systems to fix and describe "the size and character of the Project as to architectural, structural, mechanical and electrical systems, and such other elements as may be appropriate."[6] The design development documents also include outline specifications that identify major materials

---

the construction industry may develop a custom and practice that varies from the AIA standard forms. Also, regional differences exist. For example, it might be common for the owner of a project located in a particular part of the country to hire the engineers directly rather than have the architect hold subconsulting agreements with the engineers.

4. AIA Document B101–2007, *supra* note 3, §§ 3.2–3.4.

5. *Id.* § 3.2.5.

6. *Id.* § 3.3.1.

and systems and establish in general the quality of the project.[7] A significant amount of engineering services are performed during the design development phase. By the end of this phase, the architect should be confident that the project design is workable, even though the documents prepared during the design development phase are not yet a complete design and likely will be stamped "Not For Construction."

The Design Development Phase ends by delivery to the owner of the design development documents for review and approval. Once such approval is obtained, along with any authorization for changes in the project design or other project requirements, the architect proceeds with the construction documents phase.

3. *Construction Documents Phase*

Based on the approved design development documents, the architect prepares construction documents consisting of drawings and specifications setting forth in detail the quality levels of materials and systems and other requirements for construction of the Work.[8] The construction team builds the project from information contained in the construction documents.[9] The construction documents phase ends by delivery of the construction documents to the owner for review and approval.[10] Although the construction documents set forth detailed requirements for construction, the actual cost of such construction cannot be determined until potential contractors submit bids or proposals.

4. *Bidding or Negotiation Phase Services*

The architect assists the owner with obtaining and evaluating negotiated proposals or competitive bids.[11]

5. *Construction Phase Services*

Finally, the architect administers the construction contract.[12] During this phase, the architect undertakes a number of responsibilities, many of which are intended to accomplish the following:

---

7. *Id.*

8. *Id.* § 3.4.1.

9. Sometimes documents with detail sufficient to evaluate various health and safety matters are submitted to public authorities. Although such "permit documents" are adequate for purposes of issuing a building permit, further development and detail must be completed in order to produce construction documents.

10. AIA Document B101–2007, *supra* note 3, § 3.4.5.

11. *Id.* § 3.5.

12. *Id.* § 3.6.

- ❒ Provide the owner with some assurance that the work of the construction team, when complete, will comply with the requirements of the contract documents.
- ❒ Facilitate the administration of payments to the contractors.
- ❒ Review and coordinate submittals and activities pertaining to the delegation of certain aspects of the design to the construction team.
- ❒ Prepare change orders for the owner's approval.
- ❒ Determine when the construction team achieves substantial and final completion.
- ❒ Provide interpretation of the drawings and specifications.
- ❒ Attempt to resolve, and provide recommendations with respect to, disputes or claims that may arise between the owner and the construction team.

The architect's role during the construction phase is very important, particularly for larger and more complicated projects where construction necessarily involves greater coordination, evaluation, and administration. Such services cover a broad range of activities and are discussed in much greater detail in Chapter 13.

B. Basic Services versus Additional Services

The bundle of services customarily furnished by the architect is known as basic services. The architect's compensation is based on the scope of basic services. There are many different situations, however, in which an architect may need to perform services beyond those routinely classified as basic services in most standard forms of agreement between the owner and architect. Because each project is different, it is not always possible to predict at the time of contracting what services might be needed in addition to the core basic services. Most agreements between an owner and an architect, therefore, include a menu of certain additional services from which to choose.

First, the need for additional services may exist if the nature of the project eventually requires a broader scope of design services. For example, renovation projects require analysis of existing facilities; the owner may want the project to qualify for Leadership in Energy and Environmental Design (LEED) certification; or the project may involve landscape design or civil engineering services. Often, such needs are known prior to the date that the agreement between the owner and architect is executed, and the agreement may be modified to require such activities as part of basic services. However, in the event the need for such expanded services is not determined until a later date, the owner may authorize the needed services from the list of additional services set forth in the agreement.

Second, additional services may be needed if the owner initiates a change in design. As already described in Section 7.02(A), the architect develops the project design in phases. At the end of each phase, the owner has the opportunity to review the documents prepared as part of that phase and provide comments. Once the owner approves the design in its then current state, the architect has the right to rely on that approval for purposes of performing the next phase of design services. The owner is free subsequently to request a change in the design. But if such request is inconsistent with prior approvals or instructions provided by the owner, then the architect must devote additional time to modify the design documents as necessary to implement such a change. This may involve making changes to many aspects of the design in order to maintain consistency within the design documents and avoid conflicts.

The need for the architect to perform additional services also can arise from the construction process. Indeed, it is rare for a project to proceed entirely according to plan without some surprise. Events beyond the control of either party, such as a change in law or the discovery of unforeseen or concealed conditions, may require the architect to modify the design. Additional services might also be needed to respond appropriately to requests for information submitted by the construction team or to resolve problems arising out of defective work.

The distinction between basic services and additional services has practical ramifications. Most importantly, the architect is compensated for the performance of additional services in addition to compensation provided for basic services. It would be unfair to burden the architect with obligations not contemplated upon execution of the agreement between the owner and architect and to not provide additional compensation. Also, because the need and scope of additional services cannot be predicted with certainty at the time of contracting, the architect often is paid for additional services on a time and material basis. Such a compensation method may differ from that used for compensation of basic services and may not be subject to the limitations and other price protections that apply to the performance of basic services.

C. The Owner's Responsibilities

The conventional role of the parties is that the architect designs, the contractor builds, and the owner pays. There are variations on this theme, but it is still fairly accurate. The owner's typical responsibilities include the following:[13]

- Providing information regarding requirements for and limitations on the project.

13. *Id.* Art. 5.

- Periodically updating the budget for the project, including that portion allocated for the cost of the work.
- Rendering decisions in a timely manner pertaining to documents submitted by the architect in order to avoid unreasonable delay in the orderly and sequential progress of the architect's services.
- Furnishing the services of certain consultants when specialized services are requested by the architect and reasonably are required by the scope of the project.
- Furnishing tests, inspections, and reports required by law or the construction contract, such as structural, mechanical, and chemical tests, tests for air and water pollution, and tests for hazardous materials.
- Furnishing all legal, insurance, and accounting services, including auditing services, that may be reasonably necessary at any time for the project to meet the owner's needs and interests.
- Providing prompt written notice to the architect if the owner becomes aware of any fault or defect in the project.

The allocation of responsibilities in most agreements between an owner and architect reflect a philosophy that (1) the owner contributes the real property to the project and is responsible for its physical condition, as well as the procurement of services needed to evaluate or describe the property, and (2) the owner should obtain from others the design and consulting services that the architect does not customarily perform or manage.

7.03 DESIGNING TO A BUDGET

Koontz v. Thomas
333 S.C. 702, 511 S.E.2d 407 (S.C. App., 1999)[14]

HEARN, Judge:

Robert S. Koontz instituted this action against Thomas and Denziger, P.A., an architectural firm, and James G. Thomas, Jr. individually (collectively "T & D"), alleging professional negligence, negligent misrepresentation, and breach of contract in connection with an architectural contract. Koontz appeals from summary judgment in favor of T & D on all three causes of action. We affirm.

14. Internal citations, footnotes, expanded procedural history, and further discussion of negligence as applied to the facts of the case have been omitted.

In December of 1992, Koontz and T & D entered into an agreement for T & D to design Koontz's new residence in Beaufort, South Carolina. The agreement provided that (1) T & D's compensation would be 10% of the cost of construction; (2) T & D did not warrant or represent that actual construction bids or negotiated prices would not vary from any construction cost estimate T & D prepared; (3) there was no fixed upper limit of construction costs established in the contract, and the parties could agree to a limit only in a signed writing; (4) T & D's plans and specifications required Koontz's approval; and (5) the entire project would be completed within twenty four months, unless more time was needed through no fault of T & D.

With Koontz's input, T & D developed a schematic design for the home. The first plans were sent to Koontz on June 22, 1993. On June 10, 1994, after numerous designs, revisions, and modifications, Koontz approved a plan with an $800,000 construction estimate. However, Koontz believed that there was "a good chance that the construction cost could be lower than [T & D's] estimate of $800,000." Thereafter, three pre-selected contractors submitted construction bids. Contrary to Koontz's prediction, the bids ranged from $983,000 to $1.2 million. Koontz terminated the contract and demanded that T & D return all architectural fees paid through the date of termination.

T & D offered to continue working on the project, suggesting revisions and modifications that would lower construction costs, but Koontz refused the offer. Subsequently, Koontz hired a different architect to whom he paid $39,463 to design a home that was ultimately constructed for $870,000. Koontz then instituted this action against T & D, alleging professional negligence, negligent misrepresentation, and breach of contract.

On all three causes of action, Koontz made essentially the same allegations. Koontz alleged T & D negligently represented that the construction phase of the building project could be completed for $400,000, and the parties agreed to limit the project to this amount. Thus, T & D was negligent in designing a home exceeding that amount by 100%. Koontz further alleged T & D negligently represented that the preliminary architectural services necessary to the project could be completed within a reasonable time, and that T & D would charge reasonable fees for the design. T & D answered asserting affirmative defenses and counterclaimed for breach of contract and quantum meruit.

On December 27, 1996, T & D moved for summary judgment on all three of Koontz's causes of action. On June 20, 1997, the trial court granted summary judgment in favor of T & D on each cause of action. Koontz's subsequent motion for reconsideration was denied. This appeal followed.

. . .

II. Breach of Contract

A. Construction Costs

Koontz asserts the trial court erred in granting T & D summary judgment on his breach of contract claim. Specifically, Koontz alleges parol evidence should have been admissible in the breach of contract action to establish an oral agreement to limit construction costs to $400,000. We disagree.

"Where an agreement is clear and capable of legal construction, the court's only function is to interpret its lawful meaning, discover the intention of the parties as found within the agreement, and give effect to it." . . . The court must enforce an unambiguous contract according to its terms regardless of its wisdom or folly, apparent unreasonableness, or the parties' failure to guard their rights carefully. . . . It has long been held that to ascertain the intention of an instrument, the court must first look to its language, and if it is perfectly plain and capable of legal construction, then that language alone determines the instrument's force and effect.

. . .

Koontz argues that his contract with T & D is ambiguous because it is "silent as to the amount of construction costs." In support of this contention, Koontz cites *Williams & Assocs. v. Ramsey Prod. Corp.*, 19 N.C. App. 1, 198 S.E.2d 67, cert. denied, 284 N.C. 125, 199 S.E.2d 664 (1973), for the proposition that construction contracts that are silent on the issue of a maximum cost limitation fail to disclose the parties' intention as to the cost of structure contemplated. Under the facts and circumstances of this case, we decline to adopt this reasoning.

"The parol evidence rule prevents the introduction of extrinsic evidence of agreements or understandings contemporaneous with or prior to execution of a written instrument when the extrinsic evidence is to be used to contradict, vary or explain the written instrument." . . . However, if a contract is ambiguous, parol evidence is admissible to ascertain the true meaning and intent of the parties. . . . An ambiguous contract is one that can be understood in more ways than just one or is unclear because it expresses its purpose in an indefinite manner. . . .

In our view, the parties' decision not to include a maximum cost limitation in the contract does not create contractual ambiguity. Koontz's contract with T & D specifically addresses the cost of construction, stating the parties expressly declined to set a maximum cost limitation. Moreover, the parties intended to leave open the possibility of a later written agreement. We are aware of no authority, and Koontz cites none, which convinces us the parties to a construction contract may not, without creating contractual ambiguity, decline to set a fixed construction cost in the initial stages of the undertaking.

Koontz further argues that even if the contract is not ambiguous and parol evidence of a prior or contemporaneous agreement is barred, the parties subsequently modified the contract to limit the construction price to $400,000. This argument is likewise unavailing.

A written contract may be modified by oral agreement even if the contract expressly states that all changes must be in writing. . . . However, we are unable to ascertain from the record before us that the parties made such an agreement. Koontz relies upon a letter he wrote on June 7, 1993, setting a construction limit of $400,000. The record does not reveal that T & D accepted this term. Moreover, on June 10, 1994, Koontz wrote a letter indicating his approval of plans and specifications with a construction estimate of $800,000. . . . Thus, any agreement to a $400,000 construction budget made on June 7, 1993, was subsequently modified by T & D's offer and Koontz's acceptance of an $800,000 construction budget on June 10, 1994.

. . .

B. Excessive Fees

We need not address Koontz's argument that excessive construction costs led to excessive fees. The contract stipulates that T & D's fees would be 10% of the final cost of construction. Because we do not find that T & D's construction estimates were excessive or unreasonable given Koontz's approval of the $800,000 plans, it follows that any fees based on a percentage of the final construction costs must also be reasonable. . . .

. . .

C. Timeliness

We also find no merit in Koontz's claim that T & D breached the contract by failing to complete preliminary architectural services in a timely manner. Koontz points to no specific contractual provision which has been breached. Moreover, parol evidence of a prior or contemporaneous agreement is not available to establish a time limitation, and there is no evidence of a later agreement.

While the contract calls for an overall scheme of twenty-four months to complete the entire project, it is silent about when T & D would complete the design and planning phase so that construction could begin. Where a contract does not provide a time frame for performance, the court will imply a reasonable time. . . . Koontz's expert submitted an affidavit wherein he opined that "the standard of care of a reasonably prudent architect as to completing plans on this proposed project would be approximately six to nine months." In fact, T & D sent its first set of plans to Koontz on June 22, 1993, approximately six months after the parties executed their agreement. However, Koontz was still suggesting changes and modifications as late as January 27, 1994, more than one year after the parties entered into the contract.

. . . .

For the forgoing reasons, the decision of the trial court is AFFIRMED.

CONNOR and HUFF, JJ., concur.

Some measure of what a building may cost is essential to the construction process.[15] Construction professionals generally recognize that construction cost estimating is "a hazardous business."[16] Although both the importance of the

15. *See* Robert E. Johnson, The Economics of Building: A Practical Guide for the Design Professional (Wiley & Sons 1990).

16. Joseph A. Demkin, The Architect's Handbook of Professional Practice § 14.4 (13th ed. 2001); *see* Karl A. Berg, Jr., *Clairvoyance Required: The Architect's Liability for Construction Cost Estimates*, 26 Colorado Lawyer 37 (1997).

preliminary estimate and potential liability for its inaccuracy always have been understood, the significance of both has been amplified recently by unstable prices for basic construction materials.[17]

The ability of the architect to design to a budget is of obvious importance. An owner has only so much money to expend on the project, whether the source of such funds is the owner's equity, investors, or construction lenders. If the bids or proposals received from contractors exceed the owner's budget, the project may become economically unviable. Even worse, such a realization might occur only after a significant expenditure in design fees and the passage of time. That is a bleak situation for an owner, who invariably would want to know who is to blame for such excess costs.[18]

A. Contractual Provisions Regarding Estimates

The contract that the court interpreted in *Koontz* was an earlier iteration of the AIA Owner-Architect Agreement. The most current edition requires the owner to establish a budget for the project[19] and the architect to prepare a preliminary evaluation of the budget[20] and develop a design for the project that is consistent with the owner's budget,[21] including an estimate of the cost of the work.[22] The estimate is to be updated at the completion of the design development documents and again at the completion of the construction documents.[23]

The parties acknowledge that the architect's estimate is not a warranty or representation and that the architect does not control the costs of labor, materials, or equipment.[24] The architect is allowed to use contingencies in the estimate and also to control "what materials, equipment, component systems and types of construction" are to be used or allowed.[25] The estimate is only good if bidding or negotiation begins within 90 days after submittal of the construction documents; after 90 days, the owner's budget must be revised to reflect

---

17. *See* Renata A. Guidry, *The Steel Price Explosion: What Is an Owner or Contractor To Do?*, 24 The Construction Lawyer 3, 5 (2004).

18. Lawrence C. Melton & Brian A. Autry, *Beyond His Power to Build It: Who Is to Blame for the Overbudget Project?*, 25 The Construction Lawyer 1, 20 (2005).

19. AIA Document B101–2007, *supra* note 3, § 5.2 and the "Initial Information" exhibit to the document. The contract anticipates that it may be "adjusted" throughout the project. *See Id.* § 6.2.

20. *Id.* § 3.2.3.

21. *Id.* § 3.2.5.2.

22. *Id.* § 3.2.6.

23. *See id.* §§ 3.3.2 and 3.4.4.

24. *Id.*

25. *Id.* § 6.3.

level of prices in the construction market.[26] If the owner's budget is exceeded by the lowest bona fide bid or negotiated proposal, the owner may increase the budget; rebid or renegotiate; terminate the project; revise the project program, scope, or quality; or implement another mutually acceptable solution."[27] The architect's responsibility is limited to modifying the documents to meet the budget, for no additional compensation.[28]

In the event that the bids exceed the budget, however, the AIA approach provides the following:

> If the Owner chooses to proceed under Section 6.6.4, the Architect, without additional compensation, shall modify the Construction Documents as necessary to comply with the Owner's budget for the Cost of the Work at the conclusion of the Construction Documents Phase Services, or the budget as adjusted under Section 6.6.1. The Architect's modification of the Construction Documents shall be the limit of the Architect's responsibility under this Article 6.[29]

When the contract is silent on the architect's responsibility for cost overruns, or the jurisdiction has an expansive view of the extra-contractual liability, some courts have awarded damages against design professionals for producing inaccurate project cost estimates. The owner's suit against the design professional responsible for an unreasonably high project estimate is likely to be based on one or more of the following theories:

- ❒ Breach of contract arising out of any specific contractual obligation of the architect to estimate the cost of the work.
- ❒ Negligence in providing agreed estimating services, where such services fail to meet the standard of care for design professionals.[30] However, courts have allowed a broad margin of error in construction cost estimating. An expert in cost estimating generally must identify a specific act or omission as a cause of the injury.[31]

---

26. *Id.* § 6.4.

27. *Id.* § 6.6.

28. *Id.* § 6.7.

29. *Id.* § 6.7. The contract in *Koontz* interpreted previous AIA language that was triggered only when the bids exceeded a written "fixed limit of construction cost," which phase now has been replaced by the "Owner's budget."

30. 5 PHILIP L. BRUNER & PATRICK J. O'CONNOR, BRUNER & O'CONNOR ON CONSTRUCTION LAW § 17:14 (West Group 2002).

31. Pipe Welding Supply v. Haskell, Connor & Frost, 462 N.Y.S.2d 221 (1984) ("where expert pointed to no specific negligent act or omissions . . . it was improper to allow this case to go to the jury based on an inference of malpractice drawn from the discrepancy between the bids received and the cost estimate"). *See* Comstock Ins. Co. v. Thomas A. Hanson & Assoc., 550 A.2d 731 (Md. App. 1988) (negligence found from evidence of having used the wrong R.S. Means data).

- Misrepresentation based on a construction estimate, which is rare, but not unknown.
- Breach of fiduciary duty alleging that the professional is a "fiduciary" of the owner in whom the owner has placed trust, and that, as a fiduciary, the professional owes an especially high duty to the owner.[32]

Under the traditional principles applicable to breach of contract damages, no recovery can be made for underestimating the cost of a project because the resulting structure is, presumably, worth what the owner paid. This argument is sometimes referred to as "betterment." Nonetheless, if a building is "over market" as a result of the designer's error, the owner still may have suffered damages, despite the fact that the owner now possesses a building that costs what the owner was charged. An increase in construction costs for a commercial building does not result in a more valuable building unless that value is reflected in increased rental rates or revenue. A fairly typical example of this line of cases is set forth in *Kellogg v. Pizza Oven, Inc.*

Kellogg v. Pizza Oven, Inc.
157 Colo. 295, 402 P.2d 633 (Colo. 1965)[33]

DAY, Justice.

This writ of error arises out of two actions consolidated for trial in the Denver District Court.

We will refer to plaintiffs in error as the Architects and to the defendant in error as Pizza.

Pizza in one action sued the Architects, alleging that they failed to perform properly architectural services in designing and supervising the construction of a restaurant building to be occupied by Pizza. It was alleged that the cost of the building was not to exceed $62,000.00, but, because of the negligence of the Architects, it cost approximately $100,000.00, with the resultant damage to Pizza of $40,000.00. The companion action involved a mechanic's lien foreclosure in which Pizza, the Architects, and others were named the defendants and in which the Architects filed a cross claim against Pizza for $2,761.80 due for services rendered.

Trial was presented to a jury, and the verdict was returned in favor of Pizza and against the Architects in the amount of $21,489.00. A verdict in favor of the

32. Such a claim is not typical unless the architect's agreement expressly creates a fiduciary relationship.

33. Internal citations, footnotes, and expanded procedural history have been omitted.

Architects for $930.00 as the balance of their fee was also returned. Motions for new trial and for judgment notwithstanding the verdict were filed and denied.

In their writ of error the Architects assert error as follows:

1. That the evidence fails to support the verdict as no negligence was shown.
2. That the damages awarded are grossly excessive.
3. That the court failed to instruct the jury on the measure of damages.
4. That the verdicts of the jury were inconsistent in that damages were awarded to Pizza while at the same time the Architects were determined to have earned their fee and given a judgment accordingly.

Addressing ourselves to the first point in the summary of argument, we find that although the evidence was in sharp conflict, it was ample to support the verdict of the jury. The Architects knew that Pizza was not to be the owner of the building. Pizza had negotiated with the owner of the land to build the restaurant building and lease it to Pizza. The owner-lessor put a ceiling of $60,000.00 cost on the construction. The agreement between the lessor and Pizza was that if the building cost more, Pizza would be required to pay the overage.

The testimony on custom in the architect profession was that budgetary limitations are proper to impose upon an architect; that an architect must understand, in general, what the building will cost and that when he has all of the information he can estimate the construction cost. It was agreed that a 10% variation is within the norm. It was not disputed that by custom when it is discovered that a building is exceeding the cost limitations it is incumbent upon the architect to tell this fact to the one who is employing him.

There was ample testimony to establish that after construction started the Architects made no attempt to recheck their original estimate by using any of the recognized cost breakdown systems; that when they were requested to supply a detailed cost breakdown to the contractor they utilized former erroneous estimates and made no effort to recalculate their figures. There was testimony that if their figures had been rechecked they would have discovered the almost 40% excess in costs, and certain changes could have been made to meet the budgetary limitations. The Architects failed to check the various bids as they came in and approved them without regard to the total, and the building was more than half way completed before they did so. A check of the bids before the contracts on them were let would have revealed immediately the cost to be in the area of $90,000.00. The general rule upon the duty of an architect is found in 5 AM. JUR. 2D, ARCHITECTS, § 23, as follows:

> An architect who substantially underestimates, through lack of skill and care, the cost of a proposed structure, which representation is relied upon by the employer in entering in the contract and proceeding with construction, may not only forfeit his right to compensation, but may become liable to his employer for damages.

The jury was instructed that they could find for Pizza and against the Architects if the latter were 'negligent in the furnishing of the preliminary estimate and that the plaintiff reasonably relied upon such preliminary estimate; or that the defendants were negligent in their supervision of the construction work; and that such alleged negligence was the proximate cause of damages to the plaintiff.' It is our view that this is a correct statement of the law. The jury had before it evidence of careless work on the preliminary estimate, in the supervision of the various subcontracting bids, and was told of some errors in design which required additional costly installations, none of which added to the functional features of the building. As an example, a brick planter had to be built to cover exposure of a foundation wall above grade level.

The assertion that the damages were excessive and that the jury was not properly instructed as to the measure of damages must be discussed together. Counsel for the Architects did not submit an instruction on their theory on the measure of damages. The court gave the following instruction:

> Instruction No. 9
>
> You are instructed that if you find from the evidence that the plaintiff is entitled to recover damages under its second claim for relief, based upon the alleged negligence of the defendants, you will assess as the plaintiff's damages an amount which, in your considered best judgment, will reasonably and justly compensate the plaintiff for said alleged negligence. In no event, shall you return a verdict on plaintiff's second claim for relief for more than $32,768.92, the amount prayed for by the plaintiff on said second claim.
>
> . . .

Instructions such as were given in this case have been approved in similar cases in other jurisdictions such as Texas and California. It may be gleaned from these cases that if the architect fails to exercise sufficient skill and diligence in the preparation of plans, or if the plans submitted are defective, he is liable for such damages as are just.

. . .

In this case the court could have been more specific by instructing the jury that the measure of damages is the difference between the estimated costs of the building and the actual cost less the following: 10% of the estimated cost deemed to be the normal variation allowed in cost estimates, and the cost of such items as the evidence showed were caused by substantial alterations in the original plans due to instructions of Pizza. In this case the failure to give such a precise instruction is not prejudicial because the jury used the evidence and the totality of the instructions in arriving at an identical mathematical result. The actual construction was $92,768.00 and the estimated cost submitted by the Architects was $62,000.00, making a difference of $30,768.00. Subtracting from that figure $6,200.00 as 10%

of the estimated cost and the sum of $2,079.00 which it is undisputed was overage caused by Pizza, we arrive at a figure of $21,489.00, the exact amount of the jury verdict. It is, therefore, apparent that if there were to be a remand for new trial on this asserted error, even with a more precise instruction, the result, mathematically, would be a similar verdict.

The Architects call our attention to cases in which the courts have refused to allow a jury to consider the entire difference between the cost of the building and the estimated costs in forming a basis of damages. But those were actions where the ones seeking damages from the architects were also owners of the building. In such cases the courts have reasoned that even though the building costs exceeded the estimate, yet to allow the full difference would result in unjust enrichment to the owner because he still had the building. In such cases there was evidence that the increased cost also resulted in increasing the value of the building, or, in other instances, resulted in an increase in rental income by reason of the better building produced. That is not the situation in the case at bar. Although there is $92,000.00 of material and labor incorporated in the building, the owner only paid $60,000.00 for it, and Pizza was the one that had to pay the more than $21,000.00 difference. There was no evidence that the completed building was more efficient, more functional, would produce more business, or more customer revenue.

The judgment is affirmed.

SUTTON and MOORE, JJ., concur.

7.04 OWNERSHIP OF THE DESIGN DOCUMENTS

A. Statutory Rights

Absent a contractual provision to the contrary, the architect holds the copyright to the drawings and specifications that the architect prepares for a project. The statutory framework that provides such protection to architects and other design professionals was laid out in *The Yankee Candle Co., Inc. v. New England Candle Co., Inc.*:[34]

> The Copyright Act grants a copyright owner exclusive rights in a copyrighted work to authorize reproduction of the work, preparation of derivative works, and public display of the work. 17 U.S.C. §§ 106–20. To establish ownership of a valid copyright, a plaintiff must prove that "the work, when viewed as a whole, is original" and satisfy the required statutory formalities. [Citations Omitted] A work need only exhibit a "minimal degree of creativity" to qualify for copyright protection. [Citations Omitted].
>
> (1) Among the statutory formalities, the Copyright Act requires registration of a work before a plaintiff may file suit. 17 U.S.C. § 411(a). The Copyright

34. Yankee Candle Co. v. New England Candle Co., 14 F. Supp. 2d 154 (D. Mass. 1998).

Office, after accepting an application, fee, and deposit of a representative copy of the work, *see* 17 U.S.C.§ 410(c), issues a certificate of registration, which is admissible in an infringement action as "prima facie evidence of the validity of the copyright and of the facts stated in the certificate," §17 U.S.C. 410(c). "In judicial proceedings, a certificate of copyright registration constitutes prima facie evidence of copyrightability and shifts the burden to the defendant to demonstrate why the copyright is not valid." [Citations Omitted]

In passing the Copyright Act of 1976, Congress expressly indicated that architectural plans and drawings deserved copyright protection as "pictorial, graphic and sculptural work" under 17 U.S.C. §101. See H.R.Rep. No. 94-1476, *reprinted* in 1976 U.S.C.C.A.N. 5659, 5668 ("An architect's plans and drawings would, of course, be protected by copyright [law]"). Congress, however, also indicated that architectural structures built from plans possessed only minimal copyright protection:

> [T]he extent to which . . . protection would extend to the structure depicted would depend on the circumstances. Purely nonfunctional or monumental structures would be subject to full copyright protection under the bill, and the same would be true of artistic sculpture or decorative ornamentation or embellishment added to a structure. On the other hand, where the only elements of shape in an architectural design are conceptually inseparable from the utilitarian aspects of the structure, copyright protection for the design would not be available. *Id.*

After the passage of the 1976 Act, courts extended federal copyright protection to architectural plans, but not to buildings constructed from those plans. *See Robert R. Jones Assoc. v. Nino Homes,* 858 F.2d 274, 280 (6th Cir. 1988) ("one may construct a house which is identical to a house depicted in copyrights architectural plans, but one may not directly copy those plans and then use the infringing copy to construct the house"); *Donald Frederick Evans v. Continental Homes, Inc.*, 785 F.2d 897, 901 n. 7 (11th Cir. 1986) ("A builder who constructs a home substantially similar to a dwelling already constructed is not liable for copyright infringement merely based on the substantial similarity if he or she did not engage in an unauthorized copying or use of the copyrighted architectural drawings"); *Demetriades v. Kaufmann,* 680 F.Supp. 658, 665–66 (S.D.N.Y. 1988) (enjoining defendant from copying plaintiff's architectural plans or relying on infringing copies in construction but declining to enjoin construction of defendant's substantially similar house); *see also Imperial Homes Corp. v. Lamont,* 458 F.2d 895, 899 (5th cir. 1972) (holding that copying floor plans from copyrighted brochure, constitutes infringement, yet building substantially identical house not prohibited by existence of copyrighted architectural drawings for original house); *Herman Frankel Org. v. Tegman,* 367 F.Supp. 1051, 1053 (E.D. Mich. 1973) ("A person cannot, by copyrighting plans, prevent the building of a house similar to that taught by the copyrighted plans . . . A person should, however, be able to prevent another from copying copyrighted house plans and using them to build the house.")

In 1990, the Architectural Works Copyright Protection Act ("AWCPA") codified the copyright prophylaxis of architectural plans and architectural works. *See* Pub.L.No. 101-650, Tit. VII, 104 Stat. 5133 (1990)(codified in scattered sections of 17 U.S.C.). The AWCPA makes an "architectural work" an "original work of authorship" eligible for copyright protection. 17 U.S.C. § 102(a)(8). The copyright Act defines an architectural work as "the design of a building as embodied in any tangible medium of expression, including a building, architectural plans, or drawings. The work includes the overall form as well as the arrangement and composition of spaces and elements in the design but does not include individual standard features." 17 U.S.C.§101.

(2) With the passage of the AWCPA, a copyright owner may claim infringement of both the architectural plans and the structure based on such plans. *See* H.R.Rep. 101-735, *reprinted* in 1990 U.S.C.C.A.N. 6935, 6950 ("Either or both of these copyrights may be infringed and eligible separately for damages"). Courts have routinely protected modern architectural structures, such as commercial homes, that possess the minimal amount of originality that copyright law requires, as well as the plans from which owners built them. *See Johnson v. Jones,* 921 F. Supp. 1573, 1583 (E.D. Mich.1996), protecting residential home as architectural work; *Richmond Homes Management v. Raintree, Inc.,* 862 F. Supp. 1517, 1523–26 (W.D.Va. 1994) (finding residential home built from plans qualified as protected architectural work), *aff'd in part and rev'd in part,* 66 F.3d 316, 1995 WL 551274 (4th Cir. 1995); *Value Group, Inc. v. Mendham Lake Estates, L.P.,* 800 F. Supp. 1228, 1233–35(D.N.J.1992), enjoining construction of house that would infringe on developer's copyrighted single family luxury home.

The foregoing were issues in *Thomas Shine v. David M. Childs and Skidmore Owings & Merrill, LLP.*[35] Plaintiff Thomas Shine was a master's degree student at Yale School of Architecture.[36] As part of the degree program, Shine designed a skyscraper that he named "Shine 99" and later refined as "Olympic Tower."[37] Both of his designs were presented to a jury of experts assembled at Yale to critique students' work.[38] The jury applauded Shine's designs; and one juror, David Childs of the architectural firm of Skidmore Owings & Merrill, approached and complimented Shine after the event.[39] Shine's designs also appeared in a publication sent to Yale alumni, which included Childs.[40]

---

35. Thomas Shine v. David M. Childs and Skidmore Owings & Merrill, LLP, 382 F. Supp. 602 (S.D.N.Y. 2005).

36. *Id.* at 606.

37. *Id.* at 606. For a discussion of the "substantial similarity" of two buildings, *also see* Sturdza v. United Arab Emirates, 281 F.3d 1287 (D.C. Cir. 2002).

38. *Shine,* 382 F. Supp at 606.

39. *Id.* at 606–07.

40. *Id.* at 607.

Sometime after the Yale event, David Childs worked on one of the potential designs for the Freedom Tower, the building to be located at the World Trade Center site.[41] When Childs's work was published, both Shine and his Yale professors noticed a similarity between Shine's "Shine 99" and "Olympic Tower" designs and Childs's Freedom Tower design.[42] As a result, Shine commenced a lawsuit against Childs and also Skidmore Owings & Merrill claiming infringement of Shine's original skyscraper design (claiming, in other words, that Childs copied Shine's designs). In response, the defendants filed a motion for summary judgment, arguing that, as a matter of law, no similarity existed between Freedom Tower and Olympic Tower and that the Freedom Tower design did not infringe upon the Olympic Tower design.[43]

Although the court found that material issues of fact prohibited summary judgment, the court stated as follows:

> To prevail [in Shine's infringement action against Childs], plaintiff must prove (1) ownership of a valid copyright, and (2) copying of constituent elements of the work that are original. . . . To prove copying of original elements of [Shine's] work, in addition to showing originality, plaintiff must demonstrate both that defendants actually copied his works, and that such copying was illegal because there is substantial similarity between each of this works and the alleged infringing work — the Freedom Tower.[44]

In response to Shine's claims, the defendants first argued that neither Shine 99 nor Olympic Tower qualified as an architectural work under the Copyright Act because the designs had not been built. Rejecting this argument, the court stated the Copyright Act "nowhere states or implies that only designs capable of construction are worthy of protection."[45] Accordingly, the court found the two designs were protected by the Copyright Act.

The defendants next argued that Shine's designs were not "sufficiently original to warrant protection under the [Copyright Act]." The court also rejected this argument, stating that

> defendants fly high and fast over the large body of Supreme Court . . . case law on originality and copyright infringement, as well as the text of the [Copyright Act], which states that "the overall form as well as the arrangement and composition of spaces and elements in the design of an architectural work may be the subject of a valid copyright."[46]

41. *Id.*
42. *Id.*
43. *Id.* at 608.
44. *Id.*
45. *Id.*
46. *Id.* at 611.

. . .

Plaintiff need not clear a high bar in order for his architectural works to qualify as original:

> In the copyright context, originality means the work was independently created by its author, and not copied from someone else's work. The level of originality and creativity that must be shown is minimal, only an "unmistakable dash of originality need be demonstrated, high standards of uniqueness in creativity are dispensed with."[47]

B. Contractual Provisions

Standard form agreements with architects do not leave copyright issues to statutory interpretation or common law principles. Such agreements usually provide that all ownership rights in the design documents remain vested in the party who prepared the documents (i.e., the architect or its subconsultants). In addition, the drawings, specifications, and other documents prepared by the architect and its subconsultants are deemed "instruments of service," and a license is granted to the owner for use solely with respect to the project that is the subject of the contract.[48]

The terms and conditions pertaining to copyrights and use of documents set forth in most standard form agreements between owners and architects raise a number of issues. Most notably, the owner is given license to use the instruments of service that are nonexclusive and revocable.[49] The license granted to the owner also is contingent upon substantial compliance by the owner with all of its obligations set forth in the agreement, including prompt payment of all sums when due. Arguably, a failure to satisfy any one of the owner's obligations under the owner-architect agreement will extinguish the owner's license to use the construction documents and other so-called instruments of service,[50] as does any termination of the agreement.

From the standpoint of an architect, an owner retains the architect to provide professional design services. Drawings and specifications are produced as a result of those services. They are instruments of the services, but are not in and of themselves what the owner procures. This view sets up a potential conflict with many owners, who contend that upon payment, the owner should be free to use the documents however the owner sees fit. These competing portions clash. The issue most often arises when the agreement between the

47. *Id.*

48. AIA Document B101–2007, *supra* note 3, § 7.3.

49. *Id.*

50. Altman, *supra* note 1, at 6.

owner and architect is terminated and the owner needs to complete the project without the assistance of the architect, or when the owner wishes to use the drawings and specifications for another project.

Older standard form agreements contained provisions highly favorable to architects with respect to the foregoing issues.[51] Understandably, architects often viewed control over the drawings and specifications as the only leverage at their disposal to ensure payment by the owner, and their contracts gave the architect broad rights to prevent the owner from using the drawings and specifications without involvement of the architect. But, architects also harbor legitimate concerns that completion of a project, or reuse of documents on another project without the involvement of the architect, is in no one's best interests. Such action creates potential liability for the architect that was not intended at the time of contracting and disrupts the economic assumptions upon which their fee was calculated. Further, the original architect is in the best position to interpret documents or adapt them for additional projects.

More recent versions of standard form agreements attempt to balance the expectation of the owner not to be held hostage by an architect against the commercial concerns of the architect. In some current standard forms, if the owner elects to use drawings and specifications prepared by an architect without the participation of that architect, the owner is deemed to release the architect and its subconsultants from claims arising out of such use.[52] In addition, the owner agrees to indemnify and hold harmless the architect and subconsultants from claims made by third parties arising out of use of the documents by the owner without the architect.[53] While such provisions do not resolve all potential issues related to use of the documents, they reflect a compromise that allows a transaction to proceed on a reasonable basis.

7.05 PAYMENT

The AIA approach allows the parties to describe the basis of the architect's compensation.[54] Common methods include (1) hourly rates for time devoted to the services, (2) a fixed fee or stipulated sum, (3) a multiple of direct personnel expense, and (4) a percentage of construction costs. Each payment mechanism has certain advantages and disadvantages for both parties. From the owner's perspective, a stipulated sum provides the most predictability but may result in

51. AIA Document B101–2007, *supra* note 3, § 11.9, requires the owner to pay a licensing fee for continuing use of the architects' instruments of service, and such license is limited to the original project.

52. *Id.* § 7.3.1.

53. *Id.* § 7.3.1.

54. *Id.* Art. 11.

overpayment to the architect if the services ultimately require less effort than originally contemplated. From the architect's perspective, a stipulated sum may result in an underpayment if more effort is required than was anticipated when the stipulated sum was established. Hourly rates provide little predictability, although a cap on fees can be used. Multipliers of direct personnel expense present issues similar to use of hourly rates.

Payments made by the owner to the architect usually fall into three categories: (1) compensation for basic services, (2) compensation for additional services, and (3) reimbursable expenses. Compensation for the performance of basic services typically is calculated using one of the approaches described in the preceding paragraph. Because the commitment of resources for additional services often is unpredictable and the need for such services arises unexpectedly, the owner commonly pays the architect on some type of time and material basis—even if a different method of compensation applies to basic services. Finally, when performing its contractual obligations, architects will incur certain routine and sometimes administrative expenses that are not covered in the compensation for basic or additional services.[55] Owners reimburse the architect for such expenses separately, often with some modest multiplier to cover administrative costs and perhaps some profit margin for the architect.

7.06 INSURANCE

Most agreements between owners and architects specify the types and limits of insurance coverage that the architect must maintain (with the owner reimbursing the architect for the cost of any insurance beyond what the architect normally maintains).[56] Owners may want to require that the professional liability policy be kept in full force and effect for a specified period of time after substantial completion of the project because of the claims-made nature of professional liability insurance policies, which is "disappearing" coverage. For further discussion regarding insurance, a critical topic in design and construction, please see Chapter 20 of this textbook.

7.07 CONTRACTUAL LIMITATIONS OF LIABILITY

A. Disclaimers and Exculpatory Provisions

Most agreements with architects contain a number of disclaimers that limit the obligations of the architect.[57] The day of the "master builder" has long since

55. *Id.* § 11.8.1.

56. *Id.* § 2.5.

57. A number of disclaimers and exculpatory provisions can be found in AIA Document B101–2007, such as §§ 3.1.2, 3.1.5, 3.4.2, 3.6.3–6.4, 6.6.4, 6.7, and 7.3.

passed, and architects define their contractual obligations carefully. In addition to describing what is included in their services, they also define their obligations by describing what is excluded from their services. Setting forth various disclaimers and language of similar effect in agreements with architects has been a steady, but ongoing, process. This gradual erosion of the role of an architect in a project is illustrated well by the evolution of provisions in standard form agreements between an owner and an architect that describe the architect's duties to evaluate the work in progress.[58]

Typically, the architect will keep the owner informed of the progress and quality of the work. However, the architect generally only "endeavor[s] to guard the owner against defects and deficiencies" in the work to determine "in general" if the work is being performed in a manner that will comply with the requirements of the contract documents. Moreover, the architect is not required to make exhaustive or continuous on-site inspections but will visit the site only at "appropriate intervals to the stage of the contractor's operations." Finally, most architects insist that their contracts explicitly provide that the architect does not have control over or charge of, and is not responsible for, the construction means, methods, techniques, sequences, or procedures or for safety precautions and programs in connection with the work.

Such language is the direct result of common law that examined whether an architect's on-site obligations create any responsibility for the safety of workers on a construction site. A case decided by the Illinois Supreme Court is particularly instructive of how courts have looked at the issue. The plaintiffs in *Miller v. DeWitt,*[59] among other things, alleged that project architects, who had a contractual duty to evaluate work in process, also had a duty to prevent the contractor from carrying out the work in a faulty manner and, therefore, were responsible to the plaintiffs for the injuries sustained as the result of a roof collapse during construction.

The court in *Miller* affirmed the judgment in favor of the plaintiffs, finding the architect negligent in supervising the work and awarding damages for bodily injuries suffered in a job site accident.[60] Although the *Miller* court based its judgment on the architect's right to stop the work, design professionals were not inclined to take chances, and subsequent standard forms of agreement between an owner and architect (or engineer) now dilute the description of the architect's on-site responsibilities. Architects no longer "supervise" the work, but "observe" the work. Similarly, architects routinely disclaim any responsibility for construction means, methods, techniques, sequences, and

58. AIA Document B101–2007, *supra* note 3, § 3.6.2.1.

59. Miller v. DeWitt, 37 Ill. 2d 273, 226 N.E.2d 630 (1967).

60. *Id.* at 292, 226 N.E.2d at 643.

procedures, and are careful to limit other obligations where such control might expose the architect to liability for worker injuries.

The current allocation of responsibilities in most agreements between owners and architects was almost predicted by the dissenting opinion in *Miller v. DeWitt*.[61] The relevant portion of the dissent is set forth below.

Miller v. DeWitt
37 Ill. 2d 273, 226 N.E.2d 630 (1967)[62]

HOUSE, Justice (dissenting).

[The majority opinion] imposes a legal duty on an architect not only to prepare plans and specifications, but to inspect the methods employed by the contractor leading up to completion under the general inspection clause of his contract.

I cannot read into the contract a duty which is not imposed by it. The architect's contract used here is a more or less standard form generally used by architects and engineers. It provides for detailed plans and specifications, obtaining approval of various governmental agencies, issuing certificates of payment and general administration. Supervision is limited. The architect contracts to attempt prevention of defects but specifically disclaims a guarantee of the performance by the contractor.

. . .

Again, the opinion concedes that architects have no duty to specify the method used to accomplish the finished building, but the belief was stated that the architects "had the right" to insist upon a safe and adequate use of that method. True, but to parlay that "right" into a duty is neither consistent with generally accepted usage nor contemplated by the contract. Obviously, the architect did not contract to be present or represented at all phases of construction and he should not be held responsible for methods used by the contractor which may result in injury. Since there is no contractual obligation, liability is fixed by an expansion of the common law.

I find no support for such a radical departure in either this or any other jurisdiction. The cases cited for comparison are usually between contractor and architect or owner and architect, but not for liability of an architect to an employee of the contractor. The general view is stated in *Garden City Floral Co. v. Hunt* [Citations Omitted]: "To say that he (architect) must supervise the method of doing the work

61. *Supra* note 59, Dissent of Justice House, Justice Klingbiel concurring.

62. Internal citations have been omitted.

before there is full supervision would place the architect in an entirely different role from that of an architect. . . . As a matter of law the courts recognize that an architect merely supervises the results and does not dictate the methods when not controlled by the specifications."

There are sound reasons for the prevailing view that the architect's primary duty is to provide a sound completed structure in accordance with the owner's requirements, but not to dictate the methods by which the contractor attains that objective. There would be utter chaos if the contractor or his superintendent were to give an order to use his most efficient equipment and personnel, and the architect attempted to countermand and order that the work be done by another method requiring different equipment and skills. When a contractor bids a job he expects to use his equipment and the special talents and experience of his organization. If the threat existed that the details of carrying out his contract be subject to outside interference, contractors naturally would take that into consideration in fixing their bids. If the duty of architects is expanded to require that they be on the job at all times and prescribe methods of construction or be held liable for the negligence of employees of the contractor, . . . they will reflect the added burden in their supervision fees. All of this adds up to an additional and, I think, unnecessary and unwarranted financial burden upon the public without a commensurate benefit. Liability of architects as imposed here is economically unsound.

The huge construction industry in this country had functioned very well without the imposition of liability upon architects and engineers who design, but do not build, structures and other facilities. I see no justification for extending the common law to place liability on architects.

KLINGBIEL, J. joins in this dissent.

B. Waivers of Consequential Damages

Both parties to the AIA Document B101–2007 waive consequential damages under Section 8.1.3. Since, however, the architect is unlikely to suffer consequential damages from the owner's breach of the agreement, but the owner's damages may be almost exclusively consequential, the clause often is heavily negotiated by sophisticated, commercial owners.

Options are available to the owner who wishes to retain the right to recover some portion of consequential damages. For example, the owner may reserve this right to the extent there are available insurance proceeds. The parties might also agree to a limitation on consequential damages, as opposed to an outright waiver. One popular alternative is a contract provision that limits the owner's recovery to the proceeds of the architect's professional liability insurance.

C. Contractual Limitations of Liability

Valhal Corp. v. Sullivan Associates, Inc. 44 F.3d 195 (3rd Cir.1995)[63]

Before: SIRICA, NYGAARD and McKEE, Circuit Judges.

McKEE, Circuit Judge:

This dispute centers on the enforceability of a limitation of liability clause in a contract between a real estate developer (Valhal Corporation), and an architectural firm (Sullivan Associates). Valhal and Sullivan have both filed appeals from the order of the district court denying Sullivan's motion for partial summary judgment and granting Valhal's motion for partial summary judgment. The district concluded that the disputed clause was part of the contract but that it violated public policy and was therefore unenforceable. We will reverse, and dismiss for lack of jurisdiction.

I. Factual and Procedural Background

Valhal is a New York corporation which specializes in the management and development of real estate. Sullivan Associates, Inc., is a Pennsylvania corporation specializing in architectural, planning and engineering services. In March of 1989, Valhal became interested in buying a parcel of real estate located at 401 N. 21st Street in Philadelphia, Pennsylvania, known as the "Channel 57 Property." Valhal planned to build a high-rise residential tower on a portion of that property. In early June of 1989, Valhal and Sullivan discussed the possibility of Sullivan performing certain work in connection with the project, including a feasibility study.

As a result of those discussions, Sullivan forwarded a proposal to Valhal, dated June 7, 1989, detailing the services which Sullivan would perform. A document entitled "Standard Consulting Contract Terms and Conditions" was attached to the proposal and provided in part:

> . . .
>
> The OWNER agrees to limit the Design Professional's liability to the OWNER and to all construction Contractors and Subcontractors on the project, due to the Design Professional's professional negligent acts, errors or omissions, such that the total aggregate liability of each Design Professional shall not exceed $50,000 or the Design Professional's total fee for services rendered on this project.
>
> Should the OWNER find the above terms unacceptable, an equitable surcharge to absorb the Architect's increase in insurance premiums will be negotiated.
>
> . . .

63. Internal citations and footnotes omitted.

Sullivan's initial proposal provided that its services would be performed in two phases—Phase "A" and Phase "B"—and that Sullivan's total fee would be $5,000. Valhal responded by requesting that a service to be performed under Phase "B" be included under Phase "A" and by requesting that two completely new services be added to Phase "A." Sullivan agreed and a new proposal was submitted to Valhal on June 22, 1989 in which Sullivan increased its fee from $5,000 to $7,000 because of the additional work it was to perform under Phase A. This second proposal once again incorporated the Standard Contract Terms and Conditions, including the limitation of liability provision, and was again signed by Andrew Sullivan.

. . .

Sullivan responded by performing the services outlined in the July 24, 1989 proposal, and thereafter provided a written report to Valhal in which Sullivan concluded that the Channel 57 property was not burdened with any height restrictions and that it was possible to erect the tower on the property without any special governmental approvals.

Valhal thereafter entered into an Agreement of Sale for the purchase of the Channel 57 property. However, after the sale contingencies expired, Valhal learned that the property was subject to a height restriction which would be violated by its building. Nevertheless, Valhal proceeded to closing and paid the purchase price of $10.1 million. . . . Valhal then brought a diversity action against Sullivan pursuant to 28 U.S.C. § 1332 seeking damages in excess of $2,000,000 for breach of contract, negligence, gross negligence and negligent misrepresentation based upon Sullivan's failure to inform it of the height restriction.

Sullivan thereafter moved for partial summary judgment on the grounds that its liability was expressly limited to $50,000 and that the district court therefore lacked diversity jurisdiction. Valhal moved to strike the limitation of liability provision arguing that it was not a part of the contract and that even if it was, it was unenforceable. Valhal also argued that the limitation of liability clause, if enforceable, was limited only to its claim for negligence and did not apply to its breach of contract claim or to its gross negligence claim. The district court treated the parties' motions as cross-motions for summary judgment, denied Sullivan's motion, and granted Valhal's motion. The court ruled that the provision was part of the contract, but that it was against public policy as expressed in 68 Pa. Stat. Ann. § 491 (Purdons 1994) (the anti-indemnity statute) and therefore unenforceable. See Memorandum and Order, dated May 17, 1993. . . .

Thereafter, the jury returned a verdict in favor of Valhal on both the contract and negligence claims but awarded damages of $1,000,000 on the contract claim only. The jury also concluded that Sullivan was not liable for gross negligence or negligent misrepresentation. Sullivan's post-verdict motions were denied, and this appeal followed.

II. Discussion

. . .

B. The Limitation of Liability Clause Is Enforceable.

The heart of the instant controversy is Valhal's contention that the limitation of liability clause is unenforceable even if it is part of the contract with Sullivan. Valhal argues that limitation of liability provisions are disfavored in Pennsylvania and that this particular clause violates a specific public policy against an architect limiting his/her liability for damages caused by his/her own negligence.

Pennsylvania Does Not Have a General Policy Against Such Clauses.

The law recognizes different methods by which a party can limit his/her exposure to damages resulting from his/her negligent performance of a contractual obligation. An exculpatory clause immunizes a person from the consequences of his/her negligence.... Similarly, an indemnity clause holds the indemnitee harmless from liability by requiring the indemnitor to bear the cost of any damages for which the indemnitee is held liable.... The instant clause has no such consequence. The clause before us does not bar any cause of action, nor does it require someone other than Sullivan to ultimately pay for any loss caused by Sullivan's negligence. Sullivan remains liable for its own negligence and continues to be exposed to liability up to a $50,000 ceiling. Thus, the amount of liability is capped, but Sullivan still bears substantial responsibility for its actions.

. . .

Valhal asserts that exculpatory clauses, indemnity clauses and limitation of liability clauses differ only in form as the effect of each is to limit one's liability for one's own negligence. Brief of Valhal Corp. at 24–25. Valhal contends that, thus, limitation of liability clauses are disfavored in Pennsylvania and must meet stringent standards to be enforceable.

There are similarities between these types of clauses.... Indeed, the test used to determine the enforceability of exculpatory and indemnity provisions is the same. *Id.* Those clauses are disfavored and must meet certain conditions to be enforceable. First, the clause must not contravene public policy. Second, the contract must relate solely to the private affairs of the contracting parties and not include a matter of public interest. Third, each party must be a free bargaining agent. In addition, an exculpatory or indemnity clause will still not be enforced unless it is clear that the beneficiary of the clause is being relieved of liability only for his/her own acts of negligence. The clause must be construed strictly and the contract must state the intention of the parties with the greatest particularity. Furthermore, any ambiguity must be construed against the party seeking immunity, and that party also has the burden of proving each of the prerequisites to enforcement.... The district applied this test to the limitation of liability clause at issue here....

Courts have developed these limitations as reasonable conditions precedent to allowing a party to contract away responsibility for his/her negligence. It is with good reason therefore, that Pennsylvania allows such contractual provisions only where matters of public interest are not involved. One cannot contract away responsibility to the public to exercise reasonable care in performing a contract....

However, Pennsylvania appellate courts recognize that there are differences between a contract which insulates a party from liability and one which merely places a limit upon that liability.... The difference between the two clauses "is ... a real one." ... Presumably because of that difference, we find no Pennsylvania cases in which a limitation of liability clause has been disfavored or been tested by the same stringent standards developed for exculpatory, hold harmless, and indemnity clauses. Accordingly, we believe that the district court erred in applying those stringent standards to the clause before us....

Limitation of liability clauses are routinely enforced under the Uniform Commercial Code when contained in sales contracts negotiated between sophisticated parties and when no personal injury or property damage is involved. This is true whether the damages are pled in contract or tort.

...

We are persuaded that limitation of liability clauses are not disfavored under Pennsylvania law; especially when contained in contracts between informed business entities dealing at arm's length, and there has been no injury to person or property. Furthermore, such clauses are not subjected to the same stringent standards applied to exculpatory and indemnity clauses. Limitation of liability clauses are a way of allocating "unknown or undeterminable risks" ... and are a fact of everyday business and commercial life. So long as the limitation which is established is reasonable and not so drastic as to remove the incentive to perform with due care, Pennsylvania courts uphold the limitation.

Though it is possible that an agreement setting damages at a nominal level may have the practical effect of avoiding almost all culpability for wrongful action, the difference between the two concepts is nevertheless a real one. The distinction becomes more apparent in a situation which [sic] the damage level set is substantial rather than minimal, ...

The line of demarcation between the two types of agreements has significance here because of the findings needed to establish their existence. Pennsylvania permits parties to contractually relieve themselves from the consequences of negligent acts, but any agreement must spell out the intention of the parties with particularity.

... Here, Sullivan is exposed to liability which is seven times the amount of the remuneration under its contract with Valhal. Accordingly, the cap does not immunize Sullivan from the consequences for its own actions. It is a reasonable allocation of risk between two sophisticated parties and does not run afoul of the policy disfavoring clauses which effectively immunize parties from liability.

Although it could be argued that the $50,000 limitation is nominal when compared to the final verdict, we do not believe that to be the proper measure. The inquiry must be whether the cap is so minimal compared to Sullivans expected

> compensation as to negate or drastically minimize Sullivan's concern for the consequences of a breach of its contractual obligations. One cannot seriously argue that a cap which leaves Sullivan exposed to damages that are seven times its expected fee insulates Sullivan from liability.
>
> IV. Conclusion.
>
> The jurisdiction of the district court was based solely on diversity. Diversity jurisdiction requires an amount in controversy in excess of $50,000 excluding fees and costs. 28 U.S.C. § 1332(a). "This provision must be narrowly construed so as not to frustrate the congressional purpose behind it: to keep the diversity caseload of the federal courts under some modicum of control." Because we have concluded that the limitation of liability clause is an enforceable part of the contract which is the basis of this diversity action, Valhal's maximum possible recovery is $50,000. Therefore, the district court was without subject matter jurisdiction to hear this controversy. Accordingly, we will vacate the order of the district court and remand with directions to dismiss for lack of subject matter jurisdiction.

It is not surprising that many participants in the design and construction process attempt to shift project liability risks to other project participants or to an insurer. Many design and construction contracts use exculpatory provisions, indemnification, and hold harmless agreements, as well as a simple contractual allocation of responsibility and liability for particular matters to another party. Such provisions are often the source of protracted negotiations and accomplish the ultimate objective of insulating a person or entity from liability with varying degrees of success. Consequently, there has been an increase in efforts by various parties—in particular, design professionals—to include a specific contractual limitation of liability in their agreements.

Use of limitation of liability clauses by design professionals became more common in the general building sector when a trade association, then known as the Associated Soil and Foundation Engineers, developed a contractual provision that limited liability of geotechnical engineers of the owner and construction contractors to a fixed amount. Geotechnical engineers at that time were threatened by a large number of claims being filed against them for an extraordinary amount of damages, as compared to the fees earned on such jobs. As a practical matter, such concerns also affect many other design professionals.

Because the owner receives the principal benefit from the design and completion of the project, design professionals contend the owner is the more appropriate party to assume certain liability risks. Design professionals also point out that damages alleged or claimed against them are potentially enormous, whereas they receive a one-time fee that is only a small percentage of the risk that is assumed. Many owners are concerned that limiting the liability of an architect may remove an incentive for proper performance. As the court in *Valhal* concluded, however, a limitation of liability provision is not necessarily exculpation from responsibility, but rather an allocation of risk intended to reflect a proper risk/reward analysis.

7.08 TERMINATION

Many construction projects do not survive the journey from conception to completion. Sometimes an owner cannot afford what it wants, the owner's needs may change, financing for the project fails, or business plans change. Architectural libraries are filled with books that contain plans for buildings and other projects never constructed.

On other occasions, the parties part company because one or the other fails to perform its obligations. The right of the architect to terminate "for cause" is limited from a practical standpoint to failure by the owner to make payment to the architect in accordance with the agreement. At the option of the architect, nonpayment also is cause for suspension of the performance of services.[64] The architect has no liability to the owner for any delay or damage caused as a result of such suspension. This places the owner in a quandary. The architect, however, confronts a similar dilemma. Any suspension of services or termination has drastic financial implications for the owner and, therefore, the architect must consider carefully the possibility that its actions might be considered wrongful.

Many standard form agreements allow the owner to terminate the agreement for convenience and without cause.[65] There are many circumstances in which a termination for convenience is necessary or appropriate; for example, if the owner is unable to obtain necessary financing. Additionally, if the working relationship between the owner and architect breaks down or a change in personnel severely impairs the original expectations of the parties, it may be a prudent business decision simply to terminate the relationship for convenience and move on with life.

64. AIA Document B101–2007, *supra* note 3, § 9.1.

65. *Id.* § 9.5.

If the agreement between an owner and architect is terminated through the fault of the architect, some contracts provide for payment of termination expenses to the architect. The AIA approach does not include a formula for calculating termination expenses but simply provides that termination expenses are those that are "directly attributable to termination" for which the architect is not otherwise compensated, *plus* the anticipated profit on services not rendered by the architect as a consequence of termination.[66] Since the profit on unperformed work is always imprecise—and, therefore, subject to exaggeration—this portion of the termination clause is often the subject of negotiation.

66. *Id.* § 9.7.

QUESTIONS

1. During negotiations of an owner-architect agreement, the architect explains to the owner that no set of drawings and specifications is ever 100 percent complete. From prior experience, the owner knows this to be true and doesn't want to pay a premium for the architect to prepare documents that exceed the level of detail that is customary in the profession. Because of such practical concerns, the architect proposes that the owner maintain a contingency of 5 percent of the construction contract sum to pay for additional costs arising out of errors or omissions in the drawings and specifications. How do you respond if you are counsel to the owner? If you are counsel for the architect, what arguments do you make to support this position?
2. Assume that during the design development phase, the owner requests the architect to make certain modifications to the project design. Assume further that the changes do not violate any law or present any safety concerns, and that the owner is willing to pay for all additional services needed to implement the design changes. The architect objects to the desired modifications, claiming that such changes will seriously diminish the aesthetic quality of the building. If the contract is an AIA standard form or similar document, does the architect have any right to refuse to make the requested modifications? How should the owner respond to the architect?
3. An owner and architect enter into an agreement for the design of a small commercial building, using a typical AIA form of agreement. The agreement includes a project budget. The owner informs the architect that the budget cannot be exceeded by more than 10 percent, as the cost of construction is largely financed and the owner has very little ability to add equity or obtain further financing. The lowest bid from a responsible bidder comes in 30 percent over the budget. After an effort to "design down" to the budget, it becomes clear that the project is not economically viable and will need to be abandoned. What recourse does the owner have against the architect? What could the owner have done during contract negotiations to improve its position in such a circumstance?
4. Shortly after an owner retains an architect to design a commercial project, the parties discuss the program for the project. The architect encourages the owner to consider the use of "sustainable" architecture and energy efficient elements. The architect contends that such elements will pay for themselves over the useful life of the project and are simply the socially responsible thing to do. The owner likes the idea and can see certain marketing advantages for such a building. The architect's ideas, however, require use of new and untested technologies. Therefore, the architect cautions the owner that the ultimate price of the project is

unpredictable and that there likely will be a period of time after completion of construction during which the project will not function properly and will need to be optimized. The owner believes that the architect should bear those risks, as the owner still needs to rely on the architect to implement such ideas. Do you think that the owner is correct? How should an architect respond to the owner's concerns? How should such issues be addressed in the owner-architect agreement?

5. An owner retains an architect who proposes contract language that grants the owner a nonexclusive license to use the design to construct the project. Counsel for the owner rejects the language and tells the architect that the owner must own the design. In response, the architect disagrees and insists that the architect must own the design. Discuss the issues that concern the parties and suggest contract provisions that might resolve such concerns.
6. Why do many owners object to a mutual waiver of consequential damages in an owner-architect agreement? If an architect refuses to delete the waiver in an owner-architect agreement, describe what provisions an owner might try to negotiate to mitigate adverse effect caused by waiving consequential damages by the owner.

CHAPTER

8

Contractor Selection

R. HARPER HECKMAN

8.01 INTRODUCTION

Perhaps the most significant differences between construction in the private and public sectors lie in the area of procurement. Unlike private projects, public procurement is extensively regulated and generally requires that contractors be selected through a publicly announced, competitive, sealed bidding process.[1] In addition to securing for the general public the lowest price for a project, the competitive bidding process is perceived to be the best means for "giv[ing] all qualified contractors the opportunity to compete for government contracts while, at the same time, avoiding favoritism, collusion, or fraud."[2] As discussed in the section that follows, private owners enjoy considerably more freedom when procuring design and construction services.

1. 1 PHILLIP L. BRUNER & PATRICK J. O'CONNOR, JR., BRUNER AND O'CONNOR ON CONSTRUCTION LAW § 2:21, at 84 (West Group 2002) [hereinafter BRUNER & O'CONNOR].

2. *Id.* § 2:22, at 86 n.4 (quoting W. NOEL KEYES, GOVERNMENT CONTRACTS UNDER THE FEDERAL ACQUISITION REGULATION 14.1, 153 (1986)); 48 C.F.R. § 14.1 (2003).

8.02 PRIVATE PROCUREMENT GENERALLY

Procurement of construction services on private projects typically is characterized by "offer and acceptance" negotiations, which leave owners free to contract for their projects in the manner and with the contractor of their choice. Many private owners elect to contract first for the project's design and then for its construction, which is referred to as the Design-Bid-Build method of project delivery. However, there are many other means by which a private owner may contract for the construction of a project. These include:

- Design-Build
- Agency Construction Management
- At-Risk Construction Management

A detailed discussion of the various project delivery systems available to the private owner can be found in Chapter 4. A private owner also may elect to construct its project using some or all of the following pricing mechanisms:

- lump sum (also known as "fixed price")
- cost-plus-fee (with or without a "guaranteed maximum price")
- unit pricing

For more information regarding these pricing mechanisms, please refer to Chapter 9.

Private owners select their contractors in a variety of ways. Indeed, owners hire their contractors based upon similar criteria used to select their lawyers, including marketing efforts, reputation, and references. It is not uncommon for a private owner to send a request for proposal to two or more contractors, review their written submissions, and then meet with the finalists before deciding which contractor to engage. While price is an important factor in the owner's decision making, it is not the only factor.

Even when it elects to solicit competitive bids, an owner is allowed considerable freedom to negotiate privately with each bidder before deciding which bid to accept. For this reason, most competitively bid projects in the private sector are considered "closed bidding" projects, as distinguished from the "open bidding" protocol found in the public sector. In open bidding, no private negotiations take place. Instead, all bids are opened simultaneously in public and read aloud, with the contract awarded to the lowest bidder.

Nevertheless, much of the discussion concerning public bidding that follows, particularly those sections having to do with the bid advertisement and package, is applicable to the private sector as well. In reviewing these concepts, however, the reader should remember that private procurement is significantly less structured than its public counterpart. As such, private contract solicitations and proposals are often based upon incomplete project design drawings and specifications. Even under the traditional Design-Bid-Build model, private contract negotiations are dynamic processes in which indecisive owners vacillate on significant issues affecting construction scope, time, and cost. Accordingly, it is imperative that private "bidders" take care to ensure that what they ultimately contract to construct is the same project that they initially bid.

8.03 PUBLIC PROCUREMENT

A. General Considerations

Before discussing procurement issues in the public sector, we must first answer this question: "What actually is public construction?" Traditionally, the distinction was easy to make. The line between public and private projects has increasingly become blurred, however, in light of recent trends toward "privatization" of public projects and the advent of public-private partnerships.[3]

The characterization of a project as a "public" one generally involves an analysis of project (1) funding, (2) control, and (3) future ownership.[4] Privately funded, publicly created entities normally are exempt from public procurement regulations. In contrast, significant public funding of otherwise private entities—even in the form of mere tax relief—can alter the exemption.[5]

Even for projects that clearly are characterized as public, most procurement codes exempt certain types of projects from public bid laws. Examples of exemptions include:

- Small dollar exemptions, where expenditure of public funds for the project is relatively insignificant[6]
- Emergency work exemptions, where the nature of the work could not withstand the time required for traditional public procurement protocol
- "Sole source" exemptions, where competition is limited to a single bidder with proprietary rights, technology, or control over source materials[7]
- Professional services exemptions, where projects require the special skills, expertise, and services of a particular person or firm[8]

3. 1 BRUNER & O'CONNOR, *supra* note 1, § 2:25.
4. *Id.*
5. *Id.*
6. *Id.* § 2.32; 48 C.F.R. §§ 13.000 and 2.101 (F.A.R. §§ 13.000 and 2.101), setting federal small-dollar exemption at $100,000.
7. 1 BRUNER & O'CONNOR, *supra* note 1, § 2:29.
8. *Id.* § 2:30.

B. Selection of Design Professionals

During the design phase of a project, public owners may be less concerned about cost issues and more concerned with engaging those professionals who bring particular skills, expertise, leadership, and prestige to the project.[9] Accordingly, the price-based procurement policies utilized for awarding construction contracts (where the lowest bidder is awarded the contract) may not be well suited for selecting design professionals.[10] In recognition of this fact, one court has held that the traditional bidding process is not required for contracts with design professionals. The court stated:

> [T]he general rule [is] that contracts for services requiring special skills or training, such as the professions and vocations of . . . architects, . . . engineers, . . . and surveyors, are not required to be let on bids under a statutory or constitutional provision [which provides] that contracts with the state or municipality must not be entered into without first advertising for bids.[11]

Another court has noted that, because an architect requires special taste, skill, technical learning, and ability to perform its job, "it would be bad judgment to advertise and get many bids when the lowest bidder might also be the least capable and most inexperienced."[12] Thus, "the employment of a person who is highly and technically skilled in his science or profession is one which may properly be made without competitive bidding."[13] In response to these concerns, the federal government and many state governments have created statutes that either exempt or remove design professionals from the competitive bidding process and, instead, attempt to balance the goals of securing the desired design professionals, containing costs, and limiting favoritism or fraud.[14]

Under the Brooks Act, which governs the federal government's selection of architects and engineers, "[t]he policy of the Federal Government is to publicly announce all requirements for architectural and engineering services and to negotiate contracts for architectural and engineering services on the basis of demonstrated competence and qualification for the type of professional services required and at fair and reasonable prices."[15] Under this policy, the gov-

---

9. *Id.* § 2:30; FUNDAMENTALS OF CONSTRUCTION LAW 54–55 (Carina Y. Enhada et al., eds. 2001) [hereinafter FUNDAMENTALS].

10. 1 BRUNER & O'CONNOR, *supra* note 1, § 2:30.

11. Parker v. Panama City, 151 So. 2d 469, 472 (Fla. Dist. Ct. App. 1963).

12. Cobb v. Pasadena City Bd. of Ed., 285 P.2d 41, 42 (Cal. Dist. Ct. App. 1955).

13. *Id.* (citations omitted).

14. W. V. Nelson Constr. Co. v. City of Lindstrom, 565 N.W.2d 434, 435–36 (Minn. Ct. App. 1997); CONSTRUCTION LAW § 15.01[1] (Steven G. M. Stein, ed. 2006) [hereinafter CONSTRUCTION LAW].

15. 40 U.S.C.A. § 1101 (formerly 40 U.S.C.A. § 542).

ernment encourages design professionals to submit qualification and performance data for the agency's consideration.[16] The agency then "shall conduct discussions with at least [three] firms to consider anticipated concepts and compare alternative methods for furnishing services."[17] From those firms, the agency shall select three or more firms that it deems "most highly qualified to provide the services required,"[18] and then attempt to negotiate a "fair and reasonable" contract with the firm that it considers the most qualified for the job.[19] If those negotiations are unsuccessful, the agency shall then attempt to negotiate a fair and reasonable contract with the "next most qualified of the selected firms," and then the next, until the agency finally reaches an agreement for the design services.[20]

Many states follow the federal practice of advertising or publicly announcing the design project referred to as Baby Brooks Acts.[21] Similarly, many states require the awarding agency to consider a minimum number of design firms for the project.[22] Other states do not require that public agencies consider a specific number of firms, but only that they begin negotiations with the "best qualified" firm or firms.[23]

When evaluating design firms, many states specify criteria other than cost. Some states base designer selection solely on the designer's demonstrated "competence and qualifications."[24] Other states consider a larger range of factors, such as the design professional's previous performance, location, and workloads and the complexity of the proposed project.[25]

---

16. *Id.* § 1103(b) (formerly § 543).
17. *Id.* § 1103(c).
18. *Id.* § 1103(d).
19. *Id.* §§ 1104(a)(b) (formerly § 544).
20. *Id.* § 1104(b).
21. *See, e.g.*, N.C. Gen. Stat. § 143-64.31(a) (2007); 4 Pa. Code § 64.4 (2007).
22. *See, e.g.*, Cal. Gov't Code § 4527 (2007) (requiring the consideration of at least three firms); Fla. Stat. § 287.055(4) (2007) (requiring the ranking of at least three firms); Md. Code Ann. State Fin. & Proc. § 13-304(b) (2007) (requiring the consideration of at least two firms when the total cost estimate for the design services exceeds $200,000); 4 Pa. Code § 64.8 (2007) (requiring the consideration of at least three design firms); Kan. Stat. Ann. § 75-1253 (2007).
23. *See, e.g.*, N.C. Gen. Stat. § 143-64.31(a) (2007) (allowing an unspecified number of qualified firms); Okla. Stat. Ann. tit. 74, § 85.7 (2007) (noting that Oklahoma requires competitive bidding for design services costing more than $25,000 but not specifying how many firms must compete and awarding projects to the "lowest and best, or best value, bidder").
24. *See, e.g.*, La. Rev. Stat. Ann. § 38:2318.1(A) (2007); N.C. Gen. Stat. § 143-64.31(a) (2007); 4 Pa. Code § 64.5 (2007); Cal. Gov't Code § 4526 (2007) (requiring the consideration of "demonstrated competence").
25. *See, e.g.*, Fla. Stat. § 287.055(4) (2007); Kan. Stat. Ann. § 75-1257 (2007).

Some states add enhanced oversight to the design procurement process by requiring the use of third-party decision makers. For example, Pennsylvania, Maryland, and Kansas use independent selection or negotiation committees,[26] while Oklahoma requires that the State Purchasing Director approve all design agreements.[27]

Many states require their constituent agencies to "negotiate" an agreement, including a design fee, with the most qualified designer.[28] If that negotiation is unsuccessful, then they negotiate with the next most qualified firm, and then the next, until they reach an agreement.[29] In many states, the only statutory price limit on these agreements is that they must be "fair and reasonable."[30] Some states, however, have statutorily limited the designer's fee to a fixed percentage of the total project cost.[31]

C. Selection of Contractors

As stated earlier, public construction procurement traditionally has utilized the Design-Bid-Build project delivery system, in which the contract is awarded to the lowest responsive, responsible bidder under a "sealed," competitive bid process. Under this "open bidding" approach, a public owner separates the design and construction phases of the project and, before the contractor is identified or selected, obtains a complete project design from the design professional. The public owner then requests that prospective contractors bid on the project as designed. After the prospective contractors submit the bids to the public owner, the bids are opened and the contract awarded to the lowest bidder.

1. *Traditional Methodology for Selection of Contractors*

a. Advertisement for Bids

Traditional public procurement protocol begins with the public owner's advertisement for bids (also referred to as invitations or solicitations for bids).

---

26. 4 PA. CODE § 64 (2007); KAN. STAT. ANN. §§ 75-1251, 75-1253 (2007).

27. OKLA. STAT. ANN. tit. 74, § 85.7 (2007).

28. *See, e.g.*, N.C. GEN. STAT. § 143-64.31(a) (2007).

29. *Id.*

30. *See, e.g.*, CAL. GOV'T CODE §§ 4527–4528 (2007) (requiring selection and ranking of at least three firms, followed by negotiations for a "fair and reasonable" agreement). LA. REV. STAT. ANN. § 2318.1(A) (2007) (expressly forbidding the selection of design professional based on price or price-related information but selecting providers of such services "on the basis of competence and qualifications for a fair and reasonable price"); N.C. GEN. STAT. § 143-64.31(a) (2007) (selecting firms without regard to fees and then negotiating for fair and reasonable agreements).

31. KAN. STAT. ANN. §§ 75-1253–75-1263 (2007).

Advertisements for bids typically are required to be published in newspapers having general circulation in the geographic area where the project is located. Such advertisements provide prospective bidders with a description of the project sufficiently detailed to allow a reasonable opportunity to consider whether to bid upon it, and instructions on how the bidders can obtain the project drawings and specifications and other relevant documents upon which to base their bids. Other information, such as the required bid security and bonding requirements, how long the bid must be held "firm" (i.e., bid irrevocable with prices held firm) by the contractor, and any MWBE (Minority and Women Business Enterprises) requirements may be included as well. Most importantly, the announcement should include the date, time, and place for the bid submission and subsequent bid opening.

It is important that the advertisement for bids allow a prospective bidder sufficient time to formulate its bid on the project. Allowing sufficient "lead time" between bid advertisement and opening of the bids, in theory, will increase the number of contractors and subcontractors submitting bids on the project and result in lower bids.[32] More importantly, allowing adequate bid preparation time will prevent unscrupulous public officials from circumventing the requirements of the public procurement laws by giving some bidders a "head start" by sharing information about an upcoming bid that was not provided to the rest of the contracting community.

b. The Bid Package

To ensure that the submissions of all bidders are given equal consideration, the owner must be able to make a fair comparison of the bids submitted. Thus, it is essential that the owner "standardize" the information upon which competing contractors base their bids. This standardized set of owner instructions is generally referred to as the "bid package" or "project manual."

Assuming the design of the project to be bid upon has already been completed, the project drawings and specifications almost certainly will be included in the bid package. Copies of the proposed contractual terms, general conditions, and any special or supplementary conditions are also commonly included. Finally, should the owner desire to provide the potential bidders with the results of any geotechnical or other engineering reports concerning the project site, these will be included in the bid package as well.

Sometimes a public owner has not made its final decision on all items to be constructed by the successful bidder before the project is put out for bid.

32. 1 Bruner & O'Connor, *supra* note 1, §§ 2:71 and 2.2:50.

In addition, an owner may not know if the construction budget will be sufficient to cover all items on the "wish list" before the preliminary construction costs are received from the bidders. One mechanism for dealing with this uncertainty is to ask the prospective bidders to submit a "base bid" package covering those items that the owner clearly intends to construct and also one or more "alternates" that the owner is free to accept or reject before the final contract price is set. However, the use of alternates in the competitive bidding process has proven to be fertile ground for bid-related disputes on public projects. For example, what if the lowest bidder on the owner's base contract work is not the lowest bidder when its bids on the alternates are considered?

Another means for allowing an owner flexibility in the bidding process is to ask bidders to include in their bids pricing budgets, or "allowances," for certain items (e.g., floor or wall coverings and plumbing or lighting fixtures). The use of an "allowance" permits an owner to make its contract award and then later decide on whether it desires to vary the quantity or quality of that item and, therefore, its final pricing associated with the item.

Sometimes, an owner may desire to correct, clarify, or augment its bid package after it has been distributed to prospective bidders but before their bids are submitted. To do so, the owner, typically through its designer, issues "addenda." These addenda, like the other materials contained in the bid package, will be binding upon the prospective bidders. Accordingly, a public owner typically will require the prospective bidder to formally acknowledge receipt of any addenda issued.

In addition to the advertisement itself, an owner often will conduct a pre-bid conference with the prospective bidders. Such conferences allow the owner to introduce its design and ownership team, solicit questions from the bidders regarding any ambiguity in the bid package, and otherwise generate enthusiasm among the prospective contractors and subcontractors who may want to submit bids on the project. Some owners mandate attendance at these conferences; others do not. In either case, it is important that the owner not rely solely on verbal instructions provided to the prospective bidders at these meetings, but instead repeats these instructions in meeting minutes or, better yet, addenda. This is particularly important when the instructions have the effect of modifying the express terms of the other materials contained in the bid package.[33]

When construction projects are awarded by virtue of the competitive bidding system, the bid package must be sufficiently correct and complete to allow bids to be submitted without the bidders making exceptions, exclusions, or

33. *Id.* § 2:61, at n.13.

assumptions as part of the bid, which would make it difficult to compare one bid to another. However, despite efforts at accuracy, most construction projects of any size involve complex and intertwined systems of materials, facilities, and equipment that may contain ambiguities or mistakes. Project plans may include hundreds of interrelated drawings, while project specifications may include thousands of pages of technical instructions. Inevitably, these drawings and specifications will contain errors, omissions, and inconsistencies that can potentially impact the bidder's construction costs.

What, then, should a bidder do if it encounters ambiguities in the bid package that may affect its bid? Is a bidder under an obligation to obtain clarification of discrepancies in the materials that comprise the bid package? Generally speaking, bidders are required to obtain clarification of "patent" (obvious) discrepancies in the bid materials but are not held responsible for any "latent" (hidden) ambiguities contained therein.[34] Whether a discrepancy is considered patent or latent depends upon the size and complexity of the project and bid package, the obviousness of the discrepancy, and the relevant significance of the discrepancy. In the event that a discrepancy is found to be patent, a contractor who, without inquiry, makes erroneous assumptions concerning that discrepancy becomes responsible for the resulting extra costs.

The same general considerations regarding a contractor's obligation to discover discrepancies in the bid documents apply to the contractor's obligation to factor site conditions into its bid price. In conducting a pre-bid site investigation, a contractor is held to the "standard of reasonableness."[35] Generally, a bidder is entitled to rely on the accuracy of information included in the solicitation and is under no obligation to conduct its own surveys or go beyond what a simple site viewing might reveal.[36] As the U.S. District Court for the District of South Carolina has noted:

> Bidding contractors and subcontractors are not held to the knowledge and skill of geologists, geotechnical engineers, or fluid dynamic experts. They are entitled to rely on accurate and reasonably complete information supplied in the bid package and a reasonable site inspection. Beginning with *United States v. Utah, Nevada and California Stage Co.*,[37] a long line of United States Supreme Court decisions has made clear that when the government in its invitation to

---

34. *Id.* § 2:63.

35. North Slope Tech. Ltd., Inc. v. United States, 14 Cl. Ct. 242, 253 (1988).

36. Mojave Enters. v. United States, 3 Cl. Ct. 353, 358 (1983); *See* 1 Bruner & O'Connor § 2:64; Helene Curtis Indus., Inc. v. United States, 312 F.2d 774 (Ct.Cl. 1963).

37. United States v. Utah, Nevada and California Stage Co., 199 U.S. 414, 26 S.Ct. 69, 50 L.Ed. 251 (1905).

> bid undertakes to specify certain existing conditions material to the performance of the contract, it is bound by these specifications and liable for any increased cost arising from their substantial variation.[38]

But what if the information supplied to bidders, while literally accurate, leaves the bidder with an incomplete or inaccurate picture of the project or project site? In such cases, a claim under the "superior knowledge" doctrine may be made when the government "fails to provide a contractor with vital knowledge in the government's possession which bears upon the cost of the contractor's performance under the contract at issue."[39] To recover under a superior knowledge claim, the bidder must demonstrate that (1) the government possessed information that materially affected the contractor's performance costs; (2) the information was not reasonably available to the contractor; (3) any contract specification supplied misled the contractor or did not put it on notice to inquire; and (4) the government failed to provide this information.[40] It is not necessary that the contractor prove that the government's failure was intentional. Even an inadvertent failure can be actionable.[41]

In *North Slope*, for example, the U.S. Court of Claims held that an owner-provided geotechnical or "soil boring" report is "considered the most reliable reflection of subsurface conditions."[42] Thus, where the government possesses materials that qualify or expand upon the borings, this information would be "relevant and important to bidders," and the government "should have disclosed the information in the contract documents."[43]

> The government discharges its duty to disclose only when its disclosure is communicated effectively to the contractor.[44] Thus, the government does not satisfy its duty of disclosure unless it shows that the disclosure was not only made but also "heard, and understood, actually or apparently."[45] Because "[t]here is no practical way the government can know for sure that someone understands . . . [if the government] has done all it can to get the information out *loudly and clearly*, it has done all it can."[46]

---

38. W. F. Mangann v. Diamond Mfg. Co., Inc., 580 F. Supp. 1299, 1313 (D.C.S.C. 1984), *rev'd in part on other grounds*.
39. Hercules, Inc. v. United States, 24 F.3d 188, 196 (Fed. Cir. 1994).
40. *Id.*; J. F. Shea Co. v. United States, 4 Cl. Ct. 46, 53 (1983).
41. *See* Granite Constr. Co. v. United States, 24 Cl. Ct. 735, 750 (1991).
42. 14 Cl. Ct. at 254 (citations omitted).
43. *Id.*
44. PetroChem Services, Inc. v. United States, 837 F.2d 1076, 1080 (Fed. Cir. 1988).
45. *Id.*
46. *Id.* (emphasis added).

c. Bid Preparation Techniques

To prepare its bid, a contractor is required to estimate both the prices and quantities of a multitude of materials and equipment to be incorporated into the finished project. These include:

1. Material costs
2. Labor costs
3. Equipment costs
4. Subcontractor costs
5. Job site overhead or "general conditions" costs (e.g., job site trailers, temporary water and electricity, insurance, and office equipment and supplies)
6. Home office overhead expenses (e.g., rent, staff salaries, advertising costs, and professional fees)
7. Profit

On larger projects, accurately estimating the total project costs would be virtually impossible without the input of the many subcontractors and suppliers who desire to contribute to the project. Accordingly, bidders on more complex projects spend a great deal of time in the days and hours leading up to "bid day" compiling and analyzing the bids of the potential subcontractors and suppliers that the contractor may hire for the project and then determining which bids to incorporate into their own submissions to the owner.

When a subcontractor or supplier submits a subcontract price that it later cannot honor, a question arises as to who is responsible for the cost overrun: the general contractor or the subcontractor. Courts faced with deciding whether the subcontractor is "bound" by its proposal to a contractor have tended to adopt one of two distinct positions on the issue, both of which are discussed at length elsewhere.[47]

2. *Contractor Selection Criteria*

a. Duty to Award to the Lowest Bidder and Owner's Right to Rebid

As stated earlier in this chapter, an open bidding process is used on most public projects, wherein the various contractor bids are opened simultaneously and publicly at a predetermined time, with the project then awarded to the lowest bidder. Critical to this process is that the owner be required to accept

47. See Chapter 10.

that price without engaging in rebidding or post-bid opening negotiations in an attempt to further drive down project costs.

But what if the owner is not satisfied with the bids received? If the lowest bid submitted is still in excess of the funds available for the project, many states will authorize the public owner to re-advertise the project for bid at a later date. Alternatively, some states authorize the public owner to enter into negotiations with the lowest bidder to reduce the quantity or quality of work in that bidder's scope of work so as to reduce the overall project costs.[48]

Public bidding statutes routinely reserve to the owner the right to reject a bid for other compelling reasons. These statutes also typically allow a public owner to cancel an invitation whenever it is determined to be in the owner's best interests. In so doing, most states will require a showing that the public owner's actions were not undertaken arbitrarily or for purposes of evading the public procurement laws.[49]

What if the owner desires to award the project to a bidder other than the one submitting the lowest bid? Generally speaking, an owner is not required to base its award solely on price. To the contrary, a public owner must make its award to the lowest "responsive," "responsible" bidder taking into consideration factors such as quality, experience, performance, and construction duration. Issues of responsibility and responsiveness will be addressed in the sections that follow.

b. Bid Responsiveness and Waiver of Bidding Irregularities

To achieve the policy objectives underscoring competitive bidding, a bid submitted during the process must be "responsive" to the requirements stated in the owner's solicitation materials; otherwise, it will be rejected even if it is the lowest one received.[50] Responsiveness to the bid is necessary so that the bids received can be compared. One court stated it as follows:

> [a]ny material deviation from the requirements of the invitation or statutory mandates destroys any ability to compare that bid competitively on the basis of price with other bids submitted and, if permitted to be considered, would violate the public policy objectives underlying the competitive bidding procedure of allowing bidders an equal right to compete.[51]

The bid must strictly comply with mandatory requirements in the invitation or statute and should substantially comply with non-mandatory require-

---

48. N.C. GEN. STAT. § 143-129(b) (2007).
49. 1 BRUNER & O'CONNOR, *supra* note 1, §§ 2:135 and 137.
50. *Id.* § 2:74 at 181.
51. *Id.* at 184.

ments, as well.[52] A material variation will void the bid and may be defined as one that would give the bidder "a substantial advantage or benefit not enjoyed by other bidders."[53] Examples of deviations that may be material and unacceptable include bids that are incomplete or late, were "prepared without a mandatory site inspection," propose "alternate work or materials not conforming substantially to specification requirements," or "fail to include required bid security."[54]

A bid that does not conform precisely to the invitation may still be accepted if the deviation is considered to be immaterial.[55] An immaterial deviation is one that has a "negligible effect upon price, quantity, quality, or time of completion of the work to be performed."[56] Examples of nonmaterial (and therefore waivable) bid deviations include failing to initial changes or corrections, submitting a bid with changes initialed by someone other than the person who signed the bid, or failing to acknowledge an amendment to the bid schedule by submitting the bid based upon an old schedule if the changes made by the amendment do not materially affect the bid.[57] The failure to sign a bid proposal may be material, depending on whether a signature on another document is sufficiently connected to the proposal to make it a binding contract.[58]

Some awarding bodies lack the statutory discretion to determine whether a requirement is waivable.[59] For the most part, however, public agencies have broad discretion in deciding whether a bid is responsive, and courts will set aside these decisions only if they are arbitrary, capricious, or an abuse of discretion.[60] For example, the Georgia Supreme Court stated that a county school

---

52. *Id.* § 2:75 at 185.

53. Gostovich v. City of W. Richland, 75 Wash. 2d 583, 587, 452 P.2d 737, 740 (1969) (quoting Duffy v. Village of Princeton, 240 Minn. 9, 60 N.W.2d 27, 29 (1953)).

54. 1 Bruner & O'Connor, *supra* note 1, § 2.2:75, at 185 (listing thirteen "deviations that raise issues of materiality").

55. Stephanie Giggetts & Eric Surette, 64 Am. Jur. 2d *Public Works and Contracts* § 65 (Westlaw 2007).

56. 1 Bruner & O'Connor, *supra* note 1, § 2.2:91, at 206 (citing 48 C.F.R. § 14.405(d)).

57. *In re* Consol. Contracting & Eng'g, 1996 WL 639911, at *2 (Comp. Gen. Nov. 6, 1996).

58. *Compare* Farmer Constr. Ltd. v. Washington, 98 Wash. 2d 600, 605, 656 P.2d 1086, 1089 (Wash. 1983) (holding that the lack of a signature on the proposal was waivable because the bid proposal and bid bond are connected with each other by internal references, and the bond was signed), with S. J. Amoroso Constr. Co. v. United States, 981 F.2d 1073, 1076 (9th Cir. 1992) (holding that the lack of a signature on the required Certificate of Procurement Integrity was a valid reason for determining the bid nonresponsive, even though the bidder had signed other documents).

59. *See* Hamp's Constr., L.L.C. v. City of New Orleans, 924 So. 2d 104, 110–111 (La. 2006).

60. 1 Bruner & O'Connor, *supra* note 1, § 2:92, at 206.

board could waive a bid requirement, such as the contractor providing a list of all its intended subcontractors, if the board decided that the requirement was not material. Similarly, the U.S. District Court for the Ninth Circuit held that rejection of a bid as nonresponsive was proper where the required Certificate of Procurement Integrity was unsigned when submitted, particularly because the Certificate "imposes significant legal obligations on the contractor."[61] The court concluded: "If an agency decides to require a new CPI with such sealed bid, it is not the place of the court to substitute its judgment."[62]

c. Responsible Bidder

In the competitive bidding process, the "lowest responsive bidder" must also qualify as "responsible" before being awarded the contract.[63] In this regard, "[t]he sole question before the agency is whether the low bidder is 'responsible'—not whether the bidder is *most* responsible among all bidders."[64]

As with the issue of responsiveness, the awarding body usually has broad discretion in making decisions regarding bidder responsibility. These subjective decisions may be based upon "virtually any evidence reasonably related to the question of whether the contract, if awarded, will be faithfully completed in accordance with its terms—general standards such as integrity, past performance, experience, technical, financial and managerial capability, safety record, [and] licensure."[65] A bidder's "character and reputation," "energy," "fraud or unfairness in previous dealings, honesty, judgment, [and] promptness" may be considered.[66] A bidder's prior legal history and even "social responsibility" also may be valid factors in determining responsibility.[67]

The Wisconsin Court of Appeals, for example, ruled that the rejection of a bidder due to concerns about the bidder's past conduct and "performance of prior contracts" was valid because the awarding body had a rational basis for the decision and the rejection was neither arbitrary nor unreasonable.[68] Like judicial review of decisions concerning responsiveness, courts generally set

---

61. *S. J. Amoroso Constr.*, 981 F.2d, at 1076.

62. *Id.*

63. 1 Bruner & O'Connor, *supra* note 1, § 2:96, at 212–14 (discussing "public agency discretion in judging bidder 'responsibility'").

64. *Id.* § 2:94 at 210.

65. *Id.* § 2:96 at 214 (internal citations omitted).

66. *Id.* at 215 (quoting Bowen Eng'g Corp. v. W.P.M. Inc., 557 N.E.2d 1358, 1366 (Ind. Ct. App. 1990)).

67. *Id.* §§ 2:97 and 2.2:102.

68. D.M.K. Inc. v. Town of Pittsfield, 290 Wis. 2d 474, 478, 483, 711 N.W.2d 672, 674, 677 (Wis. Ct. App. 2006).

aside decisions regarding responsibility only if they are arbitrary, capricious, or an abuse of discretion.[69]

A determination that the lowest bidder is not responsible is not an abuse of discretion when the lowest bidder fails to submit requested information relevant to the owner's inquiry. In *Kinsey Contracting*, the lowest bidder did not submit the requested audited financial statement, list of pending claims, résumés of supervisors, or the name of the proposed excavation contractor.[70] The second-lowest bidder submitted the required information and was awarded the contract.[71] Although the lowest bidder argued that the term "'lowest responsible bidder' requires only that a contractor with the lowest bid have the proper license for the job, provide the necessary performance and payment bonds, and have adequate financial resources to perform the contract," the North Carolina Court of Appeals held that the award standard, as specified by statute, allowed consideration of other factors.[72] Thus, said the *Kinsey* court, "in the absence of fraud or a palpable abuse of discretion, [courts] have no power to control their action."[73]

Because a determination of responsibility based on standards that were not communicated to the bidders may be overturned as arbitrary and capricious, public agencies often include guidelines or criteria for responsibility in bid invitations.[74] Agencies that waive requirements set forth in established guidelines may still be given deference by courts, as long as the decision is "reasonable and rationally based" and the waiver is "not a material deviation from the contract."[75]

Traditionally, responsibility is determined after the bids are reviewed for responsiveness.[76] In recent years, however, prequalification programs have been used to allow public agencies to make initial determinations of responsibility

---

69. *See* Lion Raisins, Inc. v. United States, 51 Fed. Cl. 238, 247 (2001) (holding that suspending a bidder as not responsible was arbitrary and capricious when that bidder had been deemed responsible five times in the previous two years).

70. Kinsey Contracting Co., Inc. v. City of Fayetteville, 106 N.C. App. 383, 386, 416 S.E.2d 607, 609 (N.C. Ct. App. 1992).

71. *Id.* at 384, 416 S.E.2d at 608.

72. *Id.* at 385, 416 S.E.2d at 609.

73. *Id.* at 384, 416 S.E.2d at 608 (quoting Mullen v. Town of Louisburg, 225 N.C. 53, 60, 33 S.E.2d 484, 488–89 (1945)).

74. 1 Bruner & O'Connor, *supra* note 1, § 2:96, at 216 (citing Harmony Constr., Inc. v. State Dep't of Transp., 668 A.2d 746 (Del. Ch. 1995)).

75. *Id.* at 216–17 (citing Suit-Kote Corp. v. City of Binghamton Bd. of Contract & Supply, 628 N.Y.S.2d 861, 863, 216 A.D.2d 831, 833 (N.Y. App. Div. 1995)).

76. Daniel D. McMillan & Erich R. Luschei, *Prequalification of Contractors by State and Local Agencies: Legal Standard and Procedural Traps*, 27 Constr. Law 21 (Spring 2007).

before soliciting bids.[77] In some states, however, this initial screening evaluates only some of the attributes (such as financial stability and bonding capacity) used in the later responsibility determinations.[78] In these jurisdictions, while bids are accepted only from bidders who are prequalified, passing this initial screening does not imply that the bidder will be deemed responsible after bids are opened and a final "responsibility" determination is made.[79] Standards of review for prequalification decisions vary. A court may give an agency's decision no deference or substantial deference, depending on whether the agency's reason was "a legal one, a factual one, or one of mixed law and fact."[80]

d. Bid Protests

What if the low bidder on a project believes that it was unfairly rejected for reasons of responsiveness or responsibility? Conversely, what if an unsuccessful bidder believes that the low bidder should be rejected as nonresponsive or nonresponsible? In both circumstances, a challenge to a public agency's procurement process may be made by undertaking what generally is referred to as a "bid protest."

Certain threshold issues impact a protesting bidder's right to relief. These include

1. Party's standing to protest the bid
2. Waiver resulting from untimely assertion of protest
3. The proper forum in which to protest the bid[81]

A party must have standing to protest a bid. At the federal level, a protestor must be an "interested party."[82] An interested party is "an actual or prospective bidder or offeror whose direct economic interest would be affected by the award of the contract or by failure to award the contract."[83] For state and local contracts, the 2000 Model Procurement Code for State and Local Governments provides standing to "any actual or prospective bidder, offer, or contractor who

---

77. *Id.*

78. *Id.*

79. *Id.* at 25 (citing Crest Constr. Corp. v. Shelby County Bd. of Educ., 612 So. 2d 425, 430 (Ala. 1992)).

80. *Id.* at 33 (providing examples of courts giving or not giving deference to agencies).

81. 1 Bruner & O'Connor § 2:143.

82. 28 U.S.C. § 1491(b)(1) (2007).

83. 31 U.S.C. § 3551(2) (2007).

is aggrieved in connection with the solicitation or award of a contract."[84] Other states, like Nebraska, only allow bid protests to be asserted by a taxpayer or rate payer.[85]

A protestor who challenges any contract award needs to assert its claims in a timely manner.[86] Failure to do so may result in a waiver of the right to protest the award.[87] For federal contracts, different deadlines exist depending on the nature of the protest.[88] State and local contracts also are subject to the statutes and regulations of their jurisdiction, and failure to assert the claim in a timely manner could subject the protest to dismissal.

The forum in which to challenge an improper award is dictated by the type of project involved. Protests involving federal projects can be heard by the awarding agency itself, the General Accounting Office, or the U.S. Court of Federal Claims. In states that have adopted the Model Procurement Code, the timing and content of protests are mandated by statute.[89] In South Carolina, for example, a protest must be made to the Chief Procurement Officer within ten days of the award and must set forth the grounds for the protest and the relief requested.[90]

Some states, such as Florida, Illinois, and New Jersey, have a centralized agency that resolves bid protests.[91] For example, in Maryland, the bid protest must be filed with its Board of Contract Appeals within ten days from receipt of notice that the initial protest was denied by the state procurement office.[92] The Board reviews the protest, solicits comments from the procurement agency and other interested bidders, and issues a decision.[93] If the bid protestor receives an unfavorable decision from the Board, it may appeal to the state court system.[94]

Other states employ a decentralized contracting system in which each agency is responsible for resolving its own protests. For example, Ohio has different statutory protest procedures, depending upon the type of project (e.g., distinct procedures for highway department projects, for water improvement

---

84. 2000 Model Procurement Code for State and Local Governments § 9-101(1).
85. Day v. City of Beatrice, 169 Neb. 858, 101 N.W.2d 481 (1960).
86. Carlson T.V. v. City of Marble, 612 F. Supp. 669 (D. Minn. 1985).
87. 4 C.F.R. § 21.2(b) (2007).
88. *See Id.* § 21.2 (2007).
89. 1 Bruner & O'Connor § 2:146.
90. S.C. Code of Laws §11-35-4210 (2007).
91. Construction Law, *supra* note 14, § 2.03[3].
92. Md. Code Ann. State Fin. & Proc. § 15-220 (2007); Md. Regs. Code 21.10.02.10 (2007).
93. Md. Regs. Code 21.10.07.03 (2007).
94. Construction Law, *supra* note 14, § 2.03[3] State and Local Bid Protests.

projects, and for library construction).[95] In the absence of formal protest procedures at the state or local level, bid protest cases are typically heard in state court.[96]

If the protestor is successful in its protest, its remedies typically are limited to the recovery of its bid preparation costs.[97] Even where the agency has wrongly awarded a contract, the agency rarely will be required to pay lost profits to the unsuccessful bidder.[98] For this reason, many protestors seek to enjoin the owner from proceeding with its award to the competing bidder.[99] Injunctive relief may be difficult to obtain, however, because the bidder must establish that

1. the plaintiff (bidder) will suffer irreparable injury if defendant (agency) is not enjoined, including, but not limited to, absence of an adequate remedy at law;
2. the plaintiff is likely to succeed on merits;
3. the harm to the plaintiff outweighs the harm to the defendant; and
4. it would be in the public interest to enjoin the defendant.[100]

Unsuccessful bidders seeking injunctive relief may also be required to post a bond, which may be forfeited if the challenge is rejected.[101] For this reason, a bid protestor needs to be cautious when filing a claim. If it is later determined that the bidder's protest was unlawful, fraudulent, or otherwise inequitable, the competing "disappointed bidder" who ultimately loses the bid as a result of the protest may assert a common-law claim against the protestor.[102] Nevertheless, preliminary injunctions can be effective in preventing an owner from

---

95. *Id.*

96. *See* Eugene McQuillin, The Law of Municipal Corporations §§ 29.83, 29.83.05, 29.85, 29.86 (3d ed. 1949).

97. *See* Lovering-Johnson, Inc. v. City of Prior Lake, 558 N.W.2d 499, 504 (Minn. Ct. App. 1997).

98. *See* State Mechanical Contractors, Inc. v. Village of Pleasant Hill, 132 Ill. App. 3d 1027, 87 Ill. Dec. 532, 477 N.E.2d 509 (Ill. App. 1985); *but see* Clark Constr. Co. v. Pena, 930 F. Supp. 1470, 1478 (M.D. Ala. 1996).

99. *See* H. R. Johnson Constr. Co. v. Bd. of Educ., 16 Ohio Misc. 99, 241 N.E.2d 403 (Ohio Com. Pl. Ct. 1968).

100. *See* ES-KO, Inc. v. United States, 44 Fed. Cl. 429, 432 (Fed. Cl. 1999).

101. *See* NewMech Companies, Inc. v. Indep. Sch. Dist. No. 206, 558 N.W.2d 22 (Minn. Ct. App. 1997).

102. Davis & Assoc., Inc. v. Midcon, Inc., 127 N.M. 134, 138, 978 P.2d 341, 345 (Ct. App. 1999).

wrongfully awarding the project to a party other than the successful bidder or making payment on a contract that was not in compliance with statutory bid procedures.[103]

e. Bid Security and Mistakes in the Bidding Process

Absent a bid protest, the owner's selection of the lowest responsible and responsive bidder typically ends the procurement process and results in the execution of a contract with the successful bidder. But what happens if the contractor selected later refuses to execute a contract? Allowing the contractor to withdraw its bid would return the owner to "square one" in the bidding process. The owner then would face the costs and delay of repeating the bidding process and a potential price increase in the replacement bid.[104] To guard against this possibility, most public construction projects require that all bidders provide some form of bid security (generally a cash deposit or "bid bond" of at least 5 percent of the total bid price) to ensure that, if selected, the bidder will enter into a contract for the work it bid upon.[105] If the selected contractor refuses to honor its bid, the owner may retain the bid security to offset its resulting loss. This forfeiture encourages contractors to submit only a bid that they are prepared to honor if selected.

But what if a contractor's bid was made in error? In some cases, a court may allow a contractor to withdraw a mistaken bid without forfeiting its bid bond. To withdraw a bid, the contractor generally must satisfy two conditions.[106] First, the "mistake must be one of 'fact.'" Mistakes in judgment will not suffice.[107] A mistake of fact typically involves a clerical error, such as a numerical miscalculation,[108] typographical mistake,[109] or an omission of a portion of the

---

103. Clark Constr., 930 F. Supp. 1470; Pincelli v. Ohio Bridge Corp., 213 N.E.2d 356 (Ohio 1966).

104. FUNDAMENTALS, *supra* note 9, at 159.

105. *Id.*

106. *See* James Cape & Sons Co. v. Mulcahy, 285 Wis.2d 200, 225, 700 N.W.2d 243, 255 (Wis. 2005).

107. 1 BRUNER & O'CONNOR, *supra* note 1, § 2:119.

108. Vintage Constr., Inc. v. State Dept. of Transp., 713 P.2d 1213 (Alaska 1986) (relieving the construction contractor of a bid total for $249,900 more than the actual total of the unit prices listed in the bid document).

109. Kemp v. United States, 38 F. Supp. 568 (D. Md. 1941), (allowing the construction contractor to withdraw its bid of $2,953.85 when it relied on a quote it had received for a machine that stated $2,813 instead of $20,813).

project advertised for bid.[110] Second, the mistake must be substantial enough to justify depriving the owner of the benefit of its bargain. Some states even require that the mistake be such that the owner would have reason to suspect a mistake by looking at the bid document itself.[111]

f. The *Croson* and *Adarand* Decisions and Hiring Preferences

Because the construction industry has few educational and financial barriers to entry, it has long been a vehicle to effectuate change. At the same time, the construction sector historically has been dominated by white, male-owned enterprises.

In 1961, President John F. Kennedy used an executive order to introduce affirmative action into the federal procurement process.[112] In 1977, Congress furthered federal efforts to eliminate discrimination by establishing an administrative program that required all contractors to hire Minority Business Enterprises (MBE) for at least 10 percent of all public projects.[113] Three years later, the U.S. Supreme Court upheld this program in *Fullilove v. Klutznick*.[114]

As affirmative action programs gained judicial footing in the construction industry, local governments began implementing "set-aside" programs to encourage contractors to award subcontracts to minorities by requiring all city construction projects to allocate a set percentage of the project to minority-owned businesses.[115] However, in 1989, in *City of Richmond v. J.A. Croson Co.*, the Supreme Court rejected this use of affirmative action.[116]

The set-aside program in *Croson* "required prime contractors to whom the city awarded construction contracts to subcontract at least thirty percent of the dollar amount of the contract to one or more Minority Business Enterprises (MBEs)."[117] However, in the project at issue, only one MBE subcontrac-

110. James Cape & Sons Co. v. Mulcahy, 285 Wis. 2d 200, 207, 234–35, 700 N.W.2d 243, 246, 260 (allowing the contractor to withdraw a bid but only after forfeiting its bid bond, when the bid did not include a last-minute $130 increase by one of the subcontractors).

111. *See* Arcon Constr. Co. v. State, by and through Dept. of Transp., 314 N.W.2d 303 (S.D. 1982) (relieving the contractor from its bid on the grounds of unconscionability when the bid recipient "knew of the mistake prior to awarding the contract").

112. Exec. Order No. 10925 (1961) (stating that "[t]he contractor will take affirmative action to ensure that applicants are employed, and that employees are treated during employment, without regard to their race, creed, color, or national origin").

113. 42 U.S.C. § 6705(f)(2) (1977).

114. Fullilove v. Klutznick, 448 U.S. 448, 100 S. Ct. 2758 (1980) (upholding Congress's 10 percent MBE set-aside provision in the Public Works Employment Act of 1977).

115. City of Richmond v. J. A. Croson Co., 488 U.S. 469, 477, 109 S. Ct. 1842, 1849 (1989).

116. *Id.* at 511, 731.

117. *Id.* at 477–78, 713 ("The 30% set-aside did not apply to city contracts awarded to minority-owned prime contractors.").

tor expressed an interest in the project and even then at prices that were over market rates.[118] The contractor challenged the city's MBE requirements under the Equal Protection Clause of the Fourteenth Amendment.

In striking down Richmond's program, the *Croson* Court invoked a "strict scrutiny" test for evaluating affirmative action programs in the contractor selection process that asks whether the program is narrowly tailored to serve a compelling government interest and whether there is a "strong basis in evidence for its conclusion that remedial action was necessary."[119] The city of Richmond failed to satisfy this test when it did not present any direct evidentiary support for "the need for remedial action" to the Court. Thus, the Court declared the program unconstitutional.[120]

In 1995, the U.S. Supreme Court, in *Adarand Constructors, Inc. v. Pena*, affirmed its earlier ruling in *Croson* and held that federal affirmative action preference programs must also undergo the strict scrutiny standard of judicial review.[121] In the wake of *Croson* and *Adarand*, the constitutionality of Minority and Women Business Enterprises (MWBE) preference programs has faced new challenges as the parameters of permissible affirmative action programs are continuously evaluated under the Supreme Court's strict-scrutiny standard.[122] MWBE programs work to remedy "past or present discrimination," an interest courts generally find sufficiently compelling.[123] Therefore, challenges to these programs typically focus on "the adequacy of the evidence" of past discrimination and on whether the program at issue is "narrowly tailored."[124]

118. *Id.* at 482–83, 716.

119. *Id.* at 500, 509–10, 725, 730, (quoting Wygant v. Jackson Bd. of Educ., 476 U.S. 267, 277, 106 S. Ct. 1842, 1849 (1986)) (stating, "If the city of Richmond had evidence before it that nonminority contractors were systematically excluding minority businesses from subcontracting opportunities it could take action to end the discriminatory exclusion").

120. Adarand Constructors v. Slater, 228 F.3d 1147 (10th Cir. 2000). Adarand appealed the Tenth Circuit decision, asserting that the Court of Appeals misapplied the strict scrutiny standard and that the federal program was not narrowly tailored. The Supreme Court, after oral argument following its grant of certiorari, dismissed the case on the ground that certiorari had been "improvidently granted." Adarand Constructors, Inc. v. Mineta, 534 U.S. 103, 122 S.Ct. 511 (2001). The case had a nine-year history, beginning with Adarand Constructors v. Skinner, 790 F. Supp. 240 (D. Colo. 1992).

121. 115 S. Ct. 2097 (1995).

122. *See* Contractors Ass'n of E. Pa. v. City of Phila., 91 F.3d 586 (3d Cir. 1996); Concrete Gen., Inc. v. Washington Suburban Sanitary Comm'n, 779 F. Supp. 370 (D. Md. 1991); Edinger & Son v. City of Louisville, 802 F.2d 213 (6th Cir. 1986); Eng'g Contractors Ass'n of S. Fla. v. Metro. Dade Co., 122 F.3d 895 (11th Cir. 1997).

123. *Eng'g Contractors,* 122 F.3d at 907 (quoting Ensley Branch, NAACP v. Seibels, 31 F.3d 1548, 1565 (11th Cir. 1994)).

124. *Id.* at 906.

Common targets include overly broad ordinances or arbitrarily set percentages. The existence of race-neutral remedial alternatives may also be grounds for striking down a program.[125]

In evaluating preference programs, courts consider statistical studies made before and after the institution of the program, along with anecdotal evidence.[126] A general claim of societal discrimination, however, is not enough.[127] Strict scrutiny requires a "strong basis in evidence" that reveals patterns of discrimination "persistent in the local construction industry" before and after enactment—census data, statistical disparity studies, local hearings, phone surveys, and questionnaires.[128] Anecdotal evidence from the testimony of minority and nonminority contractors regarding their personal experiences with the local industry also serves to supplement the evidence.[129] Even though *Croson* did not go as far as setting forth a requirement of negating "all evidence of non-discrimination," it did set forth a more rigorous level of scrutiny.[130] The justification for MWBE programs now faces a higher standard of proof that requires clear evidence of local discriminatory practices and how the program will counteract the effects.[131]

---

125. *Contractors Ass'n,* 91 F.3d at 608–10 (invalidating a program that included subcontractors, not just prime contractors, and that lacked evidence to explain how the city arrived at the percentage set or why the city did not instead implement training and financial assistance programs).

126. Derek M. Alphran, *Proving Discrimination after* Croson *and* Adarand*: "If It Walks Like a Duck,"* 37 U.S.F. L. Rev. 887, 904, 916, 920 (2003).

127. *Croson,* 488 U.S. at 470, 109 S.Ct. at 709 ("The city's argument that it is attempting to remedy various forms of past societal discrimination . . . alleged to be responsible for the small number of minority entrepreneurs in the local contracting industry fails, since the city also lists a host of nonracial factors which would seem to face a member of any racial group seeking to establish a new business enterprise, such as deficiencies in working capital, inability to meet bonding requirements, unfamiliarity with bidding procedures, and disability caused by an inadequate track record").

128. Concrete Works of Colo., Inc. v. City & Co. of Denver, 321 F.3d 950, 990 (10th Cir. 2003), *cert. denied,* 540 U.S. 1028, 124 S. Ct. 556 (2003).

129. *Id.* at 969–70 (noting additional prerequisites, unwarranted bid rejections, slower payment schedules, higher prices for materials, additional obligations, and difficulties joining unions and obtaining lines of credit).

130. *Id.* at 991 (citation omitted).

131. George R. La Noue, *The Impact of Croson on Equal Protection Law and Policy,* 61 Alb. L. Rev. 1, 40 (1997).

Numerous MWBE programs have withstood the heightened scrutiny imposed by the *Croson* and *Adarand* Courts.[132] For example, a San Francisco ordinance satisfied the strict scrutiny standard by replacing its previous set-aside provision with "bid preferences."[133] Instead of requiring contractors to hire MWBEs for a certain amount of the contract price, the city required a 10 percent bid preference for local MWBE bidders "to be taken into account when the city purchaser calculated the low bid."[134] The set-aside program required by the Illinois Department of Transportation also withstood challenge by allowing each selected contractor to set its own MWBE participation goal for the project that would reflect the availability of MWBEs in the local market.[135] These two programs indicate that jurisdictions can meet the Supreme Court's high standards by designing affirmative action programs with built-in flexibility.

g. Alternative Procurement Methods in the Public Sector

The practice of "competitive bidding" has been widely praised for providing taxpayers with the lowest possible construction costs by fostering competition among the bidders and by encouraging each bidder to present its best price. The practice has also been recognized as providing the most effective method of protecting the integrity of the bidding process.

The practice of awarding contracts based on price alone may not always be appropriate, given the complexity of modern construction projects. As a result, private owners have long embraced alternative delivery systems and methods when appropriate for the project at issue. More recently, public owners have begun adopting some of these approaches, as well.

i. Design-Build Construction in the Public Sector As the name implies, Design-Build contracts bundle the project design and project building or construction into one contract. Design-Build contracting offers an owner several advantages over the traditional Design-Bid-Build method, which are discussed in more detail in Chapter 4. By its very nature, however, Design-Build construction contracts do not allow an owner to make the sort of "apples-to-apples"

132. Note that a number of city, state, and federal authorities use other nomenclature for the businesses that are entitled to special assistance. Many programs refer to Small Business Enterprises (SBE), firms that meet certain size criteria, and Disadvantaged Business Enterprises (DBE), firms owned by women or other groups that would typically qualify as MBEs.

133. Associated Gen. Contractors of Cal. v. City and Co. of S.F., 748 F. Supp. 1443, 1446 (N.D. Cal. 1990), *aff'd*, 950 F.2d 1401 (9th Cir. 1991).

134. *Id.* at 1445.

135. Northern Contracting, Inc. v. Illinois, 2004 WL 422704 (N.D. Ill. March 3, 2004).

comparisons that are the hallmark of the competitive bidding process. Accordingly, when awarding contracts of a Design-Build basis, public agencies typically follow procedures similar to those utilized when awarding design-only contracts. Such awards typically identify parameters of the job, advertise those parameters, invite proposals, and evaluate the relative skills of the various bidders.[136] Some public entities do consider the cost of the project, but do so by typically assigning weights to the various evaluation criteria, such as price, features, functions, time, and life cycle costs. The entities then include those relative weights in their requests for bids, so that the designers-builders can tailor their proposals accordingly.[137] Public procurement policies then either allow the public entity to select the designer-builder based on more subjective measures or require the use of published criteria and then evaluate the potential designer-builder in strict accordance with those criteria.[138]

ii. Construction Management in the Public Sector As discussed in greater detail in Chapter 4, the Construction Management project delivery method can take two distinct forms. In one context, a construction manager is used to describe a person or entity that contracts with an owner to manage, as the owner's agent, one or more of the contractors hired directly by the owner to construct the project. Such Agency Construction Management engagements, like design contracts, typically are exempted from public procurement laws,[139] but may be subject to the same sort of previously discussed regulations that govern the selection of design professionals.

In its other manifestation, "construction management" refers to a contractor who, having entered into a contract with an owner for the construction of some improvement, subcontracts all of the work on the project, reserving for itself only the work of managing the effort. Construction managers under these arrangements are generally considered to be "at risk," in that they bear financial responsibility in the event the project exceeds the agreed-upon contract sum or duration. At-Risk Construction Management arrangements generally are not exempt from public procurement laws because they are more akin to a construction contract than a design agreement.

Because At-Risk Construction Management agreements have become increasingly popular with owners, some states have expressly authorized At-Risk Con-

---

136. *See, e.g.*, CAL. PUB. CONT. CODE § 20209.7 (2007); LA. REV. STAT. ANN. § 48:250.3 (2007).

137. *See, e.g.*, CAL. PUB. CONT. CODE § 20209.7 (2007).

138. *See e.g.*, *Id.* § 20301.5 (2007); LA. REV. STAT. ANN. § 48:250.3 (2007) (noting that "the winning proposal shall be the proposal with the lowest adjusted score").

139. 2 BRUNER & O'CONNOR, *supra* note 1, § 6:66.

struction Management agreements in the public sector. In North Carolina, for example, construction managers (at risk) may be selected under the same state regulations governing the engagement of design professionals. Once selected, the construction manager (at risk) is required to conduct a traditional "open bidding" of its first-tier subcontracts, which utilizes certain prequalification criteria and then awards each subcontract to the lowest responsible, responsive bidder, taking into consideration, quality, performance, time, oversight, and other factors that might be deemed appropriate by the owner.[140]

iii. "Best Value" Awards Inherent in the competitive bid model is the possibility that the lowest bidder might not be the most qualified builder available. Moreover, the emphasis on construction costs alone ignores other long-term price considerations, such as utility costs, maintenance costs, renovation costs, or life cycle costs. Finally, the owner might not have fully anticipated or articulated its needs during the design phase of the project, and thus the design might need revision after the contract is awarded but before the project is complete.[141]

Value-based contracting in its various forms is an effort to overcome the preceding limitations. Value-based contracting allows the owner to consider factors other than initial construction costs when awarding the bid. Often those considerations will be the future costs associated with owning the building, namely: "1) excessive renovation costs due to premature obsolescence; 2) excessive repair and renovation costs due to premature deterioration; and 3) excessive operating and maintenance costs due to inadequate design and construction."[142]

At least six different types of value-based contracting have been identified:

1. Life-cycle costing, which emphasizes the future costs of owning the building
2. Sustainable design, which emphasizes long-term efficiencies and "green" technology

140. *See* N.C.GEN.STAT. § 143-128.1 (2007).

141. James Schenck, IV & Robert Lee Carson, Value-Based Procurement—*Where Is It, Where Is It Going, and Are Construction Lawyers Ready For It?*, ABA Construction Forum Fall Meeting, October 2003 at p. 14.

142. *See* Dean B. Thomson & Michael J. Kinzer, *Best Value in State Construction Contracting*, 19 CONSTRUCTION LAWYER 31 (1999) (discussing the federal government's use of "best value" procurement instead of the traditional lowest responsive, responsible bidder model for some projects). *See also* MCQUILLIN, *supra* note 96, § 37.106 (noting that competitive bidding may or may not be required by statute or ordinance).

3. Qualifications-based selection, in which the owner selects the designer or builder based upon its competence and expertise
4. Commissioning, in which the owner documents the building's operational costs and performance and compares that to specified and promised performance
5. Guaranteed cost savings contracts, where the contractor guarantees a positive return to the owner from the anticipated maintenance and operational savings
6. Privatization, in which the contractor assumes the long-term maintenance obligations on the project[143]

Regardless of the form taken, value-based contracting allows the owner to enjoy greater savings over the life of the project even though the initial construction costs might be greater than under the lowest-bidder method of contracting.

Value-based contracting is not without its detractors. First, it does little to avoid the appearance of impropriety in awarding bids.[144] Additionally, where owners consider items such as a builder's abilities and previous performance, value-based contracting can create an anticompetitive bias in favor of larger and previously employed builders.[145] A few states have attempted to address these issues by requiring that the agencies consider the "equitable distribution" of contracts when they award contracts.[146] Others have attempted to address this by requiring the awarding body to assign specific, published weights to the factors it will consider and then rigidly adhere to those published weights when awarding the project.[147]

Value-based contracting creates new challenges for the construction lawyer, particularly with respect to the long-term accountability of design professionals and builders.[148] Typically, these professionals have only been held to the standard of ordinary care and skill, have offered very short warranties, and have had their future liability limited by the statute of limitations.[149] To hold these professionals accountable in the long term, the owner and builder likely

143. Schenck & Carson, *supra* note 141, at 5–12.

144. *Id.* at 14.

145. *Id.* at 4.

146. Robert S. Barns, Michael J. Schaengold, & Rodney A. Grandon, *Best Value in Federal Construction Contracting*, 19 CONSTRUCTION LAWYER 25, 26 (1999).

147. *See, e.g.*, FLA. STAT. ANN. 287.055(4) (2007); 62 PA. CONS. STAT. § 905(e) (2007).

148. *See, e.g.*, LA. REV. STAT. ANN. § 48:250.3 (2007) (noting that "the winning proposal shall be the proposal with the lowest adjusted score").

149. ABA, *supra* note 141, at 14–15.

will have to negotiate additional specific and documented guarantees.[150] The owner and attorney will then need to keep accurate and detailed "building performance" records.[151] The design professionals and builders faced with these additional long-term liabilities will have greater risk exposure and will need longer-term insurance. That added expense will then make the projects more costly and may erode some of the financial benefits that value-based contracting originally offered.

iv. Reverse Online Bidding Techniques A reverse online auction is an Internet-based method of bidding for the supply of construction services. As in the traditional bidding process, reverse auction bidding calls for an owner to publish a set of drawings and specifications for contractors to review and bid upon. Unlike the traditional bidding process, however, where interested bidders submit their bid by hard copy, fax, or email, online bidding takes place live, in real time via the Internet with invited bidders bidding against each other. Bidders remain anonymous but receive immediate feedback on the competitiveness of their last bid and have an opportunity to lower their price if they choose. As such, reverse online auctions work something like e-Bay in reverse, with the bidding remaining open until a predetermined deadline that may be extended indefinitely for as long as lower bids continue to be submitted. Once the auction concludes, the auction service provider notifies the owner of the bidding results and then initiates the appropriate follow-up with the winning bidder to complete the formal award of the contract.

Reverse online auctions have become increasingly popular in the private sector. Recently, the federal government has begun to use them. Proponents of the reverse online auction concept believe that the process is well suited to determining market price sufficiently and fairly. For this reason, reverse online auctions have become increasingly popular with several nationwide "big box" retailers and contractors.

Others, particularly contractor groups, are vehemently opposed to the reverse online process.[152] "Critics of the net-based reversed auction bidding process question whether that process will allow sufficient communication to ferret out incomplete or uncoordinated design documents before such documents negatively impact project scheduling and costs."[153] Additionally, these groups

150. *Id.* at 17–19.

151. *Id. at* 20.

152. ABA, *supra* note 141, at 20.

153. *See* Associated General Contractors of America, White Paper on Reverse Auctions for Procurement of Construction.

caution that the nature of the auction itself, where a bidder's price is exposed to its competitors and then systematically driven downward, may encourage imprudent bidding. Some even suggest that reverse auctions do not result in any measurable savings to owners because, unlike traditional bidders, each bidder recognizes that it will have the option to provide successively lower bids as the auction progresses.[154] As a result, a bidder has little incentive to lead with its best price and subsequently may never offer its lowest price. Critics further assert that the use of reverse auctions also may lead to higher costs due to decreased competition in the marketplace. If the contractors opt out of bidding in reverse auctions because of slim profit margins or philosophical or ethical objections, owners may suffer in the long run.

As with any new process or technology, online reverse auction bidding has had its share of technical glitches that have led unsuccessful bidders to mount legal challenges to the procedural fairness of the system. The holdings in these cases highlight the importance of having unambiguous ground rules in reverse auctions.[155]

MTB Group, Inc. v. United States presented the ultimate question of whether a federal government agency is permitted to use an online reverse auction. There, a bidder objected to the Department of Housing and Urban Development's use of an online reverse auction to obtain bids for housing inspection services because it feared that the small pool of bidders would allow other contractors to identify its bid, thereby destroying bidding anonymity.[156] While recognizing that a duty of maintaining bidding anonymity is imposed upon all federal officials by law, the Court of Federal Claims nevertheless refused to grant an injunction. In doing so, the court held that the bidders waived their right to anonymity by participating in the auction, thereby consenting to the disclosure of their bids. The refusal of the Court of Federal Claims to strike down the use of online reverse auctions in *MTB Group* appears to indicate that this procurement mechanism will continue to be employed to award public as well as private contracts.

---

154. *Id.*

155. *See* Pacific Island Movers, B-287643.2 (Comp. Gen. 2001), Royal Hawaiian Movers, B-288653 (Comp. Gen. 2001).

156. MTB Group, Inc. v. United States, 65 Fed. Cl. 516, 519 (2005).

QUESTIONS

1. What if the lowest bidder on the public owner's base contract work is no longer the lowest bidder once its bids on the owner's requested alternates are considered? Must the owner select the bidder with the lowest "base bid"? Is the owner within its rights to select either bidder it chooses?
2. Would an owner be within its rights to reject a bidder as not "responsible" based upon the bidder's litigiousness? What if the bidder believes that it was well within its rights to have sued prior owners? What if it prevailed in those claims? What if the prior litigation involved the same owner awarding the current project?
3. What is the rationale for allowing mistaken bidders to withdraw their bid for clerical errors? Should similar relief be allowed for errors in judgment?

CHAPTER

9

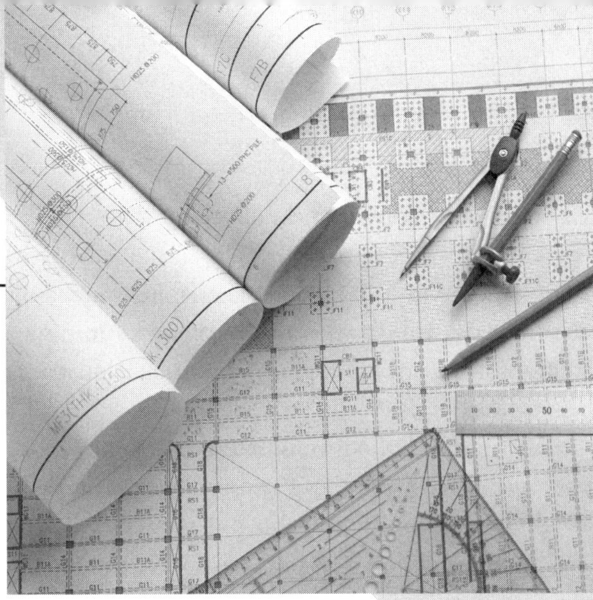

Pricing Construction Contracts

STEPHEN A. HESS

9.01 INTRODUCTION

This chapter discusses the various pricing mechanisms that owners and contractors use to fix the "contract price" through which a contractor's compensation is determined for performing the "work" required of the contractor under a prime contract.[1]

There is little recitation of case law related to this discussion. When the contract price is established by agreement between the parties, interpretation of the contract is governed by the fundamental rules of interpretation that govern all contracts, and with few exceptions those rules apply with equal force (and in many cases, were developed with reference) to construction agreements. Instead, this chapter will focus its attention on the typical terms that parties employ in pricing the work to be performed.

1. The discussion in this chapter is equally applicable to pricing contracts between general contractors and subcontractors, but the discussion focuses on the owner and contractor for the sake of simplicity.

The pricing of construction contracts goes far beyond fixing the compensation for which a contractor is willing to perform work and the owner is willing to pay for such work. Instead, the specific mechanism that the parties choose has an important role in allocating the economic risks between the owner and the contractor. In addition, many contract clauses vary depending on the type of pricing that is used.

Accordingly, the immediate goal of this chapter is to describe the different pricing mechanisms that are used by parties to construction contracts. At the same time, and as important, this chapter will endeavor to explain the extent to which the selection of a particular pricing mechanism inherently allocates risk between the parties and the manner in which parties mitigate such risks within the agreement itself or in their other commercial relationships. In addition, this chapter will identify and discuss numerous contract terms that are affected by the selection of the pricing mechanism. In completing this chapter, you should devote as much time to thinking about how the various contract terms are integrated into a cohesive unit as you think about the purpose and meaning of individual terms.

9.02 GENERAL PRICING MECHANISMS FOR CONSTRUCTION CONTRACTS

In the final analysis, pricing mechanisms can be divided into two types: those that fix the price of the work at a stipulated sum (fixed-cost) and those that allow the price to vary depending on the cost incurred by the contractor. This section discusses stipulated-sum contracts in some detail and then turns to an overview of the important considerations related to cost-plus contracts. The section continues by reviewing the manner in which these two basic types of pricing may be combined into one contract and summarizes other variations on these pricing schemes.

A. Stipulated-Sum Contracts

The simplest form of pricing is the stipulated-sum contract under which the contractor agrees to perform all of the work for a fixed figure. A typical contract provision reads:

> The Owner shall pay the Contractor the Contract Sum in current funds for the Contractor's performance of the Contract. The Contract Sum shall be Four Million Five Hundred Thousand and 00/100 Dollars ($4,500,000.00), subject to additions and deductions as provided in the Contract Documents.[2]

2. This clause is taken from AIA Document A101–2007, *Standard Form of Agreement Between Owner and Contractor Where the Basis of Payment Is a Stipulated Sum* § 4.3 (2007).

There is nothing remarkable about this language, except perhaps its simplicity. The contractor's obligation is to perform the contract, and for that work the owner is obligated to pay a set figure. The contract price (in AIA terms, the "Contract Sum") merely fixes the amount of the compensation, and other provisions in the contract deal with such important issues as the payment process and change orders.[3]

1. Scope Limitations within Contract Price

In exchange for the contract sum, the contractor is required to perform all, but no more than, the work that is identified in the contract. When the owner or circumstances attributable to the owner change in a manner that increases the amount of work that the contractor must perform, the contractor is generally entitled to an increase in the contract price. The usual mechanism through which such an increase is added is the contractor's submission (or the owner's proposal and the contractor's acceptance) of a change order. A change order takes its name from the fact that a change is being made in some term(s) of the contract, such as the scope of work, the contract sum, or the schedule.[4]

A fundamental attribute of a stipulated-sum contract is that the contractor bears all the entrepreneurial risk associated with the cost of performance, in exchange for which the contractor is exclusively entitled to reap the benefits of efficient performance. For example, if labor rates increase unpredictably and structural steel supplies experience rapid, unforeseen price spikes, the contractor may find itself in a contract whose performance will actually cost the contractor more money than the contract sum. The fact that unforeseen events cost the contractor its entire profit (and may force the contractor to incur a loss) is irrelevant to the contract sum or the enforceability of the contract. Regardless of actual costs to perform, the contractor is entitled to recover exactly the contract sum and no more.

Of course, the same principles apply if the contractor is able to work very efficiently, and if the construction costs turn out to be less than the contractor anticipated. If a contractor can perform a $4.5 million contract for out-of-pocket costs of $2.7 million, it makes a hefty profit, and the owner cannot object that the contractor is making "too much" money. In fact, the owner in a stipulated-sum contract does not usually have any means of finding out, and has no legal rights to know, how much profit the contractor built into its price, and the contractor may have good reason to keep this information from the owner.

Why are contractors willing to undertake all of this risk? At the outset, the risk for short-term contracts is not generally that great. Although there are

3. See Chapters 14 and 16, respectively.
4. Changes are discussed in more detail in Chapter 16.

price variations in materials and labor, they usually are not so dramatic as to turn a profitable job into a money-loser, even if they do shave some amount off the contractor's profit. Second, a contractor that works efficiently has the prospect of making even more than it budgets on a job, and the contractor might want to keep the entire benefit of any savings.[5] Third, a contractor can mitigate the risk of price fluctuations through advance purchases of materials or signing subcontractors to fixed price contracts.

Once the contractor binds itself to a specific stipulated sum for performing work within the scope of the contract, the only method through which it can increase the contract sum is through circumstances giving rise to a change order. For better or worse, this fact inherently gives the contractor substantial motivation to identify any of the work it is required to perform as falling outside the contractual scope of "work," as the contractor will only be entitled to additional compensation if such additional work is out of scope.

For example, civil site plans drawn in haste may fail to show precise boundaries of overlot grading, and the notes to the drawings may leave them tantalizingly ambiguous as to whether the overlot grading only encompasses the immediate building envelope or instead extends an additional 22,500 square feet to the boundaries of the owner's entire property. The contractor has some incentive to read the civil site plans narrowly, in a manner that requires it to perform very little overlot grading outside the footprint of the main building: Any additional work the contractor is required to perform is chargeable to the owner and is not within the original stipulated sum. The owner, on the other hand, will argue vociferously that the 22,500 square feet of "additional" overlot grading is within the original scope of work, and thus within the stipulated sum which the contractor is entitled to be paid.

Two important attributes of stipulated-sum contracts are evidenced by the preceding discussion. First, a pure stipulated sum contract puts all the risk of an increase in actual construction costs on the contractor and gives the contractor the sole benefit of any efficiencies or reductions in the actual costs of performing the work. Second, in stipulated-sum contracts, there is inherent tension in construing the scope of work: A contractor makes money by identifying any necessary work as falling outside the original scope of work in the contract; conversely, the owner saves money by construing as much of the work as possible as falling within the original scope of work. This tension may bleed over into the relationship between the design professional and the owner.

5. The reader may contrast this with a cost-plus contract, which is discussed later, wherein the owner generally gets most of the benefit of a contractor's efficiency in reducing job costs.

If the owner believes that it has contracted the entire scope of necessary work through the adoption of the design professional's drawings and specifications, only to find that there are substantial "extras" required to get the building the owner envisioned, the owner may try to shift the burden of the additional expense to the design professional. It is not difficult to see in this chain that disputes over scope of work may lead directly to conflicts between the design professional and the contractor.[6]

Both the owner and the contractor want to put themselves in the best financial position they can; thus, many disputes under stipulated-sum contracts involve the proper characterization of the scope of work for which the contractor is entitled to be paid the contract sum.

2. Percentage-of-Completion Payments and Potential Abuse

The payment process is discussed in more detail in a separate chapter,[7] but the pricing of lump-sum construction contracts has an important function that it is helpful to address at this juncture.

A stipulated-sum contract is usually paid over the course of time in "progress payments," which are often calculated by reference to the "percentage-of-completion" method. When this method is utilized, the contract provides that the contractor is entitled to be paid each month the total value of the contract, multiplied by the percentage of the work that is completed through that month (less retention, which is discussed in Chapter 14 on payment). When the only contract pricing is the stipulated sum itself, the determination of the percentage of work completed may be a complex and seemingly arbitrary process.

In order to add some fairness, rationality, and predictability to the percentage-of-completion method, the contractor may be required to submit with its bid a "schedule of values" that allocates the contract sum among many discrete portions of the work for purposes of more finely assessing the percentage of completion of the work. In substantial projects, the schedule of values may have hundreds of items. In even a modest project, the schedule of values includes many entries.

The use of a schedule of values, however, also offers the contractor the opportunity to take advantage of the payment process in a manner that puts the owner at risk. Assume, for example, that a contractor agrees to build a

6. Although it is beyond the scope of this chapter, the reader may wish to consider how this inherent tension plays out when the architect or design professional is put in the position of being a "neutral" administrator of the contract terms and is given the power, in the first instance, to render decisions concerning disputes between the owner and the contractor. Chapters 2 and 7 discuss the multiple roles of the design professional.

7. See Chapter 14.

medical office building for $12,000,000 on a site that requires excavation near the top of the water table, and thus the site must continually be dewatered. When offering its schedule of values, the contractor would do well to allocate $6,000,000 to "building excavation" and $500,000 per month for nine months of continuous "site dewatering." Of course, that would leave just $1,500,000 on the schedule of values for construction of the building itself. From the contractor's perspective, this is all wonderful: By the time the excavation is completed, the contractor will have earned half the contract sum. The owner, on the other hand, has given up substantial security (the unpaid contract sum) and is at substantially greater risk in the event the contractor is unable to complete the work.

As protection against this front-loading of a schedule of values, the payment provisions may void a patently front-loaded schedule of values, or give a third person (such as the architect) the responsibility for approving a schedule of values that will be used in the payment stages. These protections are important, as the failure to pay the contract sum according to the contract's payment terms may give the contractor the right to suspend or to terminate its work.

B. Cost-Plus Contracts

A cost-plus contract is one in which the contract sum is calculated as the sum of the "cost of work" *plus* some figure for the contractor's overhead and profit. A typical cost-plus contract sets out the contract sum as follows:

> The Owner shall pay the Contractor the Contract Sum in current funds for the Contractor's performance of the Contract. The Contract Sum is the actual Cost of the Work as defined in Article 7 plus the Contractor's Fee.[8]

Just as we noted the simplicity of the stipulated-sum contract language, the reader will note the apparent simplicity of the language used for the cost-plus contract. But here, the simplicity is deceptive: as the saying goes, the devil lies in the details. And in this case, the precise devil is the definition of the cost of work and, more specifically, what is and what is not included within that "cost" as opposed to being part of the contractor's assumed overhead.

Before we turn to the issue of identifying the cost of work itself, it is helpful to review how the cost-plus contract allocates economic risk. In contrast to the stipulated-sum contract, the cost-plus contract gives the owner all of the risk and all of the benefit of unanticipated changes in material and labor costs,

8. American Institute of Architects, AIA Document A103–2007, *Standard Form of Agreement Between Owner and Contractor Where the Basis of Payment Is the Cost of Work Plus a Fee without a Guaranteed Maximum Price* § 5.1 (2007) [hereinafter AIA Document A103–2007].

as well as all of the risk and benefit of the contractor's relative efficiency or inefficiency. At the same time, it has the potential of placing a substantially greater burden on the owner to monitor the course of construction from a financial standpoint, as the contractor has less financial risk at stake and less economic motivation to perform that particular work efficiently.

1. Cost of Work and Limitations Thereon

Returning to the issue of identification of costs, the following examples of potentially disputed costs reveal the necessity of a detailed definition of allowable "costs" that go into calculation of the contract sum as well as the necessity of controls on those costs. These examples are chosen to demonstrate in part that "cost of work" is not as simple a term as one might imagine, and to allow the reader to understand different mechanisms for handling cost issues.

a. Subcontracted Work

Because the contractor bears the burden of all costs in a stipulated-sum contract, it has a natural incentive to seek out the most cost-effective subcontractor that can perform the work responsibly. In a cost-plus contract, that motivation is absent, and the contractor may choose a more expensive contractor than it might otherwise choose so as to maintain a valuable professional relationship, to avoid the cost and administrative burden of pricing all the subcontract work, to increase its compensation (where it is a percentage of the total cost of work), or for other motivations or considerations that the owner does not share.

The owner may seek protection from overpriced subcontractors in a cost-plus contract. One method for providing protection is to require that the contractor bid the work out just as if it were compiling the cheapest responsible team for a stipulated-sum contract. In the alternative, the owner might retain a construction manager or utilize its design professional for purposes of monitoring the procurement of subcontractors.

This is not to suggest that most contractors are selfish or unscrupulous. One of the biggest motivations for a contractor to keep costs down is that its future in the industry may depend on it. A cost-plus contractor who develops a reputation for overcharging owners is unlikely to flourish.

b. Contractor's Equipment

The cost of work generally includes such costs as rental payments on heavy equipment. When the rental company calculates its rental rates, it assesses the diminution in value that a piece of equipment will experience over its useful rental life; the cost of maintaining, licensing, and insuring the equipment;

the cost of operating the equipment when fuels, oil, and other consumables are included in the rental; the taxes it must pay on the equipment; and such incidental items as the rental agency's overhead. These factors, together with some allowance for profit, are divided into the estimated period during which the equipment will be rented, and the result is some periodic rental rate. (As a practical matter, heavy equipment can usually be rented at hourly, daily, weekly, or monthly rates.) Owners usually do not quibble over paying these market rental fees.

On the other hand, some contractors maintain their own fleets of heavy equipment. Should they be entitled to charge the use of such equipment to the owner under a cost-plus contract, or is such equipment usually considered part of the contractor's general overhead for which it receives compensation in the "plus" part of the cost-plus formula?

There really is not much dispute that contractors who use their own equipment are normally entitled to charge for that equipment. A more difficult question is whether the rental rates for contractor-owned equipment should be calculated in a manner that allows the contractor to recover overhead and profit on the equipment itself in addition to the overhead and profit it otherwise recovers under the contract. There is no "correct" answer to a question like this; instead, what is important is that the parties agree on the method by which the contractor may charge for contractor-owned equipment. A traditional and accepted method is to append a schedule of values to the contract that specifies the exact rate that a contractor is entitled to bill to the project for its own equipment. In the absence of such a schedule, the attribution of "cost" to contractor-owner equipment can be a contentious and frustrating process for both parties.

c. Small Tool Allowance

The question in the preceding section concerned contractor-owned heavy equipment. A similar issue arises with respect to small tools. Battery-powered portable tools, for example, are not indestructible, but instead are "consumed" over the course of many projects until they must be discarded. Other more durable tools still break, wear down, or otherwise need to be replaced. Are these costs part of the normal cost of work, or are they simply part of the contractor's overhead? Again, there is no "correct" answer to this question, and as a practical matter, the answer is not all that important economically in the long run—if contractors never charged separately for small tools, their costs would simply get absorbed as part of their normal overhead, and they would start demanding higher payments for the "plus" portion of cost-plus contracts to recover consumable small tools.

As a practical matter, many cost-plus contracts have chosen to treat small tools as partially consumable items that are included in the compensable "cost" of the work.[9] That presents the question of how such tools are to be priced for cost purposes. Many contracts simply state a percentage of cost (typically 3 to 6 percent) for a small tool allowance. Others use more precise, but more ungainly mechanisms. For example, the standard AIA contract allows as a portion of the cost of work "the value of the [tool] at the time it is first used on the project site, less the value of the item when it is no longer used at the project site."[10] From an accounting, valuation, and administration perspective, it should be apparent that a stipulation as to values may cost the owner or contractor more money than it ought to cost, but will be accompanied by a substantial savings in transaction costs, i.e., the costs of determining rental value.

d. Self-Performed Work

Some general contractors perform very little of the work they are contractually obligated to complete on a project, and instead delegate all work to subcontractors. Others perform very large portions of work with their own forces. The conceptual issue arises, with respect to such self-performed work, of how the work is to be priced. Or, to put it differently, questions arise as to how to allocate the contractor's costs for performing work between recoverable "costs" of construction and non-recoverable overhead.

There is little problem with allowing a contractor to recover the cost of building materials that are incorporated into the project or the cost of laborers to execute the work. At the same time, the parties must be careful to agree as to the treatment of field personnel. When is a supervisor part of the contractor's unrecoverable overhead, and when is the supervisor appropriately considered a "cost" that the contractor incurs for which the owner should be billed? The answer is not to be found in any Restatement of the Law or in case authority—instead, it is to be found in the agreement between the parties, and in particular, in the definition of "costs" that the contractor is allowed to recover.

Is there a "typical" or "standard" answer to the question? The AIA treats supervisory and administrative personnel costs as recoverable "when they are stationed at the site with the Owner's prior approval."[11]

---

9. Some tools (or parts) are entirely consumed in a single project and can be priced at actual cost (e.g., saw blades, sanding belts).

10. AIA Document A103–2007, *supra* note 8, § 7.5.1. This formulation does not prevent the parties, as an administrative convenience, to value the diminution in value of such tools at 6 percent of their original cost.

11. *Id.* § 7.2.2.

e. Labor

One more of many issues that arise in determining the cost of the work under a cost-plus contract is the cost of labor. A contractor who pays a roofer $18 per hour in wages actually spends much more than that to keep the roofer on its staff.[12] To employ the roofer and set him out to do work, the contractor usually must pay health insurance, employment taxes, pension benefits, cafeteria plan payments, workers' compensation, insurance, and so on. These additional costs are generally referred to as the labor burden, and they are not insubstantial: Labor burdens generally run 25 to 40 percent of actual wage rates. Those familiar with the construction industry know that workers' compensation rates for certain categories of labor (e.g., roofers and steelworkers) can be greater than the actual wages paid to such workers!

There are three common ways of ensuring that the contractor recovers the actual cost of maintaining a labor force. The first is to allow the contractor to recover the actual costs of the employees by providing certified payroll accounting with the actual cost of the labor burden. This is the most accurate, most administratively expensive method of pricing labor. A second, similar method is to specify the burden as a percentage of the actual payroll that the contractor is entitled to recover. This is less precise than the first method, but much easier to administer. A third method is to schedule labor rates for different classes of workers and to allow the contractor these "burdened" rates at the stipulated values.

2. Reasonableness or Necessity of Costs Incurred

Regardless of whether the contract provides a specific mechanism for allowing the owner to participate in regulating the cost through oversight, it may contain catchall language to protect the owner to the effect that the cost of work includes only the "reasonable" and "necessary" costs incurred in performing the work. The obvious purpose of such language is to provide some protection against runaway costs where the contractor exhibits no apparent responsibility in its selection of vendors or subcontractors, or otherwise incurs costs at a rate greater than the owner anticipated. In the absence of such a clause, the owner may still have recourse to some general duty of "good faith and fair dealing," although the application of such a general contract standard to the pricing of work under a cost-plus contract can be difficult. The standard AIA formulation

---

12. In addition, in order to maintain a readily accessible workforce, the contractor may pay the roofer wages even when the roofer is idle.

simply requires that "costs shall not be at rates higher than the standard paid at the place of the project except with prior consent of the owner."[13]

Clauses such as these are expensive to litigate, however, and while they provide some theoretical protection for the owner, they do not serve as an adequate substitute for a carefully drafted definition of cost of work.

3. *Contractor's Overhead and Profit*

The "plus" portion of a cost-plus contract is the compensation the contractor receives, both for its profit and for its general "home office" overhead. As noted previously, there are often questions as to which potential cost items should be recovered separately under the contract, and which items are presumed to be part of the contractor's home office overhead. How are these "plus" figures specified?

a. Fixed Fee

One mechanism for fixing the contractor's fee is to state a flat fee for overhead and profit: "The Contract sum shall be the Cost of the Work (as defined below), plus a fee of $500,000.00 as the exclusive compensation for the Contractor's overhead and profit." When fixed fees are utilized, there is often a provision to ensure that the fixed fee is increased if there are sufficient changes on the project.

b. Percentage of Contract

Another mechanism is to pay the contractor a percentage of the cost of work: "The contract sum shall be the Cost of Work (as defined below) plus 15%, which 15% shall constitute the Contractor's exclusive compensation for overhead and profit."

Some owners are concerned that the payment of a percentage of the cost of work actually encourages a contractor to work less efficiently so as to increase its own compensation. Such fears are usually unfounded for several reasons. First, most contractors understand that the additional money they would earn by working inefficiently is not nearly worth the future business they stand to lose if they develop a reputation as a bad (inefficient) contractor. Second, a contractor who is busy with other work usually finds it more efficient to generate, to work, and to bill new business rather than to try to live off marginal increases in its compensation under a cost-plus contract. Finally, because an owner always has the ability to challenge the reasonableness of costs incurred,

13. AIA Document A103–2007, *supra* note 8, § 7.1.1.

the inefficient contractor faces some risk that it will lose the ability to collect any inefficient costs and will actually lose money on the work to the extent of excessive costs (that is, not only will it lose any profit on such work, but it will not be able to recover the cost of the work itself). That is not to say that there are no shady practices by contractors under cost-plus contracts with percentage-of-cost compensation schemes, but this concern is usually not substantial enough to justify a party's avoiding these compensation terms where they otherwise fit the parties' expectations.

4. *Estimates and Cost-Plus Contracts*

In deciding whether it wishes to use a cost-plus contract or a stipulated-sum contract, the owner may wish to have some idea of the projected cost so as to determine how much risk it is willing to take. The usual source of projected cost information is the design professional, one of whose paid tasks can be the compilation of a formal cost estimate for the work. The design professional's liability for a negligent cost estimate is discussed elsewhere in this book.[14]

The contractor may furnish an estimate for the owner's use, as well. Because the contractor and owner have divergent interests and the contractor is not generally paid for a cost estimate, it is not prudent for the owner to place too much reliance on a mere estimate. After all, if the owner wants a fixed cap, it has the alternative of entering into a stipulated-sum contract or a cost-plus contract with a guaranteed maximum price, discussed later in this chapter.

At the same time, courts may be concerned that a contractor deliberately provides an artificially low estimate for the purpose of securing the work, with the recognition that the contractor ultimately will not bear the financial burden of this estimate. As a consequence, many government contracts include clauses that cap the allowable costs at those set out in the estimate, absent advanced approval from the government. One court explained:

> The requirements of the cost limitation provision protect both the contractor and the government. The ability to obtain government approval of additional costs before incurring them protects the contractor from discovering, after performing the contract at greater cost than anticipated, that its work has been financially disastrous because the government will not pay the additional costs. It also relieves the contractor, if such advance approval is not given, from its obligation to complete the contract or to incur costs in excess of the estimate. The limitation of the government's liability to the estimated cost stated in the contract (unless additional costs are authorized) protects the government from having to pay more than it had anticipated and set aside for the contract if the contractor's costs of performance exceed the estimate.

14. See Chapter 14.

> It also gives the government the choice whether to incur additional costs for the contract or to have the contract terminated. It is therefore important that the requirements of the cost limitation provision be followed.[15]

Such clauses do not guarantee that the project will be completed at the cost estimated by the contractor, but it removes some of the incentive the contractor might have for producing a sloppy (or worse, intentionally understated) estimate.

C. Cost-Plus with Guaranteed Maximum Price

In many cases, the owner's greatest concern in a cost-plus contract is the risk of massive cost overruns, and the owner may believe that a cost-plus contract would be acceptable but for that risk. To address this concern, a third type of contract pricing has been created from cost-plus and stipulated-sum contracts that allows an owner to limit its risk, while at the same time allowing the owner to enjoy the benefits of cost-plus contract pricing. This third type of contract uses a hybrid of the two pricing mechanisms described so far and may be called Cost-Plus with a Guaranteed Maximum Price.

As the name implies, the chief pricing term is a cost-plus term in which (ideally) the allowable and excluded elements of the cost of work are set out in some detail. Such contracts contain a separate cost limitation, however, expressed in language like the following:

> The Contract Sum is guaranteed by the Contractor not to exceed four million five hundred thousand and 00/100 dollars ($4,500,000.00), subject to additions and deductions by Change Order as provided in the Contract Documents. Such maximum price is referred to in the Contract as the Guaranteed Maximum Price. Costs which would cause the Guaranteed Maximum Price to be exceeded shall be paid by the Contractor without reimbursement by the Owner.[16]

Under a clause such as this, the owner bears the risk that the cost of work will be as much as the Guaranteed Maximum Price (GMP), and the contractor bears the risk of any excess costs over the GMP.

In light of the risk of excess costs, the contractor under a contract with such a pricing term must take special care to ensure that the GMP is an acceptable figure and must effectively prepare its bid as though it were bidding a stipulated-sum contract. In short, a cost-plus with GMP pricing term operates as a

15. Advanced Materials, Inc. v. Perry, 108 F.3d 307, 310 (Fed. Cir. 1997).

16. American Institute of Architects, AIA Document A102–2007, *Standard Form of Agreement Between Owner and Contractor Where the Basis of Payment Is the Cost of Work Plus a Fee with a Guaranteed Maximum Price* § 5.2.1 (2007).

cost-plus contract up to the point that the GMP is reached and a stipulated-sum contract thereafter. The student will appreciate the operation of such contracts in the discussion of specific terms later in this chapter.

D. Unit-Price Contracts

A variation of the stipulated-sum contract is a unit-price contract in which the owner (or the contractor, where the pricing occurs in a subcontract) pays a stated price for each unit of a particular type of construction, but pays for the actual units performed, rather than paying a fixed total. For example, a contract for excavation may provide that the subcontractor will be paid $3.00 per cubic yard for the excavation and remote disposal of dirt from a building foundation. A unit-price contract, unlike a stipulated-sum contract, puts the risk of variations in estimated quantity on the owner, but gives the owner the benefit of some cost control in the form of a fixed price for each unit of work.

Owner-contractor contracts priced entirely as unit-price contracts are relatively uncommon, but unit-price contracts between general contractors and subcontractors are not unusual, and many owner-contractor agreements contain at least some elements of unit-price contracting. For example, an owner may negotiate for a fixed price (stipulated sum) for the construction of an administration building and key elements of a water treatment plant, but agree, because of uncertainty in measurement (due to the preliminary state of planning, perhaps), to pay unit costs for the excavation and placement of piping that will transport the water to and from the treatment plant. A unit-price contract requires some stipulation as to who is responsible for measuring quantities for payment purposes, as well as how disputes in measurements will be resolved.

1. *VEQ Clauses*

The use of unit prices generally puts the risk of variations in quantities on the owner, but the risk of variations in cost on the contractor. If excavation and placement of piping is priced as a unit price, the contractor suffers when the cost of pipe increases and benefits when the cost of pipe decreases.

There is another element of unit pricing where the cost estimate may vary in a manner in which the parties do not necessarily intend to allocate cost risk to the contractor, and that is in cost variations that arise from the total quantities performed under the unit price rather than from market conditions. If the contract states a unit price, it is often with some assumption by the parties about what the actual quantity will prove to be. A contractor who agrees to build approximately 600 feet of pipeline has very different unit costs from a contractor who agrees to build 60,000 feet of identical pipeline. A contractor who is willing to give a favorable unit price for 1,400 windows may find itself losing money if the design is changed mid-stream and the final building includes 350 windows.

Where the unit price is based on a common assumption or estimate of the expected quantities, the contract may contain a provision for revising the unit price for variations from the estimated quantities on which the parties relied in negotiating the unit price. A Variations in Estimated Quantities (VEQ) clause set out in the Federal Acquisition Regulation (FAR) reads as follows:

> If the quantity of a unit-priced item in this contract is an estimated quantity and the actual quantity of the unit-priced item varies more than 15 percent above or below the estimated quantity, an equitable adjustment in the contract price shall be made upon demand of either party. The equitable adjustment shall be based upon any increase or decrease in costs due solely to the variation above 115 percent or below 85 percent of the estimated quantity.[17]

The question has arisen as to whether the variation in estimated quantities outside the 85–115 percent range automatically requires repricing of the unit price. One court held that it does not, and explained its understanding of the FAR's VEQ clause as follows:

> Contract interpretation begins with the plain language of the agreement. *Gould, Inc. v. United States,* 935 F.2d 1271, 1274 (Fed. Cir. 1991). The VEQ clause requires an equitable adjustment to be based upon any increase or decrease in *costs.* The Government's preferred construction, however, requires the establishment of a new unit price for the overrun amounts by an equitable adjustment of the contractor's unit profit, even where the contractor's unit costs remained constant. This interpretation clearly contravenes the plain meaning of the VEQ clause. The express language of the VEQ clause precludes an equitable adjustment based on anything other than the contractor's costs.[18]

In this context, "equitable adjustment" in a government contract is equivalent to a change order altering the contract price of the work that is subject to the VEQ clause. The clause above does not dictate how the repricing is to be accomplished. In private contracts, such repricing is carried out in the same manner as the pricing of any other work for which the contractor or owner is entitled to a change order, and such change order pricing techniques are addressed elsewhere in this work.[19]

Although a change in the quantity of goods can be repriced by reference to the actual cost of purchase (and related bulk-quantity discounts), variations in

---

17. 48 C.F.R. § 52.212–11 (1992).
18. Foley Co. v. U.S., 11 F.3d 1032, 1034 (Fed. Cir. 1993).
19. See Chapter 16.

estimated quantities of construction elements apart from goods can have dramatic effects on a contractor's costs. Indeed, such costs can vary with actual quantities even when the cost of materials is constant. For example, a substantial portion of the cost of a discrete type of work may be mobilization and demobilization—getting equipment, materials, and workers to the site and removing equipment and excess materials at the completion of work. When the mobilization and demobilization costs are spread over the installation of 16,000 square feet of concrete masonry unit (CMU) wall, the unit cost for the wall is substantially less than the unit cost of an 800-square-foot-wall, the price of which must recover the same mobilization and demobilization costs over many fewer units. Such a reduction in CMU walls would call for an increase in the unit price of the wall, even if the cost of labor and CMU block does not change (per unit).

2. *Unbalanced Bidding*

When bids for public works are assessed on the basis of unit prices applied to estimated quantities, a contractor's superior knowledge of actual conditions (or its apprehension of an error in the public agency's estimated quantities) can allow the contractor to take advantage of the government through "unbalanced bidding."

Assume, for example, that a solicitation for a sewer installation estimates 400 linear feet of pipe and 1,200 cubic yards of excavation, and contractors are required to bid the items as unit prices with the provision that the low bidder will be the contractor whose total contract price (based on estimated quantities) is the lowest. Now suppose Abercrombie Contractors bids the pipe at \$30 per foot and the excavation at \$10 per cubic yard for a total contract bid of \$24,000 (400 × \$30, plus \$1,200 × \$10) and that both unit prices are within the range of reasonableness for that type of work. Barnard Contractors, on the other hand, bids the pipe \$5 per foot (substantially less than the wholesale cost of the pipe) and the dirt work at \$18 per yard (at the high end of excavation work). Barnard's total bid is \$23,600 (400 × \$5 per foot, plus 1,200 × \$18).

The government's estimate of dirt quantities turns out to be grossly short, and the actual quantity is 3,000 cubic yards. Although Barnard won the contract with a lowest total *projected* cost, the *actual* cost to the government of \$56,000 (400 × \$5, plus 3,000 × \$18) is much higher than the \$42,000 Abercrombie would have charged (400 × \$30, plus 3,000 × \$10). Given the fact that the unit cost Barnard bid for pipe is just a fraction of the actual cost of the pipe, one might suspect that Barnard submitted a deliberately "unbalanced bid" to exploit its superior knowledge of the deficiencies in government's estimate of excavation quantities. The government may have a remedy for such unbalanced bidding in some cases.

E. Mixed Pricing Contracts and Change Orders

There is nothing inherently wrong with using some elements of a cost-plus contract in a stipulated-sum contract or vice versa. Ultimately, pricing is a matter of risk allocation, and there is no reason that every risk on a project must be allocated in the same manner. In addition, a change order may be priced in terms very different from those of the underlying contract.

Readers will see, for example, that many stipulated-sum contracts contain clauses that allow the price to change for reasons such as differing site conditions. In such a contract, the contractor may be willing to assume all the cost risks of the project *except* the risks of unanticipated subsurface conditions. In such a case, the stipulated-sum contract is really a stipulated-sum contract with the potential variation in cost based on subsurface conditions. The unexpected construction work required by differing site conditions may itself be priced in a variety of ways. The parties may specify a unit price for particular conditions such as excavation or blasting of rock; they may price the additional work on a cost-plus basis; or they may negotiate a flat fee change order.

The careful drafter and contract administrator must ensure that the contract terms are flexible enough to accommodate these various pricing mechanisms. On small change orders, there may be too little at issue to justify the expense or administrative burden of adding sophisticated terms to the contract. On the other hand, if a contractor building an office building for a stipulated sum agrees to build an adjacent parking garage, but on a cost-plus basis, both the contractor and the owner take substantial risk by executing a simple change order adding the work for "the actual cost of the work plus 15% overhead and profit" without adding the key elements of a cost-plus contract, such as the definition of "costs" that may be charged to the owner.

9.03 SPECIFIC PRICE-RELATED TERMS UTILIZED IN CONSTRUCTION CONTRACTS

A. Pricing Mechanisms

This section identifies and discusses different pricing mechanisms that contractors, owners, and subcontractors employ in pricing work, especially under change orders. These terms are not presented for detailed analysis, as such an endeavor is beyond the scope of this book, but rather to acquaint readers with some pricing techniques that they may encounter in construction contracts.

1. *Index Pricing*

When the cost of particular materials fluctuates on the open market, the parties may agree that the cost of materials to be charged to the owner will be determined by reference to a published index. The function of such a clause is to allocate the risk of cost to the owner at then-market rates without regard to the actual cost of the materials procured by the contractor. For example, a contractor may have procured a huge quantity of steel for stockpile or through the purchase of futures and may not want to share the cost savings with the owner. After all, when the contractor made the purchase, the owner did not bear the risk that materials would become cheaper in the future. Conversely, the owner may not want to pay elevated prices for materials that the contractor purchased and stockpiled far in advance. The use of an index price to tie costs to present market conditions is one mechanism for dealing with the pricing of materials from a contractor's inventory.

2. *Time and Material Pricing*

A phrase heard with some frequency in construction circles is the pricing of a change order on a T&M basis. This refers to an agreement under which a contractor or subcontractor agrees that the cost of additional work will be based on the time and materials employed to complete the work. T&M work is essentially work that is performed on a cost-plus basis, where the cost is the cost of the labor, materials, and so on. It may or may not include an express agreement to pay some amount for the contractor's overhead and profit. To remove any ambiguity, many contracts specify that the contractor will be permitted to add a stated percentage (usually in the range of 10 to 20 percent, depending on the type of work being performed) to the T&M portion of work in a change order. T&M pricing invokes the same considerations as cost-plus contracts detailed above, but on a smaller scale that may not justify significant drafting efforts.

Time and materials pricing is often utilized as a default mechanism where the parties cannot agree to the pricing of a change order:

> B. The value of any Work covered by a Change Order or of any Claim for an adjustment in the Contract Price will be determined as follows:
>
> > 1. where the Work involved is covered by unit prices contained in the Contract Documents, by application of such unit prices . . . , or
> >
> > 2. where the Work involved is not covered by unit prices contained in the Contract Documents, by a mutually agreed lump sum (which may include an allowance for overhead and profit not necessarily in accordance with [the calculation of "cost of work"]), or

> 3. where the Work involved is not covered by unit prices contained in the Contract Documents and agreement to a lump sum is not reached . . . , on the basis of the Cost of the Work . . . plus a Contractor's fee for overhead and profit.[20]

3. *Default Pricing Through Third Party*

In the context of directed work[21] that is performed without a prior agreement as to pricing, the parties may decide in advance that some third party (the architect, for example) will determine the price of work to be performed under a change order if the parties are unable to negotiate a price. Although the use of third-party default pricing may involve some risk of favoritism, ignorance on the part of the decision maker, or other factors, in many cases the parties are willing to accept such risks as an alternative to the significant transaction costs associated with other pricing mechanisms, such as the dispute resolution techniques described next.

4. *Default Pricing Through Dispute Resolution*

Another method for default pricing—that is, pricing used where the parties cannot agree in advance—is to provide for some interim payment representing a rough value of the work (as determined, for example, by the architect, but only on an interim basis), followed by the parties undertaking some form of dispute resolution to fix the "reasonable" price of the work. The advantage of such a term is that it ensures that a neutral third party will determine the "reasonable" price of the work, although not always to the parties' satisfaction, of course. The chief drawback is that imposition of some neutral determination through the contract's dispute resolution provisions may come with substantial transaction costs. For example, a disagreement over $100,000 of value in a change order that adds work to a contract may actually cost $50,000 in attorneys' fees and arbitration expenses to resolve. Such clauses are often effective for a somewhat perverse reason—dispute resolution may not be a particularly effective way to resolve the dispute, but the transaction cost of the dispute resolution mechanism provides the parties with a strong incentive to settle the price.

---

20. EJCDC C-700, *Standard General Conditions of the Construction Contract* (2002), ¶ 12.01. *See also* AGC 200, *Standard Form of Agreement and General Conditions Between the Owner and Contractor (Lump Sum)* (2002), ¶ 8.3.1.4 (default is cost of work); American Institute of Architects, AIA Document A201–1997 *General Conditions of the Contract for Construction* § 7.3.6 (1997) (in absence of response from or agreement with contractor to proposed price, architect determines price based on "reasonable expenditures and savings" of contractor, including allowance for overhead and profit).

21. See Chapter 16.

9.04 CONTRACT TERMS RELATED TO PRICING

A. Material Price Escalation Clauses

As explained previously, in a stipulated-sum contract, the contractor bears the burden of price fluctuations in materials and labor. At the same time, the parties may recognize that the only risks the contractor anticipates bearing are those that are "normal" within the industry, even when these "normal" risks may have dramatic pricing effects. Thus, at any point in time, some building materials may be unusually expensive, and others may be unusually cheap in light of the many factors that affect pricing.

On the other hand, some unanticipated events may be completely outside the reasonable expectation of the parties, or price increases may far exceed what the parties intended to allocate as part of the risk and benefit that each accepted in the drafting of the agreement. In order to avoid a catastrophic pricing failure, the contractor may seek inclusion of a "material price escalation" clause to deal with such unusual events as worldwide shortages of particular construction materials due to natural disasters (such as Hurricane Katrina) or other causes. The necessity of such clauses lies in the fact that the common law of contracts provides little relief from unforeseen—however unfortunate—changes in the cost of an obligor's performance.[22]

In addition, known fluctuations in key construction components may counsel use of variable pricing (such as fuel cost increases) through material price escalation clauses.

B. Interim Payment Clauses

Many contracts allow the owner to "direct" that the contractor perform additional work even where there is a dispute as to the change order pricing. Such a clause allows the owner to avoid being "blackmailed" by the contractor on the price by the contractor's refusal to perform work until an outrageous price is accepted. On the other hand, the clause raises the converse problem: What if the owner refuses to agree to a reasonable price with the hope that the contractor will eventually be forced to accept a low price simply to get paid?

The solution to this potentially unfair pressure is the inclusion of an interim payment clause, under the terms of which the owner is obligated to pay

22. *Force majeure* clauses may also provide some relief to contractors in the face of disasters. Such clauses are outside the scope of this chapter.

some determinable amount for the work pending final contract pricing (and adjustment, if necessary). The interim pricing is often left in the hands of a third person, such as the architect. Through this means, the contractor is paid something for its work until the final contract price is negotiated or otherwise determined.

In the absence of an interim payment clause, the contractor (or subcontractor, as the case may be) can face substantial financial distress, as contract clauses requiring the contractor to continue work even in the face of a pricing dispute are generally enforceable, and a contractor who suspends work pending resolution of a pricing issue risks being terminated.[23]

9.05 COMPARISON OF PRICING TERMS BASED ON PRICING MECHANISM

This section discusses several attributes and clauses of construction contracts that are affected by—and affect—the pricing mechanism that is utilized. The function of this section is to give the reader some sense of the important implications of—and assumptions behind—different pricing mechanisms. Each area begins with a brief description of the characteristic of the clause at issue, and summarizes the manner in which the particular clause affects (or is affected by) choice in pricing structure. The first topic—scope of work—is set out in some detail to show the complexity of the concepts presented. Where there is only a description of the clause, it is left to the reader as an exercise to determine the importance and role of the clause depending on the pricing mechanism.

A. Scope of Work

A clause setting out the definition of the work to be performed under a contract usually incorporates detailed drawings and specifications. These drawings and specifications are not simply reference materials—they define the contractor's obligation to perform work under the contract. The scope of the work cannot change except in accordance with the change provisions in the contract. These change provisions, in turn, may include significant pricing, timing, and other terms for changes in the scope of work. At the same time, parties do not always have the luxury of a complete set of drawings and specifications when they want to begin construction.

---

23. Interim pricing may also be a requirement of a state's "Prompt Pay Act" under which owners and contractors are required to make timely payments, even in the face of disputes about pricing.

How might incomplete drawings and specifications affect pricing considerations? Precision in the scope of work under a cost-plus contract is not usually vital to the contractor.[24] After all, the owner will eventually pay for all the work that is performed, and there is thus little economic motivation to engage in heated disputes over what is within and what is outside the original scope of work. Accordingly, a cost-plus contract may be an appropriate mechanism for pricing work whose design is not complete at the time of contracting.

A stipulated sum, however, is rarely an appropriate pricing mechanism for a project whose design is not complete at the time of contracting. Because the drawings and specifications define the contractor's obligations, a contractor who quotes (and an owner who accepts) a stipulated sum on the basis of incomplete drawings undertakes a substantial risk that disputes will arise concerning whether particular tasks are inside or outside the scope of work, and therefore included (or not included) in the stipulated sum.

At the extremes—a perfect set of drawings or a mere design sketch with no details—the contractor can feel comfortable in accepting or rejecting a stipulated sum contract, respectively. The cases that end up in dispute are the borderline cases where a design is "nearly" complete at the time of contracting.

B. Changes

Contracts usually require change orders to alter the scope of work or contract time. Some contracts have magnificently detailed provisions with Byzantine procedures for effecting changes in the contract. Others have less formal change mechanisms with fewer default remedies, such as default pricing. In negotiating change-order procedures, the level of detail required in change mechanisms, and especially in the management of the change-order process during the course of construction, must take the pricing system into account.

C. Sharing of Savings

A sharing-of-savings clause is intended to give the contractor an incentive to perform efficiently by awarding the contractor a specified financial reward to the extent that the cost of construction comes in below a target amount. These cost-savings provisions are typically expressed in terms of an allocation of the cost savings by percentage to the contractor and to the owner. In what kind(s) of contract would you expect to find such a clause?

---

24. That is not to say it is unimportant to the owner, as the owner may want a bulletproof set of drawings and specifications to ensure that any cost estimates are reliable.

D. Definition of "Cost of Work"

As described previously, a clause defining the cost of work to be performed identifies specifically what costs are to be included in any cost of work calculation, as well as what costs are specifically excluded.

E. Treatment of Self-Performed Work

The contractor's right to perform work through its own forces and the compensation the contractor will receive for such work can be the subject of specific contract clauses. The discussion of cost of work set out earlier details some of the concerns that arise when a contractor performs work on a project.

F. Allowances

An allowance is a variable pricing term under which a contract price sets out an "allowance," or price of some particular element that is included in the contract sum, with the added provision that any variation from the stated allowance will result in a change in the contract price. Allowances are typically utilized for owner-finish items, the cost of which cannot be incorporated into the contract until the owner makes decisions as to the exact products to be utilized. At the same time, the owner may want some estimate of the cost of the materials or work, and the contractor thus includes a stated figure in its bid. For example, a contractor building a warehouse with an office may state an "allowance" of $8,000 for floor finishes in office spaces. If the owner chooses expensive carpet, the owner must pay the increased price. If the owner chooses unfinished floors or minimal floor coverings, it saves money.

Allowances can also be utilized as a short-form method of allowing variable pricing for materials whose cost cannot be predicted at the time of contracting. A contractor may, for example, put an "allowance" in for the log package utilized in a residential construction contract for purposes of finalizing a contract with the owner, but the parties may agree that the owner will bear the exact cost of the log package at the time that it is purchased.

There is nothing particularly challenging about the use of allowances in a construction contract, except that it is vital to understand that they fulfill very different functions from contingencies.

G. Contingencies

Contractors with substantial experience understand that when they bid jobs, they can predict with some degree of accuracy the cost of materials, labor, and so forth that go into finishing a project. They also know that most projects will experience some unforeseen occurrences that are not specifically planned in the budget. A contractor who projects job costs with no consideration for unplanned expenses undertakes substantial risk.

Over the course of many contracts, a contractor can estimate the amount by which cost overruns exceed the original estimate of identifiable costs, and thereby "plan" for unplanned expenses in future contracts. For example, a contractor who specializes in the construction of chemical treatment plants may routinely increase its cost estimates by 8 percent of the identifiable costs to account for such unplanned, but nevertheless likely, cost overruns.

Nothing in the use of a contingency as discussed to this point is troublesome. The problems arise, however, when the contractor shares its cost estimate with the owner for the purpose of allowing (for example) the owner to obtain construction financing and for the parties to formulate cost estimates for payment purposes. The owner may see the proposed contingency as a fund from which it can finance alterations in its drawings and specifications. The owner may say, for example, "We forgot to design air-conditioning into the equipment control room—let's take that out of the contingency." Theoretically, a contractor could provide a cost estimate showing no contingency so as to avoid this problem, at which point the owner's lender may impose some contingency with the understanding that no cost estimate is perfect.

As already discussed, contingencies play some role in the estimation of financing of a construction project, and they also play a role in a contractor's formulation of an acceptable stipulated-sum price. From the contractor's perspective, whether a contingency presents any concern depends on how it is employed. In a cost-plus contract where the contingency is just a cost-estimating tool for the owner, it presents no trouble—the owner pays all costs, expected or otherwise, and no dispute should arise over "ownership" of control of such contingency accounts.

When the contractor includes a contingency amount in a stipulated-sum contract, ownership and control over the contingency account should not be disputed—the contractor gets paid all of, and only, the stipulated sum, and whether the contractor internally called an expense part of flooring or part of its internal contingency is irrelevant. The only time trouble arises is when the contractor discloses its contingency to the owner for purposes of interim payments or bank-financing purposes, and the owner misunderstands the function of the contingency. Short of keeping the contingency to itself, the parties' protection against disputes is to make it clear that the owner has no interest in the control of the contingency and that it is disclosed to the owner only as a convenience to the owner.

Finally, a contingency may play a specific role in a cost-plus contract with a GMP. The owner and contractor may agree, for example, that the GMP includes up to $2,000,000 for excavation costs as part of the contractor's contingency, so that if excavation costs come to $3,000,000, the parties know how to treat those costs for purposes of calculating the shortfall or excess in relation to the GMP. Absent some stipulation as to the use of contingencies, for example, the contractor may claim that subsurface conditions entitle it to an increase in the contract sum of $3,000,000 for excavation, while the owner would contest that all excavation costs were included in the GMP.

H. Builder Rebates/Credits

Contractors often negotiate purchase agreements with vendors under the terms of which the contractor will receive a rebate on the value of materials it purchases, a credit to its "house account" for subsequent purchases, or even a price discount for timely payment of invoices. In negotiating the cost terms of a cost-plus contract, the parties should be mindful of these various considerations and agree as to their treatment. If a contractor "pays" for materials with a credit it received from prior purchases, for example, there is little doubt that the contractor should still be entitled to recoup the "cost" of the materials from the owner. On the other hand, if the contractor gets a 10 percent "rebate" in the form of cash or future discounts, should the owner be entitled to participate in the cost savings? These are questions that have no "right" answers, but rather are issues that should be addressed in the contract where the cost of work may be at issue.

I. Purpose of Budget

When a contractor creates a budget for work under a cost-plus contract, it usually serves as a planning tool only and rarely is intended (or given effect) as a cap on the contract sum. When a contract is priced as a stipulated-sum contract, on the other hand, the only "budget" that the contractor may wish to reveal is a schedule of values for use in the payment process. A contract should articulate the precise import of any construction budget that is exchanged between the parties so as to ensure that they share expectations regarding the use of the budget.

J. Audit Rights

In some contracts, owners are given the right to audit the contractor's books and records to assess the accuracy of the contractor's representations concerning costs. There is nothing unusual about audit rights, and they have their place in the cost-plus contract. They are rarely appropriate in stipulated-sum contracts. The actual costs incurred by the contractor are irrelevant to the contract sum, and the contractor may have no interest in allowing the owner to audit its books under such a contract. Similarly, the contractor may choose to limit any audit rights that it grants to those portions of a contract that are priced by reference to costs (as when, for example, the parties agree to T&M pricing for a change order to a stipulated-sum contract).

QUESTIONS

Two owner representatives call you the same day with the same project: draft a construction contract for the construction of a new project. They provide the following information:

Owner 1 is a private consortium of well-financed investors who want to get in on the wave of building in Las Vegas, where every new casino is bigger and more spectacular than the last. Speed to market is a major consideration, and like many fast-track projects on The Strip, construction will begin as the project is being designed. The casino will return several times the anticipated cost of the construction, and while the investors are not profligate spenders and will keep on eye on the construction costs, they are more concerned with early completion, as cost variations are not expected to be substantial when compared with the revenue that the project will bring in. The project is being built in an area that has ready access to building materials, and there is little prospect of unusual construction problems inherent in the Las Vegas building environment.

Owner 2 is a public wastewater treatment district that is formed by the property owners in a small community in the heart of the Rocky Mountains in Colorado. The community is growing, leach fields and septic systems are probably not a viable option for projected new housing in the coming decades, and the community will need more comprehensive wastewater treatment options in the future. Contractors will be hired from 200 miles away in Denver or Colorado Springs, and will need to complete the construction on the district's property in the mountains, where civil construction like this has not been attempted. The district has received a grant from the Colorado Department of Public Health and Environment, some federal matching funds may be available, and the district has some (albeit limited) capacity to issue bonds to finance the balance of construction. All of these sources of financing are one-time sources that probably cannot be increased during construction. Although the district is eager to get the project under way, there is no imminent need for the facilities, and construction will not proceed at an unusual pace.

Your first task is to advise the clients with respect to pricing considerations.

1. What would you tell them about the advantages and disadvantages of each of the pricing mechanisms discussed here?
2. What additional information would you solicit from each of them so as to formulate your recommendation?
3. At first blush, which do you think would be best suited to each owner?

CHAPTER

10

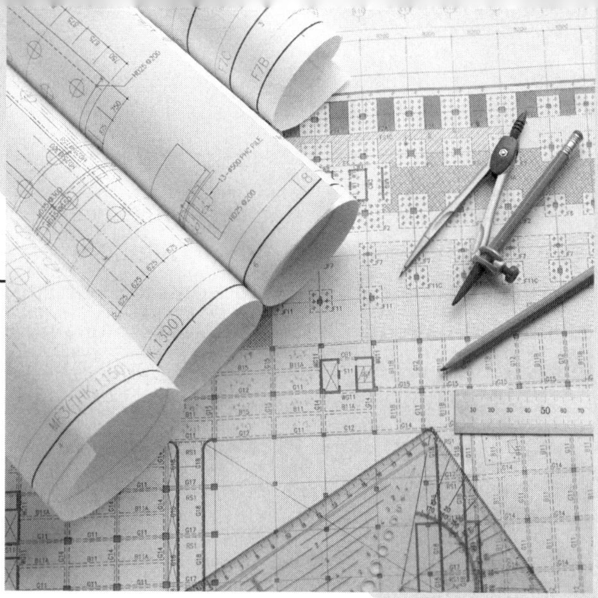

Subcontractors and Suppliers

ALLEN L. OVERCASH

10.01 SUBCONTRACTORS AND SUPPLIERS

A. Subcontractors versus Suppliers: What Is the Difference and Why Does It Matter?

It is common to think of subcontractors as those entities that furnish services such as the installation of electrical or mechanical systems to the project at the project site. It is common to think of suppliers as those entities that deliver products or goods to the project such as electrical or mechanical equipment. The distinction between the two is unclear and becoming fuzzier as traditional subcontractors increasingly furnish and even design products and goods and as more suppliers install their own products.[1]

1. The generally accepted distinction between a subcontractor who performs work at the job site and a supplier who does not is not appropriate in federal government contracting, which regards any party furnishing supplies or services for performance of a prime contract or subcontract as a subcontractor. The definition of the party has important implications for the nature of the clauses that are required to be included in its contract. Brian A. Darst, *Subcontract Incorporation by Reference and Flowdown Clauses under Federal Government Construction Projects*, CONSTRUCTION BRIEFINGS, March 2005, at 1, 8.

Why does the distinction between subcontractors and suppliers matter and what principles apply to the distinction? This section addresses those issues.

Whether the entity is a subcontractor or supplier may have serious implications for that entity's security rights. Subcontractors and suppliers are sometimes treated differently under mechanic's lien statutes and payment bonds. More importantly, the Uniform Commercial Code (UCC) commonly applies to the furnishing of goods by suppliers but not to the furnishing of services by subcontractors.

The UCC's application to a particular transaction may affect the parties' rights. These include the application of the statute of limitations,[2] the existence of special warranties,[3] the buyer's remedies,[4] buyer's notice requirements,[5] and either party's right to adequate assurance of performance.[6]

B. Subcontractors and Suppliers: Telling Them Apart

All parties are well advised to understand the nature of their transaction—whether it constitutes a subcontractor or supplier relationship. This may not be an easy determination since many transactions are mixed; that is, the same party furnishes both services and goods as part of the same agreement.

Some courts have applied a predominant factor test, analyzing "whether the predominant factor . . . is the rendition of service, with goods incidentally involved, or whether they are transactions of sale, with labor incidentally involved."[7] The fact that the product furnished by the party was fabricated

---

2. The UCC statute of limitations may be of different length than the statute applicable to furnishing services to construction projects. *Compare* Neb. U.C.C. § 2-725(1) (four years) *with* Neb. Rev. Stat. § 25-205 (five years). Further, the ability of the parties by agreement to modify the statutory period may be different. Section 2-725(1) of the Neb. U.C.C. allows the parties to reduce the period of limitation to not less than one year, but they may not extend it. Moreover, when the statute of limitations starts to run may also differ. *Compare* Neb. U.C.C. § 2-725(2) (regardless of knowledge, the cause of action accrues when the breach occurs) *with* Neb. Rev. Stat. § 25-223 (allowing extension of the statutory period if the cause of action is not discovered).

3. The UCC provides for an implied warranty of merchantability (§ 2-314) and an implied warranty of fitness for a particular purpose (§ 2-315), as well as other implied warranties that may also rise from the course of dealing (§ 2-316) and sets forth the method of creating express warranties (§ 2-313). Section 2-316 allows the parties to disclaim and exclude warranties.

4. The UCC provides buyers with certain remedies that they may exercise at various points in any transaction. *See, e.g.*, sections 2-711 (buyer's remedies in general), 2-502 (buyer's right of recovery), 2-712 ("cover"), 2-713 (contract-market damages for nondelivery), 2-714 (breach of warranty), 2-715 (incidental and consequential damages), 2-716 (buyer's right to specific performance or replevin), and 2-717 (deduction of damages from purchase price). The UCC also contains specific provisions allowing parties to limit the buyer's remedies (*see* 2-719).

5. *See* UCC §§ 2-607(3), 2-608(2).

6. *See* UCC § 2-609.

7. Mennonite Deaconess Home and Hosp., Inc. v. Gates Eng'g Co., 219 Neb. 303, 363 N.W.2d 155 (1985).

off site, moved to the project, and simply installed at the project site may be important in determining that the transaction is a supply agreement.[8]

Courts examine the entire transaction to determine its nature. An important question is whether the terms of the parties' contract more closely resemble the standard provisions of a contract for services or for the sale of goods. Therefore, the parties can shape a transaction by the provisions they insert in their contract.[9]

10.02 SELECTION OF SUBCONTRACTORS AND SUPPLIERS

A. Owner Input into Selection Process

The reality of most construction projects is that the prime contractor performs very little of the work on the project. It is generally not practical for prime contractors to maintain a staff of people who are expert in individual trades such as electrical work. It is more economical for the prime contractor to hire specialized firms that have constant experience in particular trades. The prime contractor's function is to procure and coordinate the services and supplies that are furnished to the project by others. Sometimes the owner acquires these items, but more often, the items are furnished by subcontractors and suppliers who contract directly with the prime contractor.

It is logical for the prime contractor to contract with the parties performing the work since the prime contractor is responsible for coordinating the parties' work. However, this arrangement creates some tension because the satisfactory completion of the owner's project depends largely upon the work of legal strangers over whom the owner has no direct control. Although the prime contractor is legally responsible to the owner for the work of subcontractors and suppliers, employees of the prime contractor are not expert in many of the specialty items subcontractors and suppliers furnish. In fact, the owner and/or its designer may perceive that their own expertise in the selection of certain

---

8. *See* Bonebrake v. Cox, 499 F.2d 951 (8th Cir. 1974).

9. *See* David C. Olsen & Jeffrey S. Rosenstiel, *Predicting When Construction Contracts Are Subject to Article 2 of the UCC*, 21 The Const. Lawyer 22 (Winter 2001). According to Olsen and Rosenstiel:

> [T]he contract should address issues that are customarily found in construction contracts, such as: the commencement of the work after a notice to proceed; the compliance with plans and specifications; the need to provide submittals for approval; the need to prosecute the work with adequate workers; the supervision of work by the architect or other professional; the steps that will be taken on the failure to complete the work on time; and the respective rights of the owner and contractor on default by the other.

Id. at 26.

subcontractors and suppliers exceeds that of the prime contractor. Thus, despite the traditional structure of the prime contractor contracting directly with subcontractors and suppliers, the owner still may want a voice in selecting who will perform the work.

The owner's desire to participate in the selection of subcontractors and suppliers is often reflected in the contract with the prime contractor. AIA Document A201–2007 (*General Conditions of the Contract for Construction*), for example, requires the prime contractor to furnish the names of prospective subcontractors and suppliers to the owner.[10] This starts a dialogue between the owner and the prime contractor about who will perform the work. Generally, this standard AIA form gives the owner veto power over the selection process; nevertheless, it recognizes the impracticality of allowing the owner to force an unwelcome subcontractor or supplier on the owner's prime contractor.[11]

Cost is one of the prime contractor's primary motives in selecting a subcontractor or supplier. This is particularly true if the prime contractor's contract with the owner involves a fixed price or price incentives. The owner's preferences, which are usually based on concepts of quality, may result in a higher-priced subcontractor or supplier. In such a case, the contractor may have grounds for adjusting the contract price.[12]

B. Formal Restrictions on Choosing Subcontractors and Suppliers

Often, rather than waiting to review the names of subcontractors and suppliers selected by the prime contractor, the owner desires to set the parameters of the selection process. The owner often does this as part of the invitation for bid process so that the bidders are aware of such parameters as they prepare their estimates.

On one extreme, owners actually have named a subcontractor or supplier themselves, and even negotiated the terms of the agreement with such subcontractor or supplier, with the expectation that the prime contractor is obligated to accept an assignment of that agreement. Another popular specification that a prime contractor must satisfy prescribes a "brand name or equal" product as the only product acceptable for use on the project. This approach has caused a

10. American Institute of Architects, AIA Document A201–2007, *General Conditions of the Contract for Construction* § 5.2.1 (2007) [hereinafter AIA Document A201–2007].

11. *Id.* § 5.2.2 ("The Contractor shall not contract with a proposed person or entity to whom the Owner or Architect has made reasonable and timely objection. The Contractor shall not be required to contract with anyone to whom the Contractor has made reasonable objection").

12. *Id.* § 5.2.3.

good deal of litigation in the public sector over whether particular products are to be rejected as not "equal" to those specified.[13]

In the case of public bidding, restrictions limiting the number of subcontractors and suppliers a prime contractor chooses may apply. For example, using the restrictions in the area of federal procurement, which is governed by detailed provisions of the Federal Acquisition Regulations (FARs), limits the ability of contracting officers to contract without full and open competition.[14]

C. Socioeconomic Considerations

In the instance of projects constructed for the federal government, a subcontractor or supplier may be subject to the provisions of the Buy American Act that apply to construction materials and products incorporated into the public project.[15] Many federal procurement contracts include preferences for small business concerns. Prime contractors may be required to give such small businesses the "maximum practical opportunity" to participate in the contract.[16]

These programs include preferences for subcontractors that are "small businesses," minority-owned businesses (MBEs), or disadvantaged business enterprises (DBEs).[17] Federal programs administered by the Small Business Administration include: so-called 8(a) programs designed to assist small businesses owned and controlled by socially and economically disadvantaged individuals, economically disadvantaged Indian tribes (including Alaska Native Corporations) and economically disadvantaged Native Hawaiian Organizations; small disadvantaged business programs that involve price preferences to encourage the use of so-called Small Disadvantaged Businesses (SDBs), and the HUBZone Program designed to assist small businesses located in "historically underutilized business zones." Goals also exist to award a percentage of subcontracts to women-owned businesses. Generally, companies seeking these preferences must qualify for the requisite status by application to the Small Business Administration.[18]

---

13. 1 Philip L. Bruner & Patrick J. O'Connor, Jr., Bruner & O'Connor on Construction Law § 2:46 (West Group 2002) [hereinafter Bruner & O'Connor].

14. Steven G. M. Stein, Construction Law § 2.02[4] (Matthew Bender 2007) [hereinafter Stein].

15. *See generally* John R. Tolle & Jerome H. Gress, *Buy American Act in Construction Contracting: Basic Principles and Guidelines/Edition II*, Construction Briefings, March 2003, at 1.

16. John R. Tolle, *Small Business Contracting/Edition II—Part I*, Construction Briefings, April 1999, at 1, 3.

17. 1 Bruner & O'Connor, *supra* note 13, §§ 2:54 and 2:55.

18. John R. Tolle, *Small Business Contracting/Edition II—Part II*, Construction Briefings, May 1999, at 1, 5.

Since the Supreme Court's decision in *Adarand Constructors, Inc. v. Pena,*[19] racial or ethnic classification preference programs will undergo "strict scrutiny" to determine whether such programs serve a compelling governmental interest and are narrowly tailored to further that interest.[20] Since *Adarand*, the absence of "demonstrated need" and "compelling interest" has resulted in rulings that such preferences violated the equal protection clause in the U.S. Constitution.[21]

10.03 SUBCONTRACTOR BIDDING: WHO IS BOUND BY THE BID?

A. The Bidding Process

Since subcontractors perform most of the actual construction work on many projects, the price of their work is a significant factor in the price of the total project. A large part of the cost estimate prepared by the prime contractor for the owner will be the estimate of the amounts the prime contractor will have to pay subcontractors for the work.

Theoretically, the prime contractor could, and in a few cases does, prepare its own estimate of the amount it will pay for subcontract work. But as to major items of work, the prime contractor looks to the subcontractors to tender a bid to the prime as part of the prime contractor's preparation of its overall estimate. This enables the prime contractor to make a more accurate estimate of what the subcontract work will cost. It also enables the prime contractor to take advantage of more favorable prices resulting from competition between subcontractors.

The final hours prior to the preparation of a prime contractor's estimate for a project may seem chaotic to the outside observer. Although the subcontractors and the prime contractor have studied the project for some time, final "numbers" are posted minutes before the deadline. Much of the process is verbal, with the prime contractor's final numbers varying significantly as additional subcontractor bids arrive and the manager of the prime contractor maneuvers to be the low bidder.

Added to the uncertainty introduced by changes in the subcontractor pricing is the technical issue of the scope and conditions of the subcontract work.

---

19. 115 S. Ct. 2097 (1995).

20. Richard O. Duvall, Dorn C. McGrath III, & Craig A. Holman, *Affirmative Action After Adarand*, Construction Briefings, Sept. 1995, at 1.

21. 1 Bruner & O'Connor, *supra* note 13, § 2:55.

On a complex project with many subcontractors, the specific limits of the scope of each bidder's work may vary among subcontractors vying for the same work or related work. Thus, the exact price of any particular subcontract work may be uncertain unless the record of the parties' negotiations is very precise.

B. Can the Prime Contractor Enforce the Subcontractor's Bid?

The prime contractor relies on the subcontractor's estimate for the cost of the subcontract work in framing the prime contractor's own bid. Once awarded the prime contract, the prime contractor then seeks to formalize a written agreement with the subcontractor reflecting the cost estimate that the subcontractor earlier used to prepare its own bid. But what if the subcontractor fails or refuses to honor its earlier bid by entering into a binding agreement?

The parties, as part of the bidding process, might have entered into a binding agreement that provided the subcontractor would be bound to perform its work at the bid price if the prime contractor won the project. Such an agreement would be binding on the subcontractor and could be enforced by the prime contractor.[22]

However, more often than not, no such prior agreement exists. Prime contractors receive bids from a number of competing subcontractors, and subcontractors tender bids to a number of prime contractors competing for the same project. The nature of the project makes it unlikely that the parties formally will bind themselves to each other before the contract is let by the owner.

Absent a binding bilateral agreement, what are the legal rights of the prime contractor who relied on the subcontractor's bid to prepare the prime contractor's bid? If the subcontractor walks away from its offer, does the prime contractor have any legal recourse against the subcontractor? In many jurisdictions, the prime contractor may have rights based on the legal theory of promissory estoppel. While no formal, binding, bilateral agreement exists between the parties, the actions of the prime contractor in reliance on the subcontractor's offer create a binding agreement.

The doctrine of promissory estoppel is best formalized in section 90 of the RESTATEMENT (SECOND) OF CONTRACTS, which states:[23]

> A promise on which the promissory should reasonably expect to induce action or forbearance on the part of the promisee or a third person and which does

22. *See* Bridgeport Restoration Co., Inc. v. A. Petrucci Constr. Co., 211 Conn. 230, 557 A.2d 1263 (1989) (finding an enforceable agreement to enter into subcontract if the project was awarded to the prime).

23. RESTATEMENT (SECOND) OF CONTRACTS § 90.

induce such action or forbearance is binding if injustice can be avoided only by the enforcement of the promise. The remedy granted for breach may be limited as justice requires.

On the basis of this doctrine, many courts have upheld the prime contractor's right to enforce the subcontractor's bid against it.[24] Other courts have disagreed.[25]

Assuming that the court entertains the concept of promissory estoppel as governing subcontractor bidding, what would prevent the court from enforcing the subcontractor's bid? Several facts might interfere with such a result, including the following:

1. The bid was subject to conditions or reservations, including the right to withdraw the bid.[26]
2. The bid was not a clear and definite promise but contained major ambiguities.[27]
3. The prime contractor's reliance on the bid was not reasonable because of an obvious mistake in the bid.[28]
4. The prime contractor unreasonably delayed accepting the bid.[29]
5. The prime contractor's acceptance varied the terms of the offer in such a way as to constitute a counteroffer.[30]

---

24. *See, e.g.*, Drennan v. Star Paving Co., 333 P.2d 757 (Cal. 1958).

25. *See, e.g.*, Virginia Sch. of Arts, Inc. v. Eichelbaum, 493 S.E.2d 510 (Va. 1997).

26. Neal J. Sweeney & Geoffrey Dendy, *Holding Subcontractors To Their Bids/Edition II*, CONSTRUCTION BRIEFINGS, Sept. 1999, at 1, 4; Preload Tech. v. A.B & J. Constr. Co., 696 F.2d 1080 (5th Cir. 1983); Montgomery Indus. Int'l. Inc. v. Thomas Constr. Co., 620 F.2d 91 (5th Cir. 1980), 4 CC ¶ 213; Wargo Builders, Inc. v. Douglas L. Cox Plumbing & Heating, Inc., 268 N.E.2d 597 (Ohio App. 1971); *Drennan*, 333 P.2d at 757.

27. Sweeney & Dendy, *supra* note 26, at 4; Camosy, Inc. v River Steel, Inc., 624 N.E.2d 894 (Ill. App. 1993).

28. 2 BRUNER & O'CONNOR, *supra* note 13, § 7:42. In the *Drennan* case, the subcontractor refused to honor its bid because the bid contained an error, but the court held that since the prime contractor had no reason to believe the bid was in error, it could justifiably rely upon it and "as between the subcontractor who made the bid and the general contractor who reasonably relied on it, the loss resulting from the mistake should fall on the party who caused it." *Drennan*, 333 P.2d. In the case of Sipco Serv. Marine, Inc. v. Wyatt Field Serv. Co., 857 S.W.2d 602 (Tex. App. 1993), the court upheld the reliance of the contractor as reasonable even though the subcontractor's bid was not the low bid.

29. Sweeney & Dendy, *supra* note 26, at 12; Pavel v. A. S. Johnson, 674 A.2d 521 (Md. App. 1996); *Drennan*, 333 P.2d at 757; 2 BRUNER & O'CONNOR, *supra* note 13, § 7:45.

30. Sweeney & Dendy, *supra* note 26, at 9; F.B. Reynolds v. Texarkana Constr. Co., 374 S.W.2d 818 (Ark. 1964); 2 BRUNER & O'CONNOR, *supra* note 13, § 7:43.

6. Rather than accept the bid immediately, the prime contractor engaged in bid shopping or bid chiseling.[31]

C. Can the Subcontractor Enforce Its Own Bid?

As indicated previously, many authorities are loath to side with the prime contractor because such authorities believe it would be one-sided and unfair to allow the prime contractor to enforce an agreement that the subcontractor itself cannot enforce. Absent a binding, pre-bid, bilateral agreement between the parties, what rights does the subcontractor have to force the prime contractor to award it the subcontract on the basis of its original offer?

A subcontractor may view it as unfair that a prime contractor, who used the subcontractor's bid amount in the prime contractor's own calculations of the cost of the project in order to be awarded the project, may later refuse to enter into an agreement with the subcontractor. Even though the subcontractor itself may be obligated to perform at the prime contractor's election, the subcontractor may not itself be able to enforce its own offer. Yet, this may be the likely result in a court proceeding.[32]

It is not unusual for the prime contractor, having been awarded the construction contract for the project, to negotiate further with prospective subcontractors and suppliers in order to get more favorable terms. Reducing the prices offered by a subcontractor automatically increases the prime contractor's prospective margin on the project, particularly if the prime contractor's price is fixed. Therefore, a strong motivation exists to "shop" the bid of the subcontractor by opening the subcontract work to competitors.

---

31. See discussion of bid shopping below. Some courts use this as a rationale for rejecting the prime contractor's effort to enforce the subcontractor's bid. *See Preload Technology*, 696 F.2d at 1080. The danger of bid shopping is one of the reasons the North Carolina courts will not enforce the subcontractor's bid under a theory of promissory estoppel. As stated in Home Elec. Co. of Lenoir, Inc. v. Hall, etc., 358 S.E.2d 539, 542 (N.C. 1987):

> Allowing a cause of action based on promissory estoppel in construction bidding also creates the potential for injustice. It forces the subcontractor to be bound if the general contractor uses his bid, even though the general contractor is not obligated to award the job to that subcontractor. The general contractor is still free to shop around between the time he receives the subcontractor's bid and the time he needs the goods or services, to see if he can obtain them at a lower price.

In *Sipco Serv. Marine, Inc.*, 857 S.W.2d at 602, the court held that the contractor's resolicitation of bids from all bidders did not constitute bid chiseling or bid shopping as those terms are used in the *Preload* case.

32. *See* Clark Trucking of Hope Mills, Inc. v. Lee, 109 N.C. 71, 426 S.E.2d 288 (N.C. App. 1993) (denying the subcontractor's claim based on a theory of unjust enrichment).

The practice of bid shopping has been a subject of some concern in the construction industry.[33] Many specialty subcontractors and subcontractor organizations view the practice as unfair. In addition, owners are concerned that the practice of reducing subcontract amounts by bid-shopping negotiations may reduce the quality of work on their projects.

In response to concerns over the effects of bid shopping, a number of states have enacted laws aimed at curbing the practice. Generally, these laws provide that prime contractors bidding on public projects must include in their bids a list of the principal subcontractors that will be used on the project. The purpose of these statutes is to control the quality of work and prevent subcontractor bids from being unfairly shopped.[34]

10.04 FLOW-DOWN OBLIGATIONS AND RIGHTS

A. Introduction

While the prime contractor is legally responsible for the project, its subcontractors are largely responsible for carrying out the prime contractor's responsibility. How do you then ensure that all parties will construct the project according to the owner's design? To facilitate this objective, it is normally in the interest of the prime contractor to pass on to each subcontractor the prime contractor's obligations under its contract with the owner.

The prime contractor might assign part of its contract obligations and rights to a subcontractor in order to assure the owner that the subcontractor will perform as the owner requires. But that would remove the prime contractor from its position of overall responsibility for the project. The objective is to preserve the prime contractor's management responsibility and authority but also to ensure that the project is built as planned.

The answer generally used in the industry is to include the provisions of the contract between the owner and the prime contractor in the contract between the prime contractor and the subcontractor. This is done in one of two ways. One way is to incorporate the provisions of the prime contract into the

33. Thomas P. Lambert, Comment, *Bid Shopping and Peddling in the Subcontract Construction Industry*, 18 UCLA L. Rev. 389, 398 (1970); John B. Gaides, Note, *The "Firm Offer" Problem in Construction Bids and the Need for Promissory Estoppel*, 10 Wm. & Mary L. Rev. 212 (1968).

34. Allen Holt Gwyn, *A Review of Subcontractor-Listing Statutes*, 17 The Construction Lawyer 35 (Jan. 1997). *But see* Klose v. Sequoia Union High Sch. Dist., 258 P.2d 515 (Cal. App. 1953) (holding that subcontractor's inclusion on contractor's statutory list does not create any contractual right on the part of the subcontractor).

subcontract. Another way is to include a provision in the subcontract whereby the parties agree to "flow down" to the subcontractor the prime's responsibilities under the prime contract.

B. The Nature of Flow-Down and Flow-Up Provisions

Incorporation by reference provisions is found in the major forms of subcontracts used in the industry. Article 1.1 of AIA Document 401 (*Standard Form of Agreement Between Contractor and Subcontractor*) provides that the Subcontract Documents shall include "the Prime Contract, consisting of the Agreement between the Owner and Contractor and the other Contract Documents enumerated therein."[35]

Not satisfied with the simple incorporation of the prime contract provisions into the subcontract, the popular forms also include provisions causing obligations of the prime contractor to flow down to the subcontractor. For example, Article 2 of the AIA subcontract form provides that:

> The Contractor and Subcontractor shall be mutually bound by the terms of this Agreement and, to the extent that the provisions of AIA Document A201–2007 apply to this Agreement pursuant to Section 1.2 and provisions of the Prime Contract apply to the Work of the Subcontractor, the Contractor shall assume toward the Subcontractor all obligations and responsibilities that the Owner, under such documents, assumes toward the Contractor, and the Subcontractor shall assume toward the contractor all obligations and responsibilities which the Contractor, under such documents, assumes toward the Owner and the Architect. The Contractor shall have the benefit of all rights, remedies and redress against the Subcontractor that the Owner, under such documents, has against the Contractor, and the Subcontractor shall have the benefit of all rights, remedies and redress against the Contractor that the Contractor, under such documents, has against the Owner, insofar as applicable to this Subcontract. Where a provision of such documents is inconsistent with a provision of this Agreement, this Agreement shall govern.

Note that the provision confers rights as well as imposing obligations—that is, the subcontractor inherits the rights of the prime contractor as well as the prime's obligations under its contract with the owner. This is sometimes termed a flow-up provision as well as a flow-down provision. Its purpose is to place the prime contractor in a legally neutral position between the

35. *See also* American Institute of Architects, AIA Document A401-2007, *Standard Form of Agreement Between Contractor and Subcontractor* § 16.1.2 (2007). This principle is also reflected in Article 2.3 of AGC Form 655 (*Standard Form of Agreement between Contractor and Subcontractor*). A401-2007 § 16.1.3 adds modifications to the prime contract issued prior to the subcontract.

subcontractor and the owner.[36] If the owner claims that work performed by the subcontractor does not conform to the owner's requirements, then the prime contractor can turn to the subcontractor for response to that contention.[37] On the other hand, if the subcontractor claims the owner's design information to be faulty, the subcontractor may have a right to assert that claim against the owner through the prime contractor.[38]

Flow-down provisions are commonly required of the prime contractor by owners. Article 5.3 of AIA Document A201 requires the contractor to include flow-down and flow-up provisions in its subcontracts. This evidences a policy of assuring appropriate performance on the part of all parties working on the project. Prime contractors also commonly follow this policy with their subcontractors. Section 2.2 of AIA Document A401 provides that the prime contractor may require the subcontractor to enter into similar provisions regarding assumption of rights and responsibilities with sub-subcontractors.

C. The Rights and Liabilities of Parties under Flow-Down Provisions

By their nature, flow-down provisions are general provisions applying to unspecified rights and liabilities. Therefore, when specific issues arise, the applicability of these general provisions to those issues may be in dispute. A court, for example, may enforce the flow down of a technical requirement but not enforce a general provision dealing with warranties. If the general flow-down provision is regarded as ambiguous, external evidence such as conversations between the parties may be considered to resolve the issue.

The scope of the AIA flow-down provision cited earlier is limited to the extent that the prime contract applies to the work of the subcontractor. This language follows the reasonable conclusions of some judicial authorities that, despite the broad working of a flow-down clause, the flow-down clause alone may not incorporate certain provisions of the underlying prime contract.[39] For

---

36. Note that this is not an assignment. For example, the contractor does not exit its obligations and rights vis-à-vis it and the owner. As a practical matter, the owner and the subcontractor continue to deal with each other through the contractor. (Under Section 3.2, the contractor is obligated to make certain information received from the owner available to the subcontractor.) However, Section 7.4.1 of AIA Document 401 provides that in the event of termination of the prime contract by the owner for cause, the subcontract is assigned to the owner at the owner's option.

37. Absent a flow-down provision, the prime contractor might be liable to the owner for acts of the subcontractor without adequate recourse against the subcontractor and might not be able to demand the same performance from the subcontractor as the owner demands from them. Darst, *supra* note 1, at 2, 4.

38. See following section on subcontractor pass-through claims.

39. *See* Jeff Jury, *Everything Flows Downhill—Or Does It? Flow-Down Clauses in Construction Contracts*, CONSTRUCTION BRIEFINGS, June 2007, at 1.

example, flow-down provisions have been held not to limit the security rights of the subcontractor unless the limitation is specifically set forth in the subcontract, nor to incorporate specific insurance obligations or dispute remedies.[40]

The best example of the limitations of these clauses is found in federal government contracting, the format of which is largely dictated by the extensive provisions of the Federal Acquisition Regulations. A general flow-down provision may not be enough to incorporate the special requirements of the FARs into the subcontract.[41] It is risky in government contracting to rely only upon a general flow-down clause that may not be construed to apply to all of the provisions of the prime contract and may not be sufficient to incorporate the remedial provisions of the prime contract.[42]

D. Conditional Assignment of Subcontract to the Owner

The importance of subcontractor performance to the owner becomes apparent when the prime contractor defaults on its contract with the owner. In this circumstance, it becomes critical to the owner to preserve the subcontractor's obligation to continue performance of its share of the project.

To protect the owner's rights, it is common to provide for a contingent assignment of the subcontract to the owner in the event of the prime contractor's default at the owner's election.[43] This assures the owner of some continuity in the project's performance even if the prime contractor is not able to continue.

10.05 COORDINATION OF SUBCONTRACT WORK

A. The Duty of Coordination

The problem of coordination of the work of various construction trades typically arises on a complex project where the owner employs multiple prime contractors to perform the project. Each prime contractor typically desires its work to flow smoothly in an uninterrupted sequence and schedule. This allows the prime contractor to take advantage of repetitious work performed on a "mass production" basis with attendant efficiency and savings in time and money. When a number of separate prime contractors are working in close proximity to each other with similarly tight schedules, these desires often conflict.

40. *Id.*

41. Currently there are more than 50 FAR provisions and more than 35 Department of Defense FAR Supplement flow-down provisions that are considered "mandatory flow-down clauses." Darst, *supra* note 1, at 12.

42. Darst, *supra* note 1, at 11.

43. *Supra* note 36.

At any particular time, in any particular location, who is to have precedent over the others? This conflict is continually addressed in hundreds and even thousands of individual decisions on priority, a process known as coordination. Depending upon how the decision process is handled, some prime contractors may suffer delays and inefficiencies and may seek compensation for their unanticipated expenses.

Even in the absence of definitive contract provisions, courts have held that the owner has a duty to coordinate the work of the various prime contractors that the owner hires directly. Where an owner awards multiple prime contracts, it is responsible for the coordination of the work of all the contractors and, unless otherwise provided in the contract, is liable for delays that one prime contractor causes another.[44]

Aside from the common law, this duty of coordination commonly is imposed by form construction contracts. For example, Article 6.1.3 of the AIA A201 contract form provides that: "The Owner shall provide for coordination of the activities of the Owner's own forces and of each separate contractor with the Work of the Contractor, who shall cooperate with them." This provision incorporates the owner's duty of coordination into its contract with the prime contractor.

B. Duty to Coordinate Subcontract Work

Given the strong authority supporting the owner's duty to coordinate the work of multiple prime contractors, it is no surprise that legal authority supports the prime contractor's obligation to coordinate the work of its subcontractors. Courts have found this obligation, also, to be implied by contract.

An example of court enforcement of this obligation is found in the case of *Bat Masonry Co., Inc. v. Pike-Paschen Joint Venture III*.[45] In this case, a masonry subcontractor complained about the manner in which the prime contractor made its work available. The court found that the prime contractor delayed the completion of the slab on grade work that was to precede the masonry, thus shifting the masonry to the winter months. Also, the prime contractor made the work available in small "piecemeal" segments, causing the subcontractor to demobilize. Further, the contractor resequenced the work in a manner that obstructed and interfered with the subcontractor's work.

The court in *Bat Masonry* found that these actions violated the coordination duty of the prime contractor to the subcontractor, which the court described

44. 5 Bruner & O'Connor § 15:55 at 155 and cases cited.

45. 842 F. Supp 174 (D. Maryland 1993).

as follows: "the contractor's duty is essentially that of reasonable cooperation and support in assisting the subcontractor in the completion of its portion of the work. This duty includes the duty to properly schedule and coordinate the work of subcontractors." Further, the court opined,

> [T]he prime contractor implicitly promises to provide such working conditions as may be necessary to allow its subcontractor to carry out its obligations under contract. . . . If a contractor interferes with the work of its subcontractor, it has breached its obligation and the subcontractor is entitled to recover for the resulting delays. The contractor's implied duty includes the promise not to hinder, delay or interfere with the work of the subcontractor.[46]

Prime contractors generally have been concerned over the implications of this coordination obligation since they are responsible for the entire project, and they feel they must have the right to direct the schedule and sequence of the work of all subcontractors in order to further complete the project on time. Therefore, even the standard form contracts that govern subcontractor work are relatively ambiguous about the nature of the prime contractor's coordination obligation and give the prime contractor some flexibility over scheduling subcontractor work.

The prime contractor's obligation is referenced in Article 3.1.1 of AIA Document A401 (*Standard Form of Agreement between Contractor and Subcontractor*), which provides: "The Contractor shall cooperate with the Subcontractor in scheduling and performing the Contractor's Work to avoid conflicts or interference in the Subcontractor's Work." However, Article 4.1.2 also provides that the subcontractor "shall cooperate with the Contractor in scheduling and performing the Subcontractor's Work to avoid conflict, delay in or interference with the Work of the Contractor, other subcontractors, the Owner, or separate contractors." And Article 4.1.8 provides that "the Subcontractor shall cooperate with the Contractor, other subcontractors, the Owner and separate contractors whose Work might interfere with the Subcontractor's Work." Because the clauses variously require the prime contractor to cooperate with the subcontractor and vice versa, a mutual coordination obligation is created.

C. Avoiding Coordination Obligations

The AIA provisions place some coordination burden on the subcontractor. Does this duty insulate the prime contractor from legal coordination responsibility? *Allied Fire & Safety Equipment Co. v. Dick Enterprises, Inc.*,[47] dealt

46. *Id.* at 178.
47. 886 F. Supp. 491 (E.D. Pa 1995).

with a provision that required the subcontractor to "coordinate its work with the work of other trades." This provision was held not to override the prime contractor's implied duty not to interfere with the subcontractor's work. Based on this holding, prime contractors seeking to insulate themselves from liability to subcontractors need to conceive of subcontract language stronger than simply requiring the subcontractor "to coordinate its work with the work of other trades."

The concern of prime contractors over potential liability to subcontractors for the coordination of the work of such subcontractors has fostered a number of even stronger contract provisions than those in the AIA A401 form. These include provisions giving the prime contractor absolute authority to direct the subcontractor's sequence and schedule of work during the project and obligating the subcontractor to follow those directions.

Other approaches have attempted to remove from the subcontractor any right to compensation for delays it encounters in the performance of its work. These provisions face the same hurdles that confront no-damage-for-delay clauses generally. The contract at issue in *Allied Fire* also contained an exculpatory clause under which the prime contractor was not to be held liable for the subcontractor's claims for delay. The court held that this clause was not to be given effect where the contractor "actively delayed and hindered" the performance of the contract.[48]

10.06 SUBCONTRACTOR INDEMNITY OBLIGATIONS

A. Judicial Indemnity Authority

The term "indemnity" has been defined as an assurance designed to prevent another party "from being damnified by the legal consequences of an act . . . of one of the parties or of some third person."[49] The relief from "damnification" may come from a court. Indemnity obligations may be created by common law, obligating a party who was actively negligent to indemnify a passively negligent party.[50]

It is not unusual for a party to seek a court order requiring one party to indemnify another when the first party's acts make it entirely responsible for the second party's potential liability. Some courts have applied a doctrine of

48. *Id.* at 495.

49. BLACK'S LAW DICTIONARY (5th ed. 1979).

50. Jeffrey M. Hummel & Z. Taylor Shultz, *Indemnification Principles and Restrictions on Construction Projects*, CONSTRUCTION BRIEFINGS, Aug. 2005, at 1, 2.

"implied indemnity" where a special relationship existed between the parties, and one party ultimately bears responsibility for the loss.[51]

This issue is particularly acute on construction projects, affecting subcontractors who are sharing the project site with other entities, such as other subcontractors, the prime contractor, the owner, and designers. If an accident occurs, there may be some question as to which party is responsible, and, in fact, multiple parties may share responsibility. Therefore, rather than relying upon the eventual finding of a court, the parties, through indemnity agreements, commonly seek to apportion liability before the accident happens.

B. Subcontractor Indemnity Agreements

Since subcontractors actively perform work on the project site with a number of employees, it is common for the prime contractor to request an indemnity agreement from the subcontractors. Such an agreement commonly requires the subcontractor to indemnify the prime contractor and other interested parties, such as the owner, from claims arising out of the subcontractor's work. The subcontractor's indemnity obligation consequently would indemnify the prime contractor and others against any secondary liability resulting from the negligent acts of the subcontractor or persons for whom the subcontractor is responsible.

A common version of this concept, which is widely accepted, is found in the AIA subcontract form. Article 4.6.1 of AIA Document A401 provides:

> To the fullest extent permitted by law, the Subcontractor shall indemnify and hold harmless the Owner, Contractor, Architect, Architect's consultants, and agents and employees or any of them from and against claims, damages, losses and expenses, including but not limited to attorney's fees, arising out of or resulting from performance of the Subcontractor's Work under this Subcontract, provided that any such claim, damage, loss or expense is attributable to bodily injury, sickness, disease or death, or to injury to or destruction of tangible property (other than the Work itself), but only to the extent caused by the negligent acts or omissions of the Subcontractor, the Subcontractor's Sub-subcontractors, anyone directly or indirectly employed by them or anyone for whose acts they may be liable, regardless of whether or not such claim, damage, loss or expense is caused in part by a party indemnified hereunder.

This provision captures the prime contractor's agreement to indemnify the owner.[52] Note that the AIA subcontract provision: (1) applies to losses such as

---

51. *Id.*

52. *See* AIA Document A201–2007, *supra* note 10, § 3.18.1.

bodily injury and destruction of tangible property that are commonly covered by insurance policies; (2) is limited to the extent of the subcontractor's negligence; and (3) may be held to be valid even in the face of the negligence of the indemnified party (the "indemnitee"). An additional provision provides that this remedy is not exclusive, thereby recognizing the possibility of judicially imposed indemnity.

The AIA provision is generally regarded as balanced;[53] however, the possibility of abuse in the drafting of such a clause is manifest. Since the prime contractor occupies a strong bargaining position, it might consider expanding the reach of the clause's scope to exonerate the prime contractor completely from any responsibility, even for its own negligence.

Even though a court may uphold a clause agreed to by the parties, the court may nevertheless construe the clause very strictly. If, for example, the provision would indemnify the indemnitee against its own negligence, the court may require that such intention be expressed in clear and unequivocal terms.[54] Since the courts do not favor such clauses, they must be carefully worded in order to be effective.[55] Even so, a reasonable concern exists as to whether the enforcement of an indemnity agreement designed to protect a party from the consequences of its own negligence is good public policy. This concern has prompted a number of legislatures to act on the matter, as discussed in the next section.

C. Anti-Indemnity Statutes

Because of the potential unfairness and adverse social implications of broad indemnity clauses, a number of states have enacted statutes restricting the enforceability of such clauses. Generally, these statutes have fallen into three categories:[56]

---

53. Hummel & Shultz, *supra* note 50, at 3.

54. *See* Johnson v. Bd. of County Comm'rs, 913 P.2d 119, 136 (Kan. 1996). *See also* Fretwell v. Prot. Alarm Co., 764 P.2d 149, 152–53 (Okla. 1988) (stating that it "will strictly construe an agreement which would have the result of indemnifying one against his own negligence, but where the intention to do so is unequivocally clear from an examination of the contract, such an agreement is enforceable.").

55. In fact, the Kansas Court of Appeals has recommended that drafters explicitly state the indemnitor's obligation to indemnify the indemnitee for the indemnitee's negligence. Zenda Grain & Supply Co. v. Farmland Indus., Inc., 894 P.2d 881, 888 (Kan. App. 1995).

56. Three types of anti-indemnity statutes are described in Allen Holt Gwyn and Paul E. Davis, *Fifty-State Survey of Anti-Indemnity Statutes and Related Case Law*, THE CONSTRUCTION LAWYER, Summer 2003, at 26.

1. Sole negligence laws. One type of statute restricts the enforceability of a clause that indemnifies a party against liability for a loss caused by that party's sole negligence.[57]
2. Partial negligence statutes. Certain states have enacted statutes invalidating clauses that provide for indemnification of indemnitees who are at least partially responsible through their own actions for the loss.[58]
3. Statutes barring indemnity of design professionals. Some states have enacted laws restricting the enforcement of agreements designed to indemnify design professionals from liability arising from their services.[59]

The interpretation of these anti-indemnity statutes, as well as the indemnity agreements that are potentially subject to such statutes, has given rise to substantial litigation. Since most of the statutes apply by their terms to construction projects, this may limit the nature of the relationships covered by the statutes.[60] Since the statutes differ a great deal from state to state, the decision as to which state's law to apply may be very important. This decision may be influenced by the public policy of the states involved.[61] Some statutes exclude from their scope agreements involving the procurement of insurance.[62]

10.07 PAYMENT TO SUBCONTRACTORS AND SUPPLIERS

A. The Payment Process

Generally, payments to subcontractors and suppliers are made from monies received by the prime contractor from the owner of the project. Thus, there is a third-party standing between the owner who pays for the work and the parties who performed the work and expect to receive payment. This three-party

---

57. *Id.* At the time of their writing, the authors identified 18 states that voided provisions that indemnified losses or damages arising from the indemnitee's sole negligence.

58. *Id.* At the time of their writing, the authors identified 16 states that voided provisions for losses or damages arising from the indemnitee's negligence, whether sole or concurrent.

59. *Id.* at 27. At the time of their writing, the authors identified 19 states that had passed statutes that included prescriptions in this area, some specifically voiding this type of agreement and others having general statutes that included this type of agreement.

60. Smith v. Seaboard Coast Line R.R. Co., 639 F.2d 1235 (5th Cir. 1981).

61. Hummel & Shultz, *supra* note 50, at 8.

62. MINN. STAT. ANN. § 337.05. In Seward Housing Corp. v. Conroy Bros. Co., 573 N.W.2d 364, 366 n.2, the court explicitly stated Section 337.05 allowed enforcement of a provision in a "standard subcontract agreement requiring the subcontractor to purchase liability insurance to cover any negligent acts by the [general contractor]."

relationship has created extensive contract provisions governing the handling of payments, as well as controversy regarding the rights of the parties when payments are delayed or not received at all.

The provisions for passing through payments from the owner to the subcontractor found in the AIA documents illustrate a typical system of handling such payments. Article 11 of AIA Document A401 provides for progress payments to be made by the prime contractor to the subcontractor. Progress payments of the subcontract price are based on an allocation of that price into segments of work to be performed, pursuant to a "schedule of values" (Article 11.5). The subcontractor submits pay applications each month indicating the percentage of each segment of work completed during the month (Article 11.6). The prime contractor then includes the amount of the subcontractor's application in its application to the owner (Article 11.3). Within seven working days after the prime contractor receives payment from the owner (Article 11.3), the prime contractor is to pay the subcontractor the amount of its application, less amounts retained by the owner applicable to the subcontractor's work (Article 11.7.1) and amounts withheld due to the fault of the subcontractor (Article 11.7.4).

This three-party arrangement therefore requires the subcontractor to make formal applications for payment, which are passed on to the owner for approval, who then makes payment to the prime contractor for the subcontractor's work based on the amount demonstrated by the subcontractor's application. If the subcontractor's work does not merit payment to the prime contractor, the owner will not approve the application, and the prime contractor will be justified in not making payment to the subcontractor.

If the prime contractor receives funds from the owner for the subcontractor's work, it is obligated to pass those funds on to the subcontractor. If the subcontractor is not paid, its remedies include stopping its work (Article 4.7). Courts have held that the failure of the prime contractor to pay a subcontractor may justify the subcontractor's rescission of the subcontract.[63]

B. Bearing the Risk of Owner's Failure to Pay

If the subcontractor's payment is contingent upon the prime contractor's receiving payment from the owner, there may be a serious delay in passing on the amount of retainage applicable to the subcontract work, since the bulk of the prime's retainage will not be released until substantial completion. While a subcontractor may finish its work early in the project, if its retainage pay-

63. Stein, *supra* note 14, § 5B.02[4][c].

ment must await final payment by the owner, this can delay the payment of retainage for years, something that causes a great deal of controversy in the industry.[64]

The owner may withhold payment from the prime contractor for a variety of reasons, many of which may be unrelated to the subcontractor or its performance.[65] Indeed, it is not uncommon that disputes between the prime contractor and the owner prompt the owner to withhold payment from the prime contractor even though the subcontractor's performance of work was entirely satisfactory. Even though the prime contractor does not receive money from the owner, it is arguably unfair for it to fail to pay the subcontractor if the subcontractor is not at fault.

Disputes between the prime contractor and the owner commonly affect the flow of payments on a project. In turn, this disruption of the payment process causes disputes between the prime contractor and the subcontractor over their respective rights and obligations. This situation has inspired a number of contract provisions relating to these rights and obligations as well as legal disputes over the interpretation and veracity of such provisions.

C. Risk of Payment Provisions

The risk of the owner's failure to pay is so common that the Associated General Contractors of America (AGC) has created two separate documents to deal with the problem, as follows:

- ❒ AGC Form 650 provides that if, through no fault of the subcontractor, the owner does not pay the contractor for the subcontractor's work, the contractor assumes the liability to pay the subcontractor (Article 8.2.5).
- ❒ AGC Form 655 provides that the contractor's payment to the subcontractor is conditioned upon the contractor's receipt of payment from the owner, and the subcontractor acknowledges that it is relying on the credit of the owner, not the contractor, for payment for its work (Article 8.2.5).

It is increasingly common for prime contractors to insist on including in their subcontracts a provision that the prime contractor's obligation to pay the subcontractor is contingent upon the prime contractor's receipt of payment

---

64. *See id.* § 7.04[1][d].
65. *See* AIA Document A201–2007, *supra* note 10, § 9.5.1.

from the owner. The intention of many of these provisions is to pass on to the subcontractor the risk of the owner's failure to make payment for reasons unrelated to the subcontractor's work. Nevertheless, the perception of unfairness of such a clause has created legal difficulties in its enforcement.

One legal hurdle lies in the interpretation of the clause itself. The courts have construed many clauses to apply to the timing of payment rather than a condition of payment; as a result, these clauses have not protected the prime contractor from independent liability to the subcontractor.[66] Ambiguities in such clauses are generally construed against the prime contractor seeking to enforce the clause and avoid payment.[67]

Some courts have enforced clearly written pay-if-paid clauses, which make payment by the owner an express condition of the prime's payment obligation to the subcontractor.[68] However, other courts have refused to enforce the clauses. Pay-if-paid clauses are unenforceable in the courts of California since they are inconsistent with the laws regarding subcontractor security.[69] The Iowa court has held the clause enforceable as long as nonpayment was not the fault of the prime contractor.[70]

A majority of the courts refuse to construe a pay-if-paid clause as a valid defense to nonpayment.[71] The enforceability of such clauses and the reasoning related to enforceability varies a great deal from state to state.[72]

Other states have legislated against pay-if-paid clauses, even in private contracts,[73] while other states have accepted such clauses but prevented them from interfering with subcontractor lien rights.[74]

Therefore, clauses restricting the right of subcontractors to receive payments due for their work have been faced with substantial legal barriers, and the effectiveness of such clauses is subject to substantial doubt.

---

66. *See, e.g.*, Main Elec., Ltd. v. Printz Servs. Corp., 980 P.2d 522 (Colo. 1999); Federal Ins. Co. v. I. Kruger, Inc., 829 So. 2d 732 (Ala. 2002).

67. *See* G.E.I. Recycling, Inc. v. Atl. Envtl., Inc., 821 So. 2d 431 (Fla. Dist. Ct. App. 2002); Lafayette Steel Erectors, Inc. v. Roy Anderson Corp., 71 F. Supp. 582 (S.D. Miss. 1997); Am. Drilling Serv. Co. v. City of Springfield, 614 S.W.2d 266 (Mo. Ct. App. 1981).

68. *See* Gilbane Bldg. Co. v. Brisk Waterproofing Co., 585 A.2d 248 (1991); Berkel & Co. Contractors v. Christman Co., 533 N.W.2d 838 (Mich. Ct. App. 1995).

69. *See* William R. Clarke Corp. v. Safeco Ins. Co. of Am., 938 P.2d 372 (Cal. 1997); Capitol Steel Fabricators, Inc. v. Mega Constr. Co., 68 Cal. Rptr. 2d 672 (Cal. 1998). *See also* WestFair Elec. Contractors v. Aetna Cas. & Surety Co., 661 N.E.2d 967 (N.Y. 1995).

70. Grady v. S. E. Gustafson Constr. Co., 103 N.W.2d 737 (1960).

71. Will M. Hill & Donna M. Evans, *Pay When Paid Provisions: Still a Conundrum*, 18 THE CONSTRUCTION LAWYER 16 (April 1998).

72. Robert F. Carney & Adam Cizek, *Payment Provisions in Construction Contracts and Construction Trust Fund Statutes: A Fifty-State Survey*, 24 THE CONSTRUCTION LAWYER 5 (Fall 2004).

73. *See* 6 Del. C. § 3507(e) and WISC. STAT. § 779.135 (West 2003).

74. 770 ILL. COMP. STAT. ANN. 60/21 (West 2003); KAN. STAT. ANN. § 16-1803(c) (Supp. 2005).

D. Other Methods to Secure Subcontractor Payments

To better secure payments to subcontractors and suppliers, a number of states have passed trust-fund statutes. These statutes commonly provide that monies paid to a prime contractor for work or materials furnished by subcontractors or suppliers will be held in a trust for the benefit of those subcontractors and suppliers.[75] The nature and substance of the laws vary from state to state. The laws may have a number of implications for the rights of subcontractors and others depending upon their terms. For example: (1) prime contractors who mishandle the funds may be subject to criminal or civil penalties and their officials subject to personal liability; (2) the civil remedies of subcontractors and suppliers may include expanded remedies such as attorney's fees and interest; (3) the payments held in trust may be free from the claims of the prime contractor's creditors; and (4) fiduciary responsibilities may be visited upon prime contractors which, among other things, may affect the status of the prime contractor's obligations to the funds under the bankruptcy laws.

Other measures designed to assure payment to subcontractors include a direct disbursement system (channeling payment through a title company as a disbursing agent), proposed by the American Subcontractors Association, and legislation governing federal government contracts that requires the prime contractor to pay subcontractors within a certain period of time or incur interest and possibly repay the funds to the owner.[76]

10.08 SUBCONTRACTOR CLAIMS

A. The Problem of Subcontractor Claims

In the traditional construction contracting format, an owner contracts only with a prime contractor and not with the subcontractors who will perform the work. Commonly, the prime contractor selects the subcontractors and determines the conditions of their employment. The owner regards its contract with the prime contractor as the measure of the prime contractor's responsibility and liability for performance of the project.

As a consequence of the owner's contract with the prime contractor, and the absence of a contractual relationship between the owner and the subcontractors, the owner expects the prime contractor to be responsible for the work of the subcontractors. This includes the administration of the subcontractors' work and the review and approval of any subcontractor claims. Direct dealings between the owner and the subcontractors are inconsistent with this

75. Carney & Cizek, *supra* note 72.

76. Stein, *supra* note 14, § 7.05[1][b].

arrangement and can prejudice the legal position of both the owner and the subcontractors.

On the other hand, the subcontractor is responsible for its scope of work critical to completion of the project. Only the subcontractor thoroughly understands its plan for executing its scope of work and whether and to what extent this plan is impaired by any action or omission of the owner. Particularly, with the highly specialized and technical nature of modern subcontract work, the prime contractor has much less knowledge of problems affecting the subcontract work than does the subcontractor or perhaps even the owner.

Subcontractors on construction projects can incur substantial damage from the actions of parties with whom they have no contract. For example, the subcontractor must perform changes made by the project owner or owner's representative. Delays to the prime contractor's work commonly will delay the subcontractor's work as well. Specification and design problems may adversely affect the subcontractor's scope of work. Unexpected conditions may cost the subcontractor more time and money than the prime contractor.

Like prime contractors, subcontractors base their bids on the conditions of the project represented to them at the time they price their work. When these conditions are changed, the subcontractor likely will incur extra, unanticipated costs not included in the subcontractor's bid price. Subcontractors commonly incur additional costs due to the acts or omissions of the project owner or the designer working with the owner. Naturally, the subcontractors seek compensation for such additional and unanticipated costs when they occur.

Despite the reality that the acts and omissions of owners or their agents may result in extra, unanticipated costs to the subcontractor, the only responsible party with which the subcontractor has an agreement is the prime contractor. It may be clear to the parties that the prime contractor is not the cause of the subcontractor's extra costs. If the owner or its agent is the party responsible for the change, the prime and/or the subcontractor may find it logical to pursue the owner to recover the subcontractor's extra costs. In this instance, the subcontractor and/or the prime contractor may wish to pass through the subcontractor's claim to the owner, who is the real party at fault. While the concept may be simple, finding the proper legal ground for these claims is fraught with difficulties.

B. Barriers to the Subcontractor Making Its Claim

Logical support exists for the concept of a direct contest between the subcontractor and the owner as a solution to the problem of subcontractor claims. Several authors suggest that it is at least a good management concept to permit the owner and the subcontractor to meet in the same proceeding and contest their differences.[77] As these authors point out, the primary principle standing in the way of this concept is the idea of privity.[78] There is no formal contract between the subcontractor and the owner and, therefore, the subcontractor has no contractual privity with the owner.

The subcontractor's resort to tort theories of recovery against owners traditionally has been thwarted by the Economic Loss Rule, which, in many cases, bars the suit.[79] Note that this issue may confuse some courts that are dealing with the question of whether representations and warranties of the owner to the prime contractor are passed down to the subcontractor.[80]

Given the lack of privity between the subcontractors and the owners, couldn't the parties agree to realign themselves? For example, couldn't the prime contractor assign to its subcontractor its rights against the owner relating to the subcontractor's claim? The decision in *Topco, Inc. v. State of Montana*[81] illustrates the issues raised by the application of the privity concept to situations involving prime contracts and subcontracts that differ in their treatment of the parties' rights. In *Topco*, the prime contractor entered into a contract with the State of Montana to construct a highway. The prime contract amount included a price for the clearing and grubbing work. The prime contractor then subcontracted the clearing and grubbing work at a lower price than that contained in the prime contract. The subcontractor eventually asserted a claim against the State of Montana based on the price of the work specified in the prime contract. Additionally, the prime contractor actually assigned to the subcontractor the prime's rights under the prime contract. Thus, the subcontractor was in the legal position of asserting the prime contractor's rights as its assignee. In *Topco*, the lower court held that the subcontractor's recovery was based on the subcontract price, not on the prime contract price for the clearing and grubbing. This holding was reversed on appeal on the theory that the

---

77. John W. Whelan & George H. Gnoss, *Government Contracts: Subcontractors and Privity*, 10 William & Mary L. Rev. 80 (1968).

78. *Id.* at 84.

79. *See, e.g.*, Moorman Mfg. Co. v. Nat'l Tank Co., 435 N.E.2d 443 (Ill. 1982) (holding that while a distant party may recover economic losses from an intentional misrepresentation, the rule would bar any action based on an innocent misrepresentation).

80. *But see* Murphy v. City of Springfield, 738 S.W.2d 521 (Mo. Ct. App. 1987) (allowing a drilling subcontractor to sue the owner directly after incurring economic loss due to the owner's innocent misrepresentation of the subsurface conditions).

81. 912 P.2d 805 (Mont. 1996).

prime contractor's price would govern the recovery because of the assignment to the subcontractor of the prime's rights under the prime contract. In federal contracting, assignment by the prime contractor of its claim against the government would be illegal.[82]

C. The Pass-Through System and Its Problems

In practice, a system has developed that involves both the prime contractor and the subcontractor in the assertion of the subcontractor's claims against the owner. To pass the subcontractor's claim through to the owner, the prime contractor will commence a proceeding against the owner "on behalf of" the subcontractor. In this proceeding, the prime contractor will seek to recover the extra costs incurred by the subcontractor that are caused by the owner. Thus, the prime contractor will "sponsor" the subcontractor's claim in a proceeding between the prime contractor and the owner to which the subcontractor is not a named party.

This system is predicated on the concept that the prime contractor is liable to the subcontractor for the subcontractor's claim and, therefore, the prime is "passing on" to the owner the prime contractor's liability to its subcontractor. The parties have even argued over who has the burden of proof on this issue. *See Gilbert Pac. Corp. v. Oregon*[83] in which the owner of the project took the position that the prime contractor should be required to allege that the prime contractor was liable to the subcontractor as part of the prime contractor's pleading. In *Gilbert*, the prime contractor contended that this was an affirmative defense that should be alleged by the owner. The court agreed that it was a matter of affirmative defense that was up to the owner to raise.[84]

Prime contractors go to great lengths to protect themselves from liability to subcontractors. They do this through the provisions of the subcontract agreements they draw, as well as the terms of documents they prepare during the project. If these efforts result in a legally binding release of the prime contractor's liability to the subcontractor, how can the prime later pass through any

---

82. Severin v. United States, 99 Ct. Cl. 435 (1943) (disfavoring any assignment by the subcontractor of its claim to the prime contractor, stating: "If the subcontractor did have a claim against the Government, it could not transfer that claim to another person, plaintiffs for example, since assignment of such claims is forbidden by statute." R.S. 3477; 31 U.S.C. 203).

83. 822 P.2d 729 (Or. Ct. App. 1991).

84. *See also* Kensington Corp. v. State, 253 N.W.2d 781 (Mich. Ct. App. 1997) (placing the burden of proof on the owner to show that the prime contractor was absolved from liability to the subcontractor).

claim of the subcontractor to the owner? This issue was raised in the important case of *Severin v. United States*.[85]

In *Severin*, the plaintiffs, who were the prime contractor, sued the government on behalf of their subcontractor who had incurred damages as a result of the government's action. The court questioned whether the subcontractor had released the prime contractor from liability such that the prime contractor would have no loss allowing it to bring an action. The court stated:

> If we look at plaintiffs as the real party in interest in their own suit, we encounter these facts. Plaintiffs did have a contract with the Government. That contract was breached. That breach might, if the contract had been one between private persons, have given rise to a right to win a suit, and to recover nominal damages, even if no actual damages resulted from the breach. But the futile exercise of suing merely to win a suit was not consented to by the United States when it gave its consent to be sued for its breaches of contract. *Nortz v. United States*, 294 U.S. 317, 327; *Great Lakes Construction Co. v. United States*, 95 C. Cls. 479, 502.
>
> Plaintiffs therefore had the burden of proving, not that someone suffered actual damages from the defendant's breach of contract, but that they, plaintiffs, suffered actual damages. If plaintiffs had proved that they, in the performance of their contract with the Government became liable to their subcontractor for the damages which the latter suffered, that liability, though not yet satisfied by payment, might well constitute actual damages to plaintiffs, and sustain their suit. Here, however, the proof shows the opposite. The subcontract, which is in evidence, shows that plaintiffs and the subcontractor agreed with each other as follows:
>
>> 21st. The Contractor or Subcontractor shall not in any event be held responsible for any loss, damate [sic], detention or delay caused by the Owner or any other Subcontractor upon the building; or delays in transportation, fire, strikes, lockouts, civil or military authority, or by insurrection or riot, or by any other cause beyond the control of Contractor or Subcontractor, or in any event for consequential damages.
>>
>> Thus plaintiffs, effectively so far as we are advised, protected themselves from any damage by way of liability over to the subcontractor for such breaches of contract by the Government as the one which occurred here.
>>
>> Plaintiffs must, then, so far as their claim includes items of losses suffered by their subcontractor, be merely accommodating another person who was damaged, by letting that other person use, for the

85. 99 Ct. Cl. 435 (1943).

> purposes of litigation, the name of plaintiffs, who had a contract and could properly have sued if they had been damaged. Orderly administration of justice, as well as the statute against assignment of claims, seem to us to forbid that.

However, there was a strong dissent in this case by Chief Justice Whaley, who said:

> There is no legal or equitable assignment involved. This is an action by a contractor to recover damages suffered by himself and his subcontractor, occasioned by the delay of the defendant. It is admitted that defendant's delay caused damages to both the contractor and the subcontractor. The plaintiff failed to prove the amount of his own damages but the damages suffered by the subcontractor were established by clear proof. The majority opinion admits that the subcontractor was damaged in the amount of $737.10 by allowing overhead on this amount to plaintiff.
>
> For fifty years it has been the settled doctrine of this court that a contractor could bring suit for himself and his subcontractor for losses occasioned by delay by the defendant before payment was made to the subcontractor. In innumerable cases from *Stout, Hall & Bangs v. United States,* 27 C. Cls. 385, to *Consolidated Engineering Company,* No. 43159, decided February 1, 1943 (98 C. Cls. 256), this doctrine has been uniformly followed and never been questioned.
>
> We must bear in mind that general contractors usually sublet specialized work like plumbing and electrical installations to subcontractors. The effect of the majority opinion would be to compel such subcontractors, and they are legion in numbers, to sue in their own names, which they could not do for lack of privity with the United States. This anomalous situation has never been recognized by this court in all its history. And the majority opinion cites no case in the Supreme Court in which subcontractors have been held to be assignors of claims against the United States, merely because they were unfortunate enough to be subcontractors.
>
> The subcontractor of plaintiff agreed in his contract not to hold the contractor for "loss, damage, detention or delay caused by the owner." The contractor is the plaintiff in this action. The subcontractor is not suing the contractor or the defendant. Plaintiff is suing for himself and his subcontractor for an admitted loss. The defendant was not a party to the subcontract. No consideration has been paid by the defendant for the protection given the contractor in the subcontract and without it the defendant cannot avail itself of this defense.
>
> In my judgment it is travesty of justice to allow plaintiff overhead on the losses suffered by his subcontractor and to deny recovery to plaintiff for his subcontractor of the amount admittedly due him

> from the defendant, which any court of equity would require the contractor to pay over to his subcontractor after payment to him by the defendant.

Following the concept of the majority opinion in *Severin,* what if the prime contractor and the subcontractor had liquidated the amount of the subcontractor's claim? The court, in the case of *John B. Pike & Son, Inc. v. State of New York,*[86] limited the damages recoverable by the prime to the amount that the prime had paid the subcontractor in settlement of a prior lawsuit arising out of a change order for extra work. In *Pike,* the prime had settled a lawsuit with the subcontractor for $75,000 and taken an assignment of all the subcontractor's claims against the owner. The prime's actual recovery was limited to the amount it had paid the subcontractor because the prime was to receive all the benefits of the action.

The *Severin* concept has been both criticized[87] and distinguished. In *Owens-Corning Fiberglas Corp. v. United States,*[88] the court dealt with a claim under the contract provisions rather than a claim for breach of contract. The court felt that these facts distinguished the case from *Severin*. The court stated:

> Plaintiff also argues, and quite correctly, that since this is not a suit for damages for breach of contract as present in the *Severin* cases, but rather a claim for an equitable adjustment by a prime contractor pursuing a remedy redressable under the contract terms, the *Severin* doctrine is not applicable. In *Blount Bros. Constr. v. United States,* 172 Ct. Cl. 1, 348 F.2d 471 (1965), this court held in such circumstances that "we accept the contention of plaintiff that the exculpatory clause did not affect plaintiff's liability to its subcontractor insofar as claims under the prime contract were concerned. Therefore, if the present claims are encompassed by the terms of plaintiff's contract with the Navy, then the *Severin* rule is not a bar." The Boards of Contract Appeals have held the *Severin* doctrine inapplicable to cases where the claim is one for an equitable adjustment redressable under the contract terms. See for example, *A. DuBois & Sons,* 60—2 BCA 2750; *Morrison-Knudsen Co.,* 60—2 BCA 2799, both cited by this court in the *Blount* case, *supra* at 5.

D. Liquidating Agreements

In order to avoid the *Severin* problem prior to the prime contractor's asserting a claim against the owner on behalf of a subcontractor, it is common for the prime contractor and subcontractor to enter into an agreement commonly

---

86. 647 N.Y.S.2d 654 (N.Y. Cl. 1996).
87. *See* Whelan & Gnoss, *supra* note 77.
88. 419 F.2d 439, 457, 190 Ct. Cl. 211 (1969).

known as a liquidating agreement.[89] This term comes from the fact that, although the subcontractor does not fully release the prime contractor from any liability, it "liquidates" the amount of the prime contractor's liability—generally to the amount that the prime contractor can recover from the owner on behalf of the subcontractor.

Following is an excerpt from a typical liquidating agreement:

> Subcontractor hereby agrees to release Prime Contractor from any and all liens and claims, whether arising out of tort, contract or otherwise, or of whatsoever nature, arising out of or under or relating in any way to the Subcontractor Claims, except that Subcontractor does not release Prime Contractor from the payment of any money to be paid or paid and received from or on behalf of Subcontractor Claims. It is agreed that Prime Contractor's liability with regard to the Subcontractor Claims herein reserved is limited solely to the final award and payment on Subcontractor Claims by or on behalf of Owner either through negotiation or arbitration. Upon such payment by Prime Contractor to Subcontractor of any award made and paid by or on behalf of Owner to Prime Contractor for Subcontractor Claims, if any, Prime Contractor shall be finally and fully released and discharged by Subcontractor with regard to the Subcontractor Claims. If no such award and payment is made by or on behalf of Owner on account of Subcontractor Claims, then Prime Contractor shall have no liability to Subcontractor whatsoever for the Subcontractor Claims.

The Claims Court, in *Folk Construction Co. v. United States*,[90] dealt with a release agreement that required the prime contractor to prosecute the subcontractor's claim against the government. The court held that the release was not "complete and unconditional" because the prime was obligated to cooperate in the presentation of the claim and pay the proceeds over to the subcontractor. Thus, the court held that the *Severin* doctrine did not bar the claim.

In *Bovis Lend Lease LMB, Inc. v. GCT Venture, Inc.*,[91] the court was concerned with a no-damage-for-delay provision in the subcontract agreement. Without questioning the validity of that agreement, the court referred to a later liquidating agreement entered into by the parties. The court held that the subcontractor and the prime contractor had a right to enter into this liquidating agreement without the approval of the owner and therefore, the prime's assertion of the subcontractor's claim was valid.[92]

89. The reasons for liquidating agreements and suggestions for drafting them are discussed by Carl A. Calvert & Carl F. Ingwalson, Jr., *Pass Through Claims & Liquidation Agreements*, 18 The Construction Lawyer 29 (Oct. 1998).

90. 2 Cl. Ct. 681 (1983).

91. 728 N.Y.S.2d 25 (N.Y. App. 2001).

92. *But see* Aetna Bridge Co. v. State Dept. of Transp., 795 A.2d 517 (R.I. 2002).

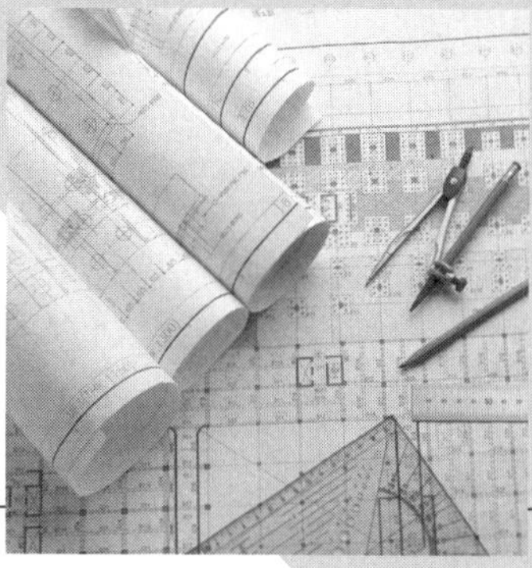

QUESTIONS

1. Company A furnishes and installs pre-built kitchen facilities for industrial plants. The principal components of the kitchen facilities are fabricated at the company's plant and then delivered to the site where they are installed by Company A together with associated mechanical and electrical hook-ups. Will Company A be treated as a subcontractor or supplier, and why might this distinction be important? What if Company A only furnishes the facilities, which are installed by another entity?
2. A subcontractor tenders a bid to a prime contractor bidding on a large construction project. The prime contractor is awarded the project and tenders a formal subcontract agreement to the subcontractor. The subcontractor refuses to execute the subcontract agreement, contending that the scope of its work was changed and the general conditions of the agreement are too onerous. Can the prime contractor require the subcontractor to execute and perform an agreement?
3. A subcontractor and prime contractor execute an AIA form of subcontract agreement that contains the normal flow-down provision. The prime contract contains an arbitration clause, but the subcontract does not. When a dispute arises over the scope of the subcontractor's work under the owner's specifications, the prime contractor asserts that the subcontractor must arbitrate rather than sue. Is the prime contractor correct?
4. Using the AIA form of subcontract agreement, the prime contractor requires the subcontractor to work overtime so that its work does not interfere with other subcontractors. Is the subcontractor obligated to follow the prime contractor's requirement? Could the prime contractor pass this cost on to the owner?
5. A subcontractor finishes its work on a project but does not receive final payment since the prime contractor's work will not be completed for some time. Must the subcontractor await payment until the entire project is complete even though its work was satisfactorily completed?
6. The prime contract provides that extra material will be handled at a price of 3X, but the prime contractor subcontracts with a subcontractor who agrees to handle it for 2X. Subcontractor now has a pass-through claim against the owner for the extra material. Can it seek compensation at the rate of 3X or is it limited to a price of 2X?

CHAPTER

11

Contract Time and Completion

ANDREW D. NESS

11.01 SIGNIFICANCE OF TIME FOR PERFORMANCE

Whether a construction contract is completed on time is one of the three most critical factors (along with the quality of the work and its price) that are likely to determine the owner's view of whether the construction contract is considered successful. Moreover, disputes between contracting parties concerning the construction process occur more frequently about the timeliness of the contractor's performance than any other single cause. The legal issues relating to time of performance, thus, are a key element of construction law.

Historically, time was not considered such a crucial element of a construction project. The uncertainties of the construction process were considered to be so great that timely completion was not deemed a material element of a construction contract, unless there was an express agreement to that effect in the contract. This rule persisted well into the 20th century. For instance, the Nebraska Supreme Court reaffirmed the principle as late as 1971.[1] If time of performance is not considered a material element of contract performance,

1. Kingery Constr. Co. v. Scherbarth Welding, Inc., 185 N.W.2d 857, 859 (Neb. 1971) (finding that time is material only if expressly so stated in the contract, because delays are foreseeable in construction).

then failure to complete the work by the agreed completion date is only a breach of an immaterial performance obligation, justifying no more than nominal damages, and certainly not sufficient to justify contract termination.

This historical common law rule does not comport with the reality of the early 21st century. In an era when firm commitments are made to host a nationally televised football game on a given date, even before ground is broken on the stadium construction, timeliness of performance can often be the most important element of the contractor's performance. No one would seriously argue that the old rule fits today's realities.[2] However, out of an abundance of caution, it is typical to specify in a construction contract that time of performance is considered material in order to avoid any doubt. This is generally accomplished by a simple statement that time is considered "of the essence," such as:

> Time limits stated in the Contract Documents are of the essence of the Contract. By executing the Agreement the Contractor confirms that the Contract Time is a reasonable period for performing the Work.[3]

11.02 CONSTRUCTION SCHEDULING—CRITICAL PATH METHOD SCHEDULES

The pace of modern life is not the only reason why timely performance is almost always a crucial element of construction contracts. Major advances in planning capabilities and scheduling techniques permit skilled construction contractors to greatly reduce, though not entirely eliminate, the uncertainties of the construction process that were considered inherent and beyond reasonable control in earlier years. The key advance in this respect is the development of network analysis systems for construction project scheduling, particularly the critical path method, or CPM. CPM was invented in the 1950s by E. I. DuPont DeNemours & Co. as a computerized program for planning and scheduling work at DuPont's chemical plants. Thanks to its adoption in the early 1960s by the U.S. Army Corps of Engineers and other military services, CPM quickly became a standard planning tool for complex construction projects. Two of the many judicial attempts to describe CPM follow:

---

2. "Anyone actively involved within the construction process knows that time is of the essence whether such is stated or not. . . . To the vast majority of entities involved in the construction process, time is an absolute implied material feature of every contract." MICHAEL S. SIMON, CONSTRUCTION CLAIMS & LIABILITY § 12.2 (1989).

3. American Institute of Architects, AIA Document A201–1997, *General Conditions of the Contract for Construction* § 8.2.1 (1997) [hereinafter AIA Document A201–1997].

> A CPM is a standard construction device used to plan the activities of a construction project in a logical, orderly sequencing manner citing durations for the different activities from the beginning of the job to the end. A CPM is created by dividing the entire project into discrete and quantifiable steps; in turn, each step is allotted an estimated time for completion. Ultimately, each step is arranged into a chronological sequence, thus revealing the anticipated length and structure of the entire construction schedule. In addition to serving as a road map for the contractors to determine when and where their work fits into the overall construction sequence, the CPM also assists contractors in assessing their hiring and material purchasing needs.[4]
>
> Essentially, the critical path method is an efficient way of organizing and scheduling a complex project which consists of numerous inter-related separate small projects. Each subproject is identified and classified as to the duration and precedence of the work. (E.g., one would not carpet an area until the flooring is down and the flooring cannot be completed until the underlying electrical and telephone conduits are installed.) The data is then analyzed, usually by computer, to determine the most efficient schedule for the entire project. Many subprojects may be performed at any time within a given period without any effect on the completion of the entire project. However, some items of work are given no leeway and must be performed on schedule; otherwise, the entire project will be delayed. These latter items of work are on the "critical path." A delay, or acceleration, of work along the critical path will affect the entire project.[5]

More succinctly, the critical path of a project can be defined as "the longest path through the network [of individual construction activities] that establishes the minimum overall project duration."[6] As such, any delay to an activity on the critical path, unless overcome by resequencing or compression (acceleration) of other critical activities, will necessarily extend (that is, delay) the overall completion of the project. By assessing the effect or lack of effect on the critical path to completion, CPM analysis can sort out inconsequential delays and problems from critical ones. This permits project managers to identify and focus their management efforts on the work activities and problems that are most important—the ones that have a high potential to ultimately cause the project delivery date to slip if something goes wrong.

A CPM schedule is thus a dynamic tool that can be used to predict the effect on the project completion date when unexpected events occur (as they always do) during the course of the project, such as an unexpected delay in the

---

4. Roberts & Schaefer Co. v. Hardaway Co., 152 F.3d 1283, 1287 (11th Cir. 1998), *reh'g denied*, 162 F.3d 1179 (11th Cir. 1998).

5. Haney v. United States, 676 F.2d 584, 595 (Ct. Cl. 1982).

6. Associated General Contractors, Construction Planning & Scheduling, 78 (1997).

delivery of a key item of equipment. The effect of the new development on the overall project schedule can be predicted by simply modifying the CPM to include the new information and re-running the overall schedule. Various alternatives for mitigating the prospective delay, such as resequencing activities or working activities in parallel using added personnel, can also be modeled and compared. This flexibility is of great utility in managing a complex project and frequently enables projects to be completed on time, despite numerous delays and unexpected events that in the pre-CPM world would have been extremely difficult to manage and nearly impossible to overcome.

11.03 LEGAL SIGNIFICANCE OF CONSTRUCTION SCHEDULES

The legal application of network analysis—using CPM as an evidentiary basis for assessing the effects of the opposing party's delays on the overall project—was quickly recognized. By the latter half of the 1960s, attempts were being made to persuade courts to admit network analysis as evidence in support of delay claims. Today, CPM analysis is the predominant means of both proving and disproving claims of delays to the work beyond the contractor's control. The traditional reluctance of courts to accept the validity of a new method of proof was overcome relatively quickly in the case of CPM analysis. Courts today routinely accept properly founded CPM analyses in resolving disputes over timely performance. Indeed, CPM expert witnesses seem ubiquitous in construction disputes involving delay, and some courts have gone so far as to hold that in the absence of a properly prepared and contemporaneously updated CPM schedule, a claim of delay can rarely be proven satisfactorily.[7]

This is not to say that courts unanimously accept specific techniques of CPM analysis, or that courts will rely on any CPM analysis that an expert witness cares to present. CPM scheduling programs can be manipulated to produce the desired result, and in any event, are only as accurate as the information on which they are based. Features added to scheduling software in recent years have only eased the task of masking dubious logic and subtle changes in a schedule analysis, making abuses harder to detect.[8] Courts that are called

---

7. Mega Constr. Co. v. United States, 29 Fed. Cl. 396, 426–35 (Fed. Cl. 1993); Hoffman Constr. Co. of Oregon v. United States, 40 Fed. Cl. 184, 198 (Fed. Cl. 1998), *aff'd in part, rev'd in part on other grounds*, 178 F.2d 131 (Fed. Cir. 1999).

8. Jon M. Wickwire & Stuart Ockman, *Use of Critical Path Method on Contract Claims—2000*, 19 CONSTRUCTION LAWYER 12 (Oct. 1999); MICHAEL T. CALLAHAN & H. MURRAY HOHNS, CONSTRUCTION SCHEDULES § 5-5 (3rd ed. 2004).

upon frequently to rule on construction delay claims have become particularly sophisticated in rooting out such abuses and not infrequently discard the views of the scheduling experts presented by one or both sides in reaching their decision.[9]

11.04 TIME OF COMMENCEMENT AND TIME FOR COMPLETION

Because of the significance of timely performance, establishing the required duration for completion, as well as determining the actual start date and the completion date, are all fertile grounds for potential disputes. Accordingly, setting out the time for completion (contract time) unambiguously in the contract, as well as establishing clear parameters for determining both the commencement and completion dates, should be key objectives for both owner and contractor.

The starting date, or time of commencement, can be a fixed date specified in the contract, or it is the date the contract is executed. More commonly, however, the contract provides for issuance of a "notice to proceed" by the owner that fixes the time of commencement. In some cases, the time of commencement will be established by a specific action, such as the date of issuance of a building permit by the local authority, or approval of the contract by the owner's board of directors.

The contract time is normally specified in terms of either calendar days, working days, or much less commonly, by establishing a fixed end date. A typical provision is as follows:

> The Contractor shall achieve Substantial Completion of the entire Work not later than ___ days from the date of commencement, subject to adjustments of this Contract Time as provided in the Contract Documents.[10]

The main benefit of using a notice to proceed and a contract time, defined as a number of days, is that if there is slippage in the project getting under way (as frequently occurs, due to waiting for construction loan approval or issuance of permits, late completion of precursor work, or any number of other reasons), then the required completion date is automatically extended by a

---

9. *See, e.g.*, Hensel Phelps Constr. Co., ASBCA No. 49270, 99-2 B.C.A. ¶ 30,531 (1999); Pathman Construction Co., ASBCA No. 23392, 85-2 B.C.A. ¶ 18,096 (1985) (finding that both experts produced unreliable CPM analyses).

10. American Institute of Architects, AIA Document A101–2007, *Standard Form of Agreement Between Owner and Contractor* § 3.3 (2007).

day for each day of delay in issuance of the notice to proceed. This generally works well for relatively brief delays in commencement, but more substantial delays in commencement tend to unduly burden the contractor. In a fixed-price environment, the contractor's costs for labor, construction materials, and subcontractor pricing all tend to increase with time, and a substantial delay in commencement can also alter the season in which particular work activities will be ongoing, with significant cost effects. For example, the cost of excavation and grading work during a wet season or during the winter will typically be significantly greater than if performed in a drier or more temperate season.

The preferred means of dealing with this problem is to limit the period during which the owner may issue the notice to proceed, allowing adjustment of the price if that limit is exceeded. Where no such period is specified, the law implies that a notice to proceed must be issued within a reasonable time, determined under the particular circumstances of the contract.[11] Several cases, however, have held that significant delays in issuing a notice to proceed did not entitle the contractor to recover damages, giving the owner wide latitude on the commencement date in the absence of a contractual time limit.[12]

The general rule also is that the contractor's execution of the contract constitutes acceptance of the time for completion allowed by the contract. The owner's specification of the time for completion does not give rise to any implied warranty that the work can reasonably be completed within that time period.[13]

11.05 WAIVER OF TIME FOR COMPLETION

Even though unambiguously established in the contract, both the time for completion and "time is of the essence" requirements can be waived by the conduct of the parties, as illustrated by the following case:

---

11. Ross Eng'g. Co. v. United States, 92 Ct. Cl. 253, 1940 WL 4077 (Ct. Cl. 1940) (finding that delay in issuing notice to proceed beyond 12 days was unreasonable because it forced the contractor to work in the winter).

12. M. A. Mortenson Co. v. United States, 843 F.2d 1360, 1362 (Fed. Cir. 1988); J. R. Youngdale Constr. Co., Inc. v. United States, 23 Cl. Ct. 460, 1991 WL 130539 (Cl. Ct. 1991), *aff'd*, 956 F.2d 1172 (Fed. Cir. 1992).

13. Willamette Crushing Co. v. Arizona Dept. of Transp., 932 P.2d 1350, 1354 (Ariz. Ct. App. 1997).

RDP ROYAL PALM HOTEL, L.P., a Florida limited partnership, by and through its general partner, PADC Hospitality Corporation I, Plaintiff, versus CLARK CONSTRUCTION GROUP, INC., a foreign corporation, Defendant

No. 04-16203, No. 05-11713

UNITED STATES COURT OF APPEALS

FOR THE ELEVENTH CIRCUIT

168 Fed. Appx. 348; 2006 U.S. App. LEXIS 3815

February 17, 2006, Decided

February 17, 2006, Filed

PRIOR HISTORY: Appeals from the United States District Court for the Southern District of Florida. D. C. Docket No. 01-03130 CV-DMM, D.C. Docket No. 01-03130 CV-TEB.

DISPOSITION: AFFIRMED.

JUDGES: Before DUBINA and KRAVITCH, Circuit Judges, and STROM*, District Judge.
* Honorable Lyle E. Strom, United States District Judge for the District of Nebraska, sitting by designation.

OPINION: PER CURIAM:
RDP Royal Palm Hotel, L.P., ("RDP") appeals an adverse judgment in favor of Clark Construction Group, Inc. ("Clark") on RDP's suit for breach of the construction contract between the parties. RDP seeks reversal of the district court's judgment denying its claims against Clark and granting the claims asserted by Clark in its counterclaim. RDP also appeals the district court's order awarding Clark damages, attorney's fees, costs, and litigation expenses. Clark cross-appeals the district court's judgment denying its claim for extended general conditions costs. In addition, Clark appeals the district court's order denying its motion seeking modification of the award of prejudgment interest. After careful review of the record, reading the parties' briefs, and after hearing oral argument, we affirm and remand the district court's judgment.

I. BACKGROUND
These cases arise from a multi-million dollar contract dispute over the construction of the Royal Palms Crowne Plaza Resort ("Resort") on Miami Beach. In May 1998, RDP, the developer and hotel owner, entered into a contract with Clark, the general contractor, for the construction of the Resort. Arquitectonica International, Inc. ("ARQ"), the lead architect, was designated as RDP's authorized agent in the contract and was required to prepare design development and construction documents. According to the contract, Clark agreed to construct the Resort for an amount not to exceed a Guaranteed Maximum Price ("GMP") of approximately

$30.4 million. The deadline for substantial completion of the Resort was 518 days from the date of commencement. RDP was permitted to require additional work from Clark within the general scope of the contract by issuing a change order or construction change directive. Under the contract, the GMP and substantial completion deadline were to be adjusted accordingly in light of change orders and construction change directives. Additionally, in the event that Clark failed to complete its work on or before the substantial completion date, RDP had a right to recover liquidated damages.

On September 28, 1998, RDP issued a formal notice to proceed to Clark, establishing the date for substantial completion as February 28, 2000. Clark commenced work but faced many setbacks during the construction. For example, Clark discovered a buried sea wall and contaminated soil which delayed excavation and sheet piling operations; RDP determined that the existing hotel, which was scheduled to be renovated, was too deteriorated for renovation and had to be demolished and reconstructed in its entirety; and dilemmas developed from the construction drawings. Consequently, hundreds of change orders and construction change directives were issued by RDP and ARQ, requiring Clark and its subcontractors to perform additional and different work from that shown on the contract construction documents.

Faced with the construction changes, Clark requested numerous extensions of the substantial completion deadline and increases to the GMP. RDP repeatedly assured Clark that the substantial completion deadline would be extended appropriately. Despite these assurances, RDP and Clark never resolved the extension requests, increased the GMP nor established a new substantial completion deadline.

Eventually, the substantial completion deadline of February 28, 2000, passed and construction continued. RDP and ARQ uninterruptedly issued hundreds of change orders and construction change directives. In turn, Clark continued to construct the Resort in accordance with those modifications, and RDP accepted Clark's continued performance. However, the constant construction modifications disrupted Clark's construction schedule, resulting in coordination problems with various subcontractors, further delays to construction, and increased construction costs. Consequently, on March 1, 2002, Clark filed a mechanic's lien in the amount of $8 million for work performed by four subcontractors.

On March 12, 2002, RDP received a temporary certificate of occupancy, allowing RDP to staff and occupy the Resort in preparation for the opening. Around this time, Clark ceased its work on the Resort because RDP discontinued paying Clark for work completed in the change orders. In turn, RDP hired another contractor to complete the Resort and ultimately opened the Resort for business on May 15, 2002.

In July 2001, RDP sued Clark for breach of contract, intentional misrepresentation/fraud in the inducement, negligent misrepresentation, violation of the Florida Deceptive and Unfair Trade Practices Act, fraudulent and wrongful lien, and tortious interference with a contract. RDP asserted that Clark caused the two year delay in the Resort's completion date. Clark filed a counterclaim against RDP for fraudulent inducement, *quantum meruit* and breach of contract, alleging that RDP caused the delay by providing incomplete and erroneous construction drawings and submitting numerous design changes. Clark also filed a third-party complaint presenting pass-through claims of three subcontractors for damages caused by RDP's breach of contract and construction delays. In another third-party complaint, Clark sued ARQ and Cornerstone Engineering ("Cornerstone"), ARQ's structural engineering consultant, for negligence, indemnity and contribution.

Following a bench trial, the district court[1] entered detailed findings of fact and conclusions of law denying all of RDP's claims against Clark.[2] The district court found that RDP failed to establish a breach of the contract based on untimely performance because it waived the February 28, 2000, substantial completion date by accepting Clark's continued performance. The district court also found that, in the absence of a contractual completion date, RDP was not entitled to recover any damages, liquidated or otherwise, from Clark and failed to prove that Clark's lien was fraudulent. In relation to Clark's counterclaim, the district court found that Clark waived any breach of the contract resulting from the construction drawings, but was entitled to fair and reasonable payment for the construction work completed. The district court awarded Clark the contract balance, damages for the pending change orders, additional general conditions costs, and a 3.75% mark-up on those costs. Notwithstanding these awards, the court denied Clark's requests for extended general conditions costs and extended home office overhead. Moreover, the district court found that Clark was entitled to reasonable subcontractor costs for its pass-through claims against RDP. Further, the district court dismissed Clark's third-party claims against ARQ and Cornerstone.

On August 4, 2004, the district court entered a final judgment denying RDP's claims against Clark in their entirety, denying Clark's claims against ARQ and Cornerstone in their entirety, granting Clark's claims against RDP in the amount of $5.5 million, and granting Clark's pass-through claims against RDP in the amount of $5.9 million.[3] Having reserved jurisdiction for the purpose of awarding interest, attorney's fees, and expenses, the district court subsequently entered an order on November 4, 2004, awarding Clark and its subcontractors over $2.4 million in attorney's fees and $1.5 million in costs and expenses. On November 17, 2004, the district court issued a supplemental judgment awarding Clark a total judgment of approximately $16.7 million with postjudgment interest.

In case number 04-16203, RDP and Clark pursue their respective appeal and cross-appeal of the supplemental judgment. Clark moved for relief from the supplemental judgment under *Fed. R. Civ. P. 60(b)*, seeking modification of the prejudgment interest award. The district court denied Clark's Rule 60(b) motion, and, in case number 05-11713, Clark appeals that denial.

II. DISCUSSION

RDP challenges the district court's findings of fact and conclusions of law, contending that the district court erred in finding that (1) RDP waived its right to liquidated damages; (2) RDP was liable to Clark on the pass-through claims; and, (3) Clark was entitled to recover from RDP an excessive amount of damages. RDP also challenges the district court's order on attorney's fees, costs, and litigation expenses. On cross-appeal, Clark challenges the district court's findings of fact and conclusions of law, arguing that the district court erroneously failed to award damages for extended general conditions costs. In the second appeal, Clark challenges the district court's order denying its Rule 60(b) motion to modify the prejudgment interest award. We address these issues in turn.

A. Liability and Damages

RDP argues that the district court erroneously found that RDP waived its claim for liquidated damages and asserts that Clark should be held liable for the delays caused by its subcontractors. RDP also argues that the district court erred in awarding Clark and its subcontractors excessive amounts of damages by failing to apportion responsibility for the total delay. RDP further argues that, in calculating damages, the district court failed to apply the contract's provision allowing Clark to recover delay-related damages for reasonable subcontractor costs attributable to RDP more than thirty days after the contractual date for substantial completion. Finally, RDP argues that the district court's finding as to the substantial completion date is incorrect, resulting in an erroneous calculation of prejudgment interest.

We review *de novo* the district court's conclusions of law and its application of the law to the facts. *Merrill Stevens Dry Dock Co. v. M/V Yeocomico II,* 329 F.3d 809, 813 (11th Cir. 2003). However, "we review the district court's factual findings for clear error." *CFTC v. Sidoti,* 178 F.3d 1132, 1135 (11th Cir. 1999). "If the district court's account of the evidence is plausible in light of the record viewed in its entirety, we must uphold the factual findings even if we would have weighed the evidence differently." *Id.* (internal quotations and citation omitted). "The district court's findings need only be plausible." *Id.* (internal quotations and citation omitted).

After reviewing the record, we conclude that the district court was correct in its conclusion that RDP waived its right to enforce the substantial completion date and "time is of the essence" provision of the contract. According to the record, RDP allowed the substantial completion date of February 28, 2000, to pass without setting a new deadline and continued issuing change orders and construction change directives requiring Clark to perform additional work. RDP's conduct in issuing hundreds of change orders and construction change directives after expiration of the substantial completion date of February 28, 2000, constituted waiver of the "time is of the essence" provision of the contract. *See Horovitz v. Levine,* 755 So. 2d 687, 688 (Fla. Dist. Ct. App. 1999) (purported "time is of the essence" proviso in the agreement was waived by the conduct of the parties subsequent to the agreed upon date of completion). In addition, RDP failed to set a new substantial completion date, thus it failed to reserve its right to enforce the liquidated damages provision for any date after February 28, 2000. *See McNeal v. Marco Bay Assocs.,* 492 So. 2d 778, 781 (Fla. Dist. Ct. App. 1986) ("When time has not been made essential to the contract or has been waived, the party entitled to insist on performance must fix a definite date in the future for performance . . . ."). Accordingly, we agree with the district court's conclusion that RDP waived its right to liquidated damages.

In addition, the district court did not err in holding RDP liable for Clark's pass-through claims and awarding damages. The record shows that some delays were caused by Clark's subcontractors, but those delays did not substantially postpone the eventual completion of the Resort. Rather, the substantial delays which severely impacted the Resort's construction (i.e., contaminated soil, incomplete and inaccurate construction documents, multiple design changes by RDP, and delays in plan approvals) were all attributable to RDP and not the fault of Clark's subcontractors. Although RDP asserts that the district court erroneously failed to apportion the total delay among the contributing parties, we find no error because RDP has not established a reasonable basis for such apportionment nor shown that apportionment was required. *See Gesco, Inc. v. Edward L. Nezelek, Inc.*, 414 So. 2d 535, 538 (Fla. Dist. Ct. App. 1982) (evidence must establish a reasonable basis for apportioning responsibility for the total delay); *Tuttle/White Constructors, Inc. v. Montgomery Elevator Co.*, 385 So. 2d 98, 100 (Fla. Dist. Ct. App. 1980) (loss resulting from breach of contract need not be segregated proportionately among the contributing factors causing the injury).

Further, since the district court's factual findings relating to the substantial completion date of March 12, 2002, and thirty-day damages grace period are plausible, we affirm the court's findings. *See Sidoti, 178 F.3d at 1135.*

[Discussion of other issues omitted.]

III. CONCLUSION

For the foregoing reasons, we affirm the district court's judgment but remand the case for the district court to correct the clerical mistakes pointed out in this opinion.

AFFIRMED

---

1. The parties consented to the jurisdiction of the United States Magistrate Judge for all purposes including trial and entry of final judgment pursuant to 28 U.S.C. § 636(c).

2. Prior to trial, the district court dismissed most of RDP's claims; only the breach of contract and fraudulent lien claims survived. On appeal, RDP does not challenge the district court's dismissal of those claims, and therefore, it has abandoned any argument on those claims. *See, e.g.,* Greenbriar, Ltd. v. City of Alabaster, 881 F.2d 1570, 1573 n.6 (11th Cir. 1989).

3. RDP appealed, and Clark cross-appealed, the final judgment. This court dismissed that appeal because the final judgment was not a final appealable order.

Waiver of the completion date generally precludes the owner both from terminating the contractor for default due to its failure to meet the completion date and from recovering damages for late completion.[14]

---

14. Baker Marine Corp. v. Weatherby Eng'g. Co., 710 S.W.2d 690, 694 (Tex. App. 1986).

In situations where the owner effectively has waived the established completion date, what then is the contractor's obligation with respect to completing the work? In the absence of any established time for completion, the contractor is responsible only for completing the work within a "reasonable time," to be determined based on the particular circumstances of the contract.[15] An owner that only is able to hold the contractor to a "reasonable" completion date, which will be judicially determined only long after the fact, is in an unenviable situation. In most cases, the owner needs to make substantial financial commitments based on the expected completion date of the project, whether this involves releasing rights to old facilities in anticipation of new ones, adding personnel to staff a new facility, such as the resort in the *Royal Palm Hotel* decision, or committing to have the new space available for tenants signing leases. Accordingly, the lack of any firm and enforceable date on which the contractor is obligated to complete its performance is likely to cause substantial practical problems.

However, the owner is not powerless in the situation in which the original completion date has been waived. Once the contractual completion date has been waived, it need not remain uncertain for the remainder of contract performance. For instance, the parties can agree on a new completion date in lieu of the original date, which becomes equally enforceable. Even absent such agreement, the owner unilaterally can establish a new completion date, which then will become as material and enforceable a term of the contract as the original completion date. However, the new date must be reasonable under the circumstances, taking into account the performance capabilities of the contractor as of the time notice of the new completion date is given.[16]

11.06 GROUNDS FOR EXTENDING THE TIME FOR COMPLETION—COMPENSABLE AND EXCUSABLE DELAYS

The time for completion, as established in the contract, is not immutable. There are several well-established grounds for extending the time for completion, and probably the majority of projects encounter situations whereby the contractor becomes entitled to some amount of time extension.

There are two keys to assessing whether a particular event amounts to a basis for extending the time for completion. First, the event must have actually caused a delay to the work and that delay must have affected the critical path

15. Drews Co., Inc. v. Ledwith-Wolfe Assocs., 371 S.E.2d 532, 533 (S.C. 1988); J. J. Brown Co. v. J. L. Simmons Co., 118 N.E.2d 781, 785 (Ill. App. Ct. 1954).

16. DeVito v. United States, 413 F.2d 1147, 1154 (Ct. Cl. 1969).

to project completion. An event that does not affect the critical path is not a delay at all, at least in terms of providing a basis for a time extension. Second, the owner must have had control over the event. An extension of the time for completion will be available to the contractor for an action or omission of the owner or other event deemed to be in the owner's control, but not for an event deemed to be in the contractor's control. Delays within the control of the owner are generally characterized as "compensable" delays because they both entitle the contractor to a time extension for the period of delay and to added compensation for the financial effects of that delay. A delay that is within the control of the contractor is generally characterized as an "inexcusable" delay, in that no relief from the required completion date or added compensation is available in such instances.

Of course, there are also causes of delay that are entirely beyond the control of either the owner or contractor. Most commonly, such delays are the result of what are often loosely called "Acts of God"—fires, floods, hurricanes, earthquakes, and the like. The expense of such delays, not caused by either party, is generally not allocated to either party and neither may recover any damages from the other for such delays. Because the contractor will not be liable for damages for its late completion due to such a delay, the net effect is that the contractor receives an extension of the time for completion to this extent, but no added compensation. Such delays beyond the control of either party are characterized as "excusable" delays.

These principles regarding time extensions for delays within the control of the owner and for delays beyond the control of either party are stated in AIA Document A201–2007 as follows:

> 8.3.1 If the Contractor is delayed at any time in the commencement or progress of the Work by an act or neglect of the Owner or Architect, or of an employee of either, or of a separate contractor employed by the Owner, or by changes ordered in the Work, or by labor disputes, fire, unusual delay in deliveries, unavoidable casualties or other causes beyond the Contractor's control, or by delay authorized by the Owner pending mediation or arbitration, or by other causes that the Architect determines may justify delay, then the Contract Time shall be extended by Change Order for such reasonable time as the Architect may determine.[17]

Federal government contracts contain a more extensive list of specific types of delays that are considered to be beyond either party's control and thus excusable. The following are all causes of delay that in most cases will be considered

17. American Institute of Architects, AIA Document A201-2007, *General Conditions of the Contract for Construction* § 8.3.1 (2007).

as arising "from unforeseeable causes beyond the control and without the fault or negligence of the contractor":

1. acts of God or of the public enemy,
2. acts of the government in either its sovereign or contractual capacity,
3. acts of another contractor in the performance of a contract with the government,
4. fires,
5. floods,
6. epidemics,
7. quarantine restrictions,
8. strikes,
9. freight embargoes,
10. unusually severe weather, or
11. delays of subcontractors or suppliers at any tier arising from unforeseeable causes beyond the control and without the fault or negligence of both the contractor and the subcontractors or suppliers.[18]

Under either the AIA documents or federal contracts, the specific examples—fire, strikes, floods, etc.—are just examples and are not exhaustive. Other delays that similarly qualify as "beyond the contractor's control" will be considered excusable delays, whether or not mentioned on the list of examples.[19]

There is more to the concept of excusable delays, however, than just lack of control over the delay-causing event. There also must be an element of unforeseeability with respect to the event. Accordingly, the particular cause of delay must not be one that the contractor would reasonably foresee under the circumstances of the contract. This reason for requiring this additional element was well explained by the U.S. Supreme Court in this 1943 decision:

UNITED STATES v. BROOKS-CALLAWAY CO.

No. 366

SUPREME COURT OF THE UNITED STATES

318 U.S. 120; 63 S. Ct. 474; 87 L. Ed. 653; 1943 U.S. LEXIS 1282

January 4, 1943, Argued

February 1, 1943, Decided

18. Federal Acquisition Regulations, 48 C.F.R. § 52.249-10.

19. AIA Document A201–1997, *supra* note 3, § 8.3.1; 48 C.F.R. § 52.249-10. 5 Philip L. Bruner & Patrick J. O'Connor, Jr., Bruner & O'Connor on Construction Law § 15:22 (West Group 2002) [hereinafter Bruner & O'Connor].

JUDGES: Stone, Roberts, Black, Reed, Frankfurter, Douglas, Murphy, Jackson, Rutledge

OPINION: MR. JUSTICE MURPHY delivered the opinion of the Court.

We are asked to decide whether the proviso to Article 9 of the Standard Form of Government Construction Contract[1] which provides that a contractor shall not be charged with liquidated damages because of delays due to unforeseeable causes beyond the control and without the fault of the contractor, including floods, requires the remission of liquidated damages for delay caused by high water found to have been customary and foreseeable by the contracting officer.

Respondent brought this suit in the Court of Claims to recover the sum of $3,900 which was deducted from the contract price as liquidated damages for delay in the completion of a contract for the construction of levees on the Mississippi River. The contract was not completed until 290 days after the date set, and liquidated damages in the amount of $5,800 (figured at the contract rate of $20 for each day of delay) were originally assessed. Respondent protested, and upon consideration the contracting officer found that respondent had been delayed a total of 278 days by high water, 183 days of which were due to conditions normally to be expected and 95 of which were unforeseeable. He recommended that liquidated damages in the amount of $1,900 (representing 95 days of unforeseeable delay at $20 per day) be remitted and that the balance of $3,900 be retained. Payment was made on this basis.

The Court of Claims held that liquidated damages should not have been assessed for any of the 278 days of delay caused by high water because the high water was a "flood" and under the proviso all floods were unforeseeable *per se*. Accordingly, it gave judgment in respondent's favor in the sum of $3,660.[2] No findings were made as to whether any of the high water was in fact foreseeable. We granted certiorari because the case presents an important question in the interpretation of the Standard Form of Government Construction Contract.

We believe that the construction adopted below is contrary to the purpose and sense of the proviso and may easily produce unreasonable results. The purpose of the proviso is to remove uncertainty and needless litigation by defining with some particularity the otherwise hazy area of unforeseeable events which might excuse nonperformance within the contract period. Thus contractors know they are not to be penalized for unexpected impediments to prompt performance, and, since their bids can be based on foreseeable and probable, rather than possible hindrances, the Government secures the benefit of lower bids and an enlarged selection of bidders.

To avoid a narrow construction of the term, "unforeseeable causes," limiting it perhaps to acts of God, the proviso sets forth some illustrations of unforeseeable interferences. These it describes as "including, but not restricted to, acts of God, or of the public enemy, acts of the Government, fires, floods, epidemics, quarantine restrictions, strikes, freight embargoes, and unusually severe weather, or delays of subcontractors due to such causes." The purpose of the proviso to protect the contractor against the unexpected, and its grammatical sense, both militate against holding that the listed events are always to be regarded as unforeseeable, no matter what the attendant circumstances are. Rather, the adjective "unforeseeable" must modify each event set out in the "including" phrase. Otherwise, absurd results are produced, as was well pointed out by Judge Madden, dissenting below:

> . . . Not every fire or quarantine or strike or freight embargo should be an excuse for delay under the proviso. The contract might be one to excavate for a building in an area where a coal mine had been on fire for years, well known to everybody, including the contractor, and where a large element of the contract price was attributable to this known difficulty. A quarantine, or freight embargo, may have been in effect for many years as a permanent policy of the controlling government. A strike may be an old and chronic one whose settlement within an early period is not expected. In any of these situations there would be no possible reason why the contractor, who of course anticipated these obstacles in his estimate of time and cost, should have his time extended because of them.
>
> The same is true of high water or "floods." The normally expected high water in a stream over the course of a year, being foreseeable, is not an "unforeseeable" cause of delay. Here plaintiff's vice-president testified that in making its bid plaintiff took into consideration the fact that there would be high water and that when there was, work on the levee would stop. . . .[3]

A logical application of the decision below would even excuse delays from the causes listed although they were within the control, or caused by the fault of the contractor, and this despite the proviso's requirement that the events be "beyond the control and without the fault or negligence of the contractor." If fire is always an excuse, a contractor is free to use inflammable materials in a tinderbox factory and escape any damages for delay due to a resulting fire. Any contractor could shut his eyes to the extremist probability that any of the listed events might occur, submit a low bid, and then take his own good time to finish the work free of the compulsion of mounting damages, thus making the time fixed for completion practically meaningless and depriving the Government of all recompense for the delay.

We intimate no opinion on whether the high water amounted to a "flood" within the meaning of the proviso. Whether high water or flood, the sense of the proviso requires it to be unforeseeable before remission of liquidated damages for delay is warranted. The contracting officer found that 183 days of delay caused by high water were due to conditions normally to be expected. No appeal appears to have been taken from his decision to the head of the department, and it is not clear whether his findings were communicated to respondent so that it might have appealed. The Court of Claims did not determine whether respondent was concluded by the findings of the contracting officer under the second proviso to Article 9[4] and not having made this threshold determination, of course made no findings itself as to foreseeability. We think these matters should be determined in the first instance by the Court of Claims. Accordingly the judgment is reversed and the cause remanded with instructions to determine whether respondent is concluded by the findings of the contracting officer, and, if not, for a finding by the court whether the 183 days of high water or any part of that time were in fact foreseeable.

REVERSED.

---

1. In general, Article 9 gives the Government the option of terminating the contractor's right to proceed, or of allowing him to proceed subject to liquidated damages if he fails to proceed with diligence or to complete the work in time. The full text of the proviso is:

> *Provided*, That the right of the contractor to proceed shall not be terminated or the contractor charged with liquidated damages because of any delays in the completion of the work due to unforeseeable causes beyond the control and without the fault or negligence of the contractor, including, but not restricted to, acts of God, or of the public enemy, acts of the Government, fires, floods, epidemics, quarantine restrictions, strikes, freight embargoes, and unusually severe weather or delays of subcontractors due to such causes: . . .

2. 97 Ct. Cl. 689.
3. 97 Ct. Cls. 701, 702.
4. The second proviso to Article 9 immediately follows the unforeseeability proviso and states:

> *Provided further,* That the contractor shall within ten days from the beginning of any such delay notify the contracting officer in writing of the causes of delay, who shall ascertain the facts and the extent of the delay, and his findings of facts thereon shall be final and conclusive on the parties thereto, subject only to appeal, within thirty days, by the contractor to the head of the department concerned, whose decision on such appeal as to the facts of delay shall be final and conclusive on the parties hereto.

There also are a number of principles that specifically address which causes of delay will be treated as within a party's control, as opposed to beyond the control of either party. Delays caused by an employee or agent of a party, or by an entity with whom the party has contracted, are considered to be within the control of that party. As such, delays by subcontractors and suppliers are considered to be within the control of the contractor and thus non-excusable. Delays by an architect retained by the owner are considered to be within the owner's control and thus compensable. The rationale for this is that a party has the ability to select those it employs and with which it contracts. Through control over that selection process and the terms of those contracts, a party has the ability to create legally enforceable obligations and rights against its agents, employees, and contractors.

Financial inability to perform, either by the contractor or its subcontractors or suppliers, also is considered within the contractor's control and not an excusable cause of delay. Events such as bankruptcy can in reality be caused by factors that as a practical matter are beyond a company's control—an unexpected spike in interest rates and late payments by usually reliable other customers.

Nevertheless, financial inability to perform is far more in the control of the company called upon to perform than it is in the other party and generally is not considered excusable. The exception is where a party's financial difficulties are directly caused by the actions of the other party, such as failure to make proper and timely payments of substantial amounts legitimately owed.[20]

Weather is another cause of delay subject to special rules. Every locale has its normal variations in weather over the course of the year, and adverse weather events like heat waves, rain, and snowstorms (in most parts of the country, at least) are certainly foreseeable. In planning and pricing its work, the contractor is expected to take into account normal weather variations at the site location during the planned period of performance. However, unusually adverse or "abnormal" weather can be a basis for excusable delay. The difficulty lies in distinguishing abnormal from normal and in measuring the effect of abnormal weather on the critical path activities. Contracts increasingly specify the means for determining if weather conditions are considered abnormal, such as by stating a number of days for each month where precipitation is to be considered "normal," or by specifying a particular methodology for determining if weather conditions are beyond "normal." The most common methodology was developed and adopted by the U.S. Army Corps of Engineers and requires a comparison of the conditions experienced on the project to a ten-year average, using the weather records from the nearest National Oceanic and Atmospheric Administration (NOAA) monitoring location.[21] Such a specified methodology avoids the potential for a statistical battle where each party chooses historical data over a period that best supports its position.

Many disputes over the effects of adverse weather involve a determination of the actual delay to the critical path due to the abnormal weather. Different types of weather can affect different types of construction activities in radically different ways, so deciding which construction activity was on the critical path when the unusually adverse conditions occurred can be outcome determinative.[22] Additionally, the contractor remains responsible for mitigating the effects of even unusual adverse weather where practicable and reasonable.

---

20. Douglas Corp., 58-1 BCA Par. 1727, 1958 WL 266 (ASBCA 1958); United States *ex rel* Taylor & Polk Constr., Inc. v. Mill Valley Constr., Inc., 29 F.3d 154, 161 (4th Cir. 1994).

21. Army Corps of Engineers Regulation, Construction Time Extension for Weather, ER 415-1-15 (31 Oct. 1989).

22. For example, several continuous days of light rain may have little to no effect on many activities, but entirely preclude exterior painting work. The critical path may be affected in such a case only if the painting was on the critical path during the unusual rain.

For example, heavy rains can be expected to preclude site grading work, and saturated soil conditions that persist after the rain ends may extend the delay period. If, however, the contractor has not taken reasonable steps to provide adequate site drainage and protect disturbed ground from the predictable effects of rainfall, the portion of the delay due to this failing is not excusable.[23]

The contract law doctrine of impossibility or commercial impracticability also is closely related to the concept of excusable delays. Commercial impracticability or "practical impossibility" excuses contract performance where the contract could only be performed at an excessive or unreasonable cost due to circumstances well beyond the contemplation of the parties when they entered into the contract.[24] To the extent performance is excused, the delay caused is excusable, and the contractor is entitled to an extension of the time for completion. Because the expense to perform must be wholly out of proportion to what was reasonably expected, and mere "hardship" does not amount to commercial impracticability, real-world examples of commercial impracticability in construction contracts are few and far between.[25] There also are situations in which the contractor essentially contracted to perform the impossible and is held to have assumed the risk of commercial impracticability, thereby making any delay due to impracticability unexcused.[26]

11.07 SUBSTANTIAL COMPLETION

A construction contract typically contains detailed drawings and specifications setting forth the construction work to be performed and the standard of quality required. The owner is entitled to the benefit of its bargain, and accordingly, the contractor is responsible for full compliance with all requirements of the contract, including all requirements of the drawings and specifications. However, the owner is not generally entitled to 100 percent completion and full compliance with the drawings and specifications by the stated contract completion date. Instead, the state of construction completion that must be reached within the time for completion is, unless otherwise defined by the

23. Titan Pac. Constr. Corp. v. United States, 17 Cl. Ct. 630, 639, 1989 WL 78828 (1989), *aff'd* 899 F.2d 1227 (Fed. Cir. 1990).

24. 5 Bruner & O'Connor, *supra* note 19, § 15:28; Iannuccillo v. Material Sand & Stone Co., 713 A.2d 1234 (R.I. 1998).

25. An illustrative real world example is Northern Corp. v. Chugach Elec. Ass'n, 518 P.2d 76 (Alaska 1974), where it was held that the use of a haul road across a frozen reservoir in winter was an implicit assumption of the contract, and the unavailability of that road due to unsuitable ice conditions two winters in a row made performance commercially impracticable.

26. Otinger v. Water Works & Sanitary Sewer Bd., 177 So. 2d 320, 323 (Ala. 1965).

contract, "substantial completion." The AIA contract definition of substantial completion is typical:

> Substantial Completion is the stage in the progress of the Work when the Work or designated portion thereof is sufficiently complete in accordance with the Contract Documents so that the Owner can occupy or utilize the Work for its intended use.[27]

This rule reflects the reality that a construction project typically involves thousands, if not hundreds of thousands, of details, and inevitably there will be finishing touches that are completed well after the point when the owner can utilize the virtually completed project for its intended purpose. Absent this rule, the contractor would remain subject to incurring damages for late completion when the project was complete but for inconsequential details, such as a missing doorknob or two.

The substantial completion principle is the construction contract equivalent of the general contract principle of "substantial performance" and has equivalent effect. A contract that has been substantially performed may not be terminated for default, and any deficiencies in performance are remedied by an award of damages representing the cost to correct or complete those deficiencies. Once the point of substantial completion is reached, the contractor has no further liability for delayed completion. In essence, the contractor cannot be said to be in material breach if it has reached substantial completion. Because of the significant consequences of reaching substantial completion, disputes as to what constitutes substantial completion in the case of a particular project are not uncommon. The following decision is illustrative of the issues arising in connection with the determination of substantial completion.

District Court of Appeal of Florida, Fourth District
J. M. BEESON COMPANY, Appellant, v. Ernesto SARTORI, Appellee

Nos. 87-1953, 87-2415 and 87-2693

Aug. 30, 1989

WARNER, Judge.

This appeal by a general contractor questions an amended final judgment in favor of the owner of a shopping center, awarding liquidated damages for delay and compensatory damages to complete construction.

27. AIA Document A201–2007, *supra* note 3, § 9.8.

[Procedural history deleted.]

The appellant and appellee entered into a construction contract in 1983. It provided for the construction of a shopping center including several anchor tenants and outparcels. The contract provided for liquidated damages of $1,000.00 per day if appellant failed to substantially complete the project within 300 days of commencement. It further provided for a bonus of $1,000.00 per day for each day the project was finished prior to the completion date. Appellant commenced construction and within six months put two anchor tenants into their spaces. The remainder of the retail spaces received their certificates of occupancy within the 300-day period, but an anchor tenant in an outparcel was not given its certificate of occupancy until over a year after commencement. Shortly thereafter, the appellant-contractor left the job because of nonpayment and filed suit. Appellee counterclaimed for breach of contract and for the liquidated damages provided in the contract. After trial, the court awarded judgment in favor of appellee. This appeal ensued.

[Discussion of unrelated issue deleted.]

Secondly, appellant contends that the trial court erred in finding that he did not substantially complete the work. In this case the contract provided that "substantial" completion occurred when "construction is sufficiently complete in accordance with the Contract Documents, so the owner can occupy or utilize the work or designated portion thereof for the use for which it is intended." The "work" under the contract "comprises the completed construction required by the contract documents." Under the contract in this case the work consisted of "construction and completion of a Shopping Center," including all of its component parts such as landscaping and paving.

The contract also provides that the date of substantial completion is to be certified by the supervising architect. On this project, however, there was no supervising architect who approved certificates for payment or certified substantial completion. In fact, the appellee's contract with his architect specifically deleted any responsibilities on behalf of the architect to supervise payments and to certify the date of substantial completion. Thus, we must determine whether or not the trial court applied the correct definition of substantial completion to the facts of this case.

We agree with appellant's contention that the contractual definition of substantial completion in this case is similar to the well established doctrine of substantial performance, and the terms are interchangeable. *See Ramada Development Co. v. Rauch,* 644 F.2d 1097 (5th Cir. 1981). The doctrine of "substantial performance" as held by this court in *Ocean Ridge Development Corp. v. Quality Plastering, Inc.,* 247 So.2d 72, 75 (Fla. 4th DCA 1971) states:

> Substantial performance is that performance of a contract which, while not full performance, is so nearly equivalent to what was bargained for that it would be unreasonable to deny the promisee the full contract price subject to the promisor's right to recover whatever damages have been

> occasioned him by the promisee's failure to render full performance. See 3A Corbin on Contracts, Section 702 et sequi.

To say that substantial performance is performance which is nearly equivalent to what was bargained for, as the case law defines the term, in essence means that the owner can use the property for the use for which it is intended. Furthermore, in defining substantial performance, one of the tests as enunciated by Corbin is the "degree of frustration of purpose":

> Extremely important factors in solving the present problem [of what is substantial performance] are the character of the performance that the plaintiff promised to render, the purposes and end that it was expected to serve in behalf of the defendant, and the extent to which the nonperformance by the plaintiff has defeated those purposes and ends, or would defeat them if the errors and omissions are corrected. Corbin on Contracts, 3A, Section 706.

Thus, substantial completion as defined in the contract is the equivalent of substantial performance under the case law and authorities hereinbefore cited.

In the instant case as of March, 1984, all phases of the shopping center had been completed sufficiently so as to obtain the certificates of occupancy from the city. Thus, the owner was capable of having tenants occupy the spaces and collecting rents thereon, and he was already collecting substantial rents on many of the tenant spaces prior to that date.

Despite the appellee's ability to rent out the shopping center spaces, the trial court found that appellant had not substantially completed the work by relying on an architect who testified that there still may not be substantial performance even though tenants were capable of occupying the premises, if the certificate of substantial completion called for in the contract had not been issued by the supervising architect. But this witness also testified that where there is no architect to give the certification, as there was none in this case, he would rely on the licensing authority and its certificate of occupancy to tell him that all work had been substantially completed under the contract. In this case, that occurred in March, 1984.[1] While this architect also testified that if a portion of the amenities such as landscaping are not fully complete, then there would be no substantial completion according to the contract, such a definition of substantial completion is contrary to both the contract and the case law, and the court erred in placing reliance on it.[2] As even the architect noted, when the owner can put tenants in possession for fixturing and can begin to collect rents, the owner begins to utilize the work for its intended purpose. When the owner was able to occupy and fixture the constructed space, the construction was substantially completed. *Fred Howland, Inc. v. Gore,* 152 Fla. 781, 13 So.2d 303 (1942). At that point, the appellant was entitled to his full contract price, less the cost to complete and the damages due to delay from the contract completion date through the date of substantial completion, and the trial court erred in determining that appellant had not substantially completed the contract.

This decision is also supported by *American Motors Inns of Fla., Inc. v. Bell Elec. Co.*, 260 So.2d 276 (Fla. 4th DCA 1972), where this court held that substantial performance is met if the omission is not so serious that it cannot be remedied by payment from the contract price. In this case, the amount due and owing under the contract was $157,392 at the time the contractor left the job. The amount required to remedy the deficiencies and unperformed work as found by the trial court was $68,939.70. Thus, there were sufficient monies still due the contractor from which the costs of completion could be deducted to satisfy the defects.

[Discussion of unrelated issue deleted.]

Consequently, we reverse the final judgment and remand for a redetermination of the date of substantial completion consistent with the principles set forth in this opinion; to calculate the amount of liquidated damages due to the appellee for failing to substantially complete within the 300-day period; and further to determine the amount due to the contractor on the contract, less the costs to complete. The other point raised for reversal is now moot. We also reverse the awards of attorney's fees and costs and remand for a redetermination after recalculation of the judgment.

REVERSED AND REMANDED.

DOWNEY and DELL, JJ., concur.

---

1. At another place in the testimony the architect conceded that utilization of the buildings may have occurred at an earlier date. Thus, we do not determine, as a matter of law, that substantial completion only occurred at the issuance of the certificate of occupancy in this case.

2. The architect witness also attempted to define substantial completion in terms of the dollar amount left to be done on the project (1/4 of 1% of the contract-price), but that too is contrary to the definition of substantial completion or performance which focuses on the owner's ability to utilize the construction.

As the *J. M. Beeson Co.* decision illustrates, obtaining a certificate of occupancy from the local jurisdiction is often closely related to substantial completion. Most localities require that a certificate of occupancy be obtained before persons other than construction workers can occupy a structure (some industrial facilities may be exempt from this requirement). Factors taken into account in issuing a certificate of occupancy typically include the degree of structural completion and elimination of (or adequate protection against) hazards such as floor openings and lack of permanent handrails. The most important factor, however, is completion of life safety systems such as fire alarms, fire sprinklers, emergency exit signage, and full, unimpeded availability of emergency exits and evacuation routes.

By definition, once a certificate of occupancy is issued, the owner may begin to move in and occupy the premises, so the close relationship to the basic test for substantial completion—that the work be sufficiently completed to be used for its intended purpose—is clear. However, the *J. M. Beeson Co.* court, consistent with most other courts, declined to hold that obtaining the certificate of occupancy automatically equated to substantial completion. Instead, the general rule is that obtaining a certificate of occupancy is relevant evidence in determining whether the work has reached substantial completion, but it is not determinative by itself. Nevertheless, the contracting parties may specify obtaining the certificate of occupancy as an objective measure of substantial completion to provide a means of reducing the potential for disputes regarding the achievement of this important milestone.

Consistent with the concept that upon substantial completion the project can begin to be used for its intended purpose, care, custody, and control of the project customarily is transferred from the contractor to the owner upon substantial completion. That is, upon substantial completion, the owner becomes responsible for security, maintenance, heat, utilities, damage to the work (unless caused by the contractor), and insuring the project.[28] The warranty period in which the contractor has continuing responsibility to correct work that does not comply with the contract requirements also normally commences upon substantial completion.[29]

Substantial completion also has a direct financial consequence because most construction contracts provide that the retention amount (amounts withheld by the owner from progress payments made earlier in the project) is to be paid to the contractor upon attaining substantial completion. However, the owner will typically continue to withhold a portion of the contract price representing the estimated value of the relatively minor work remaining as of substantial completion.[30] In that the contractor is no longer subject to termination for default or the threat of delay damages if completion of the remaining work is protracted, many owners will hold a multiple of the estimated value of the remaining work as an incentive to complete the remaining items promptly.[31] If the contractor fails to complete the outstanding items, the owner can apply the remaining contract funds to finish or correct any uncompleted work.[32]

11.08 FINAL COMPLETION

When a construction project has achieved substantial completion, there typically remain many unfinished details to be completed before the project is fully

28. *See* AIA Document A201–1997, *supra* note 3, § 9.8.4.

29. *Id.*

30. *Id.* § 9.8.5.

31. The Engineers Joint Contract Documents Committee (EJCDC) form contract provides that the withholding will be 200 percent of the estimated value of remaining work.

32. Prudential Ins. Co. v. Stratton, 685 S.W.2d 818 (Ark. Ct. App. 1985), *supplemented*, 690 S.W.2d 750 (1985).

finished. In construction industry practice, the work remaining at this point is normally tracked by creation of a "punchlist" that itemizes the unfinished work, area by area. The punchlist for a substantial project may contain thousands of items, although many of them will normally be minor work, such as areas of missing pipe insulation or places where touch-up painting is required.

In addition to unfinished physical work, there typically are a number of paperwork requirements that are not completed until after the point of substantial completion. These often include preparation of "as-built" drawings showing any variations between the construction drawings for the work and what was actually installed (for example, the routings of smaller piping lines and electrical cables are often changed in the field to avoid obstacles or constricted areas not anticipated by the construction drawings), submission to the owner of all manufacturers' warranties for equipment supplied for the project, and providing the owner with copies of the operating and maintenance instructions for all equipment supplied.

"Final completion" is the term generally utilized to refer to the point when the work is 100 percent complete, including completion of all punchlist items, and when all nonconstruction requirements of the construction contract, other than warranty responsibilities, have been performed. Final completion generally triggers the contractor's right to be paid the balance of the contract price.

Construction contracts additionally typically are written to avoid any formal acceptance of the contractor's work by the owner until the point of final completion. For example, Section 9.10.1 of AIA Document 201–1997 provides for "final inspection and acceptance" to occur only after the contractor gives notice that all work is complete and ready for such final inspection. The contract frequently will provide that inspections and tests of particular elements of the work at any earlier point by the owner are deemed to be for the owner's convenience and do not constitute acceptance of work that is not in accordance with the contract requirements. Similarly, progress payments made to the contractor throughout the construction process are deemed not to constitute acceptance of work not conforming to the contract.[33] Placing the point of formal acceptance at final completion enables the owner to continue to exercise its common law right to reject nonconforming work until the entirety of the work is fully completed. However, considering the reality that most construction work can most efficiently be corrected near the time of its installation before being covered or blocked by subsequent work, the courts have not given owners free rein to reject work at the end of the contract when the

---

33. AIA Document A201–1997, *supra* note 3, § 9.6.6 provides: "A Certificate for Payment, a progress payment, or partial or entire use or occupancy of the project by the Owner shall not constitute acceptance of Work not in accordance with the Contract Documents."

defects were actually discovered, or should have been discovered, at a much earlier stage.[34]

Finally, most construction contracts also tie a waiver of any previously unasserted claims relating to the construction process to final completion and the process of making and accepting final payment. Sections 9.10.4 and 9.10.5 of AIA Document A201–1997 address this as follows:

> 9.10.4 The making of final payment shall constitute a waiver of Claims by the Owner except those arising from:
>
> 1. liens, Claims, security interests or encumbrances arising out of the Contract and unsettled;
> 2. failure of the Work to comply with the requirements of the Contract Documents; or
> 3. terms of special warranties required by the Contract Documents.
>
> 9.10.5 Acceptance of final payment by the Contractor, a Subcontractor or material supplier shall constitute a waiver of claims by that payee except those previously made in writing and identified by that payee at the time of final Application for Payment.[35]

Accordingly, it is important for each party to set forth any claims it may have against the other party in writing prior to final payment.

11.09 DAMAGES FOR LATE COMPLETION

The basic rule of breach of contract damages is to protect the "expectation interest" of the nonbreaching party by placing that party in the same relative position as if the contract had been performed as agreed. The reach of this principle is limited by the rule that contract damages to be recoverable must either be a direct and natural result of the breach, or reasonably foreseeable as a consequence of the breach. Damages not reasonably foreseeable as a consequence of the breach are considered "remote" and are not recoverable.[36]

In the construction context, when the contractor breaches the contract by failing to substantially complete the work within the time for completion, the damages caused to the owner are those that arise naturally and directly from the delayed completion and those deemed to be reasonably foreseeable consequences of the delay. Most of these damages will be time-related and typically include the owner's cost of administering the delayed contract for the extended

---

34. Village of Pawnee v. Azzarelli Constr. Co., 539 N.E.2d 895, 904–07 (Ill App. Ct. 1989); Subsurfco, Inc. v. B-Y Water Dist., 337 N.W.2d 448, 455–56 (S.D. 1983). This topic is more fully treated in Chapter 21.

35. AIA Document A201-1997, *supra* note 3, §§ 9.10.4–5.

36. Hadley v. Baxendale, 156 Eng. Rep. 145 (1854).

period until substantial completion is reached and loss of use of the completed facility (if constructed for the owner's use) or loss of profit on resale (if intended for resale). Other recoverable damages include extended construction financing costs and cost of obtaining substitute facilities (where the owner's need must be met through use of substitute facilities in light of the unexcused delay in completion).

The simplest category of damages to prove, although typically not the largest in amount, generally is the extended cost of project administration, which may include items such as salaries of owner's inspectors and field supervision personnel, temporary power, insurance, and taxes.[37] Depending on the nature and intended use of the project, damages for loss of use can be proven via loss of rental value, lost operating profit, fair rental value, or interest on the value of the project.[38] However, a claim for loss of use frequently will raise issues of the foreseeability of the claimed use, as well as the uncertainty, and potentially speculative nature, of the lost profits or rental value, particularly for a new business. Proving loss of profit on resale or increased construction financing costs present similar or even greater complexities. Even though there are certainly situations in which owners have successfully overcome these hurdles, marshaled the needed proof, and fully recovered their actual damages for delayed completion,[39] the complexities of successfully recovering such damages are, as a practical matter, a strong incentive in favor of providing for liquidated damages for delay in lieu of actual damages.[40]

11.10 LIQUIDATED DELAY DAMAGES

The use of a contract provision providing for the payment to the owner of a liquidated sum for each day of unexcused delay in achieving substantial completion beyond the allowed time for completion has become the norm. Where liquidated delay damages are provided for, they serve as the owner's exclusive remedy for late completion (presuming the owner has allowed the contractor to complete, rather than terminating the contract), and the owner cannot

37. McDevitt & Street Co. v. Marriott Corp., 713 F. Supp. 906 (E.D. Va. 1989), *aff'd in part and rev'd in part on other grounds,* 911 F.2d 723 (4th Cir, 1990).

38. Perini Corp. v. Greate Bay Hotel & Casino, Inc., 610 A.2d 364 (N.J. 1992), *abrogated on other grounds,* Tretina Printing, Inc. v. Fitzpatrick & Assocs. Inc., 640 A.2d 788 (N.J. 1994) (awarding $14,500,000 for lost profit due to delay in renovating a casino).

39. *McDevitt & Street,* 713 F. Supp. at 906, 931.

40. Damages recoverable by the contractor for compensable delays to project completion are discussed in Chapter 24.

also recover its actual delay damages even though they exceed the liquidated amount.[41]

Liquidated damages clauses generally are enforceable so long as the liquidated sum chosen by the parties bears a reasonable relationship to the owner's likely actual damages as of the date on which the contract was entered. The limiting principle is that a contractual penalty, meaning an amount intended to punish the contractor for late completion, is unenforceable as a matter of public policy.[42] Within a wide range, however, liquidated damage amounts that can reasonably be seen as a pre-contractual estimate of owner's likely damages for delays will be enforced. Indeed, cases in recent decades finding a liquidated damages amount to be an unenforceable penalty are few and far between.[43]

To maximize the likelihood that they will be enforced, liquidated damages clauses frequently include recitals to the effect that the actual damages in the event of delayed completion would be difficult to ascertain and that the chosen amount is intended as an estimate of actual damages and not as a penalty. Such liquidated damages clauses will be generally enforced when the owner's actual damages are considerably more than the liquidated amount,[44] when they are considerably less,[45] and even when the owner arguably incurred no delay damages at all.[46] The one exception recognized in some jurisdictions is where the contractor has wrongfully abandoned the project. In those cases, the owner has been allowed to elect between recovering liquidated damages or actual damages, on the theory that the liquidated damages provision did not contemplate outright abandonment.[47]

In recent years, incentive/disincentive or "bonus/penalty" provisions have come into more frequent use, which provide for a daily bonus amount payable to the contractor for completing the project sooner than the required completion date, and a penalty amount for each day the project is completed late.

---

41. County of Dauphin v. Fidelity & Deposit Co. of Maryland, 770 F. Supp. 248, 254 (M.D. Pa. 1991), *aff'd*, 937 F.2d 596 (3d Cir. 1991).

42. Priebe & Sons v. United States, 332 U.S. 407, 413 (1947).

43. *See, e.g.*, Kingston Constructors, Inc. v. Washington Metro. Area Transit Auth., 930 F. Supp. 651, 656 (D.D.C. 1996) (finding that inclusion in liquidated damages of EPA penalties that clearly would not be assessed made liquidated damage amount a penalty).

44. Saturn Constr. Co. v. Conn. Dept. of Pub. Works, 1994 WL 590604 (Conn. Super. Ct. 1994) (enforcing $1,000 per day liquidated damages despite actual damages of $40,000 per day).

45. C. O. Falter Constr. Corp. v. City of Binghamton, 684 N.Y.S.2d 86 (1999) (finding that majority of $43,000 per day liquidated damages were based on "public inconvenience" if odor control system was not timely completed).

46. United States v. Bethlehem Steel Co., 205 U.S. 105, 121 (1907).

47. City of Elmira v. Larry Walter, Inc., 76 N.Y.2d 912, 563 N.Y.S.2d 45, 564 N.E.2d 655 (1990).

Although neither the bonus nor the penalty can appropriately be characterized as "damages," such provisions have a great deal of utility in incentivizing early completion, particularly where the daily bonus and penalty amounts are roughly equal, balanced, and not unfair to either party. Accordingly, enforcement of such a balanced incentive/disincentive arrangement should not be repugnant to public policy. At least one decision, however, has invalidated the "penalty" aspect of such a bonus/penalty arrangement as against public policy.[48]

11.11 NO-DAMAGE-FOR-DELAY CLAUSES

A no-damage-for-delay clause is a contract provision that seeks to eliminate the contractor's recovery of delay damages in the event of compensable delays to project completion and to limit the remedy for compensable delays to a time extension. No standard form contract includes such a "no-damage-for-delay" provision, and such clauses can take a wide variety of forms. One common formulation is: "No compensation shall be payable to the contractor for delays due to any cause, including delays caused by the actions or omissions of the owner, and contractor's remedy for delay shall be limited to an extension of the contract time."

No-damage-for-delay clauses are a frequent subject of litigation, and their use remains highly controversial. Some form of no-damage-for-delay clause, however, is nevertheless frequently seen in prime contracts, and even more frequently utilized in subcontracts.[49]

In accordance with the general principle of freedom of contract, the general rule remains that no-damage-for-delay clauses are enforceable in accordance with their terms.[50] However, there are numerous and important exceptions where the clause will not be enforced; in some states, the exceptions tend to swallow the general rule. To begin with, a no-damage-for-delay clause generally is considered to be a form of exculpatory clause, which is an attempt to derogate common law risk allocation in a manner tantamount to a forfeiture of sorts and, for that reason, is classed as "disfavored" and subject to strict interpretation.[51]

---

48. Milton Constr. Co. v. Alabama Highway Dept., 568 So. 2d 784, 789 (Ala. 1990).

49. An alternative clause also frequently found in subcontracts limits the subcontractor's damages for delay to amounts recovered from the owner for the benefit of the subcontractor. 5 Bruner & O'Connor on Construction Law,, *supra* note 19, § 15:75, n.17.

50. Wells Bros. Co. of New York v. United States, 254 U.S. 83, 87 (1920); Marriott Corp. v. Dasta Constr. Co., 26 F.3d 1057, 1067 n.17 (11th Cir. 1994).

51. U.S. Indus., Inc. v. Blake Constr. Co., 671 F.2d 539, 544 (D.C. Cir. 1982).

Courts frequently have applied the principle of strict construction to avoid the application of the clause in a particular set of circumstances. This tactic is made considerably easier to use by the fact that there is no generally accepted form of no-damage-for-delay clause, and, in practice, such clauses are worded in a wide variety of more or less effective ways.

There also is a sizable list of exceptions to the enforcement of no-damage-for-delay clauses that are recognized by a majority of jurisdictions, although the exceptions recognized from one state to the next can vary significantly. As stated in the leading treatise: "Courts commonly recognize four exceptions to the enforcement of these clauses; the only problem is that they are not always the same four bases. Often the courts do not even count the same way by combining and mixing various exceptions."[52] Every state will refuse to enforce a no-damage-for-delay clause in the face of fraud, misrepresentation, or bad faith by the other party. Other broadly recognized exceptions include "uncontemplated" delays, unreasonable delays (often stated as delays so unreasonable that they constitute an abandonment of the contract), and delays resulting from a breach of a fundamental obligation of the contract.[53] Many states also recognize an exception for delays caused by the "active interference" of the owner.[54] In addition to these exceptions, other cases have found that the owner waived the right to rely on the no-damages-for-delay clause. Finally, several states have enacted legislation prohibiting enforcement of no-damage-for-delay clauses, either in public construction contracts or in all construction contracts.[55]

---

52. 5 Bruner & O'Connor, *supra* note 19, § 15:75, n.38.

53. This is the list of exceptions appearing in Corinno Civetta Constr. Corp. v. City of New York, 67 N.Y.2d 297, 309, 502 N.Y.S.2d 681, 683 493 N.E.2d 905, 907 (1986), the leading case in a state generally considered as more willing than most to enforce no-damage-for-delay clauses.

54. U.S. Steel Corp. v. Missouri Pac. R.R. Co., 668 F.2d 435, 438–39 (8th Cir. 1982) (denying recovery where issuance of notice to proceed with knowledge that the work could not proceed).

55. *E.g.*, Wash. Rev. Code § 4.24.360 (2006).

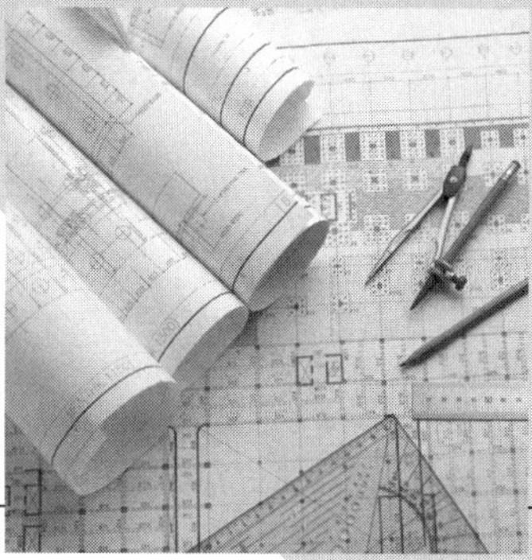

QUESTIONS

1. A hospital construction contract specifies that the contractor shall utilize a specific supplier as the "sole source" for laboratory casework. The contractor, accordingly, contracts with that supplier, but that supplier then misses the needed delivery date for the casework by six weeks and causes a critical path delay. Is this delay excusable? *See Edward M. Crough, Inc. v. Dist. of Columbia Dept. of Gen. Servs.*, 572 A.2d 457 (D.C. 1990); *Carter Steel & Fabricating Co. v. Ohio Dept. of Transp.*, 721 N.E.2d 1115 (Ohio Ct. App. 1999).
2. A liquidated damages provision states that liquidated damages will continue to accrue until the work is "fully complete in all respects." Will the contractor's liability for liquidated damages nevertheless be interpreted to conclude as of substantial completion, when the owner can use the project for its intended purpose and presumably is no longer incurring significant loss due to delayed completion, or does liquidated damages liability continue to final completion? *See Reliance Ins. Co. v. Utah Dept. of Transp.*, 858 P.2d 1363 (Utah 1993).
3. Where the contract expressly provides that liquidated damages for delay are the owner's exclusive remedy for delayed completion, and the liquidated damages are deemed an unenforceable penalty, may the owner then recover its actual delay damages, or does the "exclusive remedy" provision mean that it recovers no delay damages?
4. In many instances, some follow-on activities by others need to take place once construction is complete, before the project is actually put to its intended use. For example, in a shopping center, the completed retail spaces are typically turned over to the prospective tenants for tenant construction, fixturing, and stocking before the retail stores can be opened. In such a case, is work "sufficiently complete to be put to its intended use" when the shopping center is in condition to be opened to the public, or when it is sufficiently complete to allow tenants to begin their pre-opening activities?

CHAPTER

12

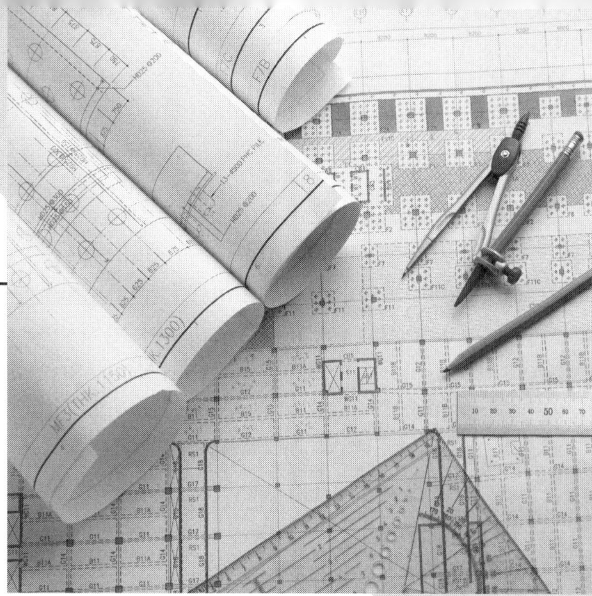

Construction Scheduling

MARK J. GROFF AND JON M. WICKWIRE

12.01 INTRODUCTION

This chapter introduces the area of construction scheduling. Time is a resource, and good scheduling practices can help ensure that parties to a construction project can meet their goals relating to budget, quality, and timely performance. This chapter complements, and should be read in conjunction with, the material in Chapter 11 on Contract Time.[1]

1. For a detailed treatment of construction scheduling principles, *see* J. O'Brien & F. Plotnick, CPM in Construction Management (6th ed. 2006); T. Glavinich, Construction Planning and Scheduling (2nd ed. 2004) [hereinafter Construction Planning and Scheduling]. For a detailed treatment of the legal aspects of construction scheduling, *see* J. Wickwire, T. Driscoll, S. Hurlbut & S. Hillman, Construction Scheduling: Preparation, Liability, and Claims (2nd ed. 2003); B. Bramble & M. Callahan, Construction Delay Claims (3rd ed. 1999). In addition, there are two major organizations dedicated to promoting professionalism and ethics in scheduling: the Project Management Institute's (PMI) College of Scheduling and the Association for the Advancement of Cost Engineering (AACE) International.

We begin with a discussion of the types of schedules generally encountered on construction projects and the requirements usually set forth in standard contract scheduling specifications. This is followed by a discussion of the baseline schedule and the schedule updating process. Finally, we discuss the major methodologies used to identify controlling project delays, who bears responsibility for such delays, and issues relating to expert scheduling testimony.

12.02 TYPES OF SCHEDULES

A. Bar Charts

Bar charts are the traditional method of planning and scheduling construction projects. They are sometimes referred to as Gantt charts, named after Henry L. Gantt, who first developed such charts in 1910.[2]

FIGURE 1
Commercial Building Schedule

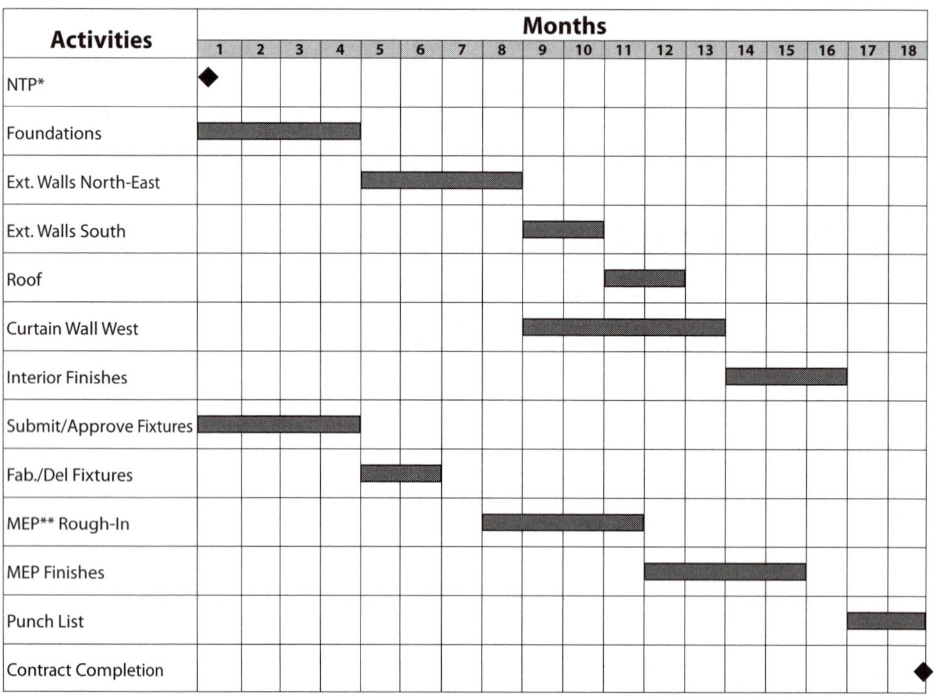

*NTP stands for "Notice to Proceed"
**MEP stands for Mechanical/Electric/Plumbing

2. Contel Advanced Systems, Inc., ASBCA No. 49075, 2004-2 B.C.A. (CCH) ¶ 32,664 (2004).

A bar chart generally comprises a horizontal scale that provides the time reference for the project, broken down into increments such as days, weeks, months, and a vertical list of the project components or work items. Bars are drawn to graphically represent the time necessary to accomplish each work item (see Figure 1).[3] Although the content of most bar charts is simple, it is possible to overlay cost and manpower curves over the time grid and the bars that are plotted for each work time (see Figures 2 and 3).

1. *Advantages and Disadvantages of Bar Charts*

Bar charts are easy to understand and are especially useful in providing less sophisticated project participants with a general understanding of the project plan. Bar charts are also often used in the field to provide short-term planning and coordination. These often take the form of what are known as "two-week

FIGURE 2
Commercial Building Schedule—Cost Loaded

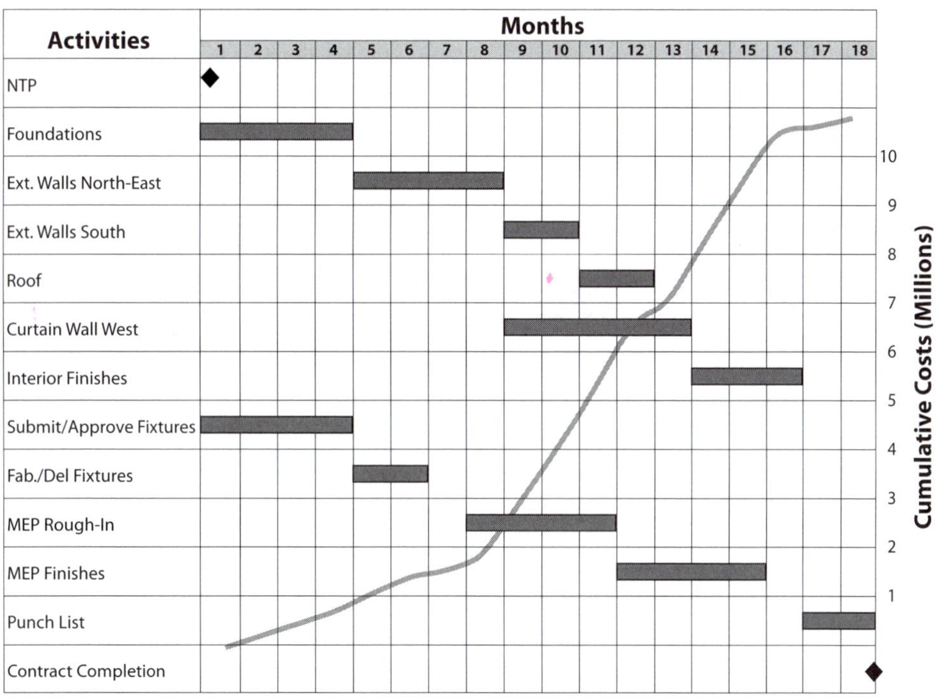

3. Figures 1–5 are based on figures prepared by the construction consulting firm TDF, LLC of Herndon, Virginia.

FIGURE 3
Commercial Building Schedule—Resource Loaded

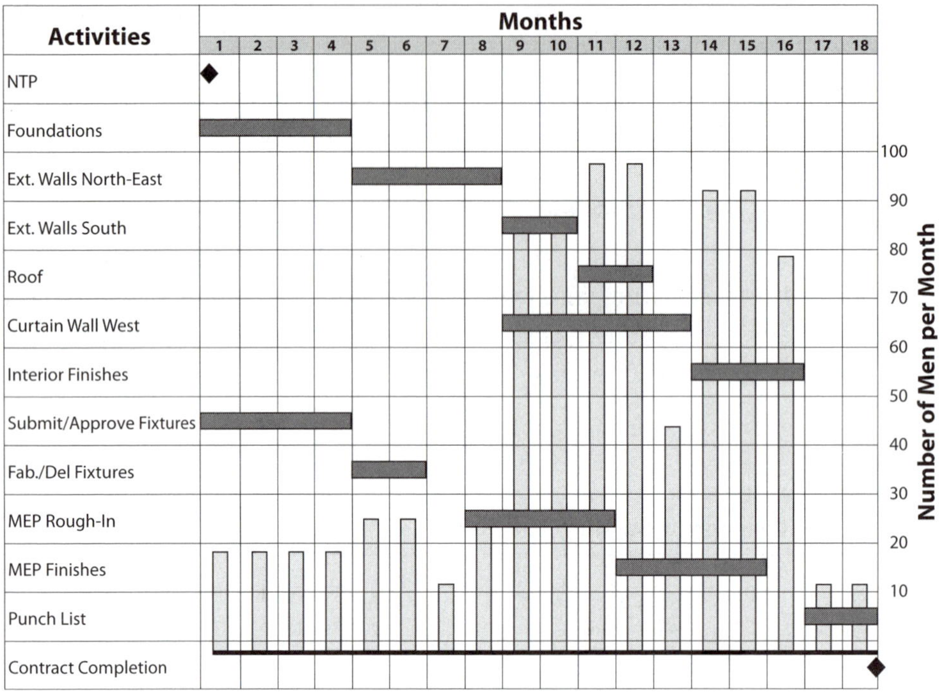

look-ahead schedules" that are distributed at weekly subcontractor coordination meetings. Bar charts, however, have significant disadvantages, which make them inappropriate to plan and schedule large, complex construction projects. The major disadvantage of bar charts is they do not adequately show the interrelationships or dependencies among activities.[4] Bar charts do not show float or contingency time (defined in the next section), nor can they show the delay impact of one bar on another. Size also limits what a bar chart can graphically represent. Not only is it difficult to manage a complex construction project with a bar chart, but many courts and federal boards of contract appeals hold that bar charts are insufficient to prove impact and delay damages on a project.[5]

---

4. Construction Planning and Scheduling, *supra* note 1, at 47.

5. Advanced Eng'g & Planning Corp., ASBCA Nos. 53366, et al., 2005-1 B.C.A. (CCH) ¶ 32,664 (2005); Bay Constr. Co., VABCA No. 5994, et al., 2002-1 B.C.A. (CCH) ¶ 31,795 (2002).

B. Critical Path Method Schedules

The critical path method (CPM) of network scheduling originated in the mid-1950s and was widely introduced in the construction field in the 1960s. CPM is a graphic presentation of the planned sequence of activities that shows the interrelationships and interdependencies of the elements composing a project. CPM techniques are used to predict project duration by analyzing which sequence of activities has the least amount of scheduling flexibility.[6] With government requirements for the use of CPM or network analysis system (NAS) techniques for major projects[7] and the perception by contractors that CPM could be an extremely useful tool for project management, the use of CPM to plan and schedule work has become the accepted standard in the construction industry. Furthermore, the federal courts and the federal boards of contract appeals increasingly require parties to a construction project to utilize CPM techniques to identify delays and disruption on projects, as well as to identify the cause and responsibilities for such delay and disruptions.[8] State court decisions in this area are not as uniform, but the modern trend is to require and/or give substantial weight to a CPM analysis of delays.[9] Furthermore, many construction contract specifications now require a CPM analysis to establish entitlement to a time extension.

1. *Basic Principles*

a. Activity Logic

Activity logic determines the sequence in which activities can or will be carried out during construction and must be carefully developed, considering physical,

---

6. *See* AACE International, Planning and Scheduling Professional Certification Glossary (2005); *see also* Haney v. United States, 676 F.2d 584 (Ct. Cl. 1982).

7. *See* J. Wickwire & M. Groff, *Drafting Schedule Specifications, Part III, Construction Management Association of America*, CM Advisor (September/October 2003).

8. Jackson Constr. Co. v. United States, 62 Fed. Cl. 84 (2004) (Contractor's claim denied where it failed to substantiate a delay to the critical path); United Pac. Ins. Co., ASBCA No.53051, 2003-2 B.C.A. (CCH) ¶ 32,267 (Contractor's claim denied where no CPM evidence presented; testimony that "any delay" on the project was "critical" held insufficient); Williams Enters, Inc. v. Strait Mfg. & Welding, Inc., 728 F.Supp. 12 (D.C. 1990), *aff'd*, 938 F.2d 230 (D.C. Cir. 1991).

9. *See, e.g.*, Granger & Sons, Inc. v. City Sch. Dist. of Albany, 296 A.D.2d 636, 744 N.Y.S.2d 567 (2002); RPR & Assoc., Inc. v. The Univ. of North Carolina, 153 N.C. App. 342, 570 S.E.2d 510 (2002). *Compare* Howard Contracting, Inc. v. G. A. MacDonald Constr. Co., 71 Cal. App. 4th 38, 83 Cal. Rptr. 590 (Cal. Ct. App. 1998) (CPM schedule not required to prove delay claim where a bar chart was used to schedule the project and where bar chart submitted at trial purportedly identified the critical path).

safety, resource, and preferential relationships among activities.[10] Three basic rules of logic should be applied to each activity:

1. What activities must be completed before the activity in question can start?
2. What activities can be performed concurrently?
3. What activities must follow the completion of the activity in question?

b. Critical Path

The critical path of a network represents the chain of interrelated activities that takes the longest time to complete from the beginning to the end. If the CPM network is time-scaled, the critical path will be the longest path in the network. The longest continuous interrelated chain of activities establishes the minimum overall project duration.[11] This will be the chain with the least amount of positive float or, if the project is behind schedule, the most amount of negative float. As discussed next, float is the amount of time an activity may slip in its start and/or completion before becoming critical.[12] The longest path is the critical path because if one of the activities along the critical path is delayed by one day and no other activity along the critical path is changed, the entire project will be delayed one day. By definition, the critical path has no float.

c. Float

The amount of float time for an activity represents the number of days by which an activity can be delayed without affecting the project's existing critical path and its forecasted completion date. Float is sometimes called slack. Float is important because it allows a contractor to shift resources from an activity with float to apply pressure to an activity on the critical path that is behind schedule and delaying the project. The float time permits the contractor subsequently to return to the noncritical activity and complete on time.

Issues often arise where one party (usually the owner) takes some action that depletes float in the schedule, thereby depriving the contractor of some scheduling flexibility and possibly transforming noncritical activities into critical activities. Unless specifically addressed upfront in the contract specifications, however, most courts and federal boards of contract appeals hold that float in a schedule is a resource that belongs to the project and is available for

---

10. Construction Planning and Scheduling, *supra* note 1, at 52.
11. AACE International, *supra* note 6.
12. *See id.*

all parties to use as long as they act in good faith.[13] This principle of shared float generally allows the use of float on a first-come, first-served basis.[14] Float values for activities may change due to current progress (or lack of progress), duration changes, or changes to milestone dates, such as the project completion date. Changes in float are the result of the dynamic nature of CPM schedules.

2. *Types of CPM*

a. Arrow Diagramming

An Arrow Diagramming Method (ADM) network graphically consists of arrows and circles that are used to describe the planned sequence and dependency relationships of work to be accomplished. The arrows represent work activities, and each activity describes a specific task to be completed. Each work activity is also assigned a time duration (see Figure 4). One exception is the dummy

FIGURE 4
Commercial Building Schedule—Arrow Diagramming Method

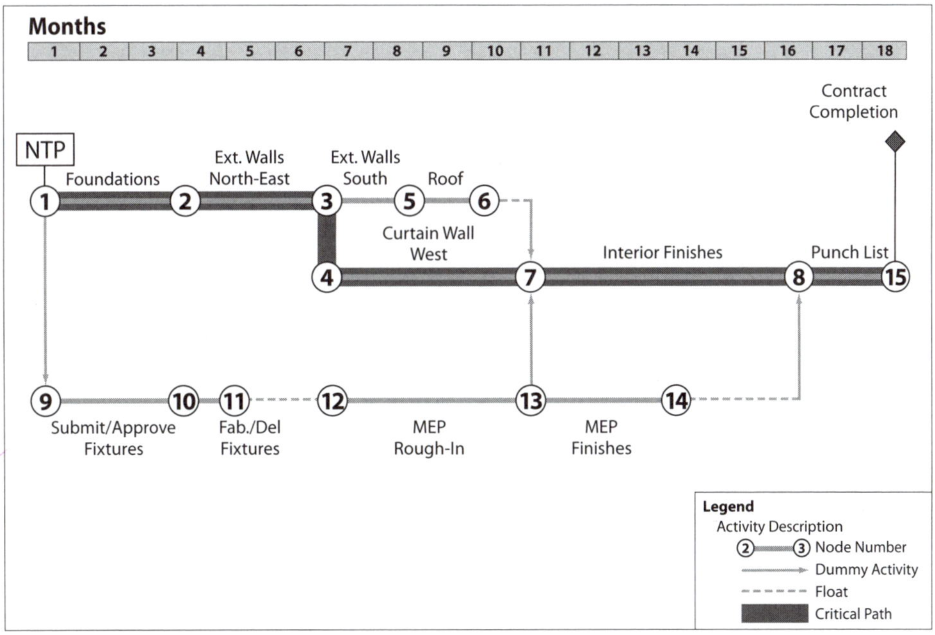

13. Weaver-Bailey Contractors, Inc. v. United States, 19 Cl. Ct 474 (1990). *See also* § 12.03(G).

14. *See, e.g.*, *AACE International Recommended Practice No. 29R-03*, Forensic Schedule Analysis § 4.3.

activity (more correctly called a "restraint"), which is used to reflect proper logic sequencing and to keep a unique event numbering system for activities. The dummy activity has a zero time dimension and does not represent a construction activity.

The circles represent the start and finish for each activity or the junction point of two or more activities. The circles are called events and are sometimes referred to as nodes. Events or nodes are points in time. An activity start event (tail) is sometimes referred to as the "i" node, and the activity finish event (head) is referred to as the "j" node. Each activity must have a unique i-j description. ADM uses only finish-to-start activity relationships, making the diagrams easy to follow; it was for many years the dominant scheduling model. Today, ADM has been almost totally replaced by Precedence Diagramming Method (PDM) networks, as most commercially available scheduling software no longer supports the ADM network model. There is a disagreement among experts and practitioners on whether the emergence of PDM as the dominant scheduling methodology has been good for the construction industry.[15] ADM principles are still relevant, however, because many scheduling specifications require the contractor to submit PDM networks, which resemble ADM's finish-to-start relationships.

b. Precedence Diagramming

In a Precedence Diagramming Method (PDM) network, activities consist of boxed figures connected by lines, which provide a sense of flow or direction (see Figure 5). The phasing of activities is accomplished by lead or lag times, sometimes called logic connectors. Dummy activities, as used in ADM, are not used in PDM. Whereas ADM uses only finish-to-start activity relationships without lag, PDM uses finish-to-start, start-to-start, and finish-to-finish activity relationships with lead or lag. These additional relationships reduce the number of activities needed to model a project and speed network development. At the same time, these tools can complicate scheduling calculations and, coupled with inexperienced schedulers and user-friendly software, can lead to schedule abuses. [16]

---

15. *See Critics Can't Find the Logic in Many of Today's CPM Schedules*, ENGINEERING NEWS RECORD, May 26, 2003 [hereinafter *Critics Can't Find the Logic*]; J. Wickwire & S. Ockman, *Use of Critical Path Method on Contract Claims*–2000, 19-4 CONSTR. LAW (1999). *See also* Donohoe Constr. Co., ASBCA No. 47,310, 98-2 B.C.A. (CCH) ¶ 30,0766, where the board noted that the contractor's as-planned schedule had a number of deficiencies because of the contractor's "imperfect" use of leads and lags in its PDM network.

16. *See* WICKWIRE ET AL., *supra* note 1, at Chapter 11.

FIGURE 5
Commercial Building Schedule—Precedence Diagramming Method

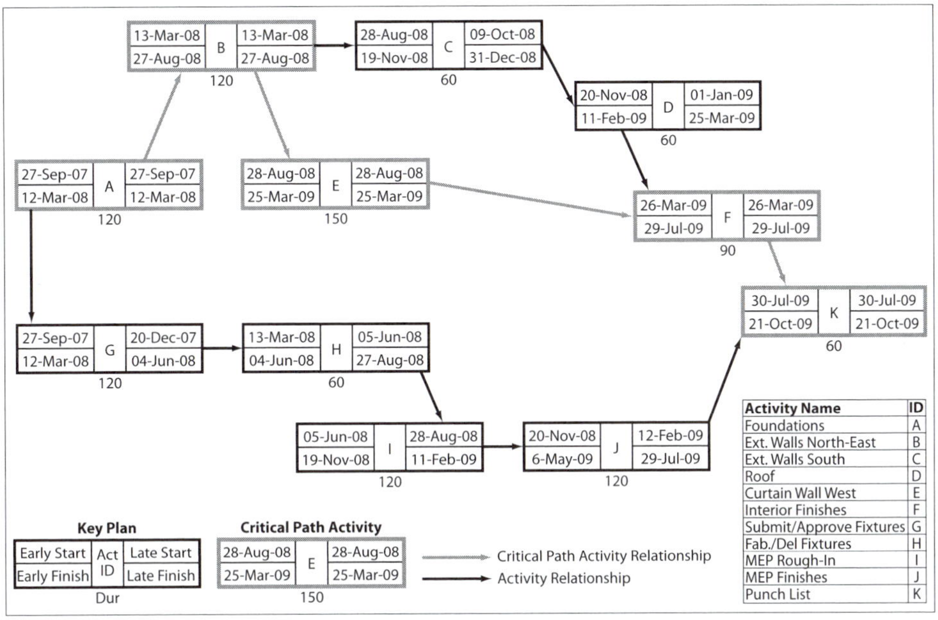

3. *Advantages and Disadvantages of CPM*

A CPM schedule can accurately forecast how long it will take to perform and complete a project and which tasks must be started and finished on time to meet the schedule. Further, a CPM schedule will show the proper relationships among all of the work tasks that are typically hard to recognize from looking at a bar chart. A CPM schedule will focus attention on troublesome aspects of the work to be performed, as well as permit management to select the most appropriate and effective methods to address such problems. A good CPM schedule will also enable project management quickly and effectively to measure the impact of proposed changes (directed or constructive) on available time, money, manpower, and equipment. Finally, a CPM schedule can assist parties in proving or defending delay claims.

A CPM schedule, however, requires the expenditure of time and money. Substantial effort is needed to maintain a good working schedule and the fact that many contract scheduling specifications are vague and ambiguous can lead to disputes. In addition, because of a lack of commitment by the parties, often CPM schedules are simply filed away and not actually used to plan and manage the project. A failure to properly manage and update the schedule can render it essentially useless, as discussed in Section 12.05 in this chapter. Finally, for smaller or less complex projects, CPM may not be worth the expense. In general, however, the advantages of using a CPM schedule usually far outweigh any disadvantages.

In the following case, the Armed Services Board of Contract Appeals provided a concise explanation of CPM's dynamic nature, including a discussion about float and proper schedule updating procedures.[17]

SANTA FE ENGINEERS, INC., ASBCA Nos. 24,578, et al., 94-2 B.C.A. (CCH) ¶ 26,872 (1994)

By Administrative Judge Coldren:

The Critical Path Method (CPM) of Scheduling

The Critical Path Method (CPM) of Scheduling is a planning technique used to determine how long a project will take to complete and to identify the most important items that need to be accomplished in order to meet the project deadline. It is a written description of the manner in which a contractor plans to complete the project on time. It tells the contractor in advance the sequence duration and parameters of dates within which specific work must be performed in order to ensure timely completion of the project. It can be used to monitor progress by measuring actual job accomplishment against the schedule as a baseline.

The CPM is based upon a graphic project model called a network which depicts all activities that must be carried out with their mutual time dependencies as well as durations in a diagram form. An activity is the basic building block of the network and is defined as a single work step that has a recognizable beginning and end, and required times for its accomplishment. Each activity is linked to another through its interdependency or logic. The logic of a network refers to the determined order in which activities are performed with the commencement of some activities logically dependent or a restraint on the completion of others. However, some activities are independent of others and can proceed concurrently. The CPM network literally builds the project on paper, defining the roles and interrelationships of the activities.

Four limiting times for each network activity are calculated. The early start of an activity is the earliest date at which it can possibly start, allowing for the times required to complete the preceding activities. The early finish of an activity is the earliest possible date at which it can be completed, and is determined by adding

17. For simplification, paragraph numbers and citations have been deleted.

the activity's duration to its early start date. The late finish of an activity is the very last date at which it can finish and still allow the project to be completed by the designated date. The late start of an activity is the latest possible date that it can be started if the project completion date is to be met. It is calculated by subtracting the activity's duration from it latest finish date.

The "float" of an activity in CPM scheduling is determined by subtracting its early start date from its late start date, although subtracting the early finish from the late finish will yield identical results. The path or paths of activities through the network with zero float is known as the "critical path." Any activity with zero float is a "critical activity" with the result that any delay in its finish dates prolongs project completion by the same amount of time.

The CPM schedule must be revised when work is added and deleted, completed and/or changes in logic are made. It must also be updated monthly to reflect work completed as well as the contractor's plans for completing the remaining work including in particular any significant changes in logic or duration. Each time updating information is entered into the network a new computation must be made because the new information may cause the critical path to shift. A CPM schedule that does not reflect what work is actually being accomplished in the field does not accurately identify the project's critical path and/or activities.

In the following case, the General Services Board of Contract Appeals also set forth an excellent primer on CPM scheduling, addressing such topics as the critical path, absolute versus preferential logic, resource leveling, acceleration, and buy-back time.[18]

UTLEY-JAMES, INC.
GSBCA No. 5370, 85-1 B.C.A. (CCH) ¶ 17,816 (1985)

By Administrative Judge Lieblich:

In critical path scheduling, each task is related to those that logically must precede and follow it. For example, in a building constructed like the McNamara Building, the concrete for the third floor slab cannot be placed until the second floor slab has been poured and the necessary support structure has been completed. The second floor slab and the supports are said to be a "restraint" on the third floor slab. The preparer of the CPM schedule must make sure to take into accounts all such restraints. . . .

18. For simplification, paragraph numbers and citations have been deleted.

In any CPM schedule there are what is called "early start" and "late start" dates. The connotations of these two terms are not the same, and the terms must be used carefully. The early start date is the earliest date by which a given activity logically can start, assuming all preceding activities are on schedule. The late start date is the latest date by which a given activity *may* start if the contractor is not to fall behind schedule. The terms "early finish" and "late finish" are parallel to "early start" and "late start," but later in time by the number of days the activity is scheduled to take. The number of days between early start and late start (or, what ought to be the same thing, between early finish and late finish) is called "float." If the schedule allows no more time to do the entire job than the entire job requires, there is no float in the job, and at any given place in the schedule there will be at least one job activity with no float. An activity with no float is said to be on the critical path. An accurate CPM schedule is a very powerful management tool for the contractor in its planning of the entire job.

Determining the critical path has an important effect on the Government's management of the contract as well. For example, a delay of fifty days on a job activity with more that fifty day's float should, in theory, have no effect on the completion of the job. The early start and early finish dates for that activity may move fifty days, but they should still be earlier than the late start and late finish dates. In contrast, a job activity on the critical path, for which there is no float, will have the same early start date and late start date (and the same date for both early finish and late finish). One day's delay to that job activity will cost one day in completion of the job unless it is somehow made up.

A job activity that is not originally in the critical path can get on the critical path as a result of a delay that uses up all of its float. If more than the original float is used up by a delay, the activity is not only on the critical path but behind it; unless the completion date for the entire job is extended, the float will become negative, i.e., the early start date will be later than the late start date. This will show that the job activity cannot be started until after it is required to be if the entire project is to adhere to the existing schedule. One of the ways to cure this problem is to change the schedule by extending the completion date for the entire project, thereby putting positive float at the end of the job that can then be distributed among the activities that have developed negative float.

The other cure for negative float is accelerating, i.e., performing activities on the critical path in less time than the schedule allows for. Almost every activity is a direct restraint on others, and the finish date for a given activity is the start date for those it directly restrains. To give a simplified example, we will treat the placing of the concrete for each floor slab as a direct restraint on the placing of the next one. If each slab is scheduled for ten days but can be poured in eight, two days can be made up on each pour. If the floor slabs are on the critical path, a ten-day delay can be made up by pouring five slabs in eight days each instead of ten (assuming that the acceleration of this activity does not put some other, unrestrained activity on the critical path in place of the slabs).

Another concept of importance to CPM scheduling is resource leveling, also called resource loading. Consider, for example, the interior finish work in an office tower like the McNamara Building. The placing of each floor slab is logically restrained by the placing of the slab beneath it, but that is not necessarily true of interior finish work. There is no reason in theory why a carpeting contractor could not come in and carpet every floor of the building on the same day. But there are a variety of reasons why that is not done, and one of the principal reasons is that it makes no sense to supply that much material and that many workers to do all the floors at once when the work can be sequenced with all sorts of other work and handled over a convenient period of time with a suitably sized workforce. To give an accurate indication of the actual planned job schedule, a CPM schedule must take resource leveling into account.

12.03 STANDARD SCHEDULE SPECIFICATIONS

With increasing frequency, construction contracts contain specifications that prescribe the form and detail of the project schedule and supporting data, the methods for adjusting and updating the schedule, and the use of the schedule for purposes of contract modifications, time extensions, and analysis of performance delays. These specifications and their interpretation must be carefully considered by contracting parties because they often result in important allocations of risk and responsibility. In practice, schedule specifications are rarely clear and concise. Experience has shown that it is difficult to establish standard schedule specifications for use by the construction industry today. Specifications may range from one paragraph to five to ten pages. Short specifications may not adequately address the major issues that might arise during construction and lengthy specifications can often be expensive and difficult to implement. Generally speaking, however, for major projects, the more detail included in the specifications, the better chance there is to eliminate misunderstandings.

A. Type of Diagram

Schedule specifications must detail clearly both the specific type of diagram and the network planning technique desired. Vague language may allow the contractor the opportunity to use bar charts, rather than CPM network analysis system diagrams, as an appropriate schedule, even though it is intended that a critical path network diagram be used on the project. For example, in *H. I. Homa Co.,*[19] the contract contained the following provision: "The schedule

19. H. I. Homa Co., ENGBCA No. PCC-41, 82-1 B.C.A. (CCH) ¶ 15,651 (1982).

shall show the order in which the contractor proposes to carry on the work, the dates on which he will start the several salient features (including procurement of materials, plant and equipment) and the contemplated days for completing same." The specifications further provided that the schedule be of a type and size acceptable to the contracting officer. Under these circumstances, the board of contract appeals determined that a CPM network was not required by the contract since a bar chart would meet the specifications. Because the government had required a CPM schedule, the board determined that the contractor was entitled to additional compensation.

B. Number of Activities

Most schedule specifications require a network showing a minimum number of activities. Such a requirement is sometimes implemented by limiting the duration of individual activities to no more than a stated number of days. It is very important to have a clear idea of the required level of detail for any network diagram. The appropriate number of activities is largely dependent upon the nature, size, and complexity of the project. If too many activities are required, the resulting diagram may not be an effective management tool for field construction. However, if too few activities are specified, the resulting activities may be so gross as to provide for a diagram that does not properly reflect the intricacies or interdependencies of various activities required to be completed on the project.

C. Resource Loading

Another extremely important requirement for an owner to include in a competent scheduling specification is one for resource loading of activities. In order to know whether a purported activity's duration has any validity, one must know its resource loading. This is because the equation that provides the activity duration requires the application of the productive resource (e.g., number of personnel, equipment) against the projected productivity of those resources. In other words, Resources/Productivity = Duration (of defined activity). Stated another way, ten workers with a productivity of two windows per day will accomplish a stated activity of twenty windows in ten days. However, if the ten workers have a productivity rate of four windows per day, the activity will only take five days. Unless the schedule clearly shows the resource loading for purported activity duration, the owner has no yardstick to determine whether the contractor is including realistic activity durations in the original planned network schedule or estimates of remaining durations in updates.[20]

D. Approval

Schedule specifications should establish whether formal approval of the schedule is required, as well as designate the party responsible for making the approval. Thus, the specifications should state whether the owner or architect/engineer or construction manager approves the CPM schedule. The specifications also should indicate the duration of the approval cycle and include the appropriate leverage to enforce the approval process. The specifications should also clearly state that the owner is entitled to fully operational electronic versions of all schedules (including updates) submitted by the contractor and electronic versions of all backup to the submitted schedules.[21]

E. Updating

Most contracts require updating of the schedule during performance. Issues addressed by the specifications include the frequency of schedule updates and what the updates must reflect about changes, delays, or modifications to the plan for the remainder of construction that may be contrary to the original network diagram. Many schedule specifications require owner approval prior to incorporating major logic changes in schedule updates, usually after a joint updating meeting.[22] Conflicts frequently arise where the specifications fail to establish monthly procedures to ensure that the historical work on the project is properly recorded and that the parties agree on the contractor's game plan for completing the work.[23]

F. Subcontractor Involvement

Most construction projects today place a great level of risk on the major trade contractors. These business entities frequently have the majority of the risk of

---

20. *See* Neal & Co. v. United States, 36 Fed. Cl. 600 (1996). *See also* Utley James, Inc., GSBCA No. 5370, 85-1 B.C.A. (CCH) ¶ 17,816 (1985), excerpted in Section 12.02 in this chapter.

21. *See* Section 12.04 in this chapter for a more detailed discussion on the submission and approval of a baseline schedule.

22. *See* Section 12.02(B)(1)(a) in this chapter for a discussion of activity logic.

23. *See* Section 12.05 in this chapter for a more detailed discussion on updating the project schedule.

labor productivity on the project and may even be operating under the framework of a total broker general contractor, where the general contractor itself performs no physical work on the project. For this reason, as well as for policy considerations of project teamwork, owners have increasingly been requiring the participation of major trade contractors in the preparation of the original CPM, as well as in the updating process.

G. Float Use and Reporting

Schedule specifications should address a number of issues concerning float, which is the amount of time that an activity may be delayed without causing delay to the project finish date. The contract should state the desired requirements for reporting float and should address the availability of the use of float to the parties and the project. Without risk-allocating provisions, neither party will be entitled to priority in use of float time. In other words, if float is available along the channel of work involved, either party (as long as it is acting in good faith) may, without being charged for a delay to the project, delay or extend work activities up to the point at which the float is exhausted.[24]

H. Major Revisions and Time Extensions

Schedule specifications should state whether the CPM or network diagrams are to be totally revised or significantly modified if there are major revisions in the project sequence. The contract should also indicate the party responsible for preparing such major revisions to the network diagram and describe the owner or construction manager's right to approve any such major revisions. At a minimum, major changes and schedule impacts should be incorporated into the schedule as promptly as possible so that the CPM network accurately reflects current conditions. Thus, contract drafters should also consider specifying the appropriate time for quantifying delays. For example, should the time-extension calculation for the delay be measured based on (a) a CPM update or diagram current as of the time of the delay, (b) the impact to the original CPM

24. *See* Weaver-Bailey Contractors, Inc. v. United States, 19 Cl. Ct. 474 (1990). In addition, many federal specifications now specifically provide that project float is a resource available to both the government and the contractor. *See* 2002 Unified Facilities Guide Specification for Network Analysis Systems § 1.10.2 (April 2002). *See also* Whitesell-Green, Inc., ASBCA Nos. 53938, et al., 2006-2 B.C.A. (CCH) ¶ 33, 323 (2006), where the board cited extensively from a Navy scheduling specification. Ironically, the board found that the parties had abandoned the use of the CPM schedule for purposes of determining time extensions. *See also* Santa Fe Eng'rs, Inc., ASBCA Nos. 24,578, et al., 94-2 B.C.A. (CCH) ¶ 26,872 (1994) (citing similar contract clauses).

plan, or (c) a "20/20" hindsight look at the project? The reference point for measurement of the impact often will have a major effect on the length of the appropriate time extension. As discussed in Section 12.06 in this chapter, the most widely accepted approach for evaluating delays is on a chronological and cumulative basis that takes into account the status and critical path(s) of the project at the time of the delay in question.[25]

I. Prohibitions on Schedule Manipulations

Use of inexperienced personnel to schedule the work and so-called user-friendly software advances that mask the logic and changes to CPM networks have led to serious scheduling abuses.[26] As a result, many owners are including specifications that specifically prohibit improper scheduling manipulations, such as the use of automatic default mechanisms to update actual start and finish dates.

12.04 BASELINE SCHEDULE

The as-planned schedule, or baseline schedule, represents the contractor's approach to executing the contract prior to the start of a project. Usually, the first step in preparing a baseline schedule requires identification of the necessary activities to perform the work in accordance with the contract drawings and specifications. Subcontractor input regarding sequences and durations should be included in the development of the baseline schedule. Once the activities have been identified, the next step is to determine the interrelationships between the activities; that is, which activities must begin before others, which activities may be performed concurrently, and which activities must be completed before others can be started. To clearly understand the relationship between activities, most schedule software programs are able to plot a logic diagram that depicts each activity's relationship with other activities. Obviously, the more complex a project, the more activities necessarily comprise the network schedule.

The next step in preparing the baseline schedule is to estimate the amount of time required to perform the scope of work represented by each activity. Many current schedule software programs allow resources to drive the duration

---

25. *See, e.g.*, SAE/American-Mid Atlantic, Inc., GSBCA Nos. 12,294, et al., 98-2 BCA (CCH) ¶ 30,084 (1998); Fru-Con Constr. Corp., ASBCA Nos. 53544, et al., 2005-1 BCA (CCH) ¶ 32,936.

26. *See Critics Can't Find the Logic*, *supra* note 15; Wickwire et al., *supra* note 1, at Chapter 11.

of specific activities. Care must be taken to consider all factors that influence the estimated activity duration, including expected weather, available manpower, equipment and material, and other constraints.

All of the sequencing and duration determinations should be made before the CPM schedule is computerized. This computerization consists of entering the activity descriptions, the activity durations, resources, and successor/predecessor activities into the CPM schedule software. The CPM schedule is completed by using a mathematical calculation performed by the software to determine the chain of interrelated activities through the network from the beginning of the project to the end. The calculation yields each activity's early and late start and finish dates, associated float, and the project's critical path. Generally, the series of activities with the least amount of float comprise the project's critical path. The project's critical path activities determine and control the project completion date.

Most contracts require the submission of the baseline schedule to the owner for formal review and approval. The owner's approval creates a rebuttable presumption of correctness or reasonableness of the schedule.[27] Because of this rebuttable presumption of the reasonableness of the schedule once an approval is in place, many owners are afraid of approving any schedule. In many cases, this type of forbearance is unsuccessful. Courts have imposed implied obligations upon owners with regard to the schedule, despite the owner's silence.[28] Furthermore, failure to approve the schedule deprives the owner of the ability to reject unreasonable drawings for performance, opens the owner up to potential claims for early completion (which may be based upon defective or unrealistic logic), and denies the parties a baseline from which to evaluate delays. If neither party is operating from an approved baseline schedule, then how will the owner or the contractor have any idea how to evaluate time extensions?

The problem of liability for approving the schedule is minimal as long as two rules are followed. First, the owner should add cautionary language to any approval, noting that the contractor remains the party responsible for development and execution of the means, method, and timing of performance reflected in the project plan with the owner. Second, the owner should review any time

---

27. Santa Fe, Inc., VABCA No. 2168, 87-3 B.C.A. (CCH) ¶ 20,104 (1987).

28. Fullerton Constr. Co., ASBCA No. 12275, 69-2 B.C.A. (CCH) ¶ 7876 (1969). *See also* G. Blindzius Contractors, Inc., ASBCA No. 37707, 90-2 BCA ¶ 22,835 (1990).

frames shown on the project plan for performance of review functions, shop drawing approvals, supply of owner-furnished equipment, and so on to make certain that the owner is not being called upon to act in a manner quicker than that called for in the contract, or in unreasonably short time frames. Presumptions of correctness as to time durations for action by the owner may be inferred when the owner accepts time approval durations shorter than those called out in the specifications.[29] There is some case law, however, that holds that even where a schedule has been approved, it will be thrown out if it contains unrealistic logic and/or durations.[30]

If the owner does not provide in the contract specifications the appropriate leverage through which to enforce the approval process, an approval standoff may occur and the construction may proceed for months or even years without an approved schedule. To help avoid this problem, the owner should expressly provide in the contract that no payment will be made on contract billings until the owner approves a schedule (absent, perhaps, some initial stipulated funding in the first month or two). If the contractor submits unreasonable or unrealistic logic, sequences, or durations (including overall project duration), the owner will then have the opportunity, through the power of the purse, to require that a reasonable schedule be submitted. The owner should also provide, by contract, that the contractor will be required to make all necessary schedule changes directed by the owner to arrive at a reasonable, realistic, and acceptable schedule.

Some public owners, such as the Los Angeles County Sanitation Districts, require the acceptance of the contractor's baseline schedule as a precondition to the issuance of the notice to proceed for the actual construction work. In other words, these types of scheduling specifications typically provide for a 60-day period, prior to notice to proceed, for the development and approval of the baseline schedule. The work on the baseline schedule is funded by a purchase order issued by the owner. To ensure contractor performance, the owner may require the contractor to post a letter of credit. If the baseline schedule is not approved within the specified 60-day time frame, the contractor is assessed liquidated damages, drawn from the letter of credit. If the contractor's baseline schedule is approved prior to 60 days, and all other conditions precedent are

29. *See* G. Blindzius Contractors, Inc., ASBCA No. 37,707, 90-2 B.C.A. (CCH) ¶ 22,835 (1990).

30. *See, e.g.*, Neal & Co. v. United States, 36 Fed. Cl. 600 (1996); Hensel Phelps Constr. Co., ASBCA No. 49270, 99-2 B.C.A. (CCH) ¶ 30,531 (1999).

satisfied, the owner extends the contract time for completion of the construction work by the amount of time saved.[31]

12.05 SCHEDULE UPDATES

The periodic review of the project schedule and the input of daily progress are termed updating. The object of updating the schedule is to determine physical progress to date, identify sequence revisions and duration changes, and thus provide a complete and accurate report of how well the actual construction progress compares with the baseline schedule or contemporaneous schedule. In addition, its purpose is to determine how all parties intend to continue the work and meet the overall schedule objective. Specifically, the project schedule should be revised to reflect the sequence actually followed, the delays encountered, and the as-built conditions, just as a set of construction drawings and specifications would be revised to reflect as-built conditions of construction work put in place.

Thus, the purpose of the updating process is threefold. First, the update should show where the project has been (or how the current point in the life of the project was reached). Next, the monthly update should show where the project is currently (as of that given point in time). Finally, the update should show the revised plan for reaching project completion. Unfortunately, the recent advent of some computer software allows the contractor to override the logic of the planned diagram (without advising the owner of changes in logic, sequence, or durations).

Even though the contractor may be properly updating the CPM for the project and transmitting updates to the owner monthly, owner representatives (because of laziness, ignorance, or the press of other duties) may simply file the document without performing any sort of analysis as to its validity. An argument can later arise as to the actual job status as of a given update. To combat this problem, many contracts require the owner and contractor to conduct joint monthly updating meetings to force a proactive and timely engagement by the parties so that they specifically agree on the progress on the various activities during the preceding month. The record of the joint updating meeting also can be a vehicle for contemporaneous quantification of the time impacts of changes and other delays. If such delays are not priced and negotiated at the time, the historical record of the joint updating (which should include both a disk that includes associated backup and a copy of the CPM diagram with historical progress to date highlighted) will provide the neces-

31. M. Fischer & R. Vakharia, *Successful Implementation of CPM Scheduling on Public Works Construction Projects*, presented at the AACE Southern California Spring Symposium, 2004.

sary factual background for the parties at a future date to discuss merit and quantum.

In *Fortec Constructors v. United States*,[32] the U.S. Court of Federal Claims recognized that control of a project is lost without employing proper updating techniques:

> Thus, if the CPM is to be used to evaluate delay on the project, it must be kept current and must reflect delays as they occur. In the instant case, the CPM was updated only once, in August of 1979. This update did not consider delays in work performed prior to the update, nor, obviously, in work that occurred after the update through the date of acceptance of the project by the Corps. . . .
>
> Reliance upon an incomplete and inaccurate CPM to substantiate denial of time extensions is clearly improper.
>
> It is difficult, if not impossible, for this Court to comprehend how the Corps could have used the CPM to evaluate Fortec's time extension claims without modifying the CPM to reflect the additional work Fortec was directed to perform—the value pit, the frame and grate, the hangar doors, and the roof ventilator. These items were all major components of the hangar project. If the Corps made such modifications to the CPM, this Court cannot understand why the Corps did not issue a formal modification (logic revision), which Fortec would have been required to include in its CPM analysis.
>
> In *Continental Consolidated Corp.*, ENGBCA Nos. 2743, 2766, 67-2 B.C.A (CCH) ¶ 6624 (1967), the Army Corps of Engineers Board of Contract Appeals held that if the CPM is used to evaluate requests for time extension it must reflect actual project conditions. In *Continental*, the contractor sought extra costs incurred due to suspensions of work and due to subsequent acceleration. . . . The Government claimed that acceleration was justified, since the CPM revealed the project was behind schedule. The board rejected the Government's reliance on the CPM, stating that:
>
> > It is essential that any changes in the work and time extensions due to the contractor be incorporated into the progress analysis concurrently with the performance of the changes or immediately after the delay and thus integrated into the periodic computer runs to reflect the effect on the critical path. Otherwise, the critical path chart produced by the computer will not reflect the current status of the work performed or the actual progress being attained.
>
> . . . Since adequate time extensions were not granted immediately upon determining the extent of the delay, the board found that the CPM's completion schedule was distorted and unreliable as a basis for denying time extensions.

32. 8 Cl. Ct. 490, 504–08 (1985).

A construction project by its very nature is dynamic. Drawings and estimates, no matter how carefully considered, are bound to change as a result of unexpected events. The project schedule must continually reflect these changes or become outdated and misleading. To be successful, a schedule must be accurate, and to be accurate, it must be kept up to date and revised on a regular basis.[33]

12.06 CPM PROOF OF DELAY CLAIMS

A. The "Classic Technique," Which Underlies CPM Analysis of Time-Related Claims

The basic technique used in evaluating contract claims with CPM is to compare the as-planned CPM schedule with the as-built CPM schedule. The technique can be summarized in the following six questions:

1. How was it planned that the project would be constructed?
2. Was the plan reasonable?
3. How did construction actually occur?
4. What are the variances, or differences, between the plan for performance and the actual performance with respect to activities, sequences, durations, manpower, and other resources?
5. What are the causes of the differences or variances between the reasonable plan and the actual performance?
6. What are the effects of the variances in sequence, duration, manpower, and so on as they relate to the costs experienced, both by the contractor and the owner for the project?

B. Methods for Proof of Delay Claims—An Overview

As noted, scheduling has become a means of both guiding performance through the contract's life and gauging the accuracy and appropriateness of

33. *See* Mergentime Corp. et al. v. Washington Metro. Area Transp. Auth., 400 F. Supp. 2d 145 (D.C. 2005) (Court criticized owner for not promptly incorporating logic changes and administering change requests but found the contractor primarily responsible given its slow responses to the owner's request for more substantiation); Lovering-Johnson, Inc., ASBCA No. 53902, 2006-1 B.C.A. (C.C.H.) ¶ 33,126 (Contractor's failure to provide proper baseline and updates rendered the schedule "useless during much of the period involved in this dispute and subject to after-the-fact manipulation and rationalization.") *Compare* Whitesell-Green, Inc., ABSCA Nos. 53928, et al., 2006-2 B.C.A. (C.C.H.) ¶ 33,323 (2006) (Parties found to have abandoned the use of the CPM schedule for the purposes of determining time extensions).

performance claims. In that regard, there are a variety of potential choices for the use of schedules to analyze project delays.[34] A listing of some of the major methodologies used today, along with a summary of advantages and disadvantages,[35] follows.

1. *Bar Chart*

Bar chart techniques continue to play a role on projects where there are only a few activities and there are linear-type relationships between those activities. This type of analysis is of limited help (without extensive supporting testimony) in proving the impact of construction delays on modern, complex construction projects with a variety of activities, complex logic relationships (predecessor, successor, concurrent), and varying resource requirements and availability. For example, in a 1998 decision, the Court of Appeals for the Federal Circuit affirmed a board decision that had criticized "the bar chart appellant provided which lacked a critical path, a favorite device with present day fact finders in contract disputes."[36] Similarly, in *Mega Constr. Co. v. United States*,[37] the Court of Federal Claims rejected the bar chart analysis of both the contractor and the government's experts. The court found that the contractor's charts had "incomprehensible overlays" and the government's chart was "virtually unreadable" and the government's expert did not know whether the chart was based on a five- or seven-day work week. Because no critical path analysis was offered by the contractor, the court only allowed those days that had been conceded by the government.

Bar charts, however, have been permitted to prove delay in circumstances in which a CPM was not used to execute the work,[38] costs of CPM analysis

---

34. The AACE International has recently identified the following 11 factors it believes should be considered in choosing an appropriate methodology: (1) contractual requirements; (2) purpose of analysis; (3) source data availability and reliability; (4) size of dispute; (5) complexity of dispute; (6) budget for schedule analysis; (7) time allowed for schedule analysis; (8) expertise of schedule analyst and resources available; (9) forum for resolution and audience; (10) legal or procedural requirements; and (11) past history/methods and what method the other side is using. *See AACE International Recommended Practice, No. 29R-03*, *supra* note 14, § 5.

35. For additional reading *see id.* and the *Delay and Disruption Protocol* issued in October 2002 by the Society of Construction Law of the United Kingdom.

36. Al Johnson Constr. Co. v. United States, 854 F.2d 467, 470 (Fed. Cir. 1988).

37. 29 Fed. Cl. 396 (1993).

38. *Compare* MES, Inc., PSBCA No. 4462, 2006-1 B.C.A. (CCH) ¶ 33,194, *mot. for recon. denied*, 2006-2 B.C.A. (CCH) ¶ 33,340 (Although contractor argued that a CPM analysis was not necessary to show delays since CPM was not a contract requirement, board stated that contractor still needed to prove its entitlement to a time extension by showing the government's actions impacted the overall completion date).

are prohibitive, and manifest delays of the owner are present. For example, in *Lamb Eng. & Constr. Co.*,[39] the board noted that although the contractor would have benefited by using CPM analysis, it found that the complexity of the project did not necessitate such analysis. Because this was coupled with evidence of serious, manifest government misconduct, the board allowed the contractor recovery on its delay claim.[40]

2. *"But For" Analysis/Collapsed As-Built*

This analysis calls for a protocol where (1) the reasonable planned schedule is compared to the as-built schedule with all delays encountered on the project; (2) the delays of the other party are removed ("collapsed") from the as-built schedule; and (3) the collapsed as-built is reviewed for anomalies representing "why hurry up and wait"[41] type of activities (rescheduled due to prior delays), and appropriate adjustments are made to remove the anomalies to arrive at an adjusted "but for" CPM analysis of project delays.

Perceived advantages of this approach are that it is based on the actual events on the project; it allows the comparison of a reasonable plan for performance with the as-built schedule. Variances between the plan and actual can be identified, and the causes of the variances and the effects, if any, can be determined. One can also compare the "but for" collapsed as-built with the plan and actual, not only to get an idea of delay effects to the critical path but also to make judgments relative to issues of lost efficiency.

There are a number of disadvantages to the collapsed as-built CPM. First, the after-the-fact approach fails to address the need to address the issue of time extensions on a real-time basis as required to address events on the project. Second, the analysis is not forward looking, chronological, or cumulative. Third, in order to collapse the schedule, the analyst is typically forced to insert after-the-fact logic ties, which may not reflect the thinking of the executor of the schedule during actual performance. Fourth, adjustments for anomalies in the adjusted schedule require experienced judgment, beyond the capability of many analysts and may be subject to dispute by experienced experts for others

39. Lamb Eng. & Constr. Co., EBCA No. C-9304172, 97-2 B.C.A. (CCH) ¶ 29,207; W. B. Constr. v. Mountains Community Hosp. Dist., 2005 Cal. App. unpub. LEXIS 5124, 12005 WL 1385571, Cal. Ct. App. 4th App. Dist., June 13, 2005) (fundamental and pervasive owner changes to plans and specifications.)

40. *See also* Hoffman Constr. Co. v. U.S., 40 Fed. Cl. 184 (1998).

41. When project delays occur and create additional float in the schedule, the contractor is not required to "hurry up and wait" and slavishly follow the original plan of performance. John Driggs Co., Inc., ENGBCA No. 4926, 87-2 B.C.A. (CCH) ¶ 19,833 at 100,388 (1987).

on the project.[42] Regardless of these disadvantages, "but for" analyses have, at times, been accepted as valid.[43]

3. Total Time Analysis

Total cost pricing, which asks the other party to pay for all losses on a project (actual costs plus markups less contract price), is subject to serious objections due to underlying assumptions.[44] For example, the approach assumes that the contractor's original bid was reasonable, that the contractor performed as efficiently on the project as anticipated in its bid, and that all the causes of the increased costs on the project were the sole responsibility of the owner.

Likewise, a total time analysis, which compares the contractor's plan with the actual as-built schedule and blames the owner for all the variances, is subject to similar attack for the underlying assumptions. These include the assumption that the original plan was reasonable, that the contractor performed at least as efficiently as planned, and that all the variances between the planned and actual are the responsibility of the owner (not due to excusable or contractor delays). There are a number of court and board decisions that have rejected the total time type of analysis.

For example, in *Morganti v. United States*,[45] the court noted that the contractor's expert "simply takes the original and extended completion dates, computes therefrom the intervening time or overrun, points to a host of individual delay incidents for which defendant was allegedly responsible and which 'contributed' to the overall extended time, and then leaps to the conclusion that the entire overrun time was attributable to defendant." The court held that a "total time" approach to proving delay was "as unsatisfactory as the 'total cost' method of proving damages," and denied the contractor's claim. Similarly, the board in *Southwest Marine, Inc.*, rejected the contractor's "disfavored total delay" theory, noting "cumulative inefficiency or impact is not proved by the issuance of numerous change orders without proof that these change

---

42. In Youngdale & Sons Constr. Co. v. United States, 27 Fed. Cl. 516 (1993), a contractor's delay claim was denied where its CPM expert testified he "did not put a lot of care" into his collapsed as-built (no delay) schedule and his as-built schedule conflicted with the project's contemporaneous record as set forth in the daily logs.

43. Fischbach & Moore Int'l Corp., ASBCA No. 18,146, 77-1 B.C.A. (CCH) ¶ 12,300 (1976); John Murphy Constr. Co., AGBCA No. 418, 79-1 B.C.A. (CCH) ¶ 13,836 (1979); Zurn Constructors, Inc. v. Castaic Lake Water Agency, 2003 Cal. App. Unpub. LEXIS 11267 (2d App. Dist, Div. 3, 2003).

44. Servidone Constr. Corp. v. United States, 931 F.2d 860 (Fed. Cir. 1991); Amelco Elec. v. City of Thousand Oaks, 27 Cal. App. 4th 228, 115 Cal. Rptr. 2d 900 (2002).

45. Morganti Nat'l, Inc. v. United States, 49 Fed. Cl. 110 (2001), *aff'd*, 36 Fed. Appx. 452 (Fed. Cir. 2002).

orders proximately caused that indirect or cumulative inefficiency or impact."[46] It should be recognized, however, that a party or an expert will rarely label its delay analysis as a total time approach. That label is usually applied by a court or board of contract appeals after they have found that the delay analysis has failed to adequately take into account any of the delays covered by the party presenting the analysis.

4. *Impacted As-Planned CPM*

This approach, which impacts the original CPM on the project solely with owner delays encountered during performance, suffers from one fatal flaw. It ignores what actually happened on the project and totally ignores excusable delays and delays by the contractor. Actual performance by all parties must be considered. Thus, hypothetical impacted as-planned network delay analyses that do not take into account actual events on the project as they evolve have been held unacceptable measures for evaluating project delays.[47]

5. *Chronological and Cumulative Approach/Time Impact Analysis*

This approach was developed to address the needs of the construction process to recognize time adjustments due the contractor in a timely manner and to provide the ability to resolve disputes prior to some exhaustive after-the-fact analysis reconstructed after the completion of the project.

This approach starts with a reasonable as-planned CPM. Next, the project is statused or updated (to take into account past performance) prior to the onset of a delay, which needs to be evaluated. In statusing the project, the location of the critical path must be identified. The original as-planned schedule is then modified from that point forward, incorporating the time impact that may or may not cause a project extension based on the location of the critical path and its relation to the delay in question. When the next delay is evaluated, the project network (as revised by the prior analysis and taking into account actual events on the project) is again statused immediately prior to the advent of the delay to locate the critical path; the second delay is then introduced into the

---

46. ASBCA No. 36854, 95-1 BCA ¶ 27,601 (1995). *See also* Santa Fe Engineers, Inc., ASBCA No. 24,578, et al., 94-2 B.C.A. (CCH) ¶ 26,872 (1994) (Board rejected total time approach because it believed some of the delays present might have been the contractor's responsibility); Webb Electric Co., ASBCA No. 54293, 2007-02 BCA (CCH) 33,717 (2007).

47. Titan Pac. Constr. Corp. v. U.S., 17 Cl. Ct. 630 (1989); Craft Machine Works, Inc., ASBCA No. 47227, 97-1 B.C.A. (CCH) ¶ 28,651 (1996) (Board rejected the impacted as-planned analysis of the contractor's expert and instead referred to CPM updates to determine whether the alleged delay was on the critical path at the time it occurred).

network to see what result it may have. The process is repeated until all delays have been analyzed.

This approach has been widely accepted and has significant merit.[48] However, care must be taken to make sure that during the period of the analysis of individual delays, other delays, which may be the responsibility of the contractor or owner, do not overtake the delay in question on the critical path. One way to avoid this problem is through the use of measuring points such as monthly updates. In this manner, the fact finder can not only look at the location of the critical path at the inception of the delay but also confirm the actual impact of delay by looking at the project status at the end of the update and the history of actual events of the project in the interim. This allows the fact finder to identify and net out concurrent delays from consideration of compensability.

6. Window Analysis

Window analysis is a variant of the chronological and cumulative approach detailed previously. This approach first looks at the status of the project immediately prior to the advent of the delay, such as the monthly update of the CPM. The actual events and delays are then introduced into a window of time, typically an update as they occurred. It is then determined which delays impacted the critical path, as well as which delays may have represented concurrent delays to the critical path. Like the chronological and cumulative approaches, this is a valuable tool and has been accepted by boards and courts. This method can be subject to criticism if the CPM expert or analyst selectively chooses or cherry-picks the "window" periods to suit the claimant's own purposes.[49]

Cogefar-Impresit U.S.A., Inc.[50] is a case that provides some lessons on the treatment of time impact and window analyses. In that case, the contractor was required to submit time impact analyses to establish entitlement to time extensions. The board found that the owner waived the right to terminate for default for delays, including the failure to submit time impact analyses in their proper form. Further, the board determined that the owner presented incorrect

---

48. *See, e.g.*, SAE/American-Mid Atlantic, Inc., GSBCA Nos. 12,294, et al., 98-2 B.C.A. (CCH) ¶ 30,084 (1998). Gulf Contracting, Inc., ASBCA Nos. 30195, et al., 89-2 B.C.A. (CCH) ¶ 21,812 (1989), *aff'd*, Gulf Contracting Co. v. U.S., 23 Cl. Ct. 525 (1991), *aff'd*, 972 F.2d 1353 (Fed. Cir. 1992) (table), *cert. den.*, 506 U.S. 999, 113 S.Ct. 598 (1992); Fru-Con Constr. Corp., ASBCA Nos. 53544, 53794, 2005-01 B.C.A. (CCH) ¶ 32,936.

49. *See* Old Dominion Elec. Coop. v. Ragnar Benson, Inc., 2006 U.S Dist. LEXIS 56145 (E. Dist. Va. 2006).

50. DOTBCA No. 2721, 97-2 B.C.A. (CCH) ¶ 29,188 (1997).

window analyses of delays because the window analyses failed to reflect actual conditions on the project at the time of the delays. This ruling was based on the principle that any evaluation of project delays must be based on a CPM schedule that is "current" as of the delay.[51]

C. The CPM Expert and the Need for Impartiality

Although the use of network analysis techniques to prove delay claims is now commonplace, a number of industry challenges remain. As the courts and boards of contract appeals become more knowledgeable about network analysis techniques, they have also become more skeptical about expert scheduling testimony, which appears biased and result driven. It is clear, therefore, that for expert scheduling testimony to be of value, it must reflect a fair and complete review of the project data and must reflect analyses that take into account major and controlling delays of all parties on the project.

The challenges facing the industry have been increased by the significant developments that have occurred in the federal arena during the past decade concerning the use and admissibility of expert testimony. These developments have effectively given the courts the power to exclude, prior to trial, "junk science" as well as other testimony that is "unreliable" for a number of reasons. Under *Daubert v. Merrell Dow Pharm., Inc.*[52] and subsequent cases, there is a clear potential to completely exclude expert testimony that may be essential to prove a cause of action for delay recovery or to deny such recovery, unless a proper groundwork has been laid by the party seeking to introduce such testimony.

Rule 702 of the Federal Rules of Evidence adopted in 1975 and amended in 2000 states:

> If scientific, technical, or other specialized knowledge will assist the trier of fact to understand the evidence or to determine fact in issue, a witness qualified as an expert by knowledge, skill, experience, training, or education may testify thereto in the form of an opinion or otherwise, if (1) the testimony is based upon sufficient facts or data, (2) the testimony is the product of reliable principles and methods, and (3) the witness has applied the principles and methods reliably to the facts of the case.

Under the Federal Rules of Evidence, the trial court must determine whether the proffered expert is reliable and fits the facts of the case. In *Daubert*, the Supreme Court provided guidance on the application of these principles. In that context, "scientific knowledge" requires that the proposed expert opin-

---

51. *Id.*
52. 509 U.S. 579 (1993).

ion be grounded in the methods and procedures of science; knowledge means more than subjective belief or unsupported speculation. "Scientific knowledge" means "any body of known facts or . . . any body of ideas inferred from such facts or accepted as truths on good grounds." The court in *Daubert* found that the trial court had the duty to act as a "gatekeeper" to determine the qualifications of the proposed expert scientific testimony *before* the trial court admits the testimony. In addition, in *Daubert*, the Court found that the gatekeeper function applied not just to novel testimony but also to more established scientific testimony.[53] Further, the 2000 amendments to Rule 702 incorporated the requirement that the trial court assure the reliability of even conventional expert testimony before admitting it into evidence. The federal principles set forth in *Daubert* are also used as guidance, if not precedent, by state courts, boards of contract appeals, and arbitration panels.

For example, in *The Sherman R. Smoot Corp.*,[54] the contractor moved to exclude the testimony and written report of the government's proposed expert witness who had prepared a CPM schedule analysis in an attempt to refute the contractor's delay claim. The contractor contended that the government's proposed witness was not qualified as an expert in construction CPM schedule analysis and could not assist the trier of fact to understand the evidence or to determine a fact in issue, pursuant to Rule 702 of the Federal Rules of Evidence. In ruling on the contractor's motion, the presiding judge at the board hearing reviewed the expert's qualifications set forth in his curriculum vitae, deposition transcript, and proposed expert report. The expert's curriculum vitae indicated that he had a Bachelor's and Master's Degree in Chemical Engineering and a Ph.D. in Industrial Engineering and industrial experience at Phillip Morris USA. The expert had also testified in his deposition that he lacked education and experience in construction, and his study and teaching of PERT/CPM analysis were in manufacturing and production, not construction. The expert further admitted in his deposition transcript that he had not prepared a construction schedule, had not used the terms "excusable" and "compensable" delay in dealings with contractors, and he was not an expert in "concurrency." After hearing this evidence, the presiding judge at the board hearing granted the contractor's motion to exclude from evidence the testimony and written report of the government's proposed expert. In so ruling, the presiding judge noted that the expert's qualifications in construction CPM generation, adjustment, and analysis were "quite thin," if not nonexistent. In the subsequent written decision on the contractor's claim, a three-member panel of the Armed

53. *Id.* at 590–93. *See also* Kumho Tire Co. v. Carmichael, 526 U.S. 137 (1997).

54. The Sherman R. Smoot Corp., ASBCA No. 52261, 03-1 B.C.A (CCH) ¶ 32,197 (2003).

Services Board of Contract Appeals affirmed the exclusion of the expert testimony and report.[55]

Of course, even when expert scheduling testimony has passed the *Daubert* test and is actually admitted in a court or board proceeding, if such testimony is not based on a fair or impartial review of the project record, it will most likely be found unpersuasive. Several board and U.S. Court of Federal Claims cases are illustrative.

In *Fraya*,[56] the government terminated a contractor for failure to make adequate progress on a renovation project in Puerto Rico. The contractor challenged the termination for default and offered the expert CPM testimony attempting to demonstrate that the contractor could, in fact, have completed the contract on time. The board rejected the expert's analysis and denied the contractor's appeal. Instead of using industry-recognized estimating guides, the expert had used what he referred to as "ways and means" of construction. This ways and means method of estimating later turned out to be nothing more than the contractor's oral estimates of the time required for various tasks. The board also rejected the expert's planned schedule because it was not resource loaded as required by the contract, did not take into account the limited building access because of security requirements, and failed to take into account the fact that the construction industry in Puerto Rico took extended holidays in December and January.

The contractor's CPM expert analysis was also rejected in *Jimenez*.[57] In this case, the contractor argued that delays resulting from the government's tardy approval of an air handling unit were compensable. The installation of the air handling units, however, was never shown as a critical path activity on any of the contractor's schedules that it had submitted during the performance of the project. Faced with that fact, the contractor hired a CPM expert. As the board noted, "not surprisingly," this expert's newly created CPM analysis showed that the delays to the air handling units accounted for the entire project delay. The board rejected the expert's opinion and noted that "such self-serving analyses, created after project completion which make adjustments to attain new and revised projected schedules, depending on theoretical contingencies, are of limited value."[58]

A CPM expert's failure to apportion government delays from concurrent contractor delays resulted in the complete rejection of a contractor's delay claim

55. 03-1 B.C.A. (CCH) ¶ 32,197 at 159,151. *See also* The Sherman R. Smoot Corp., ASBCA Nos. 52,173, et al., 03-1 B.C.A. (CCH) ¶ 32,212.

56. Fraya S.E., ASBCA No. 5222, 02-2 B.C.A. (CCH) ¶ 31,975 (2002).

57. Jimenez, Inc., VABCA No. 6351, et al., 02-2 B.C.A. (CCH) ¶ 32019 (2002).

58. 02-2 B.C.A. (CCH) ¶ 32,019 at 158,252.

in *Manuel*.[59] In this case, the contractor entered into an agreement with the Federal Aviation Administration to excavate and install concrete duct banks and manholes necessary for the installation of new fiber optic lines at the Dallas-Ft. Worth International Airport. The contractor sought over $2 million in delay-related damages based upon its allegation that the critical path of the project had been directly affected by (1) differing site conditions involving clay and rock; (2) breach of the government's implied duty to cooperate, which limited the contractor's access to the project site; and (3) additional time required for the location of unknown or misplaced utilities. Of the contractor's three claims, the court held that only the third, concerning the unknown misplaced utilities, was meritorious. The contractor's CPM expert testified that the contractor had been delayed for a total of seven and one-half months as a result of the three problems. The expert, however, could not give an opinion on the precise number of days of delay attributable to each issue. He also admitted he had based most of his analysis solely on information provided by the contractor's employees. Because the contractor could not separate the government-caused delay from the contractor's own delays, no delay damages were awarded.

12.07 CONCLUSION

CPM schedules have become widely used and accepted. Scheduling has become a means of both guiding performance through the contract's life and gauging the accuracy of and appropriateness of performance claims. The project schedule is an important management tool that enables the contractor to use personnel wisely and provide advance warning of situations threatening the profitability of a project, such as cost overruns, claims, and other difficulties. The project schedule is also an analytical device for claim recognition, preparation, and proof. If a claim does arise, the project schedule assists in determining what went wrong and who is responsible by comparing planned with actual performance.

59. Manuel Bros., Inc. v. United States, 55 Fed. Cl. 8 (2002).

QUESTIONS

1. Owner and contractor entered into a contract for the construction of a wastewater treatment facility. The contract specifications required the contractor to provide a CPM schedule, to be updated and submitted to the owner on a monthly basis. No other scheduling requirements were specified. During the course of the project, the owner took certain actions that had an impact on some of the contractor's activities, and the contractor filed a delay claim. The owner denied the claim on the basis that the activities that were impacted contained substantial float and that the overall construction schedule was not extended. The contractor admits that there was no impact to the critical path of its schedule but has nevertheless filed suit against the owner. It is the contractor's contention that it "owned" the float in its CPM schedule and that the owner's actions have adversely limited its scheduling flexibility on the project. How would a court likely decide this dispute?
2. Owner and contractor entered into a contract for the construction of a forty-story hotel and conference center. The contractor was required to submit a CPM schedule and monthly updates to the owner's construction manager. The construction manager was a recent college graduate with no experience in CPM scheduling. The construction manager was embarrassed that he could not read, analyze, or understand the contractor's CPM schedules but, because he feared he would lose his job, did not tell anyone. Instead, throughout the course of the project, the construction manager simply placed the contractor's monthly updates on the corner of his desk without reviewing or commenting upon their contents. In fact, as time went on, the construction manager would chuckle to himself every time he received a new update because placing the updates on the corner of his desk served the dual purpose of obscuring his computer screen from office visitors while he surfed the Internet and also gave the impression that he was overworked due to the volume of paperwork on the project. Shortly before substantial completion of the work, however, the contractor filed a multimillion-dollar delay and disruption claim against the owner. In defending against this claim, the owner hired an expert scheduling consultant who discovered that the contractor's monthly updates had incorrect as-built data and had been improperly manipulated to show significant owner-caused delay and to mask delays of the contractor. How could the owner have avoided this problem?

CHAPTER

13

Architect's Contract Administration

CARL J. CIRCO

13.01 INTRODUCTION

Owners often hire design professionals to provide contract administration services in addition to furnishing drawings and specifications for the project. In this context, "contract administration" refers to a wide range of activities that assist the owner in its contractual relationship with the general contractor. Those activities may include providing information to the contractor, inspecting the progress of construction, preparing documents in connection with changes to the work, helping to resolve disputes between the owner and the contractor, and many other administrative tasks. Although an owner might hire an architect, an engineer, or some other professional to provide contract administration, for convenience this chapter presumes that an architect is providing those services.

In general, the owner and the architect may agree on whatever contract administration services and procedures they think most appropriate for the project. For many projects, however, the owner retains the same architect who prepares the drawings and specifications to provide a comprehensive array of contract administration services that conform to established industry customs. The 2007 editions of AIA Documents B101[1] (especially Section 3.6) and A201[2] (especially Sections 4.2 and 15.2) illustrate one of the most common arrangements for contract administration for a commercial project using the Design-Bid-Build project delivery method. For that reason, much of this chapter discusses the architect's contract administration role in a way that reflects the contract administration provisions of these AIA documents. You may want to review these AIA provisions in conjunction with this chapter.

13.02 NATURE AND SCOPE OF ARCHITECT'S ROLE IN CONTRACT ADMINISTRATION

As you will see from the materials in this chapter, contract administration may involve many different activities and may raise several important legal issues. Depending on the owner's needs, the architect may provide such extensive services during the construction phase of a project that the architect occupies a central role throughout the entire construction process. But in other projects, the architect may be involved only in discrete administrative functions once construction begins. For still other projects, the architect may not be involved in any of the administrative aspects of the project at all.

The main purpose of this section is to familiarize you with the general concept of contract administration. Additionally, because agency law provides the underlying legal construct for certain key aspects of contract administration, the concluding subdivision in this section briefly highlights some of the agency law considerations that will be relevant to certain issues presented in later sections of this chapter.

A. Overview of Contract Administration

The contract for design services normally establishes the nature and scope of the architect's role in contract administration.[3] For most projects, nothing in the industry or the law requires the architect to have an ongoing role during construction although licensing regulations may require that certain activities

1. American Institute of Architects, AIA Document B101–2007, *Standard Form of Agreement Between Owner and Architect* (2007) [hereinafter AIA Document B101–2007].

2. American Institute of Architects, AIA Document A201–2007, *General Conditions of the Contract for Construction* (2007) [hereinafter AIA Document A201–2007].

3. *See* AIA Document B101–2007 *supra* note 1, § 3.6.1.1 (providing that the architect will provide contract administration as contemplated by AIA Document A201–2007).

relating to the project must be performed only by, or under the supervision of, a properly licensed design professional. Even though some contract administration services are customary for certain segments of the industry, the architect's role may vary greatly based on the experience, sophistication, staffing, and risk tolerance of the owner. For example, an experienced developer of commercial projects might have its own contract administration staff to perform many of the functions that an architect might provide for a less experienced owner. Or a city or other public owner might use an employee, such as the city engineer, to provide most contract administration. Other owners might limit the architect's role to a few administrative functions, such as reviewing the contractor's applications for payment and periodically inspecting the progress of the work. On a residential project, if there is any professional contract administration at all, it may be limited to a walk-through of the house before the owner makes the final payment.

An owner will also sometimes retain an architect to fill a much larger role than normal contract administration, such as serving as a construction manager or project manager. Or the owner may retain an architect to provide continuous on-site representation. The industry sometimes refers to this more intensive on-site position by the term "clerk of the works."[4] But even in these more expansive administrative arrangements, the architect's role in contract administration normally does not give the architect authority over such matters as the means and methods of construction. Indeed, the contract documents customarily reserve these matters to the contractor.[5]

In the course of typical contract administration, the architect may function in three distinct administrative capacities. First, for some purposes the architect may act as the owner's agent. For example, the contract documents may require that the contractor secure the architect's approval or review of certain construction details, such as a proposed submittal schedule or a list of proposed subcontractors.[6] On these matters, the architect frequently acts on behalf of the owner in an agency capacity. Second, for other purposes, the architect often serves as the owner's representative or consultant. This is commonly the

4. *See* Bd. of Educ. of Hudson City Sch. Dist. v. Sargent, Webster, Crenshaw & Folley, 539 N.Y.S.2d 814, 818 (N.Y. App. Div. 1989) (in which the court concluded that the evidence at trial was subject to conflicting inferences about whether the individual serving as the clerk of the works was the owner's agent or the project architect's agent).

5. *See* AIA Document B101–2007, *supra* note 1, § 3.6.1.2.

6. *See* AIA Document A201–2007, *supra* note 2, §§ 3.10.2, 5.2.

case, for example, when the architect reviews the contractor's periodic applications for payment.[7] The architect generally reviews the applications for payment for the purpose of advising the owner whether to make payment in the full amount requested. Finally, the contract between the owner and the contractor often refers to the architect some or all disagreements, disputes, and claims that arise during construction.[8] When this happens, the architect typically fulfills the role of a neutral mediator or referee between the owner and the contractor.

B. Overview of Agency Law Principles Relevant to Contract Administration

As a preliminary matter in analyzing the legal issues involved in contract administration, it is often important to determine whether, or the extent to which, the architect is serving as the owner's agent. This subdivision summarizes certain agency law principles that the reader should keep in mind when the architect acts as the owner's agent. Often, the contract documents give limited express authority to the architect to act for the owner on certain matters, and they also expressly notify the contractor that the architect has no authority to act for the owner on other matters.[9] Additionally, the circumstances may establish apparent agency authority for the architect to bind the owner. Authority may also be implied from provisions of the contract documents or from the circumstances or from a course of dealing between the parties. As already noted, however, for some contract administration purposes, the architect does not serve as the owner's agent at all, but as either a consultant to the owner or as an intermediary or referee between the owner and the contractor. For that reason, before relating the agency law principles summarized here to any particular facet of contract administration, the reader must first determine whether or not the architect is serving as the owner's agent in connection with that facet.

1. *Architect's Authority to Bind the Owner*

The scope of the architect's agency as established by the relevant contract documents determines generally the extent to which the architect may bind the owner contractually. For example, if the agreement between the owner and the contractor requires the contractor to submit the names and experience of ma-

---

7. *See* AIA Document B101–2007, *supra* note 1, § 3.6.3.

8. *See id.*, § 3.6.2.5 (unless otherwise agreed, the architect will make initial decisions on claims as contemplated by Article 15 of AIA Document A201–2007).

9. *See* AIA Document A201–2007, *supra* note 2, § 4.2.1.

jor subcontractors to the architect for a determination of whether the owner or the architect has any objections to the proposed subcontractors, the architect's approval of a subcontractor (or the architect's failure to make a timely or reasonable objection to a proposed subcontractor) probably will have the effect of waiving the owner's right to raise objections to that subcontractor. But an action by the architect that is outside the scope of the agency authority granted to the architect will generally not bind the owner. Thus, if the contract requires that both the owner and the architect must approve change orders that affect the contract price or schedule, the architect normally would not have authority to approve a change order on the owner's behalf.

In *Incorporated Town of Bono v. Universal Tank & Iron Works, Inc.*,[10] the owner's architect told the contractor that a written request for an extension of the time for performance would not be required and that the architect would "take care of the matter." Under the owner-architect agreement, the architect had neither the express nor implied authority to bind the owner to an extension of time. The court held that the architect was not a general agent of the owner but was an agent with limited authority.

2. Owner's Liability for the Architect's Acts

When the architect acts within the scope of the agency, the owner may be liable for the architect's acts and omissions. For example, if the contract documents authorize the architect to direct the contractor to perform additional work, the owner will be liable to pay the contractor for that work, even if the architect, without the contractor's knowledge, violated the owner's instructions by ordering the particular work.[11]

3. Architect's Liability for Exceeding Authority or for the Improper Exercise of Authority

An architect who acts outside the designated scope of authority or whose exercise of that authority is negligent may incur liability to the owner, the contractor, or other third parties adversely affected. This liability may be based on either a contract theory or a tort theory, depending upon the jurisdiction's enforcement of the economic loss rule.[12]

---

10. 395 S.W.2d 330 (Ark. 1965).

11. *See, e.g.*, Scribante v. Edwards, 181 P. 75 (Cal. Ct. App. 1919) (contract directed the contractor to lay a concrete floor under the direction of the owner's architect).

12. See Chapter 23.

13.03 REQUESTS FOR INFORMATION, INTERPRETATIONS, AND CLARIFICATIONS

Many technical matters relating to the drawings and specifications for a project may be referred to the architect during the construction phase through requests for information (RFIs) or may otherwise require interpretations or clarifications by the architect. Sometimes, the contract documents will establish a detailed process for securing the architect's input on these questions.[13] Most commonly, these technical questions come from the contractor or from a subcontractor who submits the inquiry through the contractor. Architects sometimes complain that contractors overuse the process and, in those situations, the architect may have a claim under the owner-architect agreement for additional compensation for providing contract administration services beyond those contemplated by the contract.[14] But contractors often argue that the RFI process is essential to uncover omissions and errors in the design documents and to avoid problems in the field.

Deficient drawings may lead to an unusual number of RFIs and may result in contractor claims for additional compensation. Although the construction contract may require the contractor to review the drawings before undertaking the work and to bring to the owner's attention errors or defects that the contractor discovers, the contractor will have a valid claim if the RFI reveals design defects that lead to delays or additional costs, especially if the defects are latent.[15] Similarly, contractor claims may arise from a failure to respond promptly or properly to RFIs. When information that the architect provides proves to be inaccurate, incomplete, or erroneous, a contractor or other participant in the project who relies on that information may assert that the architect is liable for making a fraudulent or negligent misrepresentation.[16]

Metropolitan Atlanta Rapid Transit Auth. v. Green Int'l, Inc.[17] provides an extreme example of a project plagued by requests for information stemming from design deficiencies. There, the court commented that testimony indicated that "the state of the plans created a 'design-as-you-go situation' requiring over 350 written requests for information and clarification or 'RFIs' and over 1,000

---

13. *See* AIA Document B101–2007, *supra* note 1, § 3.6.4.

14. *See id.*, § 4.3.2.

15. *See* 5 Philip L. Bruner & Patrick J. O'Connor, Jr., Bruner and O'Connor on Construction Law § 15:60 (West Group 2006) [hereinafter Bruner and O'Connor].

16. *See Hudson City*, 539 N.Y.S.2d at 814, 819; U.S. v. Rogers & Rogers, 161 F. Supp. 132 (S. D. Cal. 1958). For further information about liability for negligent misrepresentation, *see* Restatement of Torts (Second) § 552 (1977).

17. 509 S.E.2d 674 (Ga. Ct. App. 1999).

revised drawings while the project was already underway."[18] There was also evidence in that case that untimely responses to RFIs contributed to delays in construction.[19] The court upheld $2.8 million in damages that the jury awarded to the contractor on its claim against the owner.

13.04 QUALITY CONTROL

The contract documents may subject a broad array of matters to the architect's review or approval. In acting on these matters, the architect commonly acts as the owner's consultant or representative (or sometimes as the owner's agent), usually for some purpose that serves to assure the project's quality or its conformance with the drawings and specifications. The subdivisions that follow briefly discuss some of the most common situations in which the contract documents may call for the architect's involvement during the construction phase for quality control purposes.

A. Review and Approval Procedures

The contract documents may be nearly silent on the process for the architect's review or approval, or they may set out the process in great detail. In either case, disputes or claims are likely to arise if the contractor fails to seek the architect's input on a timely basis or if the architect is negligent or dilatory.

B. Approving or Objecting to Subcontractors, Manufacturers, and Suppliers

Absent restrictions stated in the contract between the contractor and the owner, the contractor has the right to assign portions of the work to others without seeking the owner's input or approval. Often, however, the contract documents give the owner or the architect (or both) the opportunity to raise reasonable objections to proposed subcontractors.[20] Additionally, the drawings and specifications or other contract documents may identify manufacturers or suppliers from which the contractor must choose to provide certain equipment and materials. Alternatively, they may require the contractor to submit to the architect information about proposed manufacturers or suppliers for the architect's approval or comment.

---

18. *Id.* at 676.

19. *Id.* at 678 (apparently the owner rather than the architect delayed in responding to the RFIs).

20. *See* AIA Document A201–2007, *supra* note 2, § 5.2.

C. Substitutions, Shop Drawings, and Other Submittals

Waggoner v. W & W Steel Co.
657 P.2d 147 (Okla. 1982)

BARNES, Vice Chief Justice:

This case presents for resolution on appeal, the following issue:

Is the architect who designed this building responsible for ensuring that the contractor employ safe methods and procedures in performing his work?

During the construction of Presbyterian Hospital in Oklahoma City, Oklahoma, a portion of the steel framework fell, resulting in the death of two workmen and the injury of another. The accident occurred on a Friday, as workers were beginning to secure a portion of the steel in the sixth, seventh and eighth floors which had been erected that day. The three men were on the structure waiting for other workers to bring up guy lines which were to be used in securing the steel before an approaching thunderstorm rolled in over the construction site. Before the task could be completed, a gust of wind hit the unsecured and unbraced steel, causing it to collapse.

In designing this building, the architects had provided for expansion joints which would allow for expansion and contraction with changing weather conditions. The expansion joint was designed so that a shelf, welded to a column, provided a seat for a beam which was held in place by "keeper angles" welded in on either side of the beam. The opposite end of the beam was secured to another column with large bolts. Unfortunately, at the time of the accident, the "keeper angles" had not been installed and the beams were not secured in any other way.

The erector had built portions of three floors rather than completing all the work on one floor at a time. As a result, the east half of the new sections were without interior columns and cross beams which would have provided lateral bracing for the outside columns. These outer columns bore the weight, not only of the steel beams and trusses which had been erected, but also of large bundles of steel decking which were placed on the beams at each floor level. The outside columns were held upright only by temporary "clip angles" provided by the fabricator to facilitate the alignment and welding of the top columns to the columns below.

After the accident, suit was brought against the owners of the hospital, the fabricator and the architect by the injured worker and representatives of the estates of the deceased workers. The case (three cases were consolidated) went to jury trial with the architect as the sole defendant, the other defendants having been released from the lawsuit by dismissals and summary judgments. After plaintiffs and defendant rested, the trial court directed a verdict for the architect. The Court of Appeals, in reversing the lower court, determined that the architect did owe a duty to the workers, in that the architect "undertook to supervise the construction project." With this, we disagree.

Architects are required to exercise ordinary professional skill and diligence and to conform to accepted architectural standards. Because an architect's undertaking does not imply a guarantee of perfect plans or results, he is liable only for failure to exercise reasonable care and professional skill in preparation and execution of plans *according to their contract. Smith v. Goff,* 325 P.2d 1061 (Okla. 1958) (emphasis added). This principal [sic] was reaffirmed in *Wills v. Black & West, Architects*, 344 P.2d 581 (Okla.1959) in which we looked to the contract between the parties to determine the responsibilities of the architect. It is stated in 5 AM.JUR.2D ARCHITECTS § 5 (1962) that:

> (t)he employment of an architect is ordinarily a matter of contract between the parties, and the terms of such employment are governed by the terms of the contract into which they entered. The architect's duties may be limited to the preparation of plans and specifications, or they may include, in addition, the supervision of construction.

At common law, privity of contract was required before an action in tort could arise from a breach of duty created by a contract. However, in *Truitt v. Diggs,* 611 P.2d 633 (Okla.1980), we indicated that in cases involving physical injury to third parties, that restriction has in many cases been eliminated or modified. Therefore, it is possible for an architect to be liable for injuries received by a person with whom he has no privity, but there can be no standard rule. The determination must be made by considering the nature of the architect's undertaking and his conduct pursuant thereto.

To do so in the present case, we must look to the contract which actually consists of several separate documents. Pertinent provisions of the General Conditions of the Contract for Construction are set forth below.

> 2.2.4 The Architect will make periodic visits to the site to familiarize himself generally with the progress and quality of the Work and to determine in general if the Work is proceeding in accordance with the Contract Documents. On the basis of his on-site observations as an architect, he will keep the Owner informed of the progress of the Work, and will endeavor to guard the Owner against defects and deficiencies in the Work of the Contractor. *The Architect will not be required to make exhaustive or continuous on-site inspections to check the quality or quantity of the Work. The Architect will not be responsible for construction means, methods, techniques, sequences or procedures, or for safety precautions and programs in connection with the Work, and he will not be responsible for the Contractor's failure to carry out the Work in accordance with the Contract Documents.* (emphasis added).
>
> 4.3 SUPERVISION AND CONSTRUCTION PROCEDURES
>
> 4.3.1 The Contractor shall supervise and direct the Work, using his best skill and attention. He shall be solely responsible for all construction

> means, methods, techniques, and sequences and procedures and for coordinating all portions of the Work under the Contract.
>
> 4.13.1 By approving and submitting Shop Drawings and Samples, the Contractor thereby represents that he has determined and verified all field measurements, field construction criteria, materials, catalog numbers and similar data, or will do so, and that he has checked and coordinated each Shop Drawing and Sample with the requirements of the Work and of the Contract Documents.
>
> 4.13.5 The Architect will review and approve Shop Drawings and Samples with reasonable promptness so as to cause no delay, but only for conformance with the design concept of the Project and with the information given in the Contract Documents. The Architect's approval of a separate item shall not indicate approval of an assembly in which the item functions.
>
> 10.2 SAFETY OF PERSONS AND PROPERTY
>
> 10.2.1 The Contractor shall take all reasonable precautions for the safety of, and shall provide all reasonable protection to prevent damage, injury or loss to:
>
> .1 all employees on the Work and all other persons who may be affected thereby; . . .
>
> 10.2.2 The Contractor shall comply with all applicable laws, ordinances, rules, regulations and lawful orders of any public authority having jurisdiction for the safety of persons or property or to protect them from damage, injury or loss. He shall erect and maintain, as required by existing conditions and progress of the Work, all reasonable safeguards for safety and protection, including posting danger signs and other warnings against hazards, promulgating safety regulations and notifying owners and users of adjacent utilities.

Appellants maintain that the trial court was in error when he directed a verdict against them because the architect owed a duty to them to exercise ordinary professional skill and diligence in preparing and approving plans for construction, and therefore the question of the alleged violation of that duty should have been submitted to the jury.

We must determine if, under the contractual provisions set out above, the architect had such a duty.

Section 4.3.1 of the General Conditions specifies that the contractor is to supervise the work, being "solely responsible for all construction means, methods, techniques, and sequences and procedures." Although the architect is to periodically visit the construction site, § 2.2.4 provides that he is not required to make "exhaustive or continuous on-site inspections to check the quality or quantity of the work." It goes on to exclude from the architect's responsibilities those which are outlined in 4.3.1 as belonging solely to the contractor. It should also be noted that the owners of the hospital employed an engineer as the project inspector.

Article 10, entitled "Protection of Persons and Property" contains provisions which require the contractor to protect workmen from injury, comply with safety regulations and laws and designate a superintendent whose job it is to prevent accidents. It is to be noted that the responsibilities for all safety precautions and programs are assigned exclusively to the contractor.

Article 4 sets forth the procedure and purpose concerning the shop drawings which appellants contend should have included specifications for temporary bracing and connections. The shop drawings, prepared by a subcontractor, were submitted to the contractor and the architects, as required by the contract. It is the architects' approval of the shop drawings, without provision for temporary connections on the expansion joints, that appellants maintain was negligence.

However, § 4.13.4 provides that the shop drawings are submitted to the *contractor* for determination and verification of "all field measurements, field construction criteria, materials, catalog number and similar data." By approving them, he represents that he has checked each shop drawing with the requirements of the contract. And, as previously noted, § 4.3.1 of the contract makes the contractor "solely responsible for all construction means, methods, techniques, sequences and procedures."

It is apparent that the shop drawings serve more than one purpose. They are submitted to the contractor for approval regarding aspects of the construction work. But, according to § 4.13.5, *they are submitted to the architects for approval "only for conformance with the design concept of the project and with the information given in the Contract Documents."*

Therefore, it was the duty of the contractor, not the architects, to see that the shop drawings included provisions for temporary connections which fall into the categories of "field construction criteria," "construction means, methods, techniques, sequences and procedures." Since it was not the responsibility of the architects, they obviously would not be negligent in failing to require temporary connections.

Because the contractor, not the architect, was required under the contract to supervise the job and employ all reasonable safety precautions, the architects cannot be held liable for injuries sustained as a result of an unsafe construction procedure. There was no question of fact for the jury. The trial court properly directed a verdict for the architects.

The judgment of the Court of Appeals is vacated and the decision of the trial court is affirmed.

IRWIN, C.J., and LAVENDER, SIMMS, DOOLIN, HARGRAVE, OPALA and WILSON, JJ., concur.

The contract documents commonly require the contractor to submit certain construction details and information for the architect's review.[21] The primary purpose of these requirements is to allow the architect to confirm that the construction will conform to the design concept embodied in the architect's drawings and specifications. For example, because certain fabrication, installation, and construction details require the input of manufacturers, suppliers, and trade contractors, the contract documents may require the contractor to prepare or secure drawings and other information relating to these details. Shop drawings are plans that furnish these special details. Other submittals may include material or product samples, models, and technical information from manufacturers, suppliers, and fabricators.

The architect usually wants to have the opportunity to review these details without taking any responsibility for them because they involve matters that are more properly within the expertise of those who will carry out the work. In other words, the architect's review does not give the architect control over construction means, methods, or techniques; it merely allows the architect to determine whether these aspects of construction appear to be consistent with the design expressed in the drawings and specifications.[22]

Additionally, the contract documents often permit the contractor to propose for the architect's approval substitutions for equipment or materials specified in the drawings and specifications. This option recognizes that the contractor and its subcontractors and suppliers may have more complete or current information about available products and equipment and may be able to suggest substitutions that meet the quality and functional requirements of the drawings and specifications while offering some advantage over the specified equipment or materials. For example, a substitution may reduce costs, avoid potential delays in delivery schedules, or facilitate the construction process in some other way.

As the *Waggoner* case illustrates, the architect's review and approval of shop drawings or other technical submittals has been a fertile source of litigation, especially when the defect created a safety hazard.[23] In reaction to cases of that nature, project architects and engineers have become exceedingly careful to avoid contractual responsibility for approving or commenting on shop

---

21. *See* AIA Document B101–2007, *supra* note 1, § 3.6.4.

22. *See id.*, § 3.6.4.2.

23. *See, e.g.*, Duncan v. Mo. Bd. for Architects, Prof'l Eng'rs & Land Surveyors, 744 S.W.2d 524 (Mo. Ct. App. 1988) (revocation of licenses of engineers who failed to detect structural error in shop drawings prepared by steel fabricator that resulted in a catastrophic collapse of pedestrian walkways in a hotel lobby); Jaeger v. Henningson, Durham & Richardson, Inc., 714 F.2d 773 (8th Cir. 1983) (architect who failed to recognize that a shop drawing called for lighter gauge steel in a stairway than that required by the construction drawings was liable for a resulting injury).

drawings and other submittals, except for the limited purpose of confirming that they conform to the specified design concepts.

D. Decisions on Aesthetic Effect

Architects often retain final authority over decisions that affect project aesthetics. In contrast to how they usually address shop drawings and submittals, the contract documents normally require the architect's approval of aesthetic details, and they may make the architect's decision on these matters final.[24]

E. Site Visits and Inspections

In many projects, one of the architect's main responsibilities during the construction phase is to visit the site periodically to observe the progress of construction in an effort to guard the owner against defects in the work.[25] The AIA form documents attempt to limit the architect's responsibility for the contractor's failure to perform the work in accordance with the contract documents,[26] but this effort has met with mixed success, as shown in the following case.

Board of Educ. of Hudson City School Dist. v. Sargent, Webster, Crenshaw & Folley
539 N.Y.S.2d 814 (N.Y. App. Div. 1989)

LEVINE, Justice.

In 1966, plaintiff, the Board of Education of the Hudson City School District (hereinafter the School District), entered into an agreement with defendant, Sargent, Webster, Crenshaw & Folley (hereinafter the Architect), on a standard form of agreement of the American Institute of Architects (hereinafter AIA) for the provision of plans and specifications and supervisory architectural services in connection with the construction of a new high school building. Under paragraph 7 (b) of the agreement, the Architect was obligated to make periodic visits to the work site "to determine in general if the work is proceeding in accordance with the Contract Documents." The Architect was also required to "keep [the School District] informed of the progress of the work, will endeavor to guard [it]against defects and deficiencies . . . and . . . may condemn work as failing to conform to the Contract Documents." The Architect was also to issue certificates of payment, thereby representing that the work was properly performed and authorizing progress

24. *See* AIA Document B101–2007, *supra* note 1, § 3.6.2.4.

25. *See id.*, § 3.6.2.

26. *Id.*, § 3.6.1.2.

and final payments by the School District to the general contractor. Paragraph 7 (b) provided, however, that the Architect was not required to make "exhaustive or continuous on-site inspections to check the quality or quantity of the work and . . . will not be responsible for the Contractors' failure to carry out the construction work in accordance with the Contract Documents." The agreement also required the Architect to assemble any written guarantees required of contractors and to make three annual inspections of the building following completion of construction and to advise the School District of any maintenance and remedial measures required.

Third-party defendant, Thompson Construction Corporation (hereinafter Thompson), was engaged as general contractor and construction commenced in the spring of 1969. Thompson contracted with Skyway Roofing Company, Inc. (hereinafter Skyway) to install the roof on the building. The specifications called for a four-ply, built-up, gravel-surfaced roof consisting of four layers of felt and coal tar pitch to assure bonding and waterproofing, with the gravel embedded in the top level of pitch. These layers were to be applied on top of a poured gypsum roof deck. The School District submitted evidence that, for a roof of this type, the layers as applied must be protected from exposure to precipitation during construction, since moisture trapped within the layers has a tendency to vaporize and expand, creating blisters in the plies of felt which may crack and cause leakages. Therefore, the specifications expressly required completion of the application of the plies of roofing in one day or, failing that, the application of a glaze coating of hot bitumen to protect the felts until application of the gravel.

The School District also proved that Skyway failed to comply with the foregoing specifications in that application of the layers was done in phases, leaving the felt plies exposed to rain and snow without protective glazing. In fact, Thompson's daily construction reports indicated that Skyway personnel shoveled snow off the partially completed roof and that this was observed by one of the Architect's representatives on the site. The Architect then engaged an outside roofing expert to inspect Skyway's installation. He concluded that the roof system was seriously out of conformity with the specifications, alluding to the phased construction but, more importantly, to another defect, i.e., the fact that the vapor barrier was not attached to the poured gypsum deck. Based upon that report, the Architect directed Thompson to remove and replace the defective roofing. Thompson refused, insisting that the roof was properly applied. Following various conferences and exchanges of correspondence, the Architect agreed to several less severe remedial measures and, once these were completed, issued certificates of payment for the roofing.

The construction of the high school was completed and a final certificate of payment was issued by the Architect in September 1972. It was much later discovered, however, that the Architect had failed to obtain the two roofing guarantees required by the specifications, i.e., a two-year guarantee from Thompson and Skyway and a manufacturer's 20-year roofing bond. Leaks in the roof developed as early as November or December 1972. Upon initially consulting the Architect about the leaks, the School District's building and grounds superintendent was told by the Architect's representative that the cause of the problem was ice buildup. Each succeeding year, the blistering and the leaks became worse and, by 1979, the School District contemplated replacement of the roof. At this point, it was informed by the Architect that the cause of the problem was the phased installation in violation of the specifications and that the guarantees required by the contract inadvertently had not been obtained.

The School District then brought the instant suit against the Architect and Thompson. Its complaint set forth five causes of action against the Architect, the first two of which were for breach of contract, alleging in the first the Architect's failure to obtain the roofing guarantees and, in the second, the Architect's failure to have informed the School District of the deviation from the specifications in the roof installation and failure to have guarded the School District against the general contractor's defective performance and to have condemned the work as failing to conform to the contract documents.

. . .

In the foregoing procedural posture, the case proceeded to trial. At the conclusion of all of the evidence, Supreme Court granted the Architect's motion for a directed verdict dismissing all but the first of the School District's causes of action. The court directed a verdict in favor of the School District on its first cause of action (the failure to obtain the roofing guarantees) and assessed damages against the Architect for $6,000, the penal sum of the roofing bond called for by the contract. Finally, the court dismissed the Architect's third-party complaint against Thompson, ruling that the evidence did not establish the Architect's claim for indemnification. These cross appeals ensued.

In our view, the School District's proof regarding its second cause of action was sufficient to withstand the motion for a directed verdict. Hence, the portion of the judgment dismissing the second cause of action should be reversed, that claim reinstated and remitted for retrial. Supreme Court's contrary ruling was based upon the provision of paragraph 7 (b) of the agreement under which the Architect was absolved from responsibility "for the Contractors' failure to carry out the construction work in accordance with the Contract Documents," reading that provision as totally exonerating the Architect from liability for construction defects. While this very clause in the standard AIA architect/owner contract has been given exculpatory effect in this State and other jurisdictions (*see, Jewish Bd. of Guardians v. Grumman Allied Indus.*, 96 AD2d 465, 467, *affd* 62 NY2d 684; *Shepard v. City of Palatka*, 414 So 2d 1077 [Fla]; *Moundsview Ind. School Dist. No. 621 v. Buetow & Assocs.*, 253 NW2d 836 [Minn]), none of the cases involved defects known by the architect during the course of construction, which he failed to apprise the owner of under the contractual duty to "keep the [School District] informed of the progress of the work." We decline to extend the application of the clause in question to an instance such as this, where the trier of facts could find that the architect was aware of the defect and failed to notify the owner of it.

First, in general, exculpatory provisions are disfavored and are narrowly construed (*see Gross v. Sweet*, 49 NY2d 102,106–107). Second, the underlying rationale for exonerating an architect under the exculpatory clause does not apply to defects known by the architect. The cases reason that an owner who has not contracted and paid for an architect to closely supervise construction should not be able to hold the architect liable for those defects in the contractor's performance which close supervision would have revealed (*see Jewish Bd. of Guardians v. Grumman Allied Indus.*, supra., at 465; *Moundsview Ind. School Dist. No. 621 v. Buetow & Assocs.*, supra., at 839). Clearly, however, the School District did bargain and agree to pay the Architect for periodic inspection and conveyance of information on the progress of the contractor's performance which the Architect acquired as the result of such on-site visits. When, as a result of periodic inspection, an architect discovers defects in the progress of the work which the owner, if notified, could have taken steps to ameliorate, the imposition of liability upon the architect for failure to notify would be based on a breach of his own contractual duty, and not as a guarantor of the contractor's performance (*see Hunt v. Ellisor & Tanner*, 739 SW2d 933, 937 [Tex]).

. . .

Alternatively, the Architect argued before Supreme Court and urges now that the School District's second cause of action failed because of the absence of expert testimony that the Architect's performance fell short of accepted professional standards of architectural practice. Such a deviation by an architect may, it is true, afford the basis of a contract cause of action as well as recovery in tort (*Sears, Roebuck & Co. v. Enco Assocs.*, 43 NY2d 389, 396). At the least, however, the claim in the second cause of action based upon the Architect's failure to notify of the known defective roof installation asserts the breach of a specific contractual undertaking by the Architect, akin to the failure to achieve a particular promised result, which does not require proof of a failure to meet professional standards of care.

. . .

We also reject the Architect's contention that the School District's proof of damages, i.e., the 1972 costs to correct by replacement of the defectively installed roof, was improper as a measure of damages. The cost to correct and replace defective construction may be a proper measure of damages for a breach of contract by an architect (*Sears, Roebuck & Co. v. Enco Assocs.*, supra.; *Hubert v. Aitken*, 2 NYS 711, 713, *order adhered to on rearg* 5 NYS839, *affd* 123 NY 655; *Schwartz v. Kuhn*, 71 Misc 149, 151; 5 Am Jur 2d, Architects, § 24, at 687–688). The testimony of the School District's roofing expert was sufficient to prima facie establish its damages on the foregoing theory.

. . .

Judgment modified, on the law, without costs, by reversing so much thereof as dismissed plaintiff's second cause of action; said cause of action reinstated and matter remitted to the Supreme Court for a new trial thereon; and, as so modified, affirmed.

WEISS, J.P., and MIKOLL, YESAWICH and HARVEY, JJ., concur.

Suffice it to say that when defective work has been installed by the contractor—especially when the contractor has subsequently become insolvent—the search for culprits often includes the architects who performed site visits. The AIA language is intended to exculpate the architects from this liability, but has met with somewhat inconsistent results, as demonstrated by *Hudson City*[27] and by the *Moundsview*[28] and *Ellison & Tanner*[29] cases cited in the opinion. Indeed, when the owner's design professional agrees to provide extensive on-site services, some courts have interpreted that obligation to create a duty to other project participants who rely on those services, occasionally including, somewhat remarkably, the contractor whose defective work caused the problem.[30]

13.05 COORDINATION

Project architects commonly retain engineering consultants to provide such design services as structural, mechanical, and electrical engineering.[31] Because these consultants work under contracts with the project architect, rather than under contracts with the owner, the architect coordinates the activities and schedules of all of these design components. Also, for some projects, in addition to retaining an architect, the owner may directly retain other design professionals for special purposes, such as interior design or audiovisual design. In those cases, unless the owner has adequate in-house construction administration capability, the owner may arrange for the architect to coordinate these specialized design activities with other project activities to which they relate.[32]

---

27. 539 N.Y.S.2d at 814.

28. Moundsview Ind. Sch. Dist. No. 621 v. Buetow & Assocs., 253 N.W.2d 836 (Minn. 1977).

29. Hunt v. Ellisor & Tanner, Inc. 739 S.W.2d 933, 937 (Tex. App. 1987). Both *Moundsview* and *Ellisor & Tanner* are discussed more thoroughly in Section 22.03(B).

30. Magnolia Constr. Co. v. Mississippi Gulf S. Eng'rs, Inc., 518 So. 2d 1194 (Miss. 1988); *Hudson City*, 539 N.Y.S.2d at 814, 818. Jurisdictions that enforce the negligence economic loss rule, discussed in Chapter 23, are less likely to follow these cases.

31. *See* AIA Document B101–2007, *supra* note 1, § 3.1. Note that the AIA standard document contemplates that the architect may provide civil engineering as an additional service for which additional compensation will be due. *See id.*, § 4.1.7.

32. *See* Metro. Atlanta Rapid Transit Auth. v. Green Int'l, Inc., 509 S.E.2d 674 (1999) (involving inadequate coordination of architectural and structural drawings.); Bd. of Educ. v. URS Co., No. 64496, 1994 WL 520862 (Ohio Ct. App. Sept. 22, 1994), *aff'd in part and rev'd in part*, 648 N.E.2d 811 (Ohio 1995) (architect that designed the main classroom building for a school district project also coordinated the design of a planetarium dome for the same project to be designed and built by an expert in dome structures, and the architect served as project manager for the entire project).

Although an architect may coordinate construction activities for a project, as well as design services, if all or most of the construction will be provided by a single general contractor and that contractor's subcontractors, it is typically the general contractor, rather than the project architect, that coordinates all of the construction activities. In some projects, however, the major participants are not all under the control or supervision of a single general contractor. For example, an owner may contract directly with multiple prime contractors who will be responsible for distinct aspects of the work.[33] Coordinating the construction activities of separate contractors working on a project imposes time demands and liability risks on the architect far in excess of normal contract administration. For those reasons, architects who provide coordination services of this type should carefully assess the additional liability risks involved.[34]

13.06 APPROVAL OF PROGRESS PAYMENTS

Contractually, the architect typically acts as the owner's consultant or representative in reviewing and making recommendations on the contractor's progress payment applications.[35] For a complete discussion of the payment process, see Chapter 14. The architect may be liable to the owner for negligence in approving payment applications. Beyond this, some cases also impose liability on the architect to others when the architect takes improper action on progress payment applications. The architect may be liable to the contractor whose payment application is wrongfully rejected and even to subcontractors, sureties, or construction lenders. Construction lenders and government funding sources frequently arrange for or require a separate or independent inspecting architect to approve progress payments to the contractor.[36]

Under what circumstances should the architect be liable to others for improperly approving a contractor's application for payment? In *Fabe v. WVP Corp.*,[37]

---

33. *See, e.g.*, RPR & Assocs. v. O'Brien/Atkins Assocs., P.A., 921 F. Supp. 1457, 1459–60 (M.D. N.C. 1995) (contractor handling general construction in a multiple prime contractor project sought damages for "items of cost arising from alleged defaults and shortcomings in the administration of the project" by both the owner and the architect).

34. *See generally* JUSTIN SWEET & MARC M. SCHNEIER, LEGAL ASPECTS OF ARCHITECTURE, ENGINEERING AND THE CONSTRUCTION PROCESS § 17.04(C) (7th ed. 2004) [hereinafter SWEET & SCHNEIER].

35. *See* AIA Document B101–2007, *supra* note 1, § 3.6.3.

36. *See, e.g.*, Sheetz, Aiken & Aiken, Inc. v. Spann, Hall, Ritchie, Inc., 512 So. 2d 99 (Ala. 1987) (holding that the developer-architect of a subsidized housing project could not recover damages on a contractual third-party beneficiary basis from the architect that the ultimate purchaser of the project hired to review and approve the contractor's payment applications).

37. 760 S.W.2d 490 (Mo. Ct. App. 1988).

an action against the project architect, the court summarily rejected a claim by the state superintendent of insurance, acting as liquidator of the contractor's insolvent surety. The superintendent of insurance alleged that the architect improperly approved the contractor's applications for payment without discovering that the contractor was not paying its subcontractors. The court based its holding on the following contractual provision, which is substantially similar to the language of a standard AIA form (emphasis by the court):

> The issuance of a Certificate for Payment shall constitute a representation by the Architect to the Owner, based on the Architect's observations at the site . . . and on the data comprising the Contractor's Application for Payment, that the Work has progressed to the point indicated; that, to the best of the Architect's knowledge, information and belief, the quality of the Work is in accordance with the Contract Documents . . . ; and that the Contractor is entitled to payment in the amount certified. However, *the issuance of a Certificate for Payment shall not be a representation that the Architect has made any examination to ascertain how and for what purpose the Contractor has used the moneys paid on account of the Contract Sum.*[38]

13.07 THE RIGHT TO REJECT WORK OR TO STOP WORK

The contract documents may authorize the architect, commonly as the owner's agent, to reject work that does not comply with the drawings and specifications.[39] They may also authorize the architect to stop construction altogether, if there are serious problems. Does the architect's authority to reject work or stop the work impose a legal duty on the architect to act or refrain from acting under any circumstances? If so, to whom is the architect liable for exercising that authority improperly or for failing to exercise it? In one case in which the contract documents gave the project engineer discretion to decide whether or not to shut down the job for adverse winter weather, the owner incurred liability to the contractor because of the engineer's improper failure to exercise the authority.[40] In several cases, the architect's authority to stop the work has been used as a basis to impose liability on the architect for worker injuries or safety violations. In reaction to these cases, design service agreements prepared by or for the protection of architects often take great care to avoid giving the architect any authority to stop the work.[41]

38. *Id.* at 491.
39. *See* AIA Document B101–2007, *supra* note 1, § 3.6.2.2.
40. *See* Bignold v. King County, 399 P.2d 611 (Wash. 1965).
41. *See* 5 Bruner & O'Connor, *supra* note 15, §§ 17:52–17:54.

13.08 ISSUING CHANGE ORDERS AND RESPONDING TO PROPOSALS AND CHANGE ORDER REQUESTS

Because Chapter 16 deals extensively with the change order process, here we will only briefly consider the architect's role in that process, which is often a dominant role. In one common arrangement, the architect reviews all proposals or requests for changes in the work, prepares written change orders, including new or revised drawings and specifications relating to the changes, and approves all change orders.[42] The architect may have authority to issue a change order unilaterally if the change involved is consistent with the agreed drawings, specifications, and other contract documents and does not require any adjustment to the price or completion schedule.[43] Ideally, before any significant change is undertaken, the parties will agree on complete drawings and specifications and other details concerning the change involved, as well as on any appropriate adjustments to the price and the completion schedule, in which case the architect will prepare a change order for the parties' signatures that includes all of that information.[44] But practical realities often dictate that work must begin on the basis of a change before a final or complete agreement is reached on any impact the change will have on the price or the schedule. In anticipation of that circumstance, contract documents often establish a procedure for the owner and the architect to order the contractor to undertake the change promptly and for any adjustments in the price or schedule to be resolved later. When that occurs, the architect may have ultimate authority to determine the adjustments, if the owner and the contractor cannot agree.[45]

Because most or all of these activities involve professional design services, the architect may incur liability for design defects, errors, or omissions in change order documentation. Additionally, an architect may be liable for failing to respond properly to proposals and change order requests or for delaying the process.

13.09 CERTIFICATION OF SUBSTANTIAL COMPLETION, FINAL COMPLETION, AND OTHER PERFORMANCE

When the project architect agrees to provide the full range of construction administration services, the architect's final administrative duties normally include determining when the contractor has achieved substantial completion and final completion.[46] Under this arrangement, the times of substantial com-

42. *See* AIA Document A201–2007, *supra* note 2, art. 7.

43. *See id.*, § 7.4.

44. *See id.*, § 7.2.

45. *See id.*, § 7.3.

46. *See* AIA Document B101–2007, *supra* note 1, § 3.6.6.

pletion and final completion are normally established by separate certificates issued by the architect.

As a condition to issuance of the certificate of substantial completion, the contract documents may provide that the architect and the contractor will inspect the project and agree on a list of items that must be corrected or completed (called a punchlist). Any portion of the contract price that the owner has retained from progress payments as security in accordance with the construction contract may be released to the contractor when the certificate of substantial completion is issued. But the architect may also determine or approve a portion of that balance of the contract price for the owner to continue to withhold until all of the punchlist work is completed. Additionally, at substantial completion, physical possession and the risk of loss typically transfer to the owner while the contractor continues to remain on site to complete the remaining work and to correct defective work.

At final completion, the architect normally issues a certificate for final payment of the balance due to the contractor. In connection with final completion, the architect may also be responsible, as the owner's consultant, to require and approve final lien waivers, operating manuals, warranty documentation, project start-up procedures and testing, performance tests to establish that the project meets performance standards or guarantees, and other project closeout steps.[47]

13.10 ARCHITECT'S ROLE IN CLAIMS PROCESS

A. Resolving Claims and Disputes

Chapter 21 addresses claims and the dispute resolution process. This chapter is concerned with these matters only with respect to the architect's often significant role in these procedures. Under one common structure, the owner and the contractor either may or must submit all or most claims to the architect for a preliminary determination or recommendation before commencing mediation, arbitration, or litigation.[48] Moreover, the contract documents often provide that

47. *See* Hudson City, 539 N.Y.S.2d at 814, 817 (N.Y. App. Div. 1989) (where the owner sued the architect for a variety of negligent administrative actions, including allegedly failing to obtain the appropriate roofing guarantees).

48. The current AIA approach is to submit most claims and disputes to the person named by the owner and the contractor to the role of "Initial Decision Maker." AIA Document A201–2007, *supra* note 2, art. 15. In turn, the AIA documents provide that unless the owner and contractor name someone else, the architect will serve in that role. *Id.*, § 15.2.1; AIA Document B101–2007, *supra* note 1, § 3.6.2.5.

the architect's decision becomes final if neither the owner nor the contractor institutes formal dispute resolution procedures.[49] Courts frequently uphold provisions of that nature.[50]

Is it a sound idea to designate a design professional retained by the owner to resolve disputes between the owner and the contractor? Because the architect is typically selected and paid solely by the owner, when a contractor challenges a decision by a project architect, courts have used varying standards to subject the architect's decision to some meaningful judicial review, such as a standard of reasonableness, honesty, good faith, or absence of serious mistake.[51] But even without that added level of judicial scrutiny, a provision giving binding dispute resolution authority to the owner's architect does not necessarily assure favorable treatment for the owner. As one leading authority noted, architects and other design professionals may be subject to professional rules of conduct that may create a professional duty to act impartially when serving as an arbitrator or mediator.[52]

Several cases uphold provisions in public construction contracts that vest binding dispute resolution authority in a design professional that is also an employee of the owner.[53]

B. Architect's Role in Contractor Termination

Contract documents often provide for the architect to advise the owner if the contractor defaults in performance.[54] The architect's advice may include a recommendation to the owner about whether or not to terminate employment of the contractor. Sometimes, the architect's approval may be a condition

49. *See* AIA Document A201–2007, *supra* note 2, § 15.2.5.

50. *See, e.g.*, Elec-Trol, Inc. v. C. J. Kern Contractors, Inc., 284 S.E.2d 119 (N.C. Ct. App. 1981).

51. *See* Sweet & Schneier, *supra* note 34, §§ 29.09–29.10.

52. *See* 5 Bruner & O'Connor, *supra* note 15, § 17:83 (2006).

53. *See, e.g.*, Westinghouse Elec. Corp. v. New York City Transit Auth., 623 N.E.2d 531 (N.Y. 1993); *see generally* Gene Ming Lee, *A Case for Fairness in Public Works Contracting*, 65 Ford. L. Rev. 1075, 1108 (1996). *But cf.* MCI Constructors, Inc. v. Hazen and Sawyer, P.C. 401 F. Supp. 2d 504, 514 (M.D.N.C. 2005) (applying "an objective standard of reasonableness based upon good faith and fair play" to a decision made in a city's favor pursuant to a construction contract that provided for disputes to be resolved by the city manager, and distinguishing the case from those in which the public employee vested with the authority to resolve disputes was a design professional).

54. *See* AIA Document A201–2007, *supra* note 2, § 14.2.2 (providing that the owner may terminate the contractor under specified circumstances if the "Initial Decision Maker" (commonly the architect) certifies that cause for termination exists).

precedent to the owner's right to terminate. In what circumstances and under what legal theories may the architect be liable in damages to the contractor for recommending or taking part in a wrongful termination of the contractor?

In *DiMaria Constr., Inc. v. Interarch*,[55] a New Jersey appellate court, following an arbitration decision holding a termination to be invalid, upheld a judgment against a designer who recommended termination. The judgment was for tortuous interference with the terminated contract. The court refused to overturn a jury verdict against the designer that implicitly determined that the designer had acted outside of the scope of its agency and with malice or bad faith. By contrast, in *Dehnert v. Arrow Sprinklers, Inc.*,[56] the Supreme Court of Wyoming reversed a jury verdict against an architect who recommended termination, stating:

> [T]he architect acting in his professional capacity typically is bound by contract to guard the interests of his principal. An architect who acts within the scope of his contractual obligations to the owner will not be liable for advising the owner to terminate a contractor's performance unless the architect acts with malice or in bad faith. Ballou v. Basic Construction Company, 407 F.2d 1137 (4th Cir. 1969); Lundgren v. Freeman, 307 F.2d 104 (9th Cir.1962); Craviolini v. Scholer & Fuller Associated Architects, 89 Ariz. 24, 357 P.2d 611 (1961); Kecko Piping Company, Inc. v. Town of Monroe, 172 Conn. 197, 374 A.2d 179 (1977). In his treaties [sic], LEGAL ASPECTS OF ARCHITECTURE, ENGINEERING AND THE CONSTRUCTION PROCESS, 2d Ed., Justin Sweet explains the concept of an architect's privilege to intervene in the contractual relationship between the owner and the contractor:
>
>> . . . [I]nterference is not wrongful if it is privileged. . . . Privilege can be created by the relationship between owner and design professional. Also, it can be created by contract clauses either giving the design professional the right to terminate or allowing the owner to terminate upon certification by the design professional that adequate cause for termination exists. The privilege is granted to enable the owner to be advised honestly without the risk of the person giving advice being taken to court.
>>
>> The privilege, whether conferred by law or by contract, must not be abused. If the interference with the contractor's rights were motivated by malice or bad faith, the design professional would be liable for any pecuniary loss caused the contractor. Section 27.10, pp. 536–537.

55. 799 A.2d 555 (N.J. Super. Ct. App. Div. 2001).
56. 705 P.2d 846 (Wyo. 1985).

> A showing of malice or bad faith, not negligence, poor judgment or inflexibility, is essential if the contractor is to recover from the architect for contract interference. Ballou v. Basic Construction Company, 407 F.2d at 1141.[57]

13.11 ARCHITECT'S IMMUNITY FOR PERFORMING QUASI-JUDICIAL FUNCTIONS

Most courts hold that the architect has immunity against liability when serving as a neutral party to resolve construction disputes or to render decisions on other questions presented by the owner or the contractor. When the architect exercises responsibility of this nature, the architect's role is sometimes characterized as a quasi-judicial or arbitral role. But if the architect is serving as the owner's agent or advisor rather than in a quasi-judicial role, the immunity may be lost. Even when the architect is acting in a quasi-judicial role, however, the immunity is not necessarily absolute. This is illustrated by *E. C. Ernst, Inc. v. Manhattan Constr. Co. of Texas*[58] where the court held that the architect lost its immunity by the nature of its actions, stating:

> The arbitrator's "quasi-judicial" immunity arises from his resemblance to a judge. The scope of his immunity should be no broader than this resemblance. The arbitrator serves as a private vehicle for the ordering of economic relationships. He is a creature of contract, paid by the parties to perform a duty, and his decision binds the parties because they make a specific, private decision to be bound. His decision is not socially momentous except to those who pay him to decide. The judge, however, is an official governmental instrumentality for resolving societal disputes. The parties submit their disputes to him through the structure of the judicial system, at mostly public expense. His decisions may be glossed with public policy considerations and fraught with the consequences of stare decisis. When in discharging his function the arbitrator resembles a judge, we protect the integrity of his decision making by guarding against his fear of being mulcted in damages. Cf. Broom v. Douglass, 175 Ala. 268, 57 So. 860 (1912). But he should be immune from liability only to the extent that his action is functionally judge-like.[59]

In *RPR & Assoc. v. O'Brien/Atkins Assoc., P.A.*,[60] an architectural firm and an engineering firm working for a public owner sought to assert sovereign immunity against claims brought by a contractor. The defendants argued that

---

57. *Id.* at 850–51.
58. 551 F.2d 1026 (5th Cir. 1977).
59. *Id.* at 1033.
60. 921 F. Supp. 1457 (M.D. N.C. 1995).

"they are clothed with the State's sovereign immunity because they are acting in the capacity as agents for the State of North Carolina."[61] The court held that the sovereign immunity defense was not available to private parties in that situation, and it also rejected the architectural firm's immunity claim based on its status as an arbiter of disputes between the contractor and the public owner. The court concluded that the claims were based on negligence in the performance of professional services. The court held that "when an architect acts not as an arbitrator, but as an architect, the arbitral immunity vanishes and the architect is susceptible to legal action."[62]

61. *Id.* at 1460.
62. *Id.* at 1462.

QUESTIONS

1. You are an attorney for the state department of roads. You are in the process of updating the department's standard bid package for highway construction. Would it be a good idea to include a provision making the department's chief engineer the final arbiter of all claims made by a contractor and all other disputes between a contractor and the department? What are the relevant considerations that you would raise with your client?
2. Contractor Termination
   a. What effect do you think cases such as *DiMaria* and *Arrow Sprinklers* have on a project architect's willingness to advise the owner candidly about the contractor's performance? Can you think of any provisions the architect's counsel might add to the owner-architect agreement to reduce the architect's risk? As counsel to the owner, would you accede to a proposal by the architect's counsel to include a provision in the owner-architect agreement by which the owner agrees to indemnify the architect against the contractor's claims of tortious interference? As counsel to a project architect, would you insist that any protective language be added to the owner-contractor agreement?
   b. Your client is the project architect for a new four-star hotel project. There have been extensive delays. Your client and the owner believe that the contractor caused most of the delays. The contractor blames the owner and your client. The owner, under advice of counsel, is pressing the architect for a recommendation to terminate the contractor for default. The owner-contractor agreement provides for all claims and disputes to be determined initially by the architect, but subject to arbitration if either party is dissatisfied with the architect's decision. The contract also provides that the owner may terminate the contract on account of contractor default if the architect certifies in writing that sufficient cause exists. What is your advice?

CHAPTER 14

Payment

DAVID A. SENTER

14.01 INTRODUCTION

Payment and cash flow issues are often the most negotiated and litigated issues on construction projects. Construction is all about the flow of funds from the lender to the owner to the contractor to the subcontractors to the suppliers. Contractors typically have to advance certain costs before payments are received from the owner. For example, contractors will typically perform work for 30 days before their first invoice is sent to the owner. The owner then has 30 days to make payment, and the owner is often entitled to withhold a portion of the payment as retainage. As a result, a contractor may have to wait 60 days or longer before receiving its first payment, which may only represent 90 percent of its billing. During that 60-day period, however, the contractor must make payment to its employees, subcontractors, and material suppliers. Contractors and subcontractors cannot "be expected to finance the operation to completion without receiving the stipulated payments on account as the work [progresses]. . . . Advance payment is a condition precedent to the contractor's obligation to proceed."[1]

Subcontractors tend to be smaller, more undercapitalized entities. While this is not always the case, they may not be as financially strong as either the general contractors for whom they work or the suppliers upon whom they rely for materials. Although they will not receive payment from the general contractor until a specified portion of the work is completed, subcontractors usually pay their field personnel and tradesmen on a weekly basis and often must pay suppliers in advance for materials received. Placed in this financial squeeze, subcontractors not uncommonly "rob Peter to pay Paul" (i.e., use payments on one project to pay bills on another).

If payment is delayed at any point in the construction chain, the contractor or subcontractor may run out of money and/or be unable to pay its subcontractors, suppliers, or tradesmen to continue working.[2] The likely result is delayed performance or nonperformance. Thus, a project's success or failure may depend upon whether the project participants are paid promptly for their work.

14.02 BASIS FOR PAYMENT

The parties' contract should clearly describe the basis and timing for payment. There are many ways to structure payment in a construction contract—stipulated sum (lump sum or fixed price), unit price, cost-plus (cost of the work plus the contractor's fee), and others.[3] The traditional pricing mechanism is fixed price. This mechanism arises out of the Design-Bid-Build method of project delivery. Under this method, an owner presents its drawings and specifications, prepared by the owner's design professional, to a contractor to price or to a group of contractors to bid. Based upon those drawings and specifications, the owner and the contractor enter into a contract to construct the project for a fixed price or a lump sum.

A unit price contract may be used where the type of work is well defined but the quantity is only estimated. If a contractor must bid a fixed price for an unknown quantity, its price will likely be high so as to include the quantity risk contingency. On the other hand, with a unit-price contract, the quantity contingency is not a significant issue, and the contractor's price likely will be lower.

---

1. Guerini Stone Co. v. P. J. Carlin Constr. Co., 248 U.S. 334, 345 39 S. Ct. 102, 106 (1919).

2. Subcontractors, as a whole, file more bankruptcies than any other entity in the construction process. *See* Carina Y. Enhada, Cheri Turnage Gatlin & Fred D. Wilshusen, eds. Fundamentals of Construction Law, ABA Construction Forum 2001, at 112 [hereinafter Fundamentals].

3. *See generally id.*, Chapter 9 (discussing pricing mechanisms in more detail).

A cost-plus or cost-reimbursement contract provides that the owner will pay or reimburse the contractor for its costs (labor, material, and equipment) and will pay the contractor a fee, which is usually a percentage of the costs (e.g., cost plus 15 percent). Such contracts should contain a clear definition of "costs," especially as the term relates to the contractor's own personnel, equipment, management, and overhead. The AIA Contract Documents, for example, deal with "cost" by defining it to include labor costs; subcontract costs; costs of material and equipment incorporated in the completion of construction; costs of other materials and equipment, temporary facilities, and related items; and miscellaneous costs.[4] Obviously, such an open-ended contract leaves the owner exposed to uncertainties and increased costs. This exposure is often managed by using a cost-plus contract coupled with a guaranteed maximum price, or a GMP. Such a contract involves a typical cost-plus arrangement up to the dollar amount of the GMP. The contractor bears the risk of any costs over the GMP, thereby reducing the owner's risk.

Given the risks involved in cost-plus contracts, AIA Document A102–2007 provides that:

> The Contractor accepts the relationship of trust and confidence established by this Agreement and covenants with the Owner to cooperate with the Architect and exercise the Contractor's skill and judgment in furthering the interests of the Owner; to furnish efficient business administration and supervision; to furnish at all times an adequate supply of workers and materials; and to perform the Work in an expeditious and economical manner consistent with the Owner's interests.[5]

Some courts have held that such language creates a fiduciary relationship between the owner and the contractor, thereby increasing the level of the contractor's duty to the owner and providing the basis for a constructive fraud claim.[6] Other courts have held that such language alone does not necessarily create a fiduciary duty.[7]

Once the price is agreed upon, the parties must also agree on how that price is to be paid. A structure is not built overnight, and the contractor is not

---

4. American Institute of Architects, AIA Document A102–2007, *Standard Form of Agreement Between Owner and Contractor Where the Basis of Payment is the Cost of the Work Plus a Fee with a Negotiated Guaranteed Maximum Price* §§ 7 and 8.

5. *Id.* at § 6.

6. *See* Jones v. J. H. Hiser Constr. Co., 60 Md. App. 671, 484 A.2d 302 (1984).

7. *See* Eastover Ridge, LLC v. Metric Constructors, Inc., 138 N.C. App. 360, 533 S.E.2d 827 (2000) (stating that, notwithstanding Article 3 of the contract, the "architect's constant, close involvement in the project belies any claim that a 'relation of trust and confidence' existed between" owner and contractor); *see generally* 2 PHILIP L. BRUNER & PATRICK J. O'CONNOR, BRUNER & O'CONNOR ON CONSTRUCTION LAW § 6:81 (West Group 2002) [hereinafter BRUNER & O'CONNOR].

expected to finance the project by waiting until the end to receive payment. As a result, the contract should provide for periodic payments as the work progresses, known as progress payments. While the parties may contract for any payment period, most progress payments are made on a monthly basis and are determined based upon the actual progress of the work.[8] Such progress payments, less retainage, are paid to the contractor at the periods or intervals required by the contract.

The contractor's periodic payment is based upon the value of the work performed during the period for which payment is made. To determine the amount of a given progress payment, the parties typically employ a "schedule of values."[9] This schedule simply sets forth how much various portions of the work are worth, including the contractor's profit or how the contract price is allocated among various portions of the work.[10] These values are determined by agreement at the beginning of the project. During the course of the project, the contractor is paid at regular intervals, usually monthly, with the amount of each payment being determined based upon the percentage of the work completed as applied against the schedule of values.

For example, if the footings value in the schedule is $50,000, and after one month the contractor has performed 40 percent of this work, then it will be paid $20,000 for its work on that schedule item for that month. In general terms, the total month's payment is the sum of the amount due for each item on the schedule, less retainage and other properly withheld amounts.[11]

A contractor's attempt to "front load" is one problem that may arise in connection with the schedule of values. That is, the contractor may attempt to assign a higher value than is appropriate for an item of work early in the project (i.e., on the front end). The contractor may "skew . . . [its] assigned values in the schedule of values so that work performed early (such as clearing and grubbing, excavation, and foundation work) is valued higher than its cost, while later work (such as painting) is undervalued."[12] The result is that the contractor gets paid more than the value of the early work and can collect its profit before the end of the project. If the contractor has been overpaid, the owner has less protection if a problem arises later in the project. Owners and

---

8. *See* American Institute of Architects, AIA Document A101–2007, *Standard Form of Agreement Between Owner and Contractor* § 5.1.1–2 (2007) [hereinafter AIA Document A101–2007]; AIA Document A201–2007, *General Conditions of the Contract for Construction* § 9.6 (2007) [hereinafter AIA Document A201–2007]; *see also infra* § 14.07.

9. *See* AIA Document A201–2007, *supra* note 8, §§ 9.2.1 and 9.3.1.

10. *See* FUNDAMENTALS, *supra* note 2, at 31.

11. *See* discussion *infra* at Sections 14.03(A) and 14.03(B) in this chapter.

12. *See* FUNDAMENTALS, *supra* note 2, at 193.

architects should carefully review a contractor's proposed schedule of values to minimize the risk of front loading.[13]

Under a unit price contract, the contractor is paid based upon the number of "units" performed during the relevant period (i.e., feet of pipe laid, cubic yards of dirt or rock excavated).

14.03 PAYMENT PROCESS

Under the so-called Doctrine of Conditions, "[t]he promise by the owner to make any payments is conditioned on the contractor's compliance with the contract documents."[14] A strict application of this doctrine would obviously result in a hardship to the contractor, as during the course of the project, the contractor would be required to perform the entire contract before it received any payment. A contractor cannot and should not be expected to finance the construction project. To alleviate this hardship, contract procedures have developed whereby the contractor can apply for and receive regular payments as the work progresses. As set forth in the previous section, the amount of each payment is based upon the value of the work performed.

A strict application of this doctrine can also cause hardship at the end of the job. For example, if an owner were entitled to withhold retainage or final payment on a multi-million-dollar project simply because of the need for some paint touch-up work, yet the owner was able to occupy the building, the owner would be unjustly enriched at the expense of the contractor.[15] The loss to the contractor would be disproportionate to the loss to the owner caused by the contractor's nonperformance.[16] To deal with such an "end of project issue," the law and certain form contract documents have developed the concepts of "substantial performance and substantial completion," which will be discussed later in this chapter.[17]

A. Applications for Payment

To be paid for its work, a contractor must submit what is known as an application for payment or a draw request. The contractor's application for payment sets forth the contractor's representation as to the status of its work,

13. *See id.*

14. Justin Sweet, Legal Aspects of Architecture, Engineering and the Construction Process § 22.06(A) (6th Ed. 2000) [hereinafter Sweet].

15. *See id.*

16. *See id.*

17. *See* discussion *infra* at Section 14.04 in this chapter.

measured against the schedule of values. The application is submitted to the project architect for review. The architect reviews the application, compares it to the work in place, and advises the owner as to whether it is appropriate to make payment. The architect's written advice is sometimes called a certificate of payment.[18] AIA Document A201–2007, § 9.4.2 provides that the architect's level of review is as follows:

> The issuance of a Certificate for Payment will constitute a representation by the Architect to the Owner, based on the Architect's evaluation of the Work and the data comprising the Application for Payment, that to the best of the Architect's knowledge, information and belief, the Work has progressed to the point indicated and that the quality of the Work is in accordance with the Contract Documents. The foregoing representations are subject to an evaluation of the Work for conformance with the Contract Documents upon Substantial Completion, to results of subsequent tests and inspections, to correction of minor deviations from the Contract Documents prior to completion and to specific qualifications expressed by the Architect. The issuance of a Certificate for Payment will further constitute a representation that the Contractor is entitled to payment in the amount certified. However, the issuance of a Certificate of Payment will not be a representation that the Architect has (1) made exhaustive or continuous on-site inspections to check the quality or quantity of the Work, (2) reviewed construction means, methods, techniques, sequences, or procedures, (3) reviewed copies of requisitions received from Subcontractors and material suppliers and other data requested by the Owner to substantiate the Contractor's right to payment, or (4) made examination to ascertain how or for what purpose the Contractor has used money previously paid on account of the Contract Sum.

The architect may decline to certify payment for a myriad of reasons—defective work, third-party claims, failure of the contractor to pay its subcontractors, or reasonable evidence that the work cannot be completed for the contract balance or within the contract time, among others.[19]

1. *The Contractor's Certification*

A contractor's submission of an application for payment is a representation made by a contractor concerning the completed work. This representation may be considered a warranty. While design professionals such as architects rarely intend to make warranties, contractors routinely make express warranties in writing, express warranties through oral representations, and even implied warranties. A typical contractor warranty clause states that:

18. The role of the architect in project administration is discussed in more detail in Chapter 13.
19. *See* AIA Document A201–2007, *supra* note 8, § 9.5.1.

> [t]he Contractor warrants to the Owner and Architect that materials and equipment furnished under the Contract will be of good quality and new unless otherwise required or permitted by the Contract Documents, that the Work will be free from defects not inherent in the quality required or permitted, and that the Work will conform to the requirement of the Contract Documents.[20]

The application for payment submitted by a subcontractor to a general contractor is a representation when it is certified or sworn to by the subcontractor.[21] A general contractor may "rely upon its [subcontractor's] representation that work has been done for which payment is now due."[22] False representations can lead to "liability for fraud, negligent misrepresentation, and representations amounting to an unfair or deceptive trade practice."[23]

2. *The Architect's Certification*

The architect's certification "constitutes a representation that the contractor is entitled to payment in the amount certified."[24] To ensure that a warranty is not created, or to limit liability, an architect may "qualify" the certification, stating that it does not constitute a representation that the architect made "continuous on-site inspections" or "reviewed construction means, methods, techniques," for example. The qualifications may limit what is and what is not covered by the representation. These qualification attempts can be seen in Section 9.4.2 of AIA Document A201–2007 quoted earlier.

Although similar to warranties in that both are representations, certifications differ from warranties. Design professionals rarely intend to make express warranties; however, issuing a certification instead of a warranty does not preclude all liability. For example, a claim of negligence can be made by an owner against an architect who over-certifies to a lender the amount of work completed because it can be reasonably foreseen that an owner may rely on the accuracy of an architect's certification.[25]

The representation made by an architect in a certification generally is not a warranty concerning the quality of the work, but it should represent an architect's "exercise [of] reasonable care and diligence in the supervision of the

---

20. *See id.* § 9.54 (*quoting* American Institute of Architects, AIA Document A201–1997, *General Conditions of the Project for Construction* § 3.5 (1997)).

21. Myers & Chapman, Inc. v. Thomas G. Evans, Inc., 374 S.E.2d 385, 390 (N.C. 1988).

22. *Id.* at 390.

23. Robert G. Byrd, *Misrepresentation in North Carolina*, 70 N.C. L. Rev. 323 (1992).

24. 3 Bruner & O'Connor, *supra* note 7, § 8:20.

25. *See* Browning v. Maurice B. Levien & Co., 262 S.E.2d 355, 358, *disc. rev. denied* 267 S.E.2d 673 (N.C. Ct. App. 1980); *see also* 2 Strong's N.C. Index *Architects* § 18 (4th 2007).

work."[26] Even if an architect does not assume the duty to "supervise" construction, it is still bound to exercise reasonable care and diligence in the performance of its professional services, whatever they may be. If a contract between an owner and an architect does not include "any special guarantees or warranties," such warranties and guarantees may be "set out in general terms such that the only implication arising from the terms is that the architect will use reasonable care in preparation of the plans and supervision of the construction of the project."[27]

In summary, inaccurate certifications related to payment, whether intentional or unintentional, may result in additional liability for contractors and architects.

B. Retainage

Almost all construction contracts provide that a percentage of payments, usually 5 to 10 percent, shall be retained until some later date, such as substantial completion of the project. The purpose of retainage is to preserve a fund of money that will be available to assure final completion and to correct defects in the work. Typically, the full percentage retained on the entire job is held until substantial completion, at which time retainage is reduced to an amount estimated as needed to cover only the incomplete or defective items described on a written punchlist. Retainage also gives the owner, lender, and sureties

26. In Palmer v. Brown, 273 P.2d 306, 317 (Cal. Dist. Ct. App. 1954), a cross-complaint (counterclaim) against the plaintiff architects included a claim "that plaintiffs fraudulently issued certificates to the contractors showing satisfactory completion of work without ascertaining whether the premises were free from liens chargeable to the contractors," and a claim "for plaintiffs' negligence in issuing certificates to the contractors without requiring proof that the premises were free from liens chargeable to the contractors and without carefully examining the work to see that it had been properly done." The appeals court reversed the judgment of the lower court, which had been in favor of the plaintiffs. The appeals court stated that the "architect's certificate as a precondition of payment was for the benefit of defendant," so that the defendant could be assured that the payments did in fact "discharge him and his property for the labor and materials for which he was paying." The plaintiff/architects could be found negligent in not requiring proper receipts, lien releases, or lien waivers from the contractor.

The court also ruled that there was sufficient evidence that the plaintiff/architects had improperly certified payment for work that had not been inspected as required by the contract. Additionally, the court explained that when work certified by an architect is faulty, the architect can be found negligent in terms of its supervision of the construction (if it undertook the duty of supervision in its contract). Although an architect's certification does not make it liable for specific defects, those defects may not have occurred if the architect had exercised reasonable care and diligence in the performance of its contractual obligations during construction.

27. Ryan v. Morgan Spear Assoc., Inc., 546 S.W.2d 678, 682 (Tex. Civ. App. 1977).

extra security or protection against other issues that might arise, such as defective work, unpaid subcontractors, or liens filed late in or after the construction process.[28]

Most construction contracts provide for 10 percent retainage to be withheld until substantial completion. Another option is to withhold 10 percent until the project is 50 percent complete and thereafter withhold no retainage.[29] This has the effect of reducing retainage to 5 percent at the end of the job. The theory is that, as the contractor completes more and more of the work, the risk that the project will not be completed or that costs to correct will reach 10 percent diminishes. Even if a contract allows for reduction in retainage, it will also likely allow the owner to reinstate full retainage in the event of a change of circumstances.

While protecting the owners, retainage can create financial difficulty for contractors by decreasing their short-term cash flow.[30] Contractors, in turn, often respond to this decrease in cash flow by transferring the retainage burden to their subcontractors and thus withholding a portion of payments to the subcontractors until the owner eventually pays retainage to the contractor. This practice effectively allows the contractors to secure financing from their subcontractors by delaying some of their payments until the end of the project. This also places a greater burden on subcontractors who do their work earlier in the construction process because those subcontractors then have to wait longer until the project is complete and the retained funds are released.

Some jurisdictions have addressed this tension between protecting the owners and financing the contractors by limiting the maximum allowable retainage, which typically ranges from 5 to 10 percent. Some jurisdictions also address this tension by requiring retention of fewer funds as the construction progresses and as adequate performance appears more likely.[31] Other jurisdictions allow contractors to pledge security or establish construction bonds in lieu of retainage.[32] Additional jurisdictions have responded by allowing the retained funds to be placed in escrow and then allowing the contractor to access the

---

28. Nat'l Union Indem. Co. v. G. E. Bass & Co., 369 F.2d 75, 77 (5th Cir. 1966); *In re* Lynch III Props. Corp., 125 B.R. 857, 863 (Bkrtcy. E.D.N.Y. 1991); FUNDAMENTALS, *supra* note 2, at 192–93.

29. *See* AIA Document A101–2007, *supra* note 8, § 5.1.8.

30. *See* Fireman's Fund Ins. Co. v. United States, 909 F.2d 495, 498 (Fed. Cir. 1988) (noting that "[i]t is true that retainage is an incentive to complete the contract, but it is equally true that contractors may need full progress payments to finance the work").

31. *See, e.g.*, ARK. CODE ANN. § 22-9-604 (allowing 10 percent retainage until the project is 50 percent complete, then not allowing any additional retainage to be withheld).

32. *See, e.g.*, IND. CODE § 36-1-12-13.1 (requiring the establishment of a construction bond or, alternatively, retainage); 3 BRUNER & O'CONNOR, *supra* note 7, § 8:18, n.5.

interest earned on those funds.[33] Finally, others have addressed the problem of contractors using retainage to secure financing from subcontractors by limiting the percentage held back from subcontractors to no more than the percentage that the owners hold back from the contractors.[34]

The tension between owner security and contractor cash flow is illustrated in *Fireman's Fund Ins. Co. v. United States*,[35] where the government, pursuant to the contract, withheld 10 percent as retainage. It released that retainage when the project was 85 percent complete, not because the contractor had completed the work, but because the contractor was having severe cash flow problems and appeared not to be paying its subcontractors. Thus, the government was concerned that the contractor would not be able to complete the job. After the government released the retainage, and thus its "security," the contractor then abandoned the contract, and the government had to procure another contractor to complete the job without the benefit of any retained funds.[36]

Retainage held by the owner can become the subject of claims by third parties. A performance bond surety who steps in to complete a contract for its principal is protected by the doctrine of equitable subrogation such that it is entitled to collect its principal's retainage.[37] In *Pearlman v. Reliance Ins. Co.*,[38] the U.S. Supreme Court held that unpaid contract balances should be paid to a surety who had stepped in to complete the project and thereby was subrogated to the rights of the debtor (its principal), the subcontractors, and the owner.[39] In so holding, the court determined that the surety's rights of subrogation were unaffected by bankruptcy, and the contract balance did not even constitute property of the bankruptcy estate. In *Grochal v. Ocean Technical Services Corp.*

---

33. TENN. CODE. ANN. § 66-11-144 (providing for retainage to be placed in escrow and for the contractor to earn the interest); 3 BRUNER & O'CONNOR, *supra* note 7, § 8:18, n.4.

34. *See, e.g.*, N.C. GEN. STAT. § 22C-4.

35. 909 F.2d at 495.

36. As is discussed in Chapter 20, the owner sometimes has the benefit of a performance bond surety to ensure completion. See also *infra* Section 14.03(C)(6), regarding overpayment.

37. Prairie State Nat'l Bank v. United States, 164 U.S. 227, 17 S. Ct. 142, 41 L. Ed. 412 (1896); *see generally*, 3 BRUNER & O'CONNOR, *supra* note 7, § 8:36.

38. 371 U.S. 132, 83 S.Ct. 232, 9 L. Ed. 2d 190 (1962).

39. *Id.* at 137 and 141, 83 S.Ct. at 235 (stating that "a surety who pays the debt of another is entitled to all the rights of the person he paid" and "the [owner] had a right to use the retained funds to pay laborers and materialmen; the laborers and materialmen had a right to be paid out of that fund; the contractor, had he completed his job and paid his laborers and materialmen, would have become entitled to the fund; and the surety, having paid the laborers and materialmen, is entitled to the benefit of all these rights to the extent necessary to reimburse it").

(In re Baltimore Marine Indus., Inc.),[40] the Fourth Circuit distinguished *Pearlman* by finding that while in the context of a general contractor's bankruptcy proceeding, a surety who has fully paid all subcontractors has subrogated the general contractor's interest in funds withheld by an owner, and an unpaid subcontractor is not subrogated to the general contractor's interest in those funds.

In the nonbankruptcy context, third-party claims to retainage are driven by state law. For example, in North Carolina a subcontractor's lien on funds has

> priority over all other interests or claims theretofore or thereafter created or suffered in the funds by the person against whose interest the lien upon funds is asserted, including, but not limited to, liens arising from garnishment, attachment, levy, judgment, assignments, security interests, and any other type of transfer, whether voluntary or involuntary.[41]

C. Owner's Payment Obligation

1. *Timing*

The terms of the parties' contract will define the timing of the owner's payment obligation. The owner's failure to make payment of undisputed amounts in a timely manner may be a material breach of the contract that would support the contractor's decision to stop work.[42] For example, Article 9.7.1 of AIA Document A201–2007 provides as follows:

> If the Architect does not issue a Certificate for Payment, through no fault of the Contractor, within seven days after receipt of the Contractor's Application for Payment, or if the Owner does not pay the Contractor within seven days after the date established in the Contract Documents the amount certified by the Architect or awarded by binding dispute resolution, then the Contractor may, upon seven additional days' written notice to the Owner and Architect, stop the Work until payment of the amount owing has been received. The Contract Time shall be extended appropriately and the Contract Sum shall be increased by the amount of the Contractor's reasonable costs of shut-down, delay and start-up, plus interest as provided for in the Contract Documents.

---

40. 476 F.3d 238, 242 (4th Cir. 2007).

41. N.C. Gen. Stat. § 44A-22.

42. *See Guerini Stone Co.*, 248 U.S. at 334, 39 S. Ct. at 102 (stating that the requirement of progress payments "must be deemed so material that a substantial failure to pay would justify the contractor in declining to proceed"); Silliman Co. v. S. Ippolito & Sons, Inc., 467 A.2d 1249 (Conn. App. Ct. 1983) (holding that a contractor's failure to make progress payments to subcontractor is breach of contract); *see also* 3 Bruner & O'Connor, *supra* note 7, § 8:17.

The contractor's decision to stop work must be reasonable and in good faith. Otherwise, the contractor may be in default under the contract despite the other party's withholding of payments.[43] Generally, if it is undisputed that the contractor is entitled to progress payments under the contract, the contractor has the right to stop work until it has been paid.[44] The RESTATEMENT (SECOND) OF CONTRACTS contains a good example of this factual scenario:

> A contracts to build a house for B for $50,000, progress payments to be made monthly in an amount equal to 85 percent of the price of the work performed during the preceding month, the balance to be paid on the architect's certificate of satisfactory completion of the house. Without justification B fails to make a $5,000 progress payment. A thereupon stops work on the house and a week goes by. A's failure to continue the work is not a breach and B has no claim against A. B's failure to make the progress payment is an uncured material failure of performance which operates as the non-occurrence of a condition of A's remaining duties of performance under the exchange.[45]

However, the requirement for progress payments "must be deemed so material that a substantial failure to pay would justify the contractor in declining to proceed."[46] Despite the harsh phrasing of this rule, courts almost always find that withholding payment is material.[47] Nonetheless, the contractor's decision to stop work must be reasonable in order to justify abandonment.[48] To determine

---

43. *See* Gramm v. Ins. Unlimited, 378 P.2d 662, 664 (Mont. 1963) (stating that "the particular facts of each case [determine] whether or not non-payment of an installment is a total breach enabling the contractor to cease work, or whether it is merely a partial breach entitling the contractor to sue for the partial breach, but not permitting him to abandon the contract").

44. *See* United States *ex rel.* Micro-King Co. v. Cmty. Science Tech., Inc., 574 F.2d 1292, 1295 n.3 (5th Cir. 1978); Macri v. U.S. for Use of John H. Maxwell & Co., 353 F.2d 804 (9th Cir. 1965) (noting that "it is well settled that an owner's failure to make progress payments on a building contract may constitute a material breach, justifying the builder's refusal to proceed with the work on the theory that the owner has rendered his performance impossible").

45. RESTATEMENT (SECOND) OF CONTRACTS § 237, illus. 1 (1981).

46. *Guerini Stone Co.*, 248 U.S. at 344, 39 S. Ct. at 106.

47. *See* Ernst v. Ohio Dept. of Adm. Serv., 69 Ohio App. 3d 330 (1990); U.S. *ex rel.* E. C. Ernst, Inc. v. Curtis T. Bedwell & Sons, Inc., 506 F. Supp. 1324 (D.C. Pa. 1981); Brady Brick & Supply Co. v. Lotito, 356 N.E.2d 1126 (Ill. App. Ct. 1976).

48. *See* Havens v. Safeway Stores, 678 P.2d 625 (Kan. 1984) (holding that minor delays of six to eight days in making monthly payments did not justify the abandonment of the construction contract).

whether the abandonment of work was reasonable, the court generally will analyze the amount of money withheld and the duration of the withholding in light of the contractual language and the factual circumstances.[49] Additionally, the court will examine the owner's reasons for withholding payment to determine whether cessation of work was reasonable.[50] Thus, a contractor should examine the language of the contract or payment schedules and the owner's reasons for withholding payment before determining whether to cease work.

In addition to the materiality requirement, a contractor must also be cognizant of whether it has waived its right to cease work when the owner has withheld payments. Waiver may occur when a contractor continues to perform under the contract, despite the other party's nonpayment.[51] If the contractor's right to cease work has been waived and it wrongfully stops work on the construction contract, the contractor may be liable for damages arising from the abandonment.[52]

2. *Mechanic's Lien*[53]

In terms of making payment, the owner must also protect itself from claims of subcontractors and suppliers. Even though subcontractors and suppliers have no direct contract with the owner, they may be entitled to secure their payment from the owner by placing a lien upon the real property and/or a lien upon any funds the owner owes the general contractor. Lien laws vary widely by state, but they offer significant protections to subcontractors and suppliers and significant risk to owners.[54]

---

49. *See id.*; Stewart v. C & C Excavating & Constr. Co., 877 F.2d 711 (8th Cir. 1989) (finding that a debt of $2,385 was an insignificant portion of the contract price and did not justify abandonment); U.S. for Use and Benefit of Aucoin Elec. Supply Co. v. Safeco Ins. Co. of America, 555 F.2d 535, 541 (5th Cir. 1977) (stating that "[e]ven when payment is not made where due, it is not sufficient reason to stop work unless it really prevents work").

50. *See* M. Glosser & Sons, Inc. v. Micco, 309 A.2d 602 (Pa. Commw. Ct. 1973) (finding that withholding payment based on belief that contractor was not paying his subcontractors was improper and constituted breach).

51. *See* Bd. of Regents of Univ. of Texas v. S & G Constr. Co., 529 S.W.2d 90 (Tex. Civ. App. 1975), *overruled on other grounds by* Fed. Sign v. Texas Southern Univ., 951 S.W.2d 401, 121 Ed. Law Rep. 394, 40 Tex. Sup. Ct. J. 676 (1997).

52. *See* Providence Washington Ins. Co. v. Beck, 255 N.E.2d 600 (Mass. 1970).

53. *See generally* Fundamentals, *supra* note 2, at 144 (excerpt).

54. For a more detailed discussion of liens, see Chapter 19.

3. *Stop Notice/Lien on Funds*

In addition to permitting liens that attach to the real property being improved, some states have adopted so-called "stop notice" statutes.[55] These statutory provisions allow an unpaid party, typically a subcontractor or materialman, to stop, "freeze," or "trap" construction funds before they are disbursed by giving a notice of claim to the person holding these funds, typically the owner or the lender who is financing the project. Some statutes impose personal liability on the fund holder if it disperses funds in spite of and after having received the stop notice. Strict compliance with timing and service requirements is required in order to enforce a stop notice.

4. *Lien Waivers*

To manage the contractor or subcontractor nonpayment risk, a prudent owner, prior to making payment, will obtain from the contractor a list of subcontractors and suppliers for the project and obtain lien waivers from the contractor and its subcontractors and suppliers. A lien waiver is a sworn statement from the contractor, subcontractor, or supplier that acknowledges receipt of payment and waives or releases claims for work performed through a stated date or up to a stated amount. Lien waivers are routinely required for a contractor or subcontractor to receive each progress payment. Sample lien waiver language is as follows:

> In addition, for and in consideration of the amounts and sums received, the undersigned hereby waives, releases and relinquishes any and all claims, rights, and causes of action whatsoever arising out of or in the course of the work performed on the aforementioned project, contract or event transpiring prior to the date hereof, excepting the right to receive payment for work performed and properly completed and retainage, if any, after the date of the above-mentioned payment application or invoice.

Releases and lien waivers "are contractual in nature and their interpretation is governed by the same rules governing interpretation of contracts."[56] "A contract which is plain and unambiguous on its face will be interpreted as a matter of law by the court."[57] Thus, lien waiver agreements entered into after

55. For a detailed discussion of the issue by jurisdiction, *see Construction and Effect of Statutes Requiring Construction Fundholder to Withhold Payments upon "Stop Notice" from Subcontractors, Materialman, or Other Person Entitled to Funds,* 4 A.L.R. 772 (5th 1992). *See also, e.g.,* TEX. PROP. CODE ANN. § 53.081; N.C. GEN. STAT. § 44A-18.

56. Chemimetals Processing, Inc. v. Schrimsher, 535 S.E.2d 594 (N.C. Ct. App. 2000).

57. Dept. of Transp. v. Idol, 440 S.E.2d 863, 864 (N.C. Ct. App. 1994).

a contract has been signed, and not as a form of consideration for signing of the contract, are generally enforceable.[58]

Some jurisdictions hold that lien waivers do not encompass retainage.[59] Other jurisdictions hold that lien waivers do encompass retainage.[60]

Sometimes owners include in their prime contracts, and contractors include in their subcontracts, a provision requiring the advance or prospective waiver of lien rights—waiver of lien rights before the contract is signed or the work done. In many states, however, prospective lien waivers are unenforceable.[61]

5. Joint Checks

Typically, payment is made directly from the owner to the contractor. In circumstances where the owner is concerned with the contractor making payment to its subcontractors or suppliers, the owner may insist on issuing a check jointly payable to the contractor and one of its subcontractors or suppliers. A joint check can only be validly negotiated if it is endorsed by both of the payees. If the contractor endorses the check, then it is in effect acknowledging payment from the owner. Similarly, the subcontractor's endorsement acknowledges payment from the contractor for the subcontractor's work. It is up to the contractor and the subcontractor to agree between themselves as to how the joint check proceeds will be divided. Both of these results inure to the benefit of the owner.

Many contracts provide that where the owner has a reasonable concern over whether the contractor will make payment to its subcontractors it may issue joint checks. Even absent contractual rights, as a practical matter, the owner can issue joint checks under such circumstances. The alternative would be no payment at all until the owner's concerns are otherwise satisfied.

58. *See* City of Houston v. R. F. Ball Constr. Co., 570 S.W.2d 75, 78 (Tex. Civ. App. 1978) (stating that "it is the general rule of the law of contracts that when an unambiguous writing has been entered into between the parties, the courts will look to the written instrument as the expression of the parties' intention since it is the parties' objective, not subjective, intent which is significant.").

59. *See* Tharp v. Keeter/Schaefer Investments, L.P., 943 S.W.2d 811 (Mo. Ct. App. 1997) (holding that a lien waiver submitted in connection with progress payments did not apply to retainage as owner knew that the subcontractor expected to recover its retainage at the end of the project).

60. *See* Kern v. City of Lawrenceburg, 625 N.E.2d 1326 (Ind. Ct. App. 1993) (lien waiver that states it applies to all labor and materials furnished through a certain date and applies to unpaid retainage and extra work is effective).

61. *See, e.g.*, Fla. Stat. § 713.20(2) 1988; N.C. Gen. Stat. § 44A-12(f); *but see* Tex. Prop. Code. Ann. § 53.160(B)(7); *see also Release on Waiver of Mechanic's Lien by General Contractor as Affecting Rights of Subcontractors or Materialmen*, 75 A.L.R.3d 505 (1977).

Oftentimes, it is the supplier who will initiate a joint check. For example, the electrical subcontract on a project may be for $150,000, with $125,000 of that amount representing the electrical materials—fixtures, conduits, switchgears, controls, and so on. The electrical material supplier to the subcontractor has a significant financial risk and may require a joint check agreement from the general contractor before it supplies materials to the electrical subcontractor.

Through a joint check agreement, the general contractor agrees that it will make checks to the subcontractor jointly payable to the supplier. If the contractor makes payments consistent with the joint check agreement, it is protected from claims by the supplier. If, however, it fails to comply (makes checks payable solely to the subcontractor), it becomes directly liable to the supplier in the event the subcontractor does not pay the supplier. Significantly, the mere execution of the joint check agreement does not make the contractor primarily liable to the supplier. It only requires the contractor to issue joint checks.

Typically, the obligation to issue joint checks is subject to any defenses the contractor might have pursuant to its contract with the subcontractor. However, the precise wording of the joint check agreement is critical because the agreement rises to the level of an enforceable contract.

A distinction should be drawn between a joint payee and an alternative payee. Payees of checks whose names are separated by "and" are joint payees, while those whose names are separated by "or" are considered alternative payees. A bank can properly cash a joint check only if both parties to the check have endorsed it, but with respect to alternative payees, only the endorsement of one of the payees is necessary.[62] The joint check rule holds "that a payee by endorsing a check is deemed to have received the full payment regardless whether this indeed is the case."[63]

However beneficial they may be, joint checks are not the panacea one might think. Problems can arise, including forged endorsements and joint checks negotiated with no endorsements. While there may be remedies for such problems, they may be time-consuming and expensive. Thus, construction participants should always carefully monitor payment and credit issues.

---

62. Glen-Gery Corp. v. Warfel Constr. Co., 734 A.2d 926 (Pa. Super. Ct. 1999); *see also* Bituminous Constr., Inc. v. Rucker Enterprises, Inc., 816 F.2d 965.

63. 3 Bruner & O'Connor, *supra* note 7, § 8:53; *see also* Post Bros. Constr. Co. v. Yoder, 569 P.2d 133 (Cal. 1977).

6. Overpayment

Premature payments and overpayments on a construction project can jeopardize the project itself and have dire consequences for those involved. As discussed earlier, contracting is a cash-flow business. The difference in timing between cash-in and cash-out becomes problematic if the contractor: (1) is undercapitalized; (2) is working on a particularly large project; (3) is working on multiple projects; or (4) is faced with a cash-flow crisis. When such situations arise, an owner may be persuaded to advance payments to a contractor in an effort to keep the contractor afloat or to keep the project moving. Such payments, however, may have unintended and costly consequences. The consequences are particularly problematic on projects involving payment and performance bonds.[64]

a. A Surety's Risks and Defenses

The overpayment risks for a surety are significant because, when a surety issues a performance bond, the surety becomes jointly and severally liable to the owner for the timely and satisfactory performance of the contractor's/obligor's responsibilities to the owner/obligee.[65] Courts repeatedly hold that an owner and contractor may not alter a bonded contract, which necessarily alters the surety's obligation, without the surety's consent. As a result, if an owner/obligee makes an unauthorized payment in violation of the underlying bonded contract (and without the surety's consent), the surety may be discharged.[66]

b. An Owner's Risks

An owner has numerous risks associated with overpayment. To begin with, if an overpaid contractor is unable to complete the project, then the owner may not have sufficient funds to hire a replacement contractor and complete the project on its own.[67] Moreover, the owner may not be entitled to the protection it expected to receive associated with the performance bond posted by the contractor. For example, a surety may claim that it is discharged due to a

---

64. Surety bonds are discussed further in Chapter 20.

65. *See* General Motors Acceptance Corp. v. Daniels, 492 A.2d 1306, 1309 (Md. 1985).

66. *See, e.g.*, Argonaut Ins. Co. v. Town of Cloverdale, Ind., 699 F.2d 417, 419 (7th Cir. 1983); *Nat'l Union Indem.*, 369 F.2d at 75, 77; McKnight v. Lange Mfg. Co., 155 S.W. 977 (Tex. App. 1913).

67. *See Fireman's Fund Ins. Co.*, 909 F.2d at 495.

material alteration in the underlying contract between the owner and general contractor if the owner has failed to follow the payment provisions recited in the contract. As a general rule, any material alterations of a contract between the principal and the obligee will discharge the surety.[68] Because the owner typically is obligated to inspect the project before rendering payment, a surety may also claim to be discharged if the owner abdicates such inspection responsibilities.[69] In addition, a surety may claim to be discharged from its obligations if the work performed by the contractor and paid for by the owner was defective.

c. An Architect's Risks

The primary risk to an architect is that if the architect negligently certified the prepayment or overpayment to the contractor, then the surety may be able to obtain reimbursement from the architect for its loss resulting from the architect's negligence.[70]

d. A Contractor's Risks

The primary risk to construction companies (and their owners) is that the sureties will pursue claims against them for indemnity. With respect to claims for indemnity, sureties will not issue a payment or performance bond on behalf of a contractor unless the contractor, and often the individual owners of the contracting company, have executed indemnity agreements. Pursuant to the surety's general agreement of indemnity (GAI), it is granted broad rights of indemnification against the contractor and other indemnitors. Accordingly, if a surety makes payment to an owner, subcontractor, or material supplier as a result of an obligation incurred under a bond, the surety will typically pursue

---

68. *See* Branch Banking and Trust Co. v. Creasy, 269 S.E.2d 117, 125 (N.C. 1980), citing Fleming v. Borden, 37 S.E. 219 (N.C. 1900); *but see* First Union Nat'l Bank of North Carolina v. King, 306 S.E.2d 508, 510 (N.C. Ct. App. 1983) (refusing to discharge a surety despite modification of the contract as between the principal and the creditor because the modification could only benefit the surety).

69. *See* Continental Ins. Co. v. City of Virginia Beach, 908 F. Supp. 341 (E.D. Va. 1995) (holding that where a contract provides for inspection of the construction project before the owner renders payment, and the owner abdicates such inspection responsibilities, the surety will be released to the extent of the harm it suffered).

70. *See* Peerless Ins. Co. v. Cerny & Assoc., Inc., 199 F. Supp. 951 (D. Minn. 1961) (holding that an architect responsible for certifying the contractor's right to an amount of payment has liability to a performance bond surety for negligent certification that results in the obligee's overpayment to the contractor).

indemnification from the contractor and from the other individual indemnitors that executed the GAI. [71]

D. Changes

Change is the only constant on a construction project, whether due to owner-directed changes, deficient or incomplete drawings and specifications, or unforeseen conditions.[72]

If the contracting parties do not agree that a given event is a change, that issue is resolved through the dispute resolution process. If, however, the parties agree that a change has occurred, then the price of the change is incorporated into the contract by way of a change order. The contract price is then increased, or decreased in the event of a deductive change order, and payment for the change is made per the contract requirements.

E. Interest

An owner's failure to make payment in a timely manner represents a material breach of the contract. In addition to its obligation to pay the contract amount due, an owner may also be obligated to pay interest on the past due amount. Typically, construction contracts set forth an interest rate to apply if payments are not timely made. Prejudgment interest is generally allowed in contract cases.[73] Interest on amounts awarded on a contract claim begins to run from the date of breach.[74] Interest will accrue at the legal rate unless a different rate is stated in the contract. Nearly every state has a "legal rate of interest" set by statute.[75]

14.04 SUBSTANTIAL COMPLETION/SUBSTANTIAL PERFORMANCE

Substantial completion is that point in the project where the work "is sufficiently completed in accordance with the Contract Documents so that the

71. *See Prairie State Nat'l Bank*, 164 U.S. at 227 (holding that a surety that completes a government contract upon the contractor's default in performance was entitled to be reimbursed out of the contract funds pursuant to its right of equitable subrogation).

72. See Chapter 16 on Changes.

73. *See, e.g.*, N.C. Gen. Stat. § 24-1; Craftique, Inc. v. Stevens & Co., 364 S.E.2d 129 (N.C. 1988).

74. *See* N.C. Gen. Stat. § 24-5(a).

75. *See id.* § 24-1 (legal rate of interest is 8 percent).

Owner can occupy or utilize the Work for its intended purpose."[76] When it believes the project is substantially complete, the contractor submits to the architect a list of items to be completed prior to final payment. The architect then inspects the project and ultimately prepares a list of incomplete or incorrect work items the contractor must correct and complete to achieve final completion. This list is known in the industry as a "punchlist." Once the architect is satisfied that the project or a designated portion thereof can be used for its intended purpose, it will issue a Certificate of Substantial Completion.[77]

Substantial completion is a milestone event for several reasons. First, it triggers the owner's obligation to make payment of retainage, but such payment is adjusted for the incomplete or incorrect work identified on the punchlist.[78] Second, it establishes the owner's responsibility for "security, heat, utilities, damage to the Work and insurance."[79] Third, warranties required by the contract commence on the date of substantial completion.[80] Fourth, the accrual of liquidated damages, if any, typically ends upon substantial completion, although contractually the parties can agree otherwise or agree to different liquidated damage amounts after substantial completion but before final completion. If the project is late, the owner likely will withhold accrued liquidated damages from the payment of retainage.

When the retainage is to be held until the construction is "substantially complete," disputes can arise as to when substantial completion occurs. Often a final survey or certification will be the event that triggers the obligation to release the funds, but certification alone may not always be dispositive of the question of whether substantial performance has occurred.[81] However, once the owner occupies the premises, substantial completion is hard to deny, whether or not the parties have obtained the necessary certifications.[82]

In contrast to contracts that provide for a retainage payment when the project is "substantially complete," courts generally are not supportive of retainage that allows for retention of the funds until the contract is "fully complete."

76. AIA Document A201–2007, *supra* note 8, § 9.8.1.

77. *See* AIA Document A201–2007, *supra* note 8, §§ 9.8.2–9.8.3.

78. *Id.* § 9.8.5.

79. *Id.* § 9.8.4.

80. *See id.*

81. *See* Casa Linda Tile & Marble Installers, Inc. v. Highlands Place 1981, Ltd., 642 So. 2d 766, 768 (Fla. Dist. Ct. App. 1994) (stating that "fail[ure] to secure a contractually required architect's certificate, alone, should not prevent" recovery of funds); Miller v. Bourgoin, 613 A.2d 292, 294–95 (Conn. App. Ct. 1992) (refusing to grant summary judgment for the builder even though a requisite certificate of occupancy had been issued); 3 Bruner & O'Connor, *supra* note 7, § 8:18, n.1.

82. 3 Bruner & O'Connor, *supra* note 7, § 8:18, n.1.

In *State v. Laconco, Inc.*,[83] a construction contract allowed the funds to be held until "all work" was complete. However, once the project was "substantially complete," the court disallowed the withholding of retainage beyond that which was necessary to fund the remaining work.

Other courts typically follow the *Laconco* court's practice, and when only a small amount of work remains, such as correcting items identified in a punchlist, those courts require that the retainage be reduced to no more than necessary to finance the remaining work.[84] Similarly, once the construction is actually finished, courts typically do not allow withholding retainage to ensure that the contractor performs warranty work.

Finally, issues also exist with respect to whether third parties can enforce retainage rights or access retainage funds. Typically, sureties and third parties who are given express rights in the construction contract can enforce the retainage provisions or access those funds. However, other third parties typically have greater difficulty in enforcing retainage rights or accessing retainage funds.

14.05 FINAL COMPLETION/FINAL PAYMENT

Once the contractor believes it has achieved final completion, it requests a final inspection and submits its final application for payment. Once the architect conducts its final inspection, and if it determines the work has been completed in accordance with the contract, it will issue a certificate of final completion.[85]

By issuing a certificate of final completion, the architect represents, among other things, that the contractor has submitted an affidavit and any other required documents, such as a release or lien waiver from subcontractors and suppliers reflecting the contractor's payment of obligations to its subcontractors and suppliers.[86] If a subcontractor or supplier refuses to furnish a release or waiver required by the owner, then the contractor can furnish a bond to indemnify the owner against any potential lien by that subcontractor or supplier. The bond allows the owner to make final payment and still be protected from potential liens.

The making of final payment can have an impact in at least two significant ways—acceptance of the work and waiver of claims.

83. 430 So. 2d 1376, 1382 (La. Ct. App. 1983).

84. *See* FUNDAMENTALS, *supra* note 2, at 132.

85. *See generally* AIA Document A201–2007, *supra* note 8, § 9.

86. *See id.* §§ 9.10.1–9.10.2.

A. Owner's Acceptance of Work/Waiver of Claims

The owner's final payment to its contractor will bar any claims for obvious defects if the payment is considered a clear intention to relinquish those claims.[87] Most standard contracts make clear that an owner's progress payment for work and final payment do not constitute acceptance of defective work.[88]

Absent contractual language specifying whether final payment constitutes acceptance, courts look to the facts surrounding the parties' discussion and the nature of the defective work. If the owner and contractor discuss the defective work, and the contractor promises to repair it, making final payment does not usually constitute a waiver by the owner; for example, in *Khoury Constr. Co., Inc. v. Earhart*,[89] the homeowners were able to pursue a claim for defective work despite having made final payment because they pointed out defects to the contractor during construction and continued to complain after taking possession.

If the defects are latent (hidden) and unknown to the owner after conducting a reasonably diligent inspection, then making final payment will not be considered a waiver.[90] If, however, the owner discovers, or should have discovered, a latent defect and fails to request that the contractor fix it, the owner could be deemed to have waived its claim.

If the work is patently (openly and obviously) defective, the owner's acceptance of the project and final payment may implicate the doctrine of waiver. The question is often whether or not the owner knows of the defect at the time of final payment. For instance, in *Eastover Corp. v. Martin Builders*,[91] the owner was a general partnership and the architect was one of the general partners. The architect knew of a patent defect, and this knowledge was imputed to the owner and the claim was barred by the doctrine of waiver.[92] Courts have also held that a defect that is discoverable through testing and inspection procedures called for by the contract may be considered patent, regardless of the owner's actual knowledge of the defect.[93]

87. *See* Stevens Constr. Corp. v. Carolina Corp., 217 N.W.2d 291 (Wis. 1974).

88. For example, Article 9 of AIA Document A201–2007 provides as follows regarding progress payments: "9.6.6. A Certificate for Payment, a progress payment, or partial or entire use or occupancy of the Project by the Owner shall not constitute acceptance of Work not in accordance with the Contract Documents." AIA Document A201–2007, *supra* note 8, § 9.6.6.

89. 382 S.E.2d 392 (Ga. Ct. App. 1989).

90. *Stevens Constr. Corp.*, 217 N.W.2d at 291 (citing Milwaukee County v. H. Neidner & Co., 263 N.W. 468 (Wis. 1936)); Fisher v. Simon, 112 N.W.2d 705 (Wis. 1961).

91. 543 So. 2d 1358 (La. Ct. App. 1989).

92. *See id.*

93. 3 Bruner & O'Connor, *supra* note 7, § 8:22; *see* Village of Pawnee v. Azzarelli Constr. Co., 539 N.E.2d 895 (Ill. App. Ct. 1989).

Regardless of whether the defect is latent or patent, any promise by the owner to pay despite deviation from an important aspect of the contract must be supported by consideration.[94] If the owner knows or should know of minor defects and makes payment, such minor defects may be waived.[95]

B. Contractor's Acceptance of Payment/Waiver of Claims

As with the owner's making of payment, the contractor's acceptance of payment, progress or final, may constitute a waiver of claims. Again, Article 9 of AIA Document A201–2007, deals with the final payment issue as follows:

> 9.10.5 Acceptance of final payment by the Contractor, a Subcontractor or material supplier shall constitute a waiver of claims by that payee except those previously made in writing and identified by that payee as unsettled at the time of final Application for Payment.[96]

An issue may arise as to whether or not particular communications are specific enough to constitute a waiver under the terms of the parties' contract, including provisions such as Article 9.10.5 of AIA Document A201–2007 set forth above. If the contractor simply reserves the right to file a claim for additional costs, but gives no specifics, the language may be considered insufficient to avoid the final payment bar to all claims.[97]

Usually, the contractor must execute an interim lien waiver to receive a progress payment.[98] Oftentimes, these lien waivers will contain an express waiver of all claims through a certain date. In most instances, such a waiver is enforceable.[99]

Absent clear language in a contract, lien waiver or other project document, the common-law doctrines of accord and satisfaction or compromise and settlement apply to whether acceptance of payment is a waiver of claims. The central question is whether the owner clearly communicated to the contractor that the payment was being tendered as full and final payment of all claims.

---

94. Sweet, *supra* note 14, § 22.06(E).

95. *See id.*

96. AIA Document A201–2007, *supra* note 8, § 9.10.5.

97. *See* Mingus Constructors, Inc. v. U.S., 812 F.2d 1387, 1389 (Fed.Cir. 1987) (contractor executed release of all claims except those with "amount(s) of which are undetermined at this time," and the court determined this language was insufficient to avoid the final payment bar of all claims).

98. *See* discussion *infra* at Section 14.03(C)(4).

99. *See id.*; *see also* Elec. Supply Co. of Durham, Inc. v. Swain Elec. Co., Inc., 403 S.E.2d 291, 297 (N.C. 1991) (court "note(s) that the use of lien waivers, used other than in anticipation of and in consideration for the awarding of a contract, may also minimize liability by contractors who deal with the owner").

Courts look both to evidence indicating a payment or billing is "final," as well as any factors indicating final payment has not yet been made. The specific factual circumstances, as well as the jurisdiction, lead to varied results in these situations. For instance, a court may find that a check offered as "final payment" under the contract and not as an offer to settle a disputed amount does not result in accord and satisfaction.[100] Alternatively, a court might hold that a situation involving a contractor who knows a check is offered in full settlement and negotiates that check would constitute accord and satisfaction.[101]

C. Liquidated Damages[102]

A liquidated damages provision provides for the assessment of a set sum against the contractor, usually on a per day basis, caused by the contractor's delay. The amount is determined or "liquidated" at the beginning of the job and is enforceable as long as it is not viewed as a penalty. A liquidated damages clause may be acceptable to the contractor in those instances where the risk of liquidated damages may be less than the risk of actual damages. Such provisions sometimes include a "bonus" if the project is completed ahead of time ("early completion bonus"). One could argue that, while these provisions encourage timely completion, they may also tend to result in poor work by forcing contractors to accelerate work efforts to complete their jobs by specified dates, which could give rise to claims for defects and damages caused by poor workmanship.

If delay damages appear to be easily calculated, or if the liquidated damage amount is grossly disproportionate to the probable loss the owner will suffer by a delay, the liquidated damages clause may be held unenforceable as a penalty.[103] Although a court may hold a contractor liable for delay damages even though the delay was caused by other contractors, the owner must prove that the contractor's delay was a substantial factor in the delay of the project.[104] If delays are experienced during the course of a project, the owner may withhold liquidated damages from the contractor's progress payments

100. *See* Peterson v. Ramsey County, 563 N.W.2d 103 (N.D. 1997).

101. *See* Malarchick v. Pierce, 264 N.W.2d 478 (N.D. 1978).

102. *See* FUNDAMENTALS, *supra* note 2, at 129.

103. *See, e.g.*, Knutton v. Cofield, 160 S.E.2d 29 (N.C. 1968) (enforcing liquidated damages clause after determining that said clause was not penal); *Contractual Provision for Per Diem Payments for Delay in Performance as One for Liquidated Damages or Penalty*, 12 A.L.R. 891 (4th 1982).

104. *See* Tuttle/White Constructors, Inc. v. Montgomery Elevator Co., 385 So.2d 98 (Fla. Dist. Ct. App. 1980).

and final payment. The AIA Contract Documents specifically allow an architect to withhold a certification of payment because of "reasonable evidence that the Work will not be completed within the Contract Time, and that the unpaid balance would not be adequate to cover actual or liquidated damages for the anticipated delay."[105]

14.06 SPECIFIC PAYMENT ISSUES

A. "Pay If Paid" versus "Pay When Paid"[106]

Pay-when-paid and pay-if-paid clauses are in vogue today, with the most onerous reading as follows: "Payment of Owner to Contractor for Subcontractor's work is a condition precedent to Contractor's duty to pay Subcontractor under this Subcontract agreement." When these clauses appear in a subcontract, they should cause a warning bell to go off in a subcontractor's head. Also known as contingent payment clauses, they mean that the general contractor does not have to pay the subcontractor until and unless the owner pays the general contractor. Thus, payment by the owner is a condition precedent to the contractor's duty to pay the subcontractor. Such clauses may be construed to shift the credit risk of nonpayment by the owner from the general contractor to the subcontractor, even though the subcontractor had no contract with the owner.

In most states, these clauses must be clear and unambiguous to be enforceable against a subcontractor.[107] If the clause is ambiguous, the general contractor must pay the subcontractor within a reasonable time.[108] However, even when payment is contingent, the general contractor must seek compensation from the owner in good faith and with due diligence, or the provision will be

105. AIA Document A201–2007, *supra* note 8, § 9.5.1.

106. *See* Fundamentals, *supra* note 2, at 131 (excerpt). Contingent payment clauses are also discussed in Chapter 11.

107. *See, e.g.*, Bentley Constr. Dev. & Eng'g, Inc. v. All-Phase Elec. and Maintenance, Inc., 562 So. 2d 800 (Fla. Dist. Ct. App. 1990); Elk and Jacobs Drywall v. Town Contractors, Inc., 229 S.E.2d 260 (S.C. 1976); Koch v. Constr. Tech., Inc., 924 S.W.2d 68 (Tenn. 1996) (containing citations to similar cases in nearly every jurisdiction); Galloway Corp. v. S. B. Ballard Constr. Co., 464 S.E.2d 349 (Va. 1995); Fla. Stat. § 713.245 (conditional payment bond permitted); *but see* Moores Bros. Co. v. Brown & Root, Inc., 207 F.3d 717 (4th Cir. 2000) (holding that a surety was not entitled to assert pay-when-paid defense of principal since provision was not incorporated into bond) (applying Virginia law).

108. *See Bentley*, 562 So.2d at 801 (determining that 90 days was a reasonable time); *Elk & Jacobs Drywall*, 229 S.E.2d at 262 (leaving determination of what constitutes a "reasonable time" up to the jury).

unenforceable.[109] Moreover, in some states, such clauses are unenforceable by statute as a matter of public policy.[110]

Even in states where "paid if paid" provisions are enforceable, courts may craft exceptions to the absolute enforceability of such clauses. First, where the condition precedent (payment by the owner to the general contractor) fails to occur because of the conduct of the general contractor, courts may apply the prevention doctrine.

> The prevention doctrine is a generally recognized principle of contract law according to which if a promissor prevents or hinders fulfillment of a condition to his performance, the condition may be waived or excused. . . . .The prevention doctrine does not require proof that the condition would have occurred "but for" the wrongful conduct of the promissor; instead it only requires that the conduct have "contributed materially" to the non-occurrence of the condition.[111]

Thus, if the conduct of the general contractor prevents payment by the owner to the general contractor, the general contractor cannot successfully defend a claim for payment by its subcontractor by arguing that the condition precedent, payment by the owner, has failed to occur.

Second, in the surety context, it is not likely that a surety can rely upon a pay-if-paid clause in its principal's contract with the owner as a defense to a payment bond claim by a subcontractor unless the surety included an express pay-if-paid condition precedent clause in its surety bond contract.[112] To allow a surety to avoid its obligation under the payment bond simply because the owner did not make payment to the general contractor "defeats the very purpose of a payment bond."[113]

B. Prompt Payment Acts

To deal with contractor to subcontractor payment abuses, the federal government and many states have enacted legislation to ensure that contractors promptly pay their subcontractors, commonly known as prompt payment acts.[114]

109. *See, e.g.*, Walker v. Chancey, 117 So. 705 (Fla. 1928).

110. *See, e.g.*, N.C. GEN. STAT. § 22C-2.

111. *See* Moore Bros. Co., 207 F.3d 717 (4th Cir. 2000) (citing RESTATEMENT (SECOND) OF CONTRACTS).

112. *Moore Bros. Co.*, 207 F.3d at 723.

113. *Id.* For further discussion with regard to surety bonds, see Chapter 20.

114. *See, e.g.*, 31 U.S.C. § 3901 (1998) (Federal Prompt Act) (applicable to those with contracts with the federal government).

These acts usually provide that the contractor will pay the subcontractor within a set number of days from the contractor's receipt of payment from the owner and that the amount of the payment will be the full amount received for the subcontractor's work and materials provided under the subcontract.[115] If a contractor does not timely pay its subcontractors, the contractor shall be liable to the subcontractor for interest on the unpaid amount.[116] Such acts do not, however, blindly require that the contractor make payment to its subcontractors. For instance, a contractor is not required to pay its subcontractor for "unsatisfactory job progress; defective construction not remedied; disputed work; third-party claims filed or reasonable evidence that claim will be filed; failure of subcontractor to make timely payments for labor, equipment, and materials; damage to contractor or another subcontractor; reasonable evidence that subcontract cannot be completed for the unpaid balance of the subcontract sum; . . ."[117] The contractor is also permitted to withhold "a reasonable amount for retainage not to exceed the initial percentage retained by the owner."[118]

C. False Claims

The Federal False Claims Act[119] was enacted in 1863 primarily to deal with privateering during the Civil War. During recent years, it has been used in connection with the construction industry to combat fraud. While there are many predicate acts, civil liability arises when one presents a "false or fraudulent claim for payment or approval" to the United States.[120] No specific proof of intent to defraud is required.[121] "Claim" is broadly defined to include "any request or demand, whether under contract or otherwise."[122] Civil damages can include a civil penalty and treble damages.[123] False claims related to payment can arise from a contractor's payment applications, claims for additional compensation, pass-through claims, and damages claims. A handful of states

115. *See, e.g.*, N.C. Gen. Stat § 22C-3 (requiring payment to be made by contractor within seven days of receipt of payment from the owner).

116. *See, e.g.*, N.C. Gen. Stat § 22C-5 (providing that interest is assessed at the rate of 1 percent per month beginning on the eighth day after contractor receives payment from owner).

117. N.C. Gen. Stat. § 22C-4.

118. *Id.*

119. 31 U.S.C. §§ 3729 et seq. Criminal liability can also arise for false claims. *See* 18 U.S.C. § 287.

120. 31 U.S.C. § 3729(a) (listing predicate acts).

121. *Id.* at § 3729(b).

122. *Id.* at § 3729(c).

123. *Id.* at § 3729(a).

have adopted their own false claims acts, which are modeled after the Federal False Claims Act.[124]

A careful review of regular payment applications and claims for additional compensation takes on special meaning on public projects. In *Al Munford Halls, Inc. v. United States*, the contractor sued the government for wrongful termination and submitted a claim for damages arising from that termination.[125] Discovery revealed that the contractor's claim included amounts previously paid by the government and costs not yet incurred by the contractor. The government thereafter asserted a False Claims Act claim by way of counterclaim, alleging that the contractor's claim was fictitious, inflated, and duplicative. The court agreed and noted as follows:

> The only issue before the court is whether plaintiff knew that its claims to the contracting officer were false or fraudulent as defined by the FCA. To prove a violation of the FCA, the government need not prove that a party intended to deceive the government. *United States v. TDC Management Corp.*, 24 F.3d 292, 298 (D.C. Cir. 1994). As stated above, the FCA requires only that the government prove that a party knowingly submitted a claim with reckless disregard to the falsity of the information. 31 U.S.C. § 3729(b); *TDC Management Corp.*, 24 F.3d at 298; *Chen-Cheng Wang ex rel. United States v. FMC Corp.*, 975 F.2d 1412, 1420 (9th Cir. 1992). However, innocent mistake or mere negligence is excused. *Chen-Cheng Wang*, 975 F.2d at 1420. For example, a party could not be held liable under the FCA for making an innocent mathematical error in calculating its claim.[126]

D. Lender Liability

As noted earlier, credit is a significant force in the construction industry. Credit begins with the construction loan provided by a financial institution. Given the depth and breadth of the construction loan documents and the level of the lender's risk, the lender often can exercise control with regard to construction payments and other matters. That control itself, however, can lead to significant risk and liability for the lender.

A construction lender's liability related to its duty to conduct inspections during the construction of a project was dealt with in *Lassiter v. Bank of North Carolina*.[127] The plaintiff homeowner/borrower complained that the lender failed to make construction inspections prior to the disbursement of funds.

124. *See, e.g.*, CAL GOV'T CODE § 1265 et seq.

125. 34 Fed. Cl. 62 (Fed. Cl. 1995), *vacated in part on other grounds*, 86 F.3d 1178 (Fed. Cir. 1996).

126. *Id.*

127. 145 N.C. App. 264, 551 S.E.2d 920 (2001).

The borrower derived such a duty from the "purpose" statement in the loan agreement, which provided that monthly draws on the construction loan "were to be made only on the basis of plaintiffs' draw requests and property inspections by defendant's inspector, 'to ensure that the loan is not drawn down below the point of construction completion.'" During the course of the project, the lender never inspected the construction project and permitted draws on the construction loan. During the construction process, the borrowers learned that the lender had advanced 61 percent of the construction loan but that the project was only 36 percent completed. The court refused to impose an inspection duty on the lender based upon the language in the purpose statement contained in the loan agreement:

> The loan agreement between plaintiffs and defendant contained no language obligating defendant to make property inspections before making or allowing a draw. Defendant was to make disbursements based on plaintiffs' request, but these payments were not contingent upon a property inspection. Purpose statements in loan agreements are permissive and do not create an affirmative duty on behalf of the lender. . . . even when a loan agreement indicates the lender will only disburse loan proceeds in proportion to the amount of construction completed, it does not require the lender to inspect the construction progress for the borrower's benefit.[128]

Under certain circumstances, the borrower's contractor may claim the benefit of or assert a lien against undisbursed loan proceeds in the hands of the lender. In *Embree Constr. Group, Inc. v. Rafcor, Inc.*, the court allowed a general contractor to assert an equitable lien on unadvanced construction loan proceeds still held by the lender.[129] The *Embree* court found that the general contractor had completed construction of the project, that the owner refused to pay the contract balance, and that the lender, while receiving the benefit of its security, a completed project, had not advanced all of the construction loan proceeds. The court looked favorably upon the contractor's claim for an equitable lien because the statutory mechanic's lien rights were of no practical value, given that those lien rights were subordinate to the prior recorded deed of trust of the construction lender. The court noted that "when a contractor's lien is subordinate to a construction loan mortgage . . . or to prior encumbrances such as a purchase money mortgage, any lien on the owner's property or its improvements is worthless when the owner is insolvent."[130] The court held

128. 146 N.C. App. at 268, 551 S.E.2d at 923; *see also* Camp v. Leonard, 133 N.C. App. 554, 515 S.E.2d 909 (1999); *but see* Rudolph v. First Southern Fed. Sav. & Loan Assoc., 414 So.2d 64 (Ala. 1982).

129. 330 N.C. 487, 411 S.E.2d 916 (1992).

130. 330 N.C. at 493, 411 S.E.2d at 921.

that the lender was unjustly enriched by virtue of its refusal to disburse the money remaining in the construction loan account, coupled with the lender's receipt of all of the security for which it bargained in the form of a completed building. Specifically, equitable relief is appropriate where the contractor has completed performance, the entire project itself is completed, and the lender receives a completed project but seeks to retain undisbursed construction loan funds.

14.07 CONCLUSION

Payment is the grease that runs the machine of a construction project. Clearly, there are many opportunities for the process to falter. Even though there must be contractual provisions in place that clearly set forth the parties' obligations to one another, a smooth process begins with parties who mutually respect one another and are committed to complying with their contractual obligations. The parties must maintain that commitment throughout the project.

QUESTIONS

1. Your client, Owner, is considering a project to dredge all of the mud, dirt, and debris from a New Orleans canal after Hurricane Katrina. Due to the effects of the hurricane, the amount of mud, dirt, and debris in the canal is unknown. Would you advise Owner to use a fixed-price contract or unit-price contract? Why?
2. Owner contracted with Contractor to build a new multi-level porch on his home for a fixed price and no progress payments using AIA Documents A101–2007 and A202–2007. After several days, Contractor submitted an application for final payment, stating that the porch was completed. Architect subsequently inspected the porch and signed a Certificate of Payment. After paying for the work pursuant to the Certificate of Payment, Owner finds that the porch supports are not properly bracketed, causing the porch to be structurally unsound. What are Owner's legal rights? Against whom? Under what causes of action? What is the likely outcome?
3. Has Owner waived its right to any claims against Contractor and Architect in Question 2?
4. What is the purpose of retainage?
5. After completing a substantial portion of work on a contract to build a house for Owner, Contractor is entitled to a progress payment of $5,000 pursuant to the terms of the contract. After Architect issues a Certificate of Payment for the progress payment, Owner informs Contractor that it currently does not have $5,000 and is attempting to obtain a new construction loan. What are Contractor's rights at this point? What additional facts would be helpful in determining Contractor's rights?
6. Because Owner was wary of Contractor's reputation for not paying its subcontractors, Owner paid Contractor and Subcontractor with a joint check addressed to both parties. Both Contractor and Subcontractor endorsed the check, and Contractor cashed it at a local bank. Contractor subsequently kept the proceeds from the check and refused to pay Subcontractor for work completed. What are Subcontractor's rights?
7. Owner contracts with Contractor to build a new house for $100,000. The contract provides for retainage of 10 percent, or $10,000. After substantial work, Architect submits a Certificate of Substantial Completion and a punchlist consisting of approximately $200 worth of minor fixes. Owner subsequently moves into the house but refuses to pay the retainage in light of the $200 worth of remaining work. Is Owner obligated to pay the retainage?
8. Federal Government contracts with Contractor to construct a new Supreme Court building in Washington, D.C. The Contract contains a liquidated damages provision of $5,000 per day. Because of a shortage in the labor force, Contractor

misses the contractual completion date by 180 days. Federal Government assesses \$900,000 in liquidated damages (\$5,000 × 180 days) against Contractor. Contractor contends that Federal Government's actual damages are only \$2,500 per day. Can Federal Government recover its liquidated damages of \$5,000 per day even if its actual damages are only \$2,500 per day?

9. Using the same facts as Question 8, Contractor submits a claim for additional compensation to the Federal Government because of costly changes requested by the Federal Government during construction. Contractor internally calculates its actual claim, but prior to submitting its claim to the Federal Government, it increases its claim by 25 percent in order to "leave room for negotiation." If Contractor submits its claim to Federal Government, is Contractor submitting a false claim?

CHAPTER

15

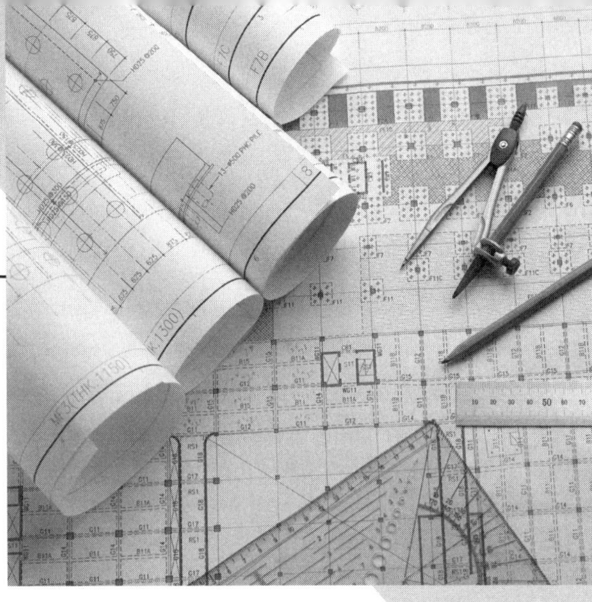

Construction Safety

WILLIAM R. ALLENSWORTH

15.01 INTRODUCTION

Construction is one of the most dangerous workplace activities, killing about 1,100 workers in the United States each year and disabling several thousands more.[1]

The prevention of construction accidents and the allocation of the responsibility for property damage and bodily injury arising from such accidents are the subjects of construction safety law. The law is hidden in a dense thicket of common law forms, obtuse regulations, complicated insurance policies, and contractual sophistry, the command of which is the stock-in-trade of a large specialty bar.

This chapter introduces the area of construction safety—more specifically, safety of workers—and the roles of the principal players. To structure the discussion, this chapter is divided into three general subdivisions that reflect the

1. Bureau of Labor Statistics, *Census of Fatal Occupational Injuries (CFOI)—Current and Revised Data* (2005), *available at* http://www.bls.gov/news.release/pdf/cfoi.pdf (last visited June 14, 2007).

most common legal issues arising in construction safety: accident prevention, compensation, and the allocation of responsibility for accidents through tort liability, insurance, or by contract.[2]

15.02 ACCIDENT PREVENTION

The prospect of tort liability for construction accidents undoubtedly encourages accident prevention programs, but this section addresses the affirmative duty of accident prevention as either a governmental or contractual requirement.

A. State Legislation

While the prevention of construction accidents has long been a subject of governmental interest, American states have enacted safety legislation only since the late 19th century. States legislating safety primarily were those with unionized workforces, with legislation often piecemeal and in response to notorious accidents, or series of accidents.[3] Most of these statutes required the general contractor to provide for workplace safety, or for a particular aspect of it, such as protection against falls. Owners either were assumed to have no responsibility or specifically were exculpated from responsibility.

Architects' and engineers' licensing boards have not been heavily involved in accident prevention—aside from a few well-publicized license suspensions arising from design or contract administration errors on notorious projects.[4] Although state licensing boards may require that a design professional produce a safe design, of more significance are state statutes and regulations requiring engineered safety systems during construction, including, in particular, trench safety and highway construction.[5]

B. Occupational Safety and Health Act

In the 1960s, Congress took up the cause of accident prevention, and President Nixon signed into law the Occupational Safety and Health Act (OSHA) in

---

2. For more information on the topic of safety in construction law, see PHILIP L. BRUNER & PATRICK J. O'CONNOR, JR., BRUNER & O'CONNOR ON CONSTRUCTION LAW §§ 5:1, 7:60, 13:14–13:15, 17:56 (West Group 2002) [hereinafter BRUNER & O'CONNOR].

3. For a well-researched history of this legislation, see MARC M. SCHNEIER, CONSTRUCTION ACCIDENT LAW: A COMPREHENSIVE GUIDE TO LEGAL LIABILITY AND INSURANCE CLAIMS 48–52 (Forum 1999) [hereinafter CONSTRUCTION ACCIDENT LAW].

4. *See, e.g.*, Duncan v. Missouri Bd. for Architects, Professional Eng'rs and Land Surveyors, 744 S.W.2d 524 (Mo. Ct. App. 1988).

5. *See, e.g.*, Texas Uniform General Conditions in State Construction Contracts, TEX. GOV'T CODE ANN. § 2166.301 (Vernon 2000).

1970.[6] OSHA set up a Safety Administration and charged the Secretary of Labor with enacting regulations to "assure safe and healthful working conditions for working men and women" and for enforcing them.[7] A number of states adopted OSHA, or portions of it, as the state statutory standards replacing local workplace safety statutes.[8]

Those regulations cover virtually every aspect of construction workplace safety in mind-numbing detail, requiring employers to institute safety programs and to ensure implementation. OSHA, however, is deliberately imprecise in its definition of "employer," and because "employer" status is necessary to be subject to OSHA, such intentional ambiguity arguably broadens the potential applicability of the act. The Safety Administration takes the position that the term includes not only the subcontractors whose employees are workers but also the prime contractors who control the subcontractors, thereby making the prime contractors responsible for complying with the OSHA regulations as well.[9]

The theory behind this broad application of OSHA was stated succinctly in *Clarkson Constr. Co. v. Occupational Safety and Health Review Comm'n*,[10] where a subcontractor backed a truck over an employee. In affirming a penalty against the prime contractor who hired the subcontractor, the court reasoned that the prime contractor was an "employer" because

> [W]e here [are not] dealing with vicarious liability of the employer to third persons in tort. Instead, we are concerned with giving meaning to an Act of Congress, which was designed to require the employer to provide a safe work place for all persons working on the premises. With that in mind, common law definitions ought not to be dispositive in interpreting the Act.
>
> . . .
>
> The Supreme Court has refused to use the traditional approach in defining employee in the context of other statutes. Moreover, instead of looking at narrow common law definitions, the Supreme Court has looked to the purpose of the Act involved in deciding what the meaning of employee should be and has

---

6. 29 U.S.C. §§ 651–678 (1994).

7. From the preamble to OSHA, 29 U.S.C 651.

8. If a state, with approval from the Secretary of Labor, adopts a comprehensive statutory safety scheme equivalent to OSHA, the enforcement of OSHA is preempted in that State. 29 U.S.C. § 667.

9. *See* OSHA Field Operations Manual, Chapter V, Section F, Multi-Employer Citation Policy (referencing 29 CFR 1926.16), *available at* http://www.osha.gov/pls/oshaweb/owadisp.show_document?p_table=DIRECTIVES&p_id=1641 (last accessed June 15, 2007); *see also* ANSI A10.33-1992, ¶¶ 2 (definitions), 5.2 (responsibilities).

10. 531 F.2d 451 (10th Cir. 1976).

> also held that it is inappropriate to use varying state common law definitions of an employee and employer in construing federal legislation.[11]
>
> The statute with which we are here dealing has a broad remedial purpose, protection of the worker from industrial injury. It has been said that, "[i]n order to accomplish this purpose, it is necessary to look to an employer who controls the working environment. Whether the employer controlling work environment [sic] is also the employer for wage or tort purposes should not be a governing factor."[12]

Design professionals performing contract administration responsibilities consistent with those contemplated by the American Institute of Architects (AIA) and National Society of Professional Engineers (NSPE) general conditions generally are not regarded as "employers" and, therefore, are usually outside the scope of OSHA.[13] Construction managers (and design professionals acting as construction managers), however, are much more likely to be viewed by OSHA as "controlling employers" if they are substantially involved in the day-to-day running of the construction site.[14]

C. OSHA in Civil Litigation

OSHA is directly enforced by its inspectors, who are empowered to levy fines and penalties on employers.[15] The extent to which OSHA can be used to estab-

---

11. *Cf.* 29 C.F.R. § 1977.5, which provides that for purposes of Section 11(c) of the Act (29 U.S.C. Sec. 660(c)) (the antidiscrimination clause) "employee" has a broad definition. The regulation states in part that: "[T]he broad remedial nature of this legislation demonstrates a clear congressional intent that the existence of an employment relationship, for purposes of section 11(c), is to be based upon economic realities rather than upon common law doctrines and concepts."

12. *Clarkson Constr. Co.*, 531 F.2d at 457 (citations omitted).

13. Skidmore, Owings & Merrill, 5 BNA OSHC 1762 (OSHRC 1977) (stating that architect who was not involved in substantial supervision of the construction was not bound by OSHA). *See also*, Reich v. Simpson, Gumpertz & Heger, Inc., 3 F.3d 1, 3 (1st Cir.1993) (holding that construction site was not a "place of employment" of the structural engineer for purposes of OSHA where it had no employees on the site on a daily or even weekly basis, did not have an office or a trailer at the site, and did not have any employees at the site on the day in question).

14. Bertrand Goldberg Assoc., 4 BNA OSHC 1587 (OSHRC 1976) (deciding that architect who was also engaged as a construction manager was bound by OSHA in its work directly overseeing the worksite); *See also* CH2M Hill, Inc. v. Herman, 192 F.3d 711, 718–21 (7th Cir. 1999) (holding that: (1) OSHA construction standards, including "substantial supervision" test did not per se exclude professionals working on or as part of a construction project, but (2) firm did not engage in construction work and, thus, was not subject to construction standards.).

15. Or, if the state has adopted an OSHA-approved version of the act, enforcement is handled by state inspectors. *See, e.g.*, OSHA Instruction CSP 01-00-002—STP 2-0.22B, Directive No. STP 2-0.22B—State Plan Policies and Procedures Manual (March 21, 2001) *available at* http://www.osha.gov/OshDoc/Directive_pdf/STP_2-0_22B.pdf (last accessed June 18, 2007).

lish a basis for a private cause of action is a much closer question, however, as this Colorado case notes:

Scott v. Matlack, Inc.
39 P.3d 1160 (Col. 2002)[16]

Justice Bender delivered the Opinion of the Court.

I. Introduction

In this case, we review and reverse the court of appeals' decision. We hold that (1) [precedent] does not preclude the admission of Occupational Safety and Health Act evidence in a negligence suit; (2) it is proper for a trial court to admit Occupational Safety and Health Act regulations as evidence of the standard of care in an industry. . . .

Plaintiffs Randy and Ann Scott sued Defendants Matlack Trucking, Inc. and Conoco Oil Company for negligence, negligence per se, and loss of consortium after Randy Scott fell from the top of a tanker while loading hot asphalt for Matlack at Conoco's Denver facility. After a two-week trial, a jury found the defendants liable on negligence and loss of consortium claims.

[W]e held [previously] that OSH Act regulations do not create a private cause of action and therefore a plaintiff could not establish a negligence per se claim by alleging a defendant violated OSH Act regulations.[1]

In this case, the court of appeals reversed the trial court, holding that this court's [previous] opinion precludes admission of evidence related to the OSH Act in a negligence action. . . . We reverse and remand this case to the court of appeals for return to the trial court to reinstate the jury award and judgment for the plaintiffs.

. . . .

III. Analysis

. . . .

1. Negligence vs. Negligence *Per Se*

As background, we start with an explanation of negligence and negligence per se. To establish a common-law negligence claim, the plaintiff must show that the defendant owed the plaintiff a legal duty to conform to a standard of care, the defendant breached that duty, the plaintiff suffered injury, and there is a causal relationship between the breach and the injury.

The defendant owes a duty to act when it is reasonably foreseeable that the failure to act will create an unreasonable risk of harm to another. A defendant's

16. Internal citations, footnotes, expanded procedural history, and further discussion of negligence as applied to the facts of the case have been omitted.

duty is based on the standard of care owed by a reasonable person in the defendant's position. When the defendant and the plaintiff are part of an industry that conforms to certain well-established safety customs, the jury may consider the customs as non-conclusive evidence of reasonable care the defendant should follow in that industry.

In contrast to negligence, negligence *per se* occurs when the defendant violates a statute adopted for the public's safety and the violation proximately causes the plaintiff's injury. To recover, the plaintiff must also demonstrate that the statute was intended to protect against the type of injury she suffered and that she is a member of the group of persons the statute was intended to protect. If the statute applies to the defendant's actions, then the statute conclusively establishes the defendant's standard of care and violation of the statute is a breach of his duty.

2. The OSH Act and Negligence

OSH Act section 653(b)(4) provides that the statute may not be used to enlarge, diminish, or affect an employee's or employer's common-law liability:

> Nothing in this Act shall be construed to supersede or in any manner affect any workmen's compensation law or to enlarge or diminish or affect in any other manner the common law or statutory rights, duties, or liabilities of employers and employees under any law with respect to injuries, diseases, or death of employees arising out of, or in the course of, employment.

[Previously], we followed a majority of courts and ruled a defendant may not be held negligent merely with proof that he violated an OSH Act regulation. We held that OSH Act section 653(b)(4) bars the use of OSH Act regulations to establish a negligence *per se* claim.

. . .

However, [we] did not address whether section 653(b)(4) precludes a plaintiff from using OSH Act regulations in a negligence action as *some* evidence of a defendant's negligence. The Scotts argue that OSH Act evidence may be used to provide non-conclusive, or some, evidence of the industry standard with which a reasonable person in the defendants' industry should comply.

Many jurisdictions that prohibit the use of OSH Act regulations to establish a negligence *per se* claim permit the use of OSH Act regulations in a negligence case as some evidence of the standard of care in the defendant's industry.

The question of whether OSH Act regulations may be admitted as some evidence of the industry standard in a negligence case turns on two issues. The first is whether the OSH Act, as a safety statute, may be borrowed as evidence of the standard of care in the defendants' industry. The second issue is whether the language of section 653(b)(4), "[n]othing in this Act shall . . . enlarge or diminish or affect in any manner the common law or statutory . . . duties, or liabilities of employers and employees," excludes OSH Act regulations from general rules regarding the borrowing of safety statutes.

We turn to the question of whether using, or borrowing, regulations or statutes as some evidence of the standard of care in an industry is permissible. Such use has been permitted in some of our cases. The court of appeals reasoned, "safety codes and standards are admissible when offered in support of expert testimony and when introduced as objective safety standards generally recognized and accepted as such in the type of industry involved."

In a negligence case in which a plaintiff seeks to borrow a safety statute as some evidence of the standard of care in defendant's industry, we adopt [this] reasoning. Therefore, to be admissible in such a situation, the safety standards must be relevant, offered in support of expert testimony, objective, and recognized and accepted in the industry involved. [Based upon its use in the case by expert witnesses to illuminate a disputed issue of fact, OSH Act regulations, if viewed simply as safety standards] meet the criteria for admissibility as evidence of standard of care in the relevant industry.

However, Matlack and Conoco argue that the usual rules for admitting a safety statute are rendered inapplicable by the express limiting language of section 653(b)(4). The defendants argue that this section restricts OSH Act regulations from being used as evidence, for any reason, in a civil suit. They argue that even borrowing the OSH Act as some evidence of industry standards affects and enlarges an employer's duty.

There is some precedential support for the defendants' position. [T]he Seventh Circuit held that workers' compensation was an injured employee's only remedy and that OSH Act regulations could not be used as evidence of "an expanded standard of care" in a negligence suit.

However, few jurisdictions excluding OSH Act evidence from a negligence suit have done so because section 653(b)(4) absolutely bars use of the OSH Act in civil cases. Instead, those jurisdictions tend to cite controlling statutes or case law particular to their jurisdiction for excluding OSH Act evidence.

Our review of authorities and case law from other jurisdictions reveals three generally accepted conclusions: (1) the OSH Act does not create a federal private cause of action and cannot be used as the sole basis of a negligence or negligence per se claim; (2) the OSH Act cannot be used to establish a state negligence per se claim because of the language in section 653(b)(4); and (3) the OSH Act may be admitted in a negligence action as some evidence of the standard of care required of defendant.

. . .

[However], the question remains; does borrowing the OSH Act as some evidence of the standard of care enlarge or affect the defendants' duty under the common law in a negligence suit?

In construing this section, we face a logical conundrum. The broad reading of section 653(b)(4) that the defendants urge would require us to conclude that borrowing OSH Act regulations to aid the jury in determining whether a defendant complied with the industry standard, "enlarges" the defendant's liability.

On the other hand, excluding OSH Act regulations might cause an equally impermissible result. If a defendant is permitted to exclude OSH Act regulations based on the defendant's proposed reading of section 653(b)(4), then the jury is left with fewer tools to determine the standard of care with which the defendant should have complied. Restricting the jury in this way would effectively "diminish" the defendant's liability. Moreover, while the language of section 653(b)(4) is expansive, there is no evidence to suggest the Congress intended for evidence of OSH Act regulations to be absolutely barred from civil cases. Also, we note that the Restatement of Torts generally supports borrowing of administrative regulations to inform the industry standard.[2]

. . .

Ultimately, we agree with . . . the Third Circuit, which concluded that "[e]vidence of an OSHA violation, in and of itself, does not 'affect' liability; it is the inferences that the trier of fact draws from the evidence that 'affect' liability." We also agree . . . that borrowing OSH Act regulations in a negligence action is not the same as applying the regulations to the case. This is consistent with Colorado precedent that allows the admission of safety regulations as some evidence of negligence in other contexts.

Thus, we hold that once a trial court determines that a defendant could owe a plaintiff a duty under Colorado law, the jury should be permitted to hear evidence of OSH Act regulations as some indication of the standard of care with which a reasonable person in the defendant's position should comply.

1. Canape v. Petersen, 897 P.2d 762 (Colo. 1995).

2. Restatement (Second) of Torts § 288B cmt. d (1965), "[T]he requirements of administrative regulations are not adopted by the court as defining a definite standard of conduct in negligence actions, but are accepted as affording relevant evidence."

The Court did not directly address whether the plaintiff had to prove that the prime contractor was subject to the regulations as a "controlling employer" to use the regulations to prove the standard of care for the prime contractor, as opposed to the subcontractor. Even though the regulations may evidence an "industry standard," does that standard apply to a prime contractor who is not a "controlling employer" under the OSHA regulations?[17]

17. *Cf.* S & H Riggers & Erectors, Inc. v. Occupational Safety & Health Comm'n, 659 F.2d 1273 (5th Cir. 1981) (holding that: (1) absent clear articulation by the Commission of circumstances in which industry practices were not controlling, due process required, in order to establish violations of OSHA regulation governing provision of personal protective equipment to employees in construction industry, showing that employers either failed to provide personal protective equipment customarily required in industry or had actual knowledge that personal protective equipment was required under circumstances of cases, and (2) substantial evidence did not support

D. Contractual Responsibility for Safety Programs

By statute, most federal construction contracts require the prime contractor to follow OSHA,[18] while both the AIA and the Engineers Joint Contract Documents Committee (EJCDC) general conditions rely on the broader catch-all of requiring compliance with all applicable laws, ordinances, rules, and regulations,[19] which presumably includes OSHA.

The form contracts between owners and prime contractors prepared by the AIA and the EJCDC, of course, are consistent with the following approach to accident prevention: namely, that responsibility is delegated almost exclusively to the prime contractor. The prime contractor is to be "solely responsible for and have control over construction means, methods, techniques, sequences, and procedures, and for coordinating all portions of the Work under the Contract,"[20] including the affirmative duty of initiating, maintaining, and supervising all safety precautions and programs, and designating a responsible member of the contractor's organization at the site whose duty shall be the prevention of accidents.[21] Most subcontracts include similar language obligating the subcontractor to fulfill such duties with respect to the subcontractor's portion of the Work, and specifically flow-down the prime contractor's safety obligations from the prime contract.[22]

---

findings by the Commission that employers, whose practices conformed to industry custom, had actual knowledge that personal protective equipment was necessary to employees under circumstances of cases, as required to establish violations of OSHA regulation governing provision of personal protective equipment to employees in construction industry).

18. The Construction Contracts Book 319–24 (Daniel S. Brennan, Richard H. Lowe, Jennifer A. Nielsen & John I. Spangler, III eds. ABA 2004).

19. American Institute of Architects, AIA Document A201–2007, *General Conditions of the Contract for Construction* § 10.2.2 (2007) [hereinafter AIA Document A201–2007]; Engineers Joint Contract Documents Committee [hereinafter EJCDC], Standard General Conditions of the Construction Contract, ¶ 6.13 (2002).

20. AIA Document A201–2007, *supra* note 19, § 3.3.1.

21. AIA Document A201–2007, *supra* note 19, §§ 10.1 (contractor shall initiate and maintain all safety precautions), 10.2.1.1 (contractor will take reasonable precautions to prevent injury to employees in the work or other persons on the project), 10.2.2 (contractor will comply with all laws bearing on the safety of persons), 10.2.3 (contractor will post warnings and take other reasonable measures), 10.2.4 (contractor will exercise utmost care when using explosives), 10.2.6 (contractor will appoint a superintendent in charge of safety), and 10.2.7 (contractor will not permit any part of the construction to be loaded so as to endanger its safety).

22. American Institute of Architects, AIA Document A401–2007, *Standard Form of Agreement Between Contractor and Subcontractor* § 4.3.1 (2007).

15.03 RECOVERY OF COMPENSATION FOR CONSTRUCTION ACCIDENTS

The contractual responsibility to provide a safe workplace does not necessarily translate into tort liability to injured workers. The primary basis of compensation for workers' injuries remains statutory workers' compensation schemes, with uncertain resort to tort liability under arcane legal principles. Workers' compensation and tort liability are discussed separately.

A. Workers' Compensation Laws

Until the last century, most construction workers were regarded as independent contractors who, with their widowed spouses and orphans, were solely responsible for themselves. If independent contractors were deemed to be "employees," then their employers had very limited common law responsibility to them, particularly since each of the following allowed a complete defense to liability: the fellow-servant rule, contributory negligence, and assumption-of-risk. With the maturing of the Industrial Revolution, including the enfranchisement of vast numbers of voting employees and the rise of their labor unions, various state legislatures began enacting workers' compensation laws.[23]

While the laws differ from state to state, they all have in common some basic tenets: the employer provides workers' compensation insurance, which provides medical and disability insurance benefits paid to the worker, regardless of fault. In return, the employer is insulated from any further liability, except under the most egregious of circumstances.[24] The resolution of workers' compensation claims is now largely in the province of administrative law, and the primary issues are the extent and severity of the injury (and, hence, the compensation for it) and whether the injury occurred in the "course and scope" of employment.[25]

---

23. For a history of the rise of workers' compensation laws, see Price V. Fishback & Shawn Everett Kantor, *The Adoption of Workers' Compensation in the United States, 1900–1930*, 41 J.L. & ECON. 305, (1998); Paul Raymond Gurtler, *The Workers' Compensation Principle: A Historical Abstract of the Nature of Workers' Compensation*, 9 HAMLINE J. PUB. L. & POL'Y 285, (1989); Richard A. Epstein, *The Historical Origins and Economic Structure of Workers' Compensation Law*, 16 GA. L. REV. 775, (1982).

24. *See, e.g.*, TEX. LABOR CODE § 408.001(b) (Vernon's 2006). Common law recovery allowed if worker's death was caused by gross negligence or intentional conduct.

25. The simplicity of these issues and the number of claimants ensured huge amounts of legal work for budding litigators. Until very recently, most of the lions of the civil trial bar cut their teeth on compensation cases, and legal lore is full of anecdotes of their trials. Ultimately, the cost of litigation became so expensive, and the results so unjust, that most legislatures have made the proceedings almost exclusively administrative.

B. Tort Liability

Workers' compensation is an essentially Faustian bargain, which generally ensures that the worker will receive less—but more certain and expedited—compensation from the employer than the worker might have been able to obtain as a tort claimant. Only workers' compensation subscribers are insulated from liability, however. The extent to which the injured worker can recover the remainder of the worker's damages from other members of the construction team, or from the owner, involves questions of duty under tort law. These questions will be examined in detail for each of the major players in any construction: the owner, the prime contractor, and the design professionals. Generally, the fact patterns in construction accident cases can be divided into two fundamental types, based upon the kind of negligence that is involved. Workers' compensation also plays a role in the ability of owners, design professionals, or other contractors who are sued to seek indemnity or contribution from the original employer. The extent to which workers' compensation insulates employers who purchase it from these types of third-party tort claims is discussed more fully in the Indemnity and Insurance section of this chapter.

1. *Active Negligence*

The first type of construction accident case involves a party whose negligence was contemporaneous with the injury (i.e., whose negligent act more or less directly caused the injury). This category, sometimes described as "active negligence," includes negligent acts like knocking a construction worker off a platform with a crane,[26] running over a construction worker with a truck,[27] or dropping tools from scaffolding onto a construction worker's head.[28] Unless the negligent actor is insulated from liability by inclusion by the workers' compensation bar,[29] such cases are resolved under traditional duty analyses, which focus on the duty of a reasonable person to avoid negligently injuring foreseeable victims.

---

26. Gonnert v. Victor At W. 53, LLC, No. 40690/04, 2006 N.Y. Slip Op. 50284(U) (N.Y. Slip Op. March 3, 2006); Nowicki v. Cannon Steel Erection Co., 711 N.E.2d 536 (Ind. Ct. App. 1999); Sulem v. B.T.R. East Greenbush Inc., 589 N.Y.S.2d 969 (N.Y. App. Div. 1992).

27. Scott v. Matlack, Inc., 39 P.3d 1160 (Col. 2002); R. E. Gaddie, Inc. v. Price, 528 S.W.2d 708 (Ky. 1975); Eades v. Capitol Material Co., Inc., 121 F.2d 72 (D.C. 1941).

28. Scully v. Otis Elevator Co., 275 N.E.2d 905 (Ill. App. Ct. 1971); Blair v. Dunham, 134 F.2d 729 (6th Cir. 1943).

29. Which determination usually is dependent upon whether the contractor or subcontractor is an "insured" under the local workers' compensation scheme, either statutorily or as an "additional insured."

2. *Liability Based Upon Control of the Workplace*

Much more difficult is the group of cases that involves indirect liability for failing to provide a safe workplace for the injured worker or for failing to remedy dangerous conditions. Generally, these claims are brought by subcontractors' employees against the prime contractor or the owner, on a theory of premises liability (i.e., a property owner or occupier has a qualified responsibility to protect others from harm caused by defects on the owner's or occupier's land). The duty analysis in these cases, therefore, is deeply rooted in the law of property and, to a lesser degree, in the law of contracts.

C. Owners

Our discussion of the various liabilities that can attach to common players in construction projects begins with the owner. The archetypal owner is the fee holder who has hired an independent contractor to build or improve a structure on the owner's property. As a general rule, the owner is not responsible for the safety of the independent contractors or for their employees, unless the owner actually hurts them[30] or retains control of their work. This law, founded upon the law of property, is enshrined in Section 414 of the RESTATEMENT OF TORTS (SECOND).[31]

Whether the owner has retained sufficient control over the independent contractor to charge the owner with responsibility for the safety of such independent contractor's employees is determined initially by reference to the general contract. Many general contracts, and all AIA and EJCDC form contracts, explicitly state that the prime contractor is solely responsible for the manner, method, and means of construction and that safety is the obligation of the prime contractor, not the owner.[32] While the general contract may allow the owner some latitude in demanding that the prime contractor execute work safely, or abide by safety programs, these rights are generally insufficient to trigger a duty to the prime contractor's employees. Occasionally, however, the

---

30. *See* "active negligence" cases at notes 26–28, *supra*.

31. RESTATEMENT (SECOND) OF TORTS § 414 (1977) (Negligence in Exercising Control Retained by Employer: One who entrusts work to an independent contractor, but who retains the control of any part of the work, is subject to liability for physical harm to others for whose safety the employer owes a duty to exercise reasonable care, which is caused by his failure to exercise his control with reasonable care).

32. AIA Document A201–2007, *supra* note 19, § 3.3; EJCDC, STANDARD GENERAL CONDITIONS OF THE CONSTRUCTION CONTRACT, *supra* note 19, ¶ 6.13.

owner nevertheless may retain so-called actual control of some aspect of the prime contractor's work sufficient to charge the owner with the duty of exercising that control responsibly.

Control of the work, however, is often a close question involving important policy issues, as the following case demonstrates:

Armenteros v. Baptist Hosp. of Miami, Inc. 714 So. 2d 518 (Fla. Dist. Ct. App. 1998)[33]

NESBITT, Judge.

Gregorio Armenteros was employed by a subcontractor hired by the general contractor building a structure on Baptist Hospital's grounds. Armenteros claimed that he was injured in a fall from defectively erected scaffolding, and sued the hospital asserting that Baptist had breached its duty of care by allowing an unsafe work condition to remain uncorrected on its premises. Baptist answered and moved for summary judgment on its affirmative defense that it owed no duty to a subcontractor's employee. Armenteros opposed summary judgment asserting that Baptist had retained the right to, and did exercise control over the day-to-day construction operations, hence, Baptist was liable as an owner/contractor. The court granted summary final judgment in Baptist's favor. We affirm.

Armenteros' accident occurred on a construction site at Baptist where a multistory addition was being added above an existing surgery center building. . . . Throughout the course of the project, Baptist continued to operate its hospital and to provide medical and surgical care to its patients. . . . To ensure the safety of these patients, the construction schedule had to be sensitive to ongoing hospital activities and the goal of maintaining quality patient care.

To this end, the construction contract stated that construction workers were limited to certain entrances and exits so that the hospital could avoid widespread contamination by construction debris and dirt. Similarly, . . . Baptist retained the right to stop work temporarily if the construction was interfering with surgical procedures. . . .

Centex-Rodgers Construction Company was hired to perform a variety of functions, including control of the construction management. . . . [A] construction contract was signed . . . between Baptist Hospital and Centex-Rodgers that specifically identified Baptist as the owner and Centex-Rodgers as the general contractor. . . .

33. Internal citations and footnotes omitted.

One of the Centex-Rodgers employees had overall responsibility to oversee all day-to-day construction activities; including overseeing all subcontractors and ensuring that they were doing their work properly and conforming [to] the contract documents and specifications.

Armenteros was injured while in the course of his employment for General Forming Corporation, a subcontractor who was responsible for certain concrete work on the project. General Forming, like all of the successful bidders/subcontractors, had its contract with Centex-Rodgers and not with Baptist.

As a general rule, one who hires an independent contractor is not liable for injuries sustained by that contractor's employees in their work. As the Supreme Court [of Florida] observed:

> the owner may be held liable if he has been actively participating in the construction to the extent that he directly influences the manner in which the work is performed. Conversely, if the owner is a passive nonparticipant, exercising no direct control over the project, he cannot be held liable. To impose liability upon an owner who is not an employer as defined by the statute, one or more specific identifiable acts of negligence, i.e., acts either negligently creating or negligently approving the dangerous condition resulting in the injury or death to the employee, must be established.

Thus, Florida case law has followed the analysis outlined in Restatement (Second) of Torts section 414, which provides:

> One who entrusts work to an independent contractor, but who retains the control of any part of the work, is subject to liability for physical harm to others for whose safety the employer owes a duty to exercise reasonable care, which is caused by his failure to exercise his control with reasonable care.

. . .

Appellants' contention is that sufficient testimony was given to make it a jury question whether Baptist had exercised enough control to be considered to have actively supervised and directed the project. However the undisputed facts demonstrate that the hospital did no more than purchase materials to utilize its tax free status, inspect the work to ensure the high standards it sought to maintain, and control schedules so as to permit surgeries to go uninterrupted and patients to go unendangered. Such limited acts, even when grouped together, will not constitute the control element necessary for a finding of liability.

Just because an owner has the right to inspect work for conformance with the contract does not change the owner from a passive nonparticipant to an active participant in the construction with the right to supervise or control the work, nor does it destroy the independent status of the contractor and render the owner liable for the contractor's negligence in performing the work by creating a dangerous condition.

Comment c to Restatement (Second) of Torts section 414 supports this analysis and provides:

> c. In order for the rule stated in this Section to apply, the employer must have retained at least some degree of control over the manner in which the work is done. It is not enough that he has merely a general right to order the work stopped or resumed, to inspect its progress or to receive reports, to make suggestions or recommendations, which need not necessarily be followed, or to prescribe alterations and deviations. Such a general right is usually reserved to employers, but it does not mean that the contractor is controlled as to his methods of work, or as to operative detail. There must be such retention of a right of supervision that the contractor is not entirely free to do the work in his own way.

The cases relied on by appellant address scenarios where the owner exercised control over day-to-day work performance. While Baptist admitted interfering with scheduling, there was no evidence that the hospital interfered with the project's actual physical construction. There is no evidence that it was supervising the project. Nor was evidence submitted to refute Baptist's contention that it completely delegated all safety inspection tasks.

Comment c, reproduced above, is directly applicable to the instant question and supports our conclusion that liability should not be imposed merely because Baptist reserved the right to inspect the work, established time schedules, or used its tax exemption status to purchase materials. A hospital/owner should not be exposed to liability merely based on its efforts to insure the safety of its patients.

Reviewing cases across the country considering this issue, the deciding question is who exercised control of the work, and thus was in the position to prevent the harm [that] occurred. . . . [T]he linchpin to a determination of an owner's liability for physical harm to others under the facts alleged is control of the methods of work and operative details. An owner could perform a number of activities and still not expose himself to liability, if the activities were not directed at manner and method of work. . . .

Accordingly, the order under review is affirmed.

D. Prime Contractors

At early common law, a prime contractor generally had no responsibility for the safety of its subcontractors' employees.[34] Relying upon principles of property law, however, courts imposed liability in egregious cases of prime contractor negligence by creating a legal fiction equating contractor with owner.

34. Prosser and Keeton on Torts § 80 at 569 (W. Page Keeton, Dan B. Dobbs, Robert E. Keeton & David G. Owen eds., West 5th ed. 1984).

In this way, the prime contractor could be treated as an "occupier" of property who had a qualified duty under premises law to protect others from harm. Prime contractors, therefore, could be liable to the employees of their independent subcontractors if the prime contractors retained control over the work and, by extension, control over the property.[35] Since a prime contractor usually is much closer to the work than the owner, the amount of control the prime contractor can retain is also much greater. Not surprisingly, close questions of retained control arise more often in actions against prime contractors than in actions against owners.

In resolving the question of control of the work, the starting place is the contract. Curiously, many courts look at this through the prism of the subcontract, rather than the general contract. The subcontract usually is replete with language devolving responsibility for safety upon the subcontractor, with the prime contractor retaining the right to insist that the subcontractor perform safely and, in many subcontracts, in accordance with the prime contractor's safety program.

Although these subcontracts attempt to absolve the prime contractor of responsibility for unsafe conditions, the prime contractor nevertheless can be found to have exercised so-called actual control over the worksite and incur liability for negligence for exercising such actual control. Actual control is a separate inquiry from contractual right to control, though courts often fuse the two. A recent Texas case, *Lee Lewis Constr. Co. v. Harrison,* illustrates the interplay of these two intersecting routes to establishing liability.

Lee Lewis Constr., Inc. v. Harrison
70 S.W.3d 778 (Tex. 2002)[36]

Justice HANKINSON delivered the opinion of the Court

Lubbock's Methodist Hospital hired LLC [Lee Lewis Construction] to remodel the eighth floor of, and add ninth and tenth floors to, its south hospital tower. As the general contractor, LLC then subcontracted the project's interior glass-glazing

35. Restatement (Second) of Torts § 414 (1977) (Negligence In Exercising Control Retained By Employer: One who entrusts work to an independent contractor, but who retains the control of any part of the work, is subject to liability for physical harm to others for whose safety the employer owes a duty to exercise reasonable care, which is caused by his failure to exercise his control with reasonable care); for exhaustive discussion of this topic, see Construction Accident Law, *supra* note 3, at 55–217.

36. Internal citations, footnotes, parts of the factual and procedural history, and discussions of proximate cause, gross negligence, and the jury charge have been omitted.

work to KK Glass. Jimmy Harrison was an employee of KK Glass. Harrison was installing thermal insulation and caulking between the window frames on the tower's tenth floor when he fell and suffered fatal injuries. Although no one witnessed Harrison's fall, the evidence is undisputed that Harrison was not using an independent lifeline that would have stopped his fall.

. . .

Our review of the evidence concerning negligence begins with duty. The parties agree that the duty in this case is governed by our well-established law concerning a general contractor's duties to a subcontractor's employees. Ordinarily, a general contractor does not owe a duty to ensure that an independent contractor performs its work in a safe manner. A duty does arise, however, if the general contractor retains some control over the manner in which the independent contractor performs its work. The general contractor's duty of care is commensurate with the control it retains over the independent contractor's work. . . .

A general contractor can retain the right to control so as to give rise to a duty of care in two ways: by contract or by actual exercise of control. We have frequently used the phrases "right of control" or "retained control" interchangeably. The distinction remains important, however, because determining what a contract says is generally a question of law for the court, while determining whether someone exercised actual control is a [sic] generally a question of fact for the jury.

Here, the trial court asked the jury, "Did LLC retain the right to control safety" on the jobsite. . . . LLC argues there is no evidence that it exercised any actual control over KK Glass employees' use of fall-protection equipment. . . . The Harrisons respond that LLC observed and expressly approved of KK Glass employees using faulty fall-protection equipment, including using a bosun's chair without an independent lifeline.

The evidence at trial supports the Harrisons' contention. At trial, Lee Lewis, LLC's owner and president, testified that he assigned to C. L. Lewis, LLC's job superintendent, "the responsibility to routinely inspect the ninth and tenth floor addition to the south tower to see to it that the subcontractors and their employees properly utilized fall protection equipment." Testimony indicated that C. L. Lewis personally witnessed and approved of the specific fall-protections systems KK Glass used. There was testimony that C. L. Lewis "definitely did approve" the lanyard system. There was also testimony that C. L. Lewis knew of and did not object to KK Glass employees using a bosun's chair without an independent lifeline. Although our law makes clear that a general contractor is not an insurer of safety on the jobsite, we agree with the Harrisons that the testimony highlighted above constitutes more than a scintilla of evidence that LLC retained the right to control fall-protection systems on the jobsite. LLC therefore had a duty of care toward Harrison commensurate with that right. Because we conclude that LLC retained the right to control fall-protection systems on the jobsite, we need not address its argument that it did not retain the right to control by contract.

. . . .

Because the evidence of LLC's negligence and gross negligence was legally sufficient to support the jury's verdict, and because any error in the charge submitted was harmless, we affirm the court of appeals' judgment.

Justice JEFFERSON, concurring.

I concur in the Court's judgment. I agree that the evidence supports liability here, but not for the reasons stated by the Court. LLC's approval of the ineffective fall-protection system is not "actual control" as defined by this Court [previously]. [L]iability must be based on more than mere acquiescence or approval. I would affirm the judgment on the ground that LLC had a contractual right to compel compliance with safety standards, actually witnessed repeated and flagrant safety violations, and approved those repeated violations even as it enforced its own standards for its own employees.

. . .

. . . LLC unquestionably retained a contractual right to enforce safety requirements. Although the Court concludes it need not reach the issue of control by contract, I believe the question of LLC's contractual right to control must be addressed.

A general contractor's promulgation and enforcement of basic safety measures should not be sufficient, in itself, to impose liability for injuries to subcontractors' employees. Our tort system should not penalize a general contractor for insisting on compliance with basic safety standards. . . . Similarly, suggesting, or even requiring, that a subcontractor meet minimal safety requirements should not amount to the sort of "control" sufficient to hold a general contractor liable for injuries to subcontractors' employees.

A different question is presented when the general contractor, who has contractual responsibility for general safety measures, permits its subcontractors to deviate routinely from the most elemental safety precautions. "[A]n employer who is aware that its contractor routinely ignores applicable federal guidelines and standard company policies related to safety may owe a duty to require corrective measures to be taken or to cancel the contract." I would affirm the judgment on this basis.

. . .

LLC is liable not merely because it adopted a general safety program or possessed a contractual right to expel subcontractors who routinely flout general safety standards, but also because LLC endorsed KK Glass's repeated use of an obviously hazardous activity.

Lee Lewis is an exceptional case because of the concurring opinion's focus on the general contract. As noted, courts tend to look at the subcontract to determine the extent of the prime contractor's control over the workplace, and to ignore the prime contract even though the AIA and EJCDC general conditions initially place the responsibility for site safety upon the prime contractor.[37] The OSHA regulations go so far as to say that the duty for worker safety is a duty that cannot be contracted away:

> (a) The prime contractor and any subcontractors may make their own arrangements with respect to obligations which might be more appropriately treated on a jobsite basis rather than individually. Thus, for example, the prime contractor and his subcontractors may wish to make an express agreement that the prime contractor or one of the subcontractors will provide all required first-aid or toilet facilities, thus relieving the subcontractors from the actual, but not any legal, responsibility (or, as the case may be, relieving the other subcontractors from this responsibility). In no case shall the prime contractor be relieved of overall responsibility for compliance with the requirements of this part [of OSHA] for all work to be performed under the contract. (b) By contracting for full performance of a contract subject to section 107 of the Act, the prime contractor assumes all obligations prescribed as employer obligations under the standards contained in this part, whether or not he subcontracts any part of the work.[38]

The terms of the prime contract generally only come into issue when the prime contractor is sued by someone other than the worker. For example, when a claim is made against an owner or a design professional, they may cross-plead against the prime contractor based on the terms of the prime contract.[39]

E. Architects and Engineers

Occasionally, an accident involving a worker occurs because of a design error. In such a case, liability is premised on professional malpractice, and modern courts have not found it difficult to find a duty of care that extends to the worker where the design professional contractually has assumed responsibility

37. AIA Document A201–2007, *supra* note 19, § 10.2.2; EJCDC, Standard General Conditions of the Construction Contract, *supra* note 19, ¶ 6.13 (2002 ed.).

38. 29 CFR § 1926.16(a); *see also* 29 CFR § 1926.20 (establishing that under even minimum standards, the prime or controlling contractor is responsible for ensuring OSHA compliance.).

39. *See, e.g.*, Secord v. Willow Ridge Stables, Inc., 690 N.Y.S.2d 375 (N.Y. App. Div. 1999, *mem. op.*).

for temporary safety precautions or for the direction of the work[40] or has a statutorily imposed duty to ensure worker safety.[41]

The most common type of case, however, seeks to impose a duty on the architect or engineer because of its responsibilities during the contract administration phase of the project. The extent of the architect's or engineer's duty to the workers is usually determined by reference to responsibilities imposed contractually on the architect or engineer. The contracts—written by design professional organizations—generally eschew any responsibility on the part of the design professionals for safety, usually with some success, as the following case demonstrates.

Peck v. Horrocks Eng'rs, Inc.
106 F.3d 949 (10th Cir., 1997)[42]

EBEL, Circuit Judge.

Appellant Mary Peck ("Peck") appeals the district court's grant of summary judgment in favor of Appellee Horrocks Engineering ("Horrocks") in a wrongful death case concerning the death of her son, Zachery Peck ("Zachery"). . . . We AFFIRM the judgment of the district court.

Zachery Peck was killed . . . when a trench wall collapsed on him during the construction of a culinary water system [for] the City of Talmage, Utah. . . . Zachery was an employee of K & P Plumbing ("K & P"), a contractor that had been selected by the [Water District] to install the pipeline on the project. Horrocks is an engineering firm, which had been hired by the Water District to design and inspect the construction and progress of the project.

. . . . On the day of the collapse, the trench reached a depth of eight feet to allow the pipe to be laid underneath a steel culvert that crossed the path of the pipeline. An employee of Horrocks was involved in the decision to lay the pipe under the culvert, as opposed to over it. Zachery was laying pipe when the trench collapsed . . . and he was killed. In violation of OSHA regulations, the trench was neither sloped nor shored [with support beams].

40. *See* Miller v. DeWitt, 226 N.E.2d 630, 639 (Ill. 1967).

41. For example, the Illinois Structural Work Act has a broad application to all persons "having charge" of the work, which has been found to include design professionals. *See* Emberton v. State Farm Mut. Auto. Ins. Co., 373 N.E.2d 1348 (Ill. 1978).

42. Internal citations, footnotes, and discussion of choice-of-law rules have been omitted.

Following the accident, Peck sued Horrocks and the Water District for wrongful death. [The District Court granted summary judgment in favor of Horrocks.] Peck appeals. . . . We review the grant of summary judgment de novo. . . .

II. ENGINEER LIABILITY

Before Horrocks can be held liable, Peck must establish that Horrocks owed a duty of care to Zachery. Whether a duty of care exists is a question of law. . . . Only after it has been determined that a duty of care exists can a jury determine whether that duty was breached.

As a general rule, an engineer with construction inspection responsibility owes no duty to an independent contractor's employees. Nonetheless, an engineer may acquire a duty of care toward an independent contractor's employees by contractually assuming a duty to maintain safety, and several courts in other states have also ruled that a duty may be created by the engineer exercising control over worker safety issues at the job site.

. . .

A. Contractual Assumption of Duty

. . .

Peck points to two provisions of the Engineering Agreement to support his claim that Horrocks assumed a duty to maintain worker safety. First, Horrocks agreed to "physically oversee . . . the work . . . to determine that the work performed [was] in accordance with the plans and specifications." Because the plans and specifications included the requirement that K & P abide by Utah OSHA excavation standards, Peck argues that Horrocks contractually agreed to assume responsibility for the supervision of safety compliance. Second, the Engineering Agreement provides that Horrocks "will comply with and assist the owner to require all contractors . . . comply with all applicable Federal, State and Local laws.

. . .

Horrocks replies that any effort by Peck to infer a duty . . . is foreclosed by Horrocks' express disavowal of any responsibility to maintain safety in both the Engineering Agreement and the Construction Contract. [T]he Engineering Agreement provides that "[Horrocks] shall not be responsible for construction means, methods, techniques, sequences or procedures, or for safety precautions and programs in connection with the work." This same provision is also found in . . . the Construction Contract. Moreover, the Construction Contract expressly allocates responsibility for safety to K & P by providing that "[K & P] will be responsible for initiating, maintaining and supervising all safety precautions and programs in connection with the Work. . . ."

There is a split of authority regarding whether an engineer who contractually assumes responsibility . . . for general supervision of a construction project thereby assumes a duty of care to supervise the safety of employees of the general contractor. . . . The minority rule imposes upon a supervising engineer a duty of care, on the logic that an "[engineers'] contractual duties of supervising work include . . . a duty to ensure that the work was performed safely." The majority rule denies contractual liability, finding that:

> [a]n architect, or other design professional, who contracts with an owner of property for the preparation of plans and specifications for the erection of a building and who agrees to be responsible for the general administration of the construction contract between the owner and the general contractor is not contractually responsible for the safety of the workmen on the job site unless such duty is specifically assumed in the contract.

[The Utah Supreme Court follows the majority rule, thus, we determine that, under Utah law, Horrocks' contractual assumption of general supervisory duties did not include an assumption of a duty to maintain safety for K & P's employees.]

B. Implied Assumption of Duty

Several courts in states other than Utah have held that even where an engineer has assumed no contractual liability for safety, the engineer may expand its liability beyond the specific provisions of the employment contract through certain acts of control at the job site. This follows the general rule of torts that "one who undertakes to render services has a duty to exercise reasonable care." That doctrine has no applicability here, even if Utah were to adopt it, because Horrocks expressly declined to undertake responsibility for worker safety on this project and there was no evidence that Horrocks in fact assumed control at the job site over issues of worker safety.

Peck argues that Horrocks assumed a duty to maintain safety precautions on the day of the accident because it was consulted about whether the trench that collapsed on Zachery should be dug underneath an interfering steel culvert. We disagree. . . . Horrocks played no role in the actual construction of the ditch [and the Utah Supreme Court] makes clear that a supervising engineer's participation in a strategic construction decision does not obligate it to ensure that all safety precautions are complied with in the implementation of that decision. . . .

Peck [also] argues that liability may be predicated upon the simple fact that Horrocks, as an engineer, had actual knowledge of a safety problem and failed to take steps to correct that problem. Although this doctrine has some limited support in other states, we do not believe Utah would impose a duty on an engineer on the basis of knowledge and inaction alone. . . . In any event, the record here is devoid of any evidence that Horrocks knew in advance of the accident that the seven foot deep trench was neither sloped nor shored, or even that a worker would be standing in the bottom of the trench.

. . .

CONCLUSION

Because Peck has failed to raise a genuine issue of material fact as to whether Horrocks breached any duty of care owed her son, Zachery, and because the district court did not abuse its discretion in disregarding Radford's affidavit as conclusory, we AFFIRM the district court's dismissal of Peck's claim for wrongful death.

The AIA and EJDCD form contracts provide even more protection for the design professional than the clause interpreted in *Peck v. Horrocks*,[43] and indeed the decisions imposing liability on architects or engineers for failing to protect workers are rare.[44]

15.04 RISK ALLOCATION AND TRANSFER

A. Risk Allocation and Transfer, Generally

Risk allocation, insurance, and indemnity issues, although conceptually separate, are often inseparable in practical application.[45]

Under negligence theories, those participants held to have a duty to the injured worker typically have their liability decided under comparative negligence principles; that is, their negligence is weighed against that of the other participants, including the plaintiff. This effects a rough allocation of the risks, assuming all of the alleged culprits are financially responsible; however, solvency

43. AIA Document A201–2007, *supra* note 19, § 4.2.2 (The Architect, as a representative of the Owner, will visit the site at intervals appropriate to the stage of the Contractor's operations . . . the Architect will not be required to make exhaustive or continuous on-site inspections to check the quality or quantity of the work. The Architect will neither have control over or charge of, nor be responsible for, the construction means, methods, techniques, sequences or procedures, or for the safety precautions and programs connected with the Work, since these are solely the Contractor's rights and responsibilities under the Contract Documents . . ."); § 4.2.6 ("The Architect has authority to reject Work that does not conform to the Contract Documents. . . . However, neither this authority of the Architect nor a decision made in good faith either to exercise or not to exercise such authority shall give rise to a duty or responsibility of the Architect to the Contractor, Subcontractors, material and equipment suppliers, their agents or employees, or other persons or entities performing portions of the Work); § 10.1.1 ("The contractor shall be responsible for initiating, maintaining, and supervising all safety precautions and programs in connection with the performance of the Contract").

44. But not unheard of: *See, e.g.*, Carvalho v. Toll Bros. & Developers, 675 A.2d 209 (N.J. 1996).

45. They are treated at greater length in Chapter 20. *See also* Construction Accident Law, *supra* note 3.

of all parties is unusual. By contrast, construction litigation often involves parties that are uninsured, under-insured, on the edge of insolvency, or protected by sovereign immunity, any of which can transfer risk to the other parties, depending upon the particular state's comparative negligence and joint liability jurisprudence.

B. Insurance

Many participants in the construction process are thinly capitalized and lack the financial wherewithal to shoulder the liability for a catastrophic injury. Consequently, many (but not all) purchase liability insurance in varying amounts to defray the risks and pay premiums based upon the degree of risk they encounter. Most commercial contracts for construction require the purchase of insurance by all the participants and provide that proof of insurance be furnished prior to commencement of the prime contractor's or subcontractor's work.[46]

Additionally, the contracts often require the other participants to cover the owner and higher-level contractors as "additional insureds" to the participants' liability policies. When this coverage is provided, the risk is borne by the first-tier and lower-tier subcontractors' insurance carriers, who are forced to build the risk into the rate structure for the contractual liability endorsements of their Commercial General Liability policies. This affects a transfer of the risk from the sophisticated owner or prime contractor to the subcontractors and their insurance carriers.

C. Indemnity

Recognizing the vagaries of the tort system, including the uncertain results in a retained control dispute, many owners and contractors seek contractually to allocate the risks of liability from construction accidents. This often is done through contractual indemnification clauses, which attempt to transfer to the employer—usually a subcontractor—even the risk of the owner's or contractor's own negligence. The perceived unfairness of these clauses has caused the courts to take a very dim view of them, especially if the language is sophistic and the clause buried in boilerplate language. The following case is typical.

46. AIA Document A201–2007, *supra* note 19, § 11.1.

Hagerman Constr. Corp. v. Long Elec. Co. 741 N.E.2d 390 (Ind. Ct. App. 2000)[47]

HOFFMAN, Senior Judge

Third Party Plaintiff-Appellant Hagerman Construction Corp. (Hagerman) appeals the trial court's denial of its motion for summary judgment. We affirm in part and reverse in part.

. . .

Hagerman was the general contractor on a construction project on the campus of Indiana University-Purdue University in Fort Wayne. Long Electric Company (Long) was a subcontractor on the project, and James Scott (Scott) was an employee of Long. In September 1997, Scott was injured when he was struck on the head by a falling light pole. Scott made a claim for workers' compensation benefits and then filed suit against Hagerman. Hagerman subsequently filed a third party action against Long based upon an indemnity clause contained in the form contract between Hagerman and Long. Hagerman filed a motion for summary judgment requesting the court to find that, pursuant to the parties' contract, Long was required to indemnify Hagerman for any losses Hagerman suffered in the Scott litigation. The trial court denied Hagerman's motion and found that Hagerman was not entitled to indemnification by Long for Hagerman's negligence. This appeal ensued.

. . .

Hagerman contends that the trial court erred in denying its motion for summary judgment. Although the parties agree that if Hagerman is solely responsible for Scott's injuries, there should be no indemnification by Long. Hagerman asserts that, based upon the indemnity clause in the parties' contract, Long is required to indemnify Hagerman for Hagerman's own negligence if Long is also at fault. Alternatively, Hagerman claims that even if the contract does not require Long to indemnify Hagerman for Hagerman's negligence, Long is liable to indemnify Hagerman for Long's own negligence.

Absent prohibitive legislation, no public policy prevents parties from contracting as they desire. For instance, in Indiana a party may contract to indemnify another for the other's own negligence. However, this may only be done if the party knowingly and willingly agrees to such indemnification. Such provisions are strictly construed and will not be held to provide indemnification unless it is so stated in clear and unequivocal terms. We disfavor indemnity clauses because we

47. Internal citations, discussion of summary judgment standards, and expanded discussion of precedent have been omitted.

are mindful that to obligate one party to pay for the negligence of another is a harsh burden that a party would not lightly accept.

This Court has followed a two-step analysis to determine whether a party has knowingly and willingly accepted this burden. First, the indemnification clause must expressly state in clear and unequivocal terms that negligence is an area of application where the indemnitor (in this case, Long) has agreed to indemnify the indemnitee (in this case, Hagerman). The second step determines to whom the indemnification clause applies. Again, in clear and unequivocal terms, the clause must state that it applies to indemnification of the indemnitee (in this case, Hagerman) by the indemnitor (in this case, Long) for the indemnitee's own negligence.

In the present case, the indemnification clause used by the parties was taken from a form agreement drafted by the American Institute of Architects and entitled "AIA Document A401, Standard Form of Agreement Between Contractor and Subcontractor, 1987 Edition." The clause provides as follows:

> 4.6 INDEMNIFICATION
>
> 4.6.1 To the fullest extent permitted by law, the Subcontractor shall indemnify and hold harmless the Owner, Contractor, Architect, Architect's consultants, and agents and employees or any of them from and against claims, damages, losses and expenses, including but not limited to attorney's fees, arising out of or resulting from performance of the Subcontractor's Work under this Subcontract, provided that such claim, damage, loss or expense is attributable to bodily injury, sickness, disease or death, or to injury to or destruction of tangible property (other than the Work itself) including loss of use resulting therefrom, but only to the extent caused in whole or in part by negligent acts or omissions of the Subcontractor, the Subcontractor's Sub-subcontractors, anyone directly or indirectly employed by them or anyone for whose acts they may be liable, regardless of whether or not such claim, damage, loss or expense is caused in part by a party indemnified hereunder. Such obligation shall not be construed to negate, abridge, or otherwise reduce other rights or obligations of indemnity which would otherwise exist as to a party or person described in this Paragraph 4.6.

Applying the two-step test set out above, we first note that this indemnification clause expressly defines negligence as an area of application in clear and unequivocal terms. The clause speaks of claims, damages, losses and expenses attributable to bodily injury, sickness, disease or death, and injury to or destruction of property, as well as negligent acts or omissions. These words, taken in this context, are the language of negligence, and, as such, clearly and unequivocally demonstrate that the indemnification clause applies to negligence.

Next, we must determine whether the indemnification clause also expressly states, in clear and unequivocal terms, that it applies to indemnify Hagerman for its own negligence. We conclude that it does not. . .

. . .

Similar to our discussion above, . . . Long asserts that the phrase "but only to the extent caused in whole or in part by negligent acts or omissions of the Subcontractor . . ." limits the scope of the liability to only those losses that are caused by the negligence of the sub-contractor or its agents. Hagerman, on the other hand, avers that Long's interpretation of the clause would be inconsistent with other language in the provision that provides for indemnity "to the fullest extent permitted by the law" and "regardless of whether or not such claim, damage, loss or expense is caused in part by a party indemnified hereunder."

We conclude that the indemnification clause does not expressly state, in clear and unequivocal terms, that it applies to indemnify Hagerman for its own negligence. The clause explicitly indemnifies Hagerman for the acts of the subcontractor, Long, and its sub-subcontractors, employees and anyone for whom it may be liable, but it does not explicitly state that Long must indemnify Hagerman for its own negligent acts. Further, the phrase "but only to the extent" clearly limits Long's obligation to indemnify Hagerman only to the extent that Long, its sub-subcontractors, employees, and anyone for whom it may be liable are negligent. Otherwise, the clause contains no clear statement that would give the contractors notice of the harsh burden that complete indemnification would impose.

. . .

Based upon the foregoing, we conclude that the trial court properly denied Hagerman's motion for summary judgment as to Long's obligation to indemnify Hagerman for Hagerman's own negligence. However, the trial court erred in denying Hagerman's motion as to Long's obligation to indemnify Hagerman for Long's negligence.

Accordingly, we affirm in part and reverse in part.

As this case shows, the AIA takes a middle of the road approach and attempts to have comparative indemnity; that is, the indemnitor only agrees to indemnify for its own negligence.[48]

The arbitrary inclusion of the indemnity clauses in adhesion contracts has led to some statutory limits on their use. In some states, architects and engineers may not be allowed to obtain indemnification from their own negligence.

48. AIA Document 201-2007, *supra* note 19, § 3.18.

For example, in Texas a design professional cannot seek indemnity from a contractor for defects in the design work.[49] This is a logical extension of Texas laws regarding professional licensure[50] of engineers and the sealing of design documents,[51] which has the stated goal of being able to "fix responsibility for work done or services or acts performed in the practice of engineering."[52]

In other states the same indemnity prohibition applies to owners and contractors. In New York, it is against public policy for an owner or contractor to bargain for indemnity from its own negligence.[53]

D. Assumption of the Risk

Finally, risk can also be allocated to the injured worker. This is usually argued through the common law assumption-of-the-risk doctrine,[54] which currently functions as a form of comparative negligence and, when appropriate, allows for a judgment to be reduced by the percentage of the plaintiff's fault in the accident.[55]

E. Workers' Compensation's Effect on Indemnity and Contribution in Third-Party Actions

If a worker's injury is caused by the combined negligence of his employer and a third party, such as a contractor or design professional, the worker is entitled to receive workers' compensations benefits from its employer, and to pursue

---

49. TEX. CIV. PRAC. & REM. CODE ANN. §§ 130.001 & 130.002 (Vernon 2002) ((a) A covenant or promise, in connection with, or collateral to a construction contract is void and unenforceable if the covenant or promise provides for a contractor who is to perform the work that is the subject of the construction contract to indemnify or hold harmless a registered architect, licensed engineer or an agent, servant, or employee of a registered architect or licensed engineer from liability for damage that: (1) is caused by or results from: (A) defects in plans, designs, or specifications prepared, approved, or used by the architect or engineer; or (B) negligence of the architect or engineer in the rendition or conduct of professional duties called for or arising out of the construction contract and the plans, designs, or specifications that are a part of the construction contract; and (2) arises from: (A) personal injury or death; (B) property injury; or (C) any other expense that arises from personal injury, death, or property injury).

50. *Id.* § 1001.301.

51. *Id.* § 1001.401.

52. *Id.* § 1001.004.

53. N.Y. GEN. OBLIG. LAW § 5-322.1 (McKinney 2001). See JUSTIN SWEET AND MARC M. SCHNEIER, LEGAL ASPECTS OF ARCHITECTURE, ENGINEERING AND THE CONSTRUCTION PROCESS § 31.05(D) (Thompson 7th ed. 2004), for a discussion of the historical background of anti-indemnity statutes, generally.

54. For a history of the assumption-of-the-risk doctrine, see CONSTRUCTION ACCIDENT LAW, *supra* note 3, at 17–19, 211–12.

55. *See generally* PROSSER AND KEETON ON TORTS § 67 (W. Page Keeton, Dan B. Dobbs, Robert E. Keeton & David G. Owen eds. West 5th ed. 1984).

tort claims against the third party. Inevitably, the third party will seek to defray its responsibility by shifting the risk back to the employer who may be primarily liable. Under a pure tort standard, this is handled through comparative negligence as discussed in Section 15.04(A). Workers' compensation, however, often changes those dynamics; employers' liability is channeled through the workers' compensation system and is removed from the tort process.

To protect the integrity of the workers' compensation scheme, many states do not allow an employer who has purchased workers' compensation to be pursued by other negligent parties for common law indemnity or contribution.[56] The general premise is that protection from common law remedies encourages employers to enroll in the workers' compensation system and that an employer who has insured his workers under the compensation scheme should not have to pay twice. It is worth noting, however, that all the protections from common law remedies asserted by third parties discussed in this section may be waived through contract, and often are.[57]

15.05 CONCLUSION

The interplay between the law of torts, contract, insurance, procedure, and administrative law is nowhere more interesting or in more frequent collision than construction accident law. The forceful exercise of political power by the various participants, in varying degrees, has affected the balance of power (and responsibility) among the participants. Nevertheless, one constant remains: Most of the financial risk of catastrophic accidents remains on the worker, who, of all the participants, is least able to shoulder it.

56. *See, e.g.*, A & B Constr., Inc. v. Atlas Roofing & Skylight Co., 867 F. Supp. 100 (D.R.I. 1994).

57. For an in-depth discussion, *see* 2 Bruner & O'Connor, *supra* note 2, § 5:88 (discussing the waiver of workers' compensation protection in standard construction contract forms).

QUESTIONS

Plaintiff is an employee of a steel frame erection subcontractor and was fatally injured when the steel frame collapsed during construction. The vertical support beams of the frames were originally supposed to be secured by an "L" brace at the foundation, over which a concrete floor would be poured. Unfortunately, an overly eager concrete subcontractor poured in the concrete floor before all the beams were properly braced. The prime contractor discovered this error soon after it happened. Under time pressure from the owner, the prime contractor proposed to the architect that the unbraced vertical beams be secured by drilling holes in the concrete and hammering in sheared straight bolts (without the horizontal part of the "L"), which would allow the work to proceed without delay. After review, the architect approved the prime contractor's shop drawings; soon after, the sheared bolts were installed and reinforcing concrete was poured around the beams. The next day, as plaintiff climbed a beam secured in this manner, the beam fell. The beam would have held if it had been secured with an "L" brace, or if the concrete had been allowed several days to fully cure before stress was placed on the beam. The prime contractor, architect, and owner were all using standard AIA contracts.[58]

1. Does the owner owe any duty to the plaintiff for putting time pressure on the prime contractor? What if the owner had approved the sheared bolt plan rather than the having the architect do it?
2. What determines if the prime contractor owes any duty to the plaintiff?
3. What determines if the architect owes any duty to the plaintiff?
4. What determines if the plaintiff's employer, the steel erection subcontractor, owes a duty to the plaintiff?
5. To what extent can OSHA rules establish the standard of care?
6. If the plaintiff's employer had purchased workers' compensation insurance, what effect would this have on the plaintiff's ability to pursue claims against other parties? What effect would it have on the allocation of liability between the potentially liable parties?
7. If the plaintiff sues the architect, will she be entitled to indemnification from the prime contractor, assuming the prime contractor had signed AIA Document A101 and the A201 General Conditions?
8. If the plaintiff had failed to use safety equipment provided on site by the employer, how would that affect the claim?

58. The facts in this scenario are based generally on Clemens v. Benzinger, 207 N.Y.S. 539 (1925).

CHAPTER

16

Changes

ALAN WINKLER

16.01 THE INEVITABILITY OF CHANGES

Contract terms dealing with changes to the work are just as important as provisions for critical concerns such as payment and time. This may seem surprising, since a changes clause only comes into play when at least one of the contracting parties contends that there has been a deviation from the work as originally designed, specified, and contemplated. However, the reality is that, notwithstanding the owner and the architect/engineer having progressed through every phase of the design process—from concept/program development to schematic designs to design development, to the issuance of "final" construction documents—every project of any significance (and many minor projects as well) will require changes to the work and, thus, to the price and time of performance as well. The most common term for the vehicle effecting a mutually agreed change is called a change order. It is the rare project indeed that is built without any change orders. By contrast, it is not unheard of to have dozens or even a hundred change orders issued on a medium-sized project, such as the construction of a school.

Changes can come from a variety of sources. The most obvious is a change in some aspect of the design of the project emanating from the owner's desire to have a finished product differing from what is depicted in the drawings and specifications. This is by no means the sole source of changes, however. Changes can also emanate from errors or omissions in the plans, the necessity to deviate from the plans due to field conditions, costs savings suggested by the contractor, or the issuance of new government regulations.

16.02 THE PURPOSE OF A CHANGES CLAUSE

Changes can take three general forms: extra work, additional work, and deleted work. Extra work has been defined as "work arising outside of and independent of the contract; something not required in its performance."[1] An example would be the owner's request to add a room to a building. Additional work has been defined as work "necessarily required in the performance of the contract, but the necessity of which arises from unanticipated conditions."[2] An example would be excavating a site to an increased depth from what was originally required due to the nature of the soil encountered. Despite these differing definitions, both can qualify under the general heading of "changes."[3]

Absent a change clause, a claim for the performance of work outside of the requirements of a contract would be treated under the principles of *quantum meruit* (i.e., an implied contract).[4] A change clause allows the claim to be treated as a matter of contractual entitlement, with the right of recovery dependent upon the eligibility of the claim under the definition in the provision and compliance with its terms.

As discussed later in this chapter, changes can be either express or constructive. In short, an express change arises when the owner directs the contractor to perform the work in a manner differing from the contract requirements. A constructive change arises when the contractor performs work in a manner differing from the contract requirements without an express directive, but in order to overcome a difficulty that would otherwise impede construction of

1. Interstate Contracting Corp. v. City of Dallas, Texas, 407 F.3d 708, 712 (5th Cir. 2005).

2. Wisch & Vaughan Constr. Co. v. Melrose Properties, 21 S.W. 3d 36, 40 (Mo. App. S.D. 2000).

3. Courts use the terms "additional work" and "extra work" interchangeably. *See, e.g.*, Dave Kolb Grading, Inc. v. Terra Venture Bridgeton Project Joint Venture, 85 F.3d 351 (8th Cir. 1996); Miller Elevator Co. v. United States, 30 Fed. Cl. 662 (Fed. Cl. 1994). The distinction has also been questioned. *See* Buckley Co. v. City of New York, 505 N.Y.S.2d 140, 143 (N.Y. App. Div. 1986).

4. Fox v. Mountain West Elect., Inc., 52 P.2d 848 (Idaho 2002); Joseph Sternberg, Inc. v. Walber 36th Street Assoc., 594 N.Y.S.2d 144 (N.Y. App. Div. 1993).

the project as designed.[5] A well-drafted change clause regulates both express and constructive changes. First, it defines the owner's right to make an express change. It also establishes the requirements for the documents memorializing the change. Additionally, it can set the parameters for the pricing of the change and state when billing can be submitted for its performance. Likewise, a change clause can set the notice and documentation requirements for a constructive change claim, as well as the pricing structure.

Perhaps the most unique feature of a change clause is that it can authorize one party, namely, the owner, to effect an amendment to a contract.[6] Ordinarily, a construction contract is a bilateral agreement, requiring the acceptance of both parties to its terms. Given the exigencies of construction, where a change must be implemented promptly in order not to delay the entire project, it is not always practical for the parties to negotiate a mutually acceptable amendment. A change clause can allow an owner to change the contract unilaterally by directing the contractor to perform changed work under a pricing formula or under an owner-imposed price, with the right to contest the price preserved.

Another important feature of a change clause is that it provides a contractual mechanism for deleting work. The deletion of work is more problematic to price than the addition of work because the actual price of the work deleted may be difficult, if not impossible, to prove.[7] Additionally, absent a change clause allowing for the deletion of work, a contractor could contend that the owner does not have the right to delete work and that any attempt to do so would constitute a breach. The change clause not only allows the owner the flexibility to delete work, but it also can provide a framework for determining the ensuing reduction in price.

Construction is not a perfect world. Changes are not always independent, since one element frequently has an impact on another. Therefore, changes can and do involve a combination of deleted work and added work. To take a simple example, an owner may decide to upgrade the specified windows. In that event, the owner would be due a credit for the originally specified windows, now deleted from the contract, and would be obligated to pay for the increased cost of the upgraded windows. The change order may have a single price, but it could only be arrived at by subtracting the price for the deduction from the price for the upgrade.

---

5. *See* Section 16.07 *infra*.

6. *See* 1 Philip L. Bruner & Patrick J. O'Connor, Jr., Bruner & O'Connor on Construction Law § 4:2, at 502 (West 2002).

7. However, additional work may impact the overall performance of the contract, thus implicating cost considerations beyond the price of the additional work itself.

16.03 THE CHANGE ORDER PROCESS

The document memorializing the owner's and the contractor's agreement to a change is typically called a change order. Although there are contracts that employ other terms for this document, the term "change order" has become standard in the construction industry, both in the public sector and in the private sector. To give three examples, "change order" is the term utilized in federal contracting[8] and in two of the most commonly used form contracts[9] in the private sector.[10]

Section 7.2.1 of AIA Document A201–2007 defines a change order as follows:

> **Change Orders**
> A Change Order is a written instrument prepared by the Architect and signed by the Owner, Contractor and Architect, stating their agreement upon all of the following:
> 1. The change in the Work;
> 2. The amount of the adjustment, if any, in the Contract Sum; and
> 3. The extent of the adjustment, if any, in the Contract Time.

Thus, although construction contracts, including the AIA forms, are typically bilateral agreements between the owner and the contractor, this AIA provision envisions a change order as a trilateral agreement among the owner, the architect, and the contractor.

Section 52.243-4 of the Federal Acquisition Regulations (FAR), on the other hand, envisions the change order as a unilateral directive from the owner (i.e., the government) to the contractor pertaining to the work itself, with the financial consequences to be addressed separately through an equitable adjustment, which modifies the contract price:

> **CHANGES**
> (a) The Contracting Officer may, at any time, without notice to the sureties, if any, by written order designated or indicated to be a change order, make changes in the work within the general scope of the contract, including changes—
> (1) In the specifications (including drawings and designs);
> (2) In the method or manner of performance of the work;

---

8. *See, e.g.*, Federal Acquisition Regulation (FAR) § 52.243-4 (the changes clause in fixed-price federal construction contracts).

9. *See* American Institute of Architects, AIA Document A201–2007, *General Conditions of the Contract for Construction* § 7.2 (2007) [hereinafter A201–2007] and Associated General Contractors of America (AGC), Document No. 200 ¶ 8.1.

10. There is nothing preventing public entities, such as townships or school districts, from utilizing these form contracts, and some do.

(3) In the Government-furnished facilities, equipment, materials, services, or site; or

(4) Directing acceleration in the performance of the work.

(b) Any other written or oral order (which, as used in this paragraph (b), includes direction, instruction, interpretation, or determination) from the Contracting Officer that causes a change shall be treated as a change order under this clause; provided, that the Contractor gives the Contracting Officer written notice stating—

(1) The date, circumstances, and source of the order; and

(2) That the Contractor regards the order as a change order.

(c) Except as provided in this clause, no order, statement, or conduct of the Contracting Officer shall be treated as a change under this clause or entitle the Contractor to an equitable adjustment.

(d) If any change under this clause causes an increase or decrease in the Contractor's cost of, or the time required for, the performance of any part of the work under this contract, whether or not changed by any such order, the Contracting Officer shall make an equitable adjustment and modify the contract in writing. However, except for an adjustment based on defective specifications, no adjustment for any change under paragraph (b) of this clause shall be made for any costs incurred more than 20 days before the Contractor gives written notice as required. In the case of defective specifications for which the Government is responsible, the equitable adjustment shall include any increased costs reasonably incurred by the Contractor in attempting to comply with the defective specifications.

(e) The Contractor must assert its right to an adjustment under this clause within 30 days after (1) receipt of a written change order under paragraph (a) of this clause, or (2) the furnishing of a written notice under paragraph (b) of this clause, by submitting to the Contracting Officer a written statement describing the general nature and amount of the proposal, unless this period is extended by the Government. The statement or proposal for adjustment may be included in the notice under paragraph (b) of this clause.

(f) No proposal by the Contractor for an equitable adjustment shall be allowed if asserted after final payment under this contract.

What is immediately apparent from a comparison between these two changes clauses is that the AIA provision seemingly does not allow for the contract to be modified unilaterally, while the FAR provision seemingly does not effect a bilateral modification. The reason for this discrepancy is the disparity in philosophy underlying the two clauses.

AIA Document A201–2007, § 7.2.1 provides for the end product of the change order process: a fully agreed, written, signed modification defining the work, price, and effect on the schedule of the change. FAR § 52.243-4, on the other hand, sets forth how the change order process is initiated by a unilateral directive. Rarely will the change order process start with an acknowledgement by the owner and the contractor (and possibly the architect) that a change is warranted or required. Rather, as the FAR section provides, the change order process will be initiated by either the owner directing a change or by the contractor claiming that an act or omission of the owner has necessitated the change.

AIA Document A201–2007 likewise makes provision for either the owner or the contractor to initiate the change. The procedure under that AIA contract document for the owner to mandate a change unilaterally is through the issuance of a construction change directive.[11] Under that procedure, the architect and the owner issue a construction change directive setting forth how the work is to be changed, and the contractor must implement the change.[12] In the meantime, the parties can negotiate a change order memorializing not only the modification to the scope of the work but also to the price and the time of performance. Another option is for the construction change directive to provide a method for adjusting the price to account for the change, with the architect resolving differences over the method for adjusting the price.[13] Undisputed amounts for the performance of the changed work must be memorialized through a change order.[14] Another way for the owner to initiate a change is for the owner to send a pricing request to the contractor. That way, the owner can decide whether to proceed with the change order based on pricing information from the contractor in advance of the performance of the changed work, rather than becoming committed to paying for the change through the issuance of a construction change directive.

The AIA conditions parallel to the FAR provision for the contractor to initiate the change are set forth in Article 15 of AIA Document A201–2007. Under subsection 15.1.2, if the contractor believes that a written interpretation or order by the architect or other reasonable grounds will cause the contractor to incur additional costs, the contractor must initiate a claim. The claim will be the assertion of the right to a change order, similar to FAR § 52.243-4(b). If the owner agrees with the contractor's assertion, a change order will be negotiated. If the owner disagrees, the contractor can implement the remaining steps in the claims process to assert the right to a change order.

There are many construction projects where neither the FAR nor an AIA form contract is used, but common to all construction contracts is the basic process of the owner initiating a change through either a directive requiring

11. AIA Document A201–2007, *supra* note 9, § 7.3.
12. *Id.* §§ 7.3.1, 7.3.5.
13. *Id.* §§ 7.3.3, 7.3.7.
14. *Id.* § 7.3.10.

the implementation of the change or a pricing request or of the contractor initiating a claim that the change order is warranted by changed circumstances.

16.04 AUTHORITY TO ISSUE CHANGES

Many issues can arise regarding changes. Among them are differences over the scope, the price, and the effect on the schedule of the change. A preliminary issue, however, concerns who directed the change in the first instance. On a medium-sized or large-scale project, there could be dozens of people in a managerial capacity at the site and in the home offices of the various participants. A general contractor may receive directives from the owner's representative, the project manager for the owner, various corporate officers in the owner's home office, the on-site representative for the architect, and various employees of the architect working on the project off the site. Likewise, a subcontractor may receive directives from the general contractor's project executive, project manager, assistant project manager, project engineer, superintendent, and assistant superintendent, not to mention the president of the general contractor. However, there is a dichotomy between the roles of some of these people with regard to assuring completion of the project versus their ability to modify the contract. Accordingly, it is important to ascertain whether the person making the directive has the authority to effect a modification.

The first source to examine for a person's authority to direct changes is, of course, the contract. In federal contracts, the person identified as the contracting officer has the right to order changes in the work.[15] In AIA contracts, the owner is obligated to designate a representative with actual authority to bind it.[16] Typically, the parties' representatives are identified in the contract,[17] and unless the scope of the contractually designated representative for the owner is specifically limited, the owner would be hard pressed to maintain that a directive to perform a change issued by its contractually designated representative was unauthorized.[18]

However, not infrequently, someone other than the contractually designated representative issues a directive for a change. In that event, resort to the examination of doctrines such as implied authority, apparent authority, or ratification

15. FAR §§ 2.101, 52.243-4, 53.301-1402.

16. AIA Document A201–2007, *supra* note 9, § 2.1.1.

17. *See, e.g.*, American Institute of Architects, AIA Document A111–2007, *Standard Form of Agreement Between Owner and Contractor Where the Basis of Payment Is the Cost of the Work Plus a Fee with a Guaranteed Maximum Price* (2007) §§ 14.3, 14.4.

18. Fraud or mistake may constitute such grounds. *See* McQuagge v. United States, 197 F. Supp. 460, 468 (W.D. La. 1961).

must be had in order to determine if the directive is valid. A dispute over authority usually arises in the context of a contractor's claim against an owner to recover for a claim arising from a constructive change (or a subcontractor's claim against the contractor). The owner may defend the claim on several grounds, including that the person who made the request did not have the authority to direct changes.[19]

This issue is treated differently on public projects than on private construction projects. On federal projects, actual authority is necessary to bind the government. This principle was decided by the Supreme Court in *Federal Crop Ins. Corp. v. Merrill*,[20] where the court held that the contractor must verify the authority of the agent or employee of the government and cannot rely on the appearance of authority:

> Whatever the form in which the Government functions, anyone entering into an arrangement with the Government takes the risk of having accurately ascertained that he who purports to act for the Government stays within the bounds of his authority. The scope of this authority may be explicitly defined by Congress or be limited by delegated legislation, properly exercised through the rule-making power. And this is so even though, as here, the agent himself may have been unaware of the limitations upon his authority.[21]

The rationale for this principle is that public funds must be zealously protected against improvident expenditures. State and local construction contracts are similarly constrained.[22]

The *Federal Crop Insurance* case precludes the use of apparent authority to validate a directive by someone other than the designated person in the contract to effect a change. However, the reality of public projects is no different than that of private projects: To get the job done, practicality intrudes upon the rigid application of the contract. To bridge this gap and to prevent gross inequities, courts have applied the principle of implied authority to government contracts. An implied delegation of actual authority can empower someone other than the contractually designated representative to direct a change. Whether

---

19. This issue may arise in other contexts, too. For instance, an owner may contend that a contractor's performance of work in a manner differing from the plans and specifications was unauthorized and defective, while the contractor may defend that assertion by claiming that it had been directed to install its work in that manner by a representative of the owner.

20. 332 U.S. 380 (1947).

21. *Id.* at 383.

22. *See, e.g.*, Butler v. Bd. of Supervisors for Hinds County, 659 So. 2d 578 (Miss. 1995); Nether Providence Twp. Sch. Auth. v. Thomas M. Durkin & Sons, Inc., 476 A.2d 904 (Pa. 1984); Contra Costa Constr. Corp. v. Daly City, 192 P. 178 (Cal. 1920).

implied authority is found in a project representative depends upon the duties that person actually performs. "Implied authority binds the Government where a Government representative without actual authority exercises an integral part of the duties assigned to that Government employee."[23] For example, where the contracting officer delegated the responsibilities for inspection and testing to the resident engineer, there was implied authority on the part of the resident engineer to direct the contractor to replace materials he considered defective, based on his interpretation of the specifications.[24]

On private projects, another substitute for actual authority is provided by the principle of apparent authority. Apparent authority is an appearance of authority to act. The person with actual authority must create the appearance of actual authority.[25] An owner will be bound by the directive of someone without actual authority where the contractor reasonably believes, on the basis of the owner's acts, that the owner had authorized that person to make the type of directive given.[26]

Like the principle of implied authority, the applicability of the principle of apparent authority depends on a factual inquiry. First, there must be a finding that the owner led the contractor to believe that someone had authority. Second, there must be a finding that the contractor's belief was reasonable, based on the information available and the circumstances of the project. Unlike implied authority, the trigger for application of apparent authority must come from the owner's words or actions, rather than from the nonauthorized person's actual function.

Yet another substitute for actual authority is ratification. Where the owner, after the fact, affirms that an unauthorized directive is valid, the owner is bound by it.[27] Ratification can be relied on even where the contractor knew that the person who issued the directive lacked the authority to do so.[28] Ratification can be based on either the owner's express acts in sanctioning the directive or the owner's conduct impliedly giving approval. An example of the latter is the owner's failure to reject changed work upon learning of its performance.[29]

---

23. Miller Elevator Co., 30 Fed. Cl. at 662, 693.
24. WRB Corp. v. United States, 193 Ct. Cl. 409 (1968).
25. Ja Din, Inc. v. L-J, Inc., 898 F. Supp. 894 (E.D. Fla. 1995).
26. Frank Sullivan Co. v. Midwest Sheet Metal Works, 335 F.2d 33 (8th Cir. 1964).
27. Audit Services, Inc. v. Francis Tindall Constr., 600 P.2d 811 (Mont. 1979).
28. Mateyka v. Schroeder, 504 N.E.2d 1289 (Ill. 1987).
29. McDevitt & Street Co. v. Marriott Corp., 713 F. Supp. 906 (E.D. Va. 1989), *aff'd in part and rev'd in part*, 911 F.2d 723 (4th Cir. 1990).

Ratification applies to public projects as well as to private projects. The principle of ratification is codified in federal contracting through the Federal Acquisition Regulations.[30] Where a contracting officer knows of an unauthorized government representative's directive to a contractor and takes no action to repudiate it, and the government receives the benefit of its performance, the government will be deemed to have ratified that directive.[31]

16.05 NOTICE REQUIREMENTS FOR CLAIMS FOR CHANGES

Most change-order clauses contain requirements for the contractor to notify the owner of a claim for extra or additional work. The notification requirements in the AIA Document A201–2007 General Conditions are contained in Article 15.[32] A contractor asserting a claim for a change must notify the owner in writing within 21 days after the occurrence of the event giving rise to the claim or within 21 days after the claimant first recognizes the claim, whichever is later.[33] Furthermore, notice must be given prior to performing the work, if a claim for additional costs will be asserted.[34] Under the AIA claims procedure, the architect or another person designated in the contract then has a period of time to review the claim and make a decision on it, which is deemed binding unless, within 30 days of the decision, the aggrieved party requests mediation or pursues the claim through arbitration or the courts, whichever dispute resolution forum is provided for by the contract.[35] In government contracts, section 52.243-4 (1987) of the Federal Acquisition Regulations contains three notification requirements for changes (see Section 16.03).

Other contracts can have more stringent requirements. For instance, a contract could require that (1) a written notice of the claim be provided within five days after the act or omission of the owner occurs; (2) written notice that a directive is being complied with under protest be given within five days after the directive is issued; (3) a verified detailed statement of the damages sustained be submitted within 30 days after the initial notification is provided; (4) time and material tickets signed by representatives of both parties be tendered on a

---

30. FAR § 1.602-3.

31. Williams v. United States, 127 F. Supp. 617 (Ct. Cl. 1978).

32. The 1997 edition of AIA Document A201 (the General Conditions) included the claims notification procedures in Article 4, defining the scope of the architect's role in administering the contract. American Institute of Architects, AIA Document A201–1997, *General Conditions of the Contract for Construction* (1997) [hereinafter AIA Document A201–1997].

33. *See id.* ¶ 15.1.2. AIA Document A201–2007 claims procedures regulate all claims, whether by the contractor or the owner and whether for a change or something else. AIA Document A201–2007, *supra* note 9.

34. AIA Document A201–1997, *supra* note 32, ¶ 15.1.4.

35. *Id.* ¶ 15.2.5, 15.2.6.1.

daily basis while the disputed directive is being carried out; and (5) the failure to comply strictly with all of the written notification requirements shall effect a waiver of the claim for the change.[36]

There has been much litigation over the enforceability of notice provisions, as well as whether compliance was achieved. Notice provisions can provide an owner with a defense to a claim apart from the substance of whether work performed was truly beyond the scope of the contract.[37] Owners reason that timely notification of a potential claim permits them to explore other less costly solutions for performing the work; to document contemporaneously the costs of the work involved in the claim; to negotiate a prompt resolution of the claim; and to adjust their budget as needed to cover the possibility that the claim will be enforced. Whether or not a notice provision will be upheld so as to bar a claim where the contractor did not comply with its requirements will depend on the wording of the clause and the jurisdiction where the claim is litigated.

In federal contracts, strict compliance with the notification requirements of the change clause will not always be required. The courts are mindful not to apply the notice requirements in an overly technical manner where the government was aware of the operative facts.[38] The degree of prejudice to the government resulting from the contractor's failure to provide proven notice is a factor employed in determining whether to excuse the insufficient notice.[39] Moreover, the 30-day notice provision in FAR § 52.243-4 (1987) does not apply to constructive changes since the change was not effected through a written change order triggering the start of the 30-day period.[40]

Many jurisdictions uphold contractual notice provisions.[41] To counter the preclusive effect of such clauses, contractors frequently make arguments that the failure to comply with notice provision should be excused through such theories as waiver or estoppel. For instance, where an owner repeatedly issued change orders to a contractor after the changed work had been performed,

---

36. *See* A.H.A. General Constr., Inc. v. New York City Housing Auth., 600 N.E.2d 368 (N.Y. Ct. App. 1998).

37. Notice provisions are not unique to construction contracts. Timely notice is required to assert a claim under an insurance policy. *See, e.g.,* Nat'l Am. Ins. Co. v. Certain Underwriters at Lloyd's London, 93 F.3d 529 (9th Cir. 1996).

38. Hoel-Steffen Constr. Co. v. United States, 456 F.2d 760 (Ct. Cl. 1972).

39. Mingus Contractors, Inc. v. United States, 812 F.2d 1387 (Fed. Cir. 1987).

40. Jo-Bar Mfg. Corp. v. United States, 535 F.2d 62 (Ct. Cl. 1976) (holding that the contractor has a "reasonable period of time" to notify the government of the claim for a constructive change).

41. *See* Clark-Fitzpatrick, Inc./Frankl Found. Co. v. Gill, 652 A.2d 440 (R.I. 1994) (holding, however, that formal notice is not required where notice was given and the owner had an opportunity to monitor the costs); Wegroup PC/Architects and Planners v. State, 885 P.2d 709 (Or. 1994); Indian River Constr. Co. v. City of Jacksonville, 390 So. 2d 1139 (Fla. Dist. App. 1977).

but without receiving the required contractual notice of the claim, the owner can be held to have waived strict compliance with the notice provisions.[42] An owner may also be estopped from maintaining that a claim for a change is barred by the failure to provide written notice where the owner's actions caused the contractor to incur the extra cost.[43]

However, some jurisdictions strictly uphold notice requirements. In New York, a contractual provision requiring a contractor to provide the owner with notice and documentation of the claim is considered a condition precedent to the right to recover on the claim.[44] Thus, to avoid the application of the condition precedent, a contractor must show that the owner's actions prevented or hindered compliance with the provision.[45] The failure to comply with a notice provision was held to bar a claim under the law of Mississippi as well.[46] Many of the cases dealing with the issue of whether a claim is barred by the failure to comply with a notice provision emanate from public works contracts. This is not surprising since courts can be protective in guarding against the expenditure of public funds. At least one court has gone so far as to hold that notice provisions are treated differently when applied to public contracts (strictly) as opposed to private construction contracts (liberally).[47]

The first step in an analysis of a notice provision is to read the provision. Does the contract state that notice must be strictly provided? Does the provision state that a claim will be waived, absent full compliance? Then, the surrounding circumstances must be examined. Was notice given in some fashion? Did the owner inform the contractor at some point in the project that change orders will not be issued unless there has been full compliance with the contractual procedures? Whether a notice provision will be applied to bar a claim or whether the failure to comply will be waived is, in many jurisdictions, a nuanced analysis based on a number of factors.

---

42. *See, e.g.*, Cardinal Dev. Co. v. Stanley Constr. Co., 497 S.E.2d 847 (Va. 1998); Joseph F. Egan, Inc. v. City of New York, 215 N.E.2d 490 (N.Y. Ct. App. 1966).

43. *See* U.S. *ex rel.* E. & R. Constr. Co. v. Guy H. James Constr. Co., 390 F. Supp. 1193 (M.D. Tenn. 1972), *aff'd*, 489 F.2d 756 (6th Cir. 1974).

44. *A.H.A. General Constr.*, 699 N.E.2d at 368.

45. *Id.*

46. *See* PYCA Indus., Inc. v. Harrison County Waste Water Mgmt. Dist., 177 F.3d 351 (5th Cir. 1999).

47. Barsotti's, Inc. v. Consolidated Edison Co. of New York, 680 N.Y.S.2d 88 (N.Y. App. Div. 1998).

16.06 ADJUSTMENTS TO CONTRACT PRICE AND COMPLETION DATE

A fully executed change order, although technically a contract modification, operates in some ways as an independent agreement. It will set forth the nature of the transaction (i.e., the scope of the work to be performed); the consideration (i.e., the price of the work); and the additional time, if any, to be allotted for the performance. The change order clauses will typically provide a framework for reaching the goal of a fully agreed, signed change order and a procedure to be followed in the event that an agreement cannot be reached as to what generally has the potential to be the most contentious aspect: the price.

As previously discussed, under the AIA contractual scheme, a construction change directive is issued to permit the work to proceed in advance of a complete agreement as to the terms of a change order. Section 7.3.3 of AIA Document A201–2007 (General Conditions) provides for four methods for determining the price of a change: (1) an agreed lump sum; (2) unit prices; (3) costs to be determined in a manner agreed to by the parties, together with a fee for the contractor; and (4) time and material costs, plus a sum to compensate the contractor for overhead and profit. As envisioned by this provision, the owner and the contractor are at least to agree in the construction change directive on one of these four methods for determining the price, if not on the price itself.

An agreed lump sum is nothing more than a mutually agreed, negotiated fixed price. It appears to be unrealistic to mandate in a construction change direction that the price will be based on an agreed lump sum before the owner and the contractor have come to an agreement on that sum. What if the owner and the contractor simply cannot agree on a price? Accordingly, the better practice is for an agreed lump sum to be listed in the construction change directive as one of two alternative pricing methods. That way, in case the parties cannot agree on a lump price, the second stipulated method can be employed to resolve the price issue.

Unit pricing is typically utilized for repetitive operations where the exact quantity of work cannot be precisely determined at the inception. A classic example is excavation, where the parties may agree that the contractor will be paid X dollars per cubic yard of soil excavated or Y dollars per cubic yard of rock excavated. The contract may already have unit prices that could apply to the changed work, or the parties can agree to unit prices specifically limited to the changed work.

The third and the fourth methods are similar. Both involve reimbursement for the contractor's actual costs of performing the changed work, plus additional compensation, whether it is denominated as a "fee" or as "overhead and profit."

The third method gives the owner and the contractor some leeway to determine those costs, while the fourth method spells out the reimbursable cost categories. Specifically, those designated cost categories in Section 7.37 of AIA Document A201–2007 are:

> Costs of labor, including social security, old age and unemployment insurance, fringe benefits required by agreement or custom, and workers' compensation insurance;
>
> Costs of materials, supplies and equipment, including costs of transportation, whether incorporated or consumed;
>
> Rental costs of machinery and equipment, exclusive of hand tools, whether rented from the Contractor or others;
>
> Costs of premiums for all bonds and insurance; permit fees, and sales, use or similar taxes related to the Work; and
>
> Additional costs of supervision and field office personnel directly attributable to the change.

These cost categories, known in the industry as time and materials or T&M, are designed to cover all of the costs of constructing the changed work. Different contract forms may have different wording for the various time and materials categories, which may affect the eligibility for payment for certain costs. Nevertheless, this general approach of payment for the actual costs for labor, materials, equipment, supplies, insurance, bonds, and supervision is widely recognized in the field. Some owners shun this method, fearing it serves to discourage efficiency, thus prolonging the work and increasing the costs. Some contractors are adverse to this method as well, believing they can achieve greater profitability through a lump sum than through the limited overhead and profit percentage stipulated in the contract. This method is not always suitable for direct application. For instance, a change cannot be priced on a pure time and materials basis where it involves both the deletion of work and additional work. Impact issues resulting from the change likewise cannot be priced on a time and materials basis. However, it is almost always made available as the method of the last resort.

Since construction projects provide fertile grounds for fomenting disputes, especially when it comes to money, a change provision may also contain a procedure for resolving disagreements as to the eligibility of an item for payment, the amount of labor or quantity of materials necessary for the work, and so on. In Section 7.37 of AIA Document A201–2007 (General Conditions), the architect determines the price adjustment on the basis of the designated time and materials categories. However, the architect's determination is not final.

The aggrieved party can challenge it in the courts. Nevertheless, the architect's determination is given deference, through the party challenging the determination having to show that it was the product of fraud or bad faith or was palpably erroneous, or a similar higher standard of proof.[48]

Even though including a method in a changes clause for determining the price for the changed work is widely accepted, it is not universal. As noted in Section 16.03, all that the FAR provides in this respect is that "the Contracting Officer shall make an equitable adjustment."[49] The equitable adjustment can be issued before or after the work is performed. If issued before, the price will be based on the contractor's proposal, together with whatever backup documentation can be produced, such as subcontractor and supplier pricing. If issued after, the preferred method is to submit the actual costs incurred in performing the changed work.[50] However, the equitable adjustment will not necessarily be in the amount of the actual costs incurred, plus overhead and profit. The measure for determining the amount of an equitable adjustment is the reasonable cost of the change work.[51] There is no presumption that the contractor's actual costs are reasonable.[52] This addresses the concern that work performed on a time and material basis may not be done efficiently.

One marked difference between determining costs in federal contracts for changed work as opposed to nonfederal contracts lies in the treatment of contractor-owned equipment. For example, assume that the changed work is to dig a foundation deeper, and the contractor uses its own backhoe to perform the work. Since it is company owned, the cost of the backhoe cannot be determined through actual invoices, as it can be for rental equipment. In nonfederal contracts, company-owned equipment is priced into change orders by multiplying the time the equipment was on the site while the changed work was performed by a rental rate published in a standard equipment manual.[53] However, in federal contracts, standard published rental rates cannot be used to price company-owned equipment.[54] Instead, a complex calculation must be made, which is based on the acquisition cost, ownership expense, and the time during which the equipment was idle on the site.[55]

---

48. *See* Sweet Co. v. White County Bridge Comm'n, 714 N.E.2d 219 (Ind. Ct. App. 1999); RaDec Constr. Inc. v. Sch. Dist., No. 17 of Douglas County, 535 N.W.2d 408 (Neb. 1995).

49. FAR ¶ 52.243-4.

50. Bruce Constr. Corp. v. United States, 324 F.2d 516 (Ct. Cl. 1963).

51. *Id.*

52. FAR ¶ 31.201-3(a).

53. *See* Felhaber Corp. v. State, 410 N.Y.S.2d 920 (N.Y. App. Div. 1978).

54. L. L. Hall Constr. Co. v. United States, 379 F.2d 559 (Ct. Cl. 1966).

55. *Id.*

Leaving the pricing of a change to be made through an equitable adjustment, without further agreed guidelines, is feasible for federal contracts where there is a contracting officer and an administrative process to challenge a contracting officer's determination. Outside of the realm of federal contracts, specifying the methods for determining the price gives a measure of certainty to the process of pricing changes.

The same principles discussed earlier with respect to pricing the changed work apply to the additional time associated with the change. If the parties agree, the change order will set forth the additional time by which the completion date is extended as a result of the changed work. Otherwise, the architect will determine the amount of time to which the contractor can take exception and challenge in a judicial proceeding.

16.07 CONSTRUCTIVE CHANGES

By contrast to a formal change, where an owner explicitly recognizes in its directive that a change order must be issued, a constructive change covers situations where the owner causes the performance of work beyond the scope of the contract but does not recognize the contractor's entitlement to a change order. A constructive change can arise out of a variety of circumstances, such as overly stringent inspection; improper rejection of work; errors, omissions, or ambiguities in the drawings or specifications; or usurping the contractor's prerogative to select the means and methods of performing the work. Thus, a constructive change arises from either the need to address a condition caused by the owner's wrongful conduct or to comply with an informal directive by the owner.[56] Boiled down to its essence, a constructive change is found where a contractor is proven right in asserting a dispute or a claim for a change order.

A classic example of a constructive change is a situation where the owner believes that certain work is encompassed by the drawings and specifications, while the contractor contends it is beyond the scope. For instance, a contract required the contractor to install several thousand new light bulbs "immediately prior to completion." The contractor believed that it could retain the light bulbs installed to provide temporary lighting during construction and replace those that had burned out, while the owner assumed that all of the light bulbs would be replaced at completion. Thus, the owner viewed the scope of the work as requiring the furnishing and installation of all new light bulbs when the building was nearly ready to be turned over to it. The court ruled that the contractor's interpretation was reasonable and consistent with the custom of

56. A. S. McGaughan Co. v. Barram, 113 F.3d1256 (Fed. Cir. 1997).

the industry. Accordingly, the owner's directive to change the light bulbs constituted a constructive change.[57] Thus, in effect, the specifications were found to be ambiguous, and the ambiguity was construed against the owner. While constructive changes often arise out of ambiguous drawings and specifications, the variations are numerous. To give another illustration, where a contract required the contractor to install fume hoods for a science laboratory, but the drawings and specifications did not contain the necessary information to install piping to vent the hoods to the building exterior, the owner's directive to install the piping was held to be a constructive change.[58]

To avoid claims for constructive change orders arising out of ambiguities in the drawings and specifications, owners, especially public owners, sometimes include a contractual provision requiring the contractor to bring ambiguities in the drawings and specifications to the owner's attention prior to bid submission, or waive any claim thereafter. This type of clause is enforceable and operates to prevent the contractor from asserting a claim for a constructive change order arising out of a patent ambiguity.[59] The distinction between a patent ambiguity and a latent ambiguity is significant. Under this type of clause, the contractor is obligated to bring obvious ambiguities to the owner's attention and ask for a clarification so that it could price the work into the bid, if necessary. However, a contractor is not required to review the architect's work and check every detail for completeness and accuracy. Thus, this type of clause would not bar a claim for a constructive change arising out of a latent ambiguity (i.e., an ambiguity not apparent on its face but that could only be reasonably discovered when that aspect of the work is addressed during construction).

A constructive change order arising out of overly stringent inspection arises extrinsically from the owner's representative's demands at the project site, rather than intrinsically from the drawings and specifications prepared before the work is performed. A contractor is required to perform the work in conformance with the drawings and specifications to the reasonable satisfaction of the owner, not to the owner's subjective satisfaction. An extreme example arose out of a bridge painting contract. The contract required that loose paint be removed by sandblasting before new paint was applied to the steel. The inspector, however, required the contractor to remove all of the old paint. Moreover, the inspector insisted that the sandblasting continue until no flecks of paint appeared when a piece of tape was stuck on the surface and pulled off. The court upheld the

57. Metric Constr., Inc. v. Nat'l Aeronautics & Space Admin., 169 F.3d 747 (Fed. Cir. 1999).

58. Julian Speer Co. v. Ohio State Univ., 680 N.E.2d 254 (Ohio Ct. Cl, 1997).

59. Newsom v. United States, 676 F.2d 647 (Ct. Cl. 1982); Arnell Constr. Corp. v. Bd. of Educ. of the City of New York, 598 N.Y.S.2d 1 (N.Y. App. Div. 1993).

contractor's claim for additional costs by virtue of having been required to perform a different job than was provided for in the contract.[60] Improper rejection of work constitutes a constructive change, independent of whether an inspector applies standards above and beyond the contract requirements and the industry norm in reviewing the adequacy of the work. An inspector may reject work as defective by making a mistake in judgment and without applying overly stringent requirements. If adequate work, as measured by the specifications, is rejected as defective, a contractor will be able to recover the costs of implementing the unnecessary corrective work as, in effect, a constructive change.[61]

Another scenario in which a constructive change may arise is when the owner improperly dictates the means and methods of work to the contractor, resulting in increased costs. Normally, the contractor has the prerogative of choosing the means and methods of performing the work to achieve the owner's design.[62] Of course, an owner could reserve the right in the contract to approve the contractor's means and methods, or it could direct that certain means and methods be utilized. However, owners rarely do so, as it could increase costs and expose the owner to additional liability.[63] Where an owner directs the contractor to perform the work in a particular way, despite not having been accorded such a right in the contract, the contractor will have a claim for a constructive change. An example occurred where the owner required the contractor to excavate in lifts of one to two feet, in order to make it easier for the owner's representatives to inspect the excavation work. Since the contract simply required the contractor to excavate to the required depth for the foundations, the costs for the more expensive "stop and start" operation were recovered by the contractor.[64]

The majority of the case law concerning changes emanates from constructive changes. There are frequently disputes over whether a change order should be issued to account for a particular condition or requirement. The changes clause sets the parameters for a recovery or a denial of the claim for a constructive change.

---

60. State v. Buckner Constr. Co., 704 S.W.2d 837 (Tex. App. 1985).

61. *See* Granite Constr. Co. v. United States, 962 F.2d 998 (Fed. Cir. 1992).

62. *See* Burns v. Black & Veatch Architects, Inc., 854 S.W.2d 450 (Mo. Ct. App. 1993).

63. By directing the means and methods of performing the work, an owner may become liable for personal injuries to third parties sustained when an accident occurs from the method directed. *See* Farabaugh v. Penn Turnpike Comm'n, 911 A.2d 1264 (Pa. 2006).

64. Luria Bros. & Co. v. United States, 369 F.2d 701 (Ct. Cl. 1966).

16.08 DUTY TO PERFORM THE CHANGED WORK

It was stated at the beginning of this chapter that a unique and essential feature of a changes clause is the ability to effectuate a unilateral change. It may, therefore, be assumed that a contractor receiving a directive to perform changed work has an obligation to execute it; if the contractor could refuse to proceed, there would not be a unilateral modification. However, the obligation to perform the changed work is not a matter of logic, it is a matter of the contract.

In the AIA Document A201–2007 (General Conditions), this obligation is stated explicitly in two places. First, the contractor agrees to continue with the performance of the contract pending a final resolution of a claim.[65] Second, the contractor agrees to proceed promptly with the change in the work upon receipt of a construction change directive.[66] The Federal Acquisition Regulations have a similar provision requiring performance pending final resolution of a claim.[67]

Thus, under this type of clause, when the contractor has asserted that a directive involves the performance of costly extra work, while the owner contends that the work is within the requirements of the contract, the contractor will have to bear the expenditure without recompense during the claims process. An owner can also abuse this provision when it does not have the funding to pay for a change or to gain leverage in price negotiations. Moreover, if the claims process results in the upholding of the owner's position, the contractor's recourse is to file a lawsuit (or an arbitration demand, if the contract provides for arbitration), which can take years to resolve. Since contractors want to avoid a situation where work is performed without payment, they, understandably, have sought ways to excuse performance of claimed extra work absent a commitment by the owner to pay for the work.

To refuse outright to perform the work runs the risk of constituting a material breach of the contract, for which the owner could hold the contractor in default.[68]

Accordingly, some contractors have attempted to file lawsuits for declaratory judgment absolving them of the responsibility to proceed with performance of the directive. The goal of such a lawsuit is to receive the prompt judicial imprimatur of the contractor's decision not to perform the work and, in

65. AIA Document A201–2007, *supra* note 9, § 15.1.3.

66. *Id.* § 7.3.5.

67. FAR § 52-233.1(i).

68. Stoeckert v. United States, 391 F.2d 639 (Ct. Cl. 1968).

that way, to prevent the owner from declaring a default. In one such example, a contractor filed an action for a declaratory judgment that another contractor on the project, and not the filing contractor, was responsible for the excavation and installation of concrete pads for underground fuel tanks. The work was indisputably shown on the drawings; it was a matter of which of the multiple prime contractors was responsible to perform that work. Under the contract, the commissioner of the public agency that had entered into the contract could determine whether work was extra, and the contractor was obligated to abide by that determination and proceed with the work but had the right to assert a claim for compensation. New York's highest court rejected this attempt to avoid performance of the work, holding that "the contractor had specifically undertaken an obligation to perform that work and postpone any claim for additional compensation until after contract completion."[69]

However, an owner's right to require a contractor to proceed with the performance of extra work is not unlimited. If a directive is so irrational and arbitrary and the consequences of compliance are so onerous and damaging as to constitute an unconscionable result, a contractor could be justified in refusing to comply with the directive.[70] Also, in the realm of public contracts, if a public official's directive is so preposterous as to suggest collusion, bad faith, or similar circumstances, the contractor could likewise be justified in refusing to perform the work.[71] These examples are clearly intended to be limited exceptions, and not the rule.

16.09 CARDINAL CHANGES

There is another exception to the rule that upon the directive of the owner, a contractor must perform work it considers extra, even absent a commitment to compensate it for the work. The exception arises when the extra work constitutes a cardinal change to the contract. A cardinal change is defined as a change to the work that goes so far beyond the scope of the original contact as to constitute a materially different undertaking.[72] The imposition by the owner of a cardinal change upon the contractor amounts to a breach of the contract.[73] A cardinal change differs from a change that is considered to be within the purview of a changes clause (although requiring performance of work outside

69. Kalisch-Jarcho, Inc. v. City of New York, 533 N.E.2d 258, 261 (N.Y. 1988).
70. *Id.* at 263.
71. *Id.* at 262.
72. Allied Materials & Equip. Co. v. United States, 569 F.2d 562, 563–64 (Ct. Cl. 1978).
73. Edwin R. Marden Corp. v. United States, 442 F.2d 364, 369 (Ct. Cl. 1971).

the requirements of the contract, of course) in another respect besides magnitude. A cardinal change can emanate from either one isolated but (obviously) extremely significant change, or from a series of changes, related or unrelated. While the number of changes alone may be grounds for invoking a cardinal change,[74] it is really the cumulative impact of the changes upon the work as a whole that is more critical to the determination.[75]

In a sense, a case involving a claim of a cardinal change is the closest construction law comes to a constitutional analysis under the First Amendment. The reason for this analogy is that, like Justice Stewart's conclusion regarding defining whether something amounts to pornography, the courts examining cardinal change claims have adopted an "I know it when I see it" standard. The Court of Claims, in one of the leading cases on cardinal changes, phrased its parameters as follows:

> There is no exact formula for determining the point at which a single change or series of changes must be considered to be beyond the scope of the contract and necessarily in breach of it. Each case must be analyzed on its own facts and in light of its own circumstances, giving just consideration to the magnitude and quality of the changes ordered and their cumulative effect upon the project as a whole [citation omitted]. The contractor cannot claim a breach of the contract if the project it ultimately constructed is essentially the same as the one it agreed in the contract to erect [citation omitted].[76]

In the final analysis, the critical factors are whether the change or changes are of such a magnitude as to radically alter the nature of the contract itself, whether by the character of the work, the quantity, the method or manner of performance, or the cost,[77] or whether the project constructed remained essentially the project that the contractor had bid for at the outset.[78]

The doctrine of cardinal change originated in the realm of government contracts.[79] It has been discussed in numerous federal decisions, and a number of states recognize this concept as well. Some states explicitly utilize the term "cardinal change,"[80] while others acknowledge the concept by description

---

74. C. Norman Peterson Co. v. Container Corp. of America, 218 Cal. Rptr. 592, 598 (1985).

75. Becho, Inc. v. United States, 47 Fed. Cl. 595, 601 (2000).

76. Wunderlich Contracting Co. v. United States, 351 F.2d 956, 966 (Ct. Cl. 1965).

77. *Becho*, 47 Fed. Ct. at 601; Peter Kiewit Sons Co. v. Summit Constr. Co., 422 F.2d 242 (8th Cir. 1969).

78. Aragona Constr. Co. v. United States, 165 Ct. Cl. 382, 390 (1964).

79. *See* Saddler v. United States, 287 F.2d 411 (Ct. Cl. 382, 390 (1961); General Contracting & Constr. Co. v. United States, 84 Ct. Cl. 570 (1937).

80. *See, e.g.*, J. A. Jones Constr. Co. v. Lehrer McGovern Bovis, Inc., 89 P.3d 1009 (Nev. 2004); Housing Auth. of City of Texarkana v. Johnson Constr. Co., 523 S.W.2d 316 (1978).

without using this term specifically.[81] Some states have refused to validate the concept of cardinal change.[82]

Since construction is, at its most basic, a search for the practical solution, it is necessary to examine the consequences of a finding of a cardinal change. A contractor that believes it has encountered a cardinal change has two choices. It may elect to refuse to perform the work and treat the contract as terminated.[83] Alternatively, the contractor can elect to proceed with the work and sue for *quantum meruit*, thereby possibly exceeding the recovery that could have been had under the contract.[84]

In the final analysis, it is the rare instance where a contractor will declare a termination on the grounds of a cardinal change. If a court ultimately rules the termination unjustified, the consequences would be so severe, both directly to its business and indirectly through its surety, that a contractor must find itself in such dire circumstances as to have little choice other than to invoke this concept in order to cease performance.

16.10 ADMINISTRATION OF CHANGES

Even after a directive is issued and the contractor notifies the owner that it views the work as change and the decision is made to proceed to perform the work, the legal consequences continue. Until payment is made, or at least until an agreement is reached regarding the price, the work remains an open issue, requiring the attention of both parties.

It would stand to reason that both the contractor and the owner have an interest in monitoring the cost of performing the claimed extra work. The contractor needs to know how much the work is costing, so that it can submit a claim that will capture all of the expenses associated with the directive. The owner, on the other hand, would want to ensure that the costs it will be presented with in the claim are accurate and do not include the cost of work already within the contract. In order to advance these interests, someone should keep track of the costs of the claimed extra work.

Many construction contracts provide that extra work for which there is either no agreement as to price or no unit price stipulated in the contract to cover the particular work operations involved are to be paid on a time and

81. *See, e.g.*, Tufano Contracting Corp. v. State, 269 N.Y.S.2d 564 (N.Y. App. Div. 1966).

82. *See, e.g.*, Gill Constr., Inc. v. 18th & Vine Auth., 157 S.W.3d 699 (Mo. App. 2004); Litton Systems, Inc. v. Frigitemp Corp., 613 F. Supp. 1377 (S.D. Miss. 1985).

83. Fuller Co. v. Brown Minneapolis Tank & Fabricating Co., 678 F. Supp. 506, 509 (E.D. Pa. 1987); *but see* Amelco Elec. v. City of Thousand Oaks, 38 P.3d 1120 (Cal. 2002) (holding that a contractor cannot recover against a public owner on the theory of abandonment of the original project).

84. *Peter Kiewit Sons Co.*, 422 F.2d at 242.

materials basis.[85] The costs typically eligible for compensation for extra work performed on a time and materials basis are labor, materials, equipment, supplies, insurance, and surety bond. A percentage for overhead and profit is then added to these costs to arrive at the final cost.

Some contracts have clauses requiring the contractor to submit time and material tickets for the owner's signature on a daily basis while the contractor is performing work it claims is extra or while the contractor is performing acknowledged extra work for which an agreement on price has not been reached in advance. An example of one such clause, in pertinent part, is as follows:

> **ARTICLE 28. PERFORMANCE OF EXTRA OR DISPUTED WORK**
> While the Contractor or his Subcontractor is performing extra work ordered by the Commissioner under Article 25 hereof (unless payment therefore is to be made by a lump sum or at unit prices previously agreed upon) or is performing disputed work or complying with a determination or order under protest in accordance with Article 27 hereof, in each such case the Contractor shall furnish the Resident Engineer daily with three (3) copies of written statements signed by the Contractor's representatives at the site showing:
>
> 1. The name and number of each workman employed on such work or engaged in complying with such determination or order, the number of hours employed thereon, and the character of the work each is doing; and
> 2. The nature and quantity of any materials, plant and equipment furnished or used in connection with the performance of such work or compliance with such determination or order, and from whom purchased or rented.
>
> A copy of such statement will be countersigned by the Resident Engineer, noting thereon any items not agreed to or questioned, and be returned to the Contractor within two (2) days after submission.[86]

A contractor's failure to comply with this type of clause by submitting daily time and material tickets may result in the dismissal of the claim.[87]

However, many widely used contracts do not have such a clause. For instance, neither the AIA form contract nor the FAR contains this type of requirement.[88] Nevertheless, even absent such a clause, the contractor should maintain records of the actual costs. The changes clause in the FAR speaks in terms of equitably adjusting the contract price for costs incurred in the performance

85. *See* AIA Document A201–2007, *supra* note 9, § 7.3.7.

86. This excerpt is from a City of New York form of construction contract.

87. *A.H.A. General Constr., Inc.*, 699 N.E.2d at 368.

88. However, under FAR § 52.243-6, a contracting officer may require the contractor to maintain segregated cost records for a change or series of related changes estimated to exceed $100,000.

of the change.[89] An equitable adjustment should be based on the contractor's actual reasonable costs.[90] Should the claim not be resolved, a contractor may find itself precluded from recovery for the change if it cannot establish the actual costs of performing the change. A claim for extra costs on a federal contract was dismissed by the Court of Claims on this very basis:

> The evidence offered consisted of the testimony of a certified public accountant, and was based upon a comparison of the cost of performance under the contract here involved and the cost of performance under a certain other contract . . . The evident [sic] on this point has been examined with great care. Without questioning the integrity or the technical skill of the witness who presented the testimony upon which plaintiff's claim for extra costs rests, the evidence is unsatisfactory and wholly unconvincing.[91]

State law may likewise bar a recovery for extra work where only estimates of the costs of performance are submitted as proof of damages, so long as the costs of performing the extra work may be segregated from the costs of performing the contract work.[92] Therefore, whether documented in the form of daily time and material records, or in some other segregated format, the actual costs of performing the extra work should be maintained, if at all possible, in the event that a negotiated change order does not result.

16.11 RELEASES—ACCORD AND SATISFACTION

Even if the owner and the contractor agree to and sign a change order, the potential for a dispute regarding the change still exists. Generally, the dispute is over scope or time or both. A change order is comparable to a contract; it is a contract to perform a particular aspect of work not provided for in the parties' original contract. The more details the change order contains as to what it covers, the clearer it will be, and the less likely there will be disputes over its terms. Conversely, the more general the language, the greater the chance that there will be an ambiguity (at least in the mind of one of the parties) and the more likely there may be a further dispute concerning a matter that should have been settled in the document.

The possibilities for disagreement as to the breadth of a signed change order run the full gamut. A time-related dispute is the easiest to conceptualize. A contractor may claim that the change order only covered the direct costs of

89. FAR § 52.243-4(d).

90. *Bruce Constr. Corp.*, 324 F.2d at 516, *superseded by rule on other grounds as stated in* Morrison Knudsen Corp. v. Fireman's Fund Ins. Co., 175 F.3d 1221, 1224, n.30 (10th Cir. 1999).

91. American Can Co. v. United States, 69 Ct. Cl. 1, 9–10 (1929).

92. Metro. Steel Indus., Inc. v. Perini Corp., 828 N.Y.S.2d 395 (N.Y. App. Div. 2007).

performing the additional work but did not compensate it for the indirect costs associated with the overall delay to the project due to the change. A scope-related dispute can be extremely technical. For example, a contractor may contend that a change order to substitute cast-in-place concrete walls for hollow masonry block walls (i.e., cinder blocks) accounts only for the cost of the concrete used to construct the walls, but not for changes to the structural framework of the building necessitated by the added weight of the concrete over the hollow masonry blocks. In such an instance, the contractor may believe that the structural ramifications of the change are a separate matter and submit only the costs strictly associated with the forming and pouring of the concrete walls. However, the owner may believe that the change order includes all work of any nature whatsoever associated with the change to concrete walls.

In such an instance, the description of the change and what the dollar value of the modification covers would be paramount. However, this sort of factual inquiry could be circumvented by an additional general provision in the change order. Where a change order states that it constitutes payment for "all costs and markups directly or indirectly attributable to the changes," a claim for additional compensation may be barred.[93] Likewise, the recitation in a change order that it "constitutes an accord and satisfaction and represents payment in full (for both time and money) for any and all costs, impact effects, and/or delays arising out of incidental to the work" will preclude a claim for additional compensation relating to the change.[94] Concomitantly, the failure of a change order to recite that it was "in full satisfaction of all claims" resulted in the denial of the government's motion to dismiss the contractor's delay claim for the indirect costs stemming from the changed work.[95]

Aside from barring a further claim on the basis of an accord and satisfaction, a change order may bar a further claim on the basis of release. While an accord and satisfaction constitutes a settlement for a particular claim, a release is broader, as it is a waiver of all claims, known or unknown. In the context of a change order, a release operates to waive all claims, known or unknown, relating to the subject matter of the change order. Whether a further claim arising out of a change order has been released depends, of course, on whether the change order includes language constituting a release. If the change order includes release language, it will operate to bar a further claim concerning the particular change or changes encompassed by the document.[96] Nevertheless, a release could be challenged as being ambiguous and, therefore, not precluding

93. King Fisher Marine Serv., Inc. v. United States, 16 Cl. Ct. 231 (1989).
94. Safeco Credit v. United States, 44 Fed. Cl. 406, 419 (1999).
95. Westchester v. United States, 28 Fed. Cl. 172 (1993).
96. Citadel Corp. v. Sun Chem. Corp., 443 S.E.2d 489 (Ga. Ct. App. 1994).

the claim being asserted.[97] Otherwise, a release in a change order would be challenged in the same manner as releases generally (i.e., on such grounds as fraud, unconscionability, or duress). Under the law of most jurisdictions, that is a high hurdle to overcome.[98]

Parties who are concerned over the possibility of waiving a claim by signing a change order sometimes try to avert that by including a reservation of rights in the document. The reservation of rights can simply recite, for example, "The Contractor reserves all rights against the Owner for damages and costs incurred due to delays arising out of this change order." Courts have upheld the right of a contractor to assert an additional claim relating to a change order when the change order included a reservation of rights agreed to by both parties.[99] On the other hand, a reservation of rights clause unilaterally inserted into a change order will not be effective to reserve a right to assert a claim that is otherwise barred by the contract.[100] Thus, a contractor could not recover additional compensation for delays resulting from a change order containing a reservation of rights where the change order stated that it constituted "full and complete compensation" for overhead as provided in the contract's changes clause and the changes clause contained a limitation on the recovery of overhead.[101]

Nevertheless, a unilateral reservation of rights clause can be effective where the additional claim is neither included in the compensation provided for by the change order, nor barred by the contract.[102] In those jurisdictions that enforce the sovereign immunity doctrine, however, a contractor's insertion of a reservation of rights clause into a change order does not entitle it to obtain compensation that was not agreed to by the governmental entity in a signed writing.[103]

---

97. *See id.*

98. *See* Strickland Tower Maintenance, Inc. v. AT&T Communications, Inc., 128 F.3d 1422 (10th Cir. 1997); Vulcan Painters, Inc. v. MCI Contractors, Inc., 41 F.3d 1457 (11th Cir. 1995); *but see* Willms Trucking Co. v. J. W. Constr. Co., 442 S.E.2d 197 (S.C. Ct. App. 1994).

99. Williams and Sons Erectors, Inc. v. South Carolina Steel Corp., 983 F.2d 1176 (2d Cir. 1993); Southland Enterprises, Inc. v. United States, 24 Ct. Cl. 596 (1991).

100. Santa Fe Eng'rs, Inc. v. United States, 801 F.2d 379 (Fed. Cir. 1986); Linda Newman Constr. Co. v. United States, 48 Fed. Cl. 231 (2000).

101. *Linda Newman Constr. Co.*, 48 Fed. Cl. at 231.

102. *Id.*; *see also* Mike M. Johnson, Inc. v. County of Spokane, 78 P.3d 161 (Wash. 2003), Noting, *in dictum*, that authority exists to uphold a reservation of rights clause inserted by a contractor into a change order.

103. C.O.B.A.D. Constr. Corp. v. Sch. Bd. of Broward County, 765 So. 2d 844 (Fla. Dist. Ct. App. 2000).

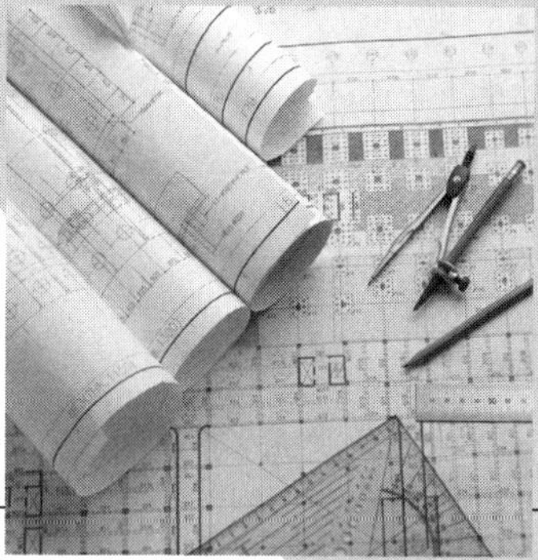

QUESTIONS

Aim To Please Construction Corp. entered into a contract with the Federal Government to build a veterans' hospital. Aim To Please then subcontracted the structural steel work for the project to Not Too Steady Erection Co. The drawings and specifications contained details on the requirements for welding and for attaching certain steel components through mechanical means. However, no bolts were specified or listed on the drawings. The structural engineer made a field visit and noticed that Not Too Steady was letting beams sit on top of columns, without bolting them together. Being alarmed at the implication, the structural engineer directed Not Too Steady's foreman, in the presence of Aim to Please's curtain wall superintendent, to bolt up all the steel that had been erected using galvanized two-inch bolts, immediately. Not Too Steady's foreman protested that the forces of gravity and the weight of the steel were sufficient to hold the beams in place until the concrete decks were poured and the curtain wall was installed, but proceeded to comply nonetheless.

In the meantime, Not Too Steady discovered that the drawings and specifications omitted to include cross bracing at certain points where trusses were to be placed, although cross bracing was specifically shown at other locations. After being informed of that by its subcontractor, Aim to Please notified the Government that cross bracing would be installed as necessary in locations where they were omitted from the design.

While all of this was happening, the Government became concerned about the possibility of a power shortage adversely affecting the hospital, once it was completed. Accordingly, the Government Contracting Officer directed Aim to Please to build a nuclear power plant at the hospital to serve as a backup source of energy.

Not Too Steady had entered into an AIA subcontract with Aim to Please, which included the A201–2007 General Conditions, without any amendments. The Government contract with Aim to Please included a clause providing that all work reasonably necessary for the complete construction of the project was included in the price; Aim to Please's subcontract with Not Too Steady did not contain such a clause. Both the prime contract and the structural steel subcontract contained a patent ambiguity provision.

1. Is Not Too Steady entitled to a change order from Aim to Please for the bolts?
2. What should Not Too Steady do to preserve its rights for furnishing and installing the bolts?
3. Is Aim To Please entitled to a change order from the Government for the bolts?
4. What should Aim To Please do to preserve its rights for furnishing and installing the bolts?
5. If it is a change, what type of change would the cross bracing be and why?
6. Is Not Too Steady entitled to a change order from Aim To Please for the cross bracing?

7. Is Aim To Please entitled to a change order from the Government for the cross bracing?
8. What options does Aim To Please have in regard to the directive to build a nuclear power plant? What are the legal ramifications to Aim To Please's contract?
9. What should Aim To Please's response be if the Government issues a change order for the cost of furnishing the bolts?
10. Do any of the individuals who issued these directives have the authority to do so?

CHAPTER

17

Differing Site Conditions

RICHARD F. SMITH, VAL S. McWHORTER, and W. STEPHEN DALE

17.01 INTRODUCTION

Many construction projects involve work below the visible surface of the earth, or rely in some part on the conditions that exist below ground. Planning and pricing the work from a contractor's perspective requires some determination of what those subsurface conditions entail and how they will impact the work. In many situations, however, neither the owner nor the contractor truly knows the precise subsurface conditions when entering into an agreement. Often the subsurface conditions that would affect the work do not manifest themselves until after the project has begun. Since the nature of these site conditions can dictate the resources needed to perform the work, and therefore the price of the contract, both owners and contractors share a keen interest in managing the risk of unknown subsurface conditions.

The phrases "differing site condition" and "changed conditions" are terms of art, which refer to unknown physical conditions encountered at the job site

The authors wish to thank Kathryn T. Muldoon for her valuable assistance in the preparation of this chapter.

that differ materially from the conditions reasonably expected.[1] For example, a differing site condition may exist where a contractor unexpectedly encounters rock rather than sand in its excavation, or a geologic formation that should not have been present in the region, or where the soil reflects a moisture content higher than expected. Regardless of whether the unexpected condition is latent or obvious, natural or manmade, the contractual question arises as to how the risk associated with unexpected conditions should be allocated between the parties and how to resolve disputes between the parties when such conditions are encountered. The Differing Site Conditions clause and other similar provisions help allocate this risk and provide a vehicle for resolving disputes between project owners and contractors relating to these unexpected conditions.

Contractors, owners, and the courts have developed several mechanisms to cope with the lack of perfect information about subsurface conditions. These mechanisms, as discussed in this chapter, typically seek a balance between a variety of competing factors including: (1) the extent of pre-bid investigations that both parties must undertake; (2) the degree of risk each party must bear; (3) the amount a contractor should include as contingency in its bid; and (4) the likelihood of unexpected problems.

As a general rule, in the absence of a specific contract clause addressing subsurface conditions, common law allocates the risk to the contractor and generally places the burden of discovering and/or overcoming unexpected subsurface conditions on the contractor. In practice, however, the bidding process rarely allows time for a contractor to conduct extensive subsurface investigations on its own, and contracts rarely provide a reimbursement tool for expensive investigations. This inability to conduct investigations forces contractors to add contingencies to their bids to cover eventualities that may never occur. This practice artificially inflates bid prices to incorporate these contingencies and causes the government or the owner to pay for contingencies, regardless of whether the adverse subsurface problems ever materialize.

To combat the practice of adding these kinds of contingencies into bid prices, standard contract clauses adopted by federal and state governments and trade organizations place a heavier contractual focus on information provided by the government or owner to the contractor. These clauses seek to reduce the contingency included by contractors in their bids by encouraging contractors to rely on information provided by owners and promising an adjustment in the contract price for material deviations from the information provided. This practice allows the government or owner to avoid paying for expensive contingencies that never occur, and allows the contractor protection against unforeseen problems that could impact project cost. In addition, it allows the

1. *See* Federal Acquisition Regulation (FAR) 52.236-2(a).

owner to evaluate bids from contractors on the same pricing basis, rather than on unknown contingencies.

To provide context for the modern clause-based approach to changed subsurface conditions, this chapter first reviews the common law heritage of the Differing Site Conditions clause and the evolution of risk allocation for subsurface conditions. The common law tradition employed general contract principles, including warranty, misrepresentation, mutual mistake, and other conceptual frameworks to allocate responsibility for price increases due to unexpected conditions. After reviewing the various common law approaches, the chapter turns to an overview of governmental and commercial contract clauses designed to address these issues as well as their application in the federal, state, and private context.

17.02 COMMON LAW

A. Contracts without a Site Indications or Risk Allocation Clause

Where fixed-price construction contracts include no special provision for subsurface conditions, the contractor normally bears the risk of unforeseen conditions. In that context, the U.S. Supreme Court summed up the allocation of risk in such circumstances saying: "Where one agrees to do, for a fixed sum, a thing possible to be performed, he will not be excused or become entitled to additional compensation, because unforeseen difficulties are encountered."[2]

This method of contract risk allocation has been echoed in decisions from state and federal courts as well as in the Federal Acquisition Regulation (FAR). The FAR declares that a firm, fixed-price contract without more

> provides for a price that is not subject to any adjustment on the basis of the contractor's cost experience in performing the contract. ***This contract type places upon the contractor maximum risk and full responsibility for all costs and resulting profit or loss.*** It provides maximum incentive for the contractor to control costs and perform effectively and imposes a minimum administrative burden upon the contracting parties.[3]

This concept of "maximum contractor risk" governed construction contracting until the advent of special risk-allocating clauses and other legal theories that allowed a contractor to recover increased costs due to unexpected subsurface conditions.

---

2. Spearin v. United States, 248 U.S. 132, 136 (1918).

3. Dalton v. Cessna Aircraft Co., 98 F.3d 1298, 1304 (Fed. Cir. 1996), discussing FAR 16.202-1 and firm fixed-price services contracts (emphasis added).

In *Simpson v. United States*,[4] for example, the contractor on a federal project agreed to construct a dry dock at the Brooklyn Navy Yard. The Navy had previously conducted soil testing at the site, but did not include that information in the solicitation or contract documents. In short, the contract documents did not contain any indications of subsurface conditions or any provision that addressed changes in those conditions. During construction, the *Simpson* contractor discovered sandy soil underneath the foundation, which caused unexpected settlement of the dock. The unforeseen soil conditions forced the contractor to incur additional costs to complete the project. The contractor submitted a request to the Navy for additional compensation to cover the costs of working in the sandy soil.

In upholding the Navy's rejection of the contractor's request, the Supreme Court in *Simpson* focused on the absence of any remedy granting provision addressing the soil conditions, finding that:

> The contract imposed upon the contractors the obligation to construct the dock according to the specifications within a designated time for an agreed price upon a site to be selected by the United States. We look in vain for any statement or agreement or even intimation that any warranty, express or implied, in favor of the contractors was entered into concerning the character of the underlying soil.[5]

Finding no contractual ground by which to release the contractor from its obligations, the Court relied on the plain language of the contract and denied the contractor any increased compensation. *Simpson* represents the trend by which the majority of jurisdictions operated prior to the introduction of the federal Differing Site Conditions clause and other similar clauses,[6] namely that the contractor must bear the risk of encountering adverse conditions.

The same risk allocation scheme persists in private commercial contracting today in the absence of contractual relief. For example, in *American Demolition,*

---

4. 172 U.S. 372 (1899).

5. *Simpson*, 172 U.S. at 380. The *Simpson* contractor also advanced an alternative theory of liability for the subsurface conditions, relying on the government's obligation to make the work site available for performance of the contract. Under this theory, the contractor contended that making the site available included an implied warranty of subsurface conditions. The Court rejected that argument, holding that making the site "available" did not encompass a broader obligation to warrant the site conditions.

6. The trend continues today in state jurisprudence where contracts lack a remedy granting provision covering unknown subsurface conditions. *See, e.g.*, American Demolition, Inc. v. Hapeville Hotel Ltd., 413 S.E.2d 749 (Ga. App. 1991) (parties expressly removed differing site conditions clause); Abbruzzese v. Miller, 1996 Ohio App. LEXIS 4304 (1996) (denying recovery for unknown subsurface conditions where contract lacked any differing site conditions clause).

Inc. v. Hapeville Hotel Ltd.,[7] the parties relied on standard AIA construction documents. In negotiation, the parties struck out the AIA standard provision allowing a contractor to recover for changed or concealed conditions at the site and did not replace the provision with any other form of relief.[8] The court held that "[t]he contract here contained no changed conditions clause, unequivocally limited the contract payment to a sum certain, and contained an inspection clause. It is clear from these provisions that the contract imposed the risk of uncertainty of subsurface conditions on [the contractor]."[9]

B. Alternate Legal Theories in the Presence of Representations or Indications

The inclusion of contract language or materials that directly address subsurface conditions works a dramatic change in the allocation of risk for subsurface problems. Even in the absence of a specific risk-allocation device, the presence of descriptions or representations in the contract or bid documents as to the nature of the subsurface conditions or to means and methods that would be suitable for addressing the work can relieve the contractor of liability for those conditions. In effecting this shift away from contractor liability, courts have relied on standard contract law devices including warranty, misrepresentation, and mutual mistake, as well as concepts specific to government contracting, such as superior knowledge.

1. Warranty

One of the legal theories on which contractors have successfully relied to recover increased costs associated with adverse conditions is the implied warranty of specifications. This well-established theory rests on the decision of the U.S. Supreme Court in *Spearin v. United States*.[10] In that decision, the Court established the maxim that "if the contractor is bound to build according to plans and specifications prepared by the owner, the contractor will not be responsible for the consequences of defects in the plans and specifications."[11] Contractors and courts have applied this concept of implied warranty of specifications to changes in subsurface conditions where the government-included specifications directed the use of particular equipment, or specific means and

7. *American Demolition, Inc.*, 413 S.E.2d at 749.
8. *See id.* at 750.
9. *Id.* at 752.
10. *Spearin*, 248 U.S. at 132. *See also* note 2 and accompanying text.
11. *Id.* at 136.

methods, or included maps or plans that represented anticipated site conditions. Courts have applied the *Spearin* doctrine as a basis for finding owner liability for these constructive "representations" of site conditions.[12]

2. *Misrepresentation*

In addition to relying upon breach of the implied warranty of specifications, contractors have also relied upon the common law theory of misrepresentation to recover increased costs associated with adverse conditions.[13] This common law approach includes both intentional misrepresentation and unintentional or negligent misrepresentation.[14] Under this family of theories, the contractor typically asserts that the government or owner made affirmative representations in the contract documents as to the nature of the work and the subsurface conditions; the contractor relied on these affirmative representations in preparing its bid; and these representations ultimately proved to be false or inaccurate. These kinds of claims include allegations akin to negligent misrepresentation,[15] (i.e., that the contract documents misled the contractor), as well as more affirmative misrepresentations evocative of claims for active concealment.[16]

The courts have almost uniformly dismissed the question of good faith or bad faith aspects of misrepresentation. Unlike a tort action, in which additional damages may be available to a victim of intentional or bad faith misrepresentation, the result in a contract action is normally the same regardless of the basis

---

12. *See, e.g.*, United States v. Atl. Dredging Co., 253 U.S. 1 (1920) (approval of contractor's plan for the work and express contract representations as to methods constituted government plans and specifications for purposes of defective specifications claim).

13. *See, e.g.*, Hollerbach v. United States, 233 U.S. 165 (1914) (dam fill material misrepresented in government plans); Christie v. United States, 237 U.S. 234 (1915) (contract borings incorrectly indicated gravel beneath lock and dam structure rather than mixture of material actually found); *Atl. Dredging Co.*, 253 U.S. at 1 (on dredging contract, direction to contractors to conduct site visit does not relieve government of liability for misrepresentations in the plans and specifications); United States v. Smith, 256 U.S. 11 (1921) (where specifications did not reveal presence of limestone, contractor awarded extra costs to work through rock); Levering & Garrigues Co. v. United States, 73 Ct. Cl. 566 (1932) (intentional misrepresentation as grounds to award increased costs).

14. *But see* Souza & McCue Constr. Co., Inc. v. Superior Court of San Benito County, 370 P.2d 338, 340 (Cal. 1962) (proof of an intentional misrepresentation adds little to the recovery since, "[t]his rule is mainly based on the theory that the furnishing of misleading plans and specifications by the public body constitutes a breach of an implied warranty of their correctness. The fact that a breach is fraudulent does not make the rule inapplicable.").

15. *See, e.g.*, *Hollerbach*, 233 U.S. at 165 (representations in the plans and specifications "must be taken as true and binding on the Government"); *Christie*, 237 U.S. at 234 (relief for misrepresentations of site conditions may be either intentional or unintentional).

16. *See, e.g.*, *Smith*, 256 U.S. at 11; *Levering & Garrigues Co.*, 73 Ct. Cl. at 566; *Accord Souza & McCue Constr. Co., Inc.*, 370 P.2d at 340.

of misrepresentation. In *Levering & Garrigues v. United States*,[17] the claimant contractor entered into a contract to construct an addition at the United States Naval Academy. As part of the claim process, the contractor learned that the Navy had long known of the existence and location of certain subsurface conditions but had intentionally failed to disclose that knowledge to bidders. In awarding damages, the Court of Claims ignored the motive and details of the withholding and focused instead on the simple fact that the government misled the contractor.[18] Thus, under this analysis, the Court looks only to the nature of the representations, inquiring as to whether that representation amounted to an affirmative statement of site conditions. With that element proven true, the contractor need only demonstrate that the actual conditions differed from that represented by the government.

3. *Superior Knowledge/Concealment*

A theory akin to misrepresentation is that of superior knowledge. Under this theory, if the government possesses special knowledge vital to the performance of a contract which the contractor does not possess and cannot reasonably acquire, the government must disclose its superior knowledge or compensate the contractor for any increased costs that result from the failure to disclose the information.[19] This legal theory shares some of the traits of misrepresentation, but focuses on concealment rather than affirmative misrepresentations of site conditions. Consequently, in a commercial or private contracting venue, the claim for relief may also proceed under the guise of the tort of concealment.

For example, in *Granite Constr. Co., Inc. v. United States*,[20] the Corps of Engineers contracted for a salmon fingerling bypass through a dam on the Columbia River. The Corps, unknown to the contractor, had encountered difficulties performing similar work on another fingerling bypass project using equipment similar to that proposed by the contractor on the current project.

---

17. 73 Ct. Cl. at 566.

18. *Id.* at 575.

19. *See* Hardeman-Monier-Hutcherson v. United States, 458 F.2d 1364, 1371–72 (1972) (unusually severe weather ruled not a changed condition, but contractor recovered increased costs based on government's failure to adequately disclose knowledge of weather issues); Helene Curtis Indus., Inc. v. United States, 312 F.2d 774, 778 (Ct. Cl. 1963) (where the government possesses "vital information which it was aware the bidders needed but would not have, [it] could not properly let them flounder on their own. Although it is not a fiduciary toward its contractors, the government—where the balance of knowledge is so clearly on its side—can no more betray a contractor into a ruinous course of action by silence than by the written or spoken word.").

20. 24 Cl. Ct. 735 (1991).

When the contractor encountered the same problems as the earlier contractor, it submitted claims to the Corps based on both superior knowledge and the Differing Site Conditions clause. Although the court denied the contractor additional compensation, it provided a detailed analysis of the factors required to demonstrate superior knowledge in the context of a differing site condition. The analysis yielded the following conclusion:

> Three elements were distilled there to find liability: the Government had information that would be material to performance of the contract; the contractor lacked that knowledge and had no reason to be on notice of it; and the Government either knew or should have known that the plaintiff was in ignorance of the relevant information. The element of misrepresentation, inadvertent or deliberate, is not directly relied upon, but may be implicit in the requirement that the Government know or have reason to know that the contractor is, in effect, stepping into a trap.[21]

The court explained the basis for its decision, stating that "in the absence of affirmative misrepresentations, the Government is not a guarantor against poor judgment with respect to methodologies selected by the contractor."[22] The formula set forth in *Granite Constr. Co.* establishes a higher evidentiary burden than that called for by many of the standard contract clauses discussed later in this chapter.

4. *Mutual Mistake*

Mutual mistake has served as an often attempted, but less often successful, means by which a contractor may recover increased costs for changed conditions. Under the general theory of mutual mistake, where parties to a contract are mistaken at the time of contract formation as to a basic assumption of the contract, and that assumption has a material bearing on the agreed exchange of performances, the adversely affected party may void the contract, unless it bears the risk of the mistake.[23] In the context of changed conditions, a contractor may argue that both parties were mistaken as to the nature of subsurface conditions. That mistake may arise from mutual assumptions drawn from information in the contract documents regarding the site, or from mutual assumptions based on conditions common to the region. This approach may present a challenge to the claimant in demonstrating that both parties labored

---

21. *Id.* at 750 (emphasis added).

22. *Id.* at 753.

23. *See* Restatement of the Law (Second) of Contracts § 152 (1981); Northrop Grumman Corp. v. United States, 47 Fed. Cl. 20, 52 (2000); Nat'l Presto Indus., Inc. v. United States, 338 F.2d 99 (Ct. Cl. 1964). *Accord* FAR 14.407-4(c).

under a mutual mistake as opposed to a unilateral mistake.[24] Nevertheless, given the proper factual situation, the claim for mutual mistake continues to exist as a viable theory of recovery.

17.03 REGULATORY AND OTHER STANDARD FRAMEWORKS FOR ADDRESSING DIFFERING SITE CONDITIONS

A contractor faced with bidding work under the common law framework faces three alternatives. First, it could simply not bid the work. Few contractors who choose this option remain in business. Second, it may bid the work based on an optimistic guess about the nature of the subsurface conditions. A poor guess, or an overly optimistic guess in this respect, could result in significant economic losses and ultimately in a lawsuit by the contractor for any of the common law claims. Worse still, the contractor may simply absorb the increased costs of performing the actual work. The financial challenges posed by the magnitude of some changed conditions could easily result in default or bankruptcy of the contractor. Third, the contractor could add a contingency to its bid to cover a range of possibilities that might occur once the contractor began work. Under this scenario, the contractor must determine the right contingency, and a poor guess could result in the same problems as option two. Alternatively, an overlarge contingency could make the contractor's bid too high and uncompetitive compared to other bids. Moreover, if the contractor does not encounter the problems contemplated by its contingency, the government or owner pays for work not performed, and the contractor receives a windfall.

This functional framework resulting from the common law remedies persisted until 1926, when the federal government began to review the financial and economic impact of these kinds of changes on bidding and performance of public contracts. As part of that review, the federal Board of Contracts and Adjustments determined that bidders on Federal contracts more often than not inserted contingencies into their bids in anticipation of problems that might arise in the work due to subsurface conditions. In addition, the government paid these hidden costs regardless of whether the contractor encountered unexpected conditions. Moreover, as later articulated by the Court of Federal Claims:

> Generally, the government is in a better position to provide the contractor with the necessary information about site conditions, given the fact that it has already taken the time to design the project, make its own borings, as well as investigate and evaluate the site prior to submitting the project for bidding.[25]

---

24. *See e.g.*, Flippin Materials Co. v. United States, 160 Ct. Cl. 357 (1963) (court rejected claim for mistake after finding contract documents actually revealed changed conditions and showed no mistake).

25. Youngdale & Sons Constr. Co. v. United States, 27 Fed. Cl. 516, 528 (1993).

As a result, the Board of Contracts and Adjustments developed a standard contracting clause to address contingencies in bids, but also provide some protection for the contractor in the event conditions differed from those expected.

A. Differing Site Conditions—FAR 52.236-2

The clause developed by the Board of Contracts and Adjustments was known as the Changed Conditions clause. The concepts and protections that gave rise to the Changed Conditions clause now appear in federal contracts, and many other contracts funded by federal contributions, under the name Differing Site Conditions clause. The federal Differing Site Conditions clause states as follows:

> **Differing Site Conditions (Apr 1984)**
>
> (a) The Contractor shall promptly, and before the conditions are disturbed, give a written notice to the Contracting Officer of—
>
> (1) Subsurface or latent physical conditions at the site which differ materially from those indicated in this contract; or
>
> (2) Unknown physical conditions at the site, of an unusual nature, which differ materially from those ordinarily encountered and generally recognized as inhering in work of the character provided for in the contract.
>
> (b) The Contracting Officer shall investigate the site conditions promptly after receiving the notice. If the conditions do materially so differ and cause an increase or decrease in the Contractor's cost of, or the time required for, performing any part of the work under this contract, whether or not changed as a result of the conditions, an equitable adjustment shall be made under this clause and the contract modified in writing accordingly.
>
> (c) No request by the Contractor for an equitable adjustment to the contract under this clause shall be allowed, unless the Contractor has given the written notice required; *provided*, that the time prescribed in paragraph (a) of this clause for giving written notice may be extended by the Contracting Officer.
>
> (d) No request by the Contractor for an equitable adjustment to the contract for differing site conditions shall be allowed if made after final payment under this contract.

The FAR mandates that this clause appear in any "fixed-price construction contract or a fixed-price dismantling, demolition, or removal of improvements contract" that exceeds the simplified acquisition threshold.[26]

26. FAR 36.502. The "simplified acquisition threshold" is defined in the FAR as the anticipated price threshold below which the federal government may use special procedures for acquiring goods and services.

In *Foster Constr. C.A. & Williams Bros. Co. v. United States*,[27] the U.S. Court of Claims articulated and reinforced the nature and purpose of the federal Differing Site Conditions clause. There, the court described what it termed the "long-standing, deliberately adopted procurement policy" behind the Differing Site Conditions clause stating:

> The starting point of the policy expressed in the changed conditions clause is the great risk, for bidders on construction projects, of adverse subsurface conditions: "no one can ever know with certainty what will be found during subsurface operations." Whenever dependable information on the subsurface is unavailable, bidders will make their own borings or, more likely, include in their bids a contingency element to cover the risk. Either alternative inflates the costs to the Government. The Government therefore often makes such borings and provides them for the use of the bidders, as part of a contract containing the standard changed conditions clause.
>
> . . .
>
> The purpose of the changed conditions clause is thus to take at least some of the gamble on subsurface conditions out of bidding. Bidders need not weigh the cost and ease of making their own borings against the risk of encountering an adverse subsurface, and they need not consider how large a contingency should be added to the bid to cover the risk. They will have no windfalls and no disasters. The Government benefits from more accurate bidding, without inflation for risks which may not eventuate. It pays for difficult subsurface work only when it is encountered and was not indicated in the logs.[28]

B. Standard Form Provisions in Commercial Agreements

The federal regulatory policy articulated in *Foster* also influenced commercial contracting, as private commercial contracts often include provisions granting relief for unexpected subsurface conditions. Many domestic organizations, including the American Institute of Architects (AIA), the Associated General Contractors of America (AGC), the Engineers' Joint Contract Documents Committee (EJCDC), and the newly developed ConsensusDOCS,[29] have developed clauses that reflect the same concepts embodied in the federal Differing Site Conditions clause. Each of these individual provisions reflects the different focus of the particular authoring organization.

27. 435 F.2d 873 (Ct. Cl. 1970).

28. *Id.* at 887.

29. ConsensusDOCS, *2007 Standard Agreement and General Conditions Between Owner and Contractor (Where the Contract Price is a Lump Sum)*, *available at* http://www.consensusdocs.org.

For example, the AIA General Conditions of the Contract for Construction have long included a variety of changed-conditions clauses. The most current version of that contained in AIA Document A201–2007 includes a clause at Section 3.7.4 entitled "Concealed or Unknown Conditions." This provision generally mirrors the purpose and language of the federal Differing Site Conditions clause, but differs primarily with respect to the role of the third-party architect. AIA Document A201–2007 (General Conditions), § 3.7.4 reads:

> 3.7.4 Concealed or Unknown Conditions.
>
> If conditions are encountered at the site that are (1) subsurface or otherwise concealed physical conditions that differ materially from those indicated in the Contract Documents or (2) unknown physical conditions of an unusual nature, that differ materially from those ordinarily found to exist and generally recognized as inherent in construction activities of the character provided for in the Contract Documents, the Contractor shall promptly provide notice to the Owner and the Architect before conditions are disturbed and in no event later than 21 days after first observance of the conditions. The Architect will promptly investigate such conditions and, if the Architect determines that they differ materially and cause an increase or decrease in the Contractor's cost of, or time required for, performance of any part of the Work, will recommend an equitable adjustment in the Contract Sum or Contract Time, or both. If the Architect determines that the conditions at the site are not materially different from those indicated in the Contract Documents and that no change in the terms of the Contract is justified, the Architect shall promptly notify the Owner and Contractor in writing, stating the reasons. If either party disputes the Architect's determination or recommendation, that party may proceed as provided in Article 15.[30]

In addition to domestic alternatives, the concepts embodied in the federal Differing Site Conditions clause have percolated into the international construction contracting community. Among the alternatives available in that market, the International Federation of Consulting Engineers (FIDIC) has prepared standard construction clauses that appear in various international contracts, both

30. As discussed below in the context of the federal framework, the AIA clause carries with it the burden that the contractor perform a reasonable site inspection. *See* American Institute of Architects, AIA Document A201–2007, *General Conditions of the Contract for Construction* § 3.2.1 (2007) [hereinafter AIA Document A201–2007] ("Execution of the Contract by the Contractor is a representation that the Contractor has visited the site, become generally familiar with local conditions under which the Work is to be performed and correlated personal observations with requirements of the Contract Documents.").

public and private.[31] These standard forms, used by entities such as the World Bank for international development projects, include an Unforeseeable Physical Conditions clause that provides cost and schedule relief for "adverse physical conditions which [the contractor] considers to have been Unforeseeable."

17.04 ELEMENTS OF A FEDERAL DIFFERING SITE CONDITIONS CLAIM

The Differing Site Conditions clause allows for contract adjustments to address two kinds of changed conditions. Those two scenarios, referred to as Type I and Type II differing site conditions, differ primarily by the nature and existence of contract indications. Using the language of the clause itself, a Type I condition involves, "Subsurface or latent physical conditions at the site which differ materially from those indicated in this contract."[32] Thus, the Type I claim depends on the difference between the contract indications or representations and the actual conditions encountered. By contrast, a Type II claim involves "[u]nknown physical conditions at the site, of an unusual nature, which differ materially from those ordinarily encountered and generally recognized as inhering in work of the character provided for in the contract."[33] Therefore, a Type II, unlike a Type I, does not depend on the contract indications at all, but instead on the unique and unusual nature of the conditions encountered.

A. Type I—Materially Different than Indicated

A claim for a Type I differing site condition turns on the difference between the actual conditions encountered by the contractor and those represented in the contract documents. As a result, the nature of the original contract indications, the reasonableness of the contractor's interpretation of those indications, and the contractor's reliance on the indications become critical parts of a claim for a Type I differing site condition. These requirements have resulted in a formal checklist necessary to prevail on a Type I claim. That checklist is as follows:

> (i) The contract documents must have affirmatively indicated or represented the subsurface conditions which form the basis of the plaintiff's claim;

---

31. *See, e.g.*, The World Bank, *Standard Bidding Documents Procurement of Works & User's Guide* ¶ 4.12 (2006) *available at* http://go.worldbank.org/HQ4HH4UILO (last accessed Aug. 31, 2007).

32. FAR 52.236-2(a)(1).

33. FAR 52.236-2(a)(2).

(ii) The contractor must have acted as a reasonably prudent contractor in interpreting the contract documents;
(iii) The contractor must have reasonably relied on the indications of subsurface conditions in the contract;
(iv) The subsurface conditions actually encountered, within the contract site area, must have differed materially from the subsurface conditions indicated in the same contract area;
(v) The actual subsurface conditions encountered must have been reasonably unforeseeable; and
(vi) The contractor's claimed excess costs must be shown to be solely attributable to the materially different subsurface conditions within the contract site.[34]

1. Indications

The contract must include affirmative representations, or indications, of the subsurface conditions. The indications may appear anywhere in the contract documents or materials outside the four corners of the contract that are expressly referred to by the contract documents.[35] Indications can be express or implied. Express indications can include practically anything that overtly represents or fairly indicates the subsurface site conditions, such as boring logs, core samples, soil test, soil samples, geotechnical reports, photographs, and field notes. Indications can also be implied from the contract documents, or from omissions in the contract documents, and can include items such as specifications on means and methods, restrictions on blasting or types of equipment, or other indirect information suggestive of certain subsurface conditions. Moreover, any type of contract indication "need not be explicit or specific, but only enough to impress or lull a reasonable bidder" into drawing reasonable conclusions regarding the site conditions.[36] Thus, in the context of subsurface conditions, "all that is required is that there be enough of an indication on the face of the contract documents for a bidder reasonably not to

---

34. *See* Weeks Dredging & Contracting v. United States, 13 Cl. Ct. 193, 218 (1987), *aff'd* 861 F.2d 728 (Fed. Cir. 1988). For a five-element version of this test, *see* Ace Constructors, Inc. v. U.S., 70 Fed. Cl. 253, 268–69 (2006).

35. For example, federal construction contracts often include FAR 52.236-4, Physical Data (APR 1984). That provision will be used any time the government decides to provide a contractor with certain types of information including surveys, borings, test pits, tunnels, or other set of data on physical conditions at the site. When so included, the materials articulated under the Physical Data clause will automatically become part of the contract indications.

36. Stock & Grove, Inc. v. United States, 204 Ct. Cl. 103, 133–34 (1974).

expect subsurface or latent physical conditions at the site differing materially from those indicated in this contract."[37]

The indications, however, must relate to physical conditions at the site. "Physical conditions" within the meaning of the clause do not generally include weather or acts of God that occur during performance of the project.[38] Physical conditions also do not encompass manmade events such as mob attacks,[39] acts of war,[40] or labor strikes.[41] The term, however, can encompass manmade or artificial conditions at the site that were present at the time of contracting.

Despite including the language "at the site," the clause does not limit the conditions to the physical limits of the work site. "At the site" can refer to off-site locations required for performance of the work, such as borrow or fill areas, referenced in the contract. Several decisions of federal panels have allowed recovery for differing site conditions that occurred at contract-specified borrow pits that existed outside the work site parameters.[42]

2. *Reasonableness*

The law requires that the contractor be reasonable in two respects: first in interpreting the indications and second in relying on them. Reasonable reliance does not impose a duty on the bidder to conduct independent subsurface tests or to have expert knowledge of the subsurface conditions at the site. Often, the

37. *Foster Constr. C.A. & Williams Bros. Co.*, 435 F.2d at 875; *see also* Pac. Alaska Contractors, Inc. v. United States, 436 F.2d 461, 469 (Ct. Cl. 1971) (asserting that express representations as to the nature of conditions are not essential to establish entitlement to an equitable adjustment for changed conditions and that all that is necessary are "indications which induced reasonable reliance by the successful bidder that subsurface conditions would be more favorable than those encountered.")

38. *See, e.g.*, Turnkey Enterprises Inc. v. United States, 597 F.2d 750, 754 (Ct. Cl. 1979) ("Generally, the government, under the standard Differing Site Conditions Article, does not assume an obligation to compensate a contractor for additional costs or losses it incurs resulting solely from weather conditions, which neither party expected or could anticipate, and not from any act or fault of the government. Weather conditions generally are considered to be acts of God. Neither party is obligated to the other for additional costs or price increases resulting solely from acts of God"). Relief for unusually severe weather, even in the presence of contract provisions that articulate expected weather conditions, arises under different federal clauses. *See, e.g.*, USACE Regulation ER 415-1-15—TIME EXTENSIONS FOR UNUSUALLY SEVERE WEATHER (OCT 89); Fortec Constructors Inc., v. United States, 8 Cl. Ct. 490 (1985).

39. *See* Cross Constr. Co., ENG BCA No. 3676, 79-1 BCA ¶ 13,707.

40. *See* Keang Nam Enter. Ltd., ASBCA No. 13747, 69-1 BCA ¶ 7705.

41. *See* Olympus Corp. v. United States, 98 F.3d 1314 (1996).

42. *See, e.g.*, L. G. Everist, Inc. v. United States, 231 Ct. Cl. 1013 (1982); *Stock & Grove, Inc.*, 204 Ct. Cl. at 103.

contract information will include highly technical facts or analysis. A contractor need only interpret and apply that technical information as a prudent contractor.[43] That scale of prudence will depend in part on the sophistication of the work and the contractor. Thus, a contractor need not predict conditions it could not have reasonably anticipated from an inspection of the contract documents, the contractor's site inspection, or the contractor's general experience.[44] In short, a bidding contractor's reliance on contract indications is reasonable, provided that reliance comports with what a reasonably prudent contractor, experienced in work of the kind called for in the solicitation, would conclude in evaluating the contract indications.[45]

3. *Differ Materially*

A contractor may satisfy the requirement that the conditions differ materially in many ways. It may demonstrate that the volume of work increased over what the contractor reasonably expected; that the equipment required to perform the work differed from that expected; that the cost of the work increased over what was expected; or any variety of approaches, provided it demonstrate a material difference in the conditions. Proof of increased cost, however, is not a prerequisite for demonstrating a material difference between the expected conditions and the actual conditions.[46] Conversely, increased cost alone is not sufficient to establish a "material difference."

---

43. *See* Peter Kiewit & Sons Co., ENG BCA No. 4861, 4967 and 4973, 85-2 BCA ¶ 18,082 (subcontractor claim denied where no evidence presented that subcontractor actually relied on contract drawings). The reasonableness analysis does not replace an analysis of actual reliance, or reliance in fact. If the contractor was reasonable in its interpretation of the contract documents, but failed to rely on them in preparing its plans or estimates for the work, it can be precluded from recovery for differing site conditions. *See, e.g.*, Kaiser Indus. Corp. v. United States, 169 Ct. Cl. 310 (1965) (the test remains that of a "prudent contractor" even though "trained geologists might debate interminably about the matter").

44. *See Youngdale & Sons Constr. Co.*, 27 Fed. Cl. at 537–38; Servidone Constr. Corp. v. United States, 19 Cl. Ct. 346, 371 (1990); J. Lawson Jones Constr. Co., Inc., ENG BCA No. 4363, 86-1 BCA ¶ 18,719 at 94,172 (1986); Maverick Diversified, Inc., ASBCA Nos. 19838, 19955, and 20091, 76-2 BCA ¶ 12,104 at 58,150 (1976); Lee R. Smith, ASBCA No. 11135, 66-2 BCA ¶ 5857 at 27,181 (1966); *Kaiser Indus. Corp.*, 169 Ct. Cl. at 324.

45. *See* Neal & Co. Inc. v. United States, 36 Fed. Cl. 600, 620 (1996) ("a contractor is not to be charged with the technical knowledge or expertise of a geologist, and is therefore 'not required to discover *hidden* subsurface conditions or those beyond the limits of an inspection appropriate to the time available'") (emphasis in original); Kinetic Builders, Inc., ASBCA No. 32627, 88-2 BCA ¶ 20,657 (no expectation that contractor should become a botanist in order to determine link between tree species and excess water at site.).

46. *See, e.g.*, Roscoe-Ajax Constr. Co. Inc. & Knickerbocker Constr. Corp. v. United States, 458 F.2d 55 (Ct. Cl. 1972) (where changed work results in decreased costs, the actual changed site condition may still exist).

4. *Reasonably Unforeseeable*

The requirement that the conditions be reasonably unforeseeable works hand in hand with the contractor's obligation to interpret and rely on contract indications reasonably. In short, a site condition is reasonably unforeseeable if the condition is not readily discoverable by a reasonable investigation of the contract documents and the site.[47] Recent decisions have expanded the basic scope of a reasonable inquiry to documents that lie outside the four corners of the contract, but that are referenced in the documents provided.[48]

5. *Causation of Increased Costs*

This final element of a claim for Type I differing site conditions calls for the contractor to demonstrate damages and causation. With respect to the quantification portion of damages, the Differing Site Conditions clause relies on the same general cost principles as an equitable adjustment under the Changes clause. The method for determining the quantum of damages for differing site conditions is therefore best expressed as follows:

> [T]he amount of the equitable adjustment to which the contractor is entitled as a result of a changed condition is "the difference between what it cost it to do the work and what it would have cost if the unforeseen conditions had not been encountered." Accordingly, expenses which the contractor would have been obliged to incur anyway had no changed condition been encountered are not recoverable as part of the equitable adjustment. The amount of an equitable adjustment for a changed condition is thus not based upon the contractor's original estimate. Similarly, where the contractor could reasonably have handled the changed condition with a less expensive method, the amount of the equitable adjustment will be appropriately adjusted.[49]

When compared to other remedy-granting provisions, however, such as the Variation in Estimated Quantities (VEQ) clause, the Differing Site Conditions clause provides broader relief since it encompasses all cost types affected by the changed conditions.[50]

---

47. *See, e.g.*, *Weeks Dredging & Contr.*, 13 Cl. Ct. at 218 (contractor responsible for site inspection and contract documents); Randa/Madison Joint Venture III v. Dahlberg, 239 F.3d 1264, 1272 (Fed. Cir. 2001) (the contractor "bears the risk associated with not examining the information that the contract explicitly referred to and made available for inspection.").

48. *See, e.g.*, *Randa/Madison Joint Venture III*, 239 F.3d at 1272.

49. *See, e.g.*, *Roscoe-Ajax Constr. Co.*, 458 F.2d at 60 (*quoting* General Contracting Corp. v. United States, 96 Ct. Cl. 255, 277 (1942)).

50. *See infra* Section 17.04(D).

A contractor must causally link any increased costs or delay in project completion to the changed conditions, and not to other events.[51] The importance of demonstrating causation was highlighted in *Weeks Dredging & Contr. v. United States*,[52] where the Claims Court could not distinguish costs attributable to differing site conditions from other contractor-caused problems. The court demanded that the contractor show a "breakdown of how many of the alleged uncompensated days are directly attributable to *each* of the differing site conditions, *independent* of the concurrent causes identified by the defendant. . . ."[53] The court opined that absent an ability to segregate increased costs caused by the differing site condition from those caused by the contractor, the contractor could not recover at all for the differing site conditions.[54] To address this problem on a practical level, contractors may use a variety of accounting methods such as discrete cost codes or special costs accounts, in an effort to isolate and to capture increased costs attributable to a differing site condition.

B. Type II—Conditions of an Unusual Nature

In contrast to a Type I condition, a Type II Differing Site Condition does not rely on contract indications, but instead arises where no contract indications have been provided.[55] To demonstrate a Type II Differing Site Condition, a contractor must show a combination of two of the following three elements:

1. The physical condition at the site was unknown; or
2. The condition was unusual and could not be reasonably anticipated by the contractor from its study of the contract documents, inspection of the site, and general experience, if any, in the contract area; and
3. The condition encountered was materially different from those ordinarily encountered and generally recognized as inhering in the work of this character.[56]

---

51. *Weeks Dredging & Contr. Inc.*, 13 Cl. Ct. at 241 (court chided contractor for being unable to "identify for the court a method whereby we can trace the impact of any *one* alleged differing site condition to the incurrence of any one identifiable excess cost" and asking the court to engage in "rank speculation" as to the causal connection of claimed costs).

52. *Id.* at 193.

53. *Id.* at 241.

54. *See id.* at 244.

55. *See generally*, John Cibinic, Jr., Ralph C. Nash, Jr. & James F. Nagle, Administration of Government Contracts 508–11 (4th ed. 2006) [hereinafter Government Contracts].

56. *See Youngdale & Sons Constr. Co.*, 27 Fed. Cl. at 527–28 (*quoting Servidone Constr. Corp.*, 19 Cl. Ct. at 360, *aff'd*, 931 F.2d 860 (Fed. Cir. 1991)).

This test creates a high hurdle for contractors seeking to recover for a Type II condition, and one with fundamentally different requirements from the Type I claim.[57] These situations are highly fact specific and allow presentation of evidence beyond that pertinent to a Type I claim. For example, while a Type I claim focuses exclusively on facts included in or referenced by the contract documents, proof of a Type II claim can extend to information discoverable from local inhabitants or other sources of information on the area.[58]

17.05 LIMITATIONS ON CLAIMS FOR DIFFERING SITE CONDITIONS

A. Duty to Investigate

In both the federal and commercial context, a contractor bears the duty to conduct a reasonable investigation of the site. While the Differing Site Conditions clause and its like have mitigated the contractor's duty to conduct extensive independent investigations of the site prior to bidding, that mitigation only exists within certain parameters. As set forth in *Foster*, the Differing Site Conditions clause creates a compromise whereby the government does not require the contractor to conduct extensive independent subsurface investigations, but in return, calls for the contractor to make reasonable use of the government-provided data. Such reasonable use of available data includes a realistic visual inspection of the work site. Thus, in the context of both Type I and Type II claims, the contractor labors under an affirmative obligation to conduct a reasonable site inspection.

Courts and boards commonly seek to balance the requirements of a site investigation against the overarching policy behind the Differing Site Conditions clause. For example, the Court of Claims in *Foster* weighed that balance in favor of the contractor, stating:

> Faithful execution of the policy requires that the promise in the changed conditions clause not be frustrated by an expansive concept of the duty of bidders to investigate the site. That duty, if not carefully limited, could force bidders to rely on their own investigations, lessen their reliance on logs in the contract and reintroduce the practice sought to be eradicated—the computation of bids

57. *See* Guy F. Atkinson Co., ENG BCA No. 4693, 87-3 BCA ¶ 19,971; Soletanche Rodio Nicholson (JV), ENG BCA Nos. 5796 and 5891, 94-1 BCA ¶ 26,472 ("When asserting a type II, differing site condition claim, the Appellant has a relatively heavy burden").

58. *See* Daymar, Inc., DOTBCA No. 77-13, 78-1 BCA ¶ 12,903 (inquiries to local excavators would have revealed general knowledge of subsurface rock conditions).

> on the basis of the bidders' own investigations, with contingency elements often substituting for investigation.[59]

The federal government and the commercial contracting industry have developed standard contract clauses to adjust in the owner's favor the struggle between providing accurate information under the Differing Site Conditions clause against the contractor's obligations to make reasonable use of that information. One means to accomplish that objective lies in site inspection provisions such as FAR 52.236-3, Site Investigation and Conditions Affecting the Work.[60] This provision requires the contractor to confirm it conducted an investigation of the site while also providing a government disclaimer of responsibility for conclusions that the contractor draws from its review of contract data.[61]

In the federal context, the level of investigation required to prevail on a Type I claim differs from that required for a Type II claim. With respect to the Type I, the law charges a contractor with a "relatively simple site viewing" and the knowledge gained from that exercise.[62] By contrast, a contractor seeking recovery for a Type II will have to satisfy a higher standard in order to demonstrate that the unknown condition was "materially different from those ordinarily encountered and generally recognized as inhering in the work of this character."

Failure to conduct an adequate investigation, without more, will not preclude recovery for a Differing Site Condition.[63] Where a contractor fails to conduct an adequate inspection, it bears the risk of all conditions that could have been discovered had the inspection been properly performed.[64] Conversely, if a condition would not have been readily apparent from a site inspection, the failure to perform one will not bar recovery. Further, where a site inspection reveals information omitted from the contract materials or in conflict with those materials, a contractor labors under a duty to inquire about the discrepancy.[65]

---

59. *Foster Constr. C.A. & Williams Bros. Co.*, 435 F.2d at 887.

60. For a similar provision in the commercial context, see AIA Document A201–2007, *supra* note 30, § 3.2.1.

61. *But see* North Slope Tech. Ltd., Inc. v. United States, 14 Cl. Ct. 242, 253 (1988) (*quoting Foster Constr. C.A. & Williams Bros. Co.*, 435 F.2d. at 888) ("[t]he duty to make a pre-bid inspection of the site does not negate the differing site conditions clause 'by putting the contractor at peril to discover hidden subsurface conditions or those beyond the limits of an inspection appropriate to the time available' ").

62. *See generally* Government Contracts, *supra* note 55, at 522; Richard J. Bednar et al., Construction Contracting 600–02 (1991).

63. *See* Vann v. United States, 190 Ct. Cl. 546 (1970).

64. *See, e.g.*, *Weeks Dredging & Contracting*, 13 Cl. Ct. at 193, *aff'd* 861 F.2d 728 (Fed. Cir. 1988).

65. *See generally*, *Fortec Constructors*, 760 F.2d at 1291 (existence of a patent ambiguity in the contract raises the duty of inquiry); Jefferson Constr. Co. v. United States, 176 Ct. Cl. 1363 (1966) (contractor has duty to inquire as to obvious discrepancies in plans).

B. Disclaimers/Exculpatory Language

In addition to site inspection clauses, the government and commercial owners have attempted to insert other exculpatory clauses designed to mitigate the risk allocation scheme embodied in the Differing Site Conditions clause.[66] Those exculpatory provisions range from disclaimers as to the accuracy of the contract data provisions that exclude data accompanying the contract from the formal contract documents and other means of avoiding responsibility for the information provided. In public contracting, courts have met these provisions with healthy skepticism and have placed limits on the scope of exculpatory provisions. In one decision, the former Corps of Engineers Board of Contract Appeals stated:

> A contractor is required to do only what he might reasonably have anticipated from the contract plans and specifications. Where the contract contains a differing site conditions clause, no matter what other clauses the owner puts in the contract, they cannot make him an insurer to perform whatever work he may encounter regardless of its nature. A contract to do a useful job for another is not to be turned into a gambling transaction.[67]

Although generally disfavored, courts may enforce exculpatory provisions that seek not to nullify information provided by the government, but that confirm an absence of information or highlight information specifically not included in the contract documents.[68] These exculpatory provisions, however, will be narrowly construed and must be explicit. In commercial contracting, courts may appear more readily disposed toward exculpatory provisions, giving them greater weight when negotiated between two commercial parties.

C. Notice

Most public and commercial contracts that allow relief for changed or unknown subsurface conditions include a notice provision. In the federal context, the Differing Site Conditions clause requires that "[t]he Contractor shall promptly, and before the conditions are disturbed, give a written notice to the Contracting

---

66. Government Contracts, *supra* note 55, at 530–32; 4 Philip L. Bruner & Patrick J. O'Connor, Bruner & O'Connor on Construction Law §14:56 (West Group 2002).

67. S. & M/Traylor Bros. (JV), ENG BCA Nos. 3878, 3904, 3943, 82-1 BCA ¶ 15,484.

68. *See* P. J. Maffei Bldg. Wrecking Corp. v. United States, 732 F.2d 913 (Fed. Cir. 1984) (where contract drawings were provided for "information only," the absence of data was not considered an indication).

officer. . . ." Likewise, the language of the AIA Concealed or Unknown Conditions provision requires that the contractor provide notice "promptly before conditions are disturbed and in no event later than 21 days after first observance of the conditions." Under both provisions, notice operates as a trigger, calling for the government, the owner, or the architect to undertake an investigation of the conditions and to evaluate how the contractor should proceed.

The failure to provide notice and the proper form of notice are frequent sources of debate between contractors and the government or commercial owners. This debate ordinarily revolves around prejudice to the government's or owner's ability to investigate the conditions and make informed decisions about how the contractor should proceed. In the federal context, a contractor's failure to follow the formal notice requirements does not, of itself, bar a contractor's claim.[69] Where a contractor provides late or otherwise deficient notice of such conditions, a court normally will deny that claim only where the deficient notice prejudices an owner's ability to investigate or mitigate the situation.[70]

Under the AIA provisions, the debate is somewhat different. The AIA Concealed or Unknown Conditions clause requires only that the contractor "promptly provide notice to the Owner and the Architect before conditions are disturbed." Unlike the FAR provision, the AIA provision does not expressly call for the contractor's notice to be written. The courts of Minnesota addressed this issue in *Robert W. Carlstrom Co., Inc. v. German Evangelical Lutheran*,[71] when the contractor provided oral notice to the owner of changed conditions impacting its roofing work. The court held that the plain language of the AIA clause did not include a writing requirement, and opined that the notice requirement in the AIA Concealed or Unknown Conditions clause was "more of an investigation tool rather than a claims provision."[72]

---

69. *See, e.g.*, Hoel-Steffen Constr. Co. v. United States, 197 Ct. Cl. 561, 573 (1972) (a "severe and narrow application of the notice requirements . . . would be out of tune with the language and purpose of the notice provisions, as well as with the court's wholesome concern that notice provisions in contract adjustment clauses not be applied too technically and illiberally where the government is quite aware of the operative facts. . . ."); Santa Fe, Inc., VABCA No. 1983, 84-3 BCA ¶ 17,538 (lack of notice merely places "upon the contractor a heavier burden in proving its case.").

70. *See, e.g.*, Miller Elevator Co. v. United States, 30 Fed. Cl. 662 (1994).

71. 662 N.W.2d 168 (Minn. Ct. App. 2003).

72. *Id.* at 173 (*quoting* Justin Sweet, Sweet on Contracts 398 (1997)).

D. Relationship to the Variation in Estimated Quantities Clauses

In federal contracts, as well as some private and state contracts, a contractor may seek recovery for increases in costs resulting from an increase in the estimated quantity of a particular item. Under the framework set forth in FAR 52.211-18, Variation in Estimated Quantity (VEQ), at bid time, a contractor submits a unit price for a certain aspect of the work based on the government's representation that a specific amount of that work will occur. Should the contractor have to perform more or less of that work within certain well-defined limits, the contractor may be entitled to additional compensation, or the government may be entitled to a credit.

To demonstrate entitlement to an equitable adjustment under the VEQ, a contractor must show: (1) that a unit-priced item varied more than 15 percent above or below an estimated quantity; (2) that the contractor incurred increased costs; and (3) those increased costs were "due solely to the variation above 115 percent or below 85 percent of the estimated quantity."[73] In determining the value of an equitable adjustment under the VEQ clause, the parties must look to the change in the contractor's costs occasioned "solely" by the quantity variation.[74]

Prior to the introduction of the VEQ clause, contractors often sought relief for material variations in estimated quantities as a differing site condition.[75] In *Perini Corp. v. United States*,[76] the Court of Claims articulated certain limitations on the circumstances under which a contractor could seek an equitable adjustment for quantity variations under the Differing Site Conditions clause as follows:

> The cases that have applied the Changed Conditions article to substantial variations from the quantity estimated in the contract have done so on the

73. FAR 52.211-18.

74. *See, e.g.*, Foley Co. v. United States, 11 F.3d 1032 (Fed. Cir. 1993) (any adjustment under the VEQ "will be confined in amount to such cost differentials as are directly attributable to a volume deviation greater than 15 percent from stated contract quantities"); Victory Constr. Co. & Paul Krummel (JV) v. United States, 206 Ct. Cl. 274 (Ct. Cl. 1975) (VEQ is not a basis to re-bid unit costs).

75. Such a practice continues today where contractors may effectively plead in the alternative. *See, e.g.*, Gulf Constr. Group, Inc., ENG BCA No. 5958, 93-3 BCA ¶ 26,174 (After overrun in timber quantities, contractor requested equitable adjustment under both provisions, recovering only under the Differing Site Conditions clause).

76. 180 Ct. Cl. 768 (1967).

> basis that such a material deviation was not reasonably foreseeable. They deal with special situations where (1) the contractor could not have verified the estimated quantities from the contract documents and his investigation of the site; (2) he had no opportunity to investigate the conditions at the site or for other reasons had the right to rely on the Government's estimate, or (3) both parties labored under a mutual mistake as to the accuracy of the Government's estimates.[77]

After the introduction of the VEQ clause, courts and boards applied that clause with logic very similar to that found in decisions relating to the Differing Site Conditions clause, focusing on the reasonableness of the contractor's plan.[78]

Following the introduction of the VEQ clause, courts and boards carved out different spheres of influence for each provision based largely on the same parameters set forth in *Perini*. For example, in *Morrison-Knudsen Co., Inc. v. United States*,[79] the soil quality of government-designated borrow pits resulted in an inability to meet the contract quantity requirements for borrow. As a result, the contracting officer directed the contractor to modify its excavation and hauling methods and to use other borrow pits, actions that resulted in significant changes in the contractor's equipment and costs. The government, in reliance on the quantity variation provision, argued that the changes in borrow material did not result in the degree of quantity change that would allow the contractor any recovery under the clause. The court rejected the government's argument, holding that the limitations imposed by the variations in quantities clause did not provide an exclusive avenue for the contractor. To do otherwise, the court opined:

> [W]ould leave a contractor without administrative recourse against any change in the contract drawings or specifications due to unforeseen field conditions which did not result in the prescribed overrun or underrun in quantities, even though the entire contemplated basis for performance of the contract may have been altered at a substantially increased expense to the contractor.[80]

Moreover, courts have determined that the Differing Site Conditions clause and the Changes clause supersede the VEQ clause where the two clauses directly

77. *Id.* at 780–81.

78. *See, e.g.*, Womack v. United States, 182 Ct. Cl. 399, 412–13 (1968) ("Assuming that the bidder acts reasonably, he is entitled to rely on Government estimates as representing honest and informed conclusions").

79. 184 Ct. Cl. 661 (1968).

80. *Id.* at 689.

compete.[81] In these situations, courts have concluded that even though the two provisions may address similar or related topics, each has its own sphere of influence. The VEQ clause addresses simply the narrow universe of cost changes due solely to a change in quantity, whereas the provisions in the broader Changes and Differing Site Conditions clauses cover any circumstance in which a change in the nature of the work results in more complicated or costly work.

17.06 REPRESENTATIVE TREATMENT OF DIFFERING SITE CONDITIONS IN STATE COURTS

A. New Jersey

The evolution of relief for differing site conditions in New Jersey tracked a process very similar to that experienced on the federal level, forcing the courts to wrestle with construction claims until the introduction of a standard contract clause. During the 1980s, the New Jersey courts tackled the issue of differing site conditions in four major cases. In each, the contract between the state and the contractor lacked a clause specifically addressing changed conditions, but each contained certain representations about the site conditions. In the first three decisions, the New Jersey courts adhered to the traditional common law rules regarding warranty, misrepresentation, and site inspection. In *Sasso Contracting Co. Inc. v. State of New Jersey,*[82] and *Golomore Associates v. New Jersey State Highway Auth.*,[83] the New Jersey Appellate Division held essentially that, "if the contract supplied the results of tests of soil or other conditions but did not purport to describe actual conditions, or if the contract included disclaimers of responsibility for descriptions, a contractor would not be justified in relying on the results without independent investigation."[84] In the third member of

81. *See, e.g.*, ThermoCor, Inc. v. United States, 35 Fed. Cl. 480, 492 (1992) ("There is no indication that the VEQ clause was to override the Changes clause nor that it was to be the exclusive means for obtaining a changes adjustment."); United Contractors v. United States, 177 Ct. Cl. 151 (1966) ("clauses of this type [VEQ] do not control when the cost of doing the extra work greatly differs from the stated unit-price because of factors not foreseen by either party. In that event, the Changed Conditions clause comes into play and overrides the Special Condition."). *But see* CEMS, Inc. v. United States, 59 Fed. Cl. 168 (2003) (where contractor failed to provide evidence of changed conditions, may be limited to recovery under the VEQ clause only).

82. 173 N.J. Super. 486 (App. Div. 1980), *cert. den.* 85 N.J. 101 (1980).

83. 173 N.J. Super. 55 (App. Div. 1980).

84. *Id.* at 58.

the trio, *Ell-Dorer Contracting Co. v. State of New Jersey,* [85] the Appellate Division reiterated that premise and reinforced a "plain language" approach to risk allocation. In doing so, the court enforced an exculpatory contract provision in which the contractor agreed to submit "no claims against the State, if in carrying out the Project he finds that the actual conditions encountered do not conform to those indicated by said borings, test excavations, and other subsurface investigations."[86]

Shortly after the *Ell-Dorer* decision, however, the New Jersey Supreme Court modified the state's approach in *P.T. & L. Constr. Co., Inc. v. State of New Jersey.*[87] In *P.T. & L.*, the state contracted with PT&L to build a highway, which required the excavation and hauling of soil across the work site, without including a Differing Site Conditions clause in the agreement. After commencement of the work, the contractor claimed that the actual soil conditions differed from those represented in the contract documents and may have actually been known to the state during bidding. The state denied liability for the additional costs associated with this condition, relying on contract terms including exculpatory provisions that called for the contractor to perform its own site inspection.

In its decision, the New Jersey Supreme Court relied on the policy statement set forth by the Federal Court of Claims in *Foster* as its touchstone and balanced that policy against a bidder's obligation to "make his own investigations of subsurface conditions prior to submitting his Proposal."[88] The court found that the New Jersey form of contracting (i.e., agreements without a Differing Site Conditions clause) held bidders to a higher standard of care than under the federal framework. If, however, "facts were known or reasonably accessible only to [the government] and were not known to or reasonably discoverable by the [contractor]," the court would reduce the burden on the contractor.[89] The court tempered its position by holding that any misrepresentations must be overt, and that "[i]nferential conclusions from contract documents, however, shall not be considered a false factual representation in the face of sufficiently clear and unambiguous disclaimers of liability by the State."[90] Thus, while the courts of New Jersey paid homage to the principles embodied in the federal Differing Site Conditions clause, the state ultimately relied on a misrepresenta-

85. 197 N.J. Super. 175, 182 (App. Div. 1984).
86. *Id.*
87. 531 A.2d 1330 (N.J. 1987).
88. *Id.* at 1339.
89. *Id.* at 1340.
90. *Id.* at 1342.

tion analysis to provide recovery to a contractor. In response to the decision in *PT&L*, several New Jersey state agencies drafted and implemented clauses that mirror the federal Differing Site Conditions clause.[91]

B. Virginia

Public contracts in Virginia also include standard site investigation and Differing Site Conditions clauses akin to those found in the FAR. These clauses address both subsurface latent conditions as well as above-ground building conditions. Similar to the federal provisions, these clauses require the contractor to conduct a site inspection and provide protection to the contractor for unanticipated differing site conditions. Virginia, however, has introduced a new concept in its standard Design-Build program that utilizes a unique "validation period" that on its face allows the contractor to conduct its own investigation and include the results in the contract documents.

In *Asphalt Roads & Materials Co. v. Virginia Dept. of Transp.*,[92] the Supreme Court of Virginia applied federal case law interpreting FAR 52.236-2 to interpret the Virginia Differing Site Conditions clause. In *Asphalt Roads*, the court addressed the interconnection between differing site conditions and variations in estimated quantities under a highway contract. In performing the work, the contractor discovered a substantially greater amount of unsuitable fill soil than indicated in the contract drawings. When responding to the contractor's request for additional compensation to bring new fill to the site, the Virginia Department of Transportation (VDOT) contended that the differing site conditions clause in the contract only applied to differences in character, not deviations from estimated quantities in the specifications. In awarding relief to the contractor, and consistent with federal case law, the *Asphalt Roads* court found that the purpose of a differing site condition clause ensures the lowest competent bid for basic services such that the owner only pays for differing conditions, including excess quantities as they occur.[93] The court further held that when a contract contains a Differing Site Conditions clause, contractors may rely upon the accuracy of the specifications and need not increase their bids to account for unforeseen conditions or otherwise increase costs by undertaking their own borings and inspections of the project while bidding for a contract.

---

91. *See, e.g.*, State of New Jersey Dept. of Transp. Standard Road and Bridge Specifications (2001) § 108.09 "Unusual Conditions."

92. 512 S.E.2d 804 (Va. 1999).

93. *Id. See also* County Redevelopment and Housing Auth. v. W. M. Schlosser Co., Inc., 41 Va. Cir. 118, 126 (1996) (reading contract as a whole, risk-shifting provision could not negate the express provision to compensate the contractor for differing site conditions).

Diverging from the federal model, however, Virginia public contracts for highway work can include unique requirements for site inspections that differ significantly from the federal clause or the AIA/commercial model. Section 4.2, Inspection of Site Conditions, of the standard 2007 Virginia Design-Build contract terms, calls for a "Validation Period," in which contractors may supplement the request for proposal (RFP) indications with their own site investigation work. More specifically, the provision includes the following pertinent language:

> 4.2.2 Design-Builder will, after the Date of Commencement, undertake such testing, inspections and investigations as may be necessary to perform its obligations under the Contract Documents, including but not limited to additional geotechnical evaluations or Hazardous Materials studies. If Design-Builder intends to conduct additional geotechnical evaluations to supplement or corroborate the information contained in the RFP Documents, it shall do so during the Scope Validation Period. Any Scope Issues that arise from such evaluations shall be treated in the manner set forth in Section 2.2.3 above. . . . [94]

Section 2.2.3 sets forth some unusual language that calls for the contractor to

1. "[Acknowledge] that the Scope Validation Period is a reasonable time to enable Design-Builder to identify Scope Issues that will materially impact Design-Builder's price or time to perform the Work";
2. "[A]ssume and accept all risks, costs, and responsibilities of any Scope Issue arising from or relating to the Contract Documents, including but not limited to conflicts within or between the RFP Documents and Proposal";
3. Warrant that the contract documents are "sufficient to enable Design-Builder to complete the design and construction of the Project without any increase in the Contract Price or extension to the Contract Time(s)"; and
4. "[E]xpressly [disclaim] any responsibility for, and Design-Builder expressly waives its right to seek any increase in the Contract Price or extension to the Contract Time(s) for, any Scope Issue associated with any of the Contract Documents."[95]

To date, the courts of Virginia have not addressed the implications of this unique language, nor its impact on the relief allowed contractors under the Differing Site Conditions clause.

---

94. *See* Section 4.2.2 of Commonwealth of Virginia Department of Transportation, *General Conditions of Contract Between Department and Design-Builder Template* (July 20, 2007) [hereinafter "Standard VDOT D/B General Conditions"], *available at* http://virginiadot.org/business/resources/DB_Template_Part4_General_Conditions_070813.pdf.

95. *See id.* at Section 2.2.3.

C. Illinois

In 1999, the Illinois legislature passed the Illinois Public Construction Contract Act (IPCCA), requiring, in part, that all public contracts for improvements in excess of $75,000 contain a Changed Conditions clause.[96] Through this act, the legislature of Illinois ensured that public contracts in Illinois contain a provision very similar to the federal Differing Site Conditions clause.

In addition to the substantive issues of differing site conditions, the courts of Illinois have wrestled with the procedural issues triggered by the Differing Site Conditions clause. For example, in *Kenny Constr. Co. of Illinois v. Metro. Sanitary Dist. of Greater Chicago*,[97] a pre-IPCCA decision, the court examined the public owner's duties under the clause and the implications of the owner's behavior after receiving notice of a differing site condition from the contractor. During construction of a sewer tunnel in Chicago, the contractor unexpectedly encountered unfavorable conditions, stopped work, and provided written notification to the district engineer of the problem along with a cost estimate for proceeding with the changed conditions. The district engineer failed to respond to the notice and threatened to default Kenny for failing to perform the work. The contractor then met with the chief engineers for the Sanitary District and received oral instruction from them to proceed with the extra work, but gave no written approval of a change to the contract.[98] The engineer represented that "[i]t would be much easier and more realistic . . . to evaluate the work after it had been completed, and check as to how much more the contractor had expended, than it would be to accept from the contractor before the work began."[99] Later, when the Sanitary District refused to compensate the contractor for its increased cost of complying with the oral direction, the contractor filed suit.[100]

The *Kenny Constr.* court found that, under the circumstances, the engineer's statements constituted an undertaking to pay for the extra work, provided the subsurface conditions actually had changed.[101] The court also concluded that, notwithstanding the contract's procedural requirements, since the contractor relied to its detriment on the representations of the chief engineer, the district was estopped from avoiding liability on the ground that the engineer did not approve the extra work in writing. The court further indicated that the Sanitary District's actions amounted to ratification, having allowed

96. 30 Ill. Comp. Stat. 557/10 (1999).
97. 288 N.E.2d 1 (Ill. 1971).
98. *Id.* at 5.
99. *Id.*
100. *Id.* at 4.
101. *Id.* at 7.

the contractor to perform the work and having accepted the benefits of that work.[102] Thus, relying on various equitable theories, the court in *Kenny Constr.* interpreted the procedural requirements of the Differing Site Conditions clause broadly in order to grant the contractor relief.

Later, in *Fru-Con Corp. and Granite Constr., JV v. Illinois,*[103] the Court of Claims of Illinois reviewed the nature of indications under the Changed Conditions clause found in the Illinois Department of Transportation (IDOT) Standard Specifications for Roads and Bridge Construction. There, a joint venture entered into a contract with IDOT to construct the substructure for a three-lane bridge and a number of piers. After construction began, the joint venture encountered numerous difficulties with river traffic and riverbed erosion. The contractor sought compensation for the work associated with overcoming these difficulties based on various theories, including misrepresentation and changed conditions.

With respect to problems caused by excess river traffic, the *Fru-Con Corp.* court found that the contract documents did not indicate or otherwise warrant that river traffic would be restricted to the navigational clearance during the bridge construction.[104] The court further found that the joint venture's site visit prior to construction should have disclosed the river traffic condition. When turning to the riverbed erosion problems, however, the court emphasized that IDOT knew or should have known of the potential for riverbed erosion.[105] The court focused on the omission of information and found that IDOT's failure to identify the potential erosion hazard qualified as an "indication" that no such condition would be encountered during construction.[106] Thus, the court held that "while it is true the contractor takes certain risks in its bid process, the [owner/government] takes certain risk when it fails to disclose information which affects the execution of the contract, whether that disclosure is intentional or not."[107]

D. California

The Court of Appeals of California first addressed the changed condition issue in *Gogo v. L.A. etc. Flood Control Dist.*,[108] where the court allowed contractor to recover on the theory that:

102. *Id.* at 8–9.
103. 50 Ill. Ct. Cl. 50 (1996).
104. *Id.* at 77.
105. *Id.* at 91.
106. *Id.*
107. *Id.* at 58.
108. 114 P.2d 65 (Cal. Ct. App. 1941).

> [W]here the plans and specifications induce a public contractor reasonably to believe that certain indicated conditions actually exist and may be relied upon in submitting a bid, he is entitled to recover the value of such extra work as was necessitated by the conditions being other than as represented.[109]

Subsequent California decisions expanded on *Gogo*, and held that there must be an affirmative misrepresentation or concealment of material facts in the drawings and specifications if the contractor is to recover. The courts indicated that without the affirmative misrepresentation requirement, contractors would be rewarded for inaccurate underbids and encouraged not to act diligently in examining specifications and drawings. For example, in *Wunderlich v. California*,[110] the Supreme Court of California denied recovery to a contractor because there was no positive misrepresentation or concealment of fact. In reaching its decision, the court commented:

> It is obvious that a governmental agency should not be put in the position of encouraging careless bidding by contractors who might anticipate that should conditions differ from optimistic expectations reflected in the bidding, the government will bear the costs of the bidder's error. . . . When there is no misrepresentation of factual matters within the state's knowledge or withholding of material information, and when both parties have equal access to information as to the nature of the tests which resulted in the state's findings, the contractor may not claim in the face of a pertinent disclaimer that the presentation of the information, or a reasonable summary thereof, amounts to a warranty of the conditions that will actually be found.[111]

Thereafter, in 1989, the California legislature enacted California Public Contract Code § 7104, which requires public works construction contracts involving excavations in the state to include a Changed Condition clause. The statute requires such contracts to include a contract provision that generally follows the language of the federal Differing Site Conditions clause. In *Condon-Johnson & Assoc., Inc. v. Sacramento Mun. Util. Dist.*,[112] the Court of Appeal of California revisited *Wunderlich* and the effect of disclaimers regarding subsurface conditions in light of Section 7104. The court held that the legislature intended the language found in Section 7104 to assist the contractor in preparing its bid and allowed a contractor to draw reasonable inferences about the actual subsurface conditions from the "indications" in the contract.[113] The court further held that the statute allocated the risk between public entity and contractor and reflected

---

109. *Id.* at 69 (citations omitted).
110. 423 P.2d 545 (Cal. 1967).
111. *Id.* at 550.
112. 149 Cal. App. 4th 1384 (2007).
113. *Id.* at 857.

the policy that "it is to a public entities' [*sic*] advantage to provide information upon which the bidder can rely in order to obtain the lowest qualified bid."[114] In that environment, the California Court of Appeal concluded that disclaimers as to the validity of contract indications violated the remedial purpose of Section 7104.[115]

17.07 CONCLUSION

Contractors and owners have addressed the fundamental problem of how to treat changes in subsurface conditions equitably, largely through express contract mechanisms. In the absence of those mechanisms, however, courts have looked to common law theories as well as theories created in interpreting the express clauses themselves to determine which party should bear the risk of these events. Administering these mechanisms has generated a body of law that assists in managing a contractor's and an owner's risk of subsurface conditions.

114. *Id.* at 857–58.
115. *Id.* at 859.

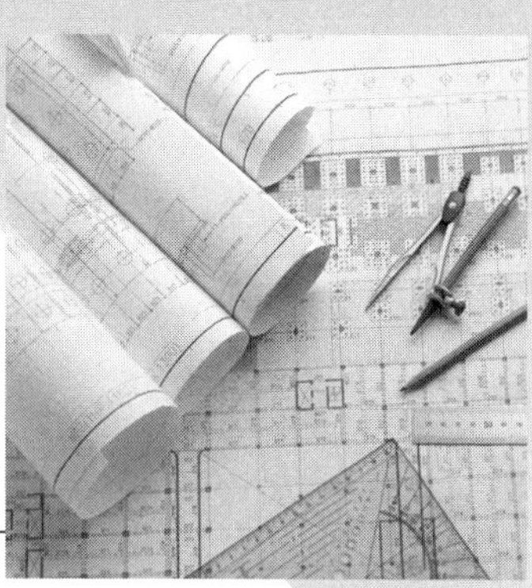

QUESTIONS

1. Your client is going to bid on a contract with the federal Differing Site Conditions clause, FAR 52.236-2. He tells you he is somewhat familiar with the property, having played on the site when he was a young boy. He remembers that there is rock around the site, even though the geotechnical report accompanying the contract does not indicate rock is present. He asks if he should go to the site, extensively walk the site, and make a written record of what he observes. He also asks what he should do if he sees rock outcroppings at the site. Should he bid the work based on his observations or the soil conditions reported in the geotechnical report? If the bidding documents provide for questions to the owner, should the contractor submit questions outlining past experience?
2. A Type I Differing Site Conditions is one where "subsurface or latent physical conditions at the site differ materially from those indicated in this contract." Does this mean that there has to be a report of some nature stating the expected subsurface or latent physical conditions, or can the conditions "indicated in this contract" be determined by reading the contract terms as a whole?
3. You have a contract with a Variation in Estimated Quantities clause that allows an adjustment in unit price if the final actual quantity is above or below 115 or 85 percent of the estimated quantity. If rock excavation significantly exceeds the estimated quantity, can you submit a claim for a differing site condition? If so, when should you give written notice to the contracting officer of your claim? How should you determine the amount of your claim? Does it make a difference if you claim a variation is estimated quantities or a differing site condition?
4. Your contract has a Differing Site Conditions clause and an extensive geotechnical report detailing the conditions at the site. The report does not mention the existence of granite, but indicates hard rock may be encountered. A clause in the contract specifically excludes a claim for encountering granite. You encounter extensive amounts of granite at the site, which requires you to change equipment and expend a significant amount of time excavating the material at great cost. Do you have a claim? What theories might you use to assert a claim?
5. You are going to bid a project that has a Differing Site Conditions clause and a geotechnical report. If you do not have the training to understand it, should you find someone in your company who has the training to review it, or should you hire an outside consultant to review the report and provide an analysis for you? How do you demonstrate reasonable reliance on the materials provided? What is the standard for a reasonable investigation? Discuss the difference in the standard for a Type I claim as compared to a Type II differing site conditions claim.

6. Shortly after you are awarded a construction contract, a hurricane causes extensive flooding on the site. Can this be claimed as a differing site condition? What arguments might you make to support your claim?
7. What types of evidence would you use to prove a differing site condition? Outline the methods you would use to document the claim or defend against a claim.
8. Is the failure to investigate the site at all an absolute defense to a differing site conditions claim? What arguments could be advanced on either side of this issue? Does it depend on the nature of the differing site conditions claimed?
9. What should be included in any notice given of a differing site conditions claim? Does the contracting officer's knowledge of the condition waive the need for notice? What arguments could the contracting officer assert to demonstrate prejudice, or other defenses, based on lack of notice?

CHAPTER

18

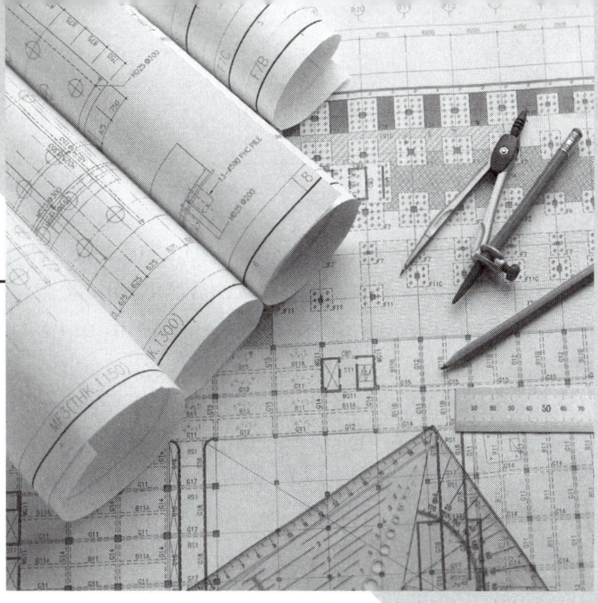

Termination and Default

STEVEN M. SIEGFRIED

18.01 INTRODUCTION

Termination of a construction contract is generally an unpleasant and sometimes unavoidable prospect for all parties involved and poses a unique set of problems. Primarily, it is not always clear whether there is a sufficient legal basis for terminating the contract. Disputes may also arise when a particular act or omission by one party constitutes a breach of a construction contract or when the terminating party is free from fault. When a party terminates a construction contract, the burden of proving that the termination was justified will lie with the terminating party.[1] Upon proof that the termination was warranted,

The author thanks Nicholas D. Siegfried, an associate in the firm of Siegfried, Rivera, Lerner, De La Torre & Sobel, PA, in Miami, for his valuable assistance in the preparation of this chapter.

1. CJP Contractors, Inc. v. U.S., 45 Fed. Cl. 343 (1999).

the burden then shifts to the terminated party to show that its default was excused.[2] The ramifications of wrongful termination also present another large concern for the parties to a construction contract. The potential liability for damages resulting from a wrongful termination is often great, as a wrongful termination will amount to a material breach of contract.

In the simplest terms, termination becomes an option after one party's performance, or lack thereof, renders performance by the other party an impossibility. In some instances, actions caused by a subcontractor or a lower-tier subcontractor may render the performance of the contractor impossible. Oftentimes, the conditions leading up to the termination are bargained-for aspects of the construction contract. By the same token, construction contracts outline the duties of the parties, the breach of which would provide grounds for termination. Construction contracts generally provide not only the grounds for which termination can be effectuated but also the requirements for the termination procedure. It is important for the terminating party to follow the requirements set forth in the construction contract or run the risk of wrongfully terminating the other party and being liable for the consequent damages.

18.02 CONTRACTUAL TERMINATION PROVISIONS

Given the potential parade of "horribles" that may result from an improper termination, as well as the unfortunate possibility of necessitating termination, it behooves all parties to a construction contract to address the prospect of termination as clearly and specifically as possible during the contractual negotiation phase. Well-drafted termination provisions go a long way in terms of limiting the potential problems associated with termination. Oftentimes, the construction contract contains separate termination provisions, each specific to the contractor and the owner.

Article 14 of the often-used AIA Document A201–2007 contains termination provisions addressing the suspension and termination of the contract by either the owner or the contractor.

A. Termination by the Contractor for Cause

One very common type of termination provision is the "for cause" provision. As its name implies, this type of provision allows a party to terminate only when there is proper cause. What constitutes proper cause is usually defined in the provision. For example, Article 14 of AIA Document A201–2007 provides:

2. *Id.*

14.1 Termination by Contractor[3]

14.1.1 The Contractor may terminate the Contract if the Work is stopped for a period of 30 consecutive days through no act or fault of the Contractor or a Subcontractor, Sub-subcontractor or their agents or employees or any other persons or entities performing portions of the Work under direct or indirect contract with the Contractor, for any of the following reasons:

1. issuance of an order of a court or other public authority having jurisdiction that requires all Work to be stopped;
2. an act of government, such as a declaration of national emergency that requires all Work to be stopped;
3. because the Architect has not issued a Certificate for Payment and has not notified the Contractor of the reason for withholding certification as provided in Subparagraph 9.4.1, or because the Owner has not made payment on a Certificate for Payment within the time stated in the Contract Documents; or
4. the Owner has failed to furnish to the Contractor promptly, upon the Contractor's request, reasonable evidence as required by Subparagraph 2.2.1.[4]

14.1.2 The Contractor may terminate the Contract, if through no act or fault of the Contractor or a Subcontractor, Sub-subcontractor or their agents or employees or any other person or entities performing portions of the Work under direct or indirect contract with the Contractor, repeated suspensions, delays or interruptions of the entire Work by the Owner as described in Paragraph 14.3 constitute in the aggregate more than 100 percent of the total number of days scheduled for completion, or 120 days in any 365-day period, whichever is less.

14.1.3 If one of the reasons described in Subparagraph 14.1.1 or 14.1.2 exists, the Contractor may, upon seven days' written notice to the Owner and Architect, terminate the Contract and recover from the Owner payment for Work executed, including reasonable overhead and profit, costs incurred by reason of such termination, and damages.[5]

14.1.4 If the Work is stopped for a period of 60 consecutive days through no act or fault of the Contractor or a Subcontractor or their agents or employees

---

3. American Institute of Architects, AIA Document A201–2007, *General Conditions of the Contract for Construction* (2007) [hereinafter AIA Document A201–2007].

4. Under this element, the contractor must make a demand pursuant to AIA Document A201–2007 § 2.2.1 and, upon not receiving a timely response, the contractor must stop the work for 30 days and thereafter terminate.

5. The seven days' notice is separate from the 30-day period during which work is stopped by the contractor.

> or any other persons performing portions of the Work under contract with the Contractor because the Owner has repeatedly failed to fulfill the Owner's obligations under the Contract Documents with respect to matters important to the progress of the Work, the Contractor may, upon seven additional days' written notice to the Owner and the Architect, terminate the Contract and recover from the Owner as provided in Subparagraph 14.1.3.

This provision requires that the contractor terminate only as a result of the specific reasons provided. The enumerated reasons are typically of concern to the contractor in terms of whether or not it will be paid and will be able to timely and properly perform under the contract. The contractor cannot terminate for reasons other than those specified. From an owner's perspective, this type of provision is beneficial because it limits the contractor's ability to terminate within the parameters set forth therein. However, it is important to note that if the contractor seeking termination has contributed to the occurrence of any of the enumerated reasons, the contractor may not terminate.[6] Under Section 14.1.4, the contractor has the ability to terminate by providing seven-days' notice after work has been stopped for a period of 60 days as a result of the owner's fault.[7] It is extremely important that a contractor terminating pursuant to a provision such as this comply with all of the notice requirements, as failure to do so may constitute improper termination of the contract.

B. Termination by the Owner for Cause

AIA Document A201–2007 also contains provisions for termination by the owner, which are instructive.

> 14.2 Termination by Owner for Cause[8]
>
> 14.2.1 The Owner may terminate the Contract if the Contractor:
>
> 1. repeatedly refuses or fails to supply enough properly skilled workers or proper materials;
> 2. fails to make payments to Subcontractors for materials or labor in accordance with the respective agreements between the Contractor and the Subcontractors;
> 3. repeatedly disregards applicable laws, statutes, ordinances, codes, rules and regulations, lawful orders of a public authority; or

6. Werner Sabo. Legal Guide to AIA Documents 398 (Aspen Publishers 14th ed. 1998).

7. *Id.* at 399. The "owner's fault" can include the owner's failure to give approvals, to provide necessary surveys, to provide access to the site, or any other duty of the owner or its agents.

8. AIA Document A201–2007, *supra* note 3.

4. otherwise is guilty of substantial breach of a provision of the Contract Documents.

14.2.2 When any of the above reasons exist, the Owner upon certification by the Initial Decision Maker that sufficient cause exists to justify such action, may without prejudice to any other rights or remedies of the Owner and after giving the Contractor and the Contractor's surety, if any, seven days' written notice, terminate employment of the Contractor, and may, subject to any prior rights of the surety:

1. Exclude the Contractor from the site and take possession of all materials, equipment, tools, and construction equipment and machinery thereon owned by the Contractor;
2. Accept assignment of subcontracts pursuant to Paragraph 5.4; and
3. Finish the Work by whatever reasonable method the Owner may deem expedient. Upon written request of the Contractor, the Owner shall furnish to the Contractor a detailed accounting of the costs incurred by the Owner in finishing the Work.

This provision mirrors the contractor's provision in that it enumerates specific reasons allowing the owner to terminate. Prior to terminating the contractor, the owner must obtain certification by the architect that sufficient cause exists to justify such action.[9] Furthermore, this provision imposes an additional requirement upon the owner to provide seven-days' written notice of the termination, not only to the contractor, but also to the surety, if there is one. By terminating under this provision, the owner does not waive its rights to alternative remedies that may include an action for damages or a demand for arbitration against the contractor.[10] Once the owner terminates the contract for any of the reasons set forth in the foregoing provision, the contractor loses its entitlement to payment until such time as the work is finished.[11] Furthermore, upon terminating the contractor, the owner can deduct all reasonable expenses incurred in completing the project after the contractor's termination.[12] If any funds are left over after such payments, however, the contractor is entitled to those amounts.[13] Likewise, if the cost to complete the contract is greater than the amount unpaid, the contractor is liable for the difference.[14] The architect must ultimately certify the amounts due under this provision to either the

9. The architect's certification should be in writing. Sabo, *supra* note 6, at 400.
10. *Id.*
11. AIA Document A201–2007, *supra* note 3, § 14.2.3.
12. Sabo, *supra* note 6, at 402; AIA Document A201–2007, *supra* note 3, § 14.2.4.
13. Sabo, *supra* note 6, at 402; AIA Document A201–2007, *supra* note 3, § 14.2.4.
14. Sabo, *supra* note 6, at 402; AIA Document A201–2007, *supra* note 3, § 14.2.4.

owner or the contractor, and payment is to be made, despite the fact that the contract has been terminated.[15]

As mentioned previously, Subsections 14.2.2.1, 14.2.2.2, and 14.2.2.3 of AIA Document A201–2007 permit the owner to take immediate action in an effort to mitigate its damages when the contractor defaults. The owner is given the right to take possession of all equipment, materials, and other items owned by the contractor. Although not provided in AIA Document A201–2007, it is presumed that the owner may use the equipment and incorporate the materials into the project until the project is completed.[16] The owner is also given the right to pick and choose which subcontracts it is willing to accept to finish the project.[17] All other subcontractors will not be allowed on the project.[18] To avoid any delays in renegotiating contracts with subcontractors, Section 5.4 of AIA Document A201–2007 grants the owner an assignment of the subcontract upon the termination of the contractor for cause. Inherent in this clause is a promise by the contractor to obtain the subcontractors' consent to these assignments.[19] Finally, without regard to whether it is the cheapest or best method, the owner is permitted to finish the project in any reasonable manner.[20]

The underlying purposes for these provisions have been described by Professor Justin Sweet as follows:

1. To provide incentive to the contractor to take away its property from the site so that the owner can efficiently bring in a successor
2. To provide the owner with material and equipment by which it can expeditiously continue the work with a successor (note the conflict with item 1)
3. To give the owner property that can be sold and the proceeds used to pay for any claim it may have against the contractor.[21]

While the policy surrounding these provisions may seem beneficial for all those involved, the practicality of all parties cooperating during this time to ensure a smooth transition is doubtful. Given the seven-day termination notice, the contractor may feel compelled to take its property away before the

15. Sabo, *supra* note 6, at 402; AIA Document A201–2007, *supra* note 3, § 14.2.4.
16. Sabo, *supra* note 6, at 401; AIA Document A201–2007, *supra* note 3, § 14.2.2.1.
17. Sabo, *supra* note 6, at 401; AIA Document A201–2007, *supra* note 3, § 14.2.2.2.
18. Sabo, *supra* note 6, at 401; AIA Document A201–2007, *supra* note 3, § 14.2.2.2.
19. Justin Sweet, Legal Aspects of Architecture, Engineering and the Construction Process, 657.
20. Sabo, *supra* note 6, at 401; AIA Document A201–2007, *supra* note 3, § 14.2.2.3.
21. Sweet, *supra* note 19 at 651.

owner does.[22] The successor contactor may want to use different materials and equipment.[23] What materials and equipment does the contractor actually own? On most construction sites, the heavy equipment is usually rented and the materials are supplied by a materialman. Thus, this clause may be of little or no value. Furthermore, in the event the contractor seeks protection in bankruptcy, the trustee will surely seek to protect its claim.[24] Notwithstanding the potential pitfalls involved with these clauses, they may, at the very least, keep the project moving toward completion during the transition process.[25]

C. Termination for Convenience

Construction contracts may also contain a provision that allows the parties, though typically only the owner, the option of terminating the contract for convenience. Essentially, pursuant to this type of provision, so long as the procedural aspects are followed, the owner can terminate without having a specific reason. These types of provisions have been upheld and can be found in many of the standard construction industry contract forms, including AIA Document A201–2007. For example, a provision allowing the owner to terminate "at any time for any reason by giving at least ten (10) days prior written notice" was held enforceable.[26]

Even though a termination for convenience provision may be a powerful sword for the owner, it should be wielded carefully. In the event an owner is faced with a defaulting contractor, should the owner terminate for cause, or should the owner simply terminate the contractor for convenience? The City of New York thought it would simply be easier to terminate a construction manager for convenience rather than for cause in the case of *Tishman Constr. Corp. v. City of New York*.[27] After being terminated for convenience, the construction manager brought suit against the city for payment under various provisions of the contract.[28] The city then counterclaimed for damages to cure the construction manager's defaults under the contract, as well as for overpayment.[29]

---

22. *Id.*
23. *Id.*
24. *Id.*
25. *Id.* at 652.
26. Avatar Dev. v. DePani Constr., Inc., 834 So. 2d 873, 875 (Fla. 4th DCA 2004).
27. Tishman Constr. Corp., Inc. v. City of New York, 228 A.D.2d 292, 643 N.Y.S.2d 589 (1st Dep't 1996).
28. *Id.* at 292.
29. *Id.* at 293.

The court found that because the city elected to terminate for convenience, it was precluded from raising its counterclaims under the contract.[30] In the event the city wished to recoup the cost of curing the construction manager's default, it should have terminated under the section of the agreement dealing with termination for cause, which allowed the city to recoup its expenses.[31] It should also be noted that the court rejected the city's argument that it was entitled to proceed with its counterclaim by claiming that its rights were without limitation under the sections in the contract entitled "All Defenses Reserved" and "No Waivers."[32] The court limited the city's counterclaim to proving that the overpayments were due to fraud or mistake.[33]

Under AIA Document A201–2007, a contractor may be entitled to lost profits where the termination is for convenience. Pursuant to paragraph 14.4.3 of AIA Document A201–2007, a contractor who is terminated for convenience is entitled to payment for work executed, costs incurred by reason of such termination, along with reasonable overhead and profit on the work not executed. The allowance for anticipated profit found in AIA Document A201 is inconsistent with the common law rule governing termination for convenience clauses, which generally excludes damages for lost profits on work not performed.[34] (Damages are discussed in further detail in Chapter 24.)

18.03 MATERIAL BREACH

Not all breaches of construction contracts are equal. It has been noted that "any contractually authorized breach for which the contract provides administrative redress cannot constitute a common-law breach of contract, such as those authorizing changes, suspension, site conditions, and termination for convenience. These convert common law breaches into contractual rights."[35] The level of a party's conduct or the impact that a breach has on the object of the contract may give rise to additional causes of action or remedies. One such type of breach is a material breach. A breach of a contract term is "material" if it is substantial enough to "reasonably compel a clear inference of an

30. *Id.*

31. *Id.*

32. *Id.*

33. *Id.*

34. Davis Sewing Mach. Co. of Delaware v. U.S., 60 Ct. Cl. 201, 217, 1925 WL 2797 (1925), *aff'd*, 63 Ct. Cl. 678, 273 U.S. 324, 47 S. Ct. 352, 71 L. Ed. 662 (1927). "[T]he right to recover prospective profits is negatived by the very insertion of a termination for convenience clause. Profits are a gain a party might have made had he been allowed to complete the contract."

35. 5 Philip L. Bruner & Patrick J. O'Connor, Bruner & O'Connor on Construction Law § 18:2 (West Group 2002) [hereinafter Bruner & O'Connor].

unwillingness or inability of one party to meet substantially the contractual future performance expectations of the other party and of the need for the other party to mitigate its damages."[36] Stated differently, "[A] 'material breach' is any breach that materially impairs the interest of the non-breaching party in the future performance of a contract."[37] If a party has materially breached the construction contract, the nonbreaching party will have grounds to terminate.

While black letter law of what constitutes a "material breach" is easy to find, applying these principles to the facts of a bitterly contested construction dispute is difficult, as the following case illustrates.

Mustang Pipeline Company, Inc. v. Driver Pipeline Company, Inc. 134 S.W.3d 195 (Texas 2004)[38]

PER CURIAM.

In this breach of contract case, both parties, Mustang Pipeline Co. (Mustang), and Driver Pipeline Co. (Driver), obtained favorable jury findings as to the other's breach of a pipeline construction contract. Mustang initially brought suit against Driver for Driver's failure to complete work in a timely manner. Driver counterclaimed for Mustang's wrongful termination of the contract. It is a fundamental principle of contract law that when one party to a contract commits a material breach of that contract, the other party is discharged or excused from further performance. Furthermore, if it is clear the parties intend that time is of the essence to a contract, timely performance is essential to a party's right to require performance by the other party.

The court of appeals, in affirming the trial court, held that Mustang could not recover damages because Mustang failed to get a jury finding that Driver's breach was material. We hold that an express jury finding on the issue of materiality is not required when "time is of the essence." Because time was of the essence under the contract and Driver failed to timely perform, Mustang was thereafter excused from performance. We reverse in part and render judgment that Driver take nothing.

In January 1997, Mustang entered into a contract under which Driver was to construct the northern 100 miles of a 200 mile pipeline from Mont Belvieu to Longview, Texas. During the bidding process, Mustang stressed that time was of the essence and that the pipeline construction had to be completed no later than April 30, 1997. In response to this time demand, Driver increased its original bid by

36. *Id.* § 18:4.

37. *Id.* § 18:1.

38. Most internal citations have been omitted as have sections discussing issues not related to this chapter. This opinion has been significantly edited for educational purposes.

approximately ten percent, but was still awarded the contract. The parties agreed on a fourteen-week schedule, working eleven hours per day, seven days a week.

Driver contends that extensive rains delayed the project soon after work began. On March 3, 1997, Driver proposed shutting down pipeline construction until the weather improved. On March 5, Driver officially requested a thirty-day extension to the agreed completion date, citing adverse weather and safety concerns as the reason for the delay. At a meeting on March 6, Driver reiterated its intent to shut-down operations and wait for drier weather, and it proposed a new 161-day construction schedule.

Fifty-eight days into the ninety-eight day construction schedule, Driver had completed only fifteen miles of pipeline and suspended operations. The parties dispute whether Driver could have accomplished more during this time by using different equipment or reassigning crews to work on drier areas. The contract includes provisions explaining how the crews and pipeline welds could be protected while working in the rain, suggesting that the parties anticipated working despite at least some inclement weather. By March 17, Driver stated that it was still not ready to resume work on the pipeline, so Mustang's project engineer certified Driver to be in default under the contract. Mustang then contracted with Sunland Corporation to finish Driver's portion of the pipeline, which it completed in September 1997.

Mustang sued Driver for breach of contract to recover the cost of completion, lost profits, and attorney's fees. Driver countersued for breach of contract, alleging that Mustang wrongfully terminated the contract, and sought damages, lost profits and attorney's fees. In their respective answers, both parties raised the other's material breach as an affirmative defense.

The case was tried to a jury.

The jury found that Driver failed to comply with the terms of the contract but also found that Mustang was not justified in terminating the contract. It awarded Mustang $2,104,601 in damages for Driver's breach of contract and awarded Driver $2,515,958 in damages for Mustang's wrongful termination.

Both parties moved for judgment notwithstanding the verdict. Mustang asked the trial court to disregard the jury's finding that Mustang wrongfully terminated the contract and the damages awarded to Driver. Mustang contended that once the jury found that Driver breached the contract, it could not also find that Mustang wrongfully terminated the contract because a breaching party cannot subsequently enforce the contract. The trial court denied Mustang's motion. Driver's motion for judgment notwithstanding the verdict asked the trial court to disregard the jury's finding that Driver breached the contract and the damages awarded to Mustang. Driver argued there was no evidence of probative force to sustain the jury's answers on either the breach of contract or damages questions. The trial court refused to disregard the jury's finding that Driver breached the contract but granted Driver's motion to disregard Mustang's damage award. Thus, the trial court rendered judgment for Driver and awarded Driver actual damages, plus attorney's fees and interest.

Both parties appealed . . . Mustang raised several issues on appeal, arguing that: (1) the trial court erred by refusing to disregard the jury's answer to the wrongful termination question as immaterial; (2) there was legally and factually insufficient evidence to support the jury's wrongful termination finding and the claim was precluded by the contract's termination for convenience provision . . .

The court of appeals affirmed the trial court's judgment. The court of appeals reasoned that the jury's finding of Driver's breach was relevant to Mustang's liability only if the breach was material. The court of appeals concluded that whether a party's breach is material is a fact question, and, absent an express jury finding, the trial court was not required to treat the breach as material as a matter of law. The court of appeals also concluded there was factually and legally sufficient evidence to support the jury's wrongful termination finding, and the contract's "termination for convenience" clause did not preclude Driver's claim because Mustang did not raise this issue in the trial court.

First we must consider whether the trial court erred in denying Mustang's motion for judgment notwithstanding the verdict. Mustang argues that the jury's answer regarding Driver's breach rendered the jury's answer regarding Mustang's wrongful termination immaterial, because once a party has breached it cannot later sue on the contract. Accordingly, Mustang contends that the trial court should have disregarded the jury's answer to the second question and granted its motion for judgment notwithstanding the verdict.

The court of appeals was correct in holding that a party is released from further obligation under the contract only if the other party materially breached. However, we disagree with the court's requirement that Mustang obtain an express jury finding that Driver's breach was material . . . The jury question regarding Driver's breach tracks the Texas Pattern Jury Charge language for the recommended breach of contract question, asking "Did Driver Pipeline Company fail to comply with the contract it had with Mustang Pipeline Company?" ("Did [defendant] fail to comply with the agreement?"). The jury answered yes.

The evidence presented at trial established that Driver's failure to comply was a material breach. The Restatement[1] lists five circumstances significant in determining whether a failure to perform is material:

(a) the extent to which the injured party will be deprived of the benefit which he reasonably expected;
(b) the extent to which the injured party can be adequately compensated for the part of that benefit of which he will be deprived;
(c) the extent to which the party failing to perform or to offer to perform will suffer forfeiture;
(d) the likelihood that the party failing to perform or to offer to perform will cure his failure, taking account of the circumstances including any reasonable assurances;
(e) the extent to which the behavior of the party failing to perform or to offer to perform comports with standards of good faith and fair dealing.

The Restatement also articulates circumstances that are significant in determining when a party's duties are discharged under a contract due to the other party's material breach. The relevant factors, in addition to those listed above, are:

(1) the extent to which it reasonably appears to the injured party that delay may prevent or hinder him in making reasonable substitute arrangements.
(2) the extent to which the agreement provides for performance without delay, but a material failure to perform or to offer to perform on a stated day does not of itself discharge the other party's remaining duties unless the circumstances, including the language of the agreement, indicate that performance or an offer to perform by that day is important.

The evidence presented established both that Driver's breach was material and that Mustang was discharged from its duties under the contract based on that breach. Evidence exists to prove, as a matter of law, that time was a material element of the contract. Driver placed an initial bid with Mustang in October 1996, but it increased its bid significantly in January 1997 after learning of the April 30 deadline. The contract specifically stated that "all time limits stated in the Contract are of the essence to the Contract," and the contract called for "100 percent completion of the pipeline system no later than April 30, 1997." In addition to the contract's emphasis on timely completion, it also required that "sufficient forces and equipment . . . be furnished at all times to adequately perform work with ample margin for emergencies and unexpected events, to carry on the work at a sufficient rate of progress to ensure completion within the time specified in the Contract."

In late February 1997, when Mustang realized Driver was behind on the construction schedule, it asked for assurances from Driver that the work would be completed by the deadline. Although Driver had workers, equipment and supervision on the job site, it accomplished only minimal pipeline construction. Mustang also requested a plan from Driver on how it would finish the project by the end of April. It recognized the unseasonably wet conditions, but offered suggestions to Driver about how it might expedite construction. The evidence indicates that Driver never provided a revised plan on how it would attempt to meet the construction schedule. When Mustang terminated the contract, it had approximately forty days before the April 30, 1997 deadline to complete construction on eighty-five miles of pipeline. Even the most ambitious of construction schedules would have great difficulty successfully meeting that deadline. At the point at which Mustang terminated the contract, there was virtually no chance that Driver would be able to cure its breach and complete the construction on time.

Based on this evidence, we hold that as a matter of law Driver committed a material breach. Mustang was thereafter discharged from its duties under the contract. Therefore, the trial court should have disregarded the jury's answer to the wrongful termination question and granted Mustang judgment notwithstanding the verdict.

> Accordingly, we grant Mustang's petition for review and, without hearing oral argument, affirm the court of appeals' judgment in part, reverse in part, and render judgment that Driver take nothing.
>
> 1. RESTATEMENT (SECOND) OF TORTS § 241 (1981).

The court made it clear that once a party has breached an agreement, the other party is excused from further performance and may properly terminate the contract. Moreover, the breaching party will have no further ability to sue on the contract.

18.04 TERMINATION BY AGREEMENT

The right of two parties to contract naturally infers that the two parties may agree to terminate the contract. In laymen's terms, the parties have "called the deal off," while in legalese, this concept is referred to as rescission, cancellation, mutual termination, or some other synonym.[39] To legally enter into such an arrangement, the parties must comply with general contracting principles: manifestations of mutual assent, consideration, a lawful purpose, and compliance with any formal requirements.[40] Problems with consideration may arise where a party has conferred a benefit on another and has not received anything in return. For example, where a contractor has been paid the full contract price, but has yet to complete performance, the owner will not have received the benefit of the bargain.[41] Notwithstanding the lack of consideration, a court could still enforce the agreement based on the principle that the owner simply waived any future obligations of the contractor or that it simply conferred a gift on the contractor.[42] Although pitfalls may exist where parties are attempting to mutually cancel a contract, courts generally enforce such agreements.[43]

18.05 NOTICE AND OPPORTUNITY TO CURE

Similar to the notice of termination provision in AIA Document A201, most standard form construction contracts provide that a notice of termination be sent to the breaching party and, if applicable, to the lender or surety. The questions now are: What is the purpose of such written notice, and what are the rights

39. SWEET, *supra* note 19 at 645.
40. *Id.*
41. *Id.*
42. *Id.*
43. *Id.*

and obligations of the parties during the notice period?[44] Professor Justin Sweet has stated that, depending on the circumstances giving rise to the termination and the language of the termination provision, the notice period can serve one of three possible functions: The notice period can be viewed as (1) an opportunity for the terminating party to "cool off" and rethink its decision to terminate;[45] (2) an opportunity for the breaching party to "lay off employees, remove equipment from the premises, cancel orders, and for similar matters"[46] and an opportunity for the nonbreaching party to make other arrangements to complete construction; or (3) a time period in which the breaching party is allowed to cure the breach. To avoid any confusion as to the purpose of the notice period, the parties should specifically set forth the rights and obligations of the parties during this time frame.

Whether a notice provision granted a contractor an opportunity to cure was discussed in the case of *New England Structures, Inc. v. Loranger* 234 N.E.2d 888 (Mass. 1968).

New England Structures, Inc. v. Loranger
234 N.E.2d 888, 893 (Mass. 1968)[47]

CUTTER, Justice.

Plaintiffs, doing business as Theodore Loranger & Sons (Loranger), the general contractor on a school project, seek to recover from New England Structures, Inc., a subcontractor (New England), damages caused by an alleged breach of the subcontract. Loranger avers that the breach made it necessary for Loranger at greater expense to engage another subcontractor to complete work on a roof deck. In a cross action, New England seeks to recover for breach of the subcontract by Loranger alleged to have taken place when Loranger terminated New England's right to proceed. The actions were consolidated for trial. A jury returned a verdict for New England in the action brought by Loranger, and a verdict for New England in the sum of $16,860.25 in the action brought by New England against Loranger. The cases are before us on Loranger's exceptions to the judge's charge.

Loranger, under date of July 11, 1961, entered into a subcontract with New England by which New England undertook to install a gypsum roof deck in a school, then being built by Loranger. New England began work on November 24,

44. *Id.* at 650.
45. *Id.* at 651.
46. New England Structures, Inc. v. Loranger, 234 N.E.2d 888, 893 (Mass. 1968).
47. Internal citations omitted. This excerpt has been heavily edited for educational purposes.

1961. On December 18, 1961, New England received a telegram from Loranger which read, 'Because of your repeated refusal or inability to provide enough properly skilled workmen to maintain satisfactory progress, we terminate your right to proceed with work at the school as of December 26, 1961, in accordance with Article 5 of our contract. We intend to complete the work with other forces and charge its costs and any additional damages resulting from your repeated delays to your account.' New England replied, 'Failure on your (Loranger') part to provide approved drawings is the cause of the delay.' The telegram also referred to various allegedly inappropriate changes in instructions.

The pertinent portions of art. 5 of the subcontract are set out in the margin. The Subcontractor agrees to furnish sufficient labor, materials, tools and equipment to maintain its work in accordance with the progress of the general construction work by the General Contractor. Should the Subcontractor fail to keep up with (such) progress then he shall work overtime with no additional compensation, if directed to do so by the General Contractor. If the Subcontractor should be adjudged a bankrupt or if he should persistently fail to supply enough properly skilled workmen or disregard instructions of the General Contractor or fail to observe or perform the provisions of the Contract, then the General Contractor may, by at least five days prior written notice to the Subcontractor without prejudice to any other rights or remedies, terminate the Subcontractor's right to proceed with the work. In such event, the General Contractor may prosecute the work to completion and the Subcontractor shall be liable to the General Contractor for any excess cost occasioned thereby (emphasis supplied).

Article 5 stated grounds on which Loranger might terminate New England's right to proceed with the subcontract.

There was conflicting evidence concerning (a) how New England had done certain work; (b) whether certain metal cross pieces (called bulb tees) had been properly "staggered" and whether joints had been welded on both sides by certified welders, as called for by the specifications; (c) whether New England had supplied an adequate number of certified welders on certain days; (d) whether and to what extent Loranger had waived certain specifications; and (e) whether New England had complied with good trade practices. The architect testified that on December 14, 1961, he had made certain complaints to New England's president. The work was completed by another company at a cost in excess of New England's bid. There was also testimony (1) that Loranger's job foreman told one of New England's welders "to do no work at the job site during the five day period following the date of Loranger's termination telegram," and (2) that, "if New England had been permitted to continue its work, it could have completed the entire subcontract within five days following the date of the termination telegram."

The trial judge ruled, as matter of law, that Loranger, by its termination telegram, confined the justification for its notice of termination to New England's "repeated refusal or inability to provide enough properly skilled workmen to maintain satisfactory progress." He then gave the following instructions: "If you should find that New England did not furnish a sufficient number of men to perform the required work under the contract within a reasonable time then you would be warranted in finding that Loranger was justified in terminating its contract; and it may recover in its suit against New England. (T)he termination cannot, as matter of law, be justified for any reason not stated in the telegram of December 18 including failure to stagger the joints of the bulb tees or failure to weld properly or any other reason, unless you find that inherent in the reasons stated in the telegram, namely, failure to provide enough skilled workmen to maintain satisfactory progress, are these aspects. Nevertheless, these allegations by Loranger of deficiency of work on the part of New England Structures may be considered by you, if you find that Loranger was justified in terminat[ing] the contract for the reason enumerated in the telegram. You may consider it or them as an element of damages sustained by Loranger." Counsel for Loranger claimed exceptions to the portion of the judge's charge quoted above in the body of this opinion.

Some authority supports the judge's ruling, in effect, that Loranger, having specified in its telegram one ground for termination of the subcontract, cannot rely in litigation upon other grounds, except to the extent that the other grounds may directly affect the first ground asserted.

Our cases somewhat more definitely require reliance or change of position based upon the assertion of the particular reason or defense before treating a person, giving one reason for his action, as estopped later to give a different reason. There it was said, "The defendant is not prevented from setting up this defense. Although he wrote respecting other reasons for declining to perform the contract, he expressly reserved different grounds for his refusal. While of course one cannot fail in good faith in presenting his reasons as to his conduct touching a controversy he is not prevented from relying upon one good defense among others urged simply because he has not always put it forward, when it does not appear that he has acted dishonestly or that the other party has been misled to his harm, or that he is estopped on any other ground."

We think Loranger is not barred from asserting grounds not mentioned in its telegram unless New England establishes that, in some manner, it relied to its detriment upon the circumstance that only one ground was so asserted. Even if some evidence tended to show such reliance, the jury did not have to believe this evidence. They should have received instructions that they might consider grounds for termination of the subcontract and defenses to New England's claim (that Loranger by the telegram had committed a breach of the subcontract), other than the ground raised in the telegram, unless they found as a fact that New England had relied to its detriment upon the fact that only one particular ground for termination was mentioned in the telegram.

As there must be a new trial, we consider whether art. 5 of the subcontract afforded New England any right during the five-day notice period to attempt to cure its default, and, in doing so, to rely on the particular ground stated in the telegram. Some evidence summarized above may suggest that such an attempt was made. Article 5 required Loranger to give "at least five days prior written notice to the Subcontractor" of termination.

If a longer notice period had been specified, one might perhaps infer that the notice period was designed to give New England an opportunity to cure its defaults. An English text writer (Hudson's, Building and Engineering Contracts, 9th ed. p. 530) says, "Where a previous warning notice of specified duration is expressly required by the contract before termination (in case of dissatisfaction), the notice should be explicit as to the grounds of dissatisfaction, so that during the time mentioned in the notice the builder may have the opportunity of removing the cause of objection." In Corbin, Contracts, § 1266, p. 66, it is said of a reserved power to terminate a contract, "If a period of notice is required, the contract remains in force and must continue to be performed according to its terms during the specified period after receipt of the notice of termination."

Whether the short five-day notice period was intended to give New England an opportunity to cure any specified breach requires interpretation of art. 5, a matter of law for the court. It would have been natural for the parties to have provided expressly that a default might be cured within the five-day period if that had been the purpose. Article 5 of this subcontract (which is part of a printed form of contract) appears to be based to some extent upon the American Institute of Architects, standard contract, Form A1, art. 22. The purpose of the notice period under this A.I.A. form is not clarified by any commentary.

Strong practical considerations support the view that as short a notice period as five days in connection with terminating a substantial building contract cannot be intended to afford opportunity to cure defaults major enough (even under art. 5) to justify termination of a contract. Such a short period suggests that its purpose is at most to give the defaulting party time to lay off employees, remove equipment from the premises, cancel orders, and for similar matters.

Although the intention of the notice provision of art. 5 is obscure, we interpret it as giving New England no period in which to cure continuing defaults, but merely as directing that New England be told when it must quit the premises and as giving it an opportunity to take steps during the five-day period to protect itself from injury. Nothing in art. 5 suggests that a termination pursuant to its provisions was not to be effective in any event at the conclusion of the five-day period, even if New England should change its conduct.

If Loranger in fact was not justified by New England's conduct in giving the termination notice, it may have subjected itself to liability for breach of the subcontract. The reason stated in the notice, however, for giving the notice cannot be advanced as the basis of any reliance by New England in action taken by it to cure defaults. After the receipt of the notice, as we interpret art. 5, New England had no further opportunity to cure defaults.

Exceptions sustained.

In the event the notice period is a cure provision or there is simply no notice period in the contract, the nonbreaching party should determine whether the breach is curable. Only where breaches are truly noncurable will a court find that a cure notice is unnecessary.[48] In the event the breach is curable, the non-breaching party must give notice sufficiently describing "the inadequate performance and must fairly advise [the breaching party] that [the nonbreaching party] considers the inadequate performance serious enough that, without prompt correction, the contract will be terminated."[49] Courts have found that mere routine correspondence, meeting agendas, or deficiency punchlists do not adequately apprise the nonbreaching party that its inadequate performance may lead to termination for cause.[50] "The remedial purpose of the cure notice is to (1) give the breaching party a reasonable opportunity to cure the identified deficiencies or to provide adequate assurances of cure; (2) present evidence that the identified deficiencies are excused or waived; (3) provide reasonable assurances that the deficiencies can be avoided or cured in the future; (4) permit the parties and the design professional of record to reach agreement upon an acceptable cure; (5) give the breaching party's performance bond surety, if any, advance notice that it may be called upon to perform its performance bond obligations; (6) permit the breaching party to mitigate its own damages; and (7) promote the informal settlement of disputes."[51]

18.06 WRONGFUL TERMINATION

A party seeking to terminate the contract should be cognizant of the potential pitfalls and liability that may lie ahead. Upon termination of the contract, the terminated party is excused from performance and may immediately sue for damages.[52] Where a party wrongfully terminates a nonbreaching party, such an act will amount to a material breach of the contract entitling the nonbreaching party to expectation damages.[53] Accordingly, the nonbreaching party should be placed in the position that would have existed had the contract been

48. Abcon Assoc., Inc. v. U.S., 44 Fed. Cl. 625 (1999) (cure notice not required where contract completion date not met and completion time had expired); L. K. Comstock & Co., Inc. v. United Engineers & Constructors Inc., 880 F.2d 219, 232, 14 Fed. R. Serv. 3d 348 (9th Cir. 1989) (cure notice was not required, because it would have been a useless gesture). *See also* Fru-Con/Fluor Daniel Joint Venture v. Corrigan Brothers, Inc., 154 S.W.3d 330 (Mo. Ct. App. E.D. 2004), transfer denied, (Jan. 26, 2005) and transfer denied, (Mar. 1, 2005) (because the parties had already mutually abandoned a subcontract, the contractor was not obliged to give the subcontractor a chance to cure any defective work because it had already waived any entitlement to back charges).

49. Blaine Economic Dev. Authority v. Royal Elec. Co., Inc., 520 N.W.2d 473, 477 (Minn. Ct. App. 1994).

50. *Id.*

51. 5 BRUNER & O'CONNOR, *supra* note 35, § 18:41.

52. Hosp. Mortgage Group v. First Prudential Dev. Corp., 411 So. 2d 181 (Fla. 1982).

53. For more on damages, see Chapter 24.

performed. Where a contractor is wrongfully terminated prior to beginning performance, it is generally entitled to "lost profits together with the reasonable cost of labor and materials incurred in good faith in partial performance of the contract."[54] In the event a contractor is terminated after performance has begun, the damage calculation is the contract price, minus costs incurred in partial performance, minus the reasonable cost of completion.[55] Although punitive damages are generally not awarded in the breach of contract context, courts have awarded punitive damages where a contractor's termination of a subcontractor was found to be unreasonable, without notice, unprofessional, and oppressive.[56]

18.07 TERMINATION AND THE ROLE OF THE SURETY

As is more thoroughly discussed in Chapter 20, upon the contractor's material breach of the contract, owner/obligee is permitted to terminate the contractor's performance and call upon the surety's performance bond to perform. In most instances, the owner will be required to formally terminate the contractor and provide proper notice to the surety of the contractor's default in order to trigger the surety's obligation under the bond. The failure of the owner to properly terminate the contractor and provide notice to the surety may prejudice its rights under the terms of the performance bond.[57]

---

54. First Atlantic Bldg. Corp. v. Neubauer Constr. Co., 352 So. 2d 103, 105 (Fla. 4th DCA 1977); Brooks v. Holsombach, 525 So. 2d 910 (Fla. 4th DCA 1988); Adams v. Dreyfus Interstate Dev. Corp., 352 So. 2d 76 (Fla. 4th DCA 1977).

55. Steven Siegfried, Introduction to Construction Law 147 (American Law Institute-American Bar Association Committee on Continuing Professional Education, 1987).

56. Cuddy Mountain Concrete Inc. v. Citadel Constr., Inc., 121 Idaho 220, 824 P.2d 151 (Idaho App. 1992).

57. *See* 4 Bruner & O'Connor, *supra* note 35, § 12.35 stating "[p]erformance bonds in the United States commonly are written as conditional instruments." "*See, e.g.*, American Institute of Architects, AIA Document A312–1984, *Performance Bond* (1984), under which the surety is obligated to perform only after (1) the obligee has notified the surety and principal of its intent to declare the principal in default, (2) the obligee has, in fact, declared the principal to be in default and has terminated the principal's rights under the contract, (3) the obligee has agreed to pay the contract balances to the surety, and (4) the obligee is not itself in default of its obligations under the contract."

QUESTIONS

Shortly before the planned completion of a medical office building, your client, the owner, received a letter from the plumbing subcontractor complaining about the contractor's failure to pay a four-month-old invoice for $7,000 (for which the contractor previously had billed the owner, who had paid it). This, coupled with the contractor's seeming inability to complete the punchlist work, was the "last straw" for the owner, who now wants to terminate the contractor "and finish the work, himself." The owner told the architect to certify the contractor's default and "get someone else in to finish the job." The architect is reluctant to issue the certification, and has asked the owner to reconsider.

1. The owner has now called to ask your advice and for you to outline the potential consequences of the proposed termination. What do you tell the owner?
2. Were the contractor's defaults "material"?
3. What remedies might the contractor have in the event that the termination later was deemed to be "wrongful"?

CHAPTER

19

Mechanic's Liens

BRUCE H. SCHOUMACHER

19.01 ORIGIN OF MECHANIC'S LIEN

A lien is a legal hold on property as security for a debt.[1] For example, Illinois statutes going back to 1825 have given a lien on real estate to ensure that payment is received to those persons who in good faith furnish material or labor for the demolition and construction of buildings or perform other improvements to real property. Thus, in the case where improvements are made to privately held property, the right of a mechanic's lien is the right to charge that property with the payment value of the material, fixtures, labor, and services that have been used to improve it. It is an encumbrance that attaches to the property in a manner similar to that of a mortgage.

Like the lien of a mortgage, a mechanic's lien is effective, not only against the owner of the premises at the time it attaches but also, if validly perfected, against purchasers, other encumbrancers, and third parties claiming any rights

The author thanks attorney Angelique Palmer and Jessica L. O'Neill, an attorney at Querrey & Harrow, Ltd. in Chicago, Illinois for their valuable assistance in the preparation of this chapter.

1. Moulding-Brownell Corp. v. E. C. Delfosse Constr. Co., 291 Ill. App. 343, 348, 9 N.E.2d 459, 461 (1st. Dist. 1937).

with reference to the real property in general. As explained in *Davis-Wellcome Mortg. v. Long Bell Lumber Co.*, there is "no fundamental difference between a statutory lien and a lien created by a mortgage. Each arises as a result of a contract with the owner of the property. Under [statutory law] a mechanic's lien is a security for a debt and is an encumbrance in the nature of a statutory mortgage founded upon consent, under which the holder parts with a valuable consideration."[2]

Mechanic's liens are purely statutory and were unknown at common law. At common law, the only lien available in favor of artisans, tradespeople, mechanics, or laborers was a lien on personal property that was limited to payment for work performed on the specific item of property.[3] It was a possessory lien but was lost when the property was surrendered to the owner.

Mechanic's liens are an American innovation and owe their origins to Thomas Jefferson and James Madison. In 1791, as an incentive for building the new Capitol in Washington, D.C., a lien to "encourage master builders to contract for the erecting and finishing of houses" was established by statute. Maryland was the first state to pass such legislation, followed by Pennsylvania in 1803.[4] Today, each state has its own particular statutory method of protecting contractors, subcontractors, and material suppliers by giving them a security interest in the improved real estate to protect their rights to payment. The statutory framework in each jurisdiction must be analyzed by the prudent practitioner to ensure perfection of the lien rights as provided by the act.[5]

A mechanic's lien statute serves three purposes. The primary purpose of mechanic's lien is to provide a security measure to ensure that payment is made for labor, material, and/or services furnished to increase the value or improve the conditions of the real property, thereby creating a benefit to the owner of the property.[6] The second purpose, which is necessitated by granting mechanic's lien rights initially, is to protect the owner, who makes payment for the construction, from double payment to *unknown* contractors for work performed.[7] A third and similar purpose is to protect innocent persons dealing

---

2. 184 Kan. 202, 206 (1959).

3. *Moulding-Brownell Corp.*, 291 Ill. App. at 343, 9 N.E. 2d at 459; Comment, *Mechanics' Liens and Surety Bonds in the Building Trades*, 68 YALE L.J. 138 (1958).

4. HAROLD SIEGAN, CASES AND COMMENTS ON MECHANICS' LIENS (1981); LAWRENCE FRIEDMAN, A HISTORY OF AMERICAN LAW (Simon & Schuster 1973).

5. *See generally* LIEN AND BOND CLAIMS IN THE 50 STATES (American Subcontractors Ass'n 1998); *e.g.*, Pembroke Villas of Broward, Inc. v. Raymundo, 447 So. 2d 324 (Fla. Dist. Ct. App. 1984).

6. ROBERT CUSHMAN, KENNETH CUSHMAN & STEPHEN B. COOK, CONSTRUCTION LITIGATION: REPRESENTING THE CONTRACTOR § 5.17, at 122 (2d ed. 1992); MECHANIC'S LIENS IN ILLINOIS § 1.5 (1990).

7. As the reader will soon come to understand, mechanic's lien statutes generally do not protect owners from the possibility of double payment for known contracts, with some notable exceptions.

with property, such as lenders and purchasers, from having their interests in real estate encumbered by liens that they were unaware of and, thus, have no reasonable opportunity to ascertain or avoid. Mechanic's lien statutes are generally comprehensive in that they define the rights and responsibilities of parties generally involved in construction contracts.[8]

19.02 STATUTORY FRAMEWORK

A mechanic's lien is a statutory remedy for certain trades, suppliers, and professions who contribute to the construction of improvements to real estate. It is a security for the amount owed for the construction of the improvement.[9] The mechanic's lien allows the contractor to recover the uncompensated value of improvements on the property on which the work was performed. The holder of a valid mechanic's lien claim can file suit to foreclose the lien, and if successful, the property can be sold by the court to pay the lien.

The mechanic's lien ensures payment of obligations created when real property is improved and prevents unjust enrichment to the detriment of those providing the services or materials that increase the value of the owner's property.[10] Even though a mechanic's lien gives contractors and subcontractors some assurance of being paid, extreme care must be used in attempting to protect and assert the rights afforded by a mechanic's lien act.

A mechanic's lien act presents an equitable balancing of the rights and duties of each of the classes of persons in a construction project, such as owners, lenders, architects, original contractors, subcontractors, material suppliers, and laborers.[11] In balancing those interests, owners are presumed to be unfamiliar with the terms of the contracts between a contractor and its subcontractors. To provide owners with the knowledge of the money due or to become due to subcontractors and to provide owners with a statutory payment defense against unknown claims of subcontractors, mechanic's lien acts can provide as follows, as exemplified by the Illinois Mechanic's Lien Act:

> It shall be the duty of the contractor to give the owner, and the duty of the owner to require of the contractor, before the owner or his agent, architect, or superintendent shall pay or cause to be paid to the contractor or to his order,

---

8. LIEN AND BOND CLAIMS IN THE 50 STATES, *supra* note 5; *see, e.g.*, Sanaghan v. Lawndale Nat'l Bank, 90 Ill. App. 2d 254, 258, 232 N.E.2d 546 (1st Dist. 1967).

9. LIEN AND BOND CLAIMS IN THE 50 STATES, *supra* note 5.

10. *Id.*

11. STEVEN M. SIEGFRIED, INTRODUCTION TO CONSTRUCTION LAW § 4.01, at 62 (1987) [hereinafter CONSTRUCTION LAW]; *see generally* Struebing Constr. Co. v. Golub-Lake Shore Place Corp., 281 Ill. App. 2d 689, 666 N.E.2d 846 (1st Dist. 1996); Alliance Steel, Inc. v. Piercy, 277 Ill. App. 3d 632, 660 N.E.2d 1341 (4th Dist. 1996); Korsunsky Krank Erickson Architects, Inc. v. Walsh, 370 N.W.2d 29 (Minn. 1985); Mel-Stu Constr. Corp. v. Melwood Constr. Corp., 101 A.D.2d 809, 475 N.Y.S.2d 459 (1984).

> any monies or other consideration due or to become due to the contractor, or make or cause to be made to the contractor, any advancement of any monies or any other consideration, a statement in writing, under oath or verified by affidavit, of the names and addresses of all parties furnishing materials and labor and of the amounts due or to become due to each.[12]

Accordingly, a mechanic's lien act presents contractors and subcontractors a means to provide owners notice of outstanding claims and benefits that they have received on a construction project. Failure of an owner to comply with the statutory requirements of a lien act can deprive an owner of the right to the statutory payment defense.[13]

Likewise, contractors and subcontractors must also adhere to the statutory framework in order to properly claim a lien. In some states, the contractor or subcontractor's failure to follow statutory guidelines will foreclose the possibility of obtaining a lien. However, many states have corresponding trust fund statutes, which also serve to protect those furnishing labor and services.[14] In general, trust fund statutes arrange a system where money obtained by the contractor is placed in trust and held to pay outstanding amounts due and claimed by those subcontractors and suppliers who have rights.[15] Trust fund statutes allow the lien claimant to make a direct claim on the fund without the adherence to the structured requirements of a mechanic's lien statute, although in some jurisdictions the failure to perfect lien rights also causes a forfeiture of trust fund rights. Statutory rights to a mechanic's lien and to a trust fund are separate and distinct.

19.03 LIEN CATEGORIES

Although each state's mechanic's lien statutes vary, two distinct lien classifications have emerged: (1) dependent or derivative claims systems and (2) independent or original claim systems.[16] Under the "dependent" or "derivative" system, the subcontractor's recovery is limited to the amount the owner owes to the contractor.[17] Thus, the subcontractor does not have a direct lien on the

12. 770 ILL. COMP. STAT. 60/5 (2007).

13. *Id.*; *In re* Romanac, 245 F. Supp. 882, 884–85 (W.D. Va. 1965), *aff'd*, 386 F.2d 225 (4th Cir. 1967).

14. *See, e.g.*, COLO. REV. STAT. § 38-22-127 (2006).

15. *See, e.g.*, WIS. STAT. § 779.02 (2006).

16. These were formerly called the "New York system" and the "Pennsylvania system," respectively, by some courts. *See, e.g.*, State v. Tabasso Homes Inc., 42 Del. 110 (1942). That nomenclature fell into disuse half a century ago, although it still appears (albeit quite infrequently) in some opinions and writings.

17. Bender v. Beverly Anne, Inc., 2002 N.D. 146 (2002).

owner's property; rather, it has a derivative lien dependent on the lien of the contractor.[18] The opposite is true for the latter system (the original or direct system) under which the subcontractor's claim is independent of the amount owed to the contractor.[19] Subcontractors operating under this system have a direct lien, and the owner, therefore, may be subject to double liability.[20] Some states have adopted a hybrid system in an attempt to eradicate the hardships imposed by either system.[21] The distinction between the two types of systems is of historical relevance and provides a basis for understanding the format of current statutory law.[22]

19.04 REQUIREMENTS FOR A LIEN

If a contractor is to assert a valid mechanic's lien claim, certain requirements must be fulfilled. First, the contractor must enter into a valid contract with the owner. Second, the contractor must furnish lienable materials or services for the real property. Third, the contractor must perform the contract. To satisfy the pre-lien requirement of performance, performance of the contract does not have to be perfect. Rather, substantial performance of the contract is enough. Substantial performance is "an honest and faithful performance of the contract in its material and substantial parts and no willful departure from or omission of the essential parts of the contract."[23] The owner's full use of the improvement can also be considered evidence of substantial completion.[24]

In addition to the three lien requirements, a lien claimant must also fall into the state's statutory parameters setting forth parties entitled to or excluded from asserting a lien.

A. Contract

To assert a contractor's mechanic's lien, (1) a contractor (2) must have an original or written contract directly with the owner *or* with a person authorized or knowingly permitted by the owner to enter into the construction contract. The contractor must (3) perform lienable work, (4) perform the contract and satisfy all contract provisions that have not been waived or excused, (5) provide a

18. Price v. Neal-Millard Co., 124 Ga. 884 (1906).

19. Collins v. Board of Trustees, 72 W. Va. 583 (1913).

20. Masten Lumber & Supply Co. v. Brown, 405 A.2d 101 (Del. 1979).

21. LeGrand v. Hubbard, 216 Ala. 164 (1927).

22. *See Bender*, 2002 N.D. at 146.

23. Bloomington Hotel Co. v. Garthwait, 227 Ill. 613, 81 N.E. 714 (1907).

24. V&V Cement Contractors, Inc. v. LaSalle Nat'l Bank, 119 Ill. App. 3d 154, 158, 456 N.E.2d 655 (1st Dist. 1983).

contractor's sworn statement or establish that the owner has waived this requirement, (6) timely record a proper claim for lien, and (7) timely file a lawsuit to foreclose the mechanic's lien.

The subcontractor's lien requirements are more extensive than those required by the contractor. Generally, for a subcontractor to assert a valid mechanic's lien, the subcontractor must (1) enter into a valid contract with a contractor or a subcontractor, (2) furnish lienable materials or services for the real property, (3) perform the contract, (4) establish valid chain of contracts between the contractor and the subcontractor, (5) provide a notice of its subcontract or work, (6) timely record its lien, and (7) timely file suit to foreclose on its lien.

Where a contract is not substantially performed by the contractor or subcontractor without fault on the part of the contractor, the contractor's lien will be limited to the reasonable value of the work performed and the materials furnished, though the contractor runs the risk of being denied the lien when it fails to substantially perform by the contractor's own fault.[25]

B. Improvement

A lien arises under the contract, specifically, when the work contracted for is an "improvement" to the property. Often, the text of a statute will define the term "improvement" explicitly.[26] Otherwise, an improvement is commonly defined as an addition to real property that "enhance[s] its value, beauty or utility" or "adapt[s] it for new or further purpose."[27] The courts decide on a case-by-case basis whether work performed constitutes an improvement.[28] Where the contracted work does not amount to an improvement, the courts will not grant a lien on the land. A number of courts focus on whether or not the improvement is "visible," while others look to the necessity or permanency of the work performed. The North Dakota court, in the case of *Nesdahl*

---

25. *See* Richmond v. Peterson, 2006 Neb. App. LEXIS 132 (Ct. App. 2006) (discussing substantial performance in relation to liens).

26. For examples, *see* Fla. Stat. § 713.01 (2006) (defining "improvement" and "improve" as part of the Construction Lien Law) and Haw. Rev. Stat. § 507-41 (2006) (specifying what qualifies as an improvement).

27. Black's Law Dictionary 757 (6th ed. 1990); *see also* Mazel v. Bain, 272 Ala. 640, 641–42 (1961) (adopting the Black's Law Dictionary definition of "improvement"); Cates v. Hunter Eng'g Co., 205 Ill. App. 3d 587, 588–89 (3rd Dist. 1990) (defining "improvement" as more than a mere repair or replacement).

28. Haz-Mat Response v. Certified Waste Servs., 259 Kan. 166, 174 (1996).

Surveying & Eng'g v. Ackerland Corp.,[29] dealt with a lien asserted by Nesdahl for surveying services. The owner Herzig claimed that this work did not amount to an improvement, and the court stated:

> In order for Nesdahl to fall within the protection of the North Dakota mechanic's lien statute, he must have improved real estate by the contribution of labor, skill, or materials. *See* NDCC § 35-27-02. Herzig argues that the surveying services provided by Nesdahl do not fall within the definition of "improve" found in section 35-27-01 of the North Dakota Century Code. We disagree with Herzig's interpretation. It is uncontested that Nesdahl performed land surveying on the real estate. The question is whether that land surveying was done in connection with [an] improvement," as required by section 35-27-01(2), NDCC. In the process of providing his surveying services, Nesdahl engaged in several activities which physically altered and affected Herzig's property. Nesdahl and his crew located the section corners, dug them up, put ties of number 5 rebar on the section corners, and then buried those ties in the ground near the brass caps which they had located. In addition to locating all existing markers, Nesdahl and his crew also drove several survey stakes into the ground. We find that these surveying acts constitute "structure[s], erection[s], . . . alteration[s], . . . [and] excavation[s] . . . placed, made, or done on real estate for its permanent benefit," and therefore constitute improvements as defined in section 35-27-01(3), NDCC. Therefore, Nesdahl's surveying was done "in connection with" improvements, satisfies the definition in 35-27-01(2), and is entitled to the protection of a mechanic's lien.

Without an improvement, a lienable interest does not exist. A lien can be granted upon the improvement itself, as well as the land on which the improvement exists. This poses a particular problem for those contractors and subcontractors who perform services that do not include an erection or construction of any sort, such as construction managers. Each court's interpretation of the term "improvement" acts to either limit or expand those persons entitled to a mechanic's lien. Likewise, the language surrounding the term "improvement" is closely analyzed to determine whether or not one is bestowed a lien. A word as seemingly insignificant as "repair" can considerably widen or narrow the scope of lien eligibility when it is included or excluded in the definition of an improvement. A mechanic's lien may be expansive enough to incorporate the entire "scheme of improvement" or so narrow as to merely include the work "directly connected" to the improvement.[30]

---

29. 507 N.W.2d 686 (N.D. 1993).

30. *See* Nolte v. Smith, 189 Cal. App. 2d 140, 149 (Ct. App. 1961) (noting the use of the statutory language, "scheme of improvement").

C. Amount of Lien

The contract between the contractor and the owner serves a dual purpose as related to liens. As you now know, the contract serves to fulfill the lien requirement that a contract exists between the parties. However, the contract also serves as evidence of the amount recoverable under the lien. A contractor, generally speaking, cannot enforce a lien for an amount greater than the contracted price. The amount liened is usually the contract price, less any amount already paid to the person imposing the lien.[31] Dependent upon the terms of the contract and the contractor's ability to prove that it performed "extras," the cost of the extra work provided may be liened as well.[32]

Where a contract is not fully but substantially performed, a contractor is entitled to lien the amount reasonable for work completed, less the amount needed to complete or correct the unfinished project. However, when the amount recoverable under a lien is not agreed upon by the parties, the actual or reasonable value of the improvements should be considered.[33]

Additionally, mechanic's lien law will often provide the contractor with the opportunity to lien the prejudgment interest incurred from the date of the lien notice, either at the contract rate or at the legal rate.[34]

Moreover, although some courts find that overhead and profits generally cannot be liened on their own,[35] other courts include both in the lien when overhead and profit are part of the contract price or part of the reasonable value of the work completed or when they need to prevent unjust enrichment. Certain courts also include industrial accident payment, contributions to union health and welfare funds, payroll taxes, depreciation of equipment, profit items anticipated by the contract, and insurance premiums, in the amount of the lien.[36]

Nevertheless, a lien cannot be used as a tool for recovering breach of contract claims. If, for example, an owner wrongfully terminates a contract, the contractor may be entitled to lost profits and overhead in the breach of contract suit, but this amount would not also be lienable.

---

31. ERB Lumber Inc. v. Gidley, 234 Mich. App. 387 (Ct. App. 1999).

32. *See* Alan Custom Homes, Inc. v. Krol, 256 Mich. App. 505 (Ct. App. 2003) (pursuing greater than the contract amount because addenda were made to the contract, making purchaser responsible for extras).

33. *See* Basic Modular Facilities v. Ehsanipour, 70 Cal. App. 4th 1480 (Ct. App. 1999) (allowing the reasonable value of the work to be liened including oral modifications).

34. Bouten Constr. Co. v. H. F. Magnuson Co., 133 Idaho 756 (1999); Scott v. Strickland, 10 Kan. App. 2d 14 (Ct. App. 1984).

35. Bangor Roofing & Sheet Metal Co. v. Robbins Plumbing Co., 151 Me. 45 (1955).

36. 53 Am. Jur. 2d, *Mechanics' Liens* § 90 (2003) (listing items for which a lien may be obtained).

After considering the contract amount, extras, interest, overhead, and lost profits, courts may find that the lien amount is overstated or that the lien is fraudulent. In fact, requesting overhead may put one at risk of losing the entire lien, if the court finds the lien has been overstated. A fraudulent lien is a lien filed by a party with the knowledge that it contains excessive amounts, contains false information, or fails to meet the requirements that would deem it a legitimate lien. Still, in most instances a minor mistake, error in claim of lien, or good faith dispute over the amount will not constitute willful exaggeration or defeat the lien.[37] Owners may allege the contractor's lien is fraudulent through an affirmative defense or in a counterclaim.[38]

When there is a question as to whether or not the lien is fraudulent, the contract amount may serve as evidence to prove a lien is overstated.[39] On the other hand, the absence of a contract, where one is required to claim a lien, may also be evidence that the lien is fraudulent.[40]

In several states, it is a statutory violation to knowingly or willfully file a fraudulent lien.[41] Furthermore, harsh penalties are often imposed on contractors who violate the statutes. Depending on the statutory language, courts may award monetary damages and attorney's fees to the victim of invalid lien.[42] Certain states may even impose a criminal sentence on fraudulent lien claimants who fail to release the improper lien.[43] Fraudulent lien law is rapidly emerging from state courts and legislatures and is currently an evolving area of law.

D. Lien Entitlement

Many parties involved in a construction project have lien rights. These rights are dependent on the contracts established between the parties and the type of work the parties perform. Most statutes include specific language designating those persons entitled to a lien. A person seeking a lien must establish that his or her role and performance is encompassed within the class of persons explicitly provided for in the statute.

---

37. *See* Fla. Stat. § 713.31(2)(b) (2008).

38. *See* Langley v. Knowles, 958 So. 2d 1149, 1150 (Fla. Dist. Ct. App. 5th Dist. 2007) (alleging the lien was fraudulent in both an affirmative defense and in a counterclaim).

39. Cordeck Sales, Inc. v. Constr. Sys., 382 Ill. App. 3d 334, 371 (1st Dist. 2008).

40. Centurion Planning Corp. v. Seabrook Venture II, 176 S.W.3d 498, 507 (Tex. Ct. App. 2004).

41. *See, e.g.*, Fla. Stat. § 713.31 (2008) and Tex. Civ. Prac. & Rem. Code Ann. § 12.002 (2006).

42. *See, e.g.*, Colo. Rev. Stat. § 33-22-128 (2008) (stating that the person who filed the overstated lien will be liable to the person against whom the lien was filed, including attorney's fees).

43. *See* Tex. Penal Code § 32.49 (2006) (criminal penalty imposed for failure to release fraudulent lien, which may apply in construction context).

While there is an immense variation among states as to who is permitted to claim a lien on property, the contractor is afforded lien rights under mechanic's lien acts. A contractor is most likely to meet the requirements for a lien, as it *contracts* directly with the owner to make an *improvement* on the property. Moreover, contractors are usually designated expressly in the statutes as having the right to a lien. As one might guess, issues more frequently arise as to those parties who do not directly contract with the owner but believe that they are entitled to a lien under a statutory definition.

A large construction project may include surveyors, architects, designers, engineers, supervisors, construction managers, material suppliers, equipment suppliers, subcontractors, and sub-subcontractors. The attorney representing these parties should look to the state's mechanic's lien act to determine if any or all of these parties come under the statute and are thereby entitled to a lien.[44] Often, state statutes extend lien entitlement to even remote subcontractors, though the more remote the contractor, the more difficult it becomes to prove entitlement.[45]

As an example, the Illinois Mechanic's Lien Act[46] provides in part that any person who performs "any services . . . as an architect, structural engineer, professional, land surveyor . . ." for an improvement has a lien against the property. Illinois's statutorily adopted policy is certainly not universal among all states, as other states have narrower statutes. The court's interpretation of a party's status, in relation to a state's statute, can preclude parties from claiming a lien on the property, even those parties who have fully completed their portion of a construction contract.

19.05 WHAT IS LIENABLE?

Almost any product or labor that can result in improving real property can be subject to a mechanic's lien. For example, in some states, an architect or engineer who only produces a drawing for a project can encumber the real property with a mechanic's lien. Also, asbestos removal, the demolition of a building, and the subsequent hauling away of debris are also lienable.[47]

The contractor or subcontractor need only show that the material was delivered to the owner, or the owner's agent, for the improvement or was delivered

44. Architects, engineers, and surveyors may be entitled to a lien by way of an individual act or statute specifically provided for those professions. For an example, *see* IND. CODE ANN. § 32-28-11-1 (2006).

45. *See* TENN. CODE ANN. § 66-11-115.

46. 770 ILL. COMP. STAT. 60/1 (2006).

47. Midwest Asbestos Removal Serv., Inc. v. Crown Cork & Seal Co., No. 90C 7010, 1993 WL 18433 *2 (N.D. Ill., January 25, 1993); Cleveland Wrecking Co. v. Central Nat'l Bank in Chicago, 216 Ill. App. 3d 279, 287, 576 N.E.2d 1055 (1st Dist. 1991).

to the place of construction for the purpose of being used in the construction project or improvement. In most circumstances, actual incorporation of the material into the improvement does not need to be proven.[48]

Examples of lienable materials could include the following: appliances affixed to the property (i.e., permanent fixtures); carpeting; electrical wiring and conduits; furnaces and pipes; lavatories; sewers, baths, and plumbing; landscaping for original planting; electrical signs and work; sanitary sewers and water mains; and storm drains. Lienable work can also include environmental services (such as excavation and removal);[49] pre-construction services by contractors (preliminary estimates, scheduling, determining project scope (can also be lienable by contractors not ultimately selected to build the project));[50] delay damages;[51] drawings for an abandoned project;[52] equipment;[53] ornamenting property;[54] and cleaning and moving.[55]

Many of the examples of lienable articles provided (e.g., carpeting) constitute a "fixture." Definitions of a fixture may vary from state to state, but generally an item is either a movable chattel or an annexed fixture. If the item is a movable chattel, it is unlikely to be lienable, unless a statute specifically

---

48. Dale R. Agthe, Annotation, *Delivery of Material to a Building Site as Sustaining Mechanic's Lien Modern Cases*, 32 A.L.R. 4th 1130 (1981); Queensboro Steel Corp. v. East Coast Mach. & Iron Works, Inc., 346 S.E.2d 248, 253 (N.C. Ct. App. 1986); Oolite Indus., Inc. v. Millman Constr. Co., 501 So.2d 655, 656 (Fla. Dist. Ct. App. 1987); 770 Ill. Comp. Stat. 60/7 (West 2005).

49. Midwest Envtl. Consulting & Remediation Serv., Inc. v. People's Bank of Bloomington, 251 Ill. App. 3d 256, 261–262, 620 N.E.2d 469 (4th Dist. 1993); Cork Plumbing Co. v. Martin Bloom Assocs., Inc., 573 S.W.2d 947, 959 (Mo. Ct. App. 1978); R. C. Freeman v. Shannon Constr., Inc., 560 S.W.2d 732, 735 (Tex. Civ. App. 1977).

50. First Nat'l Bank of Roscoe v. Rinaldi, 262 Ill. App. 3d 179, 634 N.E.2d 1204, 1207–1208 (2nd Dist. 1994); BRL Carpenters, Ltd. v. Am. Nat'l Bank & Trust Co., 126 Ill. App. 3d 137, 142, 466 N.E.2d 1166 (1st Dist. 1984).

51. Vermont Marble Co. v. Baltimore Contractors, Inc., 520 F. Supp. 922, 925 (D.D.C. 1981); *Cleveland Wrecking Co.*, 216 Ill. App. 3d at 279, 289, 576 N.E.2d at 1055.

52. Vermont Marble Co. v. Baltimore Contractors, Inc., 520 F. Supp. 922 (D.D.C. 1981); *Cleveland Wrecking Co.*, 216 Ill. App. 3d at 279, 289, 576 N.E.2d at 1055; Crowen v. Meyer, 342 Ill. 46, 50–51, 174 N.E. 55 (1930).

53. *Vermont Marble Co.*, 520 F. Supp. at 922; *Cleveland Wrecking Co.*, 216 Ill. App. 3d at 285, 576 N.E.2d at 1055. Equipment can be treated as a lienable fixture depending on (a) the physical attachment of the item to the real estate, (b) adaptation of the item to the subject property, and (c) intention to consider the item as part of the realty. Crane Erectors & Riggers, Inc. v. LaSalle Nat'l Bank, 125 Ill. App. 3d 658, 664, 466 N.E.2d 397 (2nd Dist. 1984) (overhead crane installed in warehouse found to be lienable fixture).

54. Construction Law, *supra* note 11, § 4.01, at 62 (involves altering or ornamenting property, despite the relatively short useful life of such ornaments (replacing mini blinds)); Lyons Sav. v. Gash Assoc., 279 Ill. App. 3d 742, 748, 665 N.E.2d 326 (1st Dist. 1996).

55. Construction Law, *supra* note 11, § 4.01, at 62 (if clean-up work is required due to demolition or construction); *Lyons Sav.*, 279 Ill. App. 3d at 748, 655 N.E.2d at 326; *Cleveland Wrecking Co.*, 216 Ill. App. 3d at 285, 576 N.E.2d at 1055.

provides for a lien of that item. However, if the item is permanently attached to the property, then the item will likely become a fixture and thus, lienable. Courts tend to consider three factors when determining whether or not an item is a lienable fixture: (1) whether the item is either actually or constructively annexed to the realty; (2) whether the item has been adapted to fit the use or purpose of the realty that it is affixed upon; and (3) whether it is the intent of the parties to make the article a permanent accession to the real estate.[56]

Examples of non-fixtures or other non-lienable materials and services can include electric motors, stoves, rental of earth moving equipment (although lienable in the context of a public contract); cash advances to the contractor for payroll; heat furnished during construction; landscape/maintenance service; performance bond costs; premiums on public liability and workmen's compensation insurance; and telephone service.[57]

In the case where a claim involves both lienable and non-lienable categories of work, as long as the lien claimed does not fraudulently overstate the amount due for the lienable work, the lien may still be valid.[58] The doctrine of equitable allocation of payments allows for payments to be received where lienable work is combined with non-lienable items.[59]

19.06 INTERESTS SUBJECT TO A LIEN

A mechanic's lien can attach to several different types of property interests. For example, where a contractor is building a skyscraper on the owner's land, both the property and the building may be subject to the lien of the contractor. Almost any interest that can be conveyed, assigned, or transferred is subject to a lien.[60] Consequently, a fee-simple interest in property will be subject to a lien, as well as a life estate, and in most cases, even a leasehold interest. Often issues arise as to leasehold interests where contractors attempt to lien the leasehold interest as well as the owner's interest. Several courts have encountered and decided this conflict. For example, the Kansas Appellate Court held that:

> Real property may be subject to a lien for work completed at the demand of a lessee where the lessee is regarded as an agent of the lessor. If the lessee acts for himself, no lien will attach to the property of the lessor. But where the owner rents his property to another and stipulates in the lease that improvements may be made on the property by the lessee, and the expense thereof deducted

56. *See* Cornell v. Sennes, 18 Cal. App. 3d 126, 132 (Cal. App. 2d Dist. 1971).

57. Construction Law, *supra* note 11, § 4.01, at 62; *In re* Tri-County Materials, Inc., 114 BR 160, 164 (C.D. Ill. 1990); Reno Elec. Works v. Ward, 51 Nev. 291, 297 (1929).

58. Lien and Bond Claims in the 50 States, *supra* note 5; 770 Ill. Comp. Stat. 60/7.

59. Lien and Bond Claims in the 50 States, *supra* note 5; Construction Law, *supra* note 11, § 4.01 at 62; *Crane Erectors & Riggers, Inc.*, 125 Ill. App. 3d at 658, 662, 466 N.E.2d at 397 (2d Dist. 1984).

60. Harkins v. Girouard Estates, Inc., 31 Conn. App. 485, 494 (App. Ct. 1993).

> from the rentals to be paid him, the lessee may be regarded as the agent of the owner, and those doing the work and furnishing materials for improving the property will be entitled to a lien on the interest and estate of the lessee and the owner.[61]

While the courts have been fairly open to allowing liens on leasehold interests, the same is not true for easement interests. A mechanic's lien will not attach to the easement interest or the land where the improvement is made unless an independent lien exists over the owner's property.[62] Still, there is recent case law that suggests that courts, interpreting statutory law, may modify this trend.[63]

On the whole, statutory language dictates precisely which property is subject to a lien. For instance, the Oregon construction lien statute states, in part:

> (1) The site together with the land that may be required for the convenient use and occupation of the improvement constructed on the site, to be determined by the court at the time of the foreclosure of the lien, shall also be subject to the liens created under ORS 87.010 (1), (4) and (5) if, at the time of the commencement of the improvement, the person who caused the improvement to be constructed was the owner of that site and land. If the person owned less than a fee-simple estate in the site and land, then only the interest of the person therein shall be subject to the lien.
> (2) If a lien created under ORS 87.010 (1), (4) and (5) is claimed against a unit as defined in ORS 100.005 to 100.910, the Oregon Condominium Act, the common elements appertaining to that unit are also subject to the lien.[64]

Whether property is subject to a lien depends on more than the property's ability to be transferred, conveyed, or assigned. Over time, exceptions have emerged limiting the types of property subject to a lien. One major category of property commonly excluded from mechanic's lien enforcement is public property, particularly property used for the public benefit.[65] Unless the state's statute indicates otherwise, this exception works to exclude the federal government's property, the state government's property, and the local government's property from becoming subject to a lien.[66] For those in the construction industry, this exception encompasses many potential work projects, such as city bridges, county jails, and state universities.

---

61. *In re* Mechanic's Lien Against City of Kansas City, 37 Kan. App. 2d 440 (Kan. Ct. App. 2007).

62. *Harkins,* 31 Conn. App. at 485.

63. Decker Constr. Co. v. Ohio River Pipe Line, LLC, 2007 Ohio 421, 19 (Ct. App., Franklin County, Feb. 1, 2007).

64. Or. Rev. Stat. § 87.015 (2006).

65. *See* Uniform Construction Lien Act § 202 (2002) (setting forth the standard construction lien exemption for government-owned property).

66. *Id.*

19.07 PERFECTING THE LIEN

To create or perfect a valid mechanic's lien, the contractor must take certain steps required by the statute. Some states require that the landowner or contractor file a notice of commencement.[67] Most states require the lien claimant to serve a notice of furnishing on the property owner. In California, the lien claimant must serve a notice on the landowner, contractor, and lender within 20 days of first performing work.[68] In Illinois, if the lienholder is a subcontractor, the subcontractor must serve a notice on the property owner and lender within 90 days of completing work in the event the subcontractor has not been paid.[69]

Many states require that the contractor provide the landowner with a sworn statement identifying subcontractors and the amounts to be paid to each. All states provide that the lienholder must file a claim of lien with the appropriate government office, usually the county clerk or a recorder of deeds. Some states require that the claim be filed within a specific time period. Indiana gives the lienholder 90 days after furnishing its last work on commercial property.[70]

In Illinois, to obtain a priority over other lienholders, the mechanic's lienholder must file its lien claim with the county recorder of deeds within four months of completion of its work.[71] In any event, the lien claim must be filed within two years of completion of its work or the lien claim is invalid. The statutes also require specific information to be provided in the notice of commencement, sworn affidavit, notice of furnishing, and mechanic's lien claim. If those basic requirements are not met, then the lien may not be valid.

19.08 PRIORITIES

If a contractor perfects its mechanic's lien, its lien comes prior to any interest of the owner in the property. However, there may be conflicting liens against the property. There may be other mechanic's liens, mortgage liens, liens for real estates taxes, judgment liens, and income tax liens.

The question of priority is determined by the basic scheme of the mechanic's lien act for each particular state. Some states provide that the owner or contractor must file a notice of commencement with the appropriate government agency. Those states determine priority of mechanic's liens against the claims of other lienholders from the date the notice of commencement is filed.

---

67. See Mich. Comp. Laws § 570.1108 (2007).
68. Cal. Civ. Code § 3097(a)(c)(d) (2007).
69. 770 Ill. Comp. Stat. 60/24 (2006).
70. Ind. Code § 32-28-3-3 (2007).
71. 770 Ill. Comp. 60/7 (2006).

Other states base priority solely on the date of filing the lien. Hence, if the mortgage lien is filed before the mechanic's lien, the mortgage lien has priority. Still, other states' statutes provide that priority is determined on the date the contractor started work on the improvement. The liens of third parties filed after that date come after the mechanic's lien. Illinois provides that if the mechanic's lien attaches after the mortgage or other lien is filed, then the mechanic's lien is satisfied out of the value of the improvement to the property.[72]

In most states, mechanic's lien claimants share pro rata in the proceeds of sale of property available to pay mechanic's liens. Some states provide for preference among certain classes of lienholders. For example, some jurisdictions provide that laborers who have valid mechanic's liens should be paid before any other mechanic's lienholders are paid.[73] Also, in some states the procedure of "bonding off," which is the discharge of the lien by filing a bond in the penalty of more then the amount of the liens, further complicates priority.[74] However, states like Virginia have determined that holders of "bonded off" mechanic's liens still have to establish priority of lien.[75]

Most mechanic's lien statutes have adopted the concept of "relation back." Under this concept, priority for a mechanic's lien is determined not by the date of filing of the mechanic's lien, but by some earlier date, such as the date the contract was entered into between the owner and the contractor or the date that the notice of commencement had been filed with the appropriate government office.

19.09 PUBLIC PROJECTS

Generally, mechanic's liens cannot attach or be enforced against public property, unless otherwise provided for by statute.[76] However, many states provide for subcontractors to lien funds due the contractor from the government agency, such as Wisconsin.[77] Washington provides that subcontractors may have a lien against retainage due the contractor.[78]

To protect a lien against funds due the contractor, the statutes provide that the subcontractor must give written notice of the claim to the public body, usually the clerk or secretary. The public body then must hold the funds, and

72. 770 Ill. Comp. Stat. § 60/16 (2007).

73. *See* Colo. Rev. Stat. § 38-22-108 (2007).

74. *See* Va. Code Ann. § 43-70 (describing the bonding-off procedure in Virginia); S.C. Code Ann. § 29-5-110 (setting forth the procedures for the release of a lien by bond in South Carolina).

75. York Fed. Sav. & Loan Ass'n v. William A. Hazel, Inc., 256 Va. 598 (Va. 1998).

76. F. D. Rich Co. v. United States, 417 U.S. 116 (1974).

77. Wis. Stats., § 779.15 (1).

78. Wash. Rev. Code, §§ 39.08.010 and 39.08.30 (2007).

the subcontractor usually has a specific number of days to file suit. If suit is not filed within that time period, the public entity can then release the funds to the contractor. Ordinarily, the suit is against the contractor for breach of contract, and the public entity does not have to be made a party to the suit.

19.10 WAIVER AND RELEASE

State law may allow a lien claimant to waive all or part of its right in a lien against the property after some or all work has been completed.[79] A lien waiver may be expressly written in the original contract or a subsequent contract, or it may be implicated. However, an implied lien will only be enforced if it is clear and unambiguous.[80] A partial waiver may also be recognized where properly drafted to waive lien rights on specific pieces of property or to "waive all rights acquired to date."[81]

To have a valid lien waiver, parties to the waiver must have knowledge of the rights being waived. Consideration may also be necessary for a valid lien waiver and release; however, the consideration prerequisite is not uniformly required in all states.[82] Not surprisingly, state statute also governs the form and content of a waiver and release in many instances. States such as Arizona require a very specific waiver and release form.[83] The statute requires that payment is actually made before the release becomes effective (otherwise, the waiver and release must be "conditional").

The effect of a valid lien waiver is to preclude the contractor from asserting its right to claim a lien on the owner's property. If a lien waiver agreement is reached between the contractor and the owner, a subcontractor's right to assert a lien may be affected. Where subcontractors only have a derivative suit (historically the "New York"–type system), there is greater risk that a subcontractor's lien will be unenforceable when the contractor waives its lien rights. However, under the "direct" system described previously (historically the "Pennsylvania" system), it is more likely that a subcontractor will retain its right to a lien, despite the contractor's agreement for waiver.[84]

Another form of waiver is the "no-lien" contract. No-lien contracts will, in effect, waive liens prior to the commencement of work, in anticipation of acquiring the contract. No-lien contracts are essentially a contractor's promise

79. *See* Johnathan Purver, Ill. Jur. *Property* § 18:55 (2007).

80. 36 Fla. Jur. *Mechanics' or Construction Liens* § 122.

81. Dunnell Minn. Digest, *Mechanics' Liens* § 6.02 (4th ed.).

82. *See* 53 Am. Jur. 2d, *Mechanics' Liens* § 280 (1996) (noting the two alternatives under the Uniform Construction Lien Act).

83. Ariz. Rev. Stat. § 33-1008 (2008).

84. Maurice T. Brunner, *Release or Waiver of Mechanic's Liens,* 75 A.L.R. 3d 505 (1977).

to the owner that the contractor will keep the property free from mechanic's liens or will not assert any lien against the owner's property. In doing so, the contractor waives its right to later claim an interest in the owner's property. Certain states permit such language, while others disallow such a provision.[85]

19.11 BANKRUPTCY

The owner of real property may file a petition for bankruptcy or may have a petition filed against it for a finding that the owner is insolvent under the U.S. Bankruptcy Code.[86] The Bankruptcy Code defines two major classes of creditors—secured and unsecured. Secured creditors are companies or persons that have liens against the property of the debtor, such as a mortgagee who has a lien against the real estate because of the mortgage. Unsecured creditors do not have liens against assets of the estate.

Under the Bankruptcy Code, creditors who have secured liens have priority to the proceeds of disposition of the assets subject to the lien. Thus, a mortgagee can petition the bankruptcy court to sell the real estate subject to the mortgage and pay the proceeds to the mortgagee. If the court so orders, the real estate will be sold and the proceeds paid to the mortgagee to pay the secured debt. Any sum in excess of the secured debt is paid into the bankrupt estate. If the proceeds are insufficient, the secured creditor becomes an unsecured creditor for the amount of the deficiency.

When the contractor constructs a building for an owner, the contractor is an unsecured creditor of the owner, until the contractor perfects its mechanic's lien. If the contractor perfects its mechanic's liens before the bankruptcy petition is filed, the contractor becomes a secured creditor in the bankruptcy because the law has created a lien against the real estate in its favor.

What happens if the property owner files a bankruptcy petition, but the contractor's lien has not been perfected? Under the Bankruptcy Code, when a petition for bankruptcy is filed, an automatic stay is entered by the bankruptcy court against the filing of new lawsuits against the bankrupt or continued prosecution of existing lawsuits. The automatic stay can only be lifted pursuant to the order of the bankruptcy court. Under such circumstances, despite the automatic stay, the contractor can continue to take steps to perfect its lien to make itself a secured creditor. However, the contractor cannot file suit to foreclose on the lien without leave of the bankruptcy court.[87]

85. *See* Fordeck-Kemerly Elec., Inc. v. Helmkamp, 591 N.E.2d 1035 (Ind. Ct. App. 1992); *but see,* 770 Ill. Comp. Stat. 60/1.1 (2006).

86. 11 U.S.C. § 101, *et seq* (2007).

87. 11 U.S.C. § 362 (2007).

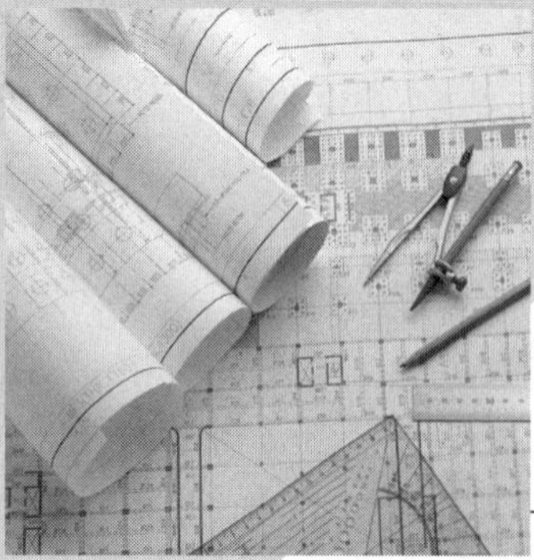

QUESTIONS

One of your longstanding clients, Buck's Construction Labor and Equipment Rental (BCLER), leases temporary skilled and unskilled construction laborers as well as construction equipment to subcontractors and contractors. It operates all across the country and has found success in this field. Although BCLER has gotten burned on a few contracts here and there (which losses it has absorbed into its pricing structure), in recent days the defaults have become more frequent, and a couple have involved fairly large contracts.

Buford, the president of BCLER, tells you that he has become disheartened by this increase in defaults and has decided to see whether the imposition of a lien program can increase his collections. He would like you to produce for him a very concise table that summarizes the lien laws for two states of your choosing whose borders are at least five hundred miles apart, after which he will craft and implement a mechanic's lien collection policy with which he can experiment.

Please create this table for Buford. It should include only the following elements, with appropriate statutory or case law citations:

- ❐ The name of the two states;
- ❐ A general citation to the lien law;
- ❐ An indication of whether an employee-leasing company may assert lien rights;
- ❐ An indication of whether an equipment lessor may assert lien rights;
- ❐ Assuming the company can assert lien rights (even if you conclude that it cannot), circumstances under which the company can lien a landlord's property;
- ❐ A statement as to when the lien statement must be recorded;
- ❐ A statement as to whether the state has a requirement that the claimant provide the parties advance notice of the intent to file a lien statement; and
- ❐ A statement as to the deadline for filing a lawsuit to foreclose the lien.

CHAPTER

20

Insurance and Bonds

DEBORAH GRIFFIN

20.01 INTRODUCTION

This chapter introduces the areas of bonds and insurance, with a particular focus on the difference between the two types of contracts and, in the case of bonds, the fundamental equitable principles that shape the rights and defenses of a surety.

20.02 THE ROLE OF INSURANCE

The role of insurance is to protect oneself against named categories of risk, shifting that risk to the insurer in exchange for payment of a premium. The one purchasing the contract of insurance is called the *insured*, who buys a policy issued by the *insurer*. In the event of a loss covered by the policy, the insurer pays or indemnifies the insured up to the amount of coverage. The insurer has no recourse against the insured except to change the pricing of the policy prospectively or to decline to renew the policy when it expires.

The business of insurance is highly regulated. Congress has made a conscious decision to leave the regulation of insurance to the states, a decision that is manifested in an express exemption from such federal statutes as the antitrust laws.[1] Even the Gramm-Leach-Bliley legislation, which regulates financial services corporations that own insurance companies, leaves the basic regulation of insurance to the states.[2]

To issue an insurance policy to an insured in a given state, the insurance company must be registered with that state's insurance commissioner. Information about what companies are registered in a particular jurisdiction is available on most state insurance commissioners' Web sites.

20.03 CONTRACTUAL INSURANCE REQUIREMENTS

Different types of insurance policies protect against different risks. Construction contracts often require the contractor, the architect, and the owner to purchase and maintain in force certain types and levels of coverage. The types of insurance most often required in connection with construction contracts or construction projects are:

For owners:	builders risk coverage property insurance flood insurance (in flood zones) general liability coverage
For design professionals:	errors and omissions coverage
For contractors:	general liability coverage workers' compensation coverage

On some larger projects, the owner can save substantial sums of money by purchasing insurance under an owner-controlled insurance program, or OCIP. With an OCIP, the insurance covers all project participants who enroll in the program (which can include contractors, subcontractors, project managers, design professionals, and the owner) against whatever risks are included. To the extent enrollment in the OCIP satisfies a party's contractual insurance obligation, that party is required to reduce its contract price by the amount of the insurance component of its price, passing the savings along to the owner.

Standard insurance requirements for a contractor are set out, by way of example, in Article 11 of AIA Document A201–2007 (General Conditions). The dollar amounts of the required coverages are stated in the contract. AIA Document A201–2007 Article 11 provides in part:

1. McCarran-Ferguson Act, 15 U.S.C. §§ 1011–1015 (2004).
2. 15 U.S.C. §§ 6801–6809 (2006).

Article 11 Insurance and Bonds

§ 11.1 CONTRACTOR'S LIABILITY INSURANCE

§ 11.1.1 The Contractor shall purchase from and maintain in a company or companies lawfully authorized to do business in the jurisdiction in which the Project is located such insurance as will protect the Contractor from claims set forth below which may arise out of or result from the Contractor's operations and completed operations under the Contract and for which the Contractor may be legally liable, whether such operations be by the Contractor or by a Subcontractor or by anyone directly or indirectly employed by any of them, or by anyone for whose acts any of them may be liable:

1. Claims under workers' compensation, disability benefit and other similar employee benefit acts that are applicable to the Work to be performed;
2. Claims for damages because of bodily injury, occupational sickness or disease, or death of the Contractor's employees;
3. Claims for damages because of bodily injury, sickness or disease, or death of any person other than the Contractor's employees;
4. Claims for damages insured by usual personal injury liability coverage;
5. Claims for damages, other than to the Work itself, because of injury to or destruction of tangible property, including loss of use resulting therefrom;
6. Claims for damages because of bodily injury, death of a person or property damage arising out of ownership, maintenance or use of a motor vehicle;
7. Claims for bodily injury or property damage arising out of completed operations; and
8. Claims involving contractual liability insurance applicable to the Contractor's obligations under Section 3.18.

. . .

§ 11.2 OWNER'S LIABILITY INSURANCE

The Owner shall be responsible for purchasing and maintaining the Owner's usual liability insurance.

§ 11.3 PROPERTY INSURANCE

§ 11.3.1 Unless otherwise provided, the Owner shall purchase and maintain, in a company or companies lawfully authorized to do business in the jurisdiction in which the Project is located, property insurance written on a builder's risk "all-risk" or equivalent policy form in the amount of the initial Contract Sum, plus value of subsequent Contract Modifications and cost of materials supplied or installed by others, comprising total value for the entire Project at the site on a replacement cost basis without optional deductibles. Such property insurance shall be maintained, unless otherwise provided in the Contract Documents or otherwise agreed in writing by all persons and entities who are beneficiaries of such insurance, until final payment has been made as provided in Section 9.10 or until no person or entity other than the Owner has an insurable interest in the property required by this Section 11.3 to be covered, whichever is later. This insurance shall include interests of the Owner, the Contractor, Subcontractors and Sub-subcontractors in the Project.

§ 11.3.1.1 Property insurance shall be on an "all-risk" or equivalent policy form and shall include, without limitation, insurance against the perils of fire (with extended coverage) and physical loss or damage including, without duplication of coverage, theft, vandalism, malicious mischief, collapse, earthquake, flood, windstorm, falsework, testing and startup, temporary buildings and debris removal including demolition occasioned by enforcement of any applicable legal requirements, and shall cover reasonable compensation for Architect's and Contractor's services and expenses required as a result of such insured loss.

. . .

§ 11.3.2 BOILER AND MACHINERY INSURANCE

The Owner shall purchase and maintain boiler and machinery insurance required by the Contract Documents or by law, which shall specifically cover such insured objects during installation and until final acceptance by the Owner; this insurance shall include interests of the Owner, Contractor, Subcontractors and Sub-subcontractors in the Work, and the Owner and Contractor shall be named insureds.

§ 11.3.3 LOSS OF USE INSURANCE

The Owner, at the Owner's option, may purchase and maintain such insurance as will insure the Owner against loss of use of the Owner's property due to fire or other hazards, however caused. The Owner waives all rights of action against the Contractor for loss of use of the Owner's property, including consequential losses due to fire or other hazards however caused.

. . .

§ 11.3.7 WAIVERS OF SUBROGATION

The Owner and Contractor waive all rights against (1) each other and any of their subcontractors, sub-subcontractors, agents and employees, each of the other, and (2) the Architect, Architect's consultants, separate contractors described in Article 6, if any, and any of their subcontractors, sub-subcontractors, agents and employees, for damages caused by fire or other causes of loss to the extent covered by property insurance obtained pursuant to this Section 11.3 or other property insurance applicable to the Work, except such rights as they have to proceeds of such insurance held by the Owner as fiduciary. The Owner or Contractor, as appropriate, shall require of the Architect, Architect's consultants, separate contractors described in Article 6, if any, and the subcontractors, sub-subcontractors, agents and employees of any of them, by appropriate agreements, written where legally required for validity, similar waivers each in favor of other parties enumerated herein. The policies shall provide such waivers of subrogation by endorsement or otherwise. A waiver of subrogation shall be effective as to a person or entity even though that person or entity would otherwise have a duty of indemnification, contractual or otherwise, did not pay the insurance premium directly or indirectly, and whether or not the person or entity had an insurable interest in the property damaged.

20.04 WORKERS' COMPENSATION INSURANCE

Workers' compensation is a system of insurance designed to provide workers who are injured in the course of their employment with a right to compensation that is awarded without regard to fault and without resort to traditional tort litigation. Most states, by statute, require all employers with a minimum number of employees to provide workers' compensation coverage. Thus, the contractual provision requiring workers' compensation coverage does not usually add a burden that the contractor does not already have by law. Rather, it makes clear which contracting party is to bear the cost of the coverage and gives the other party recourse in the event of a breach—a right not conferred by the workers' compensation statute.

The requirement of providing workers' compensation coverage can be met either by purchasing insurance from a commercial insurer or, if the employer meets certain financial criteria, through a program of self-insurance. The statutory scheme is administered in each state by an administrative agency. These agencies provide an administrative forum for the resolution of disputes about the employee's right to payment, the amount and duration of benefits, and any defenses the employer might have.

20.05 GENERAL LIABILITY INSURANCE POLICIES

Each insurance company uses its own standard form for each type of insurance. However, organizations such as the Insurance Services Office (ISO) and the American Association of Insurance Services develop standardized language that many companies adopt.

A typical commercial general liability (CGL) insurance policy consists of sections containing the declarations, insuring agreement, exclusions, conditions, and definitions. In the declarations, the policy identifies the insured and any additional insureds and states the policy period, policy territory, and the dollar limits for each type of coverage under the policy. The insuring agreement section articulates the hazards insured against, using terms as defined in the definitions section. The essence of the grant of coverage is for damage to persons or property caused by an "occurrence," which in turn is defined in such a way as to connote an event or result that is unintended and unexpected from the standpoint of the insured. There has been much litigation over what constitutes an "occurrence" so as to "trigger" the coverage under a policy in effect at that time.

To enforce the coverage, the conditions section of the policy imposes burdens on the insured, such as providing prompt notice of claims and cooperating in the defense of claims.

The exclusions section of a CGL policy defines limitations on the circumstances in which a risk that is otherwise within the scope of an insuring agreement nonetheless will not be covered. A series of the exclusions is aimed at excluding coverage for the business risks of the insured. For example, there is usually an exclusion of coverage for the cost of repairing the insured's own work product. Courts have often struggled with the application of these exclusions, as illustrated in the case that follows.

The courts generally place the burden of showing that a claim is within the scope of a grant of coverage on the insured; the burden of proving the applicability of an exclusion falls on the insurer.

Summit Custom Homes, Inc. v. Great Am. Lloyds Ins. Co.
No. 05-05-00851-CV (Tex. Civ. App., July 18, 2006)

Opinion By Justice WRIGHT

Summit Custom Homes, Inc. appeals the summary judgment granted in favor of Great American Lloyds Insurance Company . . . and the denial of its summary judgment motion. . . . Summit argues that the trial court erred in granting the summary judgment because (1) the Insurers had a duty to defend, [and] (2) the Insurers had a duty to indemnify. We affirm in part and reverse in part.

Factual and Procedural Background

Summit is a custom home builder. Great American issued Commercial General Liability ("CGL") policies to Summit from January 15, 1996 through January 15, 2000. . . . In 1996, Summit and Stephen and Helen Lazarus signed an agreement for the construction of a residence. Summit completed the Lazaruses' home that same year. In May 2003, the Lazaruses filed suit . . . complaining about construction defects in their home. They asserted claims for negligence, breach of warranty, and breach of contract claiming there were defects in the exterior finishing of the home, referred to as an "Exterior Insulating and Finishing System" or "EIFS." They further argued that the EIFS was defective and sought to have it removed and replaced. Although Summit built the home in 1996, the Lazaruses pleaded the discovery rule because EIFS-related problems were "inherently undiscoverable in that defects in EIFS are latent."

Summit later sued [Great American] seeking a declaration that [it] had breached [its] duties to defend against the suit. Thereafter, Summit filed a motion for partial summary judgment arguing that the underlying pleadings alleged "property damage" from an "occurrence" under the policies at issue. Great American also filed a joint motion for summary judgment arguing that [the plaintiff] failed to allege any facts triggering coverage under the policies. After a hearing, the trial court granted Great American's motion for summary judgment and denied Summit's motion. This appeal followed.

In its first issue, Summit contends the trial court erred in granting the Insurers' joint motion for summary judgment because the Insurers had a duty to defend them against the Lazaruses' claims. The duty to defend arises when a third party sues the insured on allegations that, if taken as true, potentially state a cause of action within the terms of the policy. [Citation omitted.] The duty to defend the lawsuit is determined by the allegations in the underlying pleadings, in light of the policy provisions, and without reference to the truth or falsity of the allegations. [Citations omitted.] This is sometimes referred to as the "eight corners" rule. [Citation omitted.] In construing the allegations of the underlying suit, the pleadings are strictly construed against the insurer, and any doubt is resolved in favor of coverage. [Citation omitted.] In considering the allegations to determine whether a liability insurer is obligated under its policy to defend, we liberally interpret the meaning of those allegations. [Citation omitted.]

Because the terms of the policy dictate whether the Insurers had a duty to defend, we must first determine which policy is at issue in this case. In *Dorchester Development Corp. v. Safeco Insurance Co.*, the issue before this Court was whether coverage existed for property damage resulting from workmanship performed during the policy period when the property damage did not manifest until after the policy period. *Dorchester Dev. Corp. v. Safeco Ins. Co.*, 737 S.W.2d 380, 383 (Tex. App.-Dallas 1987, no writ). We held that "no liability exists on the part of the insurer unless the property damage manifests itself, or becomes apparent, during the policy period." *Id.* If such damages are not manifested during the policy period, then there is no "occurrence" during the policy period. *Id.; see also State Farm & Cas. Co. v. Rodriguez*, 88 S.W.3d 313, 323 (Tex. App.-San Antonio 2002, pet. denied) (relying on *Dorchester* and holding that "property loss occurs when the injury or damages is manifested").

The First District Court of Appeals, however, has rejected the manifestation rule and applied the occurrence rule, expressly refusing to follow this Court's holding in *Dorchester. Pilgrim Enter., Inc. v. Maryland Cas. Co.*, 24 S.W.3d 488 (Tex. App.-Houston [1st Dist.] 2000, no pet.). In *Pilgrim*, the court held that an injury did not actually have to be discovered or manifested within the policy period to trigger coverage under the policy. *Id.* at 497.

Here, Summit encourages this Court to revisit its holding in *Dorchester* and apply the occurrence rule to the underlying facts of this case. We decline Summit's invitation.

Great American provided continuous coverage to Summit from 1996 to 2000. Great American's policies did not contain a specific EIFS exclusion. Summit contends that the policy provided by Great American in 1996 required it to provide a defense against the Lazarus suit because the only date alleged for the "occurrence" and "property damage" is 1996. Great American argues that because the Lazaruses pleaded the discovery rule, any property loss manifested later than 1996; therefore, the 1996 policy is inapplicable.

Despite Summit's argument to the contrary, the underlying pleadings do not establish that damage manifested in 1996. The Lazarus petition, filed in May 2003, claimed that during the construction of their home in 1996, EIFS was applied as an exterior veneer system to the home. The Lazaruses further argue that they suffered a reduction in property value, extensive damage to the home, and the need to substantially retrofit or replace the EIFS. The petition also contends that long term water penetration and retention would result in rot damage to the wooden structural elements of the house. However, missing from their factual allegations is any concrete reference to actual damages that occurred in 1996. Summit's counsel stated during oral argument that there were "no specific allegations that damages occurred in 1996, just allegations that since the house was constructed, damages have occurred." In addition to the failure to plead factual allegations of any damage manifesting in 1996, the Lazarus petition specifically invokes the discovery rule for their claims. It states the following: "Each and every claim made by the Plaintiffs herein is subject to the discovery rule because the defects of which Plaintiffs complain were latent and/or otherwise undiscoverable. The defects cause damage within the wall cavity which is not readily apparent to one examining the exterior of the EIFS surface. As a result, the named Plaintiffs would not, in the exercise of reasonable diligence, immediately perceive, or discover the defects complained of herein."

Thus, applying the "eight corners" rule, although the underlying pleadings state Summit constructed the home in 1996, it contains no allegations of actual damage during 1996. However, Great American failed to establish as a matter of law that damages did not manifest sometime in 1996, 1997, 1998, 1999, or 2000. [Citation omitted.] Therefore, summary judgment could only be appropriate regarding Great American's duty to defend if it established as a matter of law that there was no "property damage" caused by an "occurrence," thereby negating any possible coverage under the policies.

We begin by determining whether the Lazaruses' pleadings against Summit allege property damage. "Property damage" is defined in the Great American policy as "physical injury to tangible property, including all resulting loss of use of that property." Physical injury is not defined; however, the plain meaning connotes an alteration in appearance, shape, color, or in other material dimension. *Lennar Corp. v. Great American Ins. Co.*, 2006 WL 406609, *15 (Tex. App.-Houston [14th Dist.] 2006, pet. filed) (citing *Fid. & Deposit Co. of Md. v. Hartford Cas. Ins. Co.*, 215 F. Supp. 2d 1171, 1183 (D. Kan. 2002)).

Although the underlying pleadings do not explicitly describe the damage to the Lazaruses' home, they repeatedly state that the home sustained "extensive damage and additional damage is likely to occur." They further pleaded that use of EIFS causes damage within the wall cavity that is not readily apparent to one examining the exterior of the EIFS surface and that long term water penetration results in rot damage to the wooden structural elements of the house. Therefore, assuming the facts alleged are true, the Lazaruses have sufficiently pleaded physical injury to tangible property resulting in "property damage." [Citation omitted.]

Next, we must determine if the underlying pleadings establish an "occurrence" within the terms of Great American's policies. "Occurrence" means "an accident, including continuous or repeated exposure to substantially the same general harmful conditions." Great American argues that the "business risk doctrine" applies. The principle that a CGL policy does not generally cover the insured's defective construction resulting in damage to its own work is commonly known as the "business risk" doctrine. Here, the Insurers encourage us to be "mindful" of the business risk doctrine because it has vitality in Texas. This argument, as explained by Summit, overlooks the fact that coverage for some business risks is not eliminated when the damaged work, or the work out of which the damage arose, was performed by subcontractors. [Citation omitted.] Although Great American argues that the business risk exclusion precludes an "occurrence" under the policy, the Fourteenth Court of Appeals, in a recent opinion, thoroughly analyzed a similar CGL policy in another EIFS construction case involving Lennar homes and determined there was an "occurrence" under the policy triggering the insurer's duty to defend. *Lennar Corp.*, 2006 WL 406609 at *15.

Recognizing that Texas law is unsettled on whether defective construction constitutes an "occurrence," the Fourteenth Court of Appeals relied on this Court's holding in *Gehan Homes, Ltd.*, as well as other authority, to conclude that "under the standard CGL policy, negligently created, or inadvertent, defective construction resulting in damage to the insured's own work which is unintended and unexpected can constitute an 'occurrence.'" *Id.* at *4; *see also Gehan Homes, Ltd.*, 146

S.W.3d at 843 (holding that an insured builder's negligence resulting in unexpected damage can constitute an "occurrence"). [Footnote omitted.] Therefore, the Insurers' arguments that defective construction can never constitute an "occurrence" is incorrect.

Great American's policies contain business risk exclusions, specifically the "your work" exclusion, which prevents coverage for "property damage" to "your work" arising out of it or any part of it and included in the "products-completed operations hazard." [Footnote omitted.] In other words, the "your work" exclusion prevents coverage for "property damage" to the insured's work arising after a construction project is finished and in the owner's possession. *Id.* at * 8. If the analysis stopped here, then the business risk exclusion would apply, and there would be no "occurrence." *Id.* at *10 (noting that in the past the "business risk" exclusions operated collectively to preclude coverage for any damage to construction projects).

However, Great American ignores the policy's exception to the business risk exclusion. The exception provides that "[t]his exclusion does not apply if the damaged work or the work out of which the damage arises was performed on your behalf by a subcontractor." In 1986, the insurance industry incorporated the subcontractor exception in the "your work" exclusion of the CGL policy, demonstrating that insurers intended to cover some defective construction resulting in damage to the insured's work. *Id.* at *11. Further, finding no "occurrence" for defective construction that caused damage to the insured's work would render the subcontractor exception superfluous and meaningless. *Id.* Therefore, as the *Lennar* court succinctly stated: "[N]egligently created, or inadvertent, defective construction resulting in damage to the insured's own work that is unintended and unexpected can constitute an 'occurrence.' Nonetheless, the 'your work' or other 'business risk' exclusions may preclude coverage for the damage. However, in some instances, coverage will be restored if the damaged work, or the work out of which the damage arose, was performed by subcontractors." *Id.* at 12.

There is an "occurrence" if an action is intentionally taken, but negligently performed, and the damages are unexpected or unintended. [Citation omitted.] Here, Summit Homes intended to construct the Lazaruses' home using the EIFS system, which was installed and applied by subcontractors. The Lazaruses underlying petition . . . assert[s] various negligence causes of action involving the use of EIFS and the supervision of the workers applying the EIFS. Likewise, the Insurers deny that EIFS is defective; therefore, any damages were unintended because defendants believed they were using a superior cladding system that would perform well in residential construction. Therefore, following the analysis in *Lennar Corp.* and our holding in *Gehan Homes, Ltd.*, we conclude that the underlying pleadings assert an "occurrence" within the meaning of Great American's policies.

Because the underlying pleadings establish "property damage" caused by an "occurrence," Great American may have a duty to defend the Lazarus suit against Summit. As noted above, Great American has failed to establish as a matter of law when the alleged damage manifested. Thus, we cannot determine whether the damages occurred "during the policy period." Specifically, we cannot determine

> whether the 1996, 1997, 1998, 1999, or 2000 Great American policies may apply. Because a genuine issue of material fact exists, the trial court improperly granted Great American's summary judgment. Therefore, we sustain Summit's first issue regarding Great American's duty to defend and reverse and remand to the trial court for further proceedings on this issue.

20.06 BUILDERS' RISK INSURANCE POLICIES

Builders' risk insurance is intended to protect the insureds (usually the owner, general contractor, and subcontractors) from the risk of fortuitous property damage to the work under construction and to materials and equipment on the project site. Typical policies permit recovery without a showing of negligence or other fault by any person, so long as the loss was not caused by intentional misconduct or fraud by an insured.

Policies vary as to the covered causes of damage. Some policies only cover explicitly identified risks such as vandalism, earthquake, and fire. Others are broader, insuring against all causes except those explicitly excluded. The latter type of policy would cover such risks as building collapse, design defects, faulty workmanship, and freezing. The causes typically excluded from even the broadest form of builders' risk policy are government action, building code violation, and earth movement. Although faulty workmanship is a covered risk, the policy covers damage only to property other than that containing the faulty workmanship; like CGL policies, it will not cover the cost of correcting the faulty workmanship itself.

20.07 WAIVERS OF SUBROGATION

The standard contractual insurance provision quoted earlier in this chapter included a waiver of subrogation provision. Subrogation is an equitable concept whereby one person who pays for a loss or liability of another becomes entitled to assert the rights of the other against third parties, in an attempt to recoup the loss or liability paid. In the absence of a waiver of such subrogation rights, an insurer that, for example, pays workers' compensation benefits to an injured employee of a subcontractor is subrogated to that employee's rights to sue in tort anyone having tort liability for the conditions that led to the injury. Contract provisions requiring the contracting parties and their insurers to waive subrogation tend to eliminate litigation that would otherwise be instituted to realize such recoveries. The elimination of such recovery opportunities leaves the loss with the insurer that pays the loss in the first instance and, directly or indirectly, increases the insurance premiums for the party insured by that insurance carrier.

20.08 PROFESSIONAL LIABILITY INSURANCE

The insurance that a design professional is required to carry differs from that required of the contractors. Typically, design professionals must carry coverage for professional errors and omissions (E&O), that is, for damages caused by the failure of the design professional to perform in accordance with the standard of care applicable to people in that profession (which may or may not be the same as ordinary negligence). Insurers often require a significant deductible on these policies, so that the professional is responsible for the first dollars of loss up to the amount of the deductible before the insurer has to pay.

As under CGL policies, E&O policies provide coverage for the cost of defense of covered claims but differ from CGL policies by having the defense costs erode the policy limits. Such "wasting" policies may prompt an owner to require that the design professional carry higher limits to counter the risk of erosion of those limits due to defense costs.

Most E&O coverage is written on a "claims made" basis, rather than an "occurrence" basis like CGL policies. As a result, the policy for the year in which the claim is first asserted is the one—and usually the only one—that affords coverage for that claim, regardless of the policy year in which the deficient act or omission occurred or the year in which the damage was sustained. As a consequence, there are fewer disputes over which policy applies than with CGL coverage. However, once a project is done and the E&O policy year then in effect ends, coverage also ends and is not available for claims that arise in subsequent years. For that reason, owners often require in their contracts with design professionals that they maintain E&O coverage for years after a project's completion, such as for the duration of the likely statute of limitations or statute of repose.

20.09 SURETY BONDS—THE TRIPARTITE RELATIONSHIP

The purpose of a surety bond in the construction context is to provide a party to a construction contract with a guarantee by a commercial surety company that the other party to the contract will perform, and that, if it does not, the surety will respond. Put as simply as possible, a party has the status of a *surety* when, by contract, it agrees to be jointly and severally liable with another for the contractual obligations of the other. The party for whose contractual obligations the surety stands liable is called the *principal*—usually the general contractor or a subcontractor. The party to the contract to whom the surety is bound is called the *obligee*—the owner or, in the case of a bond in which the principal is a subcontractor, the general contractor. These concepts recognize a tripartite relationship, which the law has endowed with well-recognized rights and obligations.

In the construction context, the most common instruments giving rise to the surety relationship are performance bonds, payment bonds, and lien bonds. Statutes governing public construction projects of any appreciable size require that the general contractor post performance and payment bonds.[3] On private construction, an owner will often require not only a performance bond and a payment bond but also a lien bond, which protects the owner's property against the assertion of mechanics' liens.[4] Payment bonds and lien bonds, and the statutes that govern them, usually grant direct rights on the bond to those who supply labor, materials, and equipment to the bond principal. Persons entitled to make claim under payment bonds are viewed as either additional obligees or as third-party beneficiaries. The diagram in Figure 1 may help in understanding the tripartite relationship inherent in a surety bond.

FIGURE 1

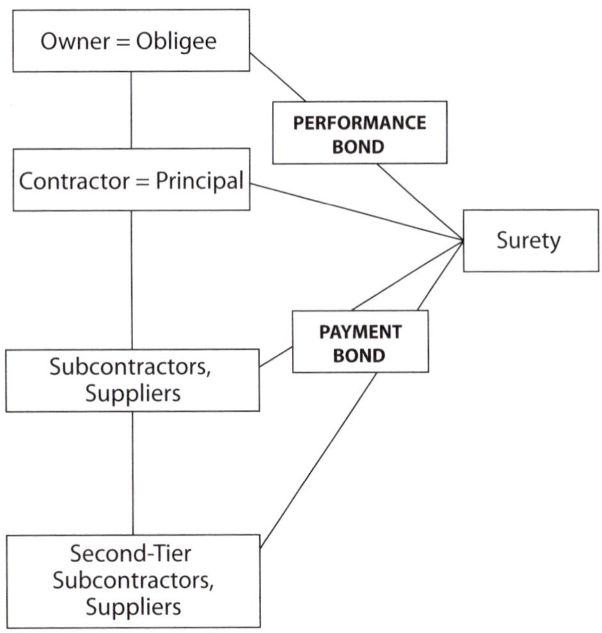

3. *See, e.g.*, Miller Act; 40 U.S.C. §§ 3131–3134 (2005) (formerly cited as 40 U.S.C. §§ 270(a)–270(f)); Mass. Gen. Laws ch. 149 § 44E(2)(E) (performance bond); Mass. Gen. Laws ch. 149 § 29 (payment bond).

4. *See, e.g.*, Mass. Gen. Laws ch. 254 § 12.

The legal and equitable principles that make up the law of suretyship have been synthesized in the RESTATEMENT (THIRD) OF SECURITY, SURETYSHIP AND GUARANTY ("Restatement"). Among these important principles are:

1. The principal and surety are jointly and severally liable to the obligee (Restatement § 1).
2. The surety is entitled to assert all defenses to liability available to its principal except for the principal's personal lack of capacity (minority, mental incapacity) and its discharge in bankruptcy (Restatement § 34).
3. The dollar amount of the surety's liability is capped by the "penal sum" of the bond, provided the surety does not breach its own bond obligations (Restatement § 73 comment b).
4. The surety's obligation is only to indemnify the obligee for its losses due to the principal's breach of the bonded contract, unless the bond specifies other obligations (See Restatement § 32).
5. The surety is entitled to be reimbursed by the principal if the surety has to pay (Restatement §§ 18, 21–26).
6. The surety is equitably subrogated to the rights of its principal against others and to the rights of the obligee and others that it pays under the bond (Restatement §§ 27–31).
7. The surety may be discharged from its obligations under the bond if the obligee impairs the surety's rights, such as by relinquishing collateral available to the obligee as security for the bonded obligation, substantially modifying the principal's obligations under the bonded contract, or permitting someone other than the principal to perform instead of the principal (Restatement §§ 37–44).
8. Surety contracts are subject to the Statute of Frauds (Restatement § 11).

20.10 OBLIGATIONS OF THE PERFORMANCE BOND SURETY

Sureties commonly meet their obligations under a performance bond in one of four ways: by financing the principal to complete its own obligations, by taking over completion of the work in lieu of the principal, by procuring a contract to complete from a new contractor and tendering the new contract to the obligee, or by cash settlement. Which of these four traditional options is actually available to a surety depends on the wording of the bond. For example, the federal Standard Form 25 Performance Bond does not spell out the surety's options, and federal regulations leave it to the government's contracting officer whether to allow the surety to arrange for completion of the work.[5] Some bond forms require that the surety complete the principal's performance if the obli-

5. *See* 48 C.F.R. § 52.249-10(a) (2006); 48 C.F.R. § 49.402-3(h)–(i) (2006).

gee so requests (a completion bond). One of the most prevalent forms is Form A312 published by the AIA; it expressly permits the surety to elect any one of the four options.

The literature is replete with advice to sureties about how to assess the relative merits of their options.[6] In general, however, where the reason for default is the contractor principal's poor financial health, the surety may opt simply to advance sufficient funds to the contractor principal to enable it to complete the job, thus electing to finance completion by the principal. When the surety arranges for the tender of a new contractor, it usually does so either through a bidding process or through negotiation, and requests the new contractor to sign an agreement to complete the contract of the defaulting principal and to supply a new set of performance and payment bonds running in favor of the obligee. The surety then pays sufficient funds to cover the difference between the contract balance on the original contract and the contract amount for the new contractor, up to the penal sum of the original bond. The option of takeover by the surety gives the surety more control over the process of completion, but carries added risk because the surety's liability might not be limited to the penal sum of the bond. The surety will satisfy its obligation through cash settlement alone when the obligee agrees either to arrange for or forego completion and the obligee's damages are readily quantifiable. This option is often used when the principal's work has not begun and other bidders for the original contract are still available, or when the principal's work is nearly complete, with a monetized punchlist to quantify the damages.

Most sophisticated developers and state agencies charged with administering large, complex projects are familiar with the surety's right to elect the option by which it will satisfy its obligations. However, owners that undertake only occasional construction projects may have difficulty accepting their lack of control over the process after a contractor's default, with dire results.

St. Paul Fire & Marine Ins. Co. v. City of Green River
93 F. Supp. 2d 1170 (D. Wyo. 2000)

ORDER GRANTING PLAINTIFF'S MOTION FOR SUMMARY JUDGMENT

BRIMMER, District Judge.

This case is about the extent of a performance bond surety's obligation to finish a construction project within the underlying construction contract's completion deadline after a default by the surety's principal. After reading the briefs, hearing oral

6. A leading source of such advice is ABA Bond Default Manual (Duncan L. Clore, Richard E. Towle, and Michael J. Sugar, Jr., eds., 3rd ed., 2005).

arguments, and being fully advised in the premises, the Court FINDS and ORDERS as follows:

BACKGROUND

. . . Westates Construction Co. ("Westates") entered a contract with the State of Wyoming Joint Powers Water Board (the "Board") for the construction of a water treatment plant in Green River, Wyoming (the "Project"). . . . Westates was to achieve substantial completion of the Project on or before December 1, 1998. Significantly, the construction contract provided that time was of the essence and that the Board would be entitled to liquidated damages in the amount of $2,500 per day for delays after the completion deadline. . . .

Prior to construction, St. Paul Fire & Marine Insurance Company ("St. Paul"), as surety, provided a performance bond to the Board, the obligee, to guarantee the principal Westates' performance under the construction contract. Paragraph 4 of the performance bond defines St. Paul's options for performance under the bond in the event of a default by Westates:

> When the Owner has satisfied the conditions of Paragraph 3 [requiring, *inter alia*, notice to the Surety, declaration of Contractor default, and Owner's agreement to pay the Surety the balance of the contract price], the Surety shall promptly and at the Surety's expense take one of the following actions:
>
> 4.1 Arrange for the Contractor, with consent of the Owner, to perform and complete the Construction Contract; or
>
> 4.2 Undertake to perform and complete the construction contract itself, through its agents or through independent contractors; or
>
> 4.3 Obtain bids or negotiated proposals from qualified contractors acceptable to the Owner for a contract for performance and completion of the Construction Contract, arrange for a contract to be prepared for execution by the Owner and the contractor selected with the Owner's concurrence, to be secured with performance and payment bonds executed by a qualified surety equivalent to the bonds issued on the Construction Contract, and pay to the Owner the amount of damages as described in Paragraph 6 in excess of the Balance of the Contract Price incurred by the Owner resulting from the Contractor's default; or
>
> 4.4 Waive its right to perform and complete, arrange for completion, or obtain a new contractor and with reasonable promptness under the circumstances:
>
> > 1. After investigation, determine the amount for which it may be liable to the Owner and, as soon as practicable after the amount is determined, tender payment therefor to the Owner; or
> > 2. Deny liability in whole or in part and notify the Owner citing reasons thereof.

. . . The Project experienced significant delays of which the causes are in dispute. On January 16, 1998, the Board terminated its contract with Westates, filed suit against Westates in state court, and made a claim against the St. Paul performance bond. St. Paul began a due diligence investigation of the Project to ascertain how it should proceed regarding its duties under the performance bond. Simultaneously, negotiations were commenced involving representatives of the Board, Westates, and St. Paul. These negotiations resulted in a compromise proposal, outlined in a term sheet, which provided, among other things, for: (1) a mutual release between the Board, Westates, and St. Paul, with claims preserved against . . . the project engineer; . . . (3) dismissal of the Board's lawsuit against Westates; (4) rescission of the Board's notice of termination; and (5) completion of the Project by St. Paul. The compromise proposal allowed St. Paul to utilize Westates personnel, though St. Paul would retain management decision making authority. The compromise proposal also called for a May 14, 1999 completion date. . . .

. . . St. Paul made presentations to the Board on February 25, 1998. At the meeting, two letters from Steven Grunsfeld of St. Paul were delivered to the Board. In the first letter, Mr. Grunsfeld indicated St. Paul's willingness to proceed under the compromise proposal, which had not been officially approved by either St. Paul or the Board. In the second letter, Mr. Grunsfeld reported St. Paul's preliminary determination that the Board's termination of Westates was wrongful, but announced St. Paul's election to complete the Project itself under Paragraph 4.2 of the performance bond, albeit under a reservation of rights.

In a February 28, 1998 letter, Dr. Joseph J. Oliver, Chairman of the Board, announced the Board's response to St. Paul's communications. Dr. Oliver wrote:

> St. Paul representatives stated [at the February 25, 1998 meeting] that St. Paul would complete the [Project] no sooner than September 24, 1999, despite a contract completion date for the Project of December 1, 1998. Accordingly, the [Board] is justifiably insecure that St. Paul will complete the Project in a timely manner.
>
> In addition, St. Paul representatives have made clear St. Paul's intention to utilize the personnel employed by Westates . . . to complete the Project . . . even though the [Board] does not consent to Westates or its personnel completing the Project. St. Paul's announced intention to use Westates to finish the Project without consent of the [Board] is an anticipatory breach of the requirements of Section 4.1 of the Performance Bond.
>
> Accordingly, please be advised that the [Board] refuses to allow St. Paul to complete [the Project] under Section 4.2 of the Performance Bond.

On March 3, 1998, St. Paul brought an action in this Court seeking a declaratory judgment that the Board's refusal to permit St. Paul to complete the Project was a material breach of the performance bond, thus exonerating St. Paul from any further obligation under the bond. . . . [T]he Board asserted counterclaims against St. Paul for breach of contract, breach of the covenant of good faith and fair dealing, tortious bad faith, and violation of Wyo. Stat. Ann. § 26-15-124(c), which allows for recovery of attorney's fees where an insurance company's refusal to pay the full amount of a claim is unreasonable or without cause. [Both parties] moved for summary judgment. . . .

ANALYSIS

Central to this dispute is whether St. Paul committed an anticipatory breach of the performance bond when it announced at the February 25, 1998 meeting that it would not complete the Project until September of 1999, well past the completion deadline in the construction contract. The Board contends that St. Paul's actions constituted an anticipatory breach because St. Paul, as surety, stepped into Westates' shoes in the construction contract and was bound by the "time is of the essence" clause contained therein. Thus, according to the Board, St. Paul's announcement that it would not meet the completion deadline constituted a repudiation of the performance bond, thereby justifying the Board's decision to treat the performance bond as repudiated and sue for damages. St. Paul, on the other hand, contends that it was not required to complete the Project within the original completion deadline, and was instead required by Paragraph 5 of the performance bond only to proceed with "reasonable promptness." In St. Paul's view, the Board's action prohibiting St. Paul from completing the Project was a material breach which excused St. Paul from any liability under the bond. Secondarily, the parties dispute whether St. Paul was allowed, under Paragraph 4 of the performance bond, to use Westates personnel without the Board's express consent.

. . . [G]eneral principles of contract interpretation apply with equal force to surety contracts. . . . [A] performance bond should "be fairly construed with a view to effect the object for which it was given and to accomplish the purpose for which it was designed." . . .

1. The Board's Claim that St. Paul Committed an Anticipatory Breach

. . . The Supreme Court of Wyoming outlined the law of anticipatory breach in *J. B. Service Court v. Wharton*:

> An anticipatory breach of contract is one committed before the time has come when there is a present duty of performance, and is the outcome of words or acts evincing an intention to refuse performance in the future. . . . In ascertaining whether an anticipatory breach of a contract has been committed by a party, it is the intention manifested by his acts and words which controls, and not his secret intention. Moreover, in order to predicate a cause of action upon an anticipatory breach, the words or conduct evidencing the breach must be unequivocal and positive in nature. . . . In order to justify the adverse party in treating the renunciation as a total breach, the refusal to perform must be of the whole contract or of a promise or obligation going to the whole consideration, and it must be distinct, unequivocal, and absolute.[1] Where a party commits an anticipatory repudiation, the other party is excused from performing under the contract and may sue immediately for damages. [Citation omitted.] . . .

St. Paul's announcement that it would not meet the December 1, 1998 completion deadline could not constitute an anticipatory repudiation unless St. Paul had an absolute duty to finish the Project by December 1, 1998. Likewise, if St. Paul was within its rights to use Westates personnel, its announced plan to do so would not be an anticipatory breach.

2. St. Paul's Obligation to Finish the Project by the Completion Deadline

Contrary to the Board's contentions, the Court concludes that, while St. Paul was liable for delay damages, it would not be in material breach if it failed to finish the Project by the completion deadline. The Board is certainly correct that, as a general rule, a performance bond surety's liability is coextensive with that of the principal. . . . [T]ime was expressly made of the essence in the underlying construction contract, and Westates was obligated to complete the Project by December 1, 1998, adjusted for extensions. Consequently, St. Paul, as Westates' surety, was likewise bound by the "time is of the essence" clause and completion deadline. [Citation omitted.] This means that, upon the principal's default, the surety is liable for damages caused by the principal's delay if the principal was liable for such damages in the underlying contract. [Citation omitted.] Here, Paragraph 6.3 of the performance bond explicitly states that St. Paul is liable for liquidated damages caused by Westates' delay. No doubt St. Paul would be in breach of the performance bond if it failed to pay liquidated damages for delays attributable to Westates. In this sense, the Board is correct that St. Paul steps into the shoes of Westates and is bound by the liquidated damages clause.

It is a far leap from this conclusion, however, to reach the Board's position that St. Paul would be in material breach of the performance bond if the Project was not completed on the construction deadline. To the contrary, the Court is convinced that the required pace of St. Paul's performance was unambiguously defined by Paragraph 5, which provides that "[i]f the Surety does not proceed with reasonable promptness, the Surety shall be deemed in default on this Bond fifteen days after receipt of an additional written notice from the Owner to the Surety demanding that the Surety perform its obligations under this Bond." . . .

Because the Court finds that the performance bond unambiguously provides that St. Paul's obligation was only to proceed with reasonable promptness, St. Paul would not have been in default under the bond if the Project was not completed by the completion deadline, so long as St. Paul did in fact proceed with reasonable promptness in completing the Project. [Additional discussion of anticipatory breach omitted.]

Far from categorically disclaiming a duty to perform under the performance bond, St. Paul promptly began a due diligence investigation, took part in negotiations with the Board's counsel, and provided the Board with a schedule for its performance under Paragraph 4.2. . . . St. Paul had a good faith basis to believe that its interpretation of the contract was correct, so that its offer of performance consistent with that interpretation was not an anticipatory breach. . . . Because St. Paul did not unequivocally renounce its obligations under the performance bond, the Board did not have the right to treat St. Paul's projected completion schedule as an anticipatory breach.

3. Use of Westates Personnel

The Board's second basis for its claim of anticipatory breach is that St. Paul planned to use Westates personnel as part of the team assembled to complete the Project. In the Board's interpretation of the performance bond, St. Paul was allowed to select one, and only one, option for performance under Paragraph 4. Paragraph 4.1, which allows St. Paul to arrange for Westates to complete the Project, requires the Board's consent. St. Paul instead announced its election to complete the project itself pursuant to Paragraph 4.2. The Board claims that, in electing to proceed under Paragraph 4.2, St. Paul was prohibited from using any Westates personnel without the Board's consent. The Board's interpretation is contrary to the unambiguous language of the performance bond, defies common sense, and is unreasonable in light of industry practices.

Turning first to Paragraph 4.2 of the performance bond, the performance option selected by St. Paul, it is clear that there are no limitations on who St. Paul could utilize to complete the Project. In proceeding under Paragraph 4.2, St. Paul assumed primary responsibility to complete the contract, and with that responsibility came the freedom to assemble the project team of its choosing. In contrast, a surety electing to proceed under Paragraph 4.1 does not assume primary responsibility for completing the contract, and the owner is required to maintain an ongoing contractual relationship with the terminated contractor. While it makes sense that the owner would have the right to object to such a "shotgun wedding" to the contractor it just terminated, it does not follow that the Board would have this right when the surety assumes primary contractual responsibility. In fact, the surety performance options contained in the performance bond are standard in the industry, and it is common practice for a surety that elects to perform the project itself to hire the principal's employees under the direction of a consultant, just as St. Paul did here. [Citation omitted.] While the Board accuses St. Paul of trying to "create a hybrid action from the two paragraphs," it is actually the Board that is attempting to graft the owner consent provision contained in Paragraph 4.1 onto the performance option actually selected by St. Paul, Paragraph 4.2. Because St. Paul had the right to utilize Westates personnel as part of its project team, St. Paul's plan to use such personnel could not constitute an anticipatory breach. In sum, neither St. Paul's proposed completion schedule nor its proposed use of Westates personnel constituted an anticipatory breach, and the Board did not have the right to terminate St. Paul, as it did in Dr. Oliver's February 28, 1998 letter.

4. The Board's Material Breach

Having determined that the Board did not have the right to terminate St. Paul, the focus now shifts to the implications of the Board's wrongful termination of St. Paul. If the Board's action constituted a material breach, St. Paul is excused from further performance under the contract. [Citation omitted.] "A material breach occurs where the covenant not performed is of such importance that the contract would not have been made without it." *Id.* Under the performance bond, St. Paul possessed four options for performance in the event of Westates' default. . . . The effect of the Board's termination of St. Paul was to divest St. Paul of its ability to minimize its liability by selecting the lowest cost option and by directing the construction or participating in the contractor selection process. Courts have consistently held that an obligee's action that deprives a surety of its ability to protect itself pursuant to performance options granted under a performance bond constitutes a material breach, which renders the bond null and void. . . . St. Paul was likewise prejudiced by the Board's actions. While the Board did properly notify St. Paul of Westates' default, it prohibited St. Paul from exercising its contractual right to perform itself or to participate in the selection of the replacement contractor. Because the Court concludes that St. Paul would not have entered the performance bond in the absence of its performance options under Paragraph 4, the Board's action depriving St. Paul of those options was a material breach, which discharged St. Paul from any further duty of performance under the bond. [Citation omitted.] . . .

For the foregoing reasons, . . . [a]ll of the Joint Powers Water Board's counterclaims against St. Paul are DISMISSED WITH PREJUDICE. . . .

1. 632 P.2d 943, 945 (Wyo.1981) (*quoting* 17 AM.JUR.2D CONTRACTS §§ 448, 450 (1964)).

20.11 PAYMENT BONDS

Payment bonds ensure that persons who supply labor, materials, and equipment to the principal are paid what the principal owes them. Payment bonds are almost universally required on public construction because mechanic's liens cannot be imposed on public land. In addition, the assurance of payment by a commercial surety is calculated to encourage competitive pricing by subcontractors and suppliers, thereby saving money for the public.

Most payment bonds, and the statutes that require them, grant direct rights of action to the suppliers of labor and materials. Such permitted claimants are not limited to those having a direct contractual relationship with the bond principal but extend to lower-tier subcontractors and suppliers as well. Statutes vary as to how remote the relationship may be and still give rights under the bond. Even where suppliers several tiers away from the bond principal have such rights, the more remote the supplier and the more fungible the materials it supplies, the greater its burden to show that its materials actually went into the project. Shipping materials directly to the job site, or supplying materials especially fabricated for the project, will usually satisfy this burden.

In addition, most payment bonds, and the statutes that require them, specify a limitation of action period much shorter than that applicable to ordinary contract claims—often one year. Lower-tier subcontractors and suppliers often have to give notice of their claims that direct subcontractors need not give. The adequacy and timeliness of notice, and the timeliness of actions on the bond, are a frequent subject of dispute. Under some bonds, the notice and limitation of action periods begin to run from the date the *claimant* last worked or supplied materials or equipment; in other instances, these periods do not begin until the *bond principal* ceases work on the project. Questions often arise as to whether a claimant's return to the project site after substantial completion—such as to perform punchlist or warranty work—restarts the notice or limitation-of-action clock.[7]

Payment bond sureties share the same attributes of the suretyship relationship as performance bond sureties. The surety may assert the principal's defenses and is subrogated to the principal's affirmative rights against the subcontractor, at least to the extent of its losses on the bonds for the project at issue. Thus, the surety may assert counterclaims against the subcontractor that, if successful, could not only reduce or cancel the payment bond claim but also could yield affirmative recovery to mitigate the surety's other losses on the project.[8]

20.12 RIGHTS AND REMEDIES OF SURETIES

An earlier section of this chapter listed various attributes of suretyship status, which included rights of equitable subrogation and the right to be reimbursed, or exonerated, by the principal. The surety's right to pursue claims of the principal against subcontractors, just discussed, provides a vivid example of the power of the right of equitable subrogation, which will be discussed further below.

In addition to these rights that inhere in the suretyship relationship, sureties routinely require a bond principal, and often its shareholders, to execute an agreement of indemnity in the surety's favor, as a condition to establishing a line of bonding credit. The typical indemnity agreement enhances the surety's equitable and common law rights by providing, for example, for an assignment of and security interest in all assets of the principal; the right to an immediate posting of collateral for the amount of reserves posted by the surety, even before it pays any claims; and an agreement that an affidavit or copies of vouchers showing payments by a surety will constitute prima facie evidence of the fact and amount of the indemnitors' liability under the agreement.

---

7. *See* David J. Oliveri, *Timeliness of Notice to Public Works Contractor on Federal Project, of Indebtedness for Labor or Materials Furnished*, 69 A.L.R. Fed. 600 at § 7 (2005); William E. Aiken, *Construction and Application of Miller Act Provision (40 U.S.C.A. § 270B(b)) Limiting Time for Suits on Payment Bond*, 10 A.L.R. Fed. 533 at §§ 4[c], [d], [e], [f] (2005).

8. *See, e.g.*, St. Paul Fire & Marine Ins. Co. v. United States, 370 F.2d 870, 872 (5th Cir. 1967); City of New York v. Aetna Casualty & Surety Co., 1997 WL 379704 at *3, *5 (S.D.N.Y. 1997); Menorah Nursing Home, Inc. v. Zukov, 548 N.Y.S.2d 702, 705 (N.Y. App. Div. 1989); Maryland Cas. Co. v. King, 381 P.2d 153, 158–59 (Okla. 1963); *see also* Am. Ins. Co. v. Ohio Bureau of Workers' Compensation, 577 N.E.2d 756, 758–79 (Ohio App. 1991). A general contractor's surety may also assert a claim against the surety for a subcontractor. *See* Argonaut Ins. Co. v. Commercial Standard Ins. Co., 380 So. 2d 1066 (Fla. App. 1980).

Returning to the right of equitable subrogation, it was noted previously that the surety is equitably subrogated to the rights of its principal against others and to the rights of the obligee and others that it pays under the bond. The simplicity of that general principle belies its breadth and its importance. Among the rights to which the surety is subrogated are:

- The principal's right to the unpaid balance of contract funds, including retainage;
- The obligee's right to make itself whole from the unpaid balance of contract funds, including retainage;
- The principal's contracts with and claims against third parties arising from the bonded contract, including subcontractors, suppliers, and design professionals;
- The obligee's rights in collateral posted by the principal;
- The rights of subcontractors and suppliers whose payment bond claims the surety pays to assert liens against the project property and other rights of direct payment from the owner; and
- The rights of the obligee and of subcontractors and suppliers whose payment bond claims the surety pays to damages from the principal on account of the principal's breach of contract.

Not only is the surety subrogated to these rights, but, in the event of a contest with another creditor claiming an interest in the same rights, the surety "stands in the shoes" of the ones to whose rights it is subrogated for purposes of determining the relative priorities among competing creditors. Under these principles, the rights a surety obtains by equitable subrogation usually defeat the competing claims of a lender holding a perfected security interest in the principal's assets, taxing authorities with statutory tax liens against the principal's assets, and, as shown in the next case, a bankruptcy trustee in the principal's bankruptcy proceeding.

Pearlman v. Reliance Ins. Co.
371 U.S. 132, 83 S.Ct. 232, 9 L. Ed. 2d 190 (1962)

Mr. Justice BLACK delivered the opinion of the Court.

This is a dispute between the trustee in bankruptcy of a government contractor and the contractor's payment bond surety over which has the superior right and title to a fund withheld by the Government out of earnings due the contractor.

The petitioner, Pearlman, is trustee of the bankrupt estate of the Dutcher Construction Corporation, which in April 1955 entered into a contract with the United

States to do work on the Government's St. Lawrence Seaway project. At the same time the respondent, Reliance Insurance Company [footnote omitted], executed two surety bonds required of the contractor by the Miller Act [then 40 U.S.C. § 270a], one to guarantee performance of the contract, the other to guarantee payment to all persons supplying labor and material for the project. [Footnote omitted.] Under the terms of the contract, which was attached to and made a part of the payment bond, the United States was authorized to retain and hold a percentage of estimated amounts due monthly until final completion and acceptance of all work covered by the contract. Before completion Dutcher had financial trouble and the United States terminated its contract by agreement. Another contractor completed the job, which was finally accepted by the Government. At this time there was left in the Government's withheld fund $87,737.35, which would have been due to be paid to Dutcher had it carried out its obligation to pay its laborers and materialmen. Since it had not met this obligation, its surety had been compelled to pay about $350,000 to discharge debts of the contractor for labor and materials. In this situation the Government was holding over $87,000 which plainly belonged to someone else, and the fund was turned over to the bankrupt's trustee, who held it on the assumption that it had been property of the bankrupt at the time of [bankruptcy] and therefore had vested in the trustee "by operation of law" under § 70 of the Bankruptcy Act. [Footnote omitted.] The surety then filed a petition in the District Court denying that the fund had vested in the trustee, alleging that it, the surety, was "the owner of said sum" of $87,737.35 "free and clear of the claims of the Trustee in Bankruptcy or any other person, firm or corporation," and seeking an order directing the trustee to pay over the fund to the surety forthwith. [Footnote omitted.] The referee in bankruptcy, relying chiefly on this Court's opinion in *United States v. Munsey Trust Co.*, [citation omitted], held that the surety had no superior rights in the fund, refused to direct payment to the surety, and accordingly ordered the surety's claim to be allowed as that of a general creditor only to share on an equality with the general run of unsecured creditors. [Footnote omitted.] The District Court vacated the referee's order and held that cases decided prior to *Munsey* had established the right of a surety under circumstances like this to be accorded priority over general creditors and that *Munsey* had not changed that rule. [Footnote omitted.] The Second Circuit affirmed. [Footnote omitted.] Other federal courts have reached a contrary result [footnote omitted], and as the question is an important and recurring one, we granted *certiorari* to decide it. [Footnote omitted.]

One argument against the surety's claim is that this controversy is governed entirely by the Bankruptcy Act and that § 64 [citation omitted], which prescribes priorities for different classes of creditors, gives no priority to a surety's claim for reimbursement. But the present dispute—who has the property interests in the fund, and how much—is not so simply solved. Ownership of property rights before bankruptcy is one thing; priority of distribution in bankruptcy of property that has passed unencumbered into a bankrupt's estate is quite another. Property interests in a fund not owned by a bankrupt at the time of [bankruptcy], whether complete or partial, legal or equitable, mortgages, liens, or simple priority of rights, are of course not a part of the bankrupt's property and do not vest in the trustee. The Bankruptcy Act simply does not authorize a trustee to distribute other people's property among a bankrupt's creditors. [Footnote omitted.] So here if the surety at the time of adjudication was, as it claimed, either the outright legal or equitable owner of this fund, or had an equitable lien or prior right to it, this property interest of the surety never became a part of the bankruptcy estate to be administered, liquidated, and distributed to general creditors of the bankrupt.... [O]ur question is not who was entitled to priority in distributions under § 64, but whether the surety had, as it claimed, ownership of, an equitable lien on, or a prior right to this fund before bankruptcy adjudication.

Since there is no statute which expressly declares that a surety does acquire a property interest in a fund like this under the circumstances here, we must seek an answer in prior judicial decisions. Some of the relevant factors in determining the question are beyond dispute. Traditionally sureties compelled to pay debts for their principal have been deemed entitled to reimbursement, even without a contractual promise such as the surety here had.[1] And probably there are few doctrines better established than that a surety who pays the debt of another is entitled to all the rights of the person he paid to enforce his right to be reimbursed.[2] This rule, widely applied in this country[3] and generally known as the right of subrogation, was relied on by the Court of Appeals in this case. It seems rather plain that at least two prior decisions of this Court have held that there is a security interest in a withheld fund like this to which the surety is subrogated, unless, as is argued, the rule laid down in those cases has been changed by passage of the Miller Act or by our holding in the *Munsey* case. Those two cases are *Prairie State Bank v. United States* [citation omitted] and *Henningsen v. United States Fid. & Guar. Co.* [citation omitted].

In the *Prairie Bank* case a surety who had been compelled to complete a government contract upon the contractor's default in performance claimed that he was entitled to be reimbursed for his expenditure out of a fund that arose from the Government's retention of 10% of the estimated value of the work done under the terms of the contract between the original contractor and the Government. That contract contained almost the same provisions for retention of the fund as the contract presently before us. The Prairie Bank, contesting the surety's claim, asserted that it had a superior equitable lien arising from moneys advanced by the bank to the contractor before the surety began to complete the work. The Court, in a well-reasoned opinion by Mr. Justice White, held that this fund materially tended to protect the surety, that its creation raised an equity in the surety's favor, that the United States was entitled to protect itself out of the fund, and that the surety, by asserting the right of subrogation, could protect itself by resort to the same securities and same remedies which had been available to the United States for its protection against the contractor. . . . The *Prairie Bank* case thus followed an already established doctrine that a surety who completes a contract has an "equitable right" to indemnification out of a retained fund such as the one claimed by the surety in the present case. The only difference in the two cases is that here the surety incurred his losses by paying debts for the contractor rather than by finishing the contract.

The *Henningsen* case, decided 12 years later in 1908, carried the *Prairie Bank* case still closer to ours. Henningsen had contracts with the United States to construct public buildings. His surety stipulated not only that the contractor would perform and construct the buildings, but also, as stated by the Court, that he would "pay promptly and in full all persons supplying labor and material in the prosecution of the work contracted for." [Footnote omitted.] Henningsen completed the buildings according to contract but failed to pay his laborers and materialmen. The surety paid. This Court applied the equitable principles declared in the *Prairie Bank* case so as to entitle the surety to the same equitable claim to the retained fund that the surety in the *Prairie Bank* case was held to have. Thus the same equitable rules as to subrogation and property interests in a retained fund were held to exist whether a surety completes a contract or whether, though not called upon to complete the contract, it pays the laborers and materialmen. These two cases therefore, together with other cases that have followed them [footnote omitted], establish the surety's right to subrogation in such a fund whether its bond be for performance or payment. Unless this rule has been changed, the surety here has a right to this retained fund. . . .

We therefore hold in accord with the established legal principles stated above that the Government had a right to use the retained fund to pay laborers and materialmen; that the laborers and materialmen had a right to be paid out of the fund; that the contractor, had he completed his job and paid his laborers and materialmen, would have become entitled to the fund; and that the surety, having paid the laborers and materialmen, is entitled to the benefit of all these rights to the extent necessary to reimburse it. [Footnote omitted.] Consequently, since the surety in this case has paid out more than the amount of the existing fund, it has a right to all of it. On this basis the judgment of the Court of Appeals is affirmed.

Affirmed.

Mr. Justice WHITE dissents.

Mr. Justice CLARK, with whom Mr. Justice DOUGLAS and Mr. Justice BRENNAN join, concurring in the result. [Opinion omitted.]

---

1. "The right of subrogation is not founded on contract. It is a creature of equity; is enforced solely for the purpose of accomplishing the ends of substantial justice; and is independent of any contractual relations between the parties." Memphis & L.R.R. Co. v. Dow, 120 U.S. 287, 301–02, 7 S.Ct. 482, 488, 489, 30 L.Ed. 595 (1887).

2. *See, e.g.*, Hampton v. Phipps, 108 U.S. 260, 263, 2 S.Ct. 622, 27 L.Ed. 719 (1883); Lidderdale's Executors v. Robinson's Executor, 12 Wheat. 594, 6 L.Ed. 740 (1827); Duncan, Fox, & Co. v. North & South Wales Bank, 6 App. Cas. 1 (H.L. 1880). *See generally* SHELDON, SUBROGATION § 11 (1882).

3. *See* cases collected in 50 AM.JUR., *Subrogation*, § 49 (1944).

20.13 SUBGUARD POLICIES

In recent years, one insurance company, Zurich North America, has offered an insurance product called Subguard, which insures against subcontractor default. It is intended as a substitute for performance and payment bonds on subcontracts and runs in favor of the general contractor. General contractors sometimes prefer Subguard insurance to subcontractor bonds because they have more control over procurement of a substitute subcontractor or other means of resolving a subcontractor default. This type of coverage is not available directly to owners and does not protect an owner against a general contractor default.

QUESTIONS

1. What are the eight corners referred to in the "eight corners rule"? What are plausible alternatives to the eight corners rule and what are the pros and cons of each?
2. What business interests of an insurer would lead it to advocate the "manifestation rule" over the "occurrence rule" in determining whether a particular policy is triggered?
3. What business or public interests are reflected in the "business risk doctrine"? What policy provisions attempt to express this doctrine?
4. Reconstruct a plausible debate that would have led the insurance industry to adopt the subcontractor exception to the "your work" exclusion in comprehensive general liability (CGL) policies in 1986.
5. What presumptions does the court use as a basis for ruling against the insurer? Which of them are rebuttable?
6. Plaintiff is an HVAC subcontractor that claims it is owed money by a general contractor on a federal building project. Plaintiff has given the Miller Act surety notice of a claim more than 90 days after completing its punchlist but during the one-year warranty period under its subcontract. Sixty days after receipt of the notice of claim, the surety denied the claim as untimely and, because the subcontractor's work was late, exposed the general contractor to liquidated damages. The general contractor did not default on the general contract and has been paid all contract funds, including retainage, by the government.
   a. What defenses might the surety have to the claim?
   b. What additional defenses might the surety have if the claimant was a sub-subcontractor, not having a direct relationship with the bond principal?
   c. If the surety is liable to the subcontractor, what recourse might the surety have against other parties to make itself whole? What steps should the surety take upon receipt of notice to protect its rights of recourse?
   d. In light of the fact that "suretyship is not insurance," *Pearlman v. Reliance*, 371 U.S. at n.19, does the surety have exposure under state unfair claims settlement practices acts, such as Mass. Gen. Law ch. 176D § 3?
   e. When will the statute of limitations run out on the subcontractor's claims?

CHAPTER

21

Dispute Resolution Processes

JAMES P. GROTON and STANLEY P. SKLAR

21.01 IMPORTANCE TO THE CONSTRUCTION INDUSTRY OF PROCESSES FOR RESOLUTION OF DISPUTES

A. Unique Dispute Resolution Needs of Construction Projects

The construction industry has used private dispute resolution techniques longer than most industries. This has been a matter of necessity because formal legal remedies are much too slow and inflexible in the context of a fast-moving activity such as a construction project. Every construction project involves a multitude of players, all with potentially conflicting interests, whose mission is to come together to plan, design, and build a completed structure, on time and within budget. Once the building process has begun, the work cannot be interrupted or delayed without serious cost consequences. In the construction business, "time is money." Accordingly, the construction industry places a premium on coordination and quick solutions to problems.

B. Historical Development of Dispute Resolution in Construction

About one hundred years ago, the industry started using a combination of two forms of private dispute resolution that were designed to keep the project moving, reduce conflict, and provide quick and expert resolution of disputes: (1) nonbinding decisions by the project architect rendered immediately upon the appearance of a problem; and (2) a prompt, informal ad hoc arbitration convened promptly at the project site to provide a quick decision on any disputed issue involved in the architect's decision that could not be resolved through negotiation. These two techniques complemented each other: the architect's awareness that its decision could quickly and easily be challenged by a prompt arbitration gave the architect an incentive to be scrupulously fair in making decisions. As a result, relatively few architects' decisions needed to be challenged, and those that were challenged were resolved quickly and expertly.

About a generation ago, for reasons that are beyond the scope of this chapter, this system of early, single-issue dispute resolution broke down, and the construction industry went through a tortured period when architects' decisions were frequently ignored, and resolution of individual disputes was deferred. Although mediation provisions began to be incorporated into standard construction contracts, as a matter of practice parties rarely invoked either mediation or arbitration during the progress of construction, causing the resolution of disputes to be routinely postponed until the end of the construction period.

Unfortunately, when unresolved disputes were allowed to accumulate and fester, their mere existence interfered with the progress and overall success of the project. Unresolved problems held up the payment of money, created uncertainty as to the outcome, impaired relationships and efficiency, and caused delays and disruption. These in turn caused added costs to the project participants and tended to generate new claims and disputes.

The postponement of dispute resolution to the end of the project also increased the difficulty, expense, delay, and uncertainty of attempting, in a court trial, arbitration, or mediation long after the fact, to reconstruct accurately all of the facts and events that occurred during the lifetime of the project that were relevant to the resolution of the accumulated disputes. This escalated the transaction costs of resolving disputes to the point where the profits of all parties to the dispute, including the nominal "winner," were often wiped out.

A couple of decades ago, construction industry leaders took a fresh look at the problems of adversarial relationships, disputes, and litigation and began to invent new techniques to prevent, control, and quickly resolve disputes. Some of the key steps in the effort to develop new dispute resolution techniques follow.

The Business Roundtable and the Construction Industry Institute investigated the basic root causes of construction problems and pioneered research into realistic risk allocation and the use of incentives, strategic alliances, and long-term "partnering" to foster cooperative relationships. The Army Corps of Engineers experimented with some very successful uses of project-specific partnering. Civil engineers, contractors, and some results-oriented public owners invented "standing neutral" dispute review boards to resolve disputes at the job site level in major tunneling and civil projects.

The main emphasis in these newer devices was on restoring concepts of reasonable dealing and teamwork into the conduct of construction projects, preventing or at least controlling the impact of problems, improving communications, fostering mutual problem solving, encouraging the earliest possible resolution of problems, and assisting disputing parties to reach early resolution of any disputes so that valuable business relationships could be preserved. The effect of these new approaches was to move the dispute resolution process "upstream," closer in time to the sources of the disputes, to deal with root causes and not just the symptoms, of disputes.

This experimental work has matured to the point where there now exist several new, well-developed techniques that have proved to be successful in curbing disputes and getting them resolved during construction.

Nevertheless, given the complexities and uncertainties of the construction process and the conflicting interests of the many participants, many disputes do not get resolved during the course of construction. In these cases, the parties continue to resort to the more traditional dispute resolution techniques of mediation, arbitration, and litigation.

21.02 THE SPECTRUM OF CONSTRUCTION INDUSTRY DISPUTE PREVENTION AND RESOLUTION TECHNIQUES

Today's construction industry members now benefit from a vast array of private dispute prevention and resolution tools that make it entirely possible that a construction project can be concluded without serious disputes ever having to be litigated, arbitrated, or even mediated.

A. The Dispute Resolution Step Chart

These tools form a spectrum or continuum that can be illustrated by the "stair step" sequential model located on the next page, which depicts dispute prevention and resolution techniques in the order in which they would normally be employed in the life of the dispute, starting first with the techniques that help most in preventing or controlling disputes and offer the greatest potential for saving money and preserving relationships. They are grouped in Figure 1 (adapted from materials developed by DART, the Dispute Avoidance and

FIGURE 1

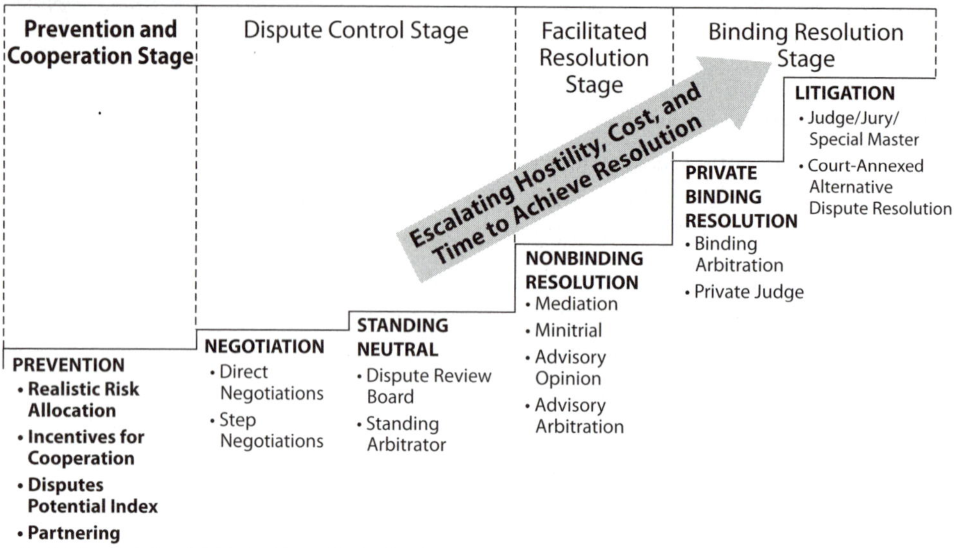

Resolution Task Force) into four successive (and escalating) stages of dispute resolution:

B. The Four Stages of Dispute Resolution

1. The Dispute Prevention and Cooperation Stage

The highest and best form of dispute resolution is prevention. One of the best ways to prevent disputes is to establish an atmosphere of cooperation, good faith, and reasonable dealing, both between the parties and among the members of each organization. An important tool in this regard is to make sure that risks are allocated realistically among the parties. Other techniques for encouraging teamwork, such as partnering, incentives for cooperation, and open communications, can help to create such an atmosphere and actually prevent some disputes from arising.

2. The Dispute Control Stage

Dealing promptly and realistically with problems and disagreements at the time they arise and before they can develop into full-fledged disputes can do much to contain and control disputes. Step negotiations, or obtaining "real-time" dispute resolution assistance from a pre-selected dispute review board or standing neutral who is already familiar with the relationship between the parties, can resolve disputes at the source, and the ready availability of these techniques can even help prevent disputes from developing.

3. The Facilitated Nonbinding Resolution Stage

If disputes have been allowed to mature to the point where positions have hardened and the parties are at an impasse, the next step is resorting to a more structured and formal facilitated dispute resolution process such as mediation, minitrial, expert's advisory opinion, or nonbinding advisory arbitration, with the assistance of one or more neutrals.

It is important to note that if the parties to a construction project are not successful in resolving all disputes during the course of construction, then the transactional costs of dealing with the disputes after the fact rise astronomically. At this point, the complexities and degree of hostility of the dispute begin to increase significantly because it becomes necessary to enlist the assistance of "strangers" to the parties' relationship—such as lawyers, outside experts, and outside neutrals—all of whom have to be educated about the background facts before they can meaningfully address the dispute. Also, with the passage of time and the dispersal of individuals who were involved in the project, the cost of reconstructing the facts becomes infinitely more difficult and expensive.

4. The Binding Resolution Stage

If all other efforts at resolution have failed, the last resort is some kind of "back stop" process for final and binding adjudication. If the parties have so agreed, that process will be arbitration; otherwise, the "default" process is court litigation. If the dispute reaches this stage, the parties have essentially lost control of the process and the outcome of the dispute; the transactional costs of processing the dispute escalate even more, and the relationship between the parties has likely been destroyed.

The various specific techniques that can be used to prevent, control, and resolve disputes will be described next in the order in which they would normally be employed in the life of the project or in the course of dealing with a developing dispute.

21.03 INDIVIDUAL CONSTRUCTION DISPUTE PREVENTION, CONTROL, AND EARLY RESOLUTION TECHNIQUES

A. Prevention and Cooperation Techniques

1. Realistic Allocation of Project Risks

Research by the Business Roundtable and the Construction Industry Institute has demonstrated that one of the most powerful ways to prevent and control disputes is to allocate project risks realistically, by assigning each potential risk to the participant in the project who is best able to manage, control, or ensure against the particular risk. It is now well recognized in the construction industry that when lawyers seek to negotiate "the best deal" for their clients, they frequently craft contract provisions that unrealistically and unfairly allocate risks to project participants who are unable to handle the risk, often creating project problems of a far greater magnitude than those the lawyers sought to solve. Realistic allocation of risks improves productivity, lowers costs, and creates better relationships among participants in the project. The result in nearly all cases will be fewer disputes and a greater chance for project success. The standard forms of agreement published by the American Institute of Architects and the Engineers Joint Contract Documents Committee are an attempt to reflect current industry consensus on sensible risk allocations.

2. *Incentives to Encourage Cooperation*

The construction industry has developed incentive plans to encourage multiple parties to work together toward a common goal. Such incentives can take many forms. For example, a general contractor or construction manager who needs to coordinate the activities of multiple subcontractors who have the potential of getting into conflict with each other can establish a "bonus pool" that, based upon the need for cooperation and the attainment of specific project goals, will be shared among all of the subcontractors. Under such a system, the bonus is payable only if *all* of the subcontractors as a group avoid conflict and meet the assigned goals—the bonus is payable either to everyone or to no one. This technique provides all participants with a common goal, places a premium on teamwork, furnishes a powerful business incentive to work cooperatively, and reduces conflicts that can occur in a common enterprise when every participant might otherwise be motivated solely by its limited perception of its own short-term interests, rather than the success of the enterprise as a whole.

3. *Partnering*

Partnering is a team-building effort developed by the construction industry in which the participants in a project establish cooperative working relationships through a mutually developed, formal strategy of commitment and communication. It can be used for long-term relationships, or on a project-specific basis.

Partnering on a long-term basis involves the development of a strategic alliance that maximizes the effectiveness of each participant's resources. The relationship is based upon trust, dedication to common goals, and understanding each other's individual expectations and values. The expected benefits from such a relationship include improved efficiencies and cost effectiveness, increased opportunity for innovation, and continual improvement of quality products and services.

When used on a project-specific basis, partnering is usually instituted at the beginning of the relationship by holding a retreat among all participants involved in the project who have leadership and management responsibilities, in which the participants, assisted by an independent facilitator, become acquainted with each other's objectives and expectations, recognize common aims, define mutual goals, develop a teamwork approach, initiate open communications, and establish nonadversarial processes for resolving potential problems. The initial retreat is followed up by periodic evaluations and meetings to ensure the communications and teamwork are continuing.

There is sometimes a misperception that partnering changes contract obligations. Partnering is strictly extracontractual, establishing a framework for the stakeholders to use for working together in a cooperative and efficient manner. It does not change any of the provisions of the contract, such as requirements for written notices of changes and differing conditions even though the parties are communicating verbally through the partnering process.

Partnering has achieved many successes. For example, the U.S. Army Corps of Engineers, which is credited with having inaugurated the use of project-specific partnering, claims that it has experienced an 85 percent reduction in construction claims and litigation by using partnering, along with other forms of alternative dispute resolution (ADR).[1]

Although the partnering process is generally accepted today as a "best management practice," it is by no means a panacea. Since it requires an "up-front" commitment to hard work and cooperation, it is not always possible to get the level of participation from all of the project stakeholders that is necessary for partnering to work. Sometimes partnering is instituted with the best of intentions, but in practice, parties may simply go through the formalities without seriously using it during the project to facilitate true communication and teamwork.

4. Open Communications and Notice Provisions

The best construction project results are obtained through good communications among project team members, so that incipient problems can be identified, brought out into the open, discussed, and solved, before they become serious. One important aspect of open communications is a contract requirement, consistently followed by all parties, that if any participant in the project needs information from another participant, or perceives that the scope of its work is being increased or impacted by any other participant, it must

---

1. Harmon, Kathleen, *Resolution of Construction Disputes: A Review of Current Methodologies*, 187 Leadership & Management in Engineering 196 (2003).

give prompt written notice to other involved parties. Careful adherence to this fundamental principle of communication, and good faith responses to such notices, will eliminate many potential misunderstandings.

This practice is encouraged by all of the standard construction industry documents. For example, the current edition of the AIA *General Conditions of the Contract for Construction*,[2] contains at least two dozen references to requirements for notices between the parties.

B. Dispute Control and "Real-Time" Early Resolution Techniques

1. Step Negotiations

Negotiation, of course, is the time-honored method by which parties try to resolve disputes through discussions and mutual agreement. There are many different techniques of negotiation. The most successful negotiations are those in which the negotiators conduct their discussions on the basis of the respective *interests* of the parties rather than the traditional approach of focusing on the *positions* of the parties. Negotiation is not only a freestanding dispute resolution technique, but it also can be a useful adjunct to every other dispute resolution technique.

A construction industry improvement on negotiation is the step negotiation procedure, a multi-tiered process that can often be used to break a deadlock. If the individuals from each organization at the lowest project level who are involved in the dispute are not able to resolve a problem promptly at their level, their immediate superiors are asked to confer and try to resolve the problem. If they fail, then the problem is to be passed on to higher management in both organizations. Because an intermediate manager is not likely to be as closely identified with the problem, the manager's interest in demonstrating to higher management that the manager has the ability to solve problems and to keep messy problems from bothering higher management is a built-in incentive to getting disputes resolved before they ever have to be referred to the highest management level.

2. Geotechnical Baseline Summary Report

One of the most common sources of disputes on a project is underground conditions, which, by their nature, are never completely known at the time the project is planned, designed, and priced. The Geotechnical Design Summary Report is a document in which the geotechnical conditions on which the designer based its design serves as a baseline by which to identify differing

2. American Institute of Architects, AIA Document A201–2007, *General Conditions of the Contract for Construction* (2007) [hereinafter AIA Document A201–2007].

site conditions that are actually experienced in the field during the course of construction. If used in connection with a "differing site conditions" contract clause in which the owner assumes the risks of unknown subsurface conditions, all competing contractors are able to bid on the same set of assumptions. Use of this device also fosters a more cooperative project climate because it demonstrates openness and candor in communications.

3. Escrow of Bid Documents

The pricing of changes in the scope of work on a project is the source of many disputes. One technique for helping to simplify this pricing process is to specify in the bidding documents that when the successful bidder is identified, that bidder's documents will be kept in an escrow where they will be accessible jointly by the contractor and owner to assist in negotiating price adjustments. Whenever any party believes that references to the original takeoffs, calculations, quotations, and other information that the successful bidder used to arrive at the bid price are relevant to review in a claim, the documents are available to be consulted by representatives of both parties. The availability of these documents is another way to foster a climate of openness and cooperation on a project.

4. Dispute Resolution Boards and Standing Neutrals

One of the most innovative and promising developments in controlling disputes between participants in a construction project is the concept of having a pre-selected or "standing" respected neutral expert serve the parties as a dispute resolver throughout the course of the project. This neutral, or a board of three disinterested, knowledgeable experts, called a dispute review board or dispute resolution board (DRB), is jointly selected by the parties early in the project relationship, is briefed on the nature of the project, is furnished with the basic construction documents, routinely receives periodic progress reports as the project progresses, and occasionally attends job site meetings with the parties simply to get a feel for the dynamics and progress of the project relationships. The dispute review board or individual neutral is expected to be available on relatively short notice to make an expert, reasoned written recommendation to the parties as to how to resolve any disputes that the parties are not able to resolve themselves.

When called upon to deal with the dispute, the neutral typically responds quickly, obtaining written position papers from the parties and then conducting a hearing (usually at or near the project site) at which the parties present their positions and respond to questions from the neutral members. The hearings are conducted informally and include decision makers and representatives of the parties with firsthand knowledge of the project and the issues in dispute.

Following the hearing, the neutral will provide a prompt written recommendation, with supporting rationale.

Although the neutral's decisions are typically not binding, experience has shown that parties have almost universally accepted these decisions as the basis for resolving the dispute, without any attempt to seek relief from any other tribunal. This result is enhanced where there is a contractual requirement that in the event of any subsequent arbitration or litigation, the decisions of third parties will be admissible in evidence. When used in accordance with the guidelines recommended by the Dispute Resolution Board Foundation or the American Arbitration Association's Dispute Review Board Procedures, this technique has been successful in establishing the following track record, according to detailed statistics maintained since 1975 by the Dispute Resolution Board Foundation:

- 60 percent of projects with a DRB have had no disputes that made it to the DRB;
- 98 percent of disputes that have been referred to a DRB for hearing have resulted in no subsequent litigation or arbitration; and
- Through the end of 2006, worldwide, over 2,000 projects with a total value of over $100 billion have used some form of DRB, with the rate of growth in use of the process now in excess of 15 percent per year.

The evaluative but nonbinding nature of the neutral review process, by providing a "dose of reality" that encourages parties to be more realistic and objective in their dealings with each other, while also giving the parties an opportunity to construct their own solutions to problems, tends to strengthen the relationship between the parties and create trust and confidence between them.

The existence of a pre-selected neutral, already familiar with the business relationship between the parties and its progress, avoids many of the initial problems and delays that are involved in selecting and appointing neutrals after a controversy has arisen. The ready availability of the neutral, the speed with which it can render decisions, and particularly the fact that this neutral will hear every dispute that occurs during the life of the relationship, all provide powerful incentives to the parties to deal with each other and the neutral in a timely and frank manner, by discouraging game-playing, dilatory tactics, and the taking of extreme and insupportable positions. In practice, the nature of this process is such that the mere existence of the neutral always results in minimizing—and often totally eliminating—the number of disputes that have to be presented to the neutral. In effect, the standing neutral serves not only as a standby dispute *resolution* technique but also as a remarkably successful dispute *prevention or control* device. Even though some expense is involved in the process of selecting, appointing, initially orienting, and periodically keeping the neutral informed about the relationship, the costs are relatively minimal, even in those rare cases where the neutral has to be called on to resolve disputes—especially when compared to the potential costs of resolving a dispute in arbitration or litigation, or even the expense of mediating a dispute after the project is completed.

The concept of the standing neutral has now been included in the new AIA form of documents. As was the case previously, under the *Standard Form of Agreement between Owner and Contractor,*[3] claims are to be initially referred to the architect as the "Initial Decision Maker" (IDM). However, the parties may appoint another person other than the architect to be the Initial Decision Maker and who will render a decision on the claim.

The nature of DRB recommendations has tended to keep the process relatively nonadversarial, with little active involvement of lawyers in the conduct of hearings. Lawyers who are highly experienced in construction have, however, been used with considerable success as members of DRB panels, particularly where issues arise that involve contract law and interpretation of complex contracts.

C. Facilitated Nonbinding Resolution of Disputes

1. Mediation

During the past twenty years, construction industry participants who have experienced difficulty in negotiating resolution of disputes have customarily turned to the use of disinterested neutrals to assist them in reaching resolution without having to resort to a binding dispute resolution process. The most common of these techniques is mediation. A neutral mediator can guide the opposing parties in a dispute through facilitated negotiations toward a settlement of their disputes, promptly and constructively.

In past practice, mediation has rarely been used to resolve disputes during the course of construction. Even though the use of dispute review boards is not yet a common practice on construction projects, anecdotal data suggest that many more disputes are typically referred to DRBs than to mediation. Reasons that have been advanced for this contrast are that the DRB is readily available to deal promptly with disputes, and the parties may be looking, not for

3. American Institute of Architects, AIA Document A101–2007, *Standard Form of Agreement Between Owner and Contractor* (2007) [hereinafter AIA Document A101–2007].

the kind of compromise that typically occurs in mediation, but rather for the expert "dose of reality" that a DRB delivers.

However, AIA document A201-2007 now expressly places mediation directly into the stream of a new in-project dispute resolution scheme. Under this scheme, most claims and disputes between the owner and the contractor must first be referred to an Initial Decision Maker (who is the project architect, or an independent neutral if the parties have so agreed), who must promptly review the claim, seek any needed additional information, and render an initial decision approving or rejecting the claim, or indicating that the IDM is unable to resolve the claim. This initial decision shall be (1) in writing; (2) state the reasons therefor; and (3) specify any change in the contract sum, the contract time, or both. This initial decision "shall be final and binding on the parties but subject to mediation." Mediation may be instituted as follows: Within 30 days from the date of the initial decision, either party may demand in writing that the other party file for mediation within 60 days after the initial decision. If such a demand is made, and the party receiving the demand fails to file for mediation within the time required, then both parties waive their rights to mediate the dispute or to pursue any binding dispute resolution proceeding with respect to the IDM's decision. Thus, this scheme requires that unless a dispute is promptly mediated, the decision of the IDM is final and binding.

Since this dispute resolution scheme is brand new as of the writing of this chapter, it remains to be seen how effective it will be in getting project disputes resolved during the course of construction (thus, in a sense, reviving the older model of in-project dispute resolution through the architect's decision, complemented by a prompt ad hoc neutral dispute resolution process).

2. *Minitrial*

A variant of mediation is the minitrial, a process in which representatives of both parties present their "best case" arguments to a panel consisting of principal executives of both parties and a disinterested neutral. Following the presentations, the executives confer and try to resolve the dispute in the light of the information they have just received. If the executives are unable to reach an agreement, the neutral steps in to try to facilitate further negotiation and reach an agreed resolution.

3. *Advisory Opinions*

Occasionally, all that disputing parties need to resolve a technical dispute is a "dose of reality." If the parties engage an independent neutral expert in the field to meet with the parties, both separately and jointly, obtain information about the issue, and render an objective expert opinion on the subject, the parties often recognize the realities of the situation and reach a negotiated settlement, sometimes without the aid of a mediator.

4. Nonbinding Summary Advisory Arbitration

A somewhat more formal process for providing a "dose of reality" to opposing parties is to engage in an informal summary arbitration in which representatives of the parties present their "best case" arguments to an arbitrator or arbitration panel, which is then asked to issue a nonbinding decision or realistic assessment of the issue. Such a ruling often leads to a consensual resolution of the dispute.

5. Binding Arbitration

The final step in any private dispute resolution process is binding arbitration pursuant to an express agreement between the parties that they will submit their cases to an independent neutral arbitrator who will issue an award or decision that, if necessary, can be enforced by a court. The arbitration is typically conducted in accordance with an accepted set of rules, such as the American Arbitration Association (AAA) Construction Arbitration Rules or the International Institute for Conflict Prevention & Resolution (CPR) Rules for Non-Administered Arbitration, or a "fast track" version of those rules. (A detailed discussion of arbitration is contained in Section 21.05 of this chapter.)

21.04 DESIGN OF DISPUTE PREVENTION AND RESOLUTION SYSTEMS

The foregoing description of currently used dispute prevention and resolution techniques has illustrated a progression of tools that can be used to prevent and resolve disputes and avoid litigation. The purpose in developing these techniques has been to move the dispute process "upstream," closer in time to the sources of disputes.

Since no single dispute prevention or resolution technique can serve all dispute resolution needs, the best practice for achieving optimum results is to include in the construction or design contract a well-planned "system" for dispute prevention, control, and final resolution. Dispute resolution system design is simply a process of analyzing the likelihood that a particular relationship or project will have disputes, predicting the kinds of disputes that might arise, and then designing a succession of techniques, filters, screens, and safety nets, using the available tools and techniques, to deal with a succession of different types of problems at different stages in the life of the project, so that problems are likely to be resolved at the earliest possible time.

One of the most important elements in the effectiveness of a dispute resolution system is to institute the system at the beginning of the project and incorporate it into the basic contract documents. Without prior agreement on a process for dealing with problems and disputes, it can be difficult, after a disagreement has developed, to get parties to come to agreement on a method of dispute resolution. A process that is already in place will provide a method for absorbing the impact of a problem; however, without a process, there is no effective way to control the outcome.

A number of contracts provide for a system that has a succession of "steps" that incorporate techniques to deal with various kinds of problems at different stages in the development of a dispute, such as techniques for prevention, techniques for control, a process for nonbinding resolution, and finally a "back stop" final resolution process. For example, on a given construction project, the participants may agree to use partnering, accompanied by a number of other devices for preventing and reducing disputes. These techniques might include

- Specific allocation of time and other resources, early in the project, to foster mutual trust as well as appropriate mutual expectations among the many players;
- Reducing conflicts from disappointed expectations or perceptions of injury;
- Explicit and realistic allocation of risks among parties;
- Incentives for cooperation built into payment structures;
- "Stepped" negotiations when problems occur, implemented immediately;
- A standing review, such as a Dispute Review Board; and
- Arbitration in the event that none of these techniques work.

21.05 PRINCIPLES OF ARBITRATION

A. Basis in Statute

Due to the increasingly complex nature of business, crowded court calendars, and the rising cost of litigation, it became apparent that parties should be entitled to select an alternative method of resolving disputes in a forum of their own choosing.

Perhaps the most compelling reason for arbitration is the need to have triers of fact who are experts in the business matters in dispute. Consequently, the doctrine that arbitration agreements may be revoked by either party has now been superseded by the Federal Arbitration Act (FAA).[4] At the present time, all states enforce agreements to arbitrate existing disputes, and 49 states, the District of Columbia, and the FAA also validate agreements to arbitrate future disputes. Both the federal and state acts make arbitration agreements irrevocable and enforceable and provide a mechanism for compelling arbitration and staying court proceedings until arbitration has been completed.

It has long been recognized that the FAA is a substantive rule that, by virtue of the Supremacy Clause of the Constitution, forecloses any state statute from undercutting the enforceability of arbitration agreements.[5] The deference that is given to arbitration and the federal policy that underlies the FAA thus supports the stay of state litigation pending the outcome of arbitration.[6] Therefore, state statutes that govern arbitration are invalid if they interfere with the policy of the FAA.[7] Even though state statutes vary, there is a model Uniform Arbitration Act.[8]

B. Contractual Nature of the Remedy

An arbitration statute, by itself, gives no right to arbitrate. In the absence of a written arbitration agreement, a party may not be required to arbitrate. Not only is it necessary to have an agreement to arbitrate in order to require arbitration, it is also appropriate that the agreement reflect the intention of the parties regarding the nature of the arbitration proceeding. Since arbitration is a creature of contract, the parties have broad leeway to prescribe in their contract agreement the nature and scope of the arbitration proceeding.

Various arbitration organizations prescribe model provisions for the parties to use in framing an arbitration agreement. The American Arbitration Association (AAA) suggests the following model provision for use in construction contracts:

> Any controversy or claim arising out of or relating to this contract, or the breach thereof, shall be settled by arbitration administered by the American Arbitration Association Rules, and judgment on the award rendered by the arbitrator(s) may be entered in any court having jurisdiction thereof.

---

4. 9 U.S.C. § 1.

5. Southland Corp. v. Keating, 465 U.S. 1 (1984).

6. 9 U.S.C. § 3.

7. David L. Threlkeld & Co., Inc. v. Metallgesellschaft Ltd. (London), 923 F.2d 245 (2nd Cir. 1991).

8. The Uniform Act is presently being revised by the National Conference of Commissioners on Uniform State Laws (NCCUSL). *See* Thomas J. Stipanowich, *What's a NCCUSL? The Uniform Law Commissioners Tackle the Uniform Arbitration Act,* The Construction Lawyer, July 1997, at 25.

By inserting this provision in their agreement, the parties have adopted the Construction Industry Arbitration Rules of the AAA, which include extensive provisions for the selection of arbitrators and the administration and hearing of arbitration cases. Parties should understand these rules before incorporating them into their contract.

Beyond the use of standard clauses, parties may wish to agree in advance on a number of aspects of the arbitration proceeding. This includes the manner of selecting the arbitrators, the location of the arbitration hearing, and the choice of the state law to be applied by the arbitrators. Consideration should also be given to the description of the scope of the arbitration delineating the subjects that the arbitrators may consider, as well as the procedures the arbitrators should follow, and the damages or other relief they may award.[9]

The power of the arbitrator to arbitrate a particular issue is derived from the agreement of the parties. A frequently contested issue is whether a particular controversy is covered by the arbitration agreement. Claims under construction contracts concern such diverse matters as extras, delays, balance due under the contract, deficiencies in performance, liquidated damages, and faulty design. The ability to arbitrate such issues depends upon the language of the agreement.

The broad arbitration clauses of the American Arbitration Association have been held to cover such typical disputes as those involving extras, delays, claims for retainage, deficiencies in performance, and liquidated damages.

The place of arbitration often may be important because of applicable procedural or substantive law bearing upon the enforceability of the arbitration clause or final award. Arbitration locale may be designated in the arbitration clause itself, either expressly or by implication. Agreement of the parties respecting locale will be enforced. The parties may mutually agree on the locale where the arbitration is to be held.

Under common law, arbitrators lack the authority to order discovery. Despite this limitation, there still are several ways to bring about discovery of relevant documents. For example, parties may always agree to voluntarily exchange information. In addition, the rule limiting discovery in aid of arbitration is inapplicable when both parties seek discovery from a third party. The Federal Arbitration Act also permits arbitrators to summon any person to appear before them or to produce any book, record, document, or paper that may be deemed material evidence in the case. When parties refuse to voluntarily exchange information, one party may air the subject before the arbitrator

9. *See* Jonathan D. Shaffer, Carl T. Hahn, and William A. Focht, *Challenges & Pitfalls in Drafting "Arbitration" Clauses,* Construction Briefings No. 97-4 (1997).

in an effort to use the negative inference that might be drawn by the arbitrator as a weapon to compel discovery. Another practical method used by imaginative counsel is to obtain a subpoena *duces tecum* to compel production of documents or things at the hearing. Many states empower arbitrators to issue subpoenas. In addition, the rules of the AAA permit limited discovery in certain cases.

Arbitrators are entitled to fees for services unless otherwise agreed. Parties to arbitration proceedings are expected to compensate an arbitrator fairly and reasonably for its services. For example, Rule R-52 of the Construction Industry Arbitration Rules of the AAA provides: "Arbitrators shall be compensated a rate consistent with the arbitrator's stated rate of compensation." The Federal Arbitration Act makes no express provision for arbitrators' fees. In contrast, the Uniform Arbitration Act 10 states: ''Unless otherwise provided in the agreement to arbitrate, the arbitrators' expenses and fees, together with other expenses, not including counsel fees, incurred in the conduct of the arbitration, shall be paid as provided in the award.''

Unlike the 1997 edition of the AIA documents, the 2007 version does not present arbitration as the sole binding method of dispute resolution. In the 2007 version, the parties are to select arbitration, litigation, or another method of resolving disputes following the decision of the "Initial Decision Maker."[10] Arbitrations are to be administered under the Construction Industry Arbitration Rules of the American Arbitration Association.[11]

C. Attacking Arbitration

Attacks on the right to arbitrate may be made either before the arbitration commences by petition to stay arbitration or in opposition to a petition to compel arbitration. Such proceedings are generally heard in the manner provided for the hearing of motions.[12] Such an attack may also be made after the arbitration award has been rendered in a proceeding to confirm or vacate such award, provided that objection has been made in the arbitration proceeding. The Uniform Arbitration Act contains a provision similar to that of the Federal Arbitration Act providing only for appeal of an order denying an application to compel arbitration.

As the right to arbitrate arises from the agreement of the parties, that right may be waived, just as any other contractual right may be waived. Since the

10. *E.g.*, AIA Document A101–2007, *supra* note 3, § 6.2.
11. *E.g.*, AIA Document A201–2007, *supra* note 2, § 15.4.
12. 9 U.S.C. § 6; Uniform Arbitration Act § 16.

parties' acts, and particularly court litigation, which is inconsistent with arbitration, may be deemed a waiver of the right to arbitrate, a contractor with a limited statutory period for filing and enforcing its mechanics' lien rights faces a difficult dilemma. Will the filing of the lien claim or suit on the lien waive arbitration? To protect its rights to arbitrate, a contractor who wishes to enforce its rights to a mechanic's lien and also arbitrate its dispute may begin by filing an arbitration demand and afterwards file its lien and complaint to foreclose. In the complaint, the contractor must specifically reserve its right to arbitrate and immediately move for a stay of the litigation pending arbitration.

D. Modification of Award by Arbitrators

In the absence of statute, an arbitrator who has rendered an award ceases to have any authority to modify the award. However, provided that application is timely made (20 days after delivery of award under the Uniform Arbitration Act), arbitrators may modify or correct an award (1) where there is an evident miscalculation of figures or an evident mistake in the description of any person, thing, or property referred to in the award; (2) where the award is imperfect in a matter of form not affecting the merits of the controversy; or (3) for the purpose of clarifying the award. As with the case of vacating awards, courts are reluctant to modify awards.[13] Parties who seek modification by the arbitrator and also wish to have the award vacated must be careful to file a motion to vacate the award within the statutory time limit, for the application to the arbitrator for modification does not extend the statutory time limit for a motion to vacate.

E. Confirmation or Vacation of Award by Court

Although the issue is rare, since parties generally seek relief in the state where the award is entered, the question sometimes arises whether an award may be vacated, modified, or confirmed in a state other than the state where the arbitration took place and the award made. In several cases, the courts have held that an arbitration award in one state is legally equivalent to the judgment of a court and may be enforced in another state.

Before the enactment of modern arbitration statutes, it was well established under common law, and remains the rule, that a court will not set aside an award for error in law or fact. Accordingly, a court will not set aside an

13. Prestige Ford v. Ford Dealer Computer Services, 324 F.3d 391 (5th Cir. 2003).

arbitrator's decision on such grounds as newly discovered evidence or that the matter could not be resolved without a party not involved in the arbitration.

The grounds for vacating an award under the Federal Arbitration Act are very limited. They include when the award is procured by corruption, fraud, or undue means; when there is evident arbitrator partiality or corruption; when the arbitrators refuse to postpone the hearing or hear pertinent evidence; or when the arbitrator exceeded its power.[14] State arbitration statutes usually include similar grounds. Certain courts have established some nonstatutory grounds for vacating arbitration awards. Some courts have considered vacating awards that involved "manifest disregard of the law" or were contrary to "public policy" or "lack[ed] due process." In general, these grounds have been very narrowly construed, and the burden born by a party seeking to overturn them is very large.[15]

On March 25, 2008, the U.S. Supreme Court decided the case of *Hall Street Assoc., L.L.C. v. Mattel, Inc.*[16] This case involved the effort of the parties to an arbitration agreement to expand the grounds of court review of the arbitration award. The parties agreed that the award could be vacated, modified, or corrected "(i) where the arbitrator's findings of fact are not supported by substantial evidence, or (ii) where the arbitrator's conclusions of law are erroneous." The court refused to allow a federal court applying the FAA to expand its review of the decision of the arbitrators past the limited statutory grounds found in Sections 9, 10, and 11 of the FAA. Therefore, the lower court was not allowed to predicate its review on the standards set forth by the parties in their contract.[17]

The Supreme Court's opinion in *Hall Street Associates* has therefore put to rest the question of whether the parties in their arbitration agreement can expand the grounds of review of a federal court considering an arbitration

---

14. 9 U.S.C. § 10.

15. *See* David E. Robbins, *Calling All Arbitrators: Reclaim Control of the Arbitration Process—The Courts Let You,* Dispute Resolution Journal, Feb.–Apr. 2005.

16. ______ U.S. ______, 128 S.Ct. 1396 (2008).

17. The Supreme Court noted that in an earlier decision [Wilko v. Swan, 346 U.S. 427, 74 S. Ct. 182 (1953)] it had used the term "manifest disregard [of the law]" but indicated that that term may have been a collective reference to the grounds set forth in the FAA as well as a new ground for review. In addition, the Supreme Court noted that it did not purport to exclude a "more searching review based on authority outside" the FAA. The Court stated: "The FAA is not the only way into court for parties wanting review of arbitration awards: they may contemplate enforcement under state statutory or common law, for example, where judicial review of different scope is arguable."

award under the FAA. Whether there is room under state law for expanded review, either by agreement of the parties or by state statutes, is not clear.

Under the Uniform Arbitration Act 12(c), where a court vacates an award for reasons other than the lack of an arbitration agreement, it may remand the case to the arbitrators for rehearing. A new panel must be appointed where fraud or similar grounds are involved. Otherwise, the remand may be to the old panel, or a new panel, at the court's discretion. A petition for vacating or modifying an award must be filed within certain narrowly prescribed time limits to be considered. Under the Federal Arbitration Act, the limit is three months after the award is filed or delivered.

QUESTIONS

1. During a partnering session, managers of the contractor and the owner agree to personally discuss any potential changes in the project in order to avoid a "letter-writing war." Later, during the project, in response to claims for extras by the contractor, the owner asserts that the contractor waived its claims by failing to give written notice as required by the contract. Has the contractor waived its claims?
2. Following the conclusion of a project constructed under the 2007 AIA documents, the contractor commences an arbitration provision over a long-standing dispute regarding extra work. The owner responds by arguing that the arbitration proceeding is premature since there has been no initial decision, nor any mediation. The contractor responds that it is clear the parties disagree over this matter and arbitration should proceed. Should it proceed? Who decides?
3. A new state statute allows the state court to vacate an award if the award is in manifest disregard of the law. After the respondent files a petition to vacate in state court, the claimant, who is a citizen of another state, endeavors to remove the action to the federal court and seek an order that the state statute is invalid. Will the claimant prevail?

CHAPTER

22

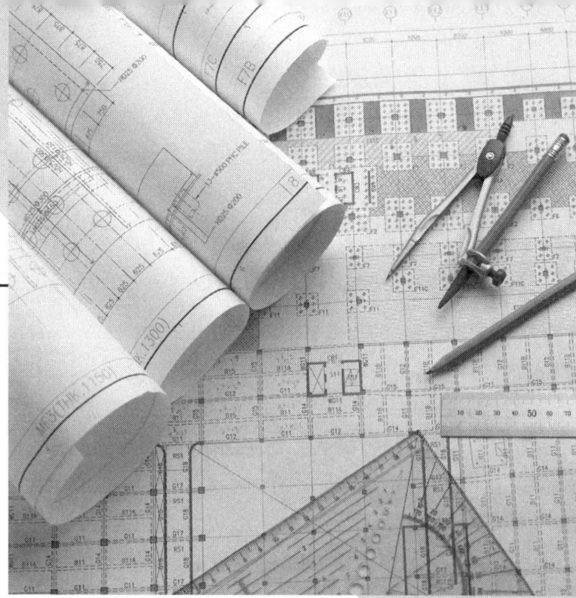

Defective Construction

RICHARD J. TYLER

Although the phrase "construction defect" or "deficiency in construction" is a convenient way to refer to some flaw in a structure that distinguishes the problem from, say, a payment dispute or an OSHA claim, the use of these shorthand phrases belies the complexity of the subject. The mission of this chapter is to bring some analytical order to the broad topic of "construction defects," identifying some of the myriad ways in which construction might be deemed "defective" or "deficient," outlining the specific legal claims under which an owner can recover for defective or deficient construction, examining some defenses that a party charged with defective or deficient construction might assert, and addressing a pair of special issues that arise in dealing with construction defects.

22.01 OVERVIEW OF THE LEGAL ANALYSIS OF DEFECTS

In this section, we provide a rough overview of construction defect analysis by reviewing the types of flaws that may qualify as defective or deficient construction, and then we provide an overview of the assignment of responsibility for such flaws.

The author thanks Michael Drew, an associate in the New Orleans office of Jones Walker; Carla Ashley, an associate in the firm's Baton Rouge office; and Mark Mercante, a partner in the Mandeville, Louisiana office of Baker, Donelson, Bearman, Caldwell & Berkowitz, PC, for their valuable assistance in the preparation of this chapter.

A. What Constitutes "Defective" or "Deficient" Construction?

1. *Types of Defects or Deficiencies*

There are many ways in which a property owner may be dissatisfied with the results of the design or building process, and the variety of dissatisfaction makes uniform definition of a construction "defect" elusive. One formulation describes a construction defect as work performed that falls below the standard promised or reasonably expected by the purchaser of the work or services. Another formulation looks at whether there is a lessening in the utility and/or value of the project. Yet a third formulation looks to the appropriateness or suitability of the work. However construed, there is general agreement that actionable defects arise out of the construction process itself and that normal wear and tear are not "defects" for purposes of construction claims.[1]

For purposes of this chapter, we will consider completed construction "defective" or "deficient" whenever it fails to satisfy an applicable contractual or legal standard that imposes requirements on the construction. The task of the next few sections will be to describe the standards that apply to construction such that a deviation renders it defective or deficient in some actionable sense.

From a broad perspective, consider the many types of deficiencies that might condemn work as defective: (1) materials and equipment furnished are not of the quality required or permitted by the contract (e.g., the pump installed does not work or has insufficient capacity); (2) the work is executed in a defective manner (e.g., paint is applied with inadequate surface preparation, resulting in premature scaling or peeling of the paint and eventual rusting of metal structural members); (3) the work, while executed correctly technically, does not conform to the requirements of the contract (e.g., the contractor applies the wrong color paint);[2] (4) the construction itself is perfectly sound but violates building codes in effect at the time of design or construction (e.g., a fire suppression system is functional but undersized); or (5) the work fails to meet professional standards for design at the time drawings were approved

---

1. *See* Richards v. Powercraft Homes, Inc., 678 P.2d 427, 430 (Ariz. 1984).

2. One treatise suggests that nonconforming work may not be defective where there is no economic consequence to the owner, such as the application of the wrong color paint. *See* 2 PHILIP L. BRUNER & PATRICK J. O'CONNOR, JR., BRUNER & O'CONNOR ON CONSTRUCTION LAW § 7:215 (West Group 2002) [hereinafter BRUNER & O'CONNOR]. *But see* American Institute of Architects, AIA Document A201–2007, *General Conditions of the Contract for Construction* § 3.5.1 (2007) [hereinafter AIA Document A201–2007] (listing circumstances under which work may be considered defective, including nonconformance with the requirements of the contract documents). We intend the phrase "defective or deficient" construction to encompass all failures to satisfy applicable standards of construction as discussed in this chapter.

(e.g., the roof is designed without taking into account snow loads and the building collapses during a heavy winter storm). Although each of these may constitute a construction "defect" or "deficiency," the legal standards under which they are condemned as flaws may be different for each of the causes listed. In assessing claims for construction defects, the first step is to determine the standard that applied to the construction at issue and the manner in which the construction failed to satisfy that standard. The variety and sources of standards are discussed throughout the chapter.

2. *Manifestations versus Defects*

As the reader approaches construction defects claims, it is vital to note that defects or deficiencies must be distinguished from their manifestations. While these terms are related—and, often, both must be corrected—a manifestation is to a defect what a symptom is to a disease. The manifestation is the obvious or apparent visible condition that serves as notice of the possibility of a defect.[3] While a defect and its manifestation may be the same thing, oftentimes they are not. For example, a crack in a newly constructed wall is the manifestation of some defect—but what defect? Consider:

1. Is the crack the result of a problem with the material used to construct the wall?
    a. Were the material components defectively manufactured?
    b. Did the design call for improper materials to be used on the wall?
    c. Was the material improperly prepared?
    d. Was the material improperly installed?
2. Is the crack the result of a problem with how the wall is attached to the rest of the structure?
    a. Did the design call for adequate reinforcement?
    b. Was the wall improperly constructed?
    c. Was the material reinforcing the wall defective?
3. Is the crack the result of a problem with how the structure's foundation was constructed?
    a. Were the cement or supports improperly manufactured?
    b. Was the foundation improperly designed?
    c. Was the foundation improperly constructed?

---

3. *See generally* Bd. of Trustees of Santa Fe Comty. Coll. v. Caudill Rowlett Scott, Inc., 461 So. 2d 239 (Fla. Dist. Ct. App. 1984). There, the plaintiff discovered leaky pipes shortly after the completion of construction. However, it was not until years after this discovery that plaintiff was able to determine that the defect causing these leaks was the contractor's decision to bury the pipes in clay, causing them to corrode.

4. Is the crack the result of a problem with the ground on which the structure was built?
   a. Will the site not support the building (i.e., unsuitable soils)?
   b. Did the design fail to consider or accommodate the soil type?
   c. Did the geotechnical engineer fail to record the soil type accurately?
   d. Did the contractor know the design was inadequate or the soils unsuitable?
5. Is the crack the result of some problem not connected with the construction (and likely not the responsibility of a party), such as an earthquake of unanticipated magnitude?
6. Is the manifestation of a purported "defect" merely a common, expected, and nonactionable result of natural forces, such as the common cracking that predictably accompanies concrete flatwork in many jurisdictions that experience dramatic climate changes?

This array of potential causes for something as "simple" as a crack in a wall reflects not only the fact that causation may be a complex issue, but that this complexity is mired in attributing responsibility to a particular party, to which topic we now turn.

B. Responsibility

The fact that a symptom such as a crack may be evidence of many discrete construction deficiencies leads immediately to a second important observation: responsibility for defective or deficient construction may be visited on any number of parties, from the owner's design professional team (including engineering consultants) to the contractor team (including subcontractors or materialmen). Indeed, the symptom under scrutiny may not evidence a *construction* defect at all, but instead may occur in circumstances in which there is no liability for the crack in the wall.

Ascertaining which party is responsible for the defect can be a complicated matter. Depending on the defect that caused the manifestation, any number of different parties may ultimately bear responsibility. For instance, if the crack was due to defective plaster supplied by the manufacturer, then the manufacturer will bear the responsibility for correcting the defect or paying damages to the owner (assuming no contractual limitations of liability). Similarly, if the crack was due to poor soil conditions that the civil engineer failed to identify in his site investigation, the defect will likely be to his account. However, the engineer's liability may be reduced, or even eliminated, if the contractor encountered the soil condition and failed to bring it to the attention of the owner and engineer. Suffice it to say, the assessment of responsibility is a fact-intensive investigation carried out within the framework of the parties' contractual obligations.

A problem such as a severely cracked wall may in fact be caused by an earthquake of unusual magnitude where the owner cannot hold another party liable for the cost of repair. The mere existence of a problem, therefore, does not compel the conclusion either that construction is defective or that someone must be liable to the owner for damages. What distinguishes a cracked wall for which there is a legal remedy from a cracked wall for which no remedy is available? In short, a remedy is available when the crack is the proximate result of some party's failure to conform its work to an applicable legal obligation governing such work. That legal obligation may arise from the voluntary agreement of the parties (such as a construction contract), it may be imposed through common law doctrines (such as warranties of fitness), or it may involve statutory and other governmental requirements (such as building codes).

The next two sections summarize the important sources of standards whose violation will render work defective or deficient and give the owner the right to recover from the party responsible for the substandard work. For simplicity of discussion, the standards have been divided between the contractor's construction standards and the design professional's liability. In reality, there are many other parties typically associated with one of those two parties who ultimately may bear responsibility for defective or deficient construction.[4]

22.02 OWNER CLAIMS AGAINST CONTRACTORS

Construction law generally recognizes three legal bases from which a defect claim may be asserted against the contractor.

A. Contract Claims

Claims that the contractor has breached the contract for the construction of the work are arguably the most common category of defect claims. Such claims typically focus on whether the contractor complied with the express or implied terms of the contract.

4. As but one example, defectively manufactured mortar may leave the manufacturer liable, as well as the subcontractor who used the mortar, as well as the contractor who hired the masonry subcontractor. For purposes of the following discussion, we discuss the legal bases of the contractor's liability to the owner alone, although downstream liability follows similar legal principles.

Construction contracts typically contain multiple provisions designed to ensure that the owner receives work that is not defective. Construction contracts usually incorporate by reference the "contract documents," which include the drawings and specifications.[5] The contractor is obligated to perform the work in accordance with the contract documents.[6] The contractor is under an obligation to review the contract documents for the purpose of facilitating construction and to report to the owner and architect any errors or omissions discovered by the contractor.[7] During the performance of the work, the contractor is obligated to correct work that is rejected by the architect or fails to conform to the requirements of the contract documents.[8] Finally, contractors are typically required to repair defective or nonconforming work discovered during the one-year period following substantial completion (commonly referred to as the contractor's "repair warranty").[9]

1. *Failure to Comply with Drawings and Specifications*

Most Design-Bid-Build construction contracts[10] incorporate specific drawings and specifications that the contractor is obligated to follow in completing its work. A contractor's primary responsibility in completing construction is to adhere to the drawings and specifications, and deviations from the drawings or specifications may render the contractor liable to the owner for such deficiencies or defects.[11]

Drawings and specifications are therefore fertile grounds for finding construction defects. Drawings and specifications generally define three aspects of the construction project: (1) the measurements, (2) the configuration, and (3) the materials to be used in completing the project. These guidelines may be general or may provide detailed instructions on how to execute the work.[12]

---

5. *See, e.g.*, AIA Document A201–2007, *supra* note 2, § 1.1.1.

6. *Id.* § 3.1.2.

7. *Id.* § 3.2.

8. *Id.* § 12.2.1.

9. *Id.* § 12.2.2.

10. Discussed in Chapter 4. *See also* Section 22.04(A), *infra*.

11. The mere deviation from the drawings and specifications itself constitutes a breach of the contractor's obligations, but inconsequential deviations may not give rise to a claim where there is no cognizable "damage" associated with the breach. *See* note 2, *supra*, discussing application of incorrect paint color.

12. *Compare* J. R. Sinnott Carpentry, Inc. v. Phillips, 443 N.E.2d 597 (Ill. App. Ct. 1982) (contract specification was only a rough drawing made by owner showing location of rooms in proposed house addition) *with* Brewer v. Custom Builders Corp., 356 N.E.2d 565 (Ill. App. Ct. 1976) (professional blueprints were incorporated into contract and showed details such as type of wood to be used in stair construction).

Each of these aspects may be subject to a separate defect claim, or all three may be part of a single claim.

2. Breach of Express Warranty

A second source of contract-based defect claims can arise out of the express warranties contained in the contract with regard to the quality of the work. Section 3.5.1 of AIA Document A201–2007 is typical:

> The Contractor warrants to the Owner and Architect that materials and equipment furnished under the Contract will be of good quality and new unless the Contract Documents require or permit otherwise. The Contractor further warrants that the Work will conform to the requirements of the Contract Documents and will be free from defects, except for those inherent in the quality of the Work the Contract Documents require or permit. Work, materials or equipment not conforming to these requirements may be considered defective. The Contractor's warranty excludes remedy for damage or defect caused by abuse, alterations to the Work not executed by the Contractor, improper or insufficient maintenance, improper operation, or normal wear and tear and normal usage.

Under this warranty, for example, the contractor assumes an obligation to use new materials, and failure to use new materials will be considered defective, even if the contract drawings did not specify whether or not the materials should be new. A breach of warranty claim, while given special nomenclature, is otherwise treated in the same fashion as any other breach of contract claim.

3. Breach of Implied Warranty

In addition to the express warranty embodied in the contract, construction law imposes a number of warranties that are implied without specific agreement of the parties.

For example, the law implies in many construction contracts a warranty that the work will be performed in a workmanlike manner.[13] The phrase "workmanlike manner" is a legal term that means that: (1) the work performed will be performed by one who has the knowledge, training, or experience necessary for the successful practice of the required trade or occupation, and (2) the work performed will be performed in a manner generally considered proficient

---

13. *See, e.g.*, McKinley v. Brandt Constr., Inc., 859 N.E.2d 572, 574 (Ohio Ct. App. 2006) ("Any contract to perform work imposes on the contractor the duty to perform the work in a workmanlike manner. 'Workmanlike manner' has been defined as the way work is customarily done by other contractors in the community.").

by those capable of judging such work.[14] This standard is objective. Courts applying the warranty of workmanlike construction will ask whether the work performed would meet the standard of work that one would expect a reasonably skilled and experienced contractor to perform in a similar situation.

Residential housing contracts often carry another implied warranty, known as the implied warranty of habitability. Generally, the elements that must be proved to support an action for breach of the implied warranty of habitability are (1) a sale, (2) by a contractor-vendor, (3) of a residence, (4) with latent defects, (5) that breach the applicable standard of habitability, and (6) cause the purchaser harm.[15] Note that this warranty will only apply to newly constructed homes and will run to the benefit of the original purchaser in many jurisdictions, as it imposes the warranty on the contractor who built the home and implies the warranty on first sale. As with the other implied warranties, this warranty is measured against an objective standard. The home need not be perfect. Rather, the test requires an inquiry as to whether a reasonable person faced with such a defect would conclude that the residence was uninhabitable. For example, a contractor's failure to properly insulate a home against cold weather may not breach the implied warranty of habitability if the home is located in Arizona. However, that same defect may constitute a breach of the warranty for a home located in Vermont.

B. Tort Claims

The availability of a tort claim against the builder depends primarily upon the jurisdiction's application of the economic loss rule, discussed in depth in Chapter 23. To the extent that the parties are in privity of contract, many states require that the claim be asserted under contract law, rather than by resort to the law of negligence, at least where the claims arise out of the parties' performance of the contract.

Of course, a contractual remedy is not available for parties not in privity of contract, and therefore an aggrieved party such as the owner may attempt to use a tort theory to state a claim against a material supplier, manufacturer, or subcontractor with whom the owner lacks contractual privity. The availability of these claims is determined by the jurisdiction's application of the economic loss rule, but many—if not most—states require the owner to allege physical damage to property outside the scope of the contract, the effect of which often is to limit the availability of these claims.

---

14. *See* Young v. Oak Leaf Builders, Inc., 626 S.E.2d 240, 242 (Ga. Ct. App. 2006).

15. *See, e.g.*, Sims v. Lewis, 374 So. 2d 298, 303 (Ala. 1979); Theis v. Heuer, 280 N.E.2d 300, 303 (Ind. 1972).

C. Statutory Claims

The final major area under which an owner can bring a claim against a contractor involves statutory rights, which some states confer on owners—particularly homeowners—under the rubric of "deceptive trade practices."

Many of these are derived from consumer protection and trade practice statutes directed at businesses taking unfair advantage of purchasers. These statutes have been invoked in construction cases when an unscrupulous contractor preys on customers through shoddy construction. Mere defective or shoddy construction, standing alone, is usually insufficient to allow the customer to utilize the powerful consumer protection statutes to obtain special relief. Rather, the shoddy construction usually must be accompanied by some egregious misconduct by the builder.

The main advantages conferred by these statutes are their broad applicability and the damages remedies available to consumers invoking their provisions. Typically, most deceptive trade practices acts include minimum statutory damages or punitive damage multipliers (such as treble damages), and permit the claimant to recover attorneys' fees that would not otherwise be available. As such, they are attractive vehicles for owners seeking to compel contractors to correct defects or to pay monetary compensation.

D. Notice Requirements

Even when work is defective, the contractor may not be required to answer for the defect or the deficiency unless it has been notified of the defect and, under some circumstances, been given an opportunity to cure the defect. For example, if the project is still under construction, the AIA General Conditions of the Contract for Construction provide that if the contractor fails to carry out the work in accordance with the contract documents, the owner is to give the contractor written notice.[16] If the contractor fails to commence curative work within 10 days after receipt of this notice, then the owner may carry out the corrective work himself.[17] The purpose of these notice requirements is to give the contractor an opportunity to cure the defect(s) in the work. Such notice serves several purposes: It allows the contractor the opportunity to avoid termination (if the defect is discovered during the performance of the contract); it permits the contractor to perform the corrective work with his own forces, presumably at lower cost; it avoids similar deficient performance in the future;

16. AIA Document A201-2007, *supra* note 2, § 2.4.

17. *Id. See also id.* § 12.2.2 (regarding notice and correction of defective work during the one-year period after substantial completion).

it promotes the informal resolution of disputes; and it assists the contractor in preserving his claim rights.

22.03 CLAIMS AGAINST DESIGN PROFESSIONALS

One treatise has noted that design professionals generally perform three functions with respect to a traditional Design-Bid-Build construction project: (1) preparation of the drawings and specifications as an independent contractor of the owner; (2) project administration as the agent of the owner; and, (3) arbiter of disputes between the owner and the contractor.[18] The design professional's failure to adequately perform these functions can adversely affect the contractor's performance of his obligations or impose obligations on the owner. For example, faulty drawings and specifications can lead to delays in the prosecution of the work and expose the owner and contractor to damage claims by each against the other. Likewise, the design professional's role in construction contract administration can create multiple potential problems for the contractor and owner if the design professional fails to perform its duties adequately.

Claims against design professionals are principally based on theories of tort and breach of contract. Contract claims are generally brought by the owner that retained the design professional's services. Although the acts that constitute a breach of contract may also give rise to a tort claim for professional negligence, tort claims against design professionals are typically brought by a contractor who claims to have suffered damage due to the acts or omissions of the design professional.

The contract between the owner and the architect defines the design professional's obligations to the owner and thus is the primary source of liability claims. The typical owner-architect agreement contains express obligations to be performed by the architect (e.g., the timely preparation of drawings and specifications and duties pertaining to the administration of the contract of construction).[19] The failure to carry out express duties constitutes a breach of the contract. Oftentimes, however, the design professional's contractual obligations are stated in subjective terms, in which event the question of contractual breach often will be determined by reference to the tort standard of care; spe-

---

18. 5 Bruner & O'Connor, *supra* note 2, § 17:4.

19. For example, American Institute of Architects, AIA Document B201–2007, *Standard Form of Architect's Services: Design and Construction Contract Administration* (2007) [hereinafter AIA Document B201–2007] requires the design professional to make site visits during construction (§ 2.6.2.1), review contractor submittals and shop drawings (§ 2.6.4.2), reject nonconforming work (§ 2.6.2.2), and interpret the contract documents for the contractor (§ 2.6.2.3).

cifically, whether the design professional acted as a reasonable design professional under like circumstances.[20]

A. Tort and Contract Claims Generally

Claims against design professionals most commonly are anchored in tort or in breach of contract. A negligence claim against a design professional is composed of the standard four elements defined in Section 22.03 in connection with claims against contractors. Similarly, breach of contract claims follow the formulation previously set forth. The precise standards that govern design professionals are discussed in more detail in Chapter 6.

B. Defective Drawings and Specifications

How do tort and contract claims typically find their way into construction defect claims against design professionals? One of the design professional's major obligations, if not the principal obligation, is to prepare drawings and specifications that will be used to construct the owner's project. The following case is a fairly typical illustration of the sort of problems that can arise from the breach of that obligation.

St. Joseph Hospital v. Corbetta Constr. Co., Inc.
316 N.E.2d 51 (Ill. App. 1974)

This involves consolidated appeals from two actions in the circuit court of Cook County. Both arose primarily out of problems flowing from the installation of Textolite plastic laminate wall paneling on the walls of the new St. Joseph Hospital during its construction, which paneling proved to have a 'flame spread' rating some 17 times the maximum permitted under the Chicago Building Code and had to be replaced with paneling complying with said Code.

. . .

The St. Joseph Hospital (the Hospital) in 1958 entered into a contract with architect Belli & Belli of Missouri (Belli) for the erection of a hospital on a new site in Chicago to replace one erected before the Great Chicago Fire of 1871, and the general construction of the hospital was undertaken by the Corbetta Construction Company (Corbetta). In April of 1965, when the building had been substantially completed, the Hospital was advised by the City Collector that its application for a license to operate the Hospital had been disapproved because the wall paneling (General

20. 5 Bruner & O'Connor, *supra* note 2, § 17:11.

Electric's "Textolite") which covered its rooms and corridors, and had been manufactured and furnished by the General Electric Company (General Electric), did not comply with a Chicago Code requirement that such paneling have a "flame spread" rating of not to exceed 15. Actually it had a rating of 255, 17 times the maximum. The city also threatened criminal action against the Hospital for operating without said license.

At this juncture, the Hospital called upon all of the parties involved to remedy the situation and withheld from Corbetta final payment of some $453,000. Although all of the parties deplored the situation, each took the position that it was not itself at fault and that only others were to blame. At about this time Corbetta advised the Hospital that it was their intention to file suit for the $453,000 so withheld by the Hospital.

Faced with the threat of two lawsuits and a complete shutdown of its operation, the Hospital, on May 14, 1965, filed a complaint for declaratory judgment against Corbetta, Belli and General Electric, setting forth the above described controversy between the parties, and attaching copies of the various contracts, subcontracts and applicable municipal ordinances. As the result of a petition for immediate relief, an order was entered in this action, permitting the Hospital to take immediate steps to remove the Textolite wall paneling and to replace it with paneling approved by the city of Chicago, all without prejudice to the rights of any of the parties.

. . .

In Section V of Belli's brief it contends, in substance, that it committed no tort and did not breach its contract with the Hospital to design and supervise the construction of the new hospital, and that it "in the normal course of such supervision ascertained that the General Electric Batten Panel System designated in its change order number G-33 would have to be replaced. This is not any unusual situation."

The evidence in this case establishes that Belli not only was legally required to know but did in fact know that the Chicago Building Code required that paneling installed in such building have a "flame spread" rating of not to exceed 15; that Belli specified General Electric's "Textolite" paneling without making any investigation whatsoever to determine whether or not it met that standard; that during the seven months it took to install the Textolite no one at Belli discovered that said material had a flame spread 17 times the maximum and could not possibly be installed upon the walls of any hospital in Chicago; that Belli therefore made a gross error in specifying it in said change order; and that it was only the insistence of the Building Inspector that a certificate be filed that uncovered the fact that no one had ever bothered to procure the consent of the Building Department to change order G-33 as issued to Corbetta.

According to Belli's brief, this was "commonplace practice" and should somehow be excused because, after the discovery of Belli's error, it refused to issue an architect's certificate certifying that Corbetta had properly completed its work, thus holding up the final payment to Corbetta. No cases are cited in support of this theory.

. . .

In 12 Vanderbilt Law Review (1959), in an article by Prof. George M. Bell entitled "Professional Negligence of Architects and Engineers," it is said, at pages 715–716, under the subtitle "Liability to the Owner for a Defective Building":

> Architects and engineers hold themselves out as competent to produce work requiring: (a) skill in the preparation of plans, drawings or designs suitable for the particular work to be executed; (b) knowledge of the materials to be used and the proper application for use; (c) knowledge of construction methods and procedures. Presumably, if the architect or engineer fails to use reasonable care to produce a satisfactory structure, he may be either sued for a breach of an implied term of his contract or in negligence. . . .

Nor can Belli avoid liability on the ground that, by refusing to issue a final certificate for payment after its error had been discovered, enough money was withheld from Corbetta to cover the loss. As we shall subsequently demonstrate, Corbetta was not itself at fault and payment to it was wrongfully withheld. The error in specifying Textolite was that of Belli; not Corbetta.

We therefore must and do hold that, although others may also be at fault, Belli, which, as architect, specified the Textolite paneling without ascertaining its flame spread rating, and permitted its installation over a period of some seven months, is liable to the Hospital for the damages proximately flowing therefrom. We therefore conclude that the jury's verdict for the Hospital and against Belli as to liability was and is correct and approve that verdict.

This court did not attempt to distinguish between professional liability based upon negligence and breach of contract, but other jurisdictions—depending upon their interpretation of the economic loss rule, discussed in Chapter 23—might be more inclined to do so. Important issues regarding allocation of comparative responsibility, including comparative negligence, the applicability of tort and contract statutes of limitation, and the assessment of consequential damages can, indeed, be dramatically different under tort and contract theories. In *St. Joseph*, none of these issues apparently was dispositive, but the court did address an interesting "betterment" issue, which often arises in design defect cases:

The defendants (putting aside for the present their several contentions that they are not liable at all) contend, inter alia, that the Hospital is not entitled to recover the extra cost of the more expensive Micarta paneling and the extra labor costs required by its more difficult installation, and that, insofar as such extra costs are included, the jury's verdict (and the court's judgment thereon) unjustly enriches the Hospital by giving it, free of charge, better and more expensive wall paneling than it had bargained for and thus puts it in a better position than it would have been had the original contracts been fully performed. We are of the considered opinion that this contention is well founded in this case.

. . .

. . . [U]nder the contracts involved, the Hospital was, insofar as wall paneling is concerned, entitled to have Belli use its best professional skill in selecting a wall paneling meeting, among other criteria, the Chicago Building Code's maximum flame spread rating of 15. There is no real doubt that Belli should originally have specified Westinghouse's Micarta asbestos wall paneling which was the only plastic laminate wall paneling which met the flame spread standard set by the Chicago Building Code.

But had Belli so complied with its contract, the Hospital would have had to pay not the relatively modest cost of the Textolite wall paneling material, and the costs of its relatively simple installation, but the greatly increased cost of the Micarta asbestos paneling, plus the greatly increased costs of its more difficult installation. And, as Sister Vincent (the administrator of the hospital during the construction, who signed the order requiring the change from "Novoply" to "Textolite" on behalf of the Hospital) testified, she would have signed a change order specifying a plastic laminate wall covering which cost more "if it was necessary to comply with the Code."

Certainly the Hospital should not receive, without paying more than it originally had agreed to pay for the Textolite, a windfall in the form of the more expensive Micarta paneling and the extra labor costs required by its more difficult installation, merely because its architect initially failed to specify it. The same applies to the door stops, solid core doors and hardware and their installation, which were necessary and were furnished in the reconstruction but were not included in Belli's original plans and specifications.

. . .

Micarta paneling . . . cost $58,000 in 1963 ($87,000 less 1/3 for paneling not removed.) On this alone the Hospital is ahead $48,000.

The Micarta, being much harder and more expensive to install than the Textolite, would have cost $80,000 to install in 1963 (the actual cost of installation of $107,750 was a 30% Increase over 1963). The Textolite cost $33,000 to install in 1963. On this, the Hospital is ahead another $47,000.

To this add door stops $4,298, solid core doors and hardware $8,811 and installation of $8,375 (which were not originally specified or installed but were added at the time the paneling was replaced) for a subtotal of $21,484.

These three items total $116,484, and we therefore will reduce the sum the Hospital is entitled to recover (by retention or otherwise) by this amount.

C. Construction Observation

Another area of significant claim exposure for design professionals arises from the design professional's observation of the contractor and the contractor's activities under the owner-contractor agreement for construction of the project.[21] As noted previously, owner-architect agreements typically impose a variety of duties on the design professional, including making site visits during construction, reviewing contractor submittals and shop drawings, rejecting nonconforming work, and interpreting the contract documents.

These obligations raise an issue as to the design professional's responsibility when a contactor's work is defective. Contractual provisions pertaining to design professional site visits and construction observation have been a fruitful source for owners seeking to expand the liability for construction defects to encompass not only contractors but also design professionals (for the defect). The extent and even the existence of this liability are treated differently in different jurisdictions.

Moundsview Indep. Sch. Dist. No. 621 v. Buetow & Assoc., Inc. 253 N.W.2d 836 (Minn. 1977)[22]

In August 1968, Moundsview retained Buetow to prepare plans and specifications for an addition to an elementary school. At the time of the execution of the agreement, Moundsview had the option of requiring Buetow to provide (1) no supervision, (2) general supervision, or (3) continuous on-site inspection of the construction project by a full-time project representative referred to as a "clerk of the works." Moundsview elected to have Buetow provide only a general supervisory function, the specific language of the contract enumerating the requirements as follows:

> The Architect shall make periodic visits to the site to familiarize himself generally with the progress and quality of the Work and to determine in general if the Work is proceeding in accordance with the Contract Documents. On the basis of his on-site observations as an Architect, he shall endeavor to guard the Owner against defects and deficiencies in the Work of the Contractor. The Architect shall not be required to make exhaustive or continuous on-site inspections to check the quality or quantity of the Work. The Architect shall not be responsible for construction means, methods, techniques, sequences or procedures, or for safety precautions

21. This topic is also addressed in Chapter 13.

22. Internal citations omitted.

> and programs in connection with the Work, and he shall not be responsible for the Contractor's failure to carry out the Work in accordance with the Contract Documents.

> The contract further provides:

> The Architect shall not be responsible for the acts or omissions of the Contractor, or any Subcontractors, or any of the Contractor's or Subcontractors' agents or employees, or any other persons performing any of the Work.

Buetow prepared plans and specifications requiring the placement of wooden plates upon the concrete walls of the building. The plates were to be fastened to the walls by attaching washers and nuts to one-half inch studs secured in cement. During the 79-week construction period, the president of Buetow made 90 visits to the construction site in performance of Buetow's general supervisory obligation.

On May 19, 1975, a severe windstorm blew a portion of the roof off the building causing damage to the addition and to other portions of the school. It was discovered that the roof had not been secured by washers and nuts to the south wall of the school as required by the plans and specifications.

Moundsview brought an action for damages caused by the roof mishap against Buetow, the general contractor, and the roofing subcontractor. In response to an interrogatory from Buetow requesting Moundsview to state all facts upon which it relied to support its allegations against Buetow, Moundsview replied:

> ... Defendant Buetow failed to properly supervise the roof construction, failed to supervise and discover the missing nuts and studs and take proper corrective action.

Thereafter, Buetow made a motion for summary judgment, basing its motion upon the affidavit of one of its officers which stated that Buetow did not observe during any of its construction site visits that the washers and nuts had not been fastened to the studs on the south wall.

...

2. Moundsview argues that Buetow breached its duty of architectural supervision by failing to discover that a contractor had failed to fasten one side of the roof to the building with washers and nuts as required by the plans and specifications. It is the general rule that the employment of an architect is a matter of contract, and consequently, he is responsible for all the duties enumerated within the contract of employment. ... An architect, as a professional, is required to perform his services with reasonable care and competence and will be liable in damages for any failure to do so. ...

Thus, consideration of whether Buetow breached a duty of supervision requires an initial examination of the contract between the parties to determine the parameters of its supervisory obligation. The argument that Buetow breached its duty to supervise would be more persuasive had Moundsview contracted for full-time project representation rather than mere general supervision. An architect's duty to inspect and supervise the construction site pursuant to a contract requiring only general supervision is not as broad as its duty when a "clerk of the works" is required. The mere fact that Buetow received additional compensation for performing the general supervisory service does not serve to expand its responsibilities to an extent equivalent to the duties of a full-time project representative. Moundsview cannot be allowed to gain the benefit of the more detailed "clerk-of-the-works" inspection service while in fact contracting and paying for only a general supervisory service.

Thus, the question of whether Buetow breached its duty to supervise the construction project is to be determined with reference to the general supervisory obligation enumerated in the contract. The contract provided that the architect ". . . shall not be responsible for the Contractor's failure to carry out the Work in accordance with the Contract Documents." When this section is read in conjunction with the section which provides that "(t)he Architect shall not be responsible for the acts or omissions of the Contractor, or any Subcontractors, or any of the Contractor's or Subcontractors' agents or employees or any other persons performing any of the Work," it is apparent that by the plain language of the contract an architect is exculpated from any liability occasioned by the acts or omissions of a contractor. The language of the contract is unambiguous. The failure of a contractor to follow the plans and specifications caused the roof mishap. By virtue of the aforementioned contractual provisions, Buetow is absolved from any liability, as a matter of law, for a contractor's failure to fasten the roof to the building with washers and nuts.

A very different result was obtained in the case of *Hunt v. Ellisor & Tanner, Inc.*[23] There, the owner brought suit against the architect for breach of contract relative to construction defects, claiming that the architect had breached its obligations to observe "the progress of the Work" and to "endeavor to guard the Owner against defects and deficiencies in the Work of the Contractor." The architect disclaimed liability, citing *Moundsview*. The court rejected the defense as follows:

We decline to follow *Moundsview*. We conclude that the language said to be exculpatory constitutes nothing other than an agreement that the architect is not the insurer or guarantor of the general contractor's obligation to carry out the work in accordance with the contract documents. We reach this conclusion because the first three sentences of paragraph 2.2.4 of the general conditions impose a nonconstruction responsibility upon the architect; to wit: to visit, to familiarize, to determine, to inform and to endeavor to guard. In short, to provide information, not to make improvements upon the job site. Therefore, we reason that the fourth sentence of paragraph 2.2.4 and the paragraph

23. 739 S.W.2d 933 (Tex. App. 1987).

paraphrasing it exist to emphasize the architect's nonconstruction responsibility and to make certain that the architect "will not be responsible for the [general] contractor's failure to carry out the work in accordance with the contract documents." In short, the provider of information to the owner does not insure or guarantee the general contractor's work. It follows, and we so hold, that the contract does not exculpate Ellisor & Tanner from liability for the general contractor's failure to carry out the work in accordance with the contract documents.[24]

D. Claims against the Design Professional by Contractors

To this point, we have focused discussion on the design professional's liability to the owner. While the acts or omissions of design professionals on construction projects can undoubtedly cause damage to the *contractor* as well by making the contractor's performance more expensive and time-consuming, in many jurisdictions the contractor is barred from pursuing claims directly against the design professional by the Economic Loss Doctrine.[25]

As noted elsewhere,[26] however, many states recognize exceptions to the Economic Loss Doctrine and permit claims against design professionals based on concepts such as the professional's knowledge of possible third-party reliance, the reasonable foreseeability that the contractor could be injured by the architect's negligence, negligent and intentional misrepresentation, and an architect/contractor relationship that is the functional equivalent of privity.

Forte Bros., Inc. v. Nat'l Amusements, Inc., et al. 525 A.2d 1301 (R.I. 1987)

The issue presented by this case is thus whether a third-party general contractor who may foreseeably be injured or suffer an economic loss proximately caused by the negligent performance of a contractual duty by an architect/site engineer has

24. The middle ground between the *Moundsview* and *Hunt* lines of cases has been to find that the architect can be liable to the owner if the design professional fails to learn of the defect due to breach of its contractual obligations to owner; *see, e.g.*, Diocese of Rochester v. R-Monde Contractors, Inc., 562 N.Y.S.2d 593 (N.Y. Sup. Ct. 1989) (failure to make timely and proper inspections); Watson, Watson, Rutland/Architects, Inc. v. Montgomery County Bd. of Educ., 559 So. 2d 168 (Ala. 1990) (failure of contractor to follow plans and specifications is known to architect during course of construction).

25. *See* 5 Bruner & O'Connor, *supra* note 2, § 17:88 n.20 (state-by-state collection of cases on applicability of economic loss doctrine and recognized exceptions thereto).

26. *See supra* note 59 and Chapter 23.

a cause of action in negligence against the architect/site engineer notwithstanding an absence of privity. The issue is novel to this jurisdiction.

. . .

A supervising architect, in the performance of its contract with the owner, is required to exercise the ability, skill and care customarily exercised by architects in similar circumstances. This duty of care extends to contractors who share an economic relationship and community of interest with the architect on a construction project. The duty is based on circumstances establishing a direct and reasonable reliance by the contractor on the contractual performance of the architect when the architect knows, or should know, of that reliance. "Altogether too much control over the contractor necessarily rests in the hands of the supervising architect for him not to be placed under a duty imposed by law to perform without negligence his functions as they affect the contractor."

We hold that Forte stated a cause of action in negligence against Allen notwithstanding the trial justice's finding that Allen was acting as agent for National. Viewing the facts in the light most favorable to Forte, as we must, it appears that Forte directly and reasonably relied on Allen's rendering of Allen's contractual duty as supervising architect/site engineer of the construction project to measure Forte's removal of mass boulders, to report the removal to National, and to authorize payment to Forte therefor. Hence, Allen owed Forte a duty to render its services professionally. In so holding, we join "an emerging majority of jurisdictions [which] have taken the position that a contractor can maintain a negligence action against an architect without direct privity of contract between the parties."

22.04 THE *SPEARIN* DOCTRINE AND CONFLICTING WARRANTIES

The taxonomy of claims set out so far should not lead the reader to believe that allocation of responsibility is always a simple matter. Although an identification of all the complexities is beyond the scope of an introductory work such as this, we will discuss two special problems that exemplify the issues that can arise in charging parties with defects.

A. The *Spearin* Doctrine

One of the fundamental doctrines that serves to allocate responsibility for construction defects is the *Spearin* Doctrine.

The most well-known project delivery system is Design-Bid-Build. Under the first step, the owner hires a design professional who transforms the owner's needs and concepts into a set of drawings and specifications for the project. These completed design documents are then presented by the owner to one or more contractors who then furnish a bid for performing the work. After evaluating these bids, the owner negotiates and executes a contract with the contractor to build the work required by the drawings and specifications. This system, however, begs the question of who is responsible when the contractor builds according to the drawings and specifications, but the work nonetheless proves to be defective.

1. Owner's Implied Warranty of Drawings and Specifications

The seminal case on the issue of responsibility for defective drawings and specifications is *United States v. Spearin*.[27] The government contracted with Spearin to build a dry dock at the Brooklyn Navy Yard in accordance with government-furnished drawings and specifications. The job site was intersected by a six-foot brick sewer, and the work required that a section of the sewer be relocated before dry dock construction began. The drawings and specifications prescribed the dimensions, material, and location of the sewer section to be relocated. Spearin fully complied with the contract requirements, and the government accepted the work as satisfactory.

Both before and after its relocation, the sewer connected with a seven-foot sewer line, located outside the project area, that emptied into a tidal basin. About a year after the relocation, a heavy downpour coincided with a high tide, forcing water from the basin up the sewer line. As a result, the relocated sewer broke in several places, and the dry dock was flooded. Upon investigation, it was discovered that the backflow was diverted into the six-foot sewer by a dam in the seven-foot sewer that was not shown on the city's sewer system drawings or on the government's drawings furnished to Spearin.

After the sewer break, Spearin notified the government that he would not resume operations unless the government assumed responsibility for the damage that had already occurred and either made changes in the sewer system or assumed responsibility for future damage. The government insisted that the responsibility for remedying existing conditions rested with the contractor. The government ultimately annulled the contract, took possession of the work and materials on the site, and completed the work using other contractors.

Spearin brought suit against the government seeking the balance due for work done under the contract and for damages resulting from the annulment of the contract. The court found in Spearin's favor, and the government appealed to the Supreme Court. The Court held that the government's breach of its implied warranty of the drawings and specifications, followed by the government's repudiation of all responsibility, past and future, justified Spearin's refusal to resume work:

27. 248 U.S. 132 (1918).

> In the case at bar, the sewer, as well as the other structures, was to be built in accordance with the plans and specifications furnished by the government. The construction of the sewer constituted as much an integral part of the contract as did the construction of any part of the dry dock proper. . . . The risk of the existing system proving adequate might have rested upon Spearin, if the contract for the dry dock had not contained the provision for relocation of the 6-foot sewer. But the insertion of the articles prescribing the character, dimensions and location of the sewer imported a warranty that if the specifications were complied with, the sewer would be adequate. This implied warranty is not overcome by the general clauses requiring the contractor to examine the site, to check up the plans, and to assume responsibility for the work until completion and acceptance. The obligation to examine the site did not impose upon him the duty of making a diligent inquiry into the history of the locality with a view to determining, at his peril, whether the sewer specifically prescribed by the government would prove adequate. The duty to check plans did not impose the obligation to pass upon their adequacy to accomplish the purpose in view. And the provision concerning contractor's responsibility cannot be construed as abridging rights arising under specific provisions of the contract.[28]

Thus, "[w]hen the Government provides specifications directing how a contract is to be performed, the Government warrants that the contractor will be able to perform the contract satisfactorily if it follows the specifications."[29]

The *Spearin* holding, known later as the *Spearin* Doctrine, has been adopted by the majority of the states.[30] Of course, this doctrine presumes that a contractor who seeks to invoke the doctrine can demonstrate that it completed construction precisely in accordance with drawings and specifications. That is, in order to establish design professional liability for defective drawings and specifications, it must be shown that the contractor reasonably relied upon the drawings and specifications (i.e., that the defects were not so patent as to provoke a duty of inquiry on the part of the contractor).[31] In addition, the contractor must also construct the work in strict compliance with the defective

---

28. *Spearin*, 248 U.S. at 136–37.

29. Hercules, Inc. v. United States, 516 U.S. 417 (1996).

30. A notable exception is the state of Texas, where the jurisprudence generally provides that the owner does not impliedly warrant the plans and specifications. Lonergan v. San Antonio Loan & Trust Co., 104 S.W. 1061 (Tex. 1907); Great Am. Ins. Co. v. North Austin Mun. Util. Dist. No. 1, 902 S.W.2d 488 (Tex. Civ. App. 1993), *rev'd in part* and *aff'd in part*, 908 S.W.2d 415 (Tex. 1995).

31. *See* 3 Bruner & O'Connor, *supra* note 2, § 9:64 (discussing implied duty to seek clarification of obvious design discrepancies). Standard form construction contracts typically place on the contractor the obligation to review the plans and specifications and promptly report any design errors or omissions noted by the contractor. *See, e.g.*, AIA Document A201–2007, *supra* note 2, § 3.2.

drawings and specifications. For example, in *Al Johnson Constr. Co. v. U.S.*,[32] the contractor was required to construct a temporary berm to stabilize coffer-cells on a spillway being constructed in connection with the Tennessee-Tombigbee Waterway project. The berm was not completed when heavy rains caused a rise in the adjacent river, and the berm sloughed away. The contractor sought to recover the extra costs incurred as a result of the failure. The Board of Contract Appeals rejected the contractor's contention that the berm would have failed even if it had been built to the specified height, relying on the testimony of government witnesses who testified that the berm, "if of full height, might or might not have failed." On appeal, the Federal Circuit affirmed, reasoning that the warranty of the drawings and specifications under the *Spearin* Doctrine only applied if the contractor complied with the drawings and specifications:

> In Justice Brandeis' opinion in *Spearin*, it is flatly stated that the implied warranty runs to contractors who have complied with the specifications.
>
> . . .
>
> Of course, this is subject to the exception that contractors can stop performing when the results of actual experience with performance to date reveal that the specifications are defective to the degree they cannot, as a matter of commercial possibility, be complied with. This was the fact situation in the *Spearin* case itself. There was no showing here that the defect of the specifications was of a kind that produced impossibility. Rather, appellant attempted to show the board that the berm would have failed anyway, even if it had met the specification completely.
>
> . . .
>
> There is nothing to show when or how the contractor became disillusioned with respect to the berm. . . . He does not show us when, or even whether, he became convinced that the berm design was defective. We think the restriction of the implied warranty to those who have fulfilled the specifications, or tried and failed to do so because of the defects themselves, has strong policy behind it that would not be served by allowing the implied warranty to run to one who has not done what he contracted to do and fails to satisfactorily explain why not. Any other exception should therefore be restricted to instances, not now foreseen, of manifest inequity, or to a deviation from the specifications shown to have been entirely irrelevant to the alleged defect.[33]

In *Ralph M. Parsons Co.*,[34] the Board explained the reasoning behind the strict compliance requirement as follows:

32. 854 F.2d 467 (Fed. Cir. 1988).

33. *Id.* at 469–70.

34. ASBCA No. 24347, 85-1 BCA ¶ 17,787; *aff'd on reconsid.*, 85-2 BCA ¶ 18,112.

We have here a threshold evidentiary problem requiring, first, that we inquire into whether or not appellant's design, indeed, was followed in the manner intended by appellant. If it was not, then the Government is hardly in a position to fault appellant's design for its difficulties. . . .

> In actions brought against architects, general rules of evidence have been applied. Where negligence of the architect consists [of] the furnishing of defective plans, specifications, and drawings, it is essential to prove that the builder substantially complied with such plans and specifications and that in consequence of such compliance damages were sustained by plaintiff.

Since this is the Government's claim, it has the burden of showing that its construction contractor "substantially complied" with appellant's plans. Once we are shown this fact, then we can further inquire whether appellant's design was negligently performed and whether for the reason of the Government's compliance therewith it incurred costs or damages.[35]

2. *Design Specifications*

In addition to the requirement that the contractor demonstrate performance of the work in strict compliance with the drawings and specifications supplied by the owner, the *Spearin* Doctrine by its very nature requires that the drawings and specifications set forth detailed and precise requirements as to materials, equipment, tolerances, or other specific information from which the contractor is not permitted to deviate. Such detailed specifications are known as "Design Specifications."[36]

An example of how the *Spearin* Doctrine can be used by a contractor defensively can be found in the case of *City of Covington v. Heard*.[37] The city sued the contractor that installed sewer pipe, which later failed. The appellate court upheld a summary judgment in favor of the contractor because the undisputed

35. 85-2 BCA at 90,936.

36. *See, e.g.*, J. L. Simmons Co. v. United States, 412 F.2d 1360, 1362 (Ct. Cl. 1969) ("The specifications, which were prepared by the defendant, are a classic example of 'design' specifications . . . . In other words, in these specifications, the defendant set forth in precise detail the materials to be employed and the manner in which the work was to be performed, and plaintiff was not privileged to deviate therefrom, but was required to follow them as one would a road map."); GAF Corp. v. United States, 19 Cl. Ct. 490, 500 n.2 (1990) ("Specifications are design when only one material or a certain composition will enable the product to meet the performance standards expressed in the specification."); Aircraft Gear Corp. v. Kaman Aerospace Corp., 856 F. Supp. 446, 452 (N.D. Ill. 1994) (design specifications "set forth precise measurements, tolerances, materials, in process and finished product tests, quality control, inspection requirements, and other specific information.").

37. 428 So. 2d 1135 (La. Ct. App. 1983).

evidence established that the contractor used the specified type of pipe, and installed it in accordance with the drawings and specifications:

> Flextran was an allowable material under the City's own specifications. The City cannot avoid responsibility for its own specifications by allowing the contractor such an option. The law is clear that where specifications call for installation of a material by brand name, and the contractor complies with the specifications by supplying and installing such brand name material, the contractor is immune from defects therein.[38]

3. *Performance Specifications*

As distinguished from design specifications, performance specifications describe an end result to be achieved by the contractor, such as requiring HVAC equipment capable of delivering a specific volume of air at a certain static pressure, but leave the determination of how to reach the result—what brand or model of equipment—to the contractor.[39] Unlike design specifications, performance specifications do not carry an implied warranty of accuracy or adequacy:

> But not all contract specifications are design specifications—some are merely performance specifications:
>
>> Design specifications explicitly state how the contract is to be performed and permit no deviations. Performance specifications, on the other hand, specify the results to be obtained, and leave it to the contractor to determine how to achieve those results.
>
> The government does not implicitly warrant performance specifications for complete accuracy or adequacy. "[T]ypical 'performance' type specifications set forth an objective or standard to be achieved, and the successful bidder is expected to exercise his ingenuity in achieving that objective or standard of performance, selecting the means and assuming a corresponding responsibility for that selection."[40]

Specifications can contain provisions that can be characterized as "design," "performance," or a hybrid of the two. To differentiate between design and performance specifications, courts generally examine the level of discretion that the specification at issue affords the contractor. "Discretion serves as the touchstone for assessing the extent of implied warranty and intended liability."[41]

---

38. *Id.*

39. *See, e.g., J. L. Simmons Co.*, 412 F.2d at 1362; Blake Constr. Co. v. United States, 987 F.2d 743, 745 (Fed. Cir. 1993) (same).

40. George Sollitt Constr. Co. v. United States, 64 Fed. Cl. 229, 296–97 (2005) (citations omitted); *see also* Stuyvesant Dredging Co. v. United States, 834 F.2d 1576 (Fed. Cir. 1987).

41. Conner Bros. Constr. Co., Inc. v. United States, 65 Fed. Cl. 657, 685 (2005).

A contractor claiming a particular specification is "design" rather than "performance" must establish the specification "do[es] not permit meaningful discretion . . . and the defective specification [is] the cause of [the] injury."[42]

4. *Conflicts between the Contractor's and the Owner's Warranties*

As previously discussed, construction contracts typically contain an express warranty in favor of the owner that materials and equipment furnished under the contract will be of good quality and new unless otherwise required or permitted by the contract documents, that the work will be free from defects not inherent in the quality required or permitted, and that the work will conform to the requirements of the contract documents.[43] The owner, on the other hand, impliedly warrants the drawings and specifications furnished to the contractor. In most cases, these warranties coexist without difficulty, but the question remains: Who bears responsibility when the materials and equipment that are expressly warranted by the contractor, but supplied under a design specification (impliedly warranted as suitable by the owner), are defective?

Rhone Poulenc Rorer Pharm., Inc. v. Newman Glass Works
112 F.3d 695 (3d Cir. 1997)

Plaintiff Turner Construction Company ("Plaintiff") appeals from an order of the district court granting Defendant Newman Glass Works' ("Defendant") motion for judgment as a matter of law pursuant to Federal Rule of Civil Procedure 50(b).

. . .

Rhone Poulenc Rorer, Inc. ("Rhone") contracted with Plaintiff to have Plaintiff install opaque spandrel glass in Rhone's headquarters and research facility. In turn, Plaintiff, as the general contractor, entered into two subcontracts with Defendant, which required Defendant to supply and to install the opaque spandrel glass that comprised the structure's curtainwall. The subcontracts, which are identical as relevant here, specify the type of glass Defendant was to install and list the three manufacturers from whom Defendant could purchase the glass. . . . In compliance with the foregoing terms, Defendant set about installing the specified spandrel glass that it had purchased from Spectrum Glass Products, Inc. ("Spectrum"), one of the three listed manufacturers. . . . Before Defendant completed the installation, the opacifier coating began to delaminate from portions of the glass. Plaintiff and its architect

42. *Id.* For a general discussion of performance specifications, *see* 5 Bruner & O'Connor, *supra* note 2, § 17.76.

43. AIA Document A201-2007, *supra* note 2, § 3.5.1.

noticed the delamination because portions of the installed glass exhibited a mottled appearance. Plaintiff demanded in writing that Defendant replace the defective glass. Defendant refused, and Plaintiff instituted this action for breach of contract.

. . .

We commence our analysis on this appeal by focusing our attention on the warranty provisions of the subcontracts. Section X of the subcontracts requires Defendant to remove and to replace all materials that Plaintiff or its architect "condemn as unsound, defective or improper." In Section 4.5.1 of the general contract, Plaintiff expressly warrants to Rhone that all work will be "free from faults and defects." Defendant, in turn, assumes this warranty toward Plaintiff in the subcontracts. Defendant argues that these express warranties are legally insufficient to support the jury's verdict because these warranties are nullified by Plaintiff's implied warranty that the specified glass was adequate for use in this building. Defendant asserts that under the Supreme Court's 1918 decision in *United States v. Spearin*, 248 U.S. 132 (1918), and its progeny, it cannot be held liable for any defects in the glass because it complied with the specifications in the subcontracts.

. . .

We are presented here with a conflict between the implied and the express warranties. We conclude that the Pennsylvania Supreme Court would find that the express warranties must prevail.

Implied warranties are generally not favored by law and are construed narrowly. This warranty of specification permits a court to allocate the risk of an inadequate specification, quite equitably, to the party that drafted the specification. Here, though, the parties have explicitly allocated to Defendant the risk that the glass would be defective. The parties are free to do so, and there is no indication of overreaching or bad faith by Plaintiff.

Defendant accurately notes that *Spearin* provides that an implied warranty of specification is not overcome by "general clauses" requiring a contractor to visit a site or to review plans. The express warranties in the subcontracts here, however, are not the sort of general clauses about which *Spearin* speaks. Section X of the subcontracts very specifically requires Defendant to remove and to replace any materials that Plaintiff or its architect deem defective. Through the incorporation of the general contract's Section 4.5.1 into the subcontracts, Defendant specifically warranted that all work would be "free from faults and defects." These duties far exceed site inspection or plan review.

. . .

It is true that here, Defendant had virtually no discretion in carrying out its contractual obligations in light of the exacting specifications in the subcontracts. But it is also true that Defendant entered into subcontracts that require it to remove and to replace any defective materials "at its own cost and expense." We conclude that the implied warranty by Plaintiff based on the specifications of the type and manufacturer of the spandrel glass must yield to Defendant's express warranties against defective materials. In consequence, the district court erred in granting Defendant's motion for judgment as a matter of law.

One might think that the allocation of responsibility to the contractor is unfair when the contractor merely carried out the drawings and specifications. The reader should bear in mind, however, that the contractor is in privity with the supplier (or the subcontractor who in turn purchased the defective materials) and through its own contract chain has the opportunity to shift the ultimate burden of the defective materials to the party responsible for the defect (in this case, presumably, the manufacturer).

22.05 TIME-BASED DEFENSES TO LIABILITY

There are many affirmative defenses to contract and tort liability, and to conclude this chapter, we will look at three: statutes of limitation, statutes of repose, and laches.

A. Statutes of Limitation

In most states, construction claims are subject to both statutes of limitation and statutes of repose. Both types of statutes specify the period of time within which a lawsuit must be commenced before the right to sue is lost. These statutes serve several important objectives. First, they protect persons from having to defend stale claims where "[t]he search for truth may be seriously impaired by the loss of evidence, whether by death or disappearance of witnesses, fading memories, disappearance of documents or otherwise."[44] Second, public policy dictates that there should come a time when persons will no longer be burdened by the possibility of liability arising from acts occurring in the past. Finally, such statutes promote the goal of certainty and finality in the administration of commercial transactions by terminating liability at a set point in time.[45]

Statutes of limitation generally require that an injured party commence suit within a specified period after the cause of action accrues. A claim for breach of contract typically accrues on the date when the contractual breach occurs, and the statute of limitations will commence to run from that date. A tort claim, on the other hand, accrues when injury or damage is sustained. Under this "accrual rule," it is entirely possible for the limitations period to expire before the injured party even knows of the claim. For example, the defective installation of piping, a breach of the contractor's obligation to perform the work in a workmanlike manner, may not become known for many years after

44. Shea v. Keuffel & Esser, 668 F. Supp. 41, 45 (D. Mass. 1986), *aff'd* 823 F.2d 543 (1st Cir. 1987); Rosenberg v. Town of North Bergen, 293 A.2d 662, 667–68 (N.J. 1972); Safeway Stores, Inc. v. Certainteed Corp., 710 S.W.2d 544 (Tex. 1986).

45. Sun Valley Water Beds v. Herm Hughes & Son, Inc., 782 P.2d 188 (Utah 1989).

the date of breach, especially if the piping is covered up by other work, such as walls. A fatal design flaw may not manifest itself for a considerable period after the design professional's negligent act. Because of the perceived unfairness of claims becoming time-barred before they are even known to exist, many jurisdictions have adopted the "Discovery Rule" (i.e., the limitations period begins to run when the nonbreaching party first becomes aware (or should have become aware) of the breach of the contract or injury).[46] The following case illustrates policy reasons behind adoption of such a rule:[47]

1000 Virginia Ltd. Partnership v. Vertecs
146 P.3d 423 (Wash. 2006)

¶ 1 In these consolidated cases the trial courts dismissed actions for breach of construction contracts as untimely. The Court of Appeals reversed in each case, holding that the discovery rule applies to determine when a cause of action accrues for breach of a construction contract and that material issues of fact exist as to when the plaintiffs discovered their causes of action. . . . The defendant construction companies challenge the Court of Appeals' holdings that the discovery rule applies. . . . We affirm the Court of Appeals.

¶ 2 1000 Virginia, acting as its own general contractor, built an apartment complex. 1000 Virginia hired subcontractor Vertecs Corporation to do the stucco work. Vertecs was not responsible for caulking, flashings, and weather protection. On December 31, 1992, the certificate of substantial completion was issued. Occasional leaks occurred, and in early 1994 1000 Virginia asked for Vertecs' presence during inspection of windows and vents. Following inspection, Vertecs concluded that caulking had failed in several areas and that some dryer vents were not connected to ductwork. Vertecs said these problems were not within the scope of its work. 1000 Virginia repaired the caulking and ductwork. Leaks continued over the next few years and, after a wet winter in 1996, 1000 Virginia noted increased cracks in the stucco and deterioration in drywall under decks. 1000 Virginia took further corrective action. Despite 1000 Virginia's efforts to resolve the problems, by the end of 1998 1000 Virginia knew substantial repairs were required to correct systemic defects that led to structural damage resulting from water intrusion and fungal problems.

---

46. Hilliard & Bartko Joint Venture v. Fedco Sys., Inc., 522 A.2d 961, 969 (Md. 1987) (holding that arbitration was timely where owner found defect within one year and then filed for arbitration within three years where construction contract provided that contractor would correct defects found within one year of completion of contract).

47. In many jurisdictions the Discovery Rule is imposed by statute rather than by common law.

¶ 3 On September 3, 2002, 1000 Virginia brought a breach of contract action against several of its subcontractors, including Vertecs, claiming that defective work led to the water intrusion and resulting damage to the apartment complex.

. . .

¶ 21 There is little to distinguish a case involving latent defects in a building and a case where a surgical instrument is left in the plaintiff's body during surgery. In both cases the plaintiff may have no way of knowing that a cause of action exists, i.e., no way of knowing the facts that show that the construction contract was breached in the first case or that a duty of care was breached in the latter. Just as a patient may lack the ability to know that a surgical sponge has been left in place which causes serious harm, so too an unsuspecting homeowner may lack the ability to know that an inferior product was used, such as a type of siding, which results in significant damage. . . .

. . .

¶ 23 Under such circumstances, it is unfair to permit a defendant to escape responsibility for shoddy construction simply because the cause of action is based on contract rather than a tort theory. In addition, it is more equitable to place the burden of loss on the party best able to prevent it, i.e., the contracting party who could avoid breaching the contract. Thus, the same fairness considerations weigh in favor of applying the discovery rule to construction contract actions where latent defects are alleged as weigh in favor of the rule in negligence actions.

¶ 24 Of course, a plaintiff cannot ignore notice of possible defects. The discovery rule requires that "when a plaintiff is placed on notice by some appreciable harm occasioned by another's wrongful conduct, the plaintiff must make further diligent inquiry to ascertain the scope of the actual harm. The plaintiff is charged with what a reasonable inquiry would have discovered." A person who has notice of facts that are sufficient to put him or her upon inquiry notice is deemed to have notice of all facts that reasonable inquiry would disclose. Whether the plaintiff should have been able to detect defects (and presumably demand that they be remedied before the structure is accepted) involves a fact question that may be resolved against the plaintiff. Thus, if a plaintiff in a construction contract case inspects as construction proceeds, voluntarily or as a matter of contractual obligation, or is placed on inquiry notice of harm during construction, these facts will bear on whether the plaintiff should have discovered the cause of action.

¶ 25 Vertecs argues, however, that this court has recognized the need for certainty and predictability in the contract setting and these concerns weigh against applying the discovery rule in cases involving actions on construction contracts. Vertecs argues that the parties to construction contracts rely on statutes of limitations and the rule that a contract action accrues at breach in allocating risks. Vertecs contends that changing accrual rules undermines the certainty and predictability of allocated risks of loss and bargained-for liability exposure. In a similar vein, amicus Associated General Contractors contends that availability and cost of insurance are significant concerns where a discovery rule of accrual applies.

> ¶ 26 Similar arguments applied and similar concerns were raised when the discovery rule was first applied in the case of negligence actions, but the benefits of applying the rule outweigh the burdens. We believe the same is true in the case of construction contracts. We hold that the discovery rule applies in the case of actions for breach of construction contracts where latent defects are alleged.
>
> . . .
>
> ¶ 43 Next, Vertecs challenges the Court of Appeals' holding that there are issues of material fact in 1000 Virginia precluding summary judgment. Vertecs contends that 1000 Virginia knew of its damage in 1994, expressly claimed it was due to Vertecs' conduct, and chose its own repair course over time. Thus, Vertecs says, 1000 Virginia should have discovered its cause of action in 1994 because there was notice of leaks and thus of the breach. However, Vertecs itself contended in 1994 that leaks were due to improper caulking and unconnected ductwork that was not within the scope of its work. Therefore, we do not agree that evidence of the leaks in 1994 is sufficient to support Vertecs' claim that as a matter of law 1000 Virginia discovered or should have discovered its cause of action against Vertecs in 1994.

As a general rule, the statute of limitations applicable to tort claims is shorter than the limitations period for contract claims.[48] In addition, even within a category of claims, the statutory limitation period can vary widely from state to state,[49] and is subject to differing judicial interpretations. Finally, certain states have adopted limitations periods specifically applicable to causes of action arising out of construction projects.[50]

There is a well-developed body of case law upholding the ability of parties to modify the statute of limitations by contract.[51] Such "private" statute of limitations may alter the limitations period by shortening it (allowing only one year to sue rather than three, for example). In addition, they may effectively shorten the legal period simply by changing the date on which a cause of

48. For example, in Louisiana, the limitations period for tort claims is one year, LA. CIV. CODE art. 3492, and the limitations period for contract claims is ten years, *id.*, art. 3499.

49. *Compare* D.C. CODE ANN. § 12-301 (three years for tort claims) *with* ALA. CODE § 6-2-33 (ten years).

50. *E.g.*, LA. CIVIL CODE art. 3500.

51. *E.g.*, College of Notre Dame of Maryland, Inc. v. Morabito Consultants, Inc., 752 A.2d 265 (Md. Ct. Spec. App. 2000) (parties may agree to a provision that modifies the statute of limitations if there is no controlling statute to the contrary, the modification is reasonable, and it is not subject to other defenses such as fraud, duress, or misrepresentation). *See also* Annotation, *Validity of Contractual Time Period, Shorter than Statute of Limitations, For Bringing Suit*, 6 A.L.R. 3rd 1197, 1207 (1966 & Electronic Supp. 2007).

action is deemed to accrue.[52] Thus, if the parties agree that all causes of action related to the construction of a project will be deemed to commence upon substantial completion (as opposed to accruing upon discovery of the claim), the practical effect could well be to cause the statute of limitations to run before it would otherwise run under the law.[53]

B. Statutes of Repose

Because of exceptions such as the Discovery Rule, statutes of limitation can nonetheless result in a participant in a construction project being subject to suit for a considerable time after project completion. For example, it would be possible for an engineer to be sued twenty years after designing a structure if the defect is not discovered until then. Statutes of repose address the issue of open-ended liability by placing an absolute time limit, from a date certain, on the bringing of claims. The repose period commences to run regardless of when the injury occurs, whether the injury has manifested itself, or whether the injured party is even aware of the injury. Once this period of time has passed, the causes of action specified in the statute are extinguished.

Virtually every jurisdiction has adopted a statute of repose for claims arising out of construction projects.[54] These statutes "demonstrate legislative recognition of the protracted and extensive vulnerability to lawsuit of architects and builders by owners and other persons. They represent a response to the inadequacy of the traditional statutes of limitations, whose time period begins upon discovery of the claim or upon occurrence of the injury."[55] Like statutes of limitation, statutes of repose vary widely from state to state[56] in terms of

---

52. The 1997 edition of AIA Document A201 provided a set of commencement dates for the running of any "applicable statute of limitations" depending on whether the act or failure to act occurred before Substantial Completion, between Substantial Completion and the Final Certificate for Payment, or after the Final Certificate for Payment. *See* American Institute of Architects, AIA Document A201-1997, *General Conditions of the Contract for Construction* § 13.7. AIA Document A201–2007 replaces this provision with a contractual statute of repose of ten years.

53. A. J. Tenwood Assocs. v. Orange Senior Citizens Housing Co., 491 A.2d 1280, 1284 (N.J. Super. Ct. App. Div. 1985).

54. Allen Holt Gwyn, *Legislative Update: Statutes of Repose*, 21 The Construction Lawyer (2001).

55. Sowders v. M.W. Kellogg Co., 663 S.W.2d 644, 647 (Tex. Ct. App. 1983).

56. *See* State by State Survey of Statutes of Repose, American Bar Association, Section of Litigation, Committee on Construction Litigation (1994).

the parties protected,[57] the date the repose period begins to run,[58] the length of time after which claims are foreclosed,[59] and the types of claims foreclosed.[60]

C. Laches

Equitable principles serve to limit a disgruntled party's right to pursue relief in many ways, one of which is through the Doctrine of Laches. The doctrine provides that an unconscionable or inexcusable delay in bringing an action can bar the action—even if it is otherwise meritorious—where the defendant suffers some undue prejudice from the delay.[61] The application of laches is somewhat limited in construction cases inasmuch as it is an equitable defense and applies only to equitable claims. Most construction claims are legal, and the Doctrine of Laches is thus not as important an affirmative defense as in other realms of the law.

Because the Doctrine of Laches is rooted in equity and the application of the doctrine requires consideration of such intangible concerns as the reasonableness of the plaintiff's action and the prejudice to the defendant, the doctrine's application as a defense to claims is not as reliable as the statute of limitations. Courts are sensitive to the uncertainty that attends application of this doctrine, however, and as a means of providing clarity and predictability to parties, many courts have adopted the rule that the "presumptive" reasonable period for bringing an equitable claim will be deemed to be the period provided by the statute of limitations for the most analogous legal claim.[62]

---

57. *Compare* LA. REV. STAT. ANN. § 9:2772 (protecting "any person performing or furnishing the design, planning, supervision, inspection, or observation of construction or the construction of an improvement to immovable property") *with* ME. REV. STAT. ANN. tit. 14 § 752-A (covering only "architects and engineers").

58. The date certain on which the repose period commences varies widely from jurisdiction. *See, e.g.*, ARK. REV. STAT. § 16-56-112 (substantial completion); FLA. STAT. ANN. § 95.11(3)(c) (date of actual possession by owner, date of issuance of certificate of occupancy, date of abandonment of construction, or date of completion or termination of the contract, whichever is later); IDAHO CODE § 5-241 (final completion); LA. REV. STAT. ANN. § 9:2772 (the date owner's acceptance of the work is filed in the mortgage office, or, if no filing is made, the date owner occupies or takes possession of the improvement, in whole or in part).

59. *Compare* ARK. REV. STAT. § 16-56-112 (five years) *with* FLA. STAT. ANN. § 95.11(3)(c) (fifteen years). Some statutes provide for an extension of the repose period if injuries occur in the latter years of the period and under other circumstances. *See, e.g.*, OKLA. STAT. ANN. tit. 15, § 110.

60. *Compare* LA. REV. STAT. ANN. § 9:2772 (precluding all actions "whether ex contractu, ex delicto, or otherwise") *with* MASS. ANN. LAWS ch. 260, § 2B (precluding only an "[a]ction of tort for damages").

61. Flaherty v. Port Auth. of Allegheny County, 299 A.2d 613, 625 n.18 (Pa. 1973) (alternate holding).

62. Indeed, in some circumstances, it is reversible error *not* to give presumptive effect to an analogous statute of limitations absent compelling reasons. Tandy Corp. v. Malone & Hyde, Inc., 769 F.2d 362 (6th Cir. 1985) (defense to trademark infringement claims).

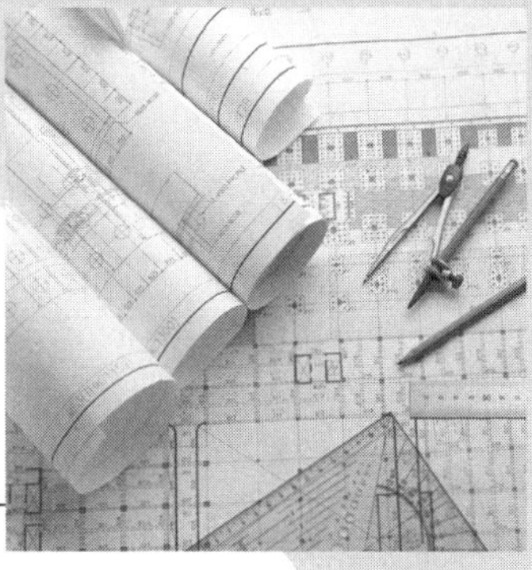

QUESTIONS

1. Subcontractor submits shop drawings for a pre-engineered metal building to Contractor who reviews them and submits them to Owner. Owner transmits the shop drawings to Designer who "reviews them for general compliance with the contract documents" and requests changes to one of the exterior column lines. Revised shop drawings are submitted, approved by the Designer, and used to fabricate the building components. While the building is being erected, it is determined that the foundation slab cannot support the weight of certain columns, which were not load-bearing in the original shop drawings but are shown as load-bearing in the revised shop drawings. Who should bear responsibility for the cost of improving the foundation to support the columns?
2. Design professionals typically claim that imposing liability on them for defective construction effectively makes the design professional the guarantor of the contractor's work. Owners respond by saying that the design professional has the duty to perform with an ordinary and reasonable degree of skill and care, which includes a duty to observe and report deviations from the drawings and specifications that should have been obvious to a skilled construction industry professional. Otherwise, the owners contend, the design professional would not be accountable to anyone for failing to make a reasonably adequate inspection so long as it made inspections, no matter how cursory. Which position is more persuasive? Does your answer change if the defective construction is not readily observable by the design professional?
3. Scholars, practitioners, and lawmakers alike debate the fundamental fairness of a rule that bars meritorious claims even before they are discovered. What policy arguments support the imposition of a statute of repose as applied to construction professionals? Why should a person injured in the collapse of a garage be barred from recovering from the party responsible for the negligent design or construction simply because more than ten years have passed since substantial completion?

4. Some states hold that, unlike statutes of limitations, statutes of repose cannot be modified by contract or waived by the parties.[63] But other states permit waiver.[64] Finally, there is a wide disparity in state law on the issue of whether a statute of repose can be equitably tolled by fraud, intentional concealment, or other actions of the putative defendant.[65] Does allowing waiver and equitable tolling defeat the very purpose for which the statute of repose was enacted?

---

63. Stone & Webster Eng'g Corp. v. Duquesne Light Co., 79 F. Supp. 2d 1, 8–9 (D. Mass. 2000) (making *Erie* guess regarding Pennsylvania law); Trax-Fax, Inc. v. Hobba, 627 S.E.2d 90, 95–96 (Ga. Ct. App. 2006) (collecting cases); Cheswold Volunteer Fire Co. v. Lambertson Constr. Co., 489 A.2d 413, 420–21 (Del. 1984) (collecting cases).

64. ESI Montgomery County, Inc. v. Montenay Int'l Corp., 899 F. Supp. 1061, 1066 (S.D.N.Y. 1995) (upholding an express waiver of period of repose in a federal securities action); First Interstate Bank of Denver v. Central Bank & Trust Co., 937 P.2d 855, 860 (Colo. Ct. App. 1997) (upholding written agreement tolling statute of repose pending a final adjudication on federal claims); One North McDowell Assoc. v. McDowell Dev. Co., 389 S.E.2d 834, 836 (N.C. Ct. App. 1990) (North Carolina construction statute of repose is subject to an express waiver by written agreement).

65. Northern Montana Hosp. v. Knight, 811 P.2d 1276 (Mont. 1991) Pfeifer v. City of Bellingham, 772 P.2d 1018 (Wash. 1989), (proof of active concealment of dangerous condition could toll statute of repose); Sharon Steel Corp. v. Workers Comp. Appeal Bd., 670 A.2d 1194, 1200 (Pa. Cmwlth. Ct. 1996) (equitable estoppel may toll operation of workers' compensation statute of repose); Curry v. Thornsberry, 98 S.W.3d 477, 482 (Ark. Ct. App. 2003), *aff'd*, 128 S.W.3d 438 (Ark. 2003) (rejecting tolling argument based on fraudulent concealment); FNB Mortgage Corp. v. Pac. Gen. Group, 76 Cal. App. 4th 1116 (3d Dist. 1999) (construction statute of repose not tolled by promises of, or attempts to repair); Monson v. Paramount Homes, Inc., 515 S.E.2d 445 (N.C. Ct. App. 1999) (same).

CHAPTER

23

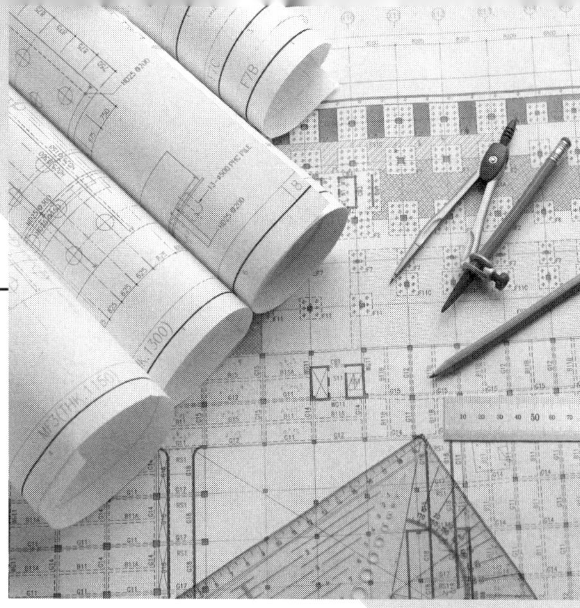

The Economic Loss Rule in Construction Law

A. HOLT GWYN

23.01 INTRODUCTION

At common law, there were bright-line distinctions between civil actions brought in contract and those brought in tort. Elements of proof were different, as were recoverable damages. Proof of a tort was generally more burdensome—requiring proof of a duty of care, foreseeability, "but for" and proximate causation, rather than simply a contract and a breach. At common law, contracting parties were free to assign various risks between them, and such assignments were generally upheld. If one commercial party was damaged by a contractual partner's act, the remedy was *ex contractu*. There was no tort recovery. This separateness survives today in Article 2 of the Uniform Commercial Code.[1] But in many other areas of commerce, the distinctions and barriers between contract and tort law can be as blurred as the colored glass in a kaleidoscope.

1. For buyers and sellers of goods, Article 2 of the Uniform Commercial Code codifies the rules for contract formation, acceptance, rejection, cover, disclaimers, damages, and consequential and incidental damage waivers.

A construction project is a monumental mix of goods and services combined to create a new and functioning edifice. Whether a building, a water-retaining structure, or a system to generate or distribute power, today's projects are often complex, requiring coordination among numerous participants who have different expertise. Specialization in the furnishing of construction goods and services permeates the construction industry. We still build with bricks and mortar, but we rely upon sophisticated components installed by specialists using delivery systems with ever more complex risk allocations. Amenities found in today's starter houses make them complex structures in comparison to the multistory office buildings of the 1930s.

As construction projects become more complex, they cost more. Owners, like all consumers, don't want to pay more.[2] This puts pressure on construction design, especially the design of components. Cost concerns now filter into the design of every construction material. And, because speed of installation is a function of overall cost, price also affects components' means and methods of application. Designers of construction components continually look for ways to make products that cost less, are quicker and easier to install, provide a more desirable appearance, and perform better than last year's models. A quantifiable improvement in any of these characteristics will increase market share. But as construction component design is revisited and revised in the name of cost, performance characteristics are often affected. When the performance criteria of components are materially altered, the performance of major building systems may be compromised. The redundancies of conservative design that were de rigueur for buildings in decades past too often play second fiddle to the latest bells and whistles of convenience.

Then, too, the increasing complexity of buildings and components has challenged the architect's role. Architects formerly were the arbiters of component design and selection. This has changed. Product design, and more recently system design, has devolved to component manufacturers. No longer is the architect the project's most knowledgeable leader. As observed by a leading Columbia University professor, by the end of the 1970s, architects came to know less and less about more and more until some were said "to know nothing about everything."[3]

---

2. Many off-the-shelf consumer goods decrease in price as availability increases. Consumers of many goods—especially newer technology goods—have become accustomed to getting more for the same price as last year, or even more for a lesser price.

3. Mario Salvadore, Why Buildings Stand Up: The Strength of Architecture (1980). No longer is the architect capable of the expertise required by the court in Hubert v. Aiken, 2 N.Y.S. 711 (C.P. 1888), *aff'd*, 25 N.E. 954 (N.Y. 1890) ("he is an expert in carpentry, cements, mortar, in the strength of materials . . . new conveniences . . .").

The construction design landscape is evolving, and the designers in the manufacturing sector are on the leading edge. Selection and coordination of building components, long the province of design professionals, are in today's world being usurped by contractors and component manufacturers. Through "value engineering" or outright design delegation,[4] architects defer to component designers. Indeed, contractor-led Design-Build arrangements often relegate the architect to backroom subcontractor status. The ability to select the component is accompanied by the responsibility to coordinate the chosen products with the building's structure and complementing systems. In this area, design professionals hand off much of their former role, if not all of their responsibility.[5]

Specialization in the construction industry has also morphed the contractual process. Virtually all projects have dozens of contractual arrangements. Specialized equipment must be furnished and installed by ever-more specialized subcontractors and suppliers. The traditional Design-Bid-Build project delivery method is but one of many available contractual vehicles. Owners can choose among construction managers, multi-prime contractors, design-builders, joint ventures, and so on—the ability to customize the construction process grows every year.[6] But there is one industry constant: there are more and more entities providing specialized goods and services. Subcontractors, suppliers, and subconsultants abound, in ever-deepening Dante-esque circles.

The monumental complexity of a construction project has strained the law of contracts. When something goes wrong on a construction site, or when the completed edifice doesn't deliver, there are usually a number of entities with a hand in the cause. Proper installation is the first place to check, but it is often not the entire story. Products that perform well in one setting may, for

---

4. The American Institute of Architects' standard general conditions of the contract between owner and contractor now explicitly allow the delegation of the architect's design responsibility to the contractor and its subcontractors. *See* American Institute of Architects, AIA A201–2007, *General Conditions of the Contract for Construction* § 3.12.10 (2007).

5. Design professionals "of record" retain liability, especially to their clients, for adequate and code-compliant design, whether the design professional performs the work or delegates it to others. Justin Sweet, Legal Aspects of Architecture, Engineering and the Construction Process § 15.10(B) (4th ed. 1989). *See also* Johnson v. Salem Title Co., 425 P.2d 519 (Or. 1967) (structural engineering requirements were non-delegable, even to a professional engineer). The result is different if the design professional never had the responsibility as part of its scope of services. *See* Aleutian Constructors v. United States, 24 Cl. Ct. 372 (1991) (specifications required contractor to design and build roof); Mudgett v. Marshall, 574 A.2d 867 (Me. 1990) (error of structural engineer hired by Design-Build contractor not imputed to owner's design professional not hired to design building).

6. The American Institute of Architects' *Standard Form of Agreement Between Owner and Architect*, AIA Document B101–2007 (2007) [hereinafter AIA Document B101–2007], provides owners a menu of architectural services from which to choose.

instance, lack the robustness to weather ambient conditions in another. As product diversification expands, the importance of selecting and coordinating construction components has increased. It is now as significant to select and coordinate as it is to install properly. When problems occur, they can be the fault of the design professional, the installer, the manufacturer, the maintenance team, or all of the above. Getting to the bottom of a construction problem often requires a detailed investigation, and a good deal of finger pointing regrettably ensues. Major responsibility may lie far down, or even outside, the contractual chain.

Contract law has adapted in several areas to market conditions. Warranties from manufacturers of certain goods (typically finished goods, specially designed equipment, and appliances) allow end users and other affected parties legal redress by direct access to the manufacturer—leapfrogging over broken links in the contractual chain.[7] There is also a federal statutory remedy for defective consumer household goods.[8] And some states have promulgated statutory warranties for residential construction that ignore which party contracts with whom, giving a remedy to end users that an ordinary contract law would not provide.[9] Outside of this sort of intervention, though, common law contract law is still the default remedy for construction disputes.

Under the common law, privity of contract is required to bring a contract action. In the historical construction setting, this worked fairly well. Until the mid-20th century, there would normally be one major construction contract—between owner and master builder, who was often responsible for the design as well as the construction. The contractor would directly employ different types of skilled labor for a job (carpenters, masons, ironworkers, concrete finishers, etc.). Today's builder is different—it separately subcontracts with subcontractors and suppliers, who often then sub-subcontract with other specialty trades. The design professionals are most often hired directly by the owner, under separate contracts. When problems develop, claims may be asserted among multiple business entities, with multiple contractual and subcontractual arrangements.

---

7. *See, e.g.*, Groppel Co. v. United States Gypsum Co., 616 S.W.2d 49 (Mo. Ct. App. 1981).

8. The Magnuson-Moss Act, 15 U.S.C.A. § 2301 et seq. (2002), provides consumers with statutory remedies for breach of a manufacturer's or vendor's written or implied warranty of a consumer product. The act also provides certain minimum requirements for written warranties. Upon establishing that the warranty was breached, the consumer may elect the remedy of refund, repair, or replacement, and may recover attorney's fees if it prevails. The act applies to personal, family, or household products, but may include "building materials" such as paneling, siding, or storm windows, when purchased in connection with remodeling a home. In one of the leading cases, it was applied to roofing materials. Muchisky v. Frederic Roofing Co., 838 S.W.2d 74 (Mo. Ct. App. 1992) (affirming verdict, and award of attorney's fees, in favor of homeowner on a re-roofing project).

9. *See, e.g.*, Fla. Stat. § 712.203 (1992) et seq.

For many of today's construction issues, common law contract causes of action can be cumbersome in the extreme. If a problem takes several years to manifest itself, if contractual responsibilities for the work are not the same up and down the chain of specialization, or if there are gaps in the scope of and responsibilities for the work, damages may occur for which there is no common law contract remedy.[10] This is a fertile field for the seeds of tort law to take root.

In many areas of commerce, tort law has insinuated itself, together with its legally imposed responsibilities and remedies. Beginning with Judge Cardozo's opinion in *MacPherson v. Buick Motor Co.*,[11] courts have imposed the duty to act reasonably as an objective standard of conduct. This duty exists independent of any agreement between parties, and applies to conduct both inside and outside a contractual relationship.[12] Tort law imposes upon all members of society certain minimum standards of behavior—including the obligation to take reasonable steps to protect others from being injured or having their possessions destroyed or damaged. If these standards are violated, tort law compensates without serious regard to risk allocation,[13] so long as the injury is foreseeable and proximately caused by the defendant's or defendants' wrong.

Tort law is more about compensation and less about risk allocation. This emphasis is reflected in the liberal rules of tort damages. A negligence victim is entitled to recover for both direct and consequential losses proximately caused by a tortfeasor's breach of duty. In contrast, the general rule of contract damages is that consequential damages are not recoverable unless they were within the contemplation of the parties at the time of contracting and became a basis for the parties' bargain.[14]

Particularly invasive into the construction setting have been the torts of negligent construction and negligent misrepresentation. The separate tort of negligent misrepresentation (normally also alleging purely economic damages) may be asserted against owners, design professionals, specialty contractors, and suppliers—any project participant who disseminates information.[15] Although these economic torts have been bemoaned by a number of commentators,[16] they have not been beset upon the construction industry alone. Many states

10. The Latin phrase is *injuria absque damno, damnum absque injuria*, or *damnum sine injuria*.

11. 111 N.E. 1050 (N.Y. 1916).

12. *See* Morse/Diesel, Inc. v. Trinity Indus., Inc., 655 F. Supp. 346, 355 (S.D.N.Y. 1987).

13. Joint tortfeasors are subject to joint and several liability.

14. *See* Hadley v. Baxendale, 156 Eng. Rep. 145 (Ex. 1854).

15. Restatement (Second) of Torts § 552 (1977), entitled "Information Negligently Supplied for the Guidance of Others."

16. *See, e.g.*, Justin Sweet, Sweet on Construction Law §§ 4.10, 11.13 (ABA 1997); Steven G. M. Stein, Paul Cottrell & Mark Friedlander, *A Blueprint for the Duties and Liabilities of Design Professionals after Moorman*, 60 Chi.-Kent L. Rev. 163 (1984).

have adopted the RESTATEMENT (SECOND) OF TORTS, allowing negligence-based causes of action for economic damages in business transactions involving appraisers,[17] building inspectors,[18] construction managers,[19] as well as lawyers,[20] and accountants, [21] despite lack of privity.

But there are certain limitations on tort damages, especially claims for purely economic damages.[22] In the construction industry, this chiefly manifests itself in the judge-made "Economic Loss Rule," prohibiting economic damages, direct or consequential, when there is no concurrent physical injury or property damage. In a way, the Economic Loss Rule is a judicial response to what some commentators have characterized as an unfair assault on privity and a cancer-like invasion by torts into the contractual setting.[23] But the economic loss rule has hardly been applied uniformly. Each year, new opinions put a new spin on the rule, making its application and exceptions a matter of locality.

23.02 ECONOMIC LOSS RULE

The Economic Loss Rule first appeared within the context of manufactured products liability.[24] Stated simply, in an action alleging negligence[25] or strict

17. Private Mortgage Inv. Servs., Inc. v. Hotel & Club Assocs., Inc., 296 F.3d 308 (4th Cir. 2002).

18. Thompson v. Waters, 526 S.E.2d 650 (N.C. 2000).

19. EH Constr. LLC v. Delor Design Group, Inc., No. 1998-CA-001476-NR, 2000 WL 339939 (Ky. Ct. App. March 31, 2000).

20. *See* Rozny v. Marnul, 250 N.E.2d 656 (Ill. 1969); Orshoski v. Krieger, No. OT-01-009, 2001 WL 1388037 (Ohio Ct. App. Nov. 9, 2001), *appeal denied,* 763 N.E.2d 1185 (Ohio 2002) ("The requirement of privity in a legal malpractice action [for negligent misrepresentation] should be put to a well-deserved burial"); Bradford Sec. Processing Services, Inc. v. Plaza Bank & Trust, 653 P.2d 188 (Okla. 1982). *But see* Bovee v. Gravel, No. 2001-347, 2002 WL 1940968 (Vt. Aug. 13, 2002).

21. United States v. Natelli, 527 F.2d 311 (2d Cir. 1975), *cert. denied,* 425 U.S. 934 (1976); Fisher v. Katz, 266 F. Supp. 180 (S.D.N.Y. 1976); Sharp v. Coopers & Lybrand, 83 F.R.D. 343 (E.D. Pa. 1974). *But see* Semida v. Rice, 863 F.2d 1156, 1160 (4th Cir. 1988); Smullian & Blumenthal, P.A. v. Futz, 762 A.2d 582 (Md. 2000).

22. Economic damages, in this instance, as opposed to bodily injury or damage to separate property.

23. *See, e.g.,* Murray H. Wright & Edward E. Nicholas, *The Collision of Tort and Contract in the Construction Industry,* 21:3 U. RICH. L. REV. 457 (1987).

24. Economic losses in negligence actions were first restricted by the holding in Seely v. White Motor Co., 403 P.2d 145 (Cal. 1965). *See also* East River S.S. Corp. v. Transamerica Delaval, Inc., 476 U.S. 858 (1986). In each of these cases, the product (a truck and a ship's turbines) injured itself, causing repair costs and lost profits, but no bodily injury or physical damage to other property or product.

25. Because many more negligence situations tend to occur in the construction industry—a simple negligence claim may exist against a trade contractor or builder not in privity, when a strict liability (or negligence *per se*) claim does not—this chapter will focus primarily upon the application of the Economic Loss Rule in negligence-based actions.

liability,[26] a plaintiff cannot recover its purely economic damages caused by a defective product that injures only itself. Economic damages represent "inadequate value, costs of repair and replacement of the defective product or consequent loss of profits without any claim of personal injury or damage to other property."[27] Economic loss also includes the "diminution in value of the product because it is inferior in quality and [because it] does not work for the general purpose for which it was manufactured and sold."[28]

The traditional inquiry concerns the type of damage caused by the defective product: Was there injury to a person or other property, or was the damage only to the product? In *Seely v. White Motor Co.*,[29] a consumer's action against a truck manufacturer, the only losses were the cost of repairing the truck and missing income due to the truck's unserviceable state. These damages were not allowed under negligence or strict liability counts, and, because they represented the plaintiff's only damages, the negligence and strict liability causes of action were dismissed.[30] Viewed in this manner, the Economic Loss Rule forms a damages restriction only—barring recovery of certain losses from an otherwise viable negligence or strict liability cause of action.[31]

A proper application of the rule occurs when there are only economic damages. If a plaintiff has sustained bodily injury or physical harm to "other property," then in both cases it may recover any economic losses, together with those physical damages.[32] All of the plaintiff's damages, of course, must be foreseeable and proximately caused by the defendant's conduct.

---

26. The Economic Loss Rule is generally also applied in actions alleging strict liability. *See Seely*, 403 P.2d at 145; Stearman v. Centex Homes, 78 Cal. App. 4th 611 (2000) (citing cases); Fieldstone Co. v. Briggs Plumbing Products, Inc., 54 Cal. App. 4th 357, 363–67 (1997) (prematurely rusted sinks damaged only themselves, strict liability claim dismissed). Strict liability actions under § 402 of the Restatement of Torts generally involve manufactured products. Those supplying services are generally not strictly liable. *See* Murphy v. E. R. Squibb & Sons, Inc., 710 P.2d 247 (Cal. 1985) (strict products liability law does not apply to services); Restatement (Third) of Torts: Products Liability § 19, subd.(b) ("services, even when provided commercially, are not products" and are governed by the law of negligence).

27. Note, *Economic Loss in Products Liability Jurisprudence*, 66 Colum. L. Rev. 917, 918 (1966).

28. Comment, *Manufacturers' Liability to Remote Purchasers for "Economic Loss" Damages—Tort or Contract?*, 114 U. Pa. L. Rev. 539, 541 (1966).

29. 403 P.2d 145 (Cal. 1965).

30. Importantly, the Supreme Court of California (Traynor, C.J.) upheld the plaintiff's claims for breach of express warranty, sustaining the jury award to the plaintiff of the purchase price of the truck and plaintiff's lost profits. *Seely*, 403 P.2d at 145.

31. *See* Casey v. Overhead Door Corp., 74 Cal. App. 4th 112, 123 (1999) (plaintiffs were not prejudiced by trial court ruling *in limine* excluding economic damages—plaintiffs could have introduced other evidence to support negligence count, if plaintiffs had any such evidence).

32. *See* Held v. Mitsubishi Aircraft Int'l, Inc., 672 F. Supp. 369, 376–77 (D. Minn. 1987) (applying Texas law). *But see* World Trade Company v. Westinghouse Elec. Corp., 682 N.Y.S.2d 385, 387

The legal and policy considerations that underpin the rule in the products liability context rest upon a manufacturer's responsibility for its goods. Defects that cause physical harm are deemed to be properly chargeable to the manufacturer. Those that affect the product's "level of performance" are not, unless the manufacturer agrees to be so charged (e.g., by giving a warranty).[33] These considerations are said to be based in the different common law concepts dividing tort and contract law. To recover in tort, there must be some showing of harm above and beyond disappointed expectations.[34]

23.03 THE SCOPE AND PURPOSE OF THE ECONOMIC LOSS RULE

The Economic Loss Rule precludes a party's recovery in negligence or strict liability for purely economic damages.[35] The principle of the Economic Loss Rule is easiest to apply in product liability cases, where the rule originated.[36] In many jurisdictions, the Economic Loss Rule has been extended to claims against all construction service providers, including design professionals, contractors, subcontractors, and suppliers, with or without privity, and whether the claims allege pure negligence, professional negligence, or negligent misrepresentation. A number of courts have held that consumer homeowners are barred from asserting negligence actions against manufacturers to recover economic damages—either to repair defective construction or for the diminution in value of their home—unless there is property damage to some other part of

---

(1998), fact that two injured workers had negligence claims against product manufacturer does not inure to the project owner's benefit when owner's only losses are replacement of defective product and consequential damages thereto.

33. *Seely*, 403 P.2d at 195.

34. *Id. See East River S.S. Corp.*, 476 U.S. at 858, 871; Nat'l Union Fire Ins. Co. v. Pratt & Whitney, 815 P.2d 601, 604 (Nev. 1991). *See also* 1 A.R. FRUMER & M. I. FRIEDMAN, PRODUCTS LIABILITY § 3.01(2)(f) (1991).

35. *See* Christopher Scott D'Angelo, *The Economic Loss Doctrine: Saving Contract Warranty from Drowning in a Sea of Tort*, 26 U. TOL. REV. 591, 609 (1995) (for a survey of jurisdictions that have adopted the economic loss rule). *See also* Pulte Home Corp. v. Osmose Wood Preserving, Inc., 60 F.3d 734 (11th Cir. 1995) (builder's claim for defective fire retardant chemical as applied to plywood sheathing barred by rule); Morris v. Osmose Wood Preserving, Inc., 667 A.2d 624 (Md. 1995) (same).

36. Commentators mostly agree that today's Economic Loss Rule was first articulated in *Seely*, 403 P.2d at 145 (*en banc*), which involved the sale, repair, and loss of income from a defective truck. There the court stated: "The distinction that the law has drawn between tort recovery for physical injuries and warranty recovery for economic loss is not arbitrary and does not rest on the 'luck' of one plaintiff in having an accident causing physical injury. The distinction rests, rather, on an understanding of the nature of the responsibility a manufacturer must undertake in distributing his products." *Seely*, 403 P.2d at 151.

the home.[37] A few courts have held that once incorporated into the structure, the entire structure itself becomes the "product."[38]

The willingness of courts and legislatures to restrict recoverable damages is a somewhat recent phenomenon. Because such statutes and decisions spring from a desire to correct injustice, it may be that the tenor of a proposed rule or statute is determined by the issue's location on the arc created by the pendulum of public opinion. If public opinion favors the contractor or subcontractor who must accept a public owner's one-sided, nonnegotiated contract, a legislature may void an onerous clause as against public policy. If the body politic believes that damages recovered by "trial lawyers" have run amok,[39] or if a court concludes that asserted losses are not the foreseeable result of a participant's conduct, damages can be limited, legislatively or judicially.

The nuances of the Economic Loss Rule make it a doctrine that is in some jurisdictions riddled with exceptions, and in others honored in the breach.[40] A decade ago, one court lamented that the concept is "stated with ease, but applied with great difficulty."[41] The same may be said today. Legal commentators and jurists cannot agree upon the reasons courts distinguish between proper and improper applications of the rule.[42] At the end of the 20th century,

37. While this has long been the position taken by general liability insurers—no property damage to other property, no coverage—recent opinions from the Supreme Courts of California, Nevada, and Alabama prohibit any economic loss claim against construction component suppliers other than a direct contract or warranty claim. These opinions foretell the upcoming battle as to the meaning of property "other than the product itself." *See* Aas v. Superior Court, 12 P.3d 1125 (Cal. 2000); Calloway v. City of Reno, 993 P. 2d 1259 (Nev. 2000) (superseded by the Nevada Legislature two years later); Keck v. Dryvit Sys., Inc., 830 So. 2d 1 (Ala. 2002).

38. *See* Wilson v. Dryvit Sys., Inc., 206 F. Supp. 2d 749 (E.D.N.C. 2002); *but see* Jimenez v. Superior Court, 58 P.3d 450 (Cal. 2002), (defective windows considered a distinct product even after installation due to the ease with which they could be removed from one house and installed in another).

39. *See, e.g.*, the present argument presented to state legislatures and Congress in support of personal injury damage caps for pain and suffering—arguments made, cynical commentators note, by insurance companies and special interest groups.

40. *See, e.g.*, Mark C. and Andrea B. Friedlander, *Malpractice and the Moorman Doctrine's "Exception of the Month,"* 86 Ill. B. J. 600 (Nov. 1998).

41. Sandarac Assoc. v. W. R. Frizzell Architects, 609 So. 2d 1349, 1352 (Fla. Dist. Ct. App. 1992), *overruled by* Moransais v. Heathman, 744 So. 2d 973 (Fla. 1999).

42. Compare the discussion and analyses of the rule within three articles within The Construction Lawyer: (1) Luther P. House, Jr., and Hubert J. Bell, Jr., *The Economic Loss Rule: A Fair Balancing of Interests,* 11:2 The Constr. Law. 1 (1991) (rule is correctly applied in the construction industry based upon the presence or lack of a duty of those sought to be held liable); (2) G. Anthony Smith, *The Continuing Decline of the 'Economic Loss Rule' in Construction Litigation,* 10:4 The Constr. Law. 1 (1990) (rule should not be applied in construction setting because of foreseeability of damage to limited class of users); (3) Steven B. Lesser, *Economic Loss Doctrine*

an astute Florida jurist observed that, in his jurisdiction, there were three distinct, but often overlapping, Economic Loss Rules:

> First, there is the products liability economic loss rule: If the defendant's product physically damages only itself, causing additional economic loss, no recovery is permitted in "tort."
>
> Second, there is the contract economic loss rule: If the parties have entered into a contract, the obligations of the contract cannot be relied upon to establish a cause of action in tort for the recovery of purely economic damages. . . . There must be a separate, "independent tort."
>
> Finally, there is the negligence economic loss rule. Common law negligence generally has not been expanded to protect economic interests in the absence of personal injury or property damage.[43]

How a jurisdiction will apply or refuse to apply these differently stated rules may depend upon how the appellate judiciary of that jurisdiction approaches the "collision" at the intersection of tort and contract law. In the absence of stare decisis as to a particular set of facts, the timbre of language used by an appellate court in similar cases can be instructive. Courts that seek to "maintain a distinction between contract and tort principles" are likely to agree with the court that held: "Economic losses from a defective building are just as *offensive* to tort law as damages sought for economic losses stemming from a defective product" (emphasis added).[44]

In contrast, if the judiciary acknowledges that a "unique relationship" exists among contractors, subcontractors, and design professionals that create "an interdependence" among project participants, then the court may be more inclined to hold noncontracting parties accountable for economic losses that foreseeably arise from the failure of a project participant to discharge its

---

and Its Impact Upon Construction Claims, 14:3 THE CONSTR. LAW. 21–26 (1994) (injured parties too often have no way to affect the hiring or contractual responsibility of, and have no contractual remedy against, project participants. To deny relief to those so harmed violates the protection and the tenets of *Marbury v. Madison*). *Compare also* the analyses of the rule by the court in Casa Clara Condominium Ass'n, Inc. v. Charley Toppino & Sons, Inc., 620 So. 2d 1244 (Fla. 1993) (high salt concrete caused rebar to rust and balconies to crumble—mere economic damage to "product") with that by the court in Borough of Lansdowne, Pa. v. Sevenson Envt. Serv., Inc., 2000 WL 1886578 (E.D. Pa. December 12, 2000) (Economic Loss Rule blocks negligence action against construction professional only when parties are in privity).

43. Woodson v. Martin, 663 So. 2d 1327, 1331 (Fla.2d DCA 1995) (Attenbernd, Jr. dissenting) *Rev'd*, 685 So.2d 1240 (Fla. 1996).

44. *Calloway,* 993 P.2d at 1259, 1265–66 (Nev. 2000). *See also East River S.S. Corp.,* 476 U.S. at 858, 866 (setting forth the often-quoted reason for the rule—to protect contract law from "drown[ing] in a sea of tort").

contractual duties in a nonnegligent manner.[45] If the appellate opinion goes on to emphasize that the Economic Loss Rule "may have some genuine, but limited, value in our damages law, . . ." and that "if the doctrine were generally applied to bar all tort claims for economic losses without accompanying personal injury or property damage, the rule would *wreak havoc on the common law of torts*"[46] (emphasis added), one can deduce at least where the court will begin its analysis.

The impact of the economic loss limitation is immense. In the construction arena, the rule eliminates many claims that otherwise would be categorized as property damage.[47] Because of the concept's draconian effect, many courts have struggled with it as applied to realty improved by the construction process.[48] A number of jurisdictions do not honor the economic loss principle in special situations involving defective construction products and services that create realty improvements, but these courts have been less than uniform in their approach.

23.04 THE ECONOMIC LOSS RULE'S APPLICATION WHERE PARTIES ARE IN PRIVITY OF CONTRACT

A. Claims of Defective Construction Products/Components (Privity)

As a child of manufactured products liability law, the Economic Loss Rule first was used in the construction context by courts considering allegations of defective construction products or equipment.[49] It follows that out of the many ways

45. *See* Caldwell v. Bechtel, 631 F.2d 989, 1000 (D.C. Cir. 1980); Davidson & Jones, Inc. v. New Hanover County, 255 S.E.2d 580 (N.C. App. 1979); United States *ex rel.* Los Angeles Testing Lab. v. Rogers & Rogers, 161 F. Supp. 132, 135–36 (S.D. Cal. 1958).

46. *Moransais,* 744 So. 2d at 973, 983 (Fla. 1999) (professional engineer's negligent inspection failed to disclose defects which caused economic loss, but which had not caused property damage or personal injury).

47. *See* Sidney R. Barrett, *Recovery of Economic Loss in Tort for Construction Defects: A Critical Analysis,* 40 S.C.L. Rev. 891 (1989). *See also* Janis K. Cheezem, *Economic Loss in the Construction Setting: Toward an Appropriate Definition of "Other Property,"* 12:2 The Constr. Law. 21, 22 (1992).

48. *See, e.g.,* discussion between majority and dissenting opinions in *Aas,* 12 P.3d at 1132–35, 1143–50.

49. *See* Sunnyslope Grading, Inc. v. Bradford, Miller, & Risberg, Inc., 437 N.W.2d 213 (Wis. 1989) (backhoe's defective parts were covered by manufacturer's warranty, but plaintiff's economic damages of additional repairs, down time, and lost profits were denied in a tort action where the defective backhoe caused no injury to another person or other property); Sacramento Regional Transit Dist. v. Grumman Flexible, 158 Cal. App. 3d 289 (1984) (repair of defective bus parts was not actionable where defect had not caused further damage).

a construction project or its participants could be economically harmed, damage due to defective equipment or to a negligently manufactured product stand the closest to the *Seely* model and constitute the scenario where the Economic Loss Rule ought to be most uniformly applied. Such has proved to be the case, especially when the product has injured "only itself."[50]

Wausau Tile, Inc. v. County Concrete Corp. 593 N.W.2d 445 (Wis. 1999)

Justice Crooks delivered the opinion of the Court.

Wausau Tile, Inc. (Wausau Tile) manufactures, sells and distributes "Terra" pavers to entities around the country. Pavers are concrete paving blocks made of cement, aggregate, water, and other materials, for use mainly in exterior walkways.

Wausau Tile contracted with Medusa Corporation (Medusa) to supply the cement for the pavers and arranged for County Concrete Corporation (County Concrete) to supply the aggregate. Wausau Tile's contract with Medusa contained warranties providing that Medusa would remedy or replace cement which did not meet particular specifications.

On April 16, 1996, Wausau Tile filed suit in Marathon County Circuit Court against Medusa, County Concrete, and their insurers, alleging breach of warranty, breach of contract, negligence, and strict liability claims. Wausau Tile claimed that several of the installed pavers had suffered "excessive expansion, deflecting, curling, cracking and/or buckling." Wausau Tile asserted that these problems were caused by alkali-silica gel reactions which resulted from high levels of alkalinity in Medusa's cement and high concentrations of silica in County Concrete's aggregate.

Wausau Tile claimed that the expansion and cracking of the pavers had led to problems and property damages which have given rise to "various claims, demands and suits against Wausau Tile." Wausau Tile alleged that it had "sustained monetary damages in remedying the property damage claims, is facing claims for personal injuries, and has suffered and will continue to suffer lost business and profits." In connection with its tort claims, Wausau Tile sought "actual and consequential damages arising from said problems and defects, including, but not limited to, costs of repair, replacement and remedy of any and all defects, complaints and resulting injuries which have arisen or will arise in the future as a result of the use of said pavers."

---

50. William K. Jones, *Product Defects Causing Commercial Loss, The Ascendancy of Contract Over Tort*, 44 U. Miami L. Rev. 731 (1990); *see* Annotation, *Strict Products Liability: Recovery for Damage to Product Alone*, 72 A.L.R. 4th 12 (1985 and Supp. 1995). *See also* Mid Continent Aircraft Corp. v. Curry County Spraying Serv., Inc., 572 S.W.2d 308, 312 (Tex. 1978) (quoting Dean Page Keeton on the difference between a dangerous condition that harms only the product and a condition that is dangerous to other persons or property).

We begin by determining whether the circuit court properly dismissed Wausau Tile's negligence and strict liability claims against Medusa as barred by the economic loss doctrine. The economic loss doctrine precludes a purchaser of a product from employing negligence or strict liability theories to recover from the product's manufacturer loss which is solely economic. Economic loss is the loss in a product's value which occurs because the product "is inferior in quality and does not work for the general purposes for which it was manufactured and sold." *Daanen & Janssen, Inc. v. Cedarapids, Inc.*, 405 N.W. 2d 842 (1998).

Economic loss may be either direct or consequential. Direct economic loss is "loss in value of the product itself." All other economic loss caused by the product defect, such as lost profits, is consequential economic loss.

The economic loss doctrine does not preclude a product purchaser's claims of personal injury or damage to property other than the product itself. Similarly, claims which allege economic loss in combination with non-economic loss are not barred by the doctrine. "In short, economic loss is damage to a product itself or monetary loss caused by the defective product, which does not cause personal injury or damage to other property."

In *Daanen,* this court identified three policies supporting the application of the economic loss doctrine to commercial transactions. First, the economic loss doctrine preserves the fundamental distinction between tort law and contract law. Second, application of the doctrine protects the parties' freedom to allocate economic risk by contract. Third, the doctrine encourages the purchaser, which is the party best situated to assess the risk of economic loss, to assume, allocate, or insure against that risk.

The first of these policies recognizes that contract law rests on bargained-for obligations, while tort law is based on legal obligations. In contract law, the parties' duties arise from the terms of their particular agreement; the goal is to hold parties to that agreement so that each receives the benefit of his or her bargain. The aim of tort law, in contrast, is to protect people from misfortunes which are unexpected and overwhelming. The law imposes tort duties upon manufacturers to protect society's interest in safety from the physical harm or personal injury which may result from defective products. Thus, where a product fails in its intended use and injures only itself, thereby causing only economic damages to the purchaser, "the reasons for imposing a tort duty are weak and those for leaving the party to its contractual remedies are strong."

In this case, the damages sought by Wausau Tile can be grouped into three categories: (1) the costs of repairing and replacing cracked, buckled or expanded pavers; (2) the costs of satisfying third parties' claims that the defective pavers either caused personal injury or damaged property adjoining the pavers, such as curbs, mortar beds and walls; and (3) lost profits and business. We consider each of these types of damages in turn.

Repair and replacement costs are typical measures of economic loss. However, it is not the measure of damages which determines whether a claim alleges solely economic loss. Physical harm to property other than the product itself may also be measured by the cost of repair or replacement of the product. Consequently, we must determine whether Wausau Tile has alleged repair or replacement costs as a measure of harm to property other than the defective product.

Wausau Tile argues that the costs of repairing and replacing the pavers do not constitute economic loss because the pavers themselves are property other than the defective product (Medusa's cement). We are not persuaded by that argument.

Damage by a defective component of an integrated system to either the system as a whole or other system components is not damage to "other property" which precludes the application of the economic loss doctrine. Comment e of the RESTATEMENT (THIRD) OF TORTS § 21 acknowledges this "integrated system" rule. It states, in part:

> A defective product that causes harm to property other than the defective product itself is governed by the rules of this Restatement. What constitutes harm to other property rather than harm to the product itself may be difficult to determine. A product that nondangerously fails to function due to a product defect has clearly caused harm only to itself. A product that fails to function and causes harm to surrounding property has clearly caused harm to other property. However, when a component part of a machine or a system destroys the rest of the machine or system, the characterization process becomes more difficult. When the product or system is deemed to be an integrated whole, courts treat such damage as harm to the product itself. When so characterized, the damage is excluded from the coverage of this Restatement. A contrary holding would require a finding of property damage in virtually every case in which a product harms itself and would prevent contractual rules from serving their legitimate function in governing commercial transactions.
>
> RESTATEMENT (THIRD) OF TORTS § 21 cmt. e (1997) (emphasis added).

In the instant case, it is undisputed that the pavers were integrated systems comprised of several component materials, including Medusa's cement. The circuit court determined that Medusa's "concrete is an indistinguishable, integral part of the pavers" which "cannot be separately identified from the finished product." We conclude that the crux of Wausau Tile's claim for repair and replacement costs is that the pavers were damaged because one or more of their ingredients was of insufficient quality and did not work for Wausau Tile's intended purpose. This is the essence of a claim for economic loss.

Second, Wausau Tile claims damages in the amount it expended, or anticipates it will expend, in remediation of third party claims of damage to property adjoining the pavers and pedestrians' possible claims of personal injury. Wausau's claims do not allege any personal injury or property damage on Wausau Tile's part. Rather, as Wausau Tile acknowledges in its brief, these claims are an attempt to recoup the commercial costs of settling the claims of third parties which resulted from the product defect. As such, the claims allege consequential economic loss.

Finally, Wausau Tile claims lost business and profits. Wausau Tile's lost business and profits are indirect losses attributable to the inferior quality of the pavers. Accordingly, they constitute economic loss which is not recoverable in tort.

We conclude that Wausau Tile's complaint alleges only economic loss. Therefore, the first policy set forth by this court in *Daanen* supports the application of the economic loss doctrine in this case. Wausau Tile's claims involve failed economic expectations, which are the province of contract law.

The second policy reason for applying the economic loss doctrine is to protect parties' freedom to allocate economic risk via contract. Allowing purchasers to elect recovery under tort theories instead of requiring them to rely on their contractual remedies "rewrites the agreement by allowing a party to recoup a benefit that was not part of the bargain." It strips sellers of the ability to protect themselves from foreseeable risk by negotiating sales agreements.

Wausau Tile and Medusa entered into a contract with a warranty which specifically addressed the suitability of the cement for use in the pavers. Wausau Tile had the opportunity to negotiate a warranty and did so. Presumably, Wausau Tile paid a price commensurate with the warranty it received. If Wausau Tile were permitted to reap the benefits of a broader warranty by recovering its damages in tort, it would receive more than it bargained for (and paid for) and Medusa would receive less than it bargained for (and was paid for). Consequently, the second policy set forth in *Daanen* also supports the application of the economic loss doctrine in this case.

The third policy reason for applying the economic loss doctrine is that the doctrine "encourages the party with the best understanding of the attendant risks of economic loss, the commercial purchaser, to assume, allocate, or insure against the risk of loss caused by a defective product." Purchasers are generally better equipped than sellers to anticipate the economic loss which a defective product could cause their particular businesses. Accordingly, courts have required purchasers to guard against foreseeable economic loss by allocating the risk by contract or by purchasing insurance. The result is a more efficient, more predictable marketplace. If tort recovery were permitted, sellers of products would be "potentially liable for unbargained-for and unexpected risks," leading eventually to higher prices for consumers.

Wausau Tile should reasonably have expected that it might receive defective or unsuitable cement. Because cement is one of the main components of pavers, Wausau Tile should also have foreseen that defective cement might produce defects in the pavers. Evidently, Wausau Tile did foresee this risk because it attempted to allocate the risk contractually with Medusa. Wausau Tile may not now turn to tort law in hopes of obtaining benefits for which it may not have bargained.

We find that the three policy reasons for applying the economic loss doctrine support the application of the doctrine in this case. Because Wausau Tile has alleged purely economic loss, the economic loss doctrine prevents Wausau Tile from maintaining its negligence and strict liability claims against Medusa.

In *Wausau Tile*, the Wisconsin Supreme Court focused on the terms of the parties' contract—the fact that the defective cement had been integrated into the pavers, and that the pavers caused no demonstrable physical injury to any persons or other property. Some courts have, in the products/privity scenario, gone even further. When the rule is considered by a court to be an anti-tort-in-commercial-transactions rule, courts have denied recovery in negligence when the parties are in privity of contract, despite damage to property "other than the product itself." In *Myrtle Beach Pipeline Corp. v. Emerson Elec. Co.*,[51] a specially made pipe leaked, causing a large fuel spill and subsequent clean-up costs. Despite damage to land adjacent to the pipeline project, tort recovery for the clean-up cost was denied on privity grounds.[52]

Beyond the simplest of fact situations, the analytical landscape for injury caused by defective construction products becomes cluttered. During the construction process, products and equipment are incorporated into systems, and many hands are laid upon them. Not all products remain in their as-purchased form once incorporated into a structure. Incorporated products may be identifiable, but to remove or repair them often injures or destroys still other products. Economic losses to the project owner and project participants can also occur when a product is improperly installed.[53] Structures may diminish in value under the stigma of nascent, but unrealized, injury due to defective components, and economic damage can be visited upon the owner or upon various project participants due to improper component-related professional services, including defective design, selection, or coordination of construction components.

B. Claims of Defective Construction Services (Including Poor Workmanship and Improper Furnishing of Materials) (Privity)

Most states agree that when parties are in privity of contract, a tort claim for negligence in the performance of that contract will not lie unless there is bodily injury or damage to property other than that which is the subject matter of the

51. 843 F. Supp. 1027 (D.S.C. 1993).

52. *Id.* at 1048–55. The court discussed the Economic Loss Rule, but was more impressed that these were "sophisticated parties" to a "negotiated" contract. The court added, even if "other property" were damaged, such damage was contemplated by the parties' contract. *Id.* at 1057. *But see* Tourist Village Motel, Inc. v. Massachusetts Engineering Co., 801 F. Supp. 903 (D. N.H. 1992) (fuel drum leak damaged plaintiff's property, Economic Loss Rule not applied).

53. Those providing construction services may be liable in negligence or implied warranty, but are generally not strictly liable for their work or for the components they install. Restatement (Third) of Torts: Products Liability, § 19, subd. (b) ("services, even when provided commercially, are not products" and are governed by the law of negligence).

contract itself. Economic damages, though they may be substantial, are deemed to be sufficiently addressed under the bargained-for rights and responsibilities set forth in the contract between the parties relating to the subject matter of the damage. Courts reason that negligent performance of a contract, without more, is a breach of contract only, and should not be confused with a tortious breach of a legal duty, which arises by law independent of the contract. This reasoning is applied to buyers and sellers, whether the contract is for the purchase of a product, the installation of the product, the leasing of equipment, or the furnishing of nonprofessional services in connection with the finished construction of a project.

Recovery in negligence for purely economic damages requires the breach of an independent tort duty, separate from the subject matter of the contract. If the plaintiff is in privity with the defendant, breach of the obligations of the contract ordinarily cannot be relied upon to support an action in negligence; the parties have already selected a contractual remedy. Failure to perform one's contract, even carelessly, generally will not support a negligence action for economic damages.[54]

Paul M. Spillman v. American Homes of Mocksville, Inc. 422 S.E.2d 740 (N.C. App. 1992)

Chief Justice Hedrick delivered the opinion of the Court.

Plaintiff Paul Spillman instituted this civil action by filing a complaint wherein he alleged negligent breach of contract arising out of plaintiff's purchase of a mobile home from defendant. The claim set forth by plaintiff was based upon the alleged improper construction and installation of the mobile home by defendant. All damage suffered by plaintiff consisted of the cost to repair the defects in the mobile home and to repair the damage to the mobile home resulting from the improper installation.

The jury returned a verdict finding that defendant had negligently performed the contract concerning "the sale and set up of the manufactured home." Defendant appeals from the trial court's denial of its motion for a directed verdict.

Defendant assigns as error the trial court's denial of its motion for a directed verdict. Defendant argues that plaintiff failed to produce evidence sufficient to submit the issue of negligent performance of the contract to the jury.

---

54. *See, e.g.*, Palmetto Linen Serv., Inc. v. U.N.X., Inc., 205 F.3d 126, 129 (4th Cir. 2000), (applying S.C. law) *quoting from* Kennedy v. Columbia Lumber and Mfg. Co., 384 S.E.2d 730, 737 (S.C. 1989) ("the 'economic loss' rule will still apply where duties are created *solely* by contract") (emphasis in original).

Plaintiff's claim of negligence is premised upon the allegation that defendant's failure to properly perform the terms of the contract between the parties resulted in damage to the mobile home which is the subject matter of the contract. Such a premise is clearly insufficient. Absent the existence of a public policy exception, as in the case of contracts involving a common carrier, innkeeper or other bailee, *see Ports Authority v. Roofing Co.,* 294 N.C. 73, 82, 240 S.E.2d 345, 350–51 (1978), a tort action does not lie against a party to a contract who simply fails to properly perform the terms of the contract, even if that failure to properly perform was due to the negligent or intentional conduct of that party, when the injury resulting from the breach is damage to the subject matter of the contract. It is the law of contract and not the law of negligence which defines the obligations and remedies of the parties in such a situation.

As the evidence presented by plaintiff does not support a claim of negligence, the trial court's denial of defendant's motion for a directed verdict as to that issue was error. The judgment entered by the District Court is reversed.

F. Craig Stuart v. Coldwell Banker Commercial Group, Inc. 745 P.2d 1284 (Wash. 1987)

UTTER, Justice.

The trial court in this case awarded damages to a homeowners' association against a builder-vendor for construction defects in common and limited common areas of a condominium complex.

Yarrowood is a condominium complex located in Bellevue. One hundred forty-four of the 155 units have private decks. Because of the steep topography and siting, there are 32 entry walkways which traverse from the parking area to provide access to its entrances. Twenty-three of these serve only one apartment each. The bulk of the areas at issue were these private decks and walkways.

In May 1975, when Phase I (67 apartments) was partially constructed, Ferguson, the initial developer, defaulted and Coldwell Banker took over. From that point, Coldwell Banker was the owner, developer, construction contractor and vendor of the apartment units at Yarrowood.

Coldwell Banker turned over management of Yarrowood to the Board of Directors of the Yarrowood Homeowners Association ("the Board") after substantially all of the units were sold. Following a short period in which Coldwell Banker acted as property manager, the Board hired a resident manager to handle complaints from homeowners, and to perform minor maintenance and repairs. The first manager served from early 1978 to 1980. During 1978 and 1979, he received complaints relating to 10 to 15 of the decks at Yarrowood. Some of the decks had damage which was plainly visible from public areas such as the parking lot for one of the buildings. By June 1979, at least one deck had so deteriorated that the homeowner asked the Board about having it rebuilt.

Currently, the state of Washington does not recognize a cause of action for negligent construction on behalf of individual homeowners. Beyond the terms expressed in the contract of sale, the only recognized duty owing from a builder-vendor of a newly completed residence to its first purchaser is that embodied in the implied warranty of habitability, which arises from the sale transaction. Plaintiff homeowners, faced with losses that are not of their own making, present a sympathetic case, and we understand the desire of the trial court to fashion a remedy. We must exercise caution, however, that we do not unduly upset the law upon which expectations are built and business is conducted.

The theory advanced by the trial court is a peculiar combination of tort and contract law, closely related to the tort law of products liability. Dean Prosser characterized the law of products liability as a "hybrid, born of the illicit intercourse of tort and contract, unique in the law." W. Prosser, *Torts* 634 (4th ed. 1971).

Initially, plaintiffs who suffered personal injuries from defective products could only reach the manufacturer under warranty theories. Courts, however, had difficulty meshing the inapposite "intricacies of the law of sale," W. Prosser, *The Assault Upon the Citadel*, 69 Yale L.J. 1099 (1960), such as notice and privity, with the plight presented by injured consumers who did not have a direct contractual relationship with the manufacturer. Accordingly, the doctrine fell prey to logical difficulties inherent in this procrustean approach, which addressed problems of unsafe conduct with a legal theory premised on receiving the benefit of a bargain.

The first cases holding manufacturers liable for injuries caused by their products in the absence of privity of contract used a negligence theory. The landmark decision in this area was *MacPherson v. Buick Motor Co.*, 217 N.Y. 382, 111 N.E. 1050 (1916) (Cardozo, J.). *MacPherson* has been accepted in every jurisdiction in the United States to provide a remedy to plaintiffs injured by manufactured items that are dangerous if negligently made. W. Prosser, *Torts* § 96, at 642–43 (4th ed. 1971). Courts reasoned that privity of contract should not stand as a bar to plaintiff consumers who suffered physical injury from a product negligently made.

As a matter of public policy, it is entirely reasonable to expect manufacturers of goods for sale to the general public to assume responsibility for the safety of their product. In *Seely v. White Motor Co.*, 63 Cal.2d 9, 45 Cal. Rptr. 17, 403 P.2d 145 (1965), Justice Traynor denied recovery to a plaintiff who had pleaded lost profits as the only damages under a negligence theory. Justice Traynor stated at page 18, 45 Cal. Rptr. 17, 403 P.2d 145:

> The distinction that the law has drawn between tort recovery for physical injuries and warranty recovery for economic loss is not arbitrary and does not rest on the "luck" of one plaintiff in having an accident causing

> physical injury. The distinction rests, rather, on an understanding of the nature of the responsibility a manufacturer must undertake in distributing his products. He can appropriately be held liable for physical injuries caused by defects by requiring his goods to match a standard of safety defined in terms of conditions that create unreasonable risks of harm. He cannot be held for the level of performance of his products in the consumer's business unless he agrees that the product was designed to meet the consumer's demands. A consumer should not be charged at the will of the manufacturer with bearing the risk of physical injury when he buys a product on the market. He can, however, be fairly charged with the risk that the product will not match his economic expectations unless the manufacturer agrees that it will. . . .

Commentators and courts faced with this issue have found it necessary to distinguish economic loss from physical harm or property damage. The distinction is usually drawn depending on the nature of the defect and the manner in which the damage occurred. Defects of quality are evidenced by internal deterioration, and designated as economic loss, while loss stemming from defects that cause accidents involving violence or collision with external objects is treated as physical injury.

The nature of the defect at Yarrowood was that the decks and walkways were not of the quality desired by the buyers. The "injury" or damage suffered was that the decks themselves deteriorated, not through accident or violent occurrence, but through exposure to the weather.

In cases such as the present one where only the defective product is damaged, the court should identify whether the particular injury amounts to economic loss or physical damage. In drawing the distinction, the determinative factor should not be the items for which damages are sought, such as repair costs. Rather, the line between tort and contract must be drawn by analyzing interrelated factors such as the nature of the defect, the type of risk, and the manner in which the injury arose. These factors bear directly on whether the safety-insurance policy of tort law or the expectation-bargain protection policy of warranty law is most applicable to the claim in question.

Imposition of tort liability upon the builder-vendors would require them to become the guarantors of the complete satisfaction of future purchasers. A builder-vendor could contract to limit liability for defects with the original purchaser and then find themselves liable for the same defects to a future purchaser with whom they had absolutely no contact. The trial court's formulation makes no distinction between subsequent purchasers who may have known nothing of the potential deterioration of the decks, and those buyers who may have been able to see the flaking observable to the naked eye from the exhibits before the court. The latter buyers may well have been able to account for the defects with an adjustment to the purchase price. As to such buyers, any recovery from the court would be a pure windfall. We decline to recognize a cause of action in negligent construction.

In each of these cases, the claimed damage was to the property which was the subject of the contract between the parties. When this is clear, and there is no personal injury or damage to "other property," only a contract action will lie.

C. Claims of Defective Design Professional/Testing Services (Including Insufficient Design and Improper Contract Administration Services) (Privity)

Professional services in the construction context are supplied by architects, engineers, and related consultants.[55] Their negligence creates a conundrum for the Economic Loss Rule. The relationship between design professionals and their owner clients is unique in the construction industry. It is a professional relationship, one in which trust and confidence are reposed. Design professionals are required to subordinate their interests to those of their clients.[56]

Negligent professional services, including design and construction contract administration services, can economically injure any construction participant, from the owner to a second-tier supplier. The harm can be direct or through principles of indemnification. The foreseeable class of economically damaged plaintiffs with legitimate negligence claims against design professionals is large, greater even perhaps than against a negligent trade contractor. The set of individuals impacted economically by a contractor's faulty work normally is limited to the users of such project. A design professional's negligence, however, can cause economic losses to owners, users, and project participants.

The scope of services undertaken by a design professional directly affects the breadth of the duty required of the professional. Once the obligation is identified, the architect or engineer must meet the standard of care for the services undertaken. To recover in negligence against a design professional, there must be proof of a breach of the standard of care owed. Similarly, most contract-based actions against design professionals also require plaintiffs to demonstrate that the professional breached the professional standard of care for the activity in question, for such is generally also the contract standard.[57] On these points, jurisdictions generally agree. Beyond this legal framework, courts part company. Who may and who may not maintain a negligence action

---

55. Hydrologists, geotechnical engineers, and geologists are examples of related professional consultants. Licensure or registration with a state governing board normally determines a consultant's status.

56. Sweet, *supra* note 16, at §2.1.

57. AIA Document B101–2007 (the latest draft) provides: "The Architect shall perform its services in accordance with generally accepted professional standards appropriate for the Project at the time services are performed." AIA Document B101–2007, *supra* note 6, § 2.2.

against a design professional for economic loss is a matter as much of local legal history as legal analysis.[58]

It is in the context of construction design professional services that the application of the economic loss rule is most difficult to reconcile. No analytical theme predominates. It is almost as if someone had written the words "privity," "special relationship," "foreseeability," "supervising architect," "no duty," and "covered by another contract" on the six sides of a die cube, and then distributed the cube to the appellate judiciary of various states to assist them in describing the policy considerations that support their interpretation of the scope of the rule. Virtually the only constant in the legal analysis among the states is that economic damages are the subject of the claim.

Still, most courts find that if an owner has reached a bargain with a design professional as to specific limitations to the scope and standard of professional services, the owner must look to that agreement for resolution of economic loss disputes.[59]

City Express, Inc. v. Express Partners
87 Hawaii 466, 959 P.2d 836 (Hawaii 1998)

Nakayama, Justice.

Express Partners filed an appeal of this action after a directed verdict was entered in favor of Architects 2, Inc. We granted certiorari to address the application of the economic loss rule to a negligence action between an owner and an architect who are in privity of contract.

City Express, Inc. entered into a joint venture agreement with developer Thomas Enomoto to construct a new warehouse suited for the purposes of City Express, a warehouse and trucking company. The business entity created by this joint venture agreement is Express Partners.

Express Partners hired Architects 2 to design the warehouse. Architects 2 prepared blueprints, and the warehouse was constructed by Dura/Constructors, Inc. The warehouse was constructed with two levels and a ramp from the first to the

58. The discussion here concerns economic losses sustained by owners, end users, and project participants, rather than instances of bodily injury and physical "property damage." The confusing issue of damage to property "other than the product itself," which pervades the discussion of the Economic Loss Rule in the manufactured product context, is recast in the professional services context, reappearing as whether the damage was contemplated by "a contract."

59. *See, e.g.*, Brushton-Moira Cent. Sch. Dist. v. Alliance Wall Corp., 600 N.Y.S.2d 511 (1993) (panels selected by architect were inappropriate for school project, held, contract action allowed, tort action dismissed); Key Int'l Mfg. v. Morse/Diesel, Inc., 142 A.D.2d 448 (N.Y. App. 1988) (owner cannot sue architect or prime contractor in negligence for economic loss since the damages sought are of the type remediable in contract).

second floor. Douglas Sonoda and Lloyd Sako, an architect and engineer respectively employed by Architects 2, testified at trial that it was their understanding, in designing the building, that the second floor would not be utilized by forklifts. Instead, the second floor would be used solely for storage of light objects. However, a portion of the blueprints prepared by Architects 2 indicates that the ramp was "for forklift use."

City Express moved into the warehouse upon its completion and began to utilize forklifts on the second floor of the structure. The use of forklifts caused the floor on the second level to crack. In an attempt to correct the problem, an additional five inches of concrete was poured. However, City Express claimed that this did not solve the problem. The additional weight caused structural damage and the floor again began cracking. City Express had to temporarily vacate the building while additional remedial work was done.

The additional remedial work could not correct the structure sufficiently to support the forklifts. Express Partners, the owner of the building, notified City Express that the type of forklifts commonly utilized in the industry could not be operated on the second floor. On March 1, 1985, City Express went out of business. The building was subsequently converted to office use.

On July 31, 1985, City Express filed a complaint against Express Partners. On January 10, 1986, Express Partners filed a third-party complaint against Architects 2. The City Express claim was settled, and Express Partners proceeded to trial against Architects 2 to recoup its damages. The sole issue at trial was the alleged professional negligence of Architects 2.

At the conclusion of Express Partners' case, Architects 2 moved for a directed verdict. The circuit court granted this motion. We granted certiorari to address an important issue of first impression in this jurisdiction, namely, the availability of economic loss damages in a negligence action against a design professional.

The economic loss rule bars recovery in tort for purely economic loss. Where, as in this case, negligent design of a building is claimed, economic loss damages are those that pertain solely to the costs related to the operation and value of the building itself. Excluded are costs for personal injuries caused by the defective design or damage to property other than the building itself. The damages claimed by Express Partners are purely economic—additional costs, lost rent, the cost of remedying the alleged building defects, and the difference between the value of the building as designed and the value it would have had if it had been properly designed.

In the context of construction litigation involving design professionals, sound policy reasons counsel against providing open-ended tort recovery to parties who have negotiated a contractual relationship. If tort and contract remedies were allowed to overlap, certainty and predictability in allocating risk would decrease and impede future business activity. The construction industry in particular would suffer, for it is in this industry that we see most clearly the importance of the precise allocation of risk as secured by contract. *The fees charged by architects, engineers, contractors, developers, vendors, and so on are founded on their expected liability exposure as bargained and provided for in the contract.*

> Parties have the freedom to contract. "Construction projects are characterized by detailed and comprehensive contracts that form the foundation of the industry's operations. Contracting parties are free to adjust their respective obligations to satisfy their mutual expectations." (citing cases)
>
> Express Partners admits that it was in privity of contract with Architects 2. However, the contract was not presented in evidence and Express Partners proceeded to trial solely on a negligence theory. Because the purely economic damages sought by Express Partners are not recoverable in negligence, the circuit court correctly granted a directed verdict in favor of Architects 2.
>
> In the context of construction litigation, where a party is in privity of contract with a design professional, economic loss damages are limited to contractual remedies, and a negligence action may not be maintained. We believe that this ruling not only encourages the parties to negotiate the limits of liability in a contractual situation, but it holds the parties to the terms of their agreement. Our holding preserves the right of design professionals to limit their exposure to liability through contract. We affirm the ruling of the circuit court.

Many owners, in order to enhance their construction budget, negotiate a lesser fee with design professionals for a reduced scope of professional services or for a limited cap on the designer's professional liability in the event of a mistake. In the *City Express* case, Express Partners made a poor bargain.

23.05 THE ECONOMIC LOSS RULE'S APPLICATION WHERE PARTIES ARE NOT IN PRIVITY OF CONTRACT

In many construction situations, those economically damaged by a defectively designed or manufactured product or by negligent professional conduct or defective construction work are not in privity of contract with (or do not have a contractual remedy against) the provider of the defective product, work, or service. Often, owners or buyers of construction services cannot influence the selection of construction materials, competently judge the quality of a professional's design selection, or evaluate the work of a trade contractor who installs the product. Nor can they bargain for a warranty from any or all of them.[60] Economic losses sustained by owners and participants due to the negligent conduct of others in the process are foreseeable. Under a strict application of the Economic Loss Rule, however, these injuries, even if proved to be proximately caused, are not recoverable in a negligence-based tort action.

60. *See* Steven B. Lesser, *Economic Loss Doctrine and Its Impact Upon Construction Claims*, 14:3 THE CONSTR. LAW. 21, 22 (1994).

In the absence of a contractual remedy against the entity causing the loss,[61] or without the ability either to close or to leap gaps in the contractual chain, no recovery will be allowed.[62]

A. Claims of Defective Construction Products/Components (No Privity)

Because the Economic Loss Rule is premised upon the type of damages suffered, most courts apply the rule in the manufactured products context whether the parties are in privity of contract or not.[63] Other courts may be less direct in referring to privity as a reason to deny tort relief, but use the fact that the affected parties have contracts (even though not between plaintiff and defendant) to restrict the type of "other property" which would avoid the application of the economic loss rule.[64] Some courts take a very strict approach to what constitutes "other property" as a way of enforcing the contractual obligations.

Pulte Home Corp. v. Osmose Wood Preserving, Inc.
60 F.3d 734, C.A.11 (Fla.), 1995

TJOFLAT, Chief Judge:

Appellant Pulte Home Corporation ("Pulte"), a nationwide homebuilder, challenges the grant of judgment notwithstanding the verdict in favor of appellee Osmose Wood Preserving, Inc. ("Osmose"), a manufacturer of chemicals that were

61. Sophisticated parties to the construction process can be very creative in crafting contract clauses that limit contractual liabilities.

62. *See* discussion in Richard E. Speidel, *Warranty Theory, Economic Loss, and the Privity Requirement, Once More Into the Void,* 67 B.U. L. Rev. 9 (1987), using as a model the outcome of Spring Motors Distrib. v. Ford Motor Co., 489 A.2d 660 (N.J. 1985) (plaintiff allowed Uniform Commercial Code (UCC) statute of limitations to pass before filing suit, negligence and strict liability claims barred by Economic Loss Rule—no recovery). Even in UCC actions, courts may find that the product loses its character as a "good" for UCC warranty purposes once it is applied upon or within a structure. *See, e.g., Keck,* 830 So. 2d at 1, 4; *but see* Johnstone, J. in dissent ("the integration of building materials . . . [as] intended by the manufacturers, suppliers, and sellers in no way detracts from the character of those materials, members, or components as products") 830 So.2d at 14. *See also* Stoney v. Franklin, 44 UCC Rep. Serv. 2d 1211 (Va. Cir. Ct. 2001) ("goods exist if they 'are movable at the time of identification to the contract of sale'").

63. *See, e.g.,* Sullivan Indus., Inc. v. Double Seal Glass Co., Inc., 480 N.W.2d 623, 628–29 (Mich. App. 1991), citing Consumer Power Co. v. Mississippi Valley Structural Steel Co., 636 F. Supp. 1100, 1106–08 (E.D. Mich. 1986).

64. *Palmetto Linen Serv., Inc.,* 205 F.3d at 126, 129 ("sophisticated parties" in privity should have contemplated type of sustained injury, thereby "rendering other property damage inseparable from the defect in the product itself").

applied to plywood Pulte used in constructing the roofs of 1876 townhouses. After the townhouses were sold, the chemicals caused the plywood to deteriorate; Pulte subsequently replaced the plywood at a cost exceeding $3,650,000. A jury found that Pulte's loss from replacing the plywood was caused by Osmose's (1) misrepresentation that the plywood would not deteriorate and (2) negligence in failing to warn Pulte that the plywood would deteriorate, and it awarded Pulte $3,750,000 in compensatory damages. Finding that Osmose's misrepresentation was made with fraudulent intent to induce Pulte to purchase Osmose-treated plywood, the jury awarded an additional $2,500,000 in punitive damages.

Following the return of the jury's verdict, but before entering judgment, the district court revisited Osmose's Fed. R. Civ. P. 50(a) motion, which had been made prior to the submission of the case to the jury, and granted Osmose judgment as a matter of law. The court acted on the theory that Pulte's tort claims were barred by the economic loss rule. We agree with the district court that the economic loss rule barred Pulte's negligence claim, but disagree that the rule precluded its fraud claim. The evidence adduced at trial, however, did not support the fraud claim. We therefore affirm the district court's judgment in full.

Pulte, one of the largest homebuilders in the United States, operates building divisions in several states including Florida and Georgia. Pulte constructs and sells between seven and ten thousand homes each year. These homes include condominiums, single family homes, and multi-family townhouses, which are sold at various prices. The multi-family townhouses consist of a row of single-family housing units that are joined by a common wall and, in part, at the roof.

During the time period relevant to this case, Osmose manufactured a chemical used by lumber suppliers to make plywood fire retardant and specified the procedures that the suppliers had to follow in treating the plywood.

FRT plywood (fire retardant treated) was primarily used as a fire wall and in the common roof areas of multi-family townhouses. Every major building code required the use of noncombustible materials, including, for example, FRT plywood, in the common roof areas of multi-family townhouses. The specific purpose of these code provisions was to ensure that builders designed and constructed multi-family housing in a manner that would prevent the spread of fire between the individual housing units.

Beginning in 1984, Pulte constructed 1876 townhouses with roofs containing FRT plywood treated with Osmose chemicals. When purchasing FRT plywood, Pulte purchasing agents did not specify a particular brand name, such as Osmose, nor did they rely on any of Osmose's promotional materials. Rather, Pulte treated FRT plywood as a "commodity," requiring only that it be cost-effective and contain the code-mandated FRT stamp.

In late 1988, after Pulte had sold the townhouses containing Osmose-treated FRT plywood, Pulte began receiving scattered reports that the plywood in its multi-family townhouses was degrading in service. At the same time, Pulte also became aware of reports in wood industry journals indicating that FRT plywood generally was unfit for use in roof systems because high temperatures and humidity levels in attic environments caused excessive reductions in the strength and structural integrity of the plywood. A normal attic environment in the United States may experience temperatures as high as 170 to 180 degrees. This extreme heat often is accompanied by excessive moisture and humidity. Of course, these temperature and humidity levels vary with the geographic location of the home.

In response to these reports, Pulte immediately notified its homeowners of the situation and warned them to stay off their roofs until Pulte could inspect them. After inspection, Pulte discovered universal degradation in the Osmose-treated FRT plywood, such that, according to Pulte, every roof containing Osmose-treated FRT plywood was "dangerously unpredictable" and unsafe. Tests Osmose conducted confirmed Pulte's findings. The tests, conducted in simulated attic environments, revealed that after 28 days of accelerated aging at 160 degrees, the Osmose-treated FRT plywood lost 72% of its strength. These test results ultimately led Osmose to cease the manufacture and sale of its fire retardant chemicals.

Pulte asked Osmose for assistance in resolving its FRT plywood dilemma, but Osmose refused. Pulte then launched an extensive remedial campaign culminating in the systematic replacement of all of the Osmose-treated FRT plywood. In replacing this plywood, Pulte was also forced to remove and replace other components of its roof systems, including untreated plywood and shingles. In total, Pulte's roof replacement project cost $3,658,200, or $1950 per roof.

On June 2, 1989, Pulte brought this diversity action against Osmose in the United States District Court for the Middle District of Florida. In its five-count complaint, which also named Lowe's Companies and Georgia Pacific Corporation as party defendants, Pulte alleged that Osmose "is engaged in the manufacture and sale of . . . fire retardant chemicals," and that Lowe's and Georgia Pacific sold Pulte the plywood after having treated it with Osmose chemicals. Pulte subsequently reached undisclosed settlement agreements with Lowe's and Georgia Pacific and dismissed its claims against those defendants.

The five counts against Osmose in Pulte's complaint alleged: (1) breach of express and implied warranties; (2) strict liability; (3) negligence; (4) "common law misrepresentation;" and (5) fraud. After the parties joined issue, Pulte voluntarily dismissed its breach of warranty claims. On February 20, 1992, the case proceeded to trial on the claims that remained. At the close of all of the evidence, Osmose moved for judgment as a matter of law pursuant to Fed. R. Civ. P. 50(a). The court denied its motion and submitted the case to the jury on Pulte's claims for strict liability, fraud, and negligence. By special verdict, the jury found in favor of Osmose on Pulte's claim of strict liability, concluding that the chemicals were not defective when they left Osmose's control. The jury also concluded that Osmose had been negligent in failing to warn Pulte that the FRT plywood would deteriorate if incorporated into Pulte's construction of the townhouse roofs (the negligence claim) and that Osmose had fraudulently induced Pulte to purchase the plywood (the misrepresentation and fraud claims). Accordingly, the jury awarded Pulte $3,750,000 in compensatory damages (on both claims) and an additional $2,500,000 in punitive damages (on the misrepresentation and fraud claims). After the jury rendered its verdict, the district court revisited Osmose's Rule 50(a) motion, and, concluding that the economic loss rule barred Pulte's negligence and fraud claims, entered judgment for Osmose. Pulte now appeals.

We conclude that Pulte is not entitled to recover on either its negligence or fraud claims. We find that Pulte's negligence claim is barred by the economic loss rule and does not meet either exception to the rule. Although we disagree with the district court that the economic loss rule, as applied in Florida, bars Pulte's claim for fraud in the inducement, we nonetheless hold that Pulte is not entitled to relief on that claim. In short, Pulte failed to establish essential elements of its claim that Osmose fraudulently induced its purchase of Osmose-treated FRT plywood. Specifically, other than the representations contained in Osmose promotional literature, Pulte presented no evidence from which the jury could conclude that Osmose made any misrepresentations or guarantees to Pulte regarding the strength of the FRT plywood treated with Osmose chemicals. Moreover, to the extent that Osmose's promotional literature misrepresented the properties of plywood treated with its chemicals, Pulte failed to establish that it actually relied on those statements in purchasing the Osmose-treated FRT plywood.

The economic loss rule "prohibits tort recovery when a product damages itself, causing economic loss, but does not cause personal injury or damage to any property other than itself." Economic loss includes "damages for inadequate value, *costs of repair and replacement of the defective product,* or consequent loss of profits—without any claim of personal injury or damage to other property." The rationale underlying the economic loss rule is that parties should protect against the risk of economic loss during contract negotiations through warranty provisions and price adjustments rather than attempt to recover under tort law after the loss occurs.

We first examine Pulte's negligence claim. Pulte alleges that Osmose negligently breached its duty to Pulte by failing to warn Pulte that FRT plywood treated with Osmose chemicals could not be used in attic environments. Stated another way, had Osmose warned Pulte (and other similarly situated purchasers) that FRT plywood treated with its chemicals "[wa]s inferior in quality and [would] not work for the general purposes for which it was manufactured and sold," Pulte would not have bought the product. Pulte's negligence claim presents precisely the type of "disappointed economic expectation" claim that Florida's strict interpretation of the economic loss rule forecloses.

Consequently, we must determine whether Pulte's claim falls within one of the two exceptions to the economic loss rule, which permit a tort claim if personal injury or damage to other property accompanies economic loss. The first exception is readily dismissed: because no individual has been physically injured as a result of the weakened plywood, Pulte's negligence claim does not fall within the personal injury exception to the economic loss rule.

To satisfy the "other property" exception, Pulte must establish damage to property aside from the FRT plywood treated with Osmose chemicals. Pulte incorporated the FRT plywood into its townhouses, such that the deterioration of the plywood destroyed the structural integrity of the roof. As a result, Pulte was forced to remove and replace the FRT plywood. In doing so, Pulte also had to remove and replace other portions of the roof, including untreated plywood and shingles. These other roof components form the basis of Pulte's "other property" argument.

In the present case, Pulte cannot establish that there was damage to other property. The product bargained for and purchased by Pulte was the Osmose-treated FRT plywood. Based on the evidence offered at trial, the FRT plywood itself was the only property damaged. Although Pulte did replace roof components other than the FRT plywood, these components were not replaced because they were *damaged.* Rather, replacing the shingles and other materials was *merely a consequence* of replacing the damaged FRT plywood. These costs simply do not trigger the "other property" exception.

Pulte failed to guard against the possibility that the Osmose-treated FRT plywood would not meet its economic expectations. Instead, Pulte chose to assume that the product would meet its specific needs. Pulte did not inquire as to the capabilities of plywood treated with Osmose fire retardant chemicals nor did Pulte demand assurances and warranties from Osmose that the products would meet its expectations. As a result, Pulte suffered a bitter and costly economic loss. Having failed to avail itself of the opportunity to mitigate the risks of potential disappointment at the time of contract negotiation, Pulte cannot now resort to the courts to save it from a bargain improvidently made.

We now turn to Pulte's fraud claim. Although the economic loss rule bars recovery for tort claims arising from breach of a contract, the doctrine does not preclude a claim for damages occasioned by an independent tort, including fraud in the inducement of a contract. Fraud in the inducement and deceit are independent torts for which compensatory and punitive damages may be recovered [citing cases]. Nevertheless, at trial, Pulte failed to prove its claim of fraud in the inducement. The grant of judgment in favor of Osmose is AFFIRMED.

The Eleventh Circuit, applying Florida law in *Pulte*, applied the Economic Loss Rule strictly against the commercial consumer of a manufactured construction product. As there was no contract between Pulte and Osmose, and no manufacturer's warranty of the chemical running to Pulte, as the purchaser, Pulte's only claim was in tort—negligence or fraud.[65] The court was impressed with Pulte's commercial sophistication and the fact that there were contracts between Pulte and the plywood suppliers, and subcontracts between the suppliers and Osmose. The court also strictly applied the "damage to other property" test.

The damage to other property analysis is fairly simple to articulate. Its application is less so. In the construction arena in particular, many if not most times the defective nature of a product can cause "physical damage." Examples include water intrusion from an improperly applied roofing membrane, and

65. Note that Pulte's fraud claim would have avoided the Economic Loss Rule if supported by the evidence.

cracked marble flooring due to a badly deflecting floor truss. Whether these damages are recoverable in negligence depends upon how a particular jurisdiction defines "product" and "other property" which is damaged by the defective product, whether the parties are in privity, and how the jurisdiction frames its desired exceptions to the rule.

If the integrated structure is the "product," the Economic Loss Rule bars recovery for any damaged part of it.[66] A significant number of courts, and the RESTATEMENT (THIRD) OF TORTS,[67] follow the broad "integrated product" analysis. If the defendant manufacturer of the allegedly defective construction product convinces the court that the product is so integrated into the structure (or a construction system within the structure) that it loses its identity,[68] then the integrated whole becomes the "product." If the structure is the "product," it cannot be "other property" for the purposes of physical damage to avoid the Economic Loss Rule against the manufacturer. On the other hand, a number of courts have not followed the strict "integrated structure as a product" analysis. These judges allow recovery against a remote manufacturer for property dam-

66. *See, e.g., Calloway,* 993 P.2d at 1267–68 (Nev. 2000) (roofing and siding defects causing water damage to interior structural components, held, damage to the "product" only); American Towers Owners v. CCI Mechanical, 930 P.2d 1182 (Utah 1996) (damage to plumbing pipes in the entire complex was damage to the "product" for purposes of the Economic Loss Rule, no recovery); Blagg v. Fred Hunt Co., 612 S.W.2d 321 (Ark. 1981) (house may be a "product" just as is an automobile); Chicago Heights Venture v. Dynamit Nobel of Am., 782 F.2d 723 (7th Cir. 1986) (defective roofing material allowed leaks damaging building, held, building is completed product); *Casa Clara Condominium Ass'n, Inc.,* 620 So. 2d at 1244 (concrete with high salt content corroded reinforcing steel in balconies causing balconies to crumble—held, no damage to "other property"); Foxcroft Townhome Owners Ass'n v. Hoffman Rosner Corp., 449 N.E.2d 125 (Ill. 1983) (damages resulting from defective siding not damage to "other property"); Dakota Gasification Co. v. Pascoe Bldg. Sys., 91 F.3d 1094, 1099 (8th Cir. 1996) (applying N. Dakota law); Detroit Edison Co. v. NABCO, Inc., 35 F.3d 236 (6th Cir. 1994) (applying Mich. Law).

67. The RESTATEMENT (THIRD) OF TORTS describes the characterization process (separate or incorporated) when products are incorporated into a system as "difficult." Nonetheless, the RESTATEMENT authors opine: "When the product or system is deemed to be an integrated whole, courts treat such damage as harm to the product itself." RESTATEMENT (THIRD) OF TORTS: PRODUCTS LIABILITY, § 21 cmt. e, 295–96 (1997). But as observed by at least one commentator and several courts, this analysis fails to recognize the duties "that arise independently of the pyramid structure of privity created by the tiers of contracts and subcontracts." Cheezem, *supra* note 47, at 22. *See also* Gilbane Bldg. Co. v. Nemours Found., 606 F. Supp. 995, 1005 (D. Del. 1985) (recognizing the interdependence of construction participants).

68. *See, e.g., Nat'l Union Fire Ins. Co.,* 815 P.2d at 601 (a defective part in the engine of a factory-assembled airplane damaged the engine and the airplane, held airplane was the integrated whole, and therefore the "product").

age to "other property" when the defective product damaged another identifiable construction component.[69]

Some defective products cases (most outside the construction arena) allow recovery of economic damages when the cause of the physical property damage to the product is from an outside source, such as fire.[70] Courts also have made distinctions when damaged property is contained within, but not integrated into, the structure.[71] In *Scott & Fetzer Co. v. Montgomery Ward & Co.*,[72] fire warning systems malfunctioned in a large warehouse occupied by several tenants. The fire spread throughout the warehouse, causing damage to the structure leased by various tenants and their personal property. The court went

---

69. *See, e.g.*, Lamb v. Georgia-Pacific Corp., 392 S.E.2d 307, 308 (Ga. App. 1990) (allegedly defective particle board installed underneath tile floor caused tile to crack, held, costs to repair damaged portion of the tile floor were recoverable as damage to "property other than the product itself," but that other economic damages would be denied); Comptech Int'l. Inc. v. Milam Commerce Park, Ltd., 753 So. 2d 1219 (Fla. 1999) (negligence claim by tenant against landlord in selection of contractor who damaged tenant's electrical systems, warehouse space and computers during renovation, held, existing electrical systems and computers were "other property"); Saratoga Fishing Co. v. J. M. Martinac & Co., 520 U.S. 875, 879 (1997) (ship's defective component damaged equipment added to the ship after the ship's initial purchase, held, added equipment was "other property"); City of La Crosse v. Schubert, Schroeder & Assoc., 240 N.W.2d 124 (Wis. 1976) (remote purchaser's negligence action against manufacturer of a roof that leaked and required replacement, held, damages to other parts of the structure were recoverable); United Air Lines, Inc. v. CEI Indus. of Illinois, Inc., 499 N.E.2d 558 (Ill. App. 1986) (water leaks in defective roof caused collapse of ceiling, held, damage to other property when damage involved "some violence or collision" and not deterioration or internal breakage, held, recovery of economic damages against manufacturer allowed). *See, e.g.*, *Jimenez*, 58 P.3d at 450 (Cal. 2002) (strict liability case against manufacturer—defective windows caused water damage to walls, floors, etc., held, "other property"); *Stearman*, 78 Cal. App. 4th at 611 (strict liability case—defective foundation construction and inadequate soil compaction caused slab movement, deformation and cracked walls, held, physical damage to "other property"); Northridge Co. v. W. R. Grace and Co., 471 N.W.2d 179 (Wis. 1991), (asbestos causes physical harm to structure, held, structure is "other property" and outside the Economic Loss Rule).

70. *See, e.g.*, Rocky Mountain Fire & Casualty Co. v. Biddulph Oldsmobile, 640 P.2d 851 (Ariz. 1982) (Winnebago destroyed by fire, held Economic Loss Rule does not apply to property damage, but would apply to claim for lost profits or other commercial losses); Gherna v. Ford Motor Co., 246 Cal. App. 2d 639, 649 (1966) (fire in the engine compartment which destroyed the entire automobile, some economic losses allowed); *but see* Lloyd Wood Coal Co. v. Clark Equip. Co., 543 So. 2d 671 (Ala. 1989) (front end loader's hydraulic hose ruptured causing a fire, but damage only to the product itself, economic losses barred).

71. *See* 2-J Corp. v. Tice, 126 F.3d 539, 544 (3d Cir. 1997) (collapsed metal building caused damage to stored inventory, held, "other property" from defective metal building, manufacturer was liable for economic losses) (applying Pa. law).

72. 493 N.E.2d 1022 (Ill. 1986).

so far as to deem the loss of the value of the tenants' leasehold and their personal items to be "not economic" because the damage was for "loss of property other than the defective product."[73]

Courts also struggle with the situation where the product is defective or potentially dangerous, but has not yet injured anyone. The Court of Judicial Appeals of Maryland (Maryland's highest court) carved out an exception to the otherwise accepted Economic Loss Rule (no recovery if the defective product has not caused physical injury) when the defect can lead to a "serious risk of creating a dangerous condition."[74] In *Whiting-Turner*, the defect was both in the product and in the construction installation services of the contractors.

Council of Co-Owners Atlantis Condominium, et al. v. Whiting-Turner Contracting Co., et al. 517 A.2d 386 (Md. 1984).

McAULIFFE, Judge.

The principal issue presented by this appeal is the extent to which tort liability should be imposed upon builders and architects for damages suffered by parties who enjoy no contractual privity with them. We hold that privity is not an absolute prerequisite to the existence of a tort duty in this type of case, and that the duty of builders and architects to use due care in the design, inspection, and construction of a building extends to those persons foreseeably subjected to the risk of personal injury because of a latent and unreasonably dangerous condition resulting from that negligence. Additionally, we hold that where the dangerous condition is discovered before it results in injury, an action in negligence will lie for the recovery of the reasonable cost of correcting the condition.

The Atlantis Condominium is a twenty-one story building containing 198 separate condominium units and located in Ocean City, Maryland. The council of unit owners as the governing regime of the condominium, and three unit owners who sue in their individual capacity, brought an action in tort against the general contractor, developer, and architects involved in the construction of the building, contending that as a result of their negligence the utility shafts and related electrical work were not constructed and installed in accordance with the plans and specifications and did not comply with the requirements of the applicable building code. They allege

73. *Id.* at 1026. The court also noted that the damage was caused by a "sudden and dangerous conflagration."

74. Council of Co-Owners Atlantis Condominium, Inc. v. Whiting-Turner Contracting Co., 517 A.2d 336, 345 (Md. 1986) (electrical utility shafts did not contain required insulation—creating a building code violation and potential hazard). Although the project was multifamily residential, the Maryland court did not create an exception to the Economic Loss Rule strictly for residential properties.

that these latent conditions create a fire hazard that "presents a threat to the safety and welfare of the owners and occupants of the Atlantis Condominium and to the personal and real property of [the] owners and occupants."

The specific deficiency alleged is the failure to construct ten vertical utility shafts with materials having a fire resistance rating of two hours. Appellants do not allege negligence in the design of the shafts, but rather a failure to construct the shafts in accordance with the design. They allege negligent construction on the part of Whiting-Turner Contracting Company, the general contractor; negligent inspection, supervision, and acceptance of the work on the part of Meyers & D'Aleo, Inc., the design and supervising architects; and negligent inspection and misrepresentation on the part of Ewing Cole Erdman Rizzio Cherry Parsky, a partnership, and Alexander Ewing, individually, who they allege were the architects employed "to inspect the building and to certify to the Building Inspection Department of Ocean City, Maryland, that the building was constructed pursuant to the approved building permit in accordance with the plans and specifications submitted with the original permit application and that the building was ready for occupancy." Additionally, Appellants allege that Colonial Mortgage Service Company ("Colonial") was a developer of the project and that it knowingly or negligently permitted the shafts to be built at variance with the plans, as well as the requirements of the building code, and negligently obtained an occupancy permit. Appellants also allege that in advertising and selling the units, the developer negligently misrepresented "the building's suitability for occupancy." Appellants seek compensatory and punitive damages from each Appellee.

The builder and the certifying architects filed demurrers, contending that the declaration failed to allege facts sufficient to show a duty owing by them to Appellants. They contended that in the absence of privity no duty could arise. Recognizing, however, that the modern trend was toward the recognition of a duty to third parties under some circumstances, they advanced the alternative argument that a duty should not be recognized in this case because Appellants had suffered only economic loss, and not personal injury or property damage. The developer also demurred, asserting as its sole challenge to the sufficiency of the count against it that there was no allegation of personal injury or property damage, and that it "cannot be liable in tort to plaintiffs for purely economic loss." The trial judge sustained the demurrers without leave to amend, concluding that Maryland law would not recognize a tort duty in the absence of privity under these circumstances, and that in any event a duty would not be recognized where only economic loss was claimed. We restrict our consideration of the sufficiency of the pleadings to those grounds specifically set forth in the demurrer.

It is now the almost universal rule that the contractor is liable to all those who may foreseeably be injured by the structure, not only when he fails to disclose dangerous conditions known to him, *but also when the work is negligently done.* This applies not only to contractors doing original work, but also to those who make repairs, or install parts, as well as supervising architects and engineers. There may be liability for negligent design, as well as for negligent construction. This trend

> was also reflected by the incorporation of § 385 into the *Restatement (Second) of Torts* (1965). One who on behalf of the possessor of land erects a structure or creates any other condition thereon is subject to liability to others upon or outside of the land for physical harm caused to them by the dangerous character of the structure or condition after his work has been accepted by the possessor, under the same rules as those determining the liability of one who as manufacturer or independent contractor makes a chattel for the use of others.
>
> In following the modern trend, we hold that privity is not an absolute prerequisite to the existence of a tort duty. The duty of the architects and the builders in this case, to use due care in the design, inspection, and construction of this condominium, extended to those persons foreseeably subjected to the risk of personal injury created, as here, by a latent and unreasonably dangerous condition resulting from their negligence.
>
> We turn to the question of whether this type of action may be maintained against a builder or architect where the risk of personal injury exists, but personal injury has not in fact resulted. We conclude that the determination of whether a duty will be imposed in this type of case should depend upon the risk generated by the negligent conduct, rather than upon the fortuitous circumstance of the nature of the resultant damage. Where the risk is of death or personal injury the action will lie for recovery of the reasonable cost of correcting the dangerous condition.

The inquiry of Maryland's highest court, drilling down into the facts, exposed one of the Economic Loss Rule's major weaknesses when applied in the construction context—that a defective product or condition may be potentially dangerous without causing physical (and, therefore, recoverable) damage.[75] Even when a defect creates an inherently dangerous situation, the Economic Loss Rule, strictly applied, bars any recovery (and, therefore, without bodily injury would prevent an action) for the repair or replacement of the defective product.[76] The Maryland court determined that focusing upon the "fortuitous circumstance of the nature of the resultant damage" was the wrong approach.[77]

---

75. In *Whiting-Turner*, the court observed: "Why should a buyer have to wait for a personal tragedy to occur in order to recover damages to remedy or repair defects?" 517 A.2d at 345.

76. One lower court has opined that only a few jurisdictions allow direct negligence actions for purely economic losses, or have recognized tort-based warranties, against *product manufacturers* of defective construction components not in privity, without a showing of physical harm or the serious risk of harm. *See Wilson,* 206 F. Supp. 2d at 749 (structure as a whole suffered only economic damages even though water intrusion allowed by defective cladding caused structural damage, held, economic loss rule applied to claim v. remote manufacturer) (citing cases). *Compare* the application of the rule to construction service providers.

77. *Id. See also Kennedy,* 384 S.E.2d at 730, 737, where the court opined: "We find that this legal framework [focusing only upon the type of damage incurred] generates difficulties. This is

Other cases have reached the opposite result. In an economic loss products case involving potentially dangerous corroding steel wall panels,[78] the U.S. Court of Appeals for the Seventh Circuit strictly applied the Economic Loss Rule to the owner's negligence count. Writing for the court, Judge Posner acknowledged that allowing recovery for bodily injuries caused by a defective product (including economic damages arising from the injuries), but barring recovery for economic damages in the absence of bodily injury caused by the same defective product, was a matter of fortuity. Still, Judge Posner observed, the court "could not recast this case as if one of the corroded wall panels had fallen and broken [the plaintiff's] foot."[79]

B. Claims of Defective Construction Services (Including Poor Workmanship and Improper Furnishing of Materials) (No Privity)

Construction involves the delegation and allocation of responsibility by contract.[80] Within the "pyramid" of this process, however, many of the participants are not in privity of contract.[81] Their defective work and products can cause economic damage to other providers, and cause other contractors to be targets of claims involving defective work. Strictly applied, the Economic Loss Rule often prohibits direct negligence actions by construction participants against

---

so because the framework's focus is on consequences, not action. . . . The framework we adopt focuses on activity, not consequence."

78. Miller v. United States Steel Corp., 902 F.2d 573 (7th Cir. 1990).

79. *Id.* at 574. Interestingly, less than two years later, in a somewhat analogous case involving general liability (GL) coverage for "property damage," Judge Posner opined that: "The incorporation of a defective product into another product inflicts *physical injury* in the relevant sense [insurance coverage] on the latter [the non-defective product] at the moment of incorporation—here, the moment when the defective Qest systems were installed in homes." (emphasis added). *See* Elger Mfg., Inc. v. Liberty Mutual Ins. Co., 972 F.2d 805, 814 (7th Cir. 1992) (declaratory judgment action brought to determine if physical injury occurred for the purposes of property damage triggering GL coverage); *see also* Roger H. Proulx & Co. v. Crest Liners, Inc., 98 Cal. App. 4th 182 (2002) (insurance broker sued in negligence for failure to procure coverage, defended on basis that the underlying claims were barred by economic loss rule, held, claimed damages raised triable issue).

80. Construction providers do not manufacture or supply "products," at least for strict liability purposes. *See* Restatement (Third) of Torts: Products Liability, § 19, subd. (b) ("services, even when provided commercially, are not products" and are therefore governed by the law of negligence).

81. For instance, the mason is a subcontractor to the general contractor, but the mason has no contract with the sitework subcontractor who has left the site a muddy mess, prohibiting the mason from making progress in constructing concrete block walls.

those principally responsible for construction defects even though the parties to the dispute are not in privity of contract.[82]

Naturally, there are exceptions, and different courts emphasize different legal concepts. The issues taken up by the courts to determine the extent of the rule's application where construction services are involved are similar, but differ slightly from those grappled with by courts concerning manufactured products.[83] If the parties to a dispute involving defective construction services are not in privity of contract, courts look to the scope of work and duties the service provider has undertaken contractually, whether damage to "other property" has occurred, and whether the defective work can be related to an independent duty. While a few courts deem lack of privity an absolute bar to any tort claim for economic losses other than for fraud,[84] most states consider the lack of privity as a starting point to discuss scope of services, damage to "other property," and the presence or absence of an independent duty.

If the plaintiff is not in privity with the provider of defective work, but the damaged work was the subject matter of the provider's contract, the rule will also generally bar recovery based on economic damages to the work, where only the work was injured.[85] On the other hand, a plaintiff may avoid the economic loss rule by demonstrating that regardless of the nature of the damage, the defendant breached a tort duty, independent of (although generally related to) the service provider's contract.[86]

---

82. As we have seen, privity will trigger the application of the Economic Loss Rule to bar negligence claims for economic losses. *See Wausau Tile*, 593 N.W.2d 445.

83. The agreements of construction service providers are, for instance, generally not governed by the UCC.

84. *See* Sensenbrenner v. Rust, Orling & Neal Architects, Inc., 374 S.E.2d 55 (Va. 1988) (homeowners' action against pool subcontractor and architect dismissed for lack of privity). This holding is consistent with the law in the Old Dominion.

85. *Calloway*, 993 P.2d at 1261–62 (townhome owners' claims against subcontractors for defective roofing and siding did not state cause of action in negligence for purely economic losses); Jardel Enterprises, Inc. v. Triconsultants, Inc., 770 P.2d 1301 (Colo. App. 1988) (restaurant owner could not bring negligence claim against subcontractor for lost profits resulting from delayed opening of restaurant due to initial construction of the restaurant at the wrong location); Nastri v. Wood Bros. Homes, Inc., 690 P.2d 158, 163–64 (Ariz. App. 1984) (despite latent structural defects, remote purchasers could not recover from builder on negligence theory, but could recover for implied-in-law warranty of habitability).

86. *See* Consol. Hardwoods, Inc. v. Alexander Concrete Constr., Inc., 811 P.2d 440, 443 (Colo. App. 1991) (owner negligence claim allowed against subcontractor because independent duty was breached); Juliano v. Gaston, 455 A.2d 523 (N.J. 1982) (owner may recover against subcontractor in negligence for economic damages).

Town of Alma v. AZCO Constr., Inc.
10 P.3d 1256 (Colo. 2000)

Justice Rice delivered the Opinion of the Court.

The Town of Alma, joined by several individual town residents, filed suit against AZCO Construction, Inc. ("AZCO"), asserting claims for breach of contract, breach of the implied warranty of sound workmanship, and negligence. The trial court dismissed Petitioners' negligence and breach of implied warranty of sound workmanship claims and a jury returned a verdict for AZCO on the breach of contract claim. Petitioners appealed the trial court's dismissal of their negligence and breach of implied warranty of sound workmanship claims and the court of appeals affirmed the dismissal of the claims. We granted Petitioners' petition for writ of certiorari to review the court of appeals' judgment affirming the trial court's dismissal of Petitioners' negligence claim. We now affirm the judgment of the court of appeals.

Petitioners' amended complaint alleged the following facts: On October 28, 1992, the town and AZCO entered into a contract for the construction of improvements to Petitioners' water distribution system. The contract called for AZCO to install new water mains, and to tie those water mains to existing water service lines which served residential properties in the town. The contract contained two separate warranty provisions. Section 29.1 of the contract, titled "GUARANTEE," provided, "[AZCO] shall guarantee all materials and equipment furnished and WORK performed for a period of one (1) year from the date of SUBSTANTIAL COMPLETION. [AZCO] warrants and guarantees . . . that the completed system is free from all defects due to faulty materials or workmanship. . . ." In addition, section 3.3.1 of the "Special Conditions" attached to the contract, titled "MAINTENANCE AND GUARANTY," provided, "[AZCO] hereby guarantees that the entire work constructed by [it] under the contract will fully meet all requirements of the contract as to quality of workmanship and materials. . . . [AZCO] hereby agrees to make at [its] own expense, any repairs or replacement made necessary by defects in materials or workmanship supplied by [it] that become evident within one year after the date of final payment. . . ."

In 1993, AZCO installed 115 flared fittings for the water service line connections. In June 1995, Petitioners discovered leaks in three water service line connections that AZCO installed pursuant to the contract. AZCO repaired those three leaks under the one-year warranty provision of the contract. In November 1995 and June 1996, additional leaks in water service line connections were discovered but AZCO refused to repair these leaks on the basis that the one-year warranty provision had expired. The leaks were repaired at the expense of individual town residents.

Petitioners filed suit against AZCO on November 27, 1996, asserting various claims, including negligence. The trial court granted AZCO's motion to dismiss the negligence claim and the case proceeded to trial on the breach of contract claim only, with the jury returning a verdict for AZCO on this claim.

Petitioners appealed the trial court's dismissal of their negligence claims. In affirming the dismissal of Petitioners' negligence claim, the court of appeals relied on the economic loss rule and stated that "[t]o hold otherwise would permit the non-breaching party to avoid the contractual limitation of remedy." Town of Alma, 985 P.2d at 57. We granted certiorari to review the dismissal of Petitioners' negligence claim.

This case presents an opportunity for us to address the status of the economic loss rule in Colorado. The rule has been applied by our court of appeals in various contexts to bar tort claims. As this is a matter of first impression, our analysis encompasses an examination of the development of the rule in other jurisdictions, as well as a discussion of the principles and rationale underlying the rule.

Limiting tort liability when a contract exists between parties is appropriate because a product's potential nonperformance can be adequately addressed by rational economic actors bargaining at arms length to shape the terms of the contract. For example, a buyer may demand additional warranties on a product while agreeing to pay a higher price, or the same buyer may choose to assume a higher level of risk that a product will not perform properly by accepting a more limited warranty in exchange for a lower product price. Limiting the availability of tort remedies in these situations holds parties to the terms of their bargain. In this way, the law serves to encourage parties to confidently allocate risks and costs during their bargaining without fear that unanticipated liability may arise in the future, effectively negating the parties' efforts to build these cost considerations into the contract. The economic loss rule thus serves to ensure predictability in commercial transactions.

The key to determining the availability of a contract or tort action lies in determining the source of the duty that forms the basis of the action. We find the following discussion by the South Carolina Supreme Court informative:

> The question, thus, is not whether the damages are physical or economic. Rather the question of whether the plaintiff may maintain an action in tort for purely economic loss turns on the determination of the source of the duty [the] plaintiff claims the defendant owed. A breach of a duty which arises under the provisions of a contract between the parties must be redressed under contract, and a tort action will not lie. A breach of a duty *arising independently* of any contract duties between the parties, however, may support a tort action. *Tommy L. Griffin Plumbing & Heating Co. v. Jordan, Jones & Goulding, Inc.,* 320 S.C. 49, 463 S.E.2d 85, 88 (1995) (emphasis added).

Determining when a contract action will lie and when a tort action will lie requires maintaining this distinction in the sources of the respective obligations. The phrase "economic loss rule" necessarily implies that the focus of the inquiry under its analysis is on the type of damages suffered by the aggrieved party.

However, the relationship between the type of damages suffered and the availability of a tort action is inexact at best. Examining the type of damages suffered may assist in determining the source of the duty underlying the action (e.g., most actions for lost profits are based on breaches of contractual duties while most actions involving physical injuries to persons are based on common law duties of care). However, some torts are expressly designed to remedy pure economic loss (*e.g.*, professional negligence, fraud, and breach of fiduciary duty). It is here that substantial confusion arises from the use of the term "economic loss rule."

This confusion can be avoided, however, by maintaining the focus on the source of the duty alleged to have been violated. For example, we have recognized that some special relationships by their nature automatically trigger an independent duty of care that supports a tort action even when the parties have entered into a contractual relationship. *See, e.g., Bebo Constr. Co. v. Mattox & O'Brien, P.C.,* 990 P.2d 78, 83 (Colo.1999) (attorney-client relationship creates independent duty of care); *Greenberg v. Perkins,* 845 P.2d 530, 534 (Colo.1993) (physician-patient relationship creates independent duty of care, as does physician's independent medical examination of non-patient); *Farmers Group, Inc. v. Trimble,* 691 P.2d 1138, 1141–42 (Colo.1984) (quasi-fiduciary nature of insurer-insured relationship creates independent duty of care). We have also recognized that certain common law claims that sound in tort and are expressly designed to remedy economic loss may exist independent of a breach of contract claim. *See Brody v. Bock,* 897 P.2d 769, 776 (Colo.1995) (common law fraud claim is based on violation of a duty independent of contract); *Keller v. A. O. Smith Harvestore Prods., Inc.,* 819 P.2d 69, 73 (Colo.1991) (negligent misrepresentation is a tort claim based "not on principles of contractual obligation but on principles of duty and reasonable conduct"). In these situations where we have recognized the existence of a duty *independent* of any contractual obligations, the economic loss rule has no application and does not bar a plaintiff's tort claim because the claim is based on a recognized independent duty of care and thus does not fall within the scope of the rule.

Turning to the facts before us, the contract in the instant case expressly assigned a duty of care to AZCO in the installation of the water system. It was this contractual duty that AZCO allegedly breached. AZCO expressly assumed the duty to guarantee its quality of workmanship and its materials when it undertook to install the water system. As such, Petitioners have failed to demonstrate that AZCO breached any duty independent of its contractual obligations.

Moreover, the town and the individual landowners are only seeking damages for the cost of repair and replacement of the water lines that were the subject of the contract. Damages for the cost of repair and replacement of property that were the subject of the contract constitute economic loss damages that must be supported by an independent duty of care to be recoverable in a negligence action. As there is no independent duty to support Petitioners' negligence claim, the economic loss rule bars this claim.

Recovery in negligence for purely economic damages to the work involved in the contract or subcontract under which the work was performed requires the breach of a separate tort duty, independent of the contract or subcontract.[87] If the plaintiff is in privity with the defendant, or the defendant's contract covers tasks that were defectively performed which caused the complained-of economic damages, the breach of the obligations of the contract ordinarily cannot be relied upon to support an action in negligence; the parties (or at least one of the parties) have already selected a contractual remedy. Failure to perform one's contract, even carelessly, generally will not support a negligence action for economic damages to the work.[88]

Economic injuries caused by the defective work but visited upon "other property" may be enough to allow a negligence action to proceed.[89] Damage to "other property" generally means property unrelated to the service provider's work. Thus, the "incorporation" and "integration" analysis concerning manufactured products also pertains to claims against and between service providers.[90] A number of courts hold that contractors and service providers both owe a tort duty to those not in privity for forseeable losses and damages

87. *See also* Olympic Products Co. v. Roof Sys., Inc., 363 S.E.2d 367 (N.C. App. 1988) (manufacturer's failure to properly inspect installation or thereafter to report improper installation deemed independent negligence, economic damages allowed); Raynor Steel Erection v. York Constr. Co., 351 S.E.2d 136 (N.C. 1986) (economic damages allowed in negligence due to general contractor's failure to properly install); Brown v. Fowler, 279 N.W.2d 907 (S.D. 1979) (remote residence purchaser has a negligence claim against builder for economic damages due to breach of separate duty. This was not an implied warranty of habitability case—which under South Dakota law extends only to the initial purchaser).

88. *See, e.g., Palmetto Linen Serv., Inc.*, 205 F.3d at 129 (applying S.C. law), *quoting from Kennedy,* 384 S.E.2d at 737 ("the 'economic loss' rule will still apply where duties are created *solely* by contract") (emphasis in original).

89. *See, e.g., Aas,* 12 P.3d at 1130 (homeowners' association and individual homeowners could not bring negligence actions against developer, contractor, or subcontractors for construction defects or diminution in value, unless property damage to "other property" could be shown. "Any construction defect can diminish the value of a house. But the difference between price paid and value received, and deviations from standards of quality that have not resulted in property damage or personal injury, are primarily the domain of contract and warranty law or the law of fraud, rather than of negligence") (citing cases).

90. *See* discussion *supra* at notes 66–71 as to the incorporation of products and services to form an "integrated product"—the structure. *But see Comptech Int'l, Inc.*, 753 So. 2d at 1219 (tenant sued landlord for negligent selection of contractor, who in renovating tenant space in warehouse, damaged electrical and computer equipment, held, contract was for a service—the renovation—and court suggested that services do not involve "other property." However, to the extent that the warehouse was the object of the contract, the wiring and computers were "other property").

proximately caused by their negligent acts.[91] These courts made no distinction between the designer and the constructor. Other courts, however, have recognized a distinction, to the benefit of the design professional.

C. Claims of Defective Design Professional Services (Including Insufficient Design and Improper Contract Administration Services) (No Privity)

2314 Lincoln Park West Condominium Ass'n v. Mann, Gin, Ebel & Frazier, Ltd. 555 N.E.2d 346 (Ill. 1990)

Justice MILLER delivered the opinion of the court:

The plaintiff, Conservatory Condominium Association, is an association of the owners of the 39 condominium units located at 2314 Lincoln Park West in Chicago. Mann, Gin, Ebel & Frazier, Ltd. (Mann) is the architectural firm that designed the project and certified its completion. The plaintiff brought the present action in Cook County seeking an award of damages for the cost of making certain repairs to the building in which the association members' condominium units are located. Plaintiff seeks recovery on a negligence theory from the architectural firm responsible for the building's design.

The plaintiff alleges that numerous defects in the design and construction of the project had become known to the unit owners by late 1986. According to the complaint, windows and glass doors were loose, the roof leaked, the heating and cooling systems and other utilities were inadequate and did not function properly, and the garage was settling. The plaintiff does not deny that the recovery requested in the present action is for what we have termed economic loss. The plaintiff is seeking an award of damages for the cost of repairing defects in the property, and there is no contention that personal injury or damage to other property has resulted. The plaintiff in the present case seeks compensation for the cost of repairing certain defects allegedly occurring in the construction of the condominium unit owners' residences.

The gravamen of the plaintiff's claim for negligence against Mann is dissatisfaction with the way in which the building was designed and constructed, and the failure of the building to meet the unit owners' expectations. As our prior decisions concerning the construction industry fully illustrate, such a claim concerns the quality, rather than the safety, of the building and thus is a matter more appropriately resolved under contract law.

91. *See* McDonough v. Whalen, 313 N.E.2d 435 (Mass. 1974) (both the designer and installer of a defective septic system owed a tort duty to property owners not in privity).

We decline to impose on Mann a duty in tort to protect the unit owners from the sort of loss asserted. The architect's responsibility originated in its contract with the original owner, and in these circumstances its duties should be measured accordingly. Recovery of the nature requested here essentially seeks damages for a difference in quality. "There is room in the market for goods of varying quality, and if the purchaser buys goods which turn out to be below its expectations, its remedy should be against the person from whom it bought the goods, based upon the contract with that person." [citing cases]

While we do not intend in the present case to determine the future application of *the* economic loss rule in all areas of professional malpractice, we must reject the plaintiff's theory that denial of the negligence claim against the present architect would signal in general the end of malpractice recovery in tort. It may be noted that in suits involving malpractice claims against health care professionals, personal injury will generally be the alleged result, and therefore the rule would not bar recovery in any event. Moreover, we believe that other professional relationships are readily distinguishable from the present case. For example, malpractice actions against attorneys have gone forward, without any suggestion that the form of recovery traditionally recognized in such actions [economic losses] would no longer be allowed. [citing cases] Moreover, those cases recognize an extra-contractual duty not only to the client but also to the group of persons the client intended to benefit. Such a duty arises from a consideration of the nature of the undertaking and the lawyer's traditional responsibilities. The same cannot be said with respect to the defendant architect in the present case.

Curiously, Illinois allows nonprivity negligence actions for economic losses against other professionals and nonprofessionals outside the construction process, including accountants,[92] attorneys,[93] and stock market indexing services,[94] but not against architects.

Ellis-Don Constr., Inc. v. HKS, Inc.
353 F. Supp. 2d 603 (M.D.N.C. 2004)

BULLOCK, District Judge.

This diversity action, filed May 27, 2004, arises out of a hospital construction project in which Ellis-Don Construction, Inc. ("Plaintiff") served as the general contractor and HKS, Inc. ("HKS"), Smith Seckman Reid, Inc. ("SSR"), and Corley Redfoot Zack, Inc. ("CRZ") (together, "Defendants") served as the design team, responsible

92. Congregation of the Passion, Holy Cross Province v. Touche Ross & Co., 636 N.E.2d 503 (1994).

93. Collins v. Reynard, 607 N.E.2d 1185 (Ill. 1992) (an attorney's duty of competence exists "without regard to the terms of any contract of employment").

94. Rosenstein v. Standard & Poor's Corp., 636 N.E.2d 665 (Ill. App. 1993).

for overall project design and management. Plaintiff alleges that Defendants performed their duties negligently, resulting in construction delays and cost overruns that forced Plaintiff to incur economic losses in excess of $1,000,000.

The University of North Carolina Hospitals, a state "public body" and owner of the construction project at issue here, contracted with HKS to design and manage the construction of the North Carolina Children's Hospital and North Carolina Women's Hospital. HKS contracted with SSR and CRZ to serve as consultants on the project, with SSR providing services relating to the mechanical, plumbing, electric, and fire protection systems, and CRZ providing architectural services, including planning and design work, administration of the construction process, and inspections. Plaintiff contracted directly with the owner to provide general contractor services for the construction of the hospitals. No contractual privity existed between Plaintiff and any of the Defendants.

In North Carolina, "in the absence of privity of contract an architect may be held liable to a general contractor and his subcontractors for economic loss resulting from breach of a common law duty of care." Such duty of care "flow[s] from the parties' working relationship." Indeed, the courts of North Carolina have stated that the "power of economic life or death" an architect holds over a contractor requires that such a duty of care be recognized, even in the absence of privity of contract.

The Defendants argue that North Carolina's economic loss doctrine has expanded beyond its traditional realm of products liability and sales of goods to preclude Plaintiff's tort claim. The court disagrees. The economic loss doctrine, first articulated by the California Supreme Court in *Seely v. White Motor Co.,* 63 Cal.2d 9, 45 Cal. Rptr. 17, 403 P.2d 145 (1965), and adopted by the United States Supreme Court in *East River S.S. Corp. v. Transamerica Delaval, Inc.,* 476 U.S. 858, 106 S. Ct. 2295, 90 L. Ed. 2d 865 (1986), was conceived of as a means by which to confine products liability in tort to damages for personal injury and injury to property other than the goods sold, and leave to contract law the question of liability for purely economic losses. This approach was, and remains, firmly rooted in traditional concepts of warranty and contract law, and operates to "keep products liability and contract law in separate spheres and to maintain a realistic limitation on damages" in products liability cases. *East River,* 476 U.S. at 871, 106 S. Ct. 2308. In North Carolina, the economic loss doctrine limits the types of damages that a purchaser of a product may recover against the seller or manufacturer of the product through a negligence action. The purchaser cannot use claims of negligence to recover "purely economic losses" caused by the product's failure to meet expectations. Instead, the purchaser must use contract law, including the Uniform Commercial Code (UCC).

The Defendants explore at length the "roots" and "flowering" of North Carolina's economic loss rule in support of their motion to dismiss, but they appear to be harvesting the wrong field in attempting to use the rule as a bar to Plaintiff's tort claim. The economic loss rule poses no obstacle to Plaintiff's claim. As noted, North Carolina's economic loss rule bars claims in tort for purely economic losses in the sale of goods covered by contract law, including the UCC. It does not limit tort actions that arise in the absence of a contract, nor is there any indication that the courts of North Carolina have expanded the rule beyond its traditional role in products liability cases.

As observed in the preceding decisions, a party not in privity with the design professional will be allowed or precluded from recovering for economic losses based on how the courts of its state apply the Economic Loss Rule. The reported cases go both ways, and the legal bases for the decisions are mixed. A number of courts find that no action for economic losses may be maintained by a third party against a construction design professional, regardless of the scope of its services or the foreseeability of harm. In the majority of these jurisdictions, the Economic Loss Rule (generally the "no duty" or "covered by another contract" analysis rather than the "privity" side) is invoked to reach this result.[95]

These courts do not dismiss actions because the third parties lack privity as such, but instead deduce that plaintiffs not in privity are not owed a duty by the design professional—thereby cutting off all economic damages.[96] To these judges, it is often at least noteworthy that the contract between the contractor and the owner contains provisions (including limitations) relating to foreseeable economic losses.[97] At least one jurisdiction, however, has not employed the Economic Loss Rule (denominated as such) in professional negligence actions, but instead has relied consistently and strictly upon privity of contract to delete a third-party cause of action for professional negligence from the state's judicial framework.[98]

---

95. *See, e.g.*, Fireman's Fund Ins. Co. v. SEC Donohue, Inc., 679 N.E.2d 1197 (Ill. 1997) (economic loss doctrine applies to engineers and architects, barring any recovery for economic losses in tort); Widett v. U.S. Fidelity and Guar. Co., 815 F.2d 885, 886–87 (2d Cir. 1987), (absent privity, architect was not liable for professional negligence to subcontractors that detrimentally relied on erroneous site plans); *American Towers Owners Ass'n, Inc.*, 930 P.2d at 1192 (same); Atherton Condominium Apartment-Owners Ass'n Bd. of Directors v. Blume Dev. Co., 799 P.2d 250, 262 (Wash. 1990) (*en banc*) (same); Floor Craft Floor Covering, Inc. v. Parma Community Gen. Hosp. Ass'n, 560 N.E.2d 206 (Ohio 1990) (same); Border Brook Terrace Condominium Assoc. v. Gladstone, 622 A.2d 1248 (N.H. 1993) (economic loss doctrine applied).

96. *See, e.g.*, Rissler & McMurray Co. v. Sheridan Area Water Supply Joint Powers Bd., 929 P.2d 1228 (Wyo. 1996) (engineer's duty under its contract with the water board did not extend to the contractor—either under a design negligence theory or a negligent misrepresentation theory); Fleischer v. Hellmuth, Obata and Kassabaum, Inc., 870 S.W.2d 832 (Mo. App. 1993) (architect had no duty to CM); *but see* Miller v. Big River Concrete, LLC, 14 S.W.3d 129, 134 (Mo. App. 2000) (engineer's negligence in conducting tests was actionable in negligence and negligent misrepresentation despite lack of privity).

97. *See also* Berschauer/Phillips Constr. Co. v. Seattle Sch. Dist. No.1, 881 P.2d 986 (Wash. 1994) (*en banc*) (limiting contractor's recovery of economic damages to remedies provided and disclaimed in owner/contractor construction contract).

98. *See* Blake Constr. Co. Inc. v. Alley, 353 S.E.2d 724, 727 (Va. 1987) and Bryant Elec. Co. v. City of Fredericksburg, 762 F.2d 1192 (4th Cir. 1985) (there is no cause of action for a contractor to recover against an engineer for economic loss in the absence of privity under Virginia law). Unlike other jurisdictions, Virginia is uniformly strict in its privity requirement, for all professionals. *See, e.g.*, Ward v. Ernst & Young, 435 S.E.2d 628 (Va. 1993) (action against CPA requires privity);

Other jurisdictions allow third parties to assert professional negligence actions for economic damages.[99] The legal basis cited in these decisions rests in the reasonable foreseeability that harm may result to a particular group from the design professional's negligent conduct.[100] This foreseeability creates a duty to use due professional care.

Some courts distinguish between design services that predate the commencement of construction (no duty to third parties for economic losses), and professional services performed during the project (which create a duty). Several of these opinions refer to a "special relationship" between an architect who has contract administration duties and the project participants.[101] California was one of the first jurisdictions to recognize that a "supervising architect" who holds the power of the purse against the contractor has power "tantamount to life or death."[102] Other jurisdictions decline to distinguish between the date or type of the allegedly negligent professional services[103] and enforce

---

Copenhaver v. Rogers, 384 S.E.2d 593 (Va. 1998) (privity required in claim against attorney). Virginia also requires privity for most indemnification actions. *See* Pulte Home Corp. v. Parex, Inc., 579 S.E.2d 188, 2003 WL 1903424 (Va., April 17, 2003).

99. *See generally* Matthew S. Steffey, *Negligence, Contract, and Architect's Liability for Economic Loss*, 82 Ky. L. J. 659, 662 (1993); Frank D. Wagner, Annotation, *Tort Liability of Project Architect for Economic Damages Suffered by Contractor*, 65 A.L.R.3d 249 (1975).

100. *See, e.g.*, Ins. Co. of N. Am. v. Town of Manchester, 17 F. Supp. 2d 81, 84 (D. Conn. 1998) (applying Conn. law) ("foreseeability is key to the determination of a cause of action in negligence").

101. *See* Craig v. Everett M. Brooks Co., 222 N.E.2d 752, 755 (Mass. 1967) (engineer was not liable for mistakes in plans, but for professional negligence in improperly placing stakes needed to locate and direct construction); A. R. Moyer, Inc. v. Graham, 285 So.2d 397 (Fla. 1973) ("supervising architect"); R. H. Macy & Co. v. Williams Tile & Terrazzo, Inc., 585 F. Supp. 175, 180 (N.D. Ga. 1984) (applying Georgia law, architect owes no tort duty to subcontractor if their relationship does not "approach" that of privity or reliance) (citations omitted).

102. *United States ex rel. Los Angeles Testing Lab.*, 161 F. Supp. at 136.

103. *See Moransais*, 744 So. 2d at 983 ("the mere existence of such a contract should not serve per se to bar an action for professional malpractice"); Stone's Throw Condominium Ass'n, Inc. v. Sand Cove Apartments, Inc., 749 So. 2d 520, 522 (Fla App. 1999) (claims based on architect's failure to design to state minimum building code and negligent misrepresentation are not barred by economic loss rule despite lack of privity); Conforti & Eisele, Inc. v. John C. Morris Assoc., 418 A.2d 1290 (N.J. Super. 1980) (contractors may sue architects for professional negligence despite lack of privity); Shoffner Indus., Inc. v. W. B. Lloyd Constr. Co., 257 S.E.2d 50, 5559 (N.C. App. 1979); Bagwell Coatings, Inc. v. Middle South Energy, Inc., 797 F.2d 1298 (5th Cir. 1986) (contractor's negligence action allowed against architect who was aware contractor would be harmed if architect did not properly perform its contractual duties to owner); Tommy L. Griffin Plumbing & Heating Co. v. Jordan, Jones & Goulding, Inc., 463 S.E.2d 85 (S.C. 1995) (engineer owed contractor duty not to negligently design or negligently supervise project and was liable to contractor absent privity for economic losses); Pritchard Bros., Inc. v. Grady Co., 428 N.W.2d 391 (Minn. 1988) (contractor may sue architect for professional negligence and for purely economic loss);

provable economic loss actions against construction design professionals for various sorts of professional conduct.[104]

23.06 THE ECONOMIC LOSS RULE AS APPLIED TO THE TORT OF NEGLIGENT MISREPRESENTATION

The tort of negligent misrepresentation is articulated in Section 552 of the Restatement (Second) of Torts. Entitled "Information Negligently Supplied for the Guidance of Others," it states:

> [O]ne who, in the course of his business, profession or employment, or in any other transaction in which he has a pecuniary interest, supplies false information for the guidance of others in business transactions, is subject to liability for pecuniary loss caused to them by their justifiable reliance upon the information, if he fails to exercise reasonable care or competence in obtaining or communicating the information.

A separate negligence-based tort of negligent misrepresentation may be asserted against any construction project participant who disseminates information. Owners, construction managers, general and trade contractors, and design professionals may be alleged to have negligently supplied inaccurate information for the guidance of others.[105] The Restatement specifically allows the tort to be maintained whether or not the injured party has a contract with the information-disseminating party.[106] In place of privity, the Restatement inserts foreseeability, requiring that the party providing the information know the class of those intended to receive and use the information,[107] and reasonable reliance by a member of that class upon the information.[108]

E. C. Ernst, Inc. v. Manhattan Constr. Co., 551 F.2d 1026, 1031–32 (5th Cir. 1977) (applying Alabama law); Donnelly Constr. Co. v. Oberg/Hunt/Gilleland, 677 P.2d 1292, 1295–97 (Ariz. 1984); Hewett-Kier Constr., Inc. v. Lemuel Ramos and Assoc., Inc., 775 So. 2d 373 (Fla. App. 2002).

104. *See, e.g.*, Dufficy & Sons, Inc. v. BRW, Inc., 2002 WL 31601023 (Colo. App. 2002) (citing cases); Mayor and City Council of Columbus v. Clark-Dietz & Assocs. Eng'rs, Inc., 550 F. Supp. 610, 623 (N.D.Miss. 1982) (architect owes duty to those parties proximately suffering economic losses as a result of architect's negligent design); Magnolia Constr. Co. v. Mississippi Gulf S. Eng'rs, Inc., 518 So. 2d 1194, 1202 (Miss. 1988) (third parties are entitled in Mississippi to rely on a design professional's contractual obligation to the owner and may recover economic losses for architect's negligence).

105. *See* Restatement (Third) of Torts § 552.

106. Restatement (Third) of Torts § 552, comment g.

107. *Id.* § 552, comment h.

108. To state a claim of negligent misrepresentation, the plaintiff (most often a contractor or subcontractor) must have reasonably relied upon the information provided and been misled by

A number of jurisdictions are less than comfortable with the breadth of the RESTATEMENT's pronouncements. To them, the RESTATEMENT's foreseeability standard is a poor substitute for contractual responsibility.[109] As noted by one court, it is "a standard that sweeps too broadly in a professional or commercial context, portending liability that is socially harmful in its potential scope and uncertainty."[110] In the construction setting in particular, courts have been uneasy assessing tort responsibility for economic damage based on the transmission of inaccurate information related to a contractual undertaking.

A number of courts believe that where there is privity of contract and only economic damages, there is no need for an additional tort of negligent misrepresentation.[111] Other courts have chosen not to allow negligent misrepresentation claims when there is a contract, any contract, which relates to the subject matter of the claimed damages.

---

that information. *See, e.g.*, Sanders Co. Plumbing & Heating, Inc. v. City of Independence, 694 S.W.2d 841 (Mo. Ct. App. 1985) (city provided contractor with inaccurate soil condition data upon which contractor relied in formulating its bid); McDevitt & Street Co. v. Marriott Corp., 713 F. Supp. 906 (E.D. Va. 1989), *aff'd in part, rev'd in part*, 911 F.2d 723 (4th Cir. 1990), *on remand*, 754 F. Supp. 513 (E.D. Va. 1991) (contractor's reliance on owner's soils reports held insufficient to support extra work claim where contractor did not perform pre-bid investigation); D. Federico Co. v. New Bedford Redevelopment Auth., 723 F.2d 122 (1st Cir. 1983); Umpqua River Navigation Co. v. Crescent City Harbor Dist., 618 F.2d 588 (9th Cir. 1980).

109. Claims against owners for faulty information were first fully articulated in American jurisprudence by a troika of opinions from the U.S. Supreme Court in the second decade of the 20th century: United States v. Atl. Dredging Co., 253 U.S. 1 (1920); Christie v. United States, 237 U.S. 234 (1915); and Hollerbach v. United States, 233 U.S. 165 (1914). For an excellent description of the effect of these three cases in the development of American construction law, *see* 4 PHILIP L. BRUNER & PATRICK J. O'CONNOR, JR., BRUNER AND O'CONNOR ON CONSTRUCTION LAW § 14:20 (West Group 2002). Although the facts in these cases involved the government owners' acts of "concealment" and "misrepresentation," accepted American tort law was not sufficiently developed at that time to sanction a separate extracontractual cause of action. And, as the parties were in privity, contract law was sufficient to allow compensatory damages for the misrepresentations. Many states today recognize both contract and tort remedies for similar conduct. Many allegedly injured contractors opt to pursue contract rather than tort claims, especially in jurisdictions slow to accept tort concepts that parallel traditional contract remedies. So long as there are no limiting contract provisions, if the contractor is in privity with the owner, there should be little if any difference in recoverable damages for an owner's negligent misrepresentation under tort or contract theories.

110. Local Joint Executive Bd. v. Stern, 651 P.2d 637, 638 (Nev. 1982).

111. *See, e.g.*, Duquesne Light Co. v. Westinghouse Elec. Corp., 66 F.3d 604 (3d Cir. 1995); City Express, Inc. v. Express Partners, 959 P.2d 836 (Haw. 1998).

Berschauer/Phillips Constr. Co. v. Seattle Sch. Dist. No. 1
881 P.2d 986 (Wash. 1994) (en banc)

GUY, Justice.

We granted review to decide whether a general contractor may recover in tort $3.8 million in economic damages for construction delays against an architect, a structural engineer and a project inspector, none of whom were in privity of contract with the general contractor. Berschauer/Phillips expended more funds than expected and was delayed in its completion of its work, and consequently was damaged monetarily.

Berschauer/Phillips requests that this court apply the Restatement (Second) of Torts § 552 (1977) to permit a general contractor not in privity of contract to bring a tort cause of action against a design professional for negligent misrepresentations. We acknowledge that § 552 provides support for the recovery of economic damages in the construction industry for negligent misrepresentations. *See* § 552, illustration 9. We also acknowledge that the tort of negligent misrepresentation is recognized in Washington.

We hold that when parties have contracted to protect against potential economic liability, as is the case in the construction industry, contract principles override the tort principles in § 552 and, thus, purely economic damages are not recoverable. There is a beneficial effect to society when contractual agreements are enforced and expectancy interests are not frustrated. In cases involving construction disputes, the contracts entered into among the various parties shall govern their economic expectations. The preservation of the contract represents the most efficient and fair manner in which to limit liability and govern economic expectations in the construction business. Berschauer/Phillips' recovery of economic damages is therefore limited to those damages recovered from the District.

In *Berschauer/Phillips,* the contractor alleged that the design professionals supplied inaccurate information, which caused the contractor to incur extra costs and delay damages in connection with the performance of the construction contract. The fact that the general contractor had a contractual remedy against the owner was dispositive for the court.[112]

112. The fact that there may have been contract provisions in the owner/general contractor agreement that severely limited the general contractor's right to recover foreseeable and substantial delay damages was apparently irrelevant to the Washington court.

Other courts limit the expanse of the RESTATEMENT's class of potential plaintiffs by making fine distinctions, on behalf of design professionals, concerning what constitutes "information in the course of one's business or profession."[113] Illinois, for instance, specifically recognizes an exception to the economic loss rule outside the construction context for damages caused by another's negligent misrepresentation as described in the RESTATEMENT (SECOND) OF TORTS § 552.[114] When presented with the case of a professional engineer whose plans for a tunnel location were inaccurate by 75 yards, causing additional costs for the contractor and damage to real property owned by others, the Illinois Supreme Court nevertheless held that the engineer was not "in the business of supplying information."[115] In a concluding remark, the court noted that the remedy for inaccurate plans could be expressed "in contract terms."[116]

Courts that allow the tort of negligent misrepresentation in the construction setting also tend to inquire rather deeply into the facts of a contested case, in order to satisfy themselves that those who incurred the economic losses were within the class of recipients whose reliance on the information, and resultant damage for inaccurate information, was foreseeable. Similar to the analysis of the "supervising architect" exception in cases of design professional negligence discussed previously, a number of courts look for a relationship between the

113. RESTATEMENT (THIRD) OF TORTS § 552.

114. *See* Moorman Mfg. Co. v. Nat'l. Tank Co., 435 N.E.2d 443, 449 (Ill. 1982), and its progeny. This exception is one of three to the economic loss rule recognized in *Moorman*. The other two are fraud and property damage from a sudden or dangerous occurrence. *Id.* at 450. *See also Rozny*, 250 N.E.2d at 660 (plaintiffs not in privity were allowed to recover economic damages against surveyor for inaccurate information on survey).

115. *SEC Donohue*, 679 N.E.2d at 1201–02 (a design professional's plans and drawings were incidental to the tangible product—the water pipe project). *See also* 2314 Lincoln Park West Condominium Ass'n v. Mann, Gin, Ebel & Frazier, Ltd., 555 N.E.2d 346 (Ill. 1990) (architect's information is incidental to a tangible product, i.e., a structure, and details on plans are normally transformed into the structure (*dicta*)); *but see* Tolan and Son, Inc. v. KLLM Architects, Inc., 719 N.E.2d 288, 298 (Ill App. 1999) (refining the "incidental" analysis to design professionals—project design architects were not "providers of information," but a consulting engineer could be).

116. *SEC Donohue*, 679 N.E.2d at 1202. Apparently, this parsing of the RESTATEMENT requirements is limited to the court's anti-tort resentment in the construction sector, and not to other commercial transactions. *See Rosenstein*, 636 N.E.2d at 665 (negligent misrepresentation claim allowed against stock market indexing service by those not in privity). Illinois also allows negligent misrepresentation claims against attorneys and CPAs. This distinction has drawn the criticism of a number of commentators, and three of the state's Supreme Court Justices. *See SEC Donohue*, 679 N.E.2d at 1202 (Heiple, J. in dissent).

parties that is "so close as to be the functional equivalent of contractual privity."[117] If it is found, the Economic Loss Rule's strictness may be avoided.[118]

Borough of Lansdowne, Pa. v. Sevenson Envtl. Serv., Inc. 2000 WL 1886578 (E.D. Pa. 2000)

Weiner, Judge.

In 1994, the EPA and Army Corps of Engineers designed a project to clean up certain buildings in the Borough that were contaminated by radiation. In 1995, the United States awarded a contract to perform construction work on the site to Sevenson. Sevenson then entered into a subcontract with Terratech to perform excavation and construction of retaining walls. During the construction of a shoring system which was designed to allow the excavation of the site to proceed, large amounts of grout filled and blocked access manholes and sewer lines owned by the Borough near the construction site. The Borough filed its first party complaint to recover damages to its property caused by the grout. During the construction project, Weston was under contract with the United States to prepare plans and review the project design. Terratech's third party complaint alleges that the plans prepared by Weston did not adequately show the location of the Borough's sewer lines which led to the infiltration of the grout.

Weston argues that Terratech fails to state a claim because Weston and Terratech lack contractual privity, and that, under Pennsylvania law, neither a general contractor nor a subcontractor has a direct right of action against the owner's architect/designer. We cannot agree. Under Pennsylvania law, which incorporates the Restatement (Second) of Torts § 552, contractual privity is not necessary to maintain a claim for negligent misrepresentation. Pennsylvania has clearly adopted § 552 as the law of negligent misrepresentation. Section 552 has supplanted the need for contractual privity as the device by which liability is limited from the world at large to those whom the actor should reasonably foresee might be harmed by his negligent provision of false information. Rather than focusing upon the existence of a contractual relationship, § 552 limits the liability of the supplier of the information to the "group of persons for whose benefit and guidance he intends to supply the information or knows that the recipient intends to supply it."

Notwithstanding the provisions of § 552, Weston argues that construction design cases are exempt from the reach of § 552 because of the economic loss doctrine; i.e. that Terratech cannot maintain a claim of negligent misrepresentation because there is no privity of contract between them. This argument misconstrues the economic loss doctrine and is not supported by the case law.

117. Ossining Union Free Sch. Dist. v. Anderson LaRocca Anderson, 539 N.E.2d 91 (N.Y. 1989).

118. *Id.* at 92.

A close reading of section 552 itself reveals that the Restatement is concerned with liability in cases where contract remedies are unavailable. Indeed, virtually all the examples provided in [§ 552] involve liability to third parties. The reason is simple: where there is privity in contract between two parties, and where the policies behind tort law are not implicated, there is no need for an additional tort of negligent misrepresentation. On the other hand, privity is not necessary here, because liability is limited to the bidders on the project, a potential class of plaintiffs that the engineer is aware may rely on his report.

Design consultants and engineers are within the scope of § 552 and may be held liable for negligent misrepresentation. This court rejects the argument that § 552 does not apply to design consultants and engineers because they are not "in the business of selling information." Design professionals such as architects and engineers are engaged for the very purpose of providing plans and specifications for a construction project. The plans and specifications they provide are an essential component of their work product. Accordingly, we conclude that § 552 has supplanted the need for contractual privity in the circumstances presented.

The Restatement (Second) of Torts § 552 concern is with foreseeability of damage to the class of recipients of information. Foreseeability supplies consistency in the identity of potential plaintiffs similar to the bargained-for protection of a contract. Foreseeability is vital to courts that uphold a plaintiff's right to sue for negligent misrepresentation regardless of privity and when harmed only economically. Thus, despite the economic loss rule, negligent misrepresentation claims have been maintained against owners,[119] builders and developers,[120] construction managers,[121] design professionals,[122] and construction product manufacturers.[123]

---

119. *See, e.g.*, D. A. Elia Constr. Corp. v. N.Y. State Thruway Auth., 734 N.Y.S.2d 295 (App. Div. 2001); *Sanders Co. Plumbing & Heating, Inc.*, 694 S.W.2d at 841 (city provided contractor with inaccurate soil condition data upon which contractor relied in formulating its bid).

120. *Council of Co-Owners Atlantis Condominium, Inc.*, 517 A.2d at 348.

121. John Martin Co. v. Morse/Diesel, Inc. 819 S.W.2d 428 (Tenn. 1991) (subcontractor was allowed to maintain negligent misrepresentation claim against CM for improper information directing sub's work).

122. Nota Constr. Corp. v. Keyes Assoc., Inc., 694 N.E.2d 401 (Mass App. 1998) (mistakes and misinformation in architect's plans could give rise to negligent misrepresentation action despite lack of privity); Guardian Constr. v. Tetra Tech, Inc., 583 A.2d 1378 (Del. Super. 1990) (plaintiffs not in privity were allowed to maintain action for negligent misrepresentation against engineer).

123. *See* Village of Cross Keys, Inc. v. U.S. Gypsum Co., 556 A.2d 1126, 1133 (Md. App. 1989) (negligent misrepresentation claim against designer/manufacturer of curtain wall system not barred by Economic Loss Rule, although recovery denied due to failure to follow manufacturer's specifications); *See also* State *ex rel.* Bronster v. United States Steel Corp., 919 P.2d 294, 302 (Haw. 1996) (Economic Loss Rule properly not applied in negligent misrepresentation case against steel manufacturer where no privity; recovery allowed).

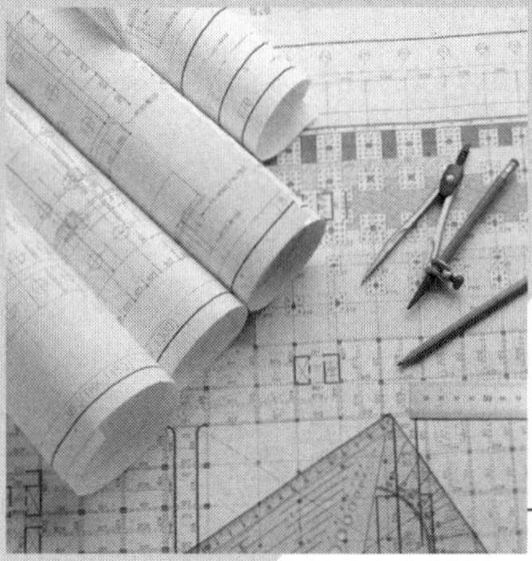

QUESTIONS

After following the state competitive bidding process, the Town of Manheister (hereinafter the "Town") entered into a contract with Delta Construction Company (hereinafter "Delta") for a multi-million-dollar construction project known as "Construction of Main Street, Manheister" (the "Main Street Project"). Prior to the execution of the general contract with Delta, the Town had entered into two contracts with Fuss & Budget, an architectural and engineering firm, pursuant to which Fuss & Budget was to provide all the design and engineering services for the Main Street Project. In addition to these services, Fuss & Budget agreed to take responsibility during the Main Street Project for utility coordination, contract administration, bid analysis, scheduling monitoring, change order processing, progress payment processing, progress meeting management, and contract document interpretation. Delta claims that Fuss & Budget had to know that Delta's performance of its contract with the Town was dependent in part upon Fuss & Budget's proper performance of its role in the Project.

Delta claims that the Main Street Project became a "boulevard of broken promises" as a result of Fuss & Budget's negligent performance of its duties. Delta also claims that the dependence of Delta's success and economic well-being on Fuss & Budget's performance of these duties made foreseeable the harm Delta would suffer as a result of Fuss & Budget's failure to exercise such care, and gave rise to a duty of care owed Delta by Fuss & Budget. Delta complains mostly that the architectural and structural drawings and specifications prepared by Fuss & Budget for the Project were not coordinated with each other, leading to many discrepancies in location and sizing of various construction components. Sorting through these issues delayed Delta, causing it to stop work while the design issues were resolved, and in some cases to remove work originally installed and replace it after the design issues were resolved by Fuss & Budget. Delta spent more than $500,000.00 over its contract price in completing the Project.

1. What causes of action may Delta bring against Fuss & Budget?
2. What are the elements Delta must prove for each cause of action to establish a prima facie case against Fuss & Budget?
3. What legal arguments would Fuss & Budget use to defend Delta's claims?
4. What documents would Fuss & Budget use to support its legal arguments?
5. Assuming Fuss & Budget was negligent in the performance of its duties in connection with the Project, what must Fuss & Budget show the court to avoid liability?
6. Describe the options available to the court as to whether this case should be allowed to go forward, and explain the analyses that a court could use to dismiss the case or allow it to proceed.

CHAPTER

24

Damages

STEPHEN A. HESS

24.01 INTRODUCTION

A. Overview of Chapter

The general principles that govern the award of damages in claims involving construction disputes are no different from those employed in other cases. At the same time, the application of these principles to construction claims provides special considerations that make the assessment of damages a matter of substantial complexity that requires attention to the particular circumstances surrounding construction contracts.

This chapter begins with a brief review of the most important measures of damages utilized in construction cases (Section 24.02). It then sets out the particular application of these rules in the calculation and proof of damages in the construction setting (Section 24.03) and discusses specific legal doctrines that serve to mitigate or bar construction claims (Section 24.04). The chapter concludes by discussing contractual and other limitations on a party's liability for construction damages (Section 24.05). The function of this chapter is not to provide an exhaustive treatment of all the intricacies of damage doctrines and the calculations of damages—it would take several books from a construction law library to cover those matters in sufficient detail for the working construction lawyer—but rather to introduce the student to the most important concepts in construction law damages.

B. "Measures of Damages" versus "Cause of Action"

This chapter is concerned with the calculation of damages once a claim is proven and does not discuss in detail the various substantive claims that a party may try to establish related to a construction project. Although a specific "cause of action" may be closely intertwined with a related "measure of damages," it is important to understand the distinction between the two concepts.[1]

A cause of action is a set of precise elements that a claimant must satisfy to establish its entitlement to relief. Once the cause of action is proven and the claimant has established its entitlement to relief, the *amount* of relief to which the aggrieved party is entitled is defined by the measure of damages. As the following discussion suggests, most causes of action are associated with a particular measure of damages.

However, some causes of action leave the claimant with the right to choose among various measures of damage with respect to *quantum meruit* below. Moreover, as the reader will see, courts may modify a presumed measure of damages (or substitute an alternative measure of damages) where the presumed measure of damages would result in injustice to one of the parties, as in the case of economic waste.

Finally, the careful reader will learn that judicial opinions are not always careful to distinguish between causes of action and measures of damages. Any person who delves into the world of "unjust enrichment" or "*quantum meruit*" will learn quickly that those phrases can be used to refer to causes of action, to measures of damages, or to both at the same time. This is not a matter of sloppy writing or muddled thinking by the courts. Rather, it is a consequence of the very close relationship between causes of action and damages, and with a little experience in this arena, an attorney can fluidly move between the analytically distinct concepts.

1. *See, e.g.*, Pepi Corp. v. Galliford, 254 S.W.3d 457, 2007 WL 441582 (Tex. Ct. App. 2007), in a dispositive issue related to competing statutes of limitations, the court held that a subcontractor's claim against a property owner was a *quantum meruit* claim founded on general principles of unjust enrichment, rather than an independent cause of action for unjust enrichment—which cause Texas recognizes.

24.02 GENERAL MEASURES OF DAMAGES

The function of this section is twofold. First, it brings together a summary of several measures of damages that students often encounter in different courses. Second, it provides some glimpse into how the common measures of damages play out in the context of construction cases.

A. Breach of Contract and Related Claims

By far the most common claims related to construction projects are those involving breach of contract.

1. *Expectation Interest (Breach of Contract)*

Generally, "a party is entitled to have what he contracts for or its equivalent,"[2] which is another way of saying that a party is entitled to the "benefit of his bargain." When a breach by the opposing party deprives the claimant of the benefit of that bargain, a court will endeavor to protect that benefit; in other words, the goal in fashioning a remedy for the breach of most contracts "is to award a sum that will put the non-breaching party in as good a position as he would have been in had the contract been performed."[3] In literature, this is generally referred to as a party's "expectation interest" in the fruit of its contract. Generally, "rights of the parties with respect to a breach of contract are fixed at the time of breach, and damages are measured as of that time."[4]

Although this single measure of damages applies to most breach of contract cases, the specific rules that apply toward calculating a party's "benefit of the bargain" may vary, depending on whether the claimant is an owner, a design professional (or construction manager), or a contractor (or subcontractor). The following discussion summarizes the typical damages that each of these parties may seek to recover. At the same time, the reader should bear in mind that the general measure of damages that applies to each is still the same—the parties all want to get the benefit of their respective bargains—but the manner of calculating that benefit differs from one party to the next.

a. **Owner's Damages**

From the perspective of understanding damage claims, the owner of a building or construction project can be understood as contracting for three specific benefits related to its project. First, the owner expects the contract to be completed for the contract price. Second, the owner expects the project to be completed

2. Charlotte v. Skidmore, Owings and Merrill, 407 S.E.2d 571, 580 (N.C. App. 1991).
3. Hernandez v. Westoak Realty & Inv., Inc., 771 S.W.2d 876, 880 (Mo. Ct. App. 1989).
4. Fairway Builders, Inc. v. Malouf Towers Rental Co., Inc., 603 P.2d 513 (Ariz. Ct. App. 1979).

within the contract time. Third, it wants the project to be of such quality as it has contracted for through the contract documents. The frustration of any of these expectations—assuming it amounts to a breach of contract—gives rise to damages. The general measure of damages for such breach, as recited previously, is based on the benefit of the bargain principle, and the fact finder must answer the question: What amount of money would put the owner in the same position it would have found itself if the contractor had finished within the budget, on time, and with a project that was of the description and quality contracted for? In determining how these damages might be calculated, it is helpful to address the three separate dimensions of performance that the owner expects.

First, the owner is entitled to get the project completed for the price specified in the contract.[5] If the contract is for a stipulated sum, this element is easier to measure. The owner is entitled to recover additional expenses when, for example, it must complete work that the contractor leaves unfinished, or the owner is forced to cure defective work. Similarly, when the contractor's breach forces the owner to spend more on design professional or construction manager fees, the owner is entitled to recover those additional fees so as to protect the benefit of its bargain. In applying the benefit of the bargain principle to the base cost element of a contract, the court merely compares the actual cost of construction to the promised cost of construction, and the owner is entitled to recover the difference. Imagine, for example, that a hospital contracts for the construction of a cardiac care wing for $14 million. As a direct result of the contractor's poor construction, the owner spends $600,000 in remedial work plus an additional $45,000 in architect fees. The owner bargained for a completed building at $14 million, and if it were compelled to pay the contractor the full contract price, it would have lost $645,000. Accordingly, the court will reduce the owner's obligation to the contractor to $13,355,000 as a means of keeping the owner whole, or protecting the owner's benefit of the bargain. Of course, the owner is not completely excused from paying the contractor. If, at the time of breach, the owner has only paid the contractor $10 million of the contract price, the owner cannot retain the balance of the contract price as a "penalty" for the breach. Instead, the owner would still be obligated to pay the contractor the $3,355,000 balance of the contract price, adjusted to reflect the owner's damages.

Second, the owner is entitled to have the project completed on time. The maxim that "time is money" applies with special force to construction projects for all parties involved. From the owner's perspective, a delay in the completion of a project has several financial ramifications. A property that produces income—through rentals or through the operation of a business—cannot pro-

5. The different mechanisms of specifying a contract price are discussed further in Chapter 9.

duce income until the project is completed, and the loss of that income is the first impact an owner feels. Moreover, many construction projects are financed through construction loans that convert to permanent financing upon substantial completion, and the interest rate for the permanent financing is often markedly less than construction financing rates because of the diminished risk upon completion of the structure. On a $10 million project where the interest rate for the construction loan is two percentage points higher than the permanent loan rate, for example, each month of delay costs the owner more than $16,667.67. In addition, a delayed project may force the owner to incur additional construction management or design professional fees. These are but a few of the damages that an owner may suffer, and even when the contractor completes the project for the contract price, the fact that the project is not completed in time may entitle the owner to some remedy measured by the benefit of the bargain principle.

Third, the owner is entitled to receive a project of the quality that is promised in the contract documents. When the construction is substandard, two ramifications may arise: The construction may need to be repaired or improved to the quality promised, and the value of the project may be diminished as a consequence of the substandard construction. When construction is substandard, the most common measure of damages to give the owner the benefit of its bargain is the cost of repair.[6]

However, where reasonable repairs may leave the project in a condition in which the fair market value is still diminished below that of the promised construction, the court may award an amount of money representing the diminution in value of the property after repair. For example, in *Northern Petrochemical Co. v. Thorsen & Thorshov, Inc.*,[7] an owner brought an action against its architect, structural engineer, and general contractor for the faulty construction and subsequent reconstruction of an industrial building. The Supreme Court of Minnesota held that "[t]he preferred measure of damages in a case such as this is to take either the cost of reconstruction in accordance with the contract, if this is possible without unreasonable economic waste, or the difference in the value of the building as contracted for and the value as actually built, if reconstruction would constitute unreasonable waste."[8]

Finally, although it is not invoked very often, an owner may simply have the right to have the offending construction removed and the property restored to the condition in which it was found prior to the construction. For example,

---

6. County Asphalt Paving Co., Inc. v. The 1861 Group, Ltd., 908 S.W.2d 184, 186 (Mo. Ct. App. 1995).

7. Northern Petrochemical Co. v. Thorsen & Thorshov, Inc., 211 N.W.2d 159, 165 (Minn. 1973).

8. *Northern Petrochemical Co.*, 211 N.W.2d at 165.

if an owner contracts for a sunroom to be added to his house and the construction is defective, the owner may be limited to recovering the cost of the room, plus demolition, plus restoration of the house to the preconstruction condition. As one court explained, "A properly constructed room addition in place of the faulty one would place the plaintiffs in a better position than they previously enjoyed. That is not in accordance with the law which requires merely that the plaintiffs be returned to the position they were in prior to the contract with [contractor]."[9] The court's concern in this case was to ensure that the owner was not improperly placed in a better position than before, which invoked the doctrine of "betterment" as discussed later.

In all but the last of these cases, the court is protecting the owner's expectation under a contract, and the measure of damages is designed to ensure that the owner gets the economic value of what it bargained for. In the last case, the owner's expectation is protected only to the extent that the court would not allow an owner to be put into a worse position than its pre-contract position. The "cost to restore," however, is really just a form of cancellation of the contract and refund of the price paid; it does not go so far as to permit the owner to obtain satisfaction of its expectations.

b. Design Professional's and Construction Manager's Damages

A design professional's or construction manager's "bargain," for purposes of measuring damages, is ordinarily the compensation to which the design professional or construction manager is entitled under the terms of the contract with the owner. When the owner fails to pay the design professional or construction manager, the measure of damages is simply the difference between the contract compensation and the compensation actually paid.

This measure is easiest to apply when the owner simply fails (or is unable) to pay a stated sum under the contract. More often, disputes arise because the design professional or construction manager performs services that are arguably outside the scope of the original contract, and the measure of damages then becomes that additional compensation to which the design professional or construction manager is entitled for the extra work. Where the contract provides specified rates for such compensation, the damage calculation is readily performed. Where the contract does not specify how compensation is to be determined, courts will typically award "reasonable" compensation. What is reasonable is simply what a court, a jury, or an arbitrator will decide is reasonable after hearing evidence regarding the additional services and the value thereof. In some cases, the design professional or construction manager may also be entitled to recover interest or attorneys' fees when permitted.

---

9. Bourgeois v. Arrow Fence Co., Inc., 592 So. 2d 445, 448 (La. Ct. App. 1991).

c. Contractor's Damages

A substantial portion of the discussion of damages in the remainder of this chapter is concerned with the calculation of damages to which a contractor is entitled. As with the previous discussion related to owners, it is helpful to understand the many dimensions of a contractor's expectations when it comes time to assess damages for breach of contract.

From a general perspective, the contractor's ultimate goal is to protect its profit interest in the contract.[10] Such a simplistic statement belies the complexity of determining damages from the contractor's perspective for breach by the owner.[11] In more complex cases, it is helpful to characterize the contractor's general expectation interest (that is, the benefit of the bargain) as the right to complete (a) the contractual scope of work in accordance with plans and specifications; (b) within the time permitted under the contract; (c) without interference from the owner, design professional, or construction manager; and (d) for payment of the negotiated price. Ultimately, the frustration of any of these expectations affects the contractor's bottom line in a manner that ultimately is measured by reference to the original contract price. At the same time, discussing each of these dimensions is instructive.

First, the contractor is entitled to complete the scope of work within the original drawings and specifications. To the extent that the scope of work is expanded by the addition of other tasks not in the original contract, the contractor is entitled to recover its costs and reasonable overhead and profit associated with the additional work. When work is simply added to the contract, the claim is relatively straightforward. More complicated analysis is required when the drawings and specifications are deficient in some respect, and the resolution of conflicts or deficiencies arguably adds to the work that the contractor is required to perform. From a conceptual standpoint, ultimately, a contractor is required to perform all *but only* the work set out in the original contract—subject to change procedures included in the contract — for the contract price.[12]

Second, the contractor is entitled to complete the contract within the time limits set out in the contract, at least to the extent the contractor's original

---

10. Similarly, when a contractor bids for work that will result in a loss to the contractor even if the owner does not breach, the contractor tries to limit its losses to those that are embodied within the contract price itself, and thus to avoid excess losses (as opposed to trying to achieve anticipated profit).

11. When the breach is simply a payment default, damages are easy to calculate. This section presumes that the breach arises from interference by the owner, delay, bad plans and specifications, or other causes resulting in more typical damages described later in this chapter.

12. This presumes that the contract is a lump-sum contract or a cost-plus contract with a guaranteed maximum price. When the contract price is a cost-plus term, the battle will less often be fought over scope of work and instead will focus on the extent to which the contractor's charges for work are reasonable.

schedule is reasonable. When the contractor's work is delayed as a consequence of problems attributable to others, the contractor will likely lose money through additional expenditures related to the delay. In protecting the benefit of the contractor's bargain, the court will allow recovery of sufficient damages to compensate the contractor for those additional expenses related to delay in completion of the work.

Third, the contractor is entitled to complete the project without unreasonable interference by the owner, by those acting on behalf of the owner, or by causes for which the owner is contractually responsible even without attribution of fault (such as extreme weather or differing site conditions). Compensable interference may cause delay, of course, for which damages are separately compensable,[13] but it can also cost the contractor economically by making the contractor's operations less efficient. This loss of efficiency is often a separately compensable element of damages when the court awards a contractor money to put it in the same position in which it would have found itself but for the interference.

Finally, the contractor expects to be paid its contract price. As set out further in Chapter 9 in this volume, the calculation of the contract price to which the contractor is entitled can be a matter of some sophistication (and ultimately, some contention).

2. *Reliance Interest—Promissory Estoppel*

The benefit of the bargain measure of damages protects a person's expectation interest in the benefits of the contract that governs the parties' relationship. At the same time, there are recurring situations in construction disputes in which a contract is never actually formed or is unenforceable, but the aggrieved party is still entitled to relief.

Circumstances may give rise to a claim (a cause of action) for "promissory estoppel" when they satisfy the following elements:

> A promise which the promisor should reasonably expect to induce action or forbearance on the part of the promisee or a third person and which does induce such action or forbearance is binding if injustice can be avoided only by enforcement of the promise. The remedy granted for breach may be limited as justice requires.[14]

The language quoted sets out the elements of a cause of action for promissory estoppel and intimates that the measure of damages should be based on "enforcement of that promise . . . limited as justice requires." From a concep-

---

13. Not all delays chargeable to the owner are compensable with an increase in the contract price. Unusually severe weather, for example, may only entitle the contractor to additional time to perform even if it causes the contractor to incur more costs to perform than it anticipated.

14. Restatement (Second) of Contracts § 90 (1982).

tual standpoint, what the rule of promissory estoppel protects is the promisee's *reliance interest* in the promise. Because promissory estoppel generally arises in the absence of an enforceable contract, it is technically not a party's expectation interest—its expectation of receiving the benefit of its bargain—that a court seeks to protect.

The measure of damages for promissory estoppel is somewhat amorphous, as it requires that the court award the promisee an amount of money to protect its reliance on the promise. This reliance interest is typically calculated as the amount of money that the promisee lost in reliance on the promise, and in many cases does not include any element of profit for the promisee. In some cases, an aggrieved party's reliance interest is very much like the calculation of damages as though a contract had been performed. Thus, for example, if a contractor justifiably relies on a steel erector's bid of $250,000 in bidding for the construction of a warehouse and the steel erector refuses to perform, thereby forcing the contractor to find a replacement at $300,000, the "reliance interest" is $50,000—the difference between the bid and the actual cost. This is precisely the measure that would be used for simple breach of the steel erector's contract had it not shown up on the job.[15] That is not always the case, however, and the measure that protects the promisee's reliance should be distinguished from the remedy for breach of contract. Moreover, as summarized by the RESTATEMENT section quoted, courts are at liberty to modify damages "as justice requires."

3. *Restitution Interest—Unjust Enrichment*

The absence of an enforceable contract between parties does not always bar recovery, as demonstrated by claims based on promissory estoppel. Even in the case of promissory estoppel, however, a court must determine that some conduct by the defendant gives rise to a cause of action on the part of the plaintiff for which a remedy is granted. There are circumstances in which a court may fashion a remedy even in the absence of culpable conduct by the defendant.

A common claim for relief that is found in construction cases—and one that is pleaded frequently as a hedge against the possibility that the fact finder will determine no enforceable contract existed—is a claim for unjust enrichment. A cause of action for unjust enrichment is established, in most jurisdictions, by proof that the claimant conferred a benefit on the defendant, which benefit the defendant appreciated (or accepted), in circumstances in which the claimant did not intend to act gratuitously, and it would be unjust to allow the defendant to retain the benefit without having to pay for it.[16]

---

15. Some courts analyze bid defaults under contract principles; others, under principles of promissory estoppel.

16. *See generally* RICHARD A. LORD, WILLISTON ON CONTRACTS § 68:5 (2007).

The value of the benefit conferred is the "reasonable value" of what it would have cost the party to obtain it from another person, or the extent to which the party's property has been increased or its other interest advanced.[17] Most courts refer to this remedy as "restitution," as it forces the benefited party to disgorge the value of the benefit, where the value is measured from the perspective of *the enriched party.* Accordingly, unjust enrichment is distinct from both breach of contract and promissory estoppel, which look to the damage suffered by the promisee. Unjust enrichment instead looks to the value conferred on the recipient and is not concerned with adequate compensation of the claiming party so much as it is with making sure the recipient does not benefit unfairly. Thus, for example, if a tenant contracts for the improvement of a building and skips out before paying the bill, the contractor may be able to recover from the owner of the building. However, the measure of damages will not be based on the contract price of the improvements or on the cost to the contractor of having relied on the tenant's promise, but rather will be the value of the improvements to the building owner if the contractor can establish entitlement to recovery.

4. *Quantum Meruit*

As a measure of damages, *quantum meruit* means "as much as he deserves" and is usually construed as the fair value of services. *Quantum meruit* is typically employed to fashion a remedy in the absence of an express, enforceable contract between two parties, but in circumstances in which a court will find that a contract is implied by law. Indeed, assertion of a *quantum meruit* claim is often deemed inconsistent with a claim for breach of an express contract as explained by one court:

> In Watson Electrical's fourth cause of action, it seeks to recover damages from the Hodgins under the theory of *quantum meruit.* "*Quantum meruit* is a measure of recovery for the reasonable value of services rendered in order to prevent unjust enrichment. It operates as an equitable remedy based upon a quasi contract or a contract implied in law. . . . An implied contract is not based on an actual agreement, and *quantum meruit* is not an appropriate remedy when there is an actual agreement between the parties. Only in the absence of an express agreement of the parties will courts impose a quasi contract or a contract implied in law in order to prevent an unjust enrichment."[18]

There are many circumstances in which *quantum meruit* is pleaded as a claim. A contractor may be terminated before the completion of its work and

17. Restatement (Second) of Contracts § 371 (1981).

18. Watson Electrical Constr. Co. v. Summit Companies, LLC, 160 N.C. App. 647, 652, 587 S.E.2d 87, 92 (N.C. App. 2003) (*quoting in part* Paul L. Whitfield, P.A. v. Gilchrist, 348 N.C. 39, 42, 497 S.E.2d 412, 414–15 (1998)).

(at least in some jurisdictions) may elect to seek the fair value of the services it rendered under a theory of *quantum meruit* rather than seeking to recover its expectation interest or the benefit of its bargain.[19] In addition, a claim for *quantum meruit* may be asserted by a *breaching* party to recover some compensation for its work even when the law will not allow it to recover breach of contract damages.[20] Third, a contract may prove to be unenforceable due to statutory limitations in circumstances in which the contractor should still be paid something for its work.[21] There are other events that may give rise to a claim for *quantum meruit*, but these are the most common.

Readers should bear in mind that courts often use *quantum meruit* to characterize a specific cause of action, the proof of whose elements will entitle the claimant to recover damages measured *in some instances* by the value of services rendered. However, when *quantum meruit* is used to refer to a specific *cause of action*, rather than to the *measure of damages* tied to the value of services rendered, a court may allow a measure of damages that is based on something other than the value of services rendered. Accordingly, when the reader crosses a case in which a party on a claim for *quantum meruit* recovers its out-of-pocket costs, such a case should be understood as invoking *quantum meruit* as a cause of action and not as a measure of damages. By keeping the difference between a cause of action and a measure of damages in mind, and by understanding that courts may use the term *quantum meruit* to refer to either, cases discussing *quantum meruit* are less troubling.

B. Tort Claims

The general measure of compensatory damages for negligence claims is similar to the measure for contract claims: It is that amount of money that will put the injured person in the same position as the injured person would have enjoyed

---

19. MCK Bldg. Assoc., Inc. v. St. Lawrence Univ., 301 A.D.2d 726, 728, 754 N.Y.S.2d 397, 399 (2003).

20. "Even a party in breach may recover in *quantum meruit* to avoid unjust enrichment so long as the breach was not willful." U.S. for Use and Benefit of Ken's Carpets Unlimited, Inc. v. Interstate Landscaping Co., Inc., 37 F.3d 1500 (6th Cir. 1994) (table). Here again, the phrase "unjust enrichment" is invoked to describe an underlying equitable principle and not a measure of damages. To confuse matters somewhat, in appropriate circumstances courts will limit a breaching party's remedy to restitution measured by principles of unjust enrichment, rather than to *quantum meruit* where the latter would actually allow a breaching party to benefit from is own malfeasance:

> If a party justifiably refuses to perform on the ground that its remaining duties of performance have been discharged by the other party's breach, the party in breach is entitled to restitution for any benefit that it has conferred by way of part performance or reliance in excess of the loss that it has caused by its own breach.

Restatement (Second) of Contracts § 374(1) (1981).

21. William Mushero, Inc. v. Hull, 667 A.2d 853, 855 (Me. 1995).

but for the negligence of the defendant. To the extent that a tort claimant can avoid the effect of the Economic Loss Rule[22] in a suit founded on breach of the duty to provide suitable drawings and specifications, or not to design a project negligently, this measure results in calculations similar or identical to those that determine breach of contract damages.[23]

24.03 CALCULATION AND PROOF OF COMPENSATORY DAMAGES

The foregoing discussion set out the common measures of damages that parties encounter in construction disputes. Once the parties establish what measure they will use to determine damages, they must turn their attention to accumulating proof of losses that support a damage claim calculated by reference to that measure. This section starts by discussing the preferred method of establishing "actual" damages through direct proof or estimation, outlines several alternative methods that may be employed when proof of actual losses is unavailable, summarizes the limitations on consequential damages in the construction setting as well as limitations on punitive damages, and concludes with some observations concerning the sophistication of the claims process in the construction industry.

A. Direct Methods of Proving Actual Damages

Courts routinely express a strong preference (in some cases, an insistence) that parties who seek remedies for breach of construction contracts provide proof of actual costs in support of their claims. The requirement of proof of actual costs should not be read as requiring precision in such proof. As noted in Section 24.03(A)(2), there are many accepted means of *estimating* losses. In principle, however, courts will generally insist on these direct methods of proof as distinct from the inferential, alternative methods discussed later.

1. *Proof of Actual Costs*

There are a few categories of additional costs that contractors incur when things do not proceed as planned on a job.

22. *See* Chapter 23 in this volume.

23. Some other tort claims have a different measure of damages. For example, the measure of damages for negligent misrepresentation is the difference in value between what the victim would have enjoyed had the facts been as represented and the value of what the victim enjoyed based upon the actual facts. Although this does not always result in a number that is different from the general calculation, it is a different measure in principle.

a. In General

Direct costs are typically those costs that can be identified specifically with and charged in whole or in part to, a particular project and a particular cause within the contract. In those cases where the contractor maintains a detailed and sophisticated cost accounting system, it may be possible to extract direct costs related to particular causes from the cost records. If the engineer changes the specification of pumps mid-project, for example, the additional compensation may be as simple as determining the difference in cost between the pumps specified and the pumps actually utilized.

Many claims are not as straightforward. If the substitution of pumps comes at a time when the pumps are on the project's critical path and there is insufficient lead time to order new pumps without delaying the project, the contractor may be entitled to recover not only the additional costs of the pump, but the results of the delay to the work as well. That delay, in turn, may be as simple as three identifiable days of idle labor and equipment. More probably, the calculation of the cost must accommodate the fact that otherwise idle crews or equipment were utilized to accelerate work on different parts of the project, perhaps diminishing the total delay, or at the very least reducing the cost of other work.

In short, the most accurate calculation of damages will always be to trace the direct consequences of a breach by one party, attribute and document a direct cost to every such consequence, and then to award as compensation the sum of those direct costs. On discrete problems in the construction process, a reasonable approximation of direct costs may be had. The accuracy and reliability of direct cost calculations diminishes as the ramifications of a particular breach widen.[24]

b. Price Increases for Labor and Material

It is not unusual for delay or disruption to a project to result in the performance of work during a later period of time at a higher cost than originally anticipated. The escalation in cost may result from scheduled increases in material costs or wage increases in a collective bargaining agreement. Escalated costs may also occur from unknown or unanticipated events (e.g., a natural disaster that increases construction material costs).

A schedule analysis may be necessary to establish that the uncertainty of the length of delay prevented the contractor from obtaining firm material, labor, or subcontract orders and to establish when such orders would have

---

24. Similarly, proof of direct costs diminishes with incompleteness, inaccuracy, or simplicity of the cost accounting system utilized by the contractor.

been placed but for the delay.[25] When changes in prices can be documented (both as to binding commitments for prices at the time work should have been performed and at the time work was actually performed), the difference in such prices may be recovered as part of the calculation of damages.

c. Idle Equipment and Related Costs

Where an owner causes delay to a project and a contractor has committed its construction equipment to the project but has been unable to use it, the contractor may be entitled to compensation for costs incurred while the equipment is idle. Equipment costs are frequently a substantial element in delay claims, and there is often a dispute regarding the rate at which the equipment should be charged.

There are different mechanisms for valuing idle equipment damages. For contractor-owned equipment, the measure is the contractor's actual cost to own the equipment (and in some cases, an element of profit for its investment in the equipment). Ownership costs include the capital cost of the equipment, depreciation, repairs, storage, taxes, and other direct costs and can be difficult to prove. However, many contracts incorporate equipment schedules for the purpose of establishing equipment costs for change orders, and these schedules can be equally applied to establish damages for idle equipment. When the contractor rents equipment, the cost attributable to idle equipment is easily established by showing an invoice from a third party and payment thereof. If the contractor's recordkeeping is not sufficiently detailed to determine actual costs, then the contractor may resort to industry-published rates that set forth the average cost of ownership or third-party rental rates.

d. Home Office Overhead

When a contractor maintains a home office for the management of work over several projects, it expects to recover the cost of that home office through its general overhead and profit on the various jobs it performs. Home office overhead includes such things as rent, utilities, salaries of management personnel, phone service, copy machines, and estimators. When one job is delayed and precludes the contractor from accepting other work, one effect is that the contractor's home office overhead is spread over fewer jobs, and thereby diminishes the contractor's ability to recover that overhead (in turn, lowering the contractor's overall profitability). Thus, if a contractor expects to bid five jobs for $1 million each and includes an allowance of $40,000 in each contract to pay for home office overhead, it expects to recover $200,000 over those five jobs. If delays on one job preclude it from performing two of those anticipated

25. *See, e.g.*, George Sollitt Const. Co. v. U.S., 64 Fed. Cl. 229, 269 (Fed. Cl. 2005).

contracts, however, it effectively loses the ability to recover $80,000 in home office overhead through other work.

Accordingly, home office overhead is recoverable as an element of delay damages when the owner is responsible for the extended performance of a contract.[26] A claimant must establish a prima facie case for recovery of home office overhead, as the general description of the circumstances of home office maintenance is not always accurate. Thus, a contractor must establish that

a. Performance of the contract was delayed due to inexcusable actions or inaction of the owner;
b. The contractor suffered actual damages as a result of the delay;
c. The nature of the delay made it impractical for the contractor to undertake the performance of other work; or
d. The nature of the delay made it impractical to reduce home office overhead costs.[27]

Once established, the claimant must then determine the proper measure of calculating home office overhead damages. As a result of delay to a project, the contractor does not realize direct billings in the same time period as anticipated. Therefore, the home office overhead is unabsorbed or underabsorbed. Unabsorbed overhead occurs when there is no extra work associated with the extended time against which the contractor can bill its overhead. Underabsorbed overhead typically occurs with disruption or partial delay where the contractor is still realizing some direct billings, but the amount of work has been severely reduced.

There are several methods for calculating an absorbed home office overhead. The most frequently used is the *Eichleay* formula. This formula is designed to calculate the daily home office overhead sustained in order to continue day-to-day operations during an owner-caused delay in a project.[28] To determine the amount of unabsorbed home office overhead, the *Eichleay* calculation requires three steps. First, the court must determine the overhead on the delayed contract by multiplying the contractor's total overhead by some ratio between the contractor's billings on the delayed contract and the contractor's total billings for the period of the delayed contract.[29] Next a daily overhead

26. *See, e.g.*, Complete Gen. Constr. Co. v. Ohio Dept. of Transp., 94 Ohio St. 3d 54, 61, 760 N.E.2d 364, 371 (Ohio 2002).

27. *See generally* George Hyman Constr. Co. v. Washington Metro. Transit Auth., 816 F.2d 753 (D.C. Cir. 1987); William Schwartzkopf & John McNamara, Calculating Construction Damages § 6.07 at 144 (2d ed. 2001 & Supp. 2007).

28. John D. Darling, *Delay of Game: One of the Thorniest Issues in Delay and Disruption Litigation is the Quantification of Actual Damages*, 27 L.A. Law. 31 (2005).

29. Reginald M. Jones, *Recovering Extended Home Office Overhead: What Is the State of Eichleay?*, 40 Procurement Lawyer 8 (2004).

rate for the delayed contract must be determined by dividing the contract overhead by the total days of contract performance.[30] Finally, the total recoverable overhead must be calculated by multiplying the daily overhead rate for the delayed contract by the total length in days of the delay.[31] Although the *Eichleay* formula is the preeminent method of calculating recoverable home office overhead, courts have developed other formulas that incorporate the same concepts.[32] For example, the Hudson formula, which is more prevalent in England and Canada, centers on the percentage of profits and overhead accounted for in the initial bid.[33]

In disputes where contractors have asserted *Eichleay*[34] or similar formulas as a means of home office overhead recovery, owners have asserted a number of defenses that administrative agency decisions have upheld. Among others, the list of defenses and mitigating factors includes (1) failure by the contractor to pursue substitute work; (2) offset of the home office overhead claimed by the contractor for additional compensation paid by the owner during the delay period pursuant to change orders; and (3) negation of the claim due to disproportionate recovery by the contractor in active periods following the delay.[35] Although these and other defenses have prevailed in occasional cases, in most situations where a contractor is able to show the existence of the aforementioned elements, *Eichleay* has become the benchmark means for recovery in owner-caused delay cases.[36]

e. Lost Profits

Lost profits are subsumed within the previously discussed damage calculations when the contractor is working under a stipulated sum contract. When the contractor is working under a cost-plus contract, the contractor's lost profit is recoverable as a separate element when establishing the "plus" portion of the contract entitlement.

Design professionals generally do not seek recovery of lost profits per se for *additional* work, but instead seek recovery based on the contractual value of their services, which may include a component for profit. On the other hand, when an owner breaches a contract with a design professional by terminating the contract impermissibly, for example, the design professional is entitled to

---

30. *Id.*

31. *Id.*

32. Patrick A. McGeehin & Carleton O. Strouss, *Learning from Eichleay: Unabsorbed Overhead Claims in State and Local Jurisdictions*, 25 PUB. CONT. L.J. 351, 358 (1996).

33. *Id.* at 359.

34. Eichleay Corp., ASBCA 5183, 60-2 BCA ¶ 2688 (1960).

35. McGeehin & Strouss, *supra* note 32, at 364.

36. *Id. at* 359.

recover the value of payments promised, less any costs the design professional avoided by not having to perform the work.[37]

Finally, an owner may be entitled to recover its lost profits for delay in completion of a project due to the fault of the general contractor or the design professional or construction manager. The recovery of lost profits for new business has never found much favor in the court inasmuch as lost future profit claims for new businesses can be regarded as speculative. Indeed, the "new business rule" in many courts bars recovery of lost profits for new businesses. The modern trend is not to prohibit the recovery of lost future profits for new businesses outright, but rather to treat the question of such recovery as a matter of evidentiary sufficiency under which the court must assess the extent to which a claim for lost future profits of a new business can survive challenges to the potentially speculative nature of such damage claims.[38] When a business owner can establish with reasonable certainty the fact of its damage and provide sufficient proof as to what its future profits would have been had the building been completed on time, it may be entitled to recover damages for lost profits.

2. *Methods of Estimating Actual Damages*

There are many cases in which a contractor cannot identify on a laborer-by-laborer basis all of the additional costs it incurs due to some defect in the owner's performance, but can establish with some reasonable degree of confidence an estimate of the losses that it suffered. In such cases, courts will permit a reasonable estimate to be substituted for proof of the cost of each actual consequence of the affected work.

For the most part, these estimates are utilized when the damages arise in the form of delays or interference with work. A thorough discussion of liability for delays and disruption can be found in Chapter 11. Here, our attention is focused on estimating damages from such delays and disruption.

a. Measured Mile

In a measured mile approach, one compares the productivity achieved on an unaffected portion of the project with the productivity achieved on an affected portion.[39] As an example, if a pipeline contractor is capable of excavating,

---

37. Other principles, such as the requirement that the design professional mitigate its damages, may alter the calculation in particular circumstances.

38. Michael Dodd & J. Duncan Findlay, 1 State-By-State Guide to Construction Contracts and Claims § 1.05[C][13] at 58 (2006 & Supp. 2007).

39. John D. Darling, *Delay of Game: One of the Thorniest Issues in Delay and Disruption Litigation Is the Quantification of Actual Damages*, 27 L.A. Lawyer 31, 36 (2005).

placing pipe, backfilling, and compacting 600 linear feet of pipe per day during unaffected portions of a construction contract but can manage only 400 feet per day in harsh winter conditions that arise when its work is improperly delayed, its loss of efficiency can be measured directly.

Although such simplistic measurements are rare, productivity is typically based on hours or dollars per unit of work achieved.[40] It is one of the preferred measures of inefficiency or loss of productivity because it uses actual historical data from the project.[41] It is critical that one prove that the affected and unaffected areas compared represent comparable work, where the only material variable is the disruption caused by the breaching party.[42] In addition, the unaffected work must be large enough to be material and to be representative of the type of work on the project as a whole. If the nonbreaching party cannot prove these criteria, the other party will attack the baseline productivity calculations for being faulty.[43]

By its very nature, the measured mile approach requires that some portion of the project be unaffected by the delay or disruption. As a result, this method may not be available where such delay or disruption has continued throughout the project.[44]

b. Industry Standards and Studies

In certain instances where the measured mile approach and the should-cost estimate may not be available (the entire project was affected and the contractor had not previously performed comparable work), it may be necessary to rely upon industry standards or studies. Such standards and studies have been developed in the industry through the Mechanical Contractors Association of America (MCAA), Business Roundtable, and RSMeans.

These studies attempt to quantify the amount by which interference in a project makes the work performed efficient. The MCAA, for example, estimates that moderately adverse weather conditions can render work 15 percent less efficient than work in normal weather conditions. A contractor can then argue, for example, that its work was rendered 15 percent less efficient when

---

40. *Id.*

41. *See id.*

42. *Id.*

43. *Id.*

44. For an extensive analysis and criticism of the measured mile approach, *see* Lee Davis, Laura Stipanowich, & Walter Bauer, P. E., *Does the "Measured Mile" Measure Up? When It Has, When It Hasn't, and What May Happen Under Daubert/Kumho*, 2007-04 CONSTRUCTION BRIEFINGS (April 2007).

it was pushed (improperly, of course) into winter conditions and its workers were not performing as well as they would have performed but for the owner's delay. The argument continues to the effect that the contractor incurred $115,000 in labor and equipment costs to complete work that it could have completed for $100,000 in normal conditions, and this additional cost is an element of damages.

Courts struggle with such methods, and there are substantial grounds on which many of these industry studies may be criticized—among them, the fact that some of the most important studies were never intended to be used to quantify damages.[45] At the same time, the grave suspicion with which these studies may be viewed is tempered by the general principle that it is better to award some damages to a party who clearly is injured than it is to award no damages so long as some reasonable degree of confidence can be established by the evidence.

c. Should-Cost Estimates

When, for whatever reason, the measured mile approach is not available, an estimate of what the cost should have been without disruption may be a reliable substitute. Whereas the measured mile approach compares performance on the same project, performance on other projects for substantially similar work may form the basis for reasonably estimating what performance on a particular project should reasonably have cost, had there been no disruption.[46] Of course, the claimant bears the burden of producing a reputable expert who invokes recognized methods for determining what a project "should" cost to build in the absence of disruption. Like the employment of industry studies, the use of should-cost estimates is viewed with some suspicion.

d. Time and Motion Studies

Where a contractor cannot provide an actual, unaffected performance period for the current job that would allow a measured mile approach or other contemporaneous jobs that would allow a should-cost estimate, the contractor may consider using a time-and-motion study. These studies are generally done by experts, and they attempt to measure the loss in both labor and equipment

---

45. For an excellent review and critique of the use of a wide variety of industry studies, see Dr. Kathleen M. J. Harmon & Bradley Cole, *Loss of Productivity Studies—Current Uses and Misuses, Part 1*, 2006-8 Construction Briefings (August 2006) and Dr. Kathleen M. J. Harmon & Bradley Cole, *Loss of Productivity Studies—Current Uses and Misuses, Part 2*, 2006-9 Construction Briefings (September 2006).

46. Dodd & Findlay, *supra* note 38, § 1.05[C][6] at 56.

efficiency by approximating an unaffected sequence and comparing it to an affected sequence.[47]

B. Alternate Methods of Proving Damages

1. *Total Cost Method*

The simplest, but also the most controversial and least accepted method, is the total cost method. The total cost method calculates damages as the difference between the total actual costs incurred, including overhead and profit, and the bid amount.[48] For example, imagine that a contractor bids $60 million to complete a hospital complex. The contractor's actual cost to complete is $75 million, and the contractor attributes all the overrun to the owner as a consequence of design deficiencies, poor administration of the contract by the architect, slow return of submittals on equipment with long lead times for order and delivery, etc. But for the owner's delays and creation of inefficiencies, the contractor argues, it would have completed the complex within its original bid.

While the simplicity of the method appeals to claimants, courts have criticized the method due to the assumption that all cost overruns are the responsibility of the project owner.[49] As such, courts disfavor use of this method, resulting in criticism and outright rejection in some courts.[50] Courts have also been suspicious of the total cost method because it assumes that the contractor's original bid was reasonable and the work could have been completed for the bid amount, absent breach of the contract. It further assumes that all of the cost overruns were caused by the conduct of the breaching party and none was caused by the contractor's own problems or inefficiencies. To protect against the inaccuracies and potential injustice inherent in the total cost method, courts have developed four elements that a claimant must establish before proceeding with the total cost method. The claimant seeking damages through the total cost method must prove:

a. The nature of the particular losses makes it impossible or highly impracticable to determine them with a reasonable degree of accuracy;
b. The contractor's bid or estimate was realistic;

---

47. *Id.* § 1.05[C][8] at 56.

48. *See generally* Stephen A. Hess, *The Total Cost Method of Proving Damages*, 2007-2 Construction Briefings (Feb. 2007).

49. Schwartzkopf & McNamara, *supra* note 27, § 1.03[C] at 14.

50. Trafalgar House Constr., Inc. v. United States, 73 Fed Cl. 675, 699–700 (2006).

c. The contractors' actual costs were reasonable; and
d. The contractor was not responsible for the cost overrun.[51]

If the claimant can establish each of the four premises, then the claimant may proceed with the calculation of damages under the total cost method.

The following excerpt from a case illustrates the typical discussions that surround these disputed issues. As the reader may imagine, opinions in construction cases tend to be very long and fact-intensive, and only a small portion of the discussion is set out here.

Baldi Bros. Constructors v. U.S. 50 Fed. Cl. 74 (2001)

MILLER, Judge.

This case is before the court after trial on damages generated by a differing site condition. Plaintiff's claim hinges on the reasonableness of its bid, which is the starting point for the damages claimed.

FACTS

On May 17, 1995, the United States Department of the Navy (the "Navy") awarded Contract No. N62470-95-C-5035 to Baldi Brothers Constructors ("plaintiff"), a small, family-owned corporation of five principals who are brothers or first cousins, based in Beaumont, California. The contract called for construction of a Multi-Purpose Training Range Project at the United States Marine Corps Base, Camp Lejeune, North Carolina.

. . .

Before deciding to bid on the project, Salvatore Pasquale ("Pat") J. Baldi, a partner and plaintiff's Operations Manager, and Michael ("Mike") V. Baldi, plaintiff's General Manager, visited the site in Jacksonville, North Carolina. . . . Based on the information available to Pat Baldi from the contract specifications and his visit to the general area, he created an earthwork estimate assuming that the dirt would be suitable for cutting and filling on-site and that the earthmoving equipment could move rapidly through the area.

. . .

During the course of construction, plaintiff encountered various subsurface conditions that it did not expect, including saturated peat marsh, super-saturated clays, subsurface water, and other unsuitable soil types. These soils were not the conditions shown on the boring logs incorporated into the contract. . . . The unexpected soil conditions required plaintiff to modify its sequence of work, which plaintiff claims increased the costs of construction. Plaintiff had intended to

51. Baldi Bros. Constructors v. U.S., 50 Fed. Cl. 74 (2001).

employ "conventional earthmoving scrapers" to perform the cut/fill work. Specifically, scrapers would be used for the excavation and embankment work in a single cut-to-fill operation, but due to the nature of the soil, plaintiff was unable to use scrapers for the earthmoving work. [The Court then found substantial differing site conditions entitling the contractor to relief.]

2. *Damages*
. . . Plaintiff claims entitlement to an equitable adjustment due to the differing site conditions for the following: 1) increased direct costs of performance; 2) overhead for both the extended field office and home office and equipment stand-by expense as costs of delay; and 3) additional costs incurred by its subcontractor Brey Electric.

1) Total-cost method
"Due to the massive and continuous changes to the earthwork activities on this [p]roject, it was not possible for [plaintiff] to discretely track the actual increased costs of performance incurred. . . ." Pl.'s Br. filed Apr. 12, 2001, at 9. Plaintiff so explains and thereupon invokes the total cost method of calculating damages, or in the alternative, the "modified" total cost method. The total cost method of damages, which provides the difference between the contract price and the actual costs, necessarily assumes that the "original bid was an accurate reflection of a reasonable cost, and that all costs in excess of that had to be attributable to the differing site condition." *Servidone Constr. Corp. v. United States,* 19 Cl. Ct. 346, 384 (1990), *aff'd,* 931 F.2d 860, 863 (Fed.Cir.1991). Trial courts are advised to use this method with caution because bidding inaccuracies can create an unrealistically low estimate of the contractor's costs, and performance inefficiencies can increase the costs incurred. *Id.,* 931 F.2d at 861–62.

. . .

The designation of a large portion of the construction site as wetlands was the equivalent of the 800-pound gorilla landing on plaintiff's plans for performing the contract. Plaintiff's testimony, largely from Pat Baldi and Debbie J. Saunders, plaintiff's Project Manager who had worked with plaintiff for over 20 years, diminished neither by defendant's cross-examination nor the Navy's unimpressive fact witnesses, detailed how the changed requirements impacted plaintiff's planned operations.

Plaintiff handily satisfies two of the four requirements: the impracticability of proving actual losses directly and plaintiff's lack of responsibility for the added costs. Before plaintiff even began its construction work on the site, it received information requiring it to change its planned operations. The Navy neither timely nor adequately responded to plaintiff's requests for information and direction concerning the soil conditions. Due to the changed site conditions, from the beginning of the project: Work was frequently halted, re-designed, re-designed again when the re-design failed, re-started in a different direction, and re-done (in the instance of a collapsing road); in addition, land required re-drainage, roads were constructed deeper and with more fill material than originally called for, and multiple large vehicles became stuck in the "muck" and "blue goo," as Pat Baldi phrased the conditions, including the off-road dump trucks that plaintiff had rented on the Navy's recommendation. In short, due to the snowball effect of the wetlands on the project plans, it would be easier for plaintiff to identify the items of contract performance that proceeded as planned, rather than the difference in costs between all aspects of the original plan and the work that the deviations occasioned.

With respect to plaintiff's responsibility for any of the additional costs that it incurred, the evidence shows that the wetlands affected the entire project site and that the conditions were exacerbated by natural causes, such as heavy rain. Plaintiff maintains that the delays and increased costs from rain do not preclude compensation for the differing site conditions. The general rule is that the risk of severe weather in a particular region is not shifted to the Government via the Differing Site Conditions clause. *Hardeman-Monier-Hutcherson, A Joint Venture,* 68-2 BCA ¶ 7,220, 1968 WL 633 (delays to pier construction caused by rough seas did not constitute changed condition). However, when severe weather interacting with an undisclosed property of the construction site delays construction, the Government has been held responsible. *D.H. Dave & Gerben Contracting Co.,* 1962 BCA ¶ 3,493, 1962 WL 939 (excessive rainfall in interaction with drainage area rendering specified performance impossible held changed condition).

2) Increased cost of performance and the reasonableness of plaintiff's bid
Plaintiff claims $819,997.00 in increased performance costs due to the differing site condition. Ms. Saunders is responsible for preparing most bids and testified regarding her preparation of the Camp Lejeune bid. After Pat and Mike Baldi returned from their Camp Lejeune pre-bid area visit and decided that plaintiff would bid on the project, Ms. Saunders sat down with the "estimators" to decide how long they felt the project would take to complete. Ms. Saunders then bid the "general conditions" based on the projected completion time, which was 15 months, or 65 weeks, or 458 days. She prepared the bid using the Lotus spreadsheet program to compile information submitted by various personnel. At trial Ms. Saunders discussed the spreadsheet with the appropriate formulas for time-dependent costs. Project staffing was based on a 65-week project duration, but certain costs were measured by months, weeks, or days.[1] Home office overhead, profit, and bond then were spread among the various line items according to a pro rata formula, whereby a certain percentage of a line item total was derived and added back into the line item. Plaintiff's final bid was $5,629,000.00, the lowest bid that the Navy received.

. . .

Pat Baldi earnestly testified that a production rate of 12,000 CY per day with a 2000-foot haul distance had been achieved in the past and that he had in fact "achieved over 26,000 [cubic] yards a day with . . . five scrapers," because plaintiff grooms its work surfaces meticulously. Although Pat Baldi has been working in this business since he was "old enough to pick up a shovel," and has been doing earthwork estimates since he was 16, his daily production rates for Camp Lejeune were optimistic, to say the least. Pat Baldi chose the 2% curve in determining the per machine production rate of 500 BCY. The Handbook describes the following conditions as warranting the 2% curve: "A hard, smooth, stabilized surfaced roadway without penetration under load, watered, maintained." While hindsight shows that the conditions at Camp Lejeune were a far cry from those warranting a 2% resistance rating, looking at plaintiff's bid prospectively satisfies the court that earnestness does not make up plaintiff's naivete in using such a low resistance rate having never before worked in that region of the country. Indeed, the photographic evidence confirms what Pat and Mike Baldi should have observed as general conditions in the area.

. . .

3) Delay costs and plaintiff's critical path

Plaintiff claims entitlement to $236,030.00, which represents the cost for 191 days of delay for both field and home office overhead. Plaintiff also claims 111 standby days for its scraper spread amounting to $253,497.63, and $24,194.40 for additional earthmoving equipment standby costs. Combined both items, plus profit and bond, total $296,561.00.

After plaintiff was awarded the contract, Ms. Saunders began preparing the schedule of prices and the project schedule. For the latter she used the bid documents and relied on discussions with the project superintendent and the estimator on their plan of action. She then input the information into the Primavera program, which calculated a project end date. The program gave a completion date four months earlier than the original end date. Ms. Saunders was concerned about submitting a schedule with such an early completion date, so she "went back into the program and made some logic changes, added some durations to items, and also took the erosion control and made that the same duration as the original contract days, so we could push the schedule out to show a completion date on the actual completion date on [sic] the contract."

As a result of Ms. Saunders' tinkering with the logic of the schedule and extending erosion control, as well as the duration of other activities, the approved project schedule does not accurately portray the critical path of the project. . . . Regardless of plaintiff's confidence that it would have completed its work well ahead of the contract completion date, the Government should not be liable when plaintiff's predictions are untested. In this case the derailment of the plan of execution occurred from day one. The court finds neither credible documentary nor testimonial evidence to support plaintiff's ambitious schedule. Thus, plaintiff, if it is entitled to any damages for compensable delays, is entitled to no more than 97 compensable delay days.

2. *Modified Total Cost Method*

The modified total cost method was developed to address some of the inherent flaws in the total cost method. Essentially, the modified total cost method reduces the amount of damages calculated under the total cost method by any errors in the bid, costs attributable to the contractor, and costs attributable to parties other than the owner.[52] The resulting damage is presumed to more accurately reflect the damage directly suffered by the claimant.[53]

In a modified total cost, the contractor scrutinizes carefully its bid and takes responsibility for any errors. The contractor also analyzes its own performance on the project and assumes responsibility for any of its own inefficiencies. This is not to say that the contractor must reduce its claim by all inefficiencies that the contractor experienced, as some normal inefficiencies would be assumed in the contractor's bid. The contractor need only assume responsibility for those problems or inefficiencies outside the norm.

3. *Jury Verdict Method*

A method of determining damages that is occasionally invoked by tribunals acting without juries is called the jury verdict method. This method is simply a shorthand name to the process by which courts estimate damages in the absence of any other acceptable means where denying any recovery would work an injustice on the aggrieved contractor.[54] It takes its name from the fact that juries are often presented with marginal damage evidence that—assuming it survives a motion for directed verdict—will be sufficient for the jury to fashion *some* remedy where the fact of the injury is very persuasive and only the quantification of damages is shaky.

As with the total cost method, the jury verdict method is generally disfavored, and where there is any cognizable means of calculating damages through traditional mechanisms—even where the trier of fact cannot be given precise numbers—such methods will be preferred to the jury verdict method. Thus, the jury verdict method may only be invoked in limited circumstances:

> This approach, most often employed when damages cannot be ascertained by any reasonable computation from actual figures, however, is not favored and may be used only when other, more exact, methods cannot be applied.

---

52. *See, e.g.*, Servidone Constr. Corp. v. United States, 931 F.2d 860, 37 Cont. Cas. Fed. (CCH) ¶ 76082 (Fed. Cir. 1991); Amertex Enterprises, Ltd. v. Uniyed States, 41 Cont. Cas. Fed. (CCH) ¶ 77047, 1995 WL 925961 (Ct. Fed. Cl. 1995), *decision aff'd on other grounds*, 108 F.3d 1392, 41 Cont. Cas. Fed. (CCH) ¶ 77048 (Fed. Cir. 1997).

53. *Baldi Bros. Constructors,* 50 Fed. Cl. at 74.

54. *See generally* Stephen A. Hess, *The Jury Verdict Method of Proving Damages*, 2007-5 Construction Briefings (May 2007). The process is generally only applicable to bench trials or to administrative proceedings—in cases before juries, the "jury verdict method" is not a separately cognizable method of calculating damages, as there is an actual jury to decide the damages. *But see* New Pueblo Constructors, Inc. v. State, 144 Ariz. 95, 696 P.2d 185 (1985) (allowing case to go to jury on express "jury verdict" method).

> Before adopting the "jury verdict method," the court must first determine three things: (1) that clear proof of injury exists; (2) that there is no more reliable method for computing damages; and (3) that the evidence is sufficient for a court to make a fair and reasonable approximation of the damages.[55]

The circumstances in which the jury verdict may be invoked are varied,[56] and the jury verdict method is generally employed by litigators only as a "last resort" or alternative basis for the calculation of damages when the damage case-in-chief bears indicia of fundamental flaws or weaknesses.

C. Consequential Damages in Construction Contracts

Consequential damages or "special damages" are damages that are caused by a breach of duty, but whose existence or impact is less readily or directly traceable to the breach than compensatory damages.[57] These damages may include lost rents, lost profits (other than profits anticipated directly from the work), extended financing costs, loss of business, and loss of bonding capacity.[58] Consequential damages may be recoverable in actions based in contract if reasonably foreseeable. They are generally not recoverable in actions based on unjust enrichment.

D. Punitive Damages

Punitive damages are intended to punish a malicious wrongdoer and to deter the wrongdoer and others from behaving in a similar fashion in the future. Punitive damages in the construction industry are rare, and most jurisdictions do not allow the recovery of punitive damages for simple breach of contract.[59] Because punitive damages are generally not available in breach of contract actions, in order to recover there must be an independent tort, outrageous or malicious conduct accompanying the breach, or a special relationship between

---

55. Dawco Constr., Inc. v. U.S., 930 F.2d 872, 37 Cont. Cas. Fed. (CCH) ¶ 76056 (Fed. Cir. 1991).

56. *See generally*, JOHN CIBNIC, JR., RALPH C. NASH, JR., & JAMES F. NAGLE, ADMINISTRATION OF GOVERNMENT CONTRACTS 704–09 (4th ed. 2006), reading some cases as utilizing the jury verdict method to adjust for lack of evidence, conflicting evidence, or lack of proof of causation, in addition to its role in approximating quantum of damage.

57. *See generally* SCHWARTZKOPF & MCNAMARA, *supra* note 27, § 1.03, at 9.

58. *See generally id.* § 1.03, at 9–10.

59. A somewhat dated but otherwise excellent survey of the general rule under which punitive damages may not be awarded for breach of contract (and a taxonomy of exceptions) may be found at Mark Pennington, *Punitive Damages for Breach of Contract: A Core Sample from the Decisions of the Last Ten Years*, 42 ARK. L. REV. 31 (1989).

the parties whose protection is important enough that the public policy of a state may be to permit punitive damages. Such relationships, however, rarely involve two parties to a construction contract.

E. The Claims Industry and Recordkeeping

The proof of claims in the construction industry—both with respect to liability and with respect to damages—often involves the review of a substantial volume of different types of construction documents beyond the basic agreement between the parties, such as drawings, specifications, daily logs, requests for information, submittals, payment applications, change orders, invoices, time and material tickets, and weather reports. Most of these documents can be understood only with experience in the construction industry. In addition, the preparation of some aspects of construction claims (especially scheduling) requires very specialized training and tools. The process of understanding what happened on a construction site in hindsight and being able to explain it to a fact finder are skills that require expertise and, in the context of litigation, expert witnesses.

An entire industry has evolved to provide this expertise. The industry does not merely provide expert witnesses for the purposes of litigation; rather, many claims companies earn their keep through the preparation of change orders, requests for adjustment, and similar claims against owners (or contractors) before the claims ever ripen into litigation.[60]

24.04 MITIGATION OF DAMAGES

A. Duty to Mitigate Damages

Even though victimized by the conduct of another party, an injured party has an affirmative obligation to mitigate its damages. That is, the injured party must take reasonable steps to minimize the damages it will suffer. For instance, where a contractor has been prevented from working on one area of the project, the contractor may have an affirmative obligation to make its best effort to continue to work by moving to another unaffected area of the project, rather than to sit idly by, accumulating additional labor and equipment costs.

60. Indeed, construction management itself has advanced to the point that there are dozens of graduate programs across the country that offer Master's and even Doctorate degrees in Construction Management or Construction Science. *See* http://www.gradschools.com (last accessed November 12, 2008).

When a party does not undertake reasonable steps to mitigate its damages, it does so at the risk of losing part of its right to recover damages. In particular, a party generally may not recover as part of its damages any money that represents damages that the claiming party could have avoided through more prudent conduct.

B. Economic Waste

While the cost of repair is the basic measure of damage for many breaches of construction contracts, this measure may not be applied where the repair cost is exorbitant, taking into account the nature of the damage, the value of the work at issue, the value of the project in general, and the diminution in value of the product in light of the damage. In other terms, where the cost of repair is manifestly excessive in light of the other elements, a court may not award the cost of repair as the measure of damages.[61] "Where such a case occurs and it is also a case where completion or repair will be inordinately expensive, the policy against economic waste suggests use of the value formula rather than the cost formula."[62] As applied, the value formula consists of the diminution in value of the structure as opposed to the cost actually required to repair the structure.[63]

As an example of economic waste, in *County of Maricopa v. Walsh and Oberg Architects, Inc.*,[64] the county obtained a judgment against its architect for the negligent preparation of drawings and specifications for an underground parking garage that was designed to be waterproof but failed to prevent water from dripping onto cars below. The county alleged that it would cost approximately $498,169 to remove landscaping and cover the top of the underground structure with a waterproof membrane. The Arizona Court of Appeals was persuaded by the fact that it cost the county far less to perform an alternate remedy that would protect the vehicles from water damage and prevent further deterioration. As such, the court of appeals recognized that it would be economic waste to award the county more than a third of a million dollars to merely remove landscaping when an alternate remedy existed.[65]

---

61. *Charlotte,* 407 S.E.2d at 581.

62. *Id.* (citing D. Dobbs, Handbook of the Law of Remedies § 12.21 (1973)).

63. *Id.* at 580.

64. County of Maricopa v. Walsh and Oberg Architects, Inc., 494 P.2d 44 (Ariz. Ct. App. 1972); *see also* Stovall v. Reliance Ins. Co., 107 P.3d 1219 (KS 2005).

65. *Walsh and Oberg Architects,* 494 P.2d at 44; *see also Stovall,* 107 P.3d at 1219.

C. Betterment

The concept of betterment is widely recognized in the construction industry.[66] As a general principle, when there is a defect in the work or an omission in the drawings, the owner should not be placed in a better position than if the error had not occurred. This concept is known as "betterment," "added value," or "added first benefit."

The courts have generally held that if the design is incomplete, the owner should pay what the omitted component would have cost if it had been included in the original design.[67] Logically, the designer should pay any additional costs to the owner, such as any "premium" associated with the change order adding the component to the project. When a design error is not discovered until during or after construction, the owner may be entitled to recover the cost of any retrofitting expense, waste, or an intervening increase in the cost of the labor or materials necessary to correct the error.[68]

From the contractor's perspective, its fundamental obligation is to build a project in accordance with the drawings and specifications. However, in some circumstances, contractors and construction managers at risk may also rely upon the defense of betterment. For example, if the owner must repair or replace defective work, the courts have generally held that it should not recover the cost of any enhancement exceeding the quality evidenced in the contract documents.[69] When the repairs do not involve any enhancement, the owner's damages may still be reduced to the extent of any extended useful life of the building component.[70]

24.05 CONTRACTUAL AND OTHER LIMITATIONS ON LIABILITY

A. Indemnification Clauses and Limitations on Liability

Two types of clauses may be found in construction agreements under the terms of which a party seeks to reduce its exposure to liability to the other party. The

66. *See* Jerome V. Bales, Shamus O'Meara, & Mark R. Azman, *The "Betterment" or Added Benefit Defense*, 26 Construction Lawyer 14 (2006).

67. *See, e.g.*, Lochrane Eng'g, Inc. v. Willingham Realgrowth Inv. Fund Ltd., 552 So. 2d 228 (Fla. App. 1989), *rev. denied*, Lochrane Eng'g, Inc. v. Willingham Realgrowth Inv. Fund Ltd., 563 So. 2d 631 (Fla. 1990).

68. *See, e.g.*, Grossman v. Sea Air Towers, Ltd., 513 So. 2d 686 (Fla. App. 1987), *rev. denied*, Grossman v. Sea Air Towers, Ltd., 520 So. 2d 584 (Fla. 1988).

69. *See, e.g.*, St. Joseph Hosp. v. Corbetta Constr. Co., Inc., 316 N.E.2d 51 (Ill. App. 1st Dist. 1974).

70. *See, e.g.*, Allied Chem. Corp. v. Van Buren Sch. Dist. No. 42, 575 S.W.2d 445 (Ark. 1979).

first type of clause—an indemnification clause—requires that one party (the indemnitor) indemnify the other party (the indemnitee) against any losses the indemnitee may suffer on the construction project. The second type of clause is a "hold harmless" or "limitation of liability" clause under the terms of which one party agrees that it will not be permitted to recover damages (or will limit the amount of damages recoverable) from the other party. [71]

Indemnification clauses largely are used to shift the burden of liability onto the party whose ultimate malfeasance results in damages to another defendant. For example, when a subcontractor's deficient construction allows the owner to recover damages from the contractor, the indemnification clause shifts the burden of this loss to the subcontractor (or its insurance carrier). When it functions in this manner, it is generally enforceable. Some indemnification clauses, however, attempt to saddle the indemnitor with all losses, even those that are caused in whole or in part by the indemnitee. Many courts and legislatures find such clauses so unfair that they have declared them unenforceable in many circumstances:[72]

> A provision in a construction contract that requires one party to the contract to indemnify, hold harmless, insure or defend the other party to the contract, including the other party's employees or agents, against liability, claims, damages, losses or expenses, including attorney fees, arising out of bodily injury to persons or damage to property caused by or resulting from, in whole or in part, the negligence, act or omission of the indemnitee, its officers, employees or agents, is void, unenforceable and against the public policy of the state.[73]

A "hold harmless" or "limitation of liability" clause, on the other hand, precludes one party from suing another for particular claims or limits the amount of damages that a culpable party may be required to pay for its malfeasance. Unless they are oppressive or operate to exclude a party from any meaningful responsibility for its own misconduct, they are generally enforceable. The conceptual difference between limitation-of-liability clauses and indemnification clauses is such that most states permit enforcement of the former, even in the face of anti-indemnification statutes.

---

71. Some courts discuss a third type of clause—an "exculpatory clause"—that completely exonerates a party from any liability, and thus operates as an extreme limitation of liability clause. Such clauses are often held unenforceable in personal contracts, although they may be given effect when the parties to a contract are businesses with sufficient bargaining power and their intent to employ an exculpatory clause is clear.

72. *See generally* Allen Holt Gwyn & Paul E. Davis, *Fifty-State Survey of Anti-Indemnity Statutes and Related Cases*, 23 Construction Law. 26 (2003).

73. N.M. Stat. Ann. § 56-7-1(A).

Fort Knox Self Storage, Inc. v. Western Technologies, Inc. 140 N.M. 233, 142 P.3d 1 (N.M. Ct. App. 2006)

FRY, Judge.

{1} Defendant Western Technologies, Inc. appeals an order awarding Plaintiff Fort Knox Self Storage, Inc. over $110,000 in damages for negligence, over $240,000 in attorney fees, and prejudgment interest. The contract between the parties, in which Western was to provide geotechnical engineering services in evaluating the subsurface conditions of a proposed building site, contained a limitation of liability clause purportedly limiting Western's liability to $50,000. The trial court refused to enforce this clause on the ground that it violated a statute prohibiting any party to a construction contract from agreeing to indemnify any entity for its own negligence. While this is the primary issue on appeal, Western also challenges the award of attorney fees and prejudgment interest.

{2} We reverse the trial court's determination that the limitation of liability clause was void and remand for entry of an order limiting the damages awarded against Western to $50,000. We affirm the award of attorney fees and prejudgment interest.

{3} Fort Knox entered into a contract with Western by which Western agreed to provide geotechnical services in connection with the site where Fort Knox intended to build a self-storage facility. Fort Knox agreed to pay Western $1,750 plus tax for its services. Western performed the agreed services, and, shortly after construction of the facility was completed, Fort Knox employees noticed damage to walls and cracks and fissures in the parking lot. Fort Knox sued, claiming that this damage resulted from the negligence of Western and other parties involved in the construction, or from breach of contract.

{4} During the course of the litigation, Western contended that a provision in its contract with Fort Knox limited its liability, if any, to $50,000. The contract provision stated:

> LIMITATION ON [WESTERN'S] LIABILITY
>
> NOTWITHSTANDING ANY OTHER PROVISION OF THIS AGREEMENT, [WESTERN'S] TOTAL AGGREGATE LIABILITY IN CONNECTION WITH THIS AGREEMENT AND WORK PERFORMED HEREUNDER SHALL BE STRICTLY LIMITED TO AN AMOUNT EQUAL TO THE GREATER OF $50,000 OR THE TOTAL CONTRACT PRICE PAID TO [WESTERN] UNDER THIS AGREEMENT (LESS DIRECT THIRD-PARTY COSTS), WHETHER SUCH LIABILITY IS ASSERTED FOR BREACH OF REPRESENTATION OR WARRANTY, UNDER ANY INDEMNITY, IN ANY OTHER RESPECT UNDER OR FOR BREACH OF CONTRACT, OR AS A LIABILITY ARISING IN TORT OR BY STATUTE. CLIENT HEREBY WAIVES AND DISCHARGES ALL PRESENT AND FUTURE CLAIMS AGAINST [WESTERN] AND (FOR ACTIONS IN SUCH CAPACITY) ITS SHAREHOLDERS, DIRECTORS, OFFICERS, AGENTS, EMPLOYEES, AND SUBCONTRACTORS, FOR ANY

> CLAIM OTHER THAN THOSE DESCRIBED IN THE PRECEDING SENTENCE OR ANY LIABILITY AMOUNT IN EXCESS OF THE AGGREGATE LIMITATION STATED IN THE PRECEDING SENTENCE.
>
> . . .

{11} Fort Knox asserts that the limitation of liability clause is unenforceable because Section 56-7-1 expressly dictates that one cannot seek to avoid responsibility for one's own negligence during the course of construction. It argues that our New Mexico Supreme Court has recognized that the provisions of this statute are broader than those in most states and prohibits agreements that attempt to indemnify in whole or in part against negligent liability. *See Sierra v. Garcia,* 106 N.M. 573, 575, 746 P.2d 1105, 1107 (1987). While we agree with Fort Knox's assessment of the holding in *Sierra,* we do not agree that Section 56-7-1, as interpreted by *Sierra,* applies in this case.

{12} We do not read Section 56-7-1 as prohibiting a limitation of liability based on one's own negligence but as prohibiting the avoidance of all liability for one's own negligence. *See Sierra,* 106 N.M. at 576, 746 P.2d at 1108 (explaining that Section 56-7-1 provides "that liability arising in whole or in part from an indemnitee's negligence . . . may not be contracted away by an indemnity agreement"). The limitation of liability clause in this case does not seek to contract away all liability for Western's negligence but seeks to limit the amount of damages Western must pay for its own negligence. The contract does not indemnify Western against its own negligence. Indeed, it provides that Western may be liable for damages, based on its own negligence, that are twenty-eight times higher than the amount of the contract.

{13} Our view is supported by *Valhal Corp. v. Sullivan Assocs., Inc.,* 44 F.3d 195 (3rd Cir.1995), which involved strikingly similar facts. The contract in *Valhal* contained a limitation of liability clause stating:

> The OWNER agrees to limit the Design Professional's liability to the OWNER and to all construction Contractors and Subcontractors on the project, due to the Design Professional's professional negligent acts, errors or omissions, such that the total aggregate liability of each Design Professional shall not exceed $50,000 or the Design Professional's total fee for services rendered on this project. Should the OWNER find the above terms unacceptable, an equitable surcharge to absorb the Architect's increase in insurance premiums will be negotiated.

Id. at 198 (internal quotation marks omitted). The lower court determined that this provision violated the public policy behind Pennsylvania's anti-indemnification statute. *Id.* at 199 n. 2.

{14} The appellate court reversed, treating the provision as a limitation of liability clause, and rejected the contention that it was unenforceable under Pennsylvania law. The court first distinguished a limitation of liability clause from two other methods of limiting exposure to damages for the negligent performance of a contract. *Id.* at 202. One method is an exculpatory clause, which immunizes a person from the consequences of his or her own negligence. *Id.* Another method is an indemnity clause, which holds the indemnitee harmless by requiring the indemnitor to bear the cost of damages for which the indemnitee may be liable. *Id.* The court then concluded that the clause at issue bore no relation to either of these two methods. The clause did not bar any cause of action or require someone other than the beneficiary of the clause to pay for any loss caused by the beneficiary's negligence. *Id.* The court held that the beneficiary remained liable for its own negligence and was exposed to liability up to a maximum of $50,000. *Id.* The court stated, "Thus, the amount of liability is capped, but [the beneficiary] still bears substantial responsibility for its actions." *id.; cf. Berlangieri v. Running Elk Corp.,* 2003-nMSC-024, ¶ 1, 134 N.M. 341, 76 P.3d 1098 (holding that an exculpatory release was unenforceable as contrary to public policy).

{15} We recognize, as did the court in *Valhal,* the similarities between exculpatory clauses, indemnity clauses, and limitation of liability clauses. *Valhal,* 44 F.3d at 202. But we also recognize, as did the court in *Valhal,* that there is a significant difference between contracts that insulate a party from any and all liability and those that simply limit liability. *Id.* The *Valhal* court noted that clauses limiting liability "are a fact of everyday business and commercial life" and that such clauses are enforceable as long as they are "reasonable and not so drastic as to remove the incentive to perform with due care." *Id.* at 204. Because the contract exposed the architectural firm in *Valhal* to damages that were seven times the amount of remuneration it would receive under the contract, the court determined that the $50,000 cap did not immunize it from the consequences of its own actions but was a reasonable allocation of risk. *Id.*

{16} Fort Knox argues that the large difference between its total damages (over $1,000,000) and the contract limitation ($50,000) violates public policy. Fort Knox also appears to argue that Western is "immunized" in the manner of an indemnity clause for any damages over $50,000. We are not persuaded, however, that the correct measure of whether a cap is so small as to render the clause unenforceable is the difference between the damages suffered and the cap. The court in *Valhal* was equally unpersuaded by this argument. Although the court recognized that the $50,000 cap could be considered nominal when compared to the more than $2,000,000 in damages being sought, the court rejected this measure of reasonableness. *Id.* at 204. The court stated that the relevant inquiry must be whether the cap is so minimal compared to the expected compensation as to negate or drastically minimize concern for liability for one's actions. *Id.* In this case, the cap of $50,000 is twenty-eight times the amount of remuneration Western received under the contract, which was $1,750 plus tax. Similar to the court's determination in *Valhal,* where the cap was only seven times the expected fee, *id.,* we conclude the cap in this case leaves Western exposed to substantial damages and does not negate Western's liability.

B. Waivers of Consequential/Incidental Damages

When a party breaches an agreement, the opposing party loses the immediate benefit of its bargain, and the breach may cause other, less direct damages such as lost future profits. These less direct damages—proximately caused by the breach but not directly implied by the aggrieved party's benefit of the bargain—are "consequential damages." As mentioned in Section 24.03(C), these damages are limited to those reasonably within the contemplation of the parties at the time of their agreement. Many construction contracts—including, notably, the AIA family of documents—contain a contractual waiver of consequential damages:

> 15.1.6 CLAIMS FOR CONSEQUENTIAL DAMAGES
>
> The Contractor and Owner waive Claims against each other for consequential damages arising out of or relating to this Contract. This mutual waiver includes
>
> 1. damages incurred by the Owner for rental expenses, for losses of use, income, profit, financing, business and reputation, and for loss of management or employee productivity or of the services of such persons; and
> 2. damages incurred by the Contractor for principal office expenses including the compensation of personnel stationed there, for losses of financing, business and reputation, and for loss of profit except anticipated profit arising directly from the Work.
>
> This mutual waiver is applicable, without limitation, to all consequential damages due to either party's termination in accordance with Article 14. Nothing contained in this Section 15.1.6 shall be deemed to preclude an award of liquidated damages, when applicable, in accordance with the requirements of the Contract Documents.[74]

The intent of the clause is to limit consequential damages, generally, and to eliminate the damages for home office overhead, specifically.[75]

C. No Damages for Delay

Owners will quite commonly insert liquidated damage clauses providing for specific compensation for contractor-caused delays and will also frequently include a cause precluding recovery of damages for delay caused by the owner. Such clauses will normally provide that a contractor's sole remedy for delay,

74. American Institute of Architects, AIA Document A201–2007, *General Conditions of the Contract for Construction* § 15.1.6 (2007).

75. *See* discussions of home office overhead [*Eichleay*] damages at Section 24.03(A)(1)(d).

regardless of the cause, is for additional time and that no additional compensation will be paid.

No-damages-for-delay clauses are generally enforceable. Courts, however, will narrowly construe them and scrutinize them under equitable principles. This has resulted in the development of several exceptions regarding the enforceability of no-damage-for-delay clauses.[76] Such clauses generally are not enforceable if the cause of the delay is

1. The bad faith of the party seeking to enforce the claim;
2. So long in duration that it amounts to abandonment of the contract;
3. Beyond the contemplation of the parties at the time of contract formation; or
4. The result of active interference of the parties seeking to enforce the clause.[77]

The allocation of risk for delay in construction contracts (and in the common law rules that define remedies) is of vital importance to contractors, owners, and design professionals. In many cases, delay damages can mount far in excess of other damages.

---

76. *See, e.g.*, Richard Gary Thomas and Fred Wilshusen, *How to Beat A No-Damages-for-Delay Clause*, 9 Const. Lawyer 17 (1989).

77. John E. Green Plumbing & Heating Co., Inc. v. Turner Constr. Co., 500 F. Supp. 910 (E.D. Mich. 1980), *aff'd*, 742 F.2d 965 (7th Cir. 2004); this topic is more fully explored in Chapter 11.

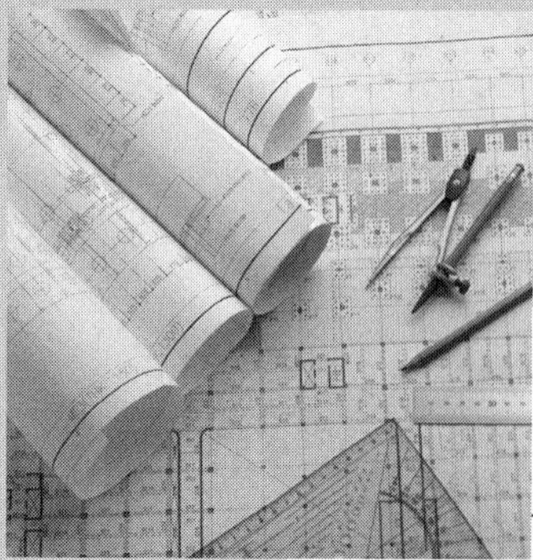

QUESTIONS

1. If a contractor works in a state that allows an election between *quantum meruit* and benefit of the bargain as a measure of damages, in what circumstances would the contractor choose to employ *quantum meruit*?
2. Suppose that you represent the government in the *Baldi* case discussed in this chapter, and are mapping out your litigation strategy. What kind of evidence would you try to discover to fend off use of the total cost method under the facts presented? The *Baldi* court mentioned one indication that the contractor's bid was unreasonably low. What other facts or expert testimony might help your case?

Table of Cases

Aas v. Superior Court, 12 P.3d 1125 (Cal. 2000) 623 n.37, 654 n.89
Abbruzzese v. Miller, 1996 Ohio App. LEXIS 4304 (1996) 462 n.6
Abcon Assoc., Inc. v. U.S., 44 Fed. Cl. 625 (1999) 510 n.48
A & B Constr., Inc. v. Atlas Roofing & Skylight Co., 867 F. Supp. 100 (D.R.I. 1994) 429 n.56
Ace Constructors, Inc. v. U.S., 70 Fed. Cl. 253 (2006) 472 n.34
Adams v. Dreyfus Interstate Dev. Corp., 352 So. 2d 76 (Fla. 4th DCA 1977) 511 n.54
Adarand Constructors, Inc. v. Mineta, 534 U.S. 103, 122 S.Ct. 511 (2001) 209 n.120
Adarand Constructors, Inc. v. Pena, 115 S. Ct. 2097 (1995) 209–210, 252
Adarand Constructors v. Skinner, 790 F. Supp. 240 (D. Colo. 1992) 209 n.120
Adarand Constructors v. Slater, 228 F.3d 1147 (10th Cir. 2000) 209 n.120
Advanced Eng'g & Planning Corp., ASBCA Nos. 53366, et al., 2005-1 B.C.A. (CCH) ¶ 32,664 (2005) 314 n.5
Advanced Materials, Inc. v. Perry, 108 F.3d 307 (Fed. Cir. 1997) 231 n.15
Advertising and Policy Committee of the Avis Rent A Car System v. Avis Rent A Car System, 780 S.W.2d 391 (Tex. App.–Houston [14th Dist.] 1989), writ granted, *judgment vacated,* 796 S.W.2d 707 (Tex. 1990) 116
Aetna Bridge Co. v. State Dept. of Transp., 795 A.2d 517 (R.I. 2002) 276 n.92
A.H.A. General Constr., Inc. v. New York City Housing Auth., 600 N.E.2d 368 (N.Y. Ct. App. 1998) 441 n.36, 442 n.44–45, 453 n.87
Aircraft Gear Corp. v. Kaman Aerospace Corp., 856 F. Supp. 446 (N.D. Ill. 1994) 603 n.36
A. J. Tenwood Assocs. v. Orange Senior Citizens Housing Co., 491 A.2d 1280 (N.J. Super. Ct. App. Div. 1985) 611 n.53
Alan Custom Homes, Inc. v. Krol, 256 Mich. App. 505 (Ct. App. 2003) 520 n.32
Aleutian Constructors v. United States, 24 Cl. Ct. 372 (1991) 617 n.5
Al Johnson Constr. Co. v. United States, 854 F.2d 467 (Fed. Cir. 1988) 333 n.36, 602
Alliance Steel, Inc. v. Piercy, 277 Ill. App. 3d 632 N.E.2d 1341 (4th Dist. 1996) 515 n.11
Allied Chemical Corp. v. Am. Independent Oil Co., 623 S.W.2d 760 (Tex. App.–Houston [1st Dist.] 1981) 122
Allied Chemical Corp. v. Van Buren Sch. Dist. No. 42, 575 S.W.2d 445 (Ark. 1979) 695 n.70

Allied Contractors, Inc. v. United States, 381 F.2d 995 (Ct. Cl. 1967) 48 n.42
Allied Fire & Safety Equipment Co. v. Dick Enterprises, Inc., 886 F. Supp. 491 (E.D. Pa 1995) 261–262
Allied Materials & Equip. Co. v. United States, 569 F.2d 562 (Ct. Cl. 1978) 450 n.72
Alma, Town of, v. AZCO Constr., Inc. 10 P.3d 1256 (Colo. 2000) 651–653
Al Munford Halls, Inc. v. United States, 34 Fed. Cl. 62 (Fed. Cl. 1995), *vacated in part on other grounds,* 86 F.3d 1178 (Fed. Cir. 1996) 396
Alstom Power, Inc. v. Balcke-Durr, Inc., 849 A.2d 804 (Conn. 2004) 115 n.29, 116 n.37, 38
Altrutech, Inc. v. Hooper Holmes, Inc., 6 F. Supp. 2d 1269 (D. Kan. 1998) 115 n.27–28
Amelco Elec. v. City of Thousand Oaks, 27 Cal. App. 4th 228, 115 Cal. Rptr. 2d 900 (2002) 335 n.44
Amelco Elec. v. City of Thousand Oaks, 38 P.3d 1120 (Cal. 2002) 452 n.83
American Can Co. v. United States, 69 Ct. Cl. 1 (1929) 454 n.91
American Demolition, Inc. v. Hapeville Hotel Ltd., 413 S.E.2d 749 (Ga. App. 1991) 462 n.6, 463 n.7
American Drilling Serv. Co. v. City of Springfield, 614 S.W.2d 266 (Mo. Ct. App. 1981) 268 n.67
American Ins. Co. v. Ohio Bureau of Workers' Compensation, 577 N.E.2d 756 (Ohio App. 1991) 552 n.8
American Mach. & Tool v. Strite–Anderson Mfg., 353 N.W.2d 592 (Minn. App. 1934) 121 n.76
American Motors Inns of Fla., Inc. v. Bell Elec. Co., 260 So.2d 276 (Fla. 4th DCA 1972) 301
American Ship Building Co. v. United States, 654, F.2d 75 (Ct. Cl. 1981) 49 n.47
American Towers Owners v. CCI Mechanical, 930 P.2d 1182 (Utah 1996) 644 n.66, 658 n.95
Amertex Enterprises, Ltd. v. United States, 41 Cont. Cas. Fed. (CCH) ¶ 77047, 1995 WL 925961 (Ct. Fed. Cl. 1995), *decision aff'd on other grounds,* 108 F.3d 1392, 41 Cont. Cas. Fed. (CCH) ¶ 77048 (Fed. Cir. 1997) 691 n.52
A. R. Moyer, Inc. v. Graham, 285 So.2d 397 (Fla. 1973) 659 n.101
Aragona Constr. Co. v. United States, 165 Ct. Cl. 382 (1964) 451 n.78
Arcon Constr. Co. v. State, by and through Dept. of Transp., 314 N.W.2d 303 (S.D. 1982) 208 n.111
Argonaut Ins. Co. v. Commercial Standard Ins. Co., 380 So. 2d 1066 (Fla. App. 1980) 552 n.8
Argonaut Ins. Co. v. Town of Cloverdale, Ind., 699 F.2d 417 (7th Cir. 1983) 385 n.66
Armenteros v. Baptist Hosp. of Miami, Inc. 714 So. 2d 518 (Fla. Dist. Ct. App. 1998) 413–415
Arnell Constr. Corp. v. Bd. of Educ. of the City of New York, 598 N.Y.S.2d 1 (N.Y. App. Div. 1993) 447 n.59
A. S. McGaughan Co. v. Barram, 113 F.3d 1256 (Fed. Cir. 1997) 446 n.56
Asphalt Roads & Materials Co. v. Virginia Dept. of Transp., 512 S.E.2d 804 (Va. 1999) 485
Associated Gen. Contractors of Cal. v. City and Co. of S.F., 748 F. Supp. 1443 (N.D. Cal. 1990), *aff'd,* 950 F.2d 1401 (9th Cir. 1991) 211 n.133–134
Atherton Condominium Apartment-Owners Ass'n Bd. of Directors v. Blume Dev. Co., 799 P.2d 250 (Wash. 1990) 658 n.95

Atlantic Dredging Co.; United States v., 253 U.S. 1 (1920) 464 n.12, 13, 661 n.109
Atlantic Mutual Ins. Co. v. Metron Eng'g and Constr. Co., 83 F.3d 897 (7th Cir. 1996) 112 n.15
Aucoin Elec. Supply Co., U.S. for Use and Benefit of, v. Safeco Ins. Co. of America, 555 F.2d 535 (5th Cir. 1977) 381 n.49
Audit Services, Inc. v. Francis Tindall Constr., 600 P.2d 811 (Mont. 1979) 439 n.27
Austin Co. v. Vaughn Bldg. Corp., 643 S.W.2d 113 (Tex. 1982) 119
Avatar Dev. v. DePani Constr., Inc., 834 So. 2d 873 (Fla. 4th DCA 2004) 499 n.26
A.W. Wendell & Sons, Inc. v. Qazi, 626 N.E.2d 280 (Ill. App. Ct. 1993) 115 n.31

Bagwell Coatings, Inc. v. Middle South Energy, Inc., 797 F.2d 1298 (5th Cir. 1986) 659 n.103
Baker Marine Corp. v. Weatherby Eng'g. Co., 710 S.W.2d 690 (Tex. App. 1986) 289 n.14
Baldi Bros. Constructors v. U.S., 50 Fed. Cl. 74 (2001) 687–690, 691 n.53
Ballou v. Basic Construction Company, 407 F.2d 1137 (4th Cir. 1969) 365, 366
Baltimore Marine Indus., Inc., *In re,* 476 F.3d 238 (4th Cir. 2007) 378–379
Bangor Roofing & Sheet Metal Co. v. Robbins Plumbing Co., 151 Me. 45 (1955) 520 n.35
Barsotti's, Inc. v. Consolidated Edison Co. of New York, 680 N.Y.S.2d 88 (N.Y. App. Div. 1998) 442 n.47
Basic Modular Facilities v. Ehsanipour, 70 Cal. App. 4th 1480 (Ct. App. 1999) 520 n.33
Bat Masonry Co., Inc. v. Pike-Paschen Joint Venture III, 842 F. Supp 174 (D. Maryland 1993) 260–261
Bay Constr. Co., VABCA No. 5994, et al., 2002-1 B.C.A. (CCH) ¶ 31,795 (2002) 314 n.5
Beacon Constr. Co. v. United States, 314 F.2d 501 (1963) 48 n.44
Bebo Constr. Co. v. Mattox & O'Brien, P.C., 990 P.2d 78 (Colo. 1999) 653
Becho, Inc. v. United States, 47 Fed. Cl. 595 (2000) 451 n.75, 77
Bender v. Beverly Anne, Inc., 2002 N.D. 146 (2002) 516 n.17, 517 n.22
Bentley Constr. Dev. & Eng'g, Inc. v. All-Phase Elec. and Maintenance, Inc., 562 So. 2d 800 (Fla. Dist. Ct. App. 1990) 393 n.107, 108
Berkel & Co. Contractors v. Christman Co., 533 N.W.2d 838 (Mich. Ct. App. 1995) 268 n.68
Berlangieri v. Running Elk Corp., 2003-nMSC-024, ¶ 1, 134 N.M. 341, 76 P.3d 1098, 699
Berschauer/Phillips Constr. Co. v. Seattle Sch. Dist. No. 1, 881 P.2d 986 (Wash. 1994) 658 n.97, 662
Bertrand Goldberg Assoc., 4 BNA OSHC 1587 (OSHRC 1976) 404 n.14
Bethlehem Steel Co.; United States v., 205 U.S. 105 (1907) 306 n.46
Big Chief Drilling Co. v. U.S., 26 Ct. Cl. 1276, 1992 U.S. Cl. Ct. Lexis 454 (1992) 45 n.32
Bignold v. King County, 399 P.2d 611 (Wash. 1965) 361 n.40
Bituminous Constr., Inc. v. Rucker Enterprises, Inc., 816 F.2d 965, 384 n.62
Blagg v. Fred Hunt Co., 612 S.W.2d 321 (Ark. 1981) 644 n.66
Blaine Economic Dev. Authority v. Royal Elec. Co., Inc., 520 N.W.2d 473 (Minn. Ct. App. 1994) 510 n.49–50
Blair v. Dunham, 134 F.2d 729 (6th Cir. 1943) 411 n.28

Blake Constr. Co. Inc. v. Alley, 353 S.E.2d 724 (Va. 1987) 658 n.98
Blake Constr. Co., Inc. v. C. G. Coakley Co., Inc., 431 A.2d 569 (D.C. 1981) 13 n.34
Blake Constr. Co. v. United States, 987 F.2d 743 (Fed. Cir. 1993) 604 n.39
Bloomington Hotel Co. v. Garthwait, 227 Ill. 613, 81 N.E. 714 (1907) 517 n.23
Blount Bros. Constr. v. United States, 172 Ct. Cl. 1, 348 F.2d 471 (1965) 275
Board of Educ. of Hudson City Sch. Dist. v. Sargent, Webster, Crenshaw & Folley, 539 N.Y.S.2d 814 (N.Y. App. Div. 1989) 345 n.4, 348 n.16, 355–358, 359, 363 n.47
Board of Educ. v. URS Co., No. 64496, 1994 WL 520862 (Ohio Ct. App. Sept. 22, 1994), *aff'd in part and rev'd in part,* 648 N.E.2d 811 (Ohio 1995) 359 n.32
Board of Regents of Univ. of Texas v. S & G Constr. Co., 529 S.W.2d 90 (Tex. Civ. App. 1975), overruled on other grounds by Fed. Sign v. Texas Southern Univ., 951 S.W.2d 401, 121 ED. LAW REP. 394, 40 Tex. Sup. Ct. J. 676 (1997) 120, 381 n.51
Board of Trustees of Santa Fe Comty. Coll. v. Caudill Rowlett Scott, Inc., 461 So. 2d 239 (Fla. Dist. Ct. App. 1984) 583 n.3
Bonebrake v. Cox, 499 F.2d 951 (8th Cir. 1974) 249 n.8
Bono, Incorporated Town of, v. Universal Tank & Iron Works, Inc., 395 S.W.2d 330 (Ark. 1965) 347
Border Brook Terrace Condominium Assoc. v. Gladstone, 622 A.2d 1248 (N.H. 1993) 658 n.95
Borg-Warner Acceptance Corp. v. Tascosa Nat'l. Bank, 784 S.W.2d 129 (Tex. App.–Amarillo 1990) 119 n.61
Bourgeois v. Arrow Fence Co., Inc., 592 So. 2d 445 (La. Ct. App. 1991) 672 n.9
Bouten Constr. Co. v. H. F. Magnuson Co., 133 Idaho 756 (1999) 520 n.34
Bovee v. Gravel, No. 2001-347, 2002 WL 1940968 (Vt. Aug. 13, 2002) 620 n.20
Bovis Lend Lease LMB, Inc. v. GCT Venture, Inc., 728 N.Y.S.2d 25 (N.Y. App. 2001) 276
Bowen Eng'g Corp. v. W.P.M. Inc., 557 N.E.2d 1358 (Ind. Ct. App. 1990) 202 n.66
Bowman v. Walnut Mountain Property Owners Ass'n, 251 Ga. App. 91, 553 S.E.2d 389 (2001) 38 n.8
Bradford Sec. Processing Services, Inc. v. Plaza Bank & Trust, 653 P.2d 188 (Okla. 1982) 620 n.20
Brady Brick & Supply Co. v. Lotito, 356 N.E.2d 1126 (Ill. App. Ct. 1976) 380 n.47
Branch Banking and Trust Co. v. Creasy, 269 S.E.2d 117 (N.C. 1980) 386 n.68
Branstetter v. Cox, 496 P.2d 1345 (Kan. 1972) 116 n.37
Brewer v. Custom Builders Corp., 356 N.E.2d 565 (Ill. App. Ct. 1976) 586 n.12
Bridgeport Restoration Co., Inc. v. A. Petrucci Constr. Co., 211 Conn. 230, 557 A.2d 1263 (1989) 253 n.22
BRL Carpenters, Ltd. v. Am. Nat'l Bank & Trust Co., 126 Ill. App. 3d 137, 466 N.E.2d 1166 (1st Dist. 1984) 523 n.50
Brody v. Bock, 897 P.2d 769 (Colo. 1995) 653
Bronster, State *ex rel.*, v. United States Steel Corp., 919 P.2d 294 (Haw. 1996) 665 n.123
Brooks-Callaway Co.; United States v., 318 U.S. 120; 63 S. Ct. 474; 87 L. Ed. 653; 1943 U.S. LEXIS 1282, 292–295
Brooks v. Holsombach, 525 So. 2d 910 (Fla. 4th DCA 1988) 511 n.54
Browning v. Maurice B. Levien & Co., 262 S.E.2d 355, *disc. rev. denied*, 267 S.E.2d 673 (N.C. Ct. App. 1980) 375 n.25

Brown v. Fowler, 279 N.W.2d 907 (S.D. 1979) 654 n.87
Brown v. Royal Maccabees Life Ins. Co., 137 F.3d 1236 (10th Cir. 1998) 114 n.25
Bruce Constr. Corp. v. United States, 324 F.2d 516 (Ct. Cl. 1963) 445 n.50–51, 454 n.90
Brushton-Moira Cent. Sch. Dist. v. Alliance Wall Corp., 600 N.Y.S.2d 511 (1993) 636 n.59
Bryant Elec. Co. v. City of Fredericksburg, 762 F.2d 1192 (4th Cir. 1985) 658 n.98
Buckley Co. v. City of New York, 505 N.Y.S.2d 140 (N.Y. App. Div. 1986) 432 n.3
Buckner Constr. Co.; State v., 704 S.W.2d 837 (Tex. App. 1985) 448 n.60
Burns v. Black & Veatch Architects, Inc., 854 S.W.2d 450 (Mo. Ct. App. 1993) 448 n.62
Burran v. Dambold, 422 F.2d 133 (10th Cir. 1970) 147 n.26
Butler v. Bd. of Supervisors for Hinds County, 659 So. 2d 578 (Miss. 1995) 438 n.22

C. Norman Peterson Co. v. Container Corp. of America, 218 Cal. Rptr. 592 (1985) 451 n.74
Caddell Constr. Co., Inc. v. United States, 78 Fed. Cl. 406, 48–49
Cadral Corp. v. Solomon, Cordwell, Buenz & Assoc., Inc., 147 Ill. App. 3d 466, 497 N.E.2d 1285 (1st Dist. 1986) 43
Caldwell v. Bechtel, 631 F.2d 989 (D.C. Cir. 1980) 625 n.45
Calloway v. City of Reno, 993 P. 2d 1259 (Nev. 2000) 623 n.37, 624 n.44, 644 n.66, 650 n.85
Canape v. Petersen, 897 P.2d 762 (Colo. 1995) 408 n.1
Capitol Steel Fabricators, Inc. v. Mega Constr. Co., 68 Cal. Rptr. 2d 672 (Cal. 1998) 268 n.69
Cardinal Dev. Co. v. Stanley Constr. Co., 497 S.E.2d 847 (Va. 1998) 442 n.42
Carlson T.V. v. City of Marble, 612 F. Supp. 669 (D. Minn. 1985) 205 n.86
Carter Steel & Fabricating Co. v. Ohio Dept. of Transp., 721 N.E.2d 1115 (Ohio Ct. App. 1999) 309
Carter v. Deitz, 556 So. 2d 842 (La. Ct. App. 4th Cir. 1990) 140 n.20
Carter v. State, 259 Ga. App. 798, 578 S.E.2d 508 (2003) 36 n.1, 51 n.55
Carvalho v. Toll Bros. & Developers, 675 A.2d 209 (N.J. 1996) 423 n.44
Casa Clara Condominium Ass'n, Inc. v. Charley Toppino & Sons, Inc., 620 So. 2d 1244 (Fla. 1993) 624 n.42, 644 n.66
Casa Linda Tile & Marble Installers, Inc. v. Highlands Place 1981, Ltd., 642 So. 2d 766 (Fla. Dist. Ct. App. 1994) 388 n.81
Casey v. Overhead Door Corp., 74 Cal. App. 4th 112 (1999) 621 n.31
Cates v. Hunter Eng'g Co., 205 Ill. App. 3d 587 (3rd Dist. 1990) 518 n.27
Cave Hill Corp. v. Hiers, 570 S.E.2d 790 (Va. 2002) 115 n.33
CEMS, Inc. v. United States, 59 Fed. Cl. 168 (2003) 483 n.81
Centurion Planning Corp. v. Seabrook Venture II, 176 S.W.3d 498 (Tex. Ct. App. 2004) 521 n.40
CFTC v. Sidoti, 178 F.3d 1132 (11th Cir. 1999) 288, 289
Chaney Bldg. Co. v. City of Tucson, 716 P.2d 28 (Ariz. 1986) 140 n.20
Charlotte v. Skidmore, Owings and Merrill, 407 S.E.2d 571 (N.C. App. 1991) 669 n.2, 694 n.61
Chemimetals Processing, Inc. v. Schrimsher, 535 S.E.2d 594 (N.C. Ct. App. 2000) 382 n.56

Chen-Cheng Wang *ex rel.* United States v. FMC Corp., 975 F.2d 1412 (9th Cir. 1992) 396
Chevron Oil Co. v. E. D. Walton Constr. Co., 517 F.2d 1119 (5th Cir. 1975) 119 n.60
Chicago Heights Venture v. Dynamit Nobel of Am., 782 F.2d 723 (7th Cir. 1986) 644 n.66
Chicago R.I. & P. Ry. Co. v. Md. Cas. Co., 75 F.2d 596 (8th Cir. 1935) 124 n.91, 94
CH2M Hill, Inc. v. Herman, 192 F.3d 711 (7th Cir. 1999) 404 n.14
Christie v. United States, 237 U.S. 234 (1915) 44, 464 n.13, 15, 661 n.109
Citadel Corp. v. Sun Chem. Corp., 443 S.E.2d 489 (Ga. Ct. App. 1994) 455 n.96
City Express, Inc. v. Express Partners, 959 P.2d 836 (Haw. 1998) 636–638, 661 n.111
City v. *See Name of opposing party*
CJP Contractors, Inc. v. U.S., 45 Fed. Cl. 343 (1999) 493 n.1
Clark Constr. Co. v. Pena, 930 F. Supp. 1470 (M.D. Ala. 1996) 206 n.98, 207 n.103
Clark-Fitzpatrick, Inc./Frankl Found. Co. v. Gill, 652 A.2d 440 (R.I. 1994) 441 n.41
Clarkson Constr. Co. v. Occupational Safety and Health Review Comm'n, 531 F.2d 451 (10th Cir. 1976) 403–404
Clark Trucking of Hope Mills, Inc. v. Lee, 109 N.C. 71, 426 S.E.2d 288 (N.C. App. 1993) 255 n.32
Clemens v. Benzinger, 207 N.Y.S. 539 (1925) 430 n.58
Cleveland Wrecking Co. v. Central Nat'l Bank in Chicago, 216 Ill. App. 3d 279, 576 N.E.2d 1055 (1st Dist. 1991) 522 n.47, 523 n.51, 52, 53, 55
C. O. Falter Constr. Corp. v. City of Binghamton, 684 N.Y.S.2d 86 (1999) 306 n.45
C.O.B.A.D. Constr. Corp. v. Sch. Bd. of Broward County, 765 So. 2d 844 (Fla. Dist. Ct. App. 2000) 456 n.103
Cobb v. Pasadena City Bd. of Ed., 285 P.2d 41 (Cal. Dist. Ct. App. 1955) 192 n.12–13
Cogefar-Impresit U.S.A., Inc., DOTBCA No. 2721, 97-2 B.C.A. (CCH) ¶ 29,188 (1997) 337–338
College of Notre Dame of Maryland, Inc. v. Morabito Consultants, Inc., 752 A.2d 265 (Md. Ct. Spec. App. 2000) 610 n.51
Collins v. Board of Trustees, 72 W. Va. 583 (1913) 517 n.19
Collins v. Reynard, 607 N.E.2d 1185 (Ill. 1992) 656 n.93
Columbia Gas Transmission Corp. v. New Ulm Gas, Ltd., 940 S.W.2d 587 (Tex. 1996) 117 n.44
Columbus, City of, v. Barngrower, 250 Ga. App. 589, 552 S.E.2d 536 (2001) 37 n.7
Commercial Union Ins. Co. v. Martinez, 635 S.W.2d 611 (Tex. App.–Dallas 1982) 116 n.39
Complete Gen. Constr. Co. v. Ohio Dept. of Transp., 94 Ohio St. 3d 54, 760 N.E.2d 364 (Ohio 2002) 681 n.26
Comptech Int'l. Inc. v. Milam Commerce Park, Ltd., 753 So. 2d 1219 (Fla. 1999) 645 n.69, 654 n.90
Comstock Ins. Co. v. Thomas A. Hanson & Assoc., 550 A.2d 731 (Md. App. 1988) 166 n.31
Concrete Gen., Inc. v. Washington Suburban Sanitary Comm'n, 779 F. Supp. 370 (D. Md. 1991) 209 n.122
Concrete Works of Colo., Inc. v. City & Co. of Denver, 321 F.3d 950 (10th Cir. 2003), *cert. denied,* 540 U.S. 1028, 124 S. Ct. 556 (2003) 210 n.128–130

Condon-Johnson & Assoc., Inc. v. Sacramento Mun. Util. Dist., 149 Cal. App. 4th 1384 (2007) 489–490
Conforti & Eisele, Inc. v. John C. Morris Assoc., 418 A.2d 1290 (N.J. Super. 1980) 659 n.103
Congregation of the Passion, Holy Cross Province v. Touche Ross & Co., 636 N.E.2d 503 (1994) 656 n.92
Connelly v. Paul, 731 S.W.2d 657 (Tex. App.–Houston [1st Dist.] 1987) 121 n.74
Connor Bros. Constr. Co. v. United States, 65 Fed. Cl. 657 (Fed. Cir. 2005) 49 n.48, 604 n.41
Conrad Bros., Inc., Appeal of, 1996 WL 675811 (B.C.A. 1996), *aff'd* 135 F.3d 778 (Fed. Cir. 1998) 47 n.38
Consolidated Contracting & Eng'g, *In re,* 1996 WL 639911, at *2 (Comp. Gen. Nov. 6, 1996) 201 n.57
Consolidated Engineering Company, No. 43159, decided February 1, 1943 (98 C. Cls. 256) 274
Consolidated Hardwoods, Inc. v. Alexander Concrete Constr., Inc., 811 P.2d 440 (Colo. App. 1991) 650 n.86
Consumer Power Co. v. Mississippi Valley Structural Steel Co., 636 F. Supp. 1100 (E.D. Mich. 1986) 639 n.63
Contel Advanced Systems, Inc., ASBCA No. 49075, 2004-2 B.C.A. (CCH) ¶ 32,664 (2004) 312 n.2
Continental Consolidated Corp., ENGBCA Nos. 2743, 2766, 67-2 B.C.A (CCH) ¶ 6624 (1967) 331
Continental Ins. Co. v. City of Virginia Beach, 908 F. Supp. 341 (E.D. Va. 1995) 386 n.69
Contra Costa Constr. Corp. v. Daly City, 192 P. 178 (Cal. 1920) 438 n.22
Contractors Ass'n of E. Pa. v. City of Phila., 91 F.3d 586 (3d Cir. 1996) 209 n.122, 210 n.125
Coombs v. Beede, 36 A. 104 (Me. 1896) 141
Copenhaver v. Rogers, 384 S.E.2d 593 (Va. 1998) 659 n.98
Cordeck Sales, Inc. v. Constr. Sys., 2008 Ill. App. LEXIS 272 (Ill. App. Ct. 1st Dist. 2008) 521 n.39
Corinno Civetta Constr. Corp. v. City of New York, 67 N.Y.2d 297, 502 N.Y.S.2d 681, 493 N.E.2d 905 (1986) 308 n.53
Cork Plumbing Co. v. Martin Bloom Assocs., Inc., 573 S.W.2d 947 (Mo. Ct. App. 1978) 523 n.49
Cornell v. Sennes, 18 Cal. App. 3d 126 (Cal. App. 2d Dist. 1971) 524 n.56
Corpus Christi Bank & Trust v. Smith, 525 S.W.2d 501 (Tex. 1975) 124 n.90
Council of Co-Owners Atlantis Condominium, Inc. v. Whiting-Turner Contracting Co., 517 A.2d 336 (Md. 1986) 646–648, 665 n.120
County Asphalt Paving Co., Inc. v. The 1861 Group, Ltd., 908 S.W.2d 184 (Mo. Ct. App. 1995) 671 n.6
County Redevelopment and Housing Auth. v. W. M. Schlosser Co., Inc., 41 Va. Cir. 118 (1996) 485 n.93
Covington, City of, v. Heard, 428 So. 2d 1135 (La. Ct. App. 1983) 603–604
Craftique, Inc. v. Stevens & Co., 364 S.E.2d 129 (N.C. 1988) 387 n.73

Craft Machine Works, Inc., ASBCA No. 47227, 97-1 B.C.A. (CCH) ¶ 28,651 (1996) 336 n.47
Craig v. Everett M. Brooks Co., 222 N.E.2d 752 (Mass. 1967) 659 n.101
Crane Erectors & Riggers, Inc. v. LaSalle Nat'l Bank, 125 Ill. App. 3d 658, 466 N.E.2d 397 (2nd Dist. 1984) 523 n.53, 524 n.59
Craviolini v. Scholer & Fuller Associated Architects, 89 Ariz. 24, 357 P.2d 611 (1961) 365
Crest Constr. Corp. v. Shelby County Bd. of Educ., 612 So. 2d 425 (Ala. 1992) 204 n.79
Cross Constr. Co., ENG BCA No. 3676, 79-1 BCA ¶ 13,707, 473 n.39
Crowen v. Meyer, 342 Ill. 46, 174 N.E. 55 (1930) 523 n.52
Cuddy Mountain Concrete Inc. v. Citadel Constr., Inc., 121 Idaho 220, 824 P.2d 151 (Idaho App. 1992) 511 n.56

D. Federico Co. v. New Bedford Redevelopment Auth., 723 F.2d 122 (1st Cir. 1983) 661 n.108
D. A. Elia Constr. Corp. v. N.Y. State Thruway Auth., 734 N.Y.S.2d 295 (App. Div. 2001) 665 n.119
Daanen & Janssen, Inc. v. Cedarapids, Inc., 405 N.W.2d 842 (1998) 627
Dakota Gasification Co. v. Pascoe Bldg. Sys., 91 F.3d 1094 (8th Cir. 1996) 644 n.66
Dalton v. Cessna Aircraft Co., 98 F.3d 1298 (Fed. Cir. 1996) 461 n.3
Daubert v. Merrell Dow Pharm., Inc., 509 U.S. 579 (1993) 338–340
Dauphin, County of, v. Fidelity & Deposit Co. of Maryland, 770 F. Supp. 248 (M.D. Pa. 1991), *aff'd*, 937 F.2d 596 (3d Cir. 1991) 306 n.41
Dave Kolb Grading, Inc. v. Terra Venture Bridgeton Project Joint Venture, 85 F.3d 351 (8th Cir. 1996) 432 n.3
David L. Threlkeld & Co., Inc. v. Metallgesellschaft Ltd. (London), 923 F.2d 245 (2nd Cir. 1991) 573 n.7
Davidson & Jones, Inc. v. New Hanover County, 255 S.E.2d 580 (N.C. App. 1979) 625 n.45
Davis & Assoc., Inc. v. Midcon, Inc., 127 N.M. 134, 978 P.2d 341 (Ct. App. 1999) 206 n.102
Davis Sewing Mach. Co. of Delaware v. U.S., 60 Ct. Cl. 201, 1925 WL 2797 (1925), *aff'd*, 63 Ct. Cl. 678, 273 U.S. 324, 47 S. Ct. 352, 71 L. Ed. 662 (1927) 500 n.34
Davis-Wellcome Mortg. v. Long Bell Lumber Co., 184 Kan. 202 (1959) 514
Dawco Constr., Inc. v. U.S., 930 F.2d 872, 37 Contr. Cas. Fed. (CCH) ¶ 76056 (Fed. Cir. 1991) 692 n.55
Daymar, Inc., DOTBCA No. 77-13, 78-1 BCA ¶ 12,903, 477 n.58
Day v. City of Beatrice, 169 Neb. 858, 101 N.W.2d 481 (1960) 205 n.85
Day v. United States, 245 U.S. 159, 44
D. C. McClean, Inc. v. Arlington County, 249 Va. 131, 452 S.E.2d 659 (1995) 37 n.5
Decker Constr. Co. v. Ohio River Pipe Line, LLC, 2007 Ohio 421, 19 (Ct. App., Franklin County, Feb. 1, 2007) 525 n.63
Dehnert v. Arrow Sprinklers, Inc., 705 P.2d 846 (Wyo. 1985) 365
Demetriades v. Kaufmann, 680 F. Supp. 658 (S.D.N.Y. 1988) 171
Department of Transp. v. Idol, 440 S.E.2d 863 (N.C. Ct. App. 1994) 382 n.57
Dermott v. Jones, 69 U.S. 1, 17 L. Ed. 762 (1864) 5 n.9, 44
Detroit Edison Co. v. NABCO, Inc., 35 F.3d 236 (6th Cir. 1994) 644 n.66

Devaney v. New Haven Board of Zoning Appeals, 132 Conn. 537, 45 A.2d 828 (1946) 38 n.9
DeVito v. United States, 413 F.2d 1147 (Ct. Cl. 1969) 290 n.16
DiMaria Constr., Inc. v. Interarch, 799 A.2d 555 (N.J. Super. Ct. App. Div. 2001) 365
Diocese of Rochester v. R-Monde Contractors, Inc., 562 N.Y.S.2d 593 (N.Y. Sup. Ct. 1989) 598 n.24
D.M.K. Inc. v. Town of Pittsfield, 290 Wis. 2d 474, 711 N.W.2d 672 (Wis. Ct. App. 2006) 202 n.68
Donald Frederick Evans v. Continental Homes, Inc., 785 F.2d 897 (11th Cir. 1986) 171
Donnelly Constr. Co. v. Oberg/Hunt/Gilleland, 677 P.2d 1292 (Ariz. 1984) 660 n.103
Donovan v. City of Deadwood, 538 N.W.2d 790 (S.D. 1995) 139 n.18
Dorchester. Pilgrim Enter., Inc. v. Maryland Cas. Co., 24 S.W.3d 488 (Tex. App.–Houston [1st Dist.] 2000, no pet.) 538
Dorchester Dev. Corp. v. Safeco Ins. Co., 737 S.W.2d 380, 383 (Tex. App.–Dallas 1987, no writ) 537–538
Douglas Corp., 58-1 BCA Par. 1727, 1958 WL 266 (ASBCA 1958) 296 n.20
Drennan v. Star Paving Co., 333 P.2d 757 (Cal. 1958) 254 n.24, 26, 28, 29
Drews Co., Inc. v. Ledwith-Wolfe Assocs., 371 S.E.2d 532 (S.C. 1988) 290 n.15
Dufficy & Sons, Inc. v. BRW, Inc., 2002 WL 31601023 (Colo. App. 2002) 660 n.104
Duffy v. Village of Princeton, 240 Minn. 9, 60 N.W.2d 27 (1953) 201 n.53
Dugan & Meyer Constr. Co., Inc. v. Ohio Dep't of Admin. Services, 113 Ohio St. 3d 864 N.E. 2d 68 (2007) 46–47
Duncan, Fox, & Co. v. North & South Wales Bank, 6 App. Cas. 1 (H.L. 1880) 556 n.2
Duncan v. Missouri Bd. for Architects, Professional Eng'rs and Land Surveyors, 744 S.W.2d 524 (Mo. Ct. App. 1988) 354 n.23, 402 n.4
Duquesne Light Co. v. Westinghouse Elec. Corp., 66 F.3d 604 (3d Cir. 1995) 661 n.111

Eades v. Capitol Material Co., Inc., 121 F.2d 72 (D.C. 1941) 411 n.27
Eastover Corp. v. Martin Builders, 543 So. 2d 1358 (La. Ct. App. 1989) 390
Eastover Ridge, LLC v. Metric Constructors, Inc., 138 N.C. App. 360, 533 S.E.2d 827 (2000) 371 n.7
East River S.S. Corp. v. Transamerica Delaval, Inc., 476 U.S. 858 (1986) 620 n.24, 622 n.34, 624 n.44, 657
E. C. Ernst, Inc., United States *ex rel,* v. Curtis T. Bedwell & Sons, Inc., 506 F. Supp. 1324 (D.C. Pa. 1981) 380 n.47
E. C. Ernst, Inc. v. Manhattan Constr. Co., 387 F. Supp. 1001 (S.D. Ala. 1974) 13 n.32
E. C. Ernst, Inc. v. Manhattan Constr. Co., 551 F.2d 1026 (5th Cir. 1977) 366, 660 n.103
Economy Forms Corp. v. Williams Bros. Constr. Co., 754 S.W.2d 451 (Tex. App.–Houston [14th Dist.] 1988) 124 n.90
Edinger & Son v. City of Louisville, 802 F.2d 213 (6th Cir. 1986) 209 n.122
Edward M. Crough, Inc. v. Dist. of Columbia Dept. of Gen. Servs., 572 A.2d 457 (D.C. 1990) 309
Edwin R. Marden Corp. v. United States, 442 F.2d 364 (Ct. Cl. 1971) 450 n.73
EH Constr. LLC v. Delor Design Group, Inc., No. 1998-CA-001476-NR, 2000 WL 339939 (Ky. Ct. App. March 31, 2000) 620 n.19
Eichleay Corp., ASBCA 5183, 60-2 BCA ¶ 2688 (1960) 681–682

Electrical Supply Co. of Durham, Inc. v. Swain Elec. Co., Inc., 403 S.E.2d 291 (N.C. 1991) 391 n.99
Elec-Trol, Inc. v. C. J. Kern Contractors, Inc., 284 S.E.2d 119 (N.C. Ct. App. 1981) 364 n.50
Elger Mfg., Inc. v. Liberty Mutual Ins. Co., 972 F.2d 805 (7th Cir. 1992) 649 n.79
Elk and Jacobs Drywall v. Town Contractors, Inc., 229 S.E.2d 260 (S.C. 1976) 393 n.107, 108
Ell-Dorer Contracting Co. v. State of New Jersey, 197 N.J. Super. 175 (App. Div. 1984) 484
Ellerbe Becket Inc.; United States v., 976 F. Supp. 1262 (D. Minn. 1997) 138 n.13
Ellis-Don Constr., Inc. v. HKS, Inc. 353 F. Supp. 2d 603 (M.D.N.C. 2004) 656–657
Elmira, City of, v. Larry Walter, Inc., 76 N.Y.2d 912, 563 N.Y.S.2d 45, 564 N.E.2d 655 (1990) 306 n.47
Emberton v. State Farm Mut. Auto. Ins. Co., 373 N.E.2d 1348 (Ill. 1978) 420 n.41
Embree Constr. Group, Inc. v. Rafcor, Inc., 330 N.C. 487, 411 S.E.2d 916 (1992) 397–398
Engineering Contractors Ass'n of S. Fla. v. Metro. Dade Co., 122 F.3d 895 (11th Cir. 1997) 209 n.122, 123
Ensley Branch, NAACP v. Seibels, 31 F.3d 1548 (11th Cir. 1994) 209 n.123
E. & R. Constr. Co., U.S. *ex rel.* v. Guy H. James Constr. Co., 390 F. Supp. 1193 (M.D. Tenn. 1972), *aff'd,* 489 F.2d 756 (6th Cir. 1974) 442 n.43
ERB Lumber Inc. v. Gidley, 234 Mich. App. 387 (Ct. App. 1999) 520 n.31
Erlich v. Menezes, 21 Cal. 4th 543, 87 Cal. Rptr. 2d 886, 981 P.2d 978 (1999) 10 n.26
Ernst v. Ohio Dept. of Adm. Serv., 69 Ohio App. 3d 330 (1990) 380 n.47
ES-KO, Inc. v. United States, 44 Fed. Cl. 429 (Fed. Cl. 1999) 206 n.100

F. Craig Stuart v. Coldwell Banker Commercial Group, Inc. 745 P.2d 1284 (Wash. 1987) 632–634
Fabe v. WVP Corp., 760 S.W.2d 490 (Mo. Ct. App. 1988) 360–361
Fairway Builders, Inc. v. Malouf Towers Rental Co., Inc., 603 P.2d 513 (Ariz. Ct. App. 1979) 669 n.4
Farabaugh v. Penn Turnpike Comm'n, 911 A.2d 1264 (Pa. 2006) 448 n.63
Farmer Constr. Ltd. v. Washington, 98 Wash. 2d 600 P.2d 1086 (Wash. 1983) 201 n.58
Farmers Group, Inc. v. Trimble, 691 P.2d 1138 (Colo. 1984) 653
F. B. Reynolds v. Texarkana Constr. Co., 374 S.W.2d 818 (Ark. 1964) 254 n.30
F. D. Rich Co. v. United States, 417 U.S. 116 (1974) 527 n.76
Federal Crop Ins. Corp. v. Merrill, 332 U.S. 380 (1947) 438
Federal Ins. Co. v. I. Kruger, Inc., 829 So. 2d 732 (Ala. 2002) 268 n.66
Federal Sign v. Tex. So. Univ., 951 S.W.2d 401 (Tex. 1997) 120 n.66, 381 n.51
Felhaber Corp. v. State, 410 N.Y.S.2d 920 (N.Y. App. Div. 1978) 445 n.53
Fidelity and Deposit Co. v. City of Sheboygan Falls, 713 F.2d 1261 (7th Cir. 1983) 47 n.35
Fidelity & Deposit Co. of Md. v. Hartford Cas. Ins. Co., 215 F. Supp. 2d 1171 (D. Kan. 2002) 539
Fieldstone Co. v. Briggs Plumbing Products, Inc., 54 Cal. App. 4th 357 (1997) 621 n.26
Filbert v. Philadelphia, 181 Pa. St. 530, 44

Filippi v. Filippi, 818 A.2d 608 (R.I. 2003) 115 n.34
Fireman's Fund Ins. Co. v. SEC Donohue, Inc., 679 N.E.2d 1197 (Ill. 1997) 658 n.95, 663 n.115, 116
Fireman's Fund Ins. Co. v. United States, 909 F.2d 495 (Fed. Cir. 1988) 377 n.30, 378, 385 n.67
First Atlantic Bldg. Corp. v. Neubauer Constr. Co., 352 So. 2d 103 (Fla. 4th DCA 1977) 511 n.54
First Nat'l Bank of Roscoe v. Rinaldi, 262 Ill. App. 3d 179, 634 N.E.2d 1204 (2nd Dist. 1994) 523 n.50
First Union Nat'l Bank of North Carolina v. King, 306 S.E.2d 508, (N.C. Ct. App. 1983) 386 n.68
Fischbach & Moore Int'l Corp., ASBCA No. 18,146, 77-1 B.C.A. (CCH) ¶ 12,300 (1976) 335 n.43
Fisher v. Katz, 266 F. Supp. 180 (S.D.N.Y. 1976) 620 n.21
Fisher v. Simon, 112 N.W.2d 705 (Wis. 1961) 390 n.90
Flaherty v. Port Auth. of Allegheny County, 299 A.2d 613 (Pa. 1973) 612 n.61
Fleischer v. Hellmuth, Obata and Kassabaum, Inc., 870 S.W.2d 832 (Mo. App. 1993) 658 n.96
Fleming v. Borden, 37 S.E. 219 (N.C. 1900) 386 n.68
Flippin Materials Co. v. United States, 160 Ct. Cl. 357 (1963) 467 n.24
Floor Craft Floor Covering, Inc. v. Parma Community Gen. Hosp. Ass'n, 560 N.E.2d 206 (Ohio 1990) 658 n.95
Foley Co. v. United States, 11 F.3d 1032 (Fed. Cir. 1993) 233 n.18, 481 n.74
Folk Construction Co. v. United States, 2 Cl. Ct. 681 (1983) 276
Fordeck-Kemerly Elec., Inc. v. Helmkamp, 591 N.E.2d 1035 (Ind. Ct. App. 1992) 529 n.85
Forte Bros., Inc. v. Nat'l Amusements, Inc., et al. 525 A.2d 1301 (R.I. 1987) 598–599
Fortec Constructors v. United States, 8 Cl. Ct. 490 (1985) 331
Fort Knox Self Storage, Inc. v. Western Technologies, Inc. 140 N.M. 233, 142 P.3d 1 (N.M. Ct. App. 2006) 697–699
Foster Constr. C.A. & Williams Bros. Co. v. United States, 435 F.2d 873 (Ct. Cl. 1970) 469, 473 n.37, 478 n.59, 61, 484
Foxcroft Townhome Owners Ass'n v. Hoffman Rosner Corp., 449 N.E.2d 125 (Ill. 1983) 644 n.66
Fox v. Mountain West Elect., Inc., 52 P.2d 848 (Idaho 2002) 432 n.4
Frank Sullivan Co. v. Midwest Sheet Metal Works, 335 F.2d 33 (8th Cir. 1964) 439 n.26
Fraya S.E., ASBCA No. 5222, 02-2 B.C.A. (CCH) ¶ 31,975 (2002) 340
Fred Howland, Inc. v. Gore, 152 Fla. 781, 13 So. 2d 303 (1942) 300
Fretwell v. Prot. Alarm Co., 764 P.2d 149 (Okla. 1988) 264 n.54
Fru-Con Constr. Corp., ASBCA Nos. 53544, et al., 2005-1 BCA (CCH) ¶ 32,936, 327 n.25, 337 n.48
Fru-Con Corp. and Granite Constr., JV v. Illinois, 50 Ill. Ct. Cl. 50 (1996) 488
Fru-Con/Fluor Daniel Joint Venture v. Corrigan Brothers, Inc., 154 S.W.3d 330 (Mo. Ct. App. E.D. 2004) 510 n.48

Fuller Co. v. Brown Minneapolis Tank & Fabricating Co., 678 F. Supp. 506 (E.D. Pa. 1987) 452 n.83
Fullerton Constr. Co., ASBCA No. 12275, 69-2 B.C.A. (CCH) ¶ 7876 (1969) 328 n.28
Fullilove v. Klutznick, 448 U.S. 448, 100 S. Ct. 2758 (1980) 208

G. Blindzius Contractors, Inc., ASBCA No. 37707, 90-2 BCA ¶ 22,835 (1990) 328 n.28, 329 n.29
GAF Corp. v. United States, 19 Cl. Ct. 490 (1990) 603 n.36
Galloway Corp. v. S. B. Ballard Constr. Co., 464 S.E.2d 349 (Va. 1995) 393 n.107
Gardner-Zenkes Co. v. State, 109 N.E.2d 729, 790 P.2d 1010 (1990) 48 n.43
Gehan Homes, Ltd. v. Employers Mutual Casualty Co., 146 S.W.3d 833 (Tex. App.–Dallas 2004) 539–540
G.E.I. Recycling, Inc. v. Atl. Envtl., Inc., 821 So. 2d 431 (Fla. Dist. Ct. App. 2002) 268 n.67
Geller v. Brownstone Condominium Ass'n, 82 Ill. App. 3d 334, 402 N.E.2d 807 (1985) 36 n.3
General Contracting & Constr. Co. v. United States, 84 Ct. Cl. 570 (1937) 451 n.79
General Contracting Corp. v. United States, 96 Ct. Cl. 255 (1942) 475 n.49
General Corrosion Servs. Corp. v. K Way Equip. Co., 631 S.W.2d 578 (Tex. App.–Tyler 1982) 120 n.70
General Motors Acceptance Corp. v. Daniels, 492 A.2d 1306 (Md. 1985) 385 n.65
George Hyman Constr. Co. v. Washington Metro. Transit Auth., 816 F.2d 753 (D.C. Cir. 1987) 681 n.27
George Sollitt Constr. Co. v. United States, 64 Fed. Cl. 229 (2005) 604 n.40, 680 n.25
George v. El Paso County Water Control & Imp. Dist. No. 1, 332 S.W.2d 144 (Tex. Civ. App.–El Paso 1960) 114 n.24, 122 n.81
Gesco, Inc. v. Edward L. Nezelek, Inc., 414 So. 2d 535 (Fla. Dist. Ct. App. 1982) 289
Gherna v. Ford Motor Co., 246 Cal. App. 2d 639 (1966) 645 n.70
Gilbane Bldg. Co. v. Brisk Waterproofing Co., 585 A.2d 248 (1991) 268 n.68
Gilbane Bldg. Co. v. Nemours Found., 606 F. Supp. 995 (D. Del. 1985) 644 n.67
Gilbert Pac. Corp. v. Oregon, 822 P.2d 729 (Or. Ct. App. 1991) 272
Gill Constr., Inc. v. 18th & Vine Auth., 157 S.W.3d 699 (Mo. App. 2004) 452 n.82
Gillum v. Republic Health Corp., 778 S.W.2d 558 (Tex. App.–Dallas 1989) 100 n.7
Glen-Gery Corp. v. Warfel Constr. Co., 734 A.2d 926 (Pa. Super. Ct. 1999) 384 n.62
Gogo v. L.A. etc. Flood Control Dist., 114 P.2d 65 (Cal. Ct. App. 1941) 488–489
Golomore Associates v. New Jersey State Highway Auth., 173 N.J. Super. 55 (App. Div. 1980) 483
Gonnert v. Victor At W. 53, LLC, No. 40690/04, 2006 N.Y. Slip Op. 50284(U) (N.Y. Slip Op. March 3, 2006) 411 n.26
Gostovich v. City of W. Richland, 75 Wash. 2d 583, 452 P.2d 737(1969) 201 n.53
Gould, Inc. v. United States, 935 F.2d 1271 (Fed. Cir. 1991) 233
Grady v. S. E. Gustafson Constr. Co., 103 N.W.2d 737 (1960) 268 n.69
Graham Constr. Co., Inc. v. Earl, 362 Ark. 220, 2005 WL 1041149, at *3 (2005) 48 n.42
Gramm v. Ins. Unlimited, 378 P.2d 662 (Mont. 1963) 380 n.43
Granger & Sons, Inc. v. City Sch. Dist. of Albany, 296 A.D.2d 636, 744 N.Y.S.2d 567 (2002) 315 n.9
Granite Constr. Co., Inc. v. United States, 24 Cl. Ct. 735 (1991) 198 n.41, 465–466

Granite Constr. Co. v. United States, 962 F.2d 998 (Fed. Cir. 1992) 448 n.61
Gravely v. Providence Partnership, 549 F.2d 958 (4th Cir. 1977) 151 n.30
Great Am. Ins. Co. v. North Austin Mun. Util. Dist. No. 1, 902 S.W.2d 488 (Tex. Civ. App. 1993), *rev'd in part* and *aff'd in part,* 908 S.W.2d 415 (Tex. 1995) 601 n.30
Great Lakes Construction Co. v. United States, 95 C. Cls. 479, 273
Greenberg v. Perkins, 845 P.2d 530 (Colo.1993) 653
Greenbriar, Ltd. v. City of Alabaster, 881 F.2d 1570 (11th Cir. 1989) 289
Green Constr. Co. v. Kansas Power & Light Co., 1 F.3d 1005 (10th Cir. 1993) 49 n.50
Grochal v. Ocean Technical Services Corp., 476 F.3d 238 (4th Cir. 2007) 378–379
Groppel Co. v. United States Gypsum Co., 616 S.W.2d 49 (Mo. Ct. App. 1981) 618 n.7
Grossman v. Sea Air Towers, Ltd., 513 So. 2d 686 (Fla. App. 1987), *rev. denied,* Grossman v. Sea Air Towers, Ltd., 520 So. 2d 584 (Fla. 1988) 695 n.68
Gross v. Sweet, 49 N.Y.2d 102, 358
Guardian Constr. v. Tetra Tech, Inc., 583 A.2d 1378 (Del. Super. 1990) 665 n.122
Guerini Stone Co. v. P. J. Carlin Constr. Co., 248 U.S. 334 39 S. Ct. 102 (1919) 370 n.1, 379 n.42, 380 n.46
Gulf Constr. Co., v. Self, 676 S.W.2d 624 (Tex. App.–Corpus Christi 1984) 119 n.60
Gulf Constr. Group, Inc., ENG BCA No. 5958, 93-3 BCA ¶ 26,174, 481 n.75
Gulf Contracting Co. v. U.S., 23 Cl. Ct. 525 (1991), *aff'd,* 972 F.2d 1353 (Fed. Cir. 1992) (table), *cert. den.,* 506 U.S. 999, 113 S.Ct. 598 (1992) 337 n.48
Guy F. Atkinson Co., ENG BCA No. 4693, 87-3 BCA ¶ 19,971, 477 n.57

Hadley v. Baxendale, 156 Eng. Rep. 145 (1854) 304 n.36, 619 n.14
Hagerman Constr. Corp. v. Long Elec. Co. 741 N.E.2d 390 (Ind. Ct. App. 2000) 425–427
Hall Street Assoc., L.L.C. v. Mattel, Inc., ______ U.S. ______, 128 S. Ct. 1396 (2008) 577–578
Hamp's Constr., L.L.C. v. City of New Orleans, 924 So. 2d 104 (La. 2006) 201 n.59
Hampton v. Phipps, 108 U.S. 260, 2 S. Ct. 622, 27 L. Ed. 719 (1883) 556 n.2
Haney v. United States, 676 F.2d 584 (Ct. Cl. 1982) 281 n.5, 315 n.6
Harco Energy, Inc. v. Re-Entry People, Inc., 23 S.W.3d 389 (Tex. App.–Amarillo 2000) 98 n.2
Hardeman-Monier-Hutcherson v. United States, 458 F.2d 1364 (1972) 465 n.19
Harkins v. Girouard Estates, Inc., 31 Conn. App. 485 (App. Ct. 1993) 524 n.60, 525 n.62
Harmony Constr., Inc. v. State Dep't of Transp., 668 A.2d 746 (Del. Ch. 1995) 203 n.74
Harrison v. Facade, Inc., 355 S.W.2d 543 (Tex. Civ. App.–Dallas 1962) 121 n.78
Harrison v. Williams Dental Group, P.C., 140 S.W.3d 912 (Tex. App.–Dallas 2004) 99 n.4
Harris v. Phillips Pipe Line Co., 517 S.W.2d 361 (Tex. Civ. App.–Austin 1974) 119 n.60
Hathaway v. Gen. Mills, Inc., 711 S.W.2d 227 (Tex. 1986) 115 n.32
Havens v. Safeway Stores, 678 P.2d 625 (Kan. 1984) 380 n.48
Haws & Garrett General Contractors, Inc. v. Gorbett Bros. Welding Co., 480 S.W.2d 607 (Tex. 1972) 100 n.6, 8, 101
Haz-Mat Response v. Certified Waste Servs., 259 Kan. 166 (1996) 518 n.28
Held v. Mitsubishi Aircraft Int'l, Inc., 672 F. Supp. 369 (D. Minn. 1987) 621 n.32

Helene Curtis Indus., Inc. v. United States, 312 F.2d 774 (Ct. Cl. 1963) 197 n.36, 465 n.19
Hensel Phelps Constr. Co., ASBCA No. 49270, 99-2 B.C.A. (CCH) ¶ 30,531 (1999) 283 n.9, 329 n.30
Hercules, Inc. v. United States, 24 F.3d 188 (Fed. Cir. 1994) 198 n.39–40
Hercules, Inc. v. United States, 516 U.S. 417 (1996) 601 n.29
Herman Frankel Org. v. Tegman, 367 F. Supp. 1051 (E.D. Mich. 1973) 171
Hernandez v. Westoak Realty & Inv., Inc., 771 S.W.2d 876 (Mo. Ct. App. 1989) 669 n.3
Hewett-Kier Constr., Inc. v. Lemuel Ramos and Assoc., Inc., 775 So. 2d 373 (Fla. App. 2002) 660 n.103
H. I. Homa Co., ENGBCA No. PCC-41, 82-1 B.C.A. (CCH) ¶ 15,651 (1982) 323–324
Hicks v. Smith, 330 S.W.2d 641 (Tex. Civ. App.–Fort Worth 1959) 114 n.23
Hilliard & Bartko Joint Venture v. Fedco Sys., Inc., 522 A.2d 961 (Md. 1987) 608 n.46
Hill v. Polar Pantries, 64 S.E.2d 885 (S.C. 1951) 148
Hoel-Steffen Constr. Co. v. United States, 456 F.2d 760 (Ct. Cl. 1972) 441 n.38
Hoffman Constr. Co. of Oregon v. United States, 40 Fed. Cl. 184 (Fed. Cl. 1998), *aff'd in part, rev'd in part on other grounds,* 178 F.2d 131 (Fed. Cir. 1999) 282 n.7, 334 n.40
Hoffman v. United States, 166 Ct. Cl. 39, 340 F.2d 645 (1964) 36 n.2
Hollerbach v. United States, 233 U.S. 165 (1914) 44, 464 n.13, 15, 661 n.109
Home Elec. Co. of Lenoir, Inc. v. Hall, etc., 358 S.E.2d 539 (N.C. 1987) 255 n.31
Horovitz v. Levine, 755 So. 2d 687 (Fla. Dist. Ct. App. 1999) 288
Hospital Mortgage Group v. First Prudential Dev. Corp., 411 So. 2d 181 (Fla. 1982) 510 n.52
Housing Auth. of the City of Texarkana v. Johnson Constr. Co., 264 Ark. 523 S.W.2d 316 (1978) 48 n.42
Houston, City of, v. R. F. Ball Constr. Co., 570 S.W.2d 75 (Tex. Civ. App. 1978) 383 n.58
Howard Contracting, Inc. v. G. A. MacDonald Constr. Co., 71 Cal. App. 4th 38, 83 Cal. Rptr. 590 (Cal. Ct. App. 1998) 315 n.9
H. R. Johnson Constr. Co. v. Bd. of Educ., 16 Ohio Misc. 99, 241 N.E.2d 403 (Ohio Com. Pl. Ct. 1968) 206 n.99
Hubbard v. Shankle, 138 S.W.3d 474 (Tex. App.–Fort Worth 2004) 98 n.3
Hubert v. Aiken, 15 Daly 237, 19 N.Y.ST. REP. 914, 2 N.Y.S. 711 (1888) 141–142
Hubert v. Aiken, 2 N.Y.S. 711 (C.P. 1888), *aff'd,* 25 N.E. 954 (N.Y. 1890) 616 n.3
Hubert v. Aitken, 2 NYS 711, *order adhered to on rearg,* 5 NYS839, *aff'd* 123 NY 655, 358
Hunt v. Ellisor & Tanner, 739 S.W.2d 933 (Tex. App. 1987) 358, 359, 597–598

Imperial Homes Corp. v. Lamont, 458 F.2d 895 (5th Cir. 1972) 171
Indian River Constr. Co. v. City of Jacksonville, 390 So. 2d 1139 (Fla. Dist. App. 1977) 441 n.41
In re. See Name of party
Insurance Co. of N. Am. v. Town of Manchester, 17 F. Supp. 2d 81, 84 (D. Conn. 1998) 659 n.100
Interstate Contracting Corp. v. City of Dallas, Texas, 407 F.3d 708 (5th Cir. 2005) 432 n.1

J. Lawson Jones Constr. Co., Inc., ENG BCA No. 4363, 86-1 BCA ¶ 18,719 at 94,172 (1986) 474 n.44
J. A. Jones Constr. Co. v. Lehrer McGovern Bovis, Inc., 89 P.3d 1009 (Nev. 2004) 451 n.80
Jackson Constr. Co. v. United States, 62 Fed. Cl. 84 (2004) 315 n.8
Ja Din, Inc. v. L-J, Inc., 898 F. Supp. 894 (E.D. Fla. 1995) 439 n.25
Jaeger v. Henningson, Durham & Richardson, Inc., 714 F.2d 773 (8th Cir. 1983) 354 n.23
James Cape & Sons Co. v. Mulcahy, 285 Wis. 2d 200, 700 N.W.2d 243 (Wis. 2005) 207 n.106, 208 n.110
Jardel Enterprises, Inc. v. Triconsultants, Inc., 770 P.2d 1301 (Colo. App. 1988) 650 n.85
Jefferson Constr. Co. v. United States, 176 Ct. Cl. 1363 (1966) 478 n.65
Jewish Bd. of Guardians v. Grumman Allied Indus., 96 A.D.2d 465, *aff'd* 62 N.Y.2d 684, 357, 358
Jimenez, Inc., VABCA No. 6351, et al., 02-2 B.C.A. (CCH) ¶ 32019 (2002) 340–341
Jimenez v. Superior Court, 58 P.3d 450 (Cal. 2002) 623 n.38, 645 n.69
J. J. Brown Co. v. J. L. Simmons Co., 118 N.E.2d 781 (Ill. App. Ct. 1954) 290 n.15
J. L. Simmons Co. v. United States, 412 F.2d 1360 (Ct. Cl. 1969) 603 n.36, 604 n.39
J. M. Beeson Company v. Ernesto Sartori, Nos. 87-1953, 87-2415 and 87-2693, 298–301
Jo-Bar Mfg. Corp. v. United States, 535 F.2d 62 (Ct. Cl. 1976) 441 n.40
Johanson v. Huizenga Holdings, Inc., 963 F. Supp 1175 (S.D. Fla. 1997) 138 n.13
John B. Pike & Son, Inc. v. State of New York, 647 N.Y.S.2d 654 (N.Y. Cl. 1996) 275
John Driggs Co., Inc., ENGBCA No. 4926, 87-2 B.C.A. (CCH) ¶ 19,833 at 100,388 (1987) 334 n.41
John E. Green Plumbing & Heating Co., Inc. v. Turner Constr. Co., 500 F. Supp. 910 (E.D. Mich. 1980), *aff'd,* 742 F.2d 965 (7th Cir. 2004) 701 n.77
John H. Maxwell & Co., 353 F.2d 804 (9th Cir. 1965) 380 n.44
John Lonberg, et al. v. Saborn Theaters, Inc., et al., 259 F.3d 1029 (9th Cir. 2001) 138 n.15
John Martin Co. v. Morse/Diesel, Inc. 819 S.W.2d 428 (Tenn. 1991) 665 n.121
John Murphy Constr. Co., AGBCA No. 418, 79-1 B.C.A. (CCH) ¶ 13,836 (1979) 335 n.43
Johnson v. Bd. of County Comm'rs, 913 P.2d 119 (Kan. 1996) 264 n.54
Johnson v. Jones, 921 F. Supp. 1573 (E.D. Mich.1996) 172
Johnson v. Salem Title Co., 425 P.2d 519 (Or. 1967) 617 n.5
Jones v. J. H. Hiser Constr. Co., 60 Md. App. 671, 484 A.2d 302 (1984) 371 n.6
Jordan v. Bd. of Educ. of Taylor's Falls, 39 Minn. 298, 39 N.W. 801 (1888) 5 n.11
Joseph F. Egan, Inc. v. City of New York, 215 N.E.2d 490 (N.Y. Ct. App. 1966) 442 n.42
Joseph Sternberg, Inc. v. Walber 36th Street Assoc., 594 N.Y.S.2d 144 (N.Y. App. Div. 1993) 432 n.4
J. R. Sinnott Carpentry, Inc. v. Phillips, 443 N.E.2d 597 (Ill. App. Ct. 1982) 586 n.12
J. R. Youngdale Constr. Co., Inc. v. United States, 23 Cl. Ct. 460, 1991 WL 130539 (Cl. Ct. 1991), *aff'd,* 956 F.2d 1172 (Fed. Cir. 1992) 284 n.12
Juliano v. Gaston, 455 A.2d 523 (N.J. 1982) 650 n.86
Julian Speer Co. v. Ohio State Univ., 680 N.E.2d 254 (Ohio Ct. Cl, 1997) 447 n.58

Kaiser Indus. Corp. v. United States, 169 Ct. Cl. 310 (1965) 474 n.43, 44
Kalisch-Jarcho, Inc. v. City of New York, 533 N.E.2d 258 (N.Y. 1988) 450 n.69–71
Keang Nam Enter. Ltd., ASBCA No. 13747, 69-1 BCA ¶ 7705, 473 n.40
Kecko Piping Co., Inc. v. Town of Monroe, 172 Conn. 197, 374 A.2d 179 (1977) 365
Keck v. Dryvit Sys., Inc., 830 So. 2d 1 (Ala. 2002) 623 n.37, 639 n.62
Keller v. A. O. Smith Harvestore Prods., Inc., 819 P.2d 69 (Colo. 1991) 653
Kellogg v. Pizza Oven, Inc., 157 Colo. 295, 402 P.2d 633 (Colo. 1965) 167–170
Kemp v. United States, 38 F. Supp. 568 (D. Md. 1941) 207 n.109
Kennedy v. Columbia Lumber and Mfg. Co., 384 S.E.2d 730 (S.C. 1989) 631 n.54, 648–649 n.77, 654 n.88
Kenny Constr. Co. of Illinois v. Metro. Sanitary Dist. of Greater Chicago, 288 N.E.2d 1 (Ill. 1971) 487–488
Ken's Carpets Unlimited, Inc., [U.S. for Use and Benefit of,] v. Interstate Landscaping Co., Inc., 37 F.3d 1500 (6th Cir. 1994) 677 n.20
Kensington Corp. v. State, 253 N.W.2d 781 (Mich. Ct. App. 1997) 272 n.84
Kern v. City of Lawrenceburg, 625 N.E.2d 1326 (Ind. Ct. App. 1993) 383 n.60
Key Int'l Mfg. v. Morse/Diesel, Inc., 142 A.D.2d 448 (N.Y. App. 1988) 636 n.59
Khoury Constr. Co., Inc. v. Earhart, 382 S.E.2d 392 (Ga. Ct. App. 1989) 390
Kincaid v. Landing Dev. Corp., 344 S.E.2d 869 (S.C. App. 1986) 146 n.25
Kinetic Builders, Inc., ASBCA No. 32627, 88-2 BCA ¶ 20,657, 474 n.45
Kingery Constr. Co. v. Scherbarth Welding, Inc., 185 N.W.2d 857 (Neb. 1971) 279 n.1
King Fisher Marine Serv., Inc. v. United States, 16 Cl. Ct. 231 (1989) 455 n.93
Kingston Constructors, Inc. v. Washington Metro. Area Transit Auth., 930 F. Supp. 651 (D.D.C. 1996) 306 n.43
Kinsey Contracting Co., Inc. v. City of Fayetteville, 106 N.C. App. 383, 416 S.E.2d 607 (N.C. Ct. App. 1992) 203
Klose v. Sequoia Union High Sch. Dist., 258 P.2d 515 (Cal. App. 1953) 256 n.34
Knutton v. Cofield, 160 S.E.2d 29 (N.C. 1968) 392 n.103
Koch v. Constr. Tech., Inc., 924 S.W.2d 68 (Tenn. 1996) 393 n.107
Koontz v. Thomas 333 S.C. 702, 511 S.E.2d 407 (S.C. App., 1999) 161–164
Korsunsky Krank Erickson Architects, Inc. v. Walsh, 370 N.W.2d 29 (Minn. 1985) 515 n.11

Laconco, Inc.; State v., 430 So. 2d 1376 (La. Ct. App. 1983) 389
La Crosse, City of, v. Schubert, Schroeder & Assoc., 240 N.W.2d 124 (Wis. 1976) 645 n.69
Lafayette Steel Erectors, Inc. v. Roy Anderson Corp., 71 F. Supp. 582 (S.D. Miss. 1997) 268 n.67
Lamb v. Georgia-Pacific Corp., 392 S.E.2d 307 (Ga. App. 1990) 645 n.69
Lamb Eng. & Constr. Co., EBCA No. C-9304172, 97-2 B.C.A. (CCH) ¶ 29,207, 334
Langley v. Knowles, 958 So. 2d 1149 (Fla. Dist. Ct. App. 5th Dist. 2007) 521 n.38
Lansdowne, Pa., Borough of, v. Sevenson Envt. Serv., Inc., 2000 WL 1886578 (E.D. Pa. December 12, 2000) 624 n.42, 664–665
Lassiter v. Bank of North Carolina, 145 N.C. App. 264, 551 S.E.2d 920 (2001) 396–397
Lee Lewis Constr., Inc. v. Harrison, 70 S.W.3d 778 (Tex. 2002) 416–418
Lee R. Smith, ASBCA No. 11135, 66-2 BCA ¶ 5857 at 27,181 (1966) 474 n.44

LeGrand v. Hubbard, 216 Ala. 164 (1927) 517 n.20
Lennar Corp. v. Great American Ins. Co., 2006 WL 406609, *15 (Tex. App.–Houston [14th Dist.]) 539, 540
Levering & Garrigues Co. v. United States, 73 Ct. Cl. 566 (1932) 464 n.13, 16, 465
L. G. Everist, Inc. v. United States, 231 Ct. Cl. 1013 (1982) 473 n.42
Lidderdale's Executors v. Robinson's Executor, 12 Wheat. 594, 6 L.Ed. 740 (1827) 556 n.2
Lidia Hernandez v. Home Depot, U.S.A. et al., 2007 U.S. Dist. LEXIS 57395, 124 n.94
Linda Newman Constr. Co. v. United States, 48 Fed. Cl. 231 (2000) 456 n.100, 101
Lion Raisins, Inc. v. United States, 51 Fed. Cl. 238 (2001) 203 n.69
Litton Systems, Inc. v. Frigitemp Corp., 613 F. Supp. 1377 (S.D. Miss. 1985) 452 n.82
L. K. Comstock & Co., Inc. v. United Engineers & Constructors Inc., 880 F.2d 219, 14 Fed. R. Serv. 3d 348 (9th Cir. 1989) 510 n.48
L. L. Hall Constr. Co. v. United States, 379 F.2d 559 (Ct. Cl. 1966) 445 n.54–55
Lloyd Wood Coal Co. v. Clark Equip. Co., 543 So. 2d 671 (Ala. 1989) 645 n.70
Local Joint Executive Bd. v. Stern, 651 P.2d 637 (Nev. 1982) 661 n.110
Lochrane Eng'g, Inc. v. Willingham Realgrowth Inv. Fund Ltd., 552 So. 2d 228 (Fla. App. 1989), *rev. denied,* Lochrane Eng'g, Inc. v. Willingham Realgrowth Inv. Fund Ltd., 563 So. 2d 631 (Fla. 1990) 695 n.67
Lonergan v. San Antonio Loan & Trust Co., 104 S.W. 1061 (Tex. 1907) 601 n.30
Los Angeles Testing Lab., United States *ex rel.,* v. Rogers & Rogers, 161 F. Supp. 132 (S.D. Cal. 1958) 625 n.45, 659 n.102
Lovering-Johnson, Inc., ASBCA No. 53902, 2006-1 B.C.A. (C.C.H.) ¶ 33,126, 332 n.33
Lovering-Johnson, Inc. v. City of Prior Lake, 558 N.W.2d 499 (Minn. Ct. App. 1997) 206 n.97
Lundgren v. Freeman, 307 F.2d 104 (9th Cir. 1962) 365
Luria Bros. & Co. v. United States, 369 F.2d 701 (Ct. Cl. 1966) 448 n.64
Lynch III Props. Corp., *In re,* 125 B.R. 857 (Bkrtcy. E.D.N.Y. 1991) 377 n.28
Lyons Sav. v. Gash Assoc., 279 Ill. App. 3d 742, 665 N.E.2d 326 (1st Dist. 1996) 523 n.54, 55

M. Glosser & Sons, Inc. v. Micco, 309 A.2d 602 (Pa. Commw. Ct. 1973) 381 n.50
M. A. Mortenson Co. v. United States, 843 F.2d 1360 (Fed. Cir. 1988) 284 n.12
MacKnight Flintic Stone Co. v. The Mayor, 160 N.Y. 72, 44
MacPherson v. Buick Motor Co., 217 N.Y. 382, 111 N.E. 1050 (1916) 619, 633
Magnolia Constr. Co. v. Mississippi Gulf S. Eng'rs, Inc., 518 So. 2d 1194 (Miss. 1988) 359 n.30, 660 n.104
Main Elec., Ltd. v. Printz Servs. Corp., 980 P.2d 522 (Colo. 1999) 268 n.66
Malarchick v. Pierce, 264 N.W.2d 478 (N.D. 1978) 392 n.101
Mann v. Clowser, 59 S.E.2d 78 (Va. 1950) 48 n.42
Manuel Bros., Inc. v. United States, 55 Fed. Cl. 8 (2002) 341
Manzo v. Ford, 731 S.W.2d 673 (Tex. App.–Houston [14th Dist.] 1987, no writ) 118 n.55, 57, 120 n.70
Maricopa, County of, v. Walsh and Oberg Architects, Inc., 494 P.2d 44 (Ariz. Ct. App. 1972) 694
Mark V, Inc. v. Mellekas, 845 P.2d 1232 (N.M. 1993) 113 n.18
Marriott Corp. v. Dasta Constr. Co., 26 F.3d 1057 (11th Cir. 1994) 307 n.50

Martin K. Eby Constr. Co., Inc. v. Jacksonville Transp. Auth., 436 F. Supp. 2d 1276 (M.D. Fla. 2005) 45 n.31, 47 n.35
Maryland Cas. Co. v. King, 381 P.2d 153 (Okla. 1963) 552 n.8
Masten Lumber & Supply Co. v. Brown, 405 A.2d 101 (Del. 1979) 517 n.20
Mateyka v. Schroeder, 504 N.E.2d 1289 (Ill. 1987) 439 n.28
Maverick Diversified, Inc., ASBCA Nos. 19838, 19955, and 20091, 76-2 BCA ¶ 12,104 at 58,150 (1976) 474 n.44
Mayor and City Council of Columbus v. Clark-Dietz & Assocs. Eng'rs, Inc., 550 F. Supp. 610 (N.D.Miss. 1982) 660 n.104
Mazel v. Bain, 272 Ala. 640 (1961) 518 n.27
McConnell v. Corona City Water Co., 149 Cal. 60, 85 P. 929 (1906) 47 n.39
McDevitt & Street Co. v. Marriott Corp., 713 F. Supp. 906 (E.D. Va. 1989), *aff'd in part, rev'd in part,* 911 F.2d 723 (4th Cir. 1990), *on remand,* 754 F. Supp. 513 (E.D. Va. 1991) 305 n.37, 39, 661 n.108
McDevitt & Street Co. v. Marriott Corp., 713 F. Supp. 906 (E.D. Va. 1989), *aff'd in part and rev'd in part,* 911 F.2d 723 (4th Cir. 1990) 439 n.29
McDonough v. Whalen, 313 N.E.2d 435 (Mass. 1974) 655 n.91
MCI Constructors, Inc. v. Hazen and Sawyer, P.C. 401 F. Supp. 2d 504 (M.D.N.C. 2005) 364 n.53
McJunkin Corp. v. City of Orangeburg, 238 F.2d 528 (4th Cir. 1956) 117 n.45–47
MCK Bldg. Assoc., Inc. v. St. Lawrence Univ., 301 A.D.2d 726, 754 N.Y.S.2d 397 (2003) 677 n.19
McKee By and Through McKee v. City of Pleasanton, 242 Kan. 649, 750 P.2d 1007 (Kan. 1988) 143–146
McKeever v. Or. Mortgage Co., 198 P. 752 (Mont. 1921) 124 n.91
McKinley v. Brandt Constr., Inc., 859 N.E.2d 572 (Ohio Ct. App. 2006) 587 n.13
McKnight v. Lange Mfg. Co., 155 S.W. 977 (Tex. App. 1913) 385 n.66
McNeal v. Marco Bay Assocs., 492 So. 2d 778 (Fla. Dist. Ct. App. 1986) 288
McQuagge v. United States, 197 F. Supp. 460 (W.D. La. 1961) 437 n.18
Mechanic's Lien Against City of Kansas City, *In re,* 37 Kan. App. 2d 440 (Kan. Ct. App. 2007) 525 n.61
Mega Constr. Co. v. United States, 29 Fed. Cl. 396 (Fed. Cl. 1993) 282 n.7, 333
Mel-Stu Constr. Corp. v. Melwood Constr. Corp., 101 A.D.2d 809, 475 N.Y.S.2d 459 (1984) 515 n.11
Memphis & L.R.R. Co. v. Dow, 120 U.S. 287, 7 S. Ct. 482, 30 L.Ed. 595 (1887) 556 n.1
Mennonite Deaconess Home and Hosp., Inc. v. Gates Eng'g Co., 219 Neb. 303 N.W.2d 155 (1985) 248 n.7
Menorah Nursing Home, Inc. v. Zukov, 548 N.Y.S.2d 702 (N.Y. App. Div. 1989) 552 n.8
Mergentime Corp. et al. v. Washington Metro. Area Transp. Auth., 400 F. Supp. 2d 145 (D.C. 2005) 332 n.33
Merrill Stevens Dry Dock Co. v. M/V Yeocomico II, 329 F.3d 809 (11th Cir. 2003) 288
MES, Inc., PSBCA No. 4462, 2006-1 B.C.A. (CCH) ¶ 33,194, *mot. for recon. denied,* 2006-2 B.C.A. (CCH) ¶ 33,340, 333 n.38
Metric Constr., Inc. v. Nat'l Aeronautics & Space Admin., 169 F.3d 747 (Fed. Cir. 1999) 447 n.57

Metropolitan Atlanta Rapid Transit Auth. v. Green Int'l, Inc., 509 S.E.2d 674 (Ga. Ct. App. 1999) 348–349, 359 n.32
Metropolitan Steel Indus., Inc. v. Perini Corp., 828 N.Y.S.2d 395 (N.Y. App. Div. 2007) 454 n.92
Micro-King Co., United States *ex rel,* v. Cmty. Science Tech., Inc., 574 F.2d 1292 (5th Cir. 1978) 380 n.44
Mid Continent Aircraft Corp. v. Curry County Spraying Serv., Inc., 572 S.W.2d 308, 312 (Tex. 1978) 626 n.50
Midwest Asbestos Removal Serv., Inc. v. Crown Cork & Seal Co., No. 90C 7010, 1993 WL 18433 (N.D. Ill., January 25, 1993) 522 n.47
Midwest Envtl. Consulting & Remediation Serv., Inc. v. People's Bank of Bloomington, 251 Ill. App. 3d 256, 620 N.E.2d 469 (4th Dist. 1993) 523 n.49
Mike M. Johnson, Inc. v. County of Spokane, 78 P.3d 161 (Wash. 2003) 456 n.102
Miller v. Big River Concrete, LLC, 14 S.W.3d 129 (Mo. App. 2000) 658 n.96
Miller v. Bourgoin, 613 A.2d 292 (Conn. App. Ct. 1992) 388 n.81
Miller v. DeWitt, 37 Ill. 2d 273, 226 N.E.2d 630 (1967) 177–179, 420 n.40
Miller v. United States Steel Corp., 902 F.2d 573 (7th Cir. 1990) 649 n.78–79
Miller Elevator Co. v. United States, 30 Fed. Cl. 662 (Fed. Cl. 1994) 432 n.3, 439 n.23
Milton Constr. Co. v. Alabama Highway Dept., 568 So. 2d 784 (Ala. 1990) 307 n.48
Milwaukee County v. H. Neidner & Co., 263 N.W. 468 (Wis. 1936) 390 n.90
Mingus Constructors, Inc. v. United States, 812 F.2d 1387 (Fed. Cir. 1987) 391 n.97, 441 n.39
Mississippi Meadows, Inc. v. Hodson, 299 N.E.2d 359 (3d Dist. 1973) 151 n.30
Missler v. Anne Arundel County, 314 A.2d 451 (Md. 1974) 117 n.42–43
Mojave Enters. v. United States, 3 Cl. Ct. 353 (1983) 197 n.36
Money v. Fort Hayes State Univ. Endowment Ass'n., 64 P.3d 458 (Kan. App. 2003) 116 n.35
Montgomery Indus. Int'l. Inc. v. Thomas Constr. Co., 620 F.2d 91 (5th Cir. 1980) 254 n.26
Montgomery Ward & Co., Inc. v. Dalton, 665 S.W.2d 507 (Tex. App.–El Paso 1983) 122–123
Montrose Contracting Co. v. County of Westchester, 80 F.2d 841 (2d Cir. 1936) 47 n.39
Moores Bros. Co. v. Brown & Root, Inc., 207 F.3d 717 (4th Cir. 2000) 393 n.107, 394 n.11, 12
Moorman Mfg. Co. v. Nat'l Tank Co., 435 N.E.2d 443 (Ill. 1982) 271 n.79, 663 n.114
Moransais v. Heathman, 744 So. 2d 973 (Fla. 1999) 623 n.41, 625 n.46, 659 n.103
Morganti Nat'l, Inc. v. United States, 49 Fed. Cl. 110 (2001), *aff'd,* 36 Fed. Appx. 452 (Fed. Cir. 2002) 335
Morrison-Knudsen Co., Inc. v. United States, 184 Ct. Cl. 661 (1968) 482
Morrison Knudsen Corp. v. Fireman's Fund Ins. Co., 175 F.3d 1221 (10th Cir. 1999) 454 n.90
Morris v. Osmose Wood Preserving, Inc., 667 A.2d 624 (Md. 1995) 622 n.35
Morse/Diesel, Inc. v. Trinity Indus., Inc., 655 F. Supp. 346 (S.D.N.Y. 1987) 619 n.12
Moses H. Cone Memorial Hosp. v. Mercury Constr. Corp., 460 U.S. 1, 103 S. Ct. 927, 74 L. Ed. 2d 765 (1983) 9 n.24

Moulding-Brownell Corp. v. E. C. Delfosse Constr. Co., 291 Ill. App. 343, 9 N.E.2d 459 (1st. Dist. 1937) 513 n.1, 514 n.3
Mounds View, City of, v. Walijarvi, 263 N.W.2d 420 (Minn. 1978) 150 n.29
Moundsview Ind. School Dist. No. 621 v. Buetow & Assocs., 253 N.W.2d 836 (Minn. 1977) 357, 358, 359, 595–597
MTB Group, Inc. v. United States, 65 Fed. Cl. 516 (2005) 216
Muchisky v. Frederic Roofing Co., 838 S.W.2d 74 (Mo. Ct. App. 1992) 618 n.8
Mudgett v. Marshall, 574 A.2d 867 (Me. 1990) 617 n.5
Mullen v. Town of Louisburg, 225 N.C. 53, 33 S.E.2d 484 (1945) 203 n.73
Murphy v. City of Springfield, 738 S.W.2d 521 (Mo. Ct. App. 1987) 271 n.80
Murphy v. E. R. Squibb & Sons, Inc., 710 P.2d 247 (Cal. 1985) 621 n.26
Mustang Pipeline Company, Inc. v. Driver Pipeline Company, Inc. 134 S.W.3d 195 (Texas 2004) 501–505
Myers & Chapman, Inc. v. Thomas G. Evans, Inc., 374 S.E.2d 385 (N.C. 1988) 375 n.21
Myrtle Beach Pipeline Corp. v. Emerson Elec. Co., 843 F. Supp. 1027 (D.S.C. 1993) 630

Nastri v. Wood Bros. Homes, Inc., 690 P.2d 158 (Ariz. App. 1984) 650 n.85
Natelli; United States v., 527 F.2d 311 (2d Cir. 1975), *cert. denied*, 425 U.S. 934 (1976) 620 n.21
National Am. Ins. Co. v. Certain Underwriters at Lloyd's London, 93 F.3d 529 (9th Cir. 1996) 441 n.37
National Presto Indus., Inc. v. United States, 338 F.2d 99 (Ct. Cl. 1964) 466 n.23
National Union Fire Ins. Co. of Pittsburgh, PA v. CBI Indus., Inc., 907 S.W.2d 517 (Tex. 1995) 117
National Union Fire Ins. Co. v. Pratt & Whitney, 815 P.2d 601 (Nev. 1991) 622 n.34, 644 n.68
National Union Indem. Co. v. G. E. Bass & Co., 369 F.2d 75 (5th Cir. 1966) 377 n.28, 385 n.66
Neal & Co. Inc. v. United States, 36 Fed. Cl. 600 (1996) 325 n.20, 329 n.30, 474 n.45
Neece v. AAA. Realty Co., 322 S.W.2d 597 (1959) 114 n.23
Nelson Co., *In re*, 959 F.2d 1260 (3d Cir. 1992) 114 n.20
Nelson v. Commonwealth, 368 S.E.2d 239 (N.E. 1988) 140 n.20
Nesdahl Surveying & Eng'g v. Ackerland Corp., 507 N.W.2d 686 (N.D. 1993) 518–519
Netherland v. Winner, 662 S.W.2d 786 (Tex. App.-Houston [14th Dist.] 1983) 119 n.60
Nether Providence Twp. Sch. Auth. v. Thomas M. Durkin & Sons, Inc., 476 A.2d 904 (Pa. 1984) 438 n.22
New England Structures, Inc. v. Loranger, 234 N.E.2d 888 (Mass. 1968) 506–509
NewMech Companies, Inc. v. Indep. Sch. Dist. No. 206, 558 N.W.2d 22 (Minn. Ct. App. 1997) 206 n.101
New Pueblo Constructors, Inc. v. State, 144 Ariz. 95, 696 P.2d 185 (1985) 691 n.54
Newsom v. United States, 676 F.2d 647 (Ct. Cl. 1982) 447 n.59
New York, City of, v. Aetna Casualty & Surety Co., 1997 WL 379704 at *3 (S.D.N.Y. 1997) 552 n.8
Nolte v. Hudson Nav. Co., 16 F.2d 182 (9th Cir. 1926) 120 n.71, 121 n.74
Nolte v. Smith, 189 Cal. App. 2d 140 (Ct. App. 1961) 519 n.30

Northern Contracting, Inc. v. Illinois, 2004 WL 422704 (N.D. Ill. March 3, 2004) 211 n.135
Northern Corp. v. Chugach Elec. Ass'n, 518 P.2d 76 (Alaska 1974) 297 n.25
Northern Petrochemical Co. v. Thorsen & Thorshov, Inc., 211 N.W.2d 159 (Minn. 1973) 671
Northridge Co. v. W. R. Grace and Co., 471 N.W.2d 179 (Wis. 1991) 645 n.69
Northrop Grumman Corp. v. United States, 47 Fed. Cl. 20 (2000) 466 n.23
North Slope Tech. Ltd., Inc. v. United States, 14 Cl. Ct. 242 (1988) 197 n.35, 198, 478 n.61
Nortz v. United States, 294 U.S. 317 (1935) 273
Nota Constr. Corp. v. Keyes Assoc., Inc., 694 N.E.2d 401 (Mass App. 1998) 665 n.122
Nowicki v. Cannon Steel Erection Co., 711 N.E.2d 536 (Ind. Ct. App. 1999) 411 n.26

Ocean Ridge Dev. Corp. v. Quality Plastering, Inc., 247 So. 2d 72 (Fla. 4th DCA 1971) 299–300
Ohio Crane Co. v. Hicks, 110 Ohio St. 168, 143, N.E. 388 (1924) 46
Old Dominion Elec. Coop. v. Ragnar Benson, Inc., 2006 U.S Dist. LEXIS 56145 (E. Dist. Va. 2006) 337 n.49
Olympic Products Co. v. Roof Sys., Inc., 363 S.E.2d 367 (N.C. App. 1988) 654 n.87
Olympus Corp. v. United States, 98 F.3d 1314 (1996) 473 n.41
1000 Virginia Ltd. Partnership v. Vertecs 146 P.3d 423 (Wash. 2006) 608–610
Orshoski v. Krieger, No. OT-01-009, 2001 WL 1388037 (Ohio Ct. App. Nov. 9, 2001), *appeal denied,* 763 N.E.2d 1185 (Ohio 2002) 620 n.20
Ossining Union Free Sch. Dist. v. Anderson LaRocca Anderson, 539 N.E.2d 91 (N.Y. 1989) 664 n.117–118
Otinger v. Water Works & Sanitary Sewer Bd., 177 So. 2d 320 (Ala. 1965) 297 n.26
Owens-Corning Fiberglas Corp. v. United States, 419 F.2d 439, 190 Ct. Cl. 211 (1969) 275

Pacific Alaska Contractors, Inc. v. United States, 436 F.2d 461 (Ct. Cl. 1971) 473 n.37
Pacific Island Movers, B-287643.2 (Comp. Gen. 2001) 216 n.155
Palmer v. Brown, 273 P.2d 306 (Cal. Dist. Ct. App. 1954) 376 n.26
Palmetto Linen Serv., Inc. v. U.N.X., Inc., 205 F.3d 126 (4th Cir. 2000) 631 n.54, 639 n.64, 654 n.88
Paralyzed Veterans of Am. v. Ellerbe Becket Architects & Eng'rs, 945 F. Supp. 1 (D.D.C. 1996) 138 n.14
Parker v. Panama City, 151 So. 2d 469 (Fla. Dist. Ct. App. 1963) 192 n.11
Pathman Construction Co., ASBCA No. 23392, 85-2 B.C.A. ¶ 18,096 (1985) 283 n.9
Paul Hardeman, Inc. v. Arkansas Power & Light Co., 380 F. Supp. 298 (E.D. Ark. 1974) 11 n.30
Paul L. Whitfield, P.A. v. Gilchrist, 348 N.C. 39, 497 S.E.2d 412 (1998) 676 n.18
Paul M. Spillman v. American Homes of Mocksville, Inc. 422 S.E.2d 740 (N.C. App. 1992) 631–632
Pavel v. A. S. Johnson, 674 A.2d 521 (Md. App. 1996) 254 n.29
Pawnee, Village of, v. Azzarelli Constr. Co., 539 N.E.2d 895 (Ill App. Ct. 1989) 304 n.34
Paxton v. Alameda County, 259 P.2d 934 (Cal. App. 1st Dist. 1953) 151 n.30

PCL Constr. Services., Inc. v. United States, 47 Fed. Cl. 745 (2000) 48 n.40, 42
Pearlman v. Reliance Ins. Co., 371 U.S. 132, 83 S.Ct. 232, 9 L. Ed. 2d 190 (1962) 378–379, 553–556
Peck v. Horrocks Eng'rs, Inc. 106 F.3d 949 (10th Cir., 1997) 420–423
Pecos & N.T. Ry. Co. v. Amarillo St. Ry. Co., 171 S.W. 1103 (Tex. Civ. App.–Amarillo 1914, no writ) 150
Peerless Ins. Co. v. Cerny & Assoc., Inc., 199 F. Supp. 951 (D. Minn. 1961) 386 n.70
Pembroke Villas of Broward, Inc. v. Raymundo, 447 So. 2d 324 (Fla. Dist. Ct. App. 1984) 514 n.5
Penn Cent. Transp. Co. v. City of New York, 438 U.S. 104, 98 S. Ct. 2646 (1978) 139 n.19
Pennsylvania Dep't. of Transp. v. Jones D. Morrisey, Inc., 682 A.2d 9 (Pa. Comm. Ct. 1996) 45 n.31
Pepi Corp. v. Galliford, 254 S.W.3d 457, 2007 WL 441582 (Tex. Ct. App. 2007) 668 n.1
Perini Corp. v. Greate Bay Hotel & Casino, Inc., 610 A.2d 364 (N.J. 1992), *abrogated on other grounds,* Tretina Printing, Inc. v. Fitzpatrick & Assocs. Inc., 640 A.2d 788 (N.J. 1994) 305 n.38
Perini Corp. v. United States, 180 Ct. Cl. 768 (1967) 481–482
Peter Kiewit & Sons Co., ENG BCA No. 4861 85-2 BCA ¶ 18,082, 474 n.43
Peter Kiewit Sons Co. v. Summit Constr. Co., 422 F.2d 242 (8th Cir. 1969) 451 n.77, 452 n.84
Peterson v. Ramsey County, 563 N.W.2d 103 (N.D. 1997) 392 n.100
PetroChem Services, Inc. v. United States, 837 F.2d 1076 (Fed. Cir. 1988) 198 n.44–46
PGA Tour, Inc. v. Martin, 531 U.S. 1049, 121 S.Ct. 1879, 149 L. Ed. 2d 904 (2001) 138 n.12
Phoenix Bridge Co. v. United States, 211 U.S. 188, 44
Pincelli v. Ohio Bridge Corp., 213 N.E.2d 356 (Ohio 1966) 207 n.103
Pipe Welding Supply v. Haskell, Connor & Frost, 462 N.Y.S.2d 221 (1984) 166 n.31
P. J. Maffei Bldg. Wrecking Corp. v. United States, 732 F.2d 913 (Fed. Cir. 1984) 479 n.68
Ports Authority v. Roofing Co., 294 N.C. 73, 240 S.E.2d 345 (1978) 632
Post Bros. Constr. Co. v. Yoder, 569 P.2d 133 (Cal. 1977) 384 n.63
Prairie State Nat'l Bank v. United States, 164 U.S. 227, 17 S. Ct. 142, 41 L. Ed. 412 (1896) 378 n.37, 387 n.71, 555–556
Preload Tech. v. A.B & J. Constr. Co., 696 F.2d 1080 (5th Cir. 1983) 254 n.26, 255 n.31
Prestige Ford v. Ford Dealer Computer Services, 324 F.3d 391 (5th Cir. 2003) 576 n.13
Price v. Neal-Millard Co., 124 Ga. 884 (1906) 517 n.18
Priebe & Sons v. United States, 332 U.S. 407 (1947) 306 n.42
Pritchard Bros., Inc. v. Grady Co., 428 N.W.2d 391 (Minn. 1988) 659 n.103
Private Mortgage Inv. Servs., Inc. v. Hotel & Club Assocs., Inc., 296 F.3d 308 (4th Cir. 2002) 620 n.17
Providence Washington Ins. Co. v. Beck, 255 N.E.2d 600 (Mass. 1970) 381 n.52
Prudential Ins. Co. v. Stratton, 685 S.W.2d 818 (Ark. Ct. App. 1985), *supplemented,* 690 S.W.2d 750 (1985) 302 n.32
P.T. & L. Constr. Co., Inc. v. State of New Jersey, 531 A.2d 1330 (N.J. 1987) 484

Pulte Home Corp. v. Osmose Wood Preserving, Inc., 60 F.3d 734 (11th Cir. 1995) 622 n.35, 639–643

Pulte Home Corp. v. Parex, Inc., 579 S.E.2d 188, 2003 WL 1903424 (Va., April 17, 2003) 659 n.98

PYCA Indus., Inc. v. Harrison County Waste Water Mgmt. Dist., 177 F.3d 351 (5th Cir. 1999) 442 n.46

Queensboro Steel Corp. v. East Coast Mach. & Iron Works, Inc., 346 S.E.2d 248 (N.C. Ct. App. 1986) 523 n.48

RaDec Constr. Inc. v. Sch. Dist., No. 17 of Douglas County, 535 N.W.2d 408 (Neb. 1995) 445 n.48

Ralph M. Parsons Co., ASBCA No. 24347, 85-1 BCA ¶ 17,787; *aff'd on reconsid.*, 85-2 BCA ¶ 18,112, 602–603

Ramada Dev. Co. v. Rauch, 644 F.2d 1097 (5th Cir. 1981) 299

Randa/Madison Joint Venture III v. Dahlberg, 239 F.3d 1264 (Fed. Cir. 2001) 475 n.47, 48

Raymond v. Baehr, 163 N.W.2d 51 (Minn. 1968) 146 n.25

Raynor Steel Erection v. York Constr. Co., 351 S.E.2d 136 (N.C. 1986) 654 n.87

R. B. Wright Constr. Co. Through Rembrant, Inc. v. United States, 919 F.2d 1569 (Fed. Cir. 1990) 49 n.48

R. C. Freeman v. Shannon Constr., Inc., 560 S.W.2d 732 (Tex. Civ. App. 1977) 523 n.49

RDP Royal Palm Hotel, L.P. v. Clark Constr. Group, Inc., 168 Fed. Appx. 348; 2006 U.S. App. LEXIS 3815, 285–289, 290

R. E. Gaddie, Inc. v. Price, 528 S.W.2d 708 (Ky. 1975) 411 n.27

Reich v. Simpson, Gumpertz & Heger, Inc., 3 F.3d 1 (1st Cir. 1993) 404 n.13

Reilly v. Rangers Mgmt., Inc., 727 S.W.2d 527 (Tex. 1987) 114

Reliance Ins. Co. v. Utah Dept. of Transp., 858 P.2d 1363 (Utah 1993) 309

Reno Elec. Works v. Ward, 51 Nev. 291 (1929) 524 n.57

Ressler v. Nielsen, 76 N.W.2d 157 (N.D. 1956) 151 n.30

R. H. Macy & Co. v. Williams Tile & Terrazzo, Inc., 585 F. Supp. 175, 180 (N.D. Ga. 1984) 659 n.101

Rhone Poulenc Rorer Pharm., Inc. v. Newman Glass Works, 112 F.3d 695 (3d Cir. 1997) 47 n.37, 49, 605–606

Richards v. Powercraft Homes, Inc., 678 P.2d 427 (Ariz. 1984) 582 n.1

Richmond, City of, v. J. A. Croson Co., 488 U.S. 469, 109 S. Ct. 1842 (1989) 208–209, 210, 211

Richmond Homes Mgmt. v. Raintree, Inc., 862 F. Supp. 1517 (W.D.Va. 994), *aff'd in part and rev'd in part,* 66 F.3d 316, 1995 WL 551274 (4th Cir. 1995) 172

Richmond v. Peterson, 2006 Neb. App. LEXIS 132 (Ct. App. 2006) 518 n.25

Rissler & McMurray Co. v. Sheridan Area Water Supply Joint Powers Bd., 929 P.2d 1228 (Wyo. 1996) 658 n.96

Ritchies Food Distrib., Inc. v. Refrigerated Constr. Serv., 2004 WL 957659 (Ohio Ct. App. 2004) 112 n.15

Robert R. Jones Assoc. v. Nino Homes, 858 F.2d 274 (6th Cir. 1988) 171

Roberts & Schaefer Co. v. Hardaway Co., 152 F.3d 1283 (11th Cir. 1998), *reh'g denied,* 162 F.3d 1179 (11th Cir. 1998) 281 n.4
Robert W. Carlstrom Co., Inc. v. German Evangelical Lutheran, 662 N.W.2d 168 (Minn. Ct. App. 2003) 480 n.71
Rocky Mountain Fire & Casualty Co. v. Biddulph Oldsmobile, 640 P.2d 851 (Ariz. 1982) 645 n.70
Roger H. Proulx & Co. v. Crest Liners, Inc., 98 Cal. App. 4th 182 (2002) 649 n.79
Rogers & Rogers; United States v., 161 F. Supp. 132 (S. D. Cal. 1958) 348 n.16
Romanac, *In re,* 245 F. Supp. 882 (W.D. Va. 1965), *aff'd,* 386 F.2d 225 (4th Cir. 1967) 516 n.13
Roscoe-Ajax Constr. Co. Inc. & Knickerbocker Constr. Corp. v. United States, 458 F.2d 55 (Ct. Cl. 1972) 474 n.46, 475 n.49
Rosenberg v. Town of North Bergen, 293 A.2d 662 (N.J. 1972) 607 n.44
Rosenstein v. Standard & Poor's Corp., 636 N.E.2d 665 (Ill. App. 1993) 656 n.94, 663 n.116
Rosen v. Bureau of Prof'l and Occupational Affairs, State Architects Licensure Bd., 763 A.2d 962 (Pa. Commonwealth 2000) 136 n.4
Ross Eng'g. Co. v. United States, 92 Ct. Cl. 253, 1940 WL 4077 (Ct. Cl. 1940) 284 n.11
Royal Hawaiian Movers, B-288653 (Comp. Gen. 2001) 216 n.155
Rozny v. Marnul, 250 N.E.2d 656 (Ill. 1969) 620 n.20, 663 n.114
R & P Enters. v. LaGuarta, Gavrel & Kirk, Inc., 596 S.W.2d 517 (Tex. 1980) 114 n.19, 25
RPR & Assoc., Inc. v. The Univ. of North Carolina, 153 N.C. App. 342, 570 S.E.2d 510 (2002) 315 n.9
RPR & Assoc. v. O'Brien/Atkins Assoc., P.A., 921 F. Supp. 1457 (M.D. N.C. 1995) 360 n.33, 366–367
Ryan v. Morgan Spear Assoc., Inc., 546 S.W.2d 678 (Tex. Civ. App. 1977) 148–150, 151 n.30, 376 n.27

S. & M./Traylor Bros. (JV), ENG BCA Nos. 3878, et al., 82-1 BCA ¶ 15,484, 479 n.67
Sacramento Regional Transit Dist. v. Grumman Flexible, 158 Cal. App. 3d 289 (1984) 625 n.49
Saddler v. United States, 287 F.2d 411 (Ct. Cl. 382, 390 (1961)) 451 n.79
SAE/American-Mid Atlantic, Inc., GSBCA Nos. 12,294, et al., 98-2 BCA (CCH) ¶ 30,084 (1998) 327 n.25, 337 n.48
Safeco Credit v. United States, 44 Fed. Cl. 406 (1999) 455 n.94
Safeway Stores, Inc. v. Certainteed Corp., 710 S.W.2d 544 (Tex. 1986) 607 n.44
Sandarac Assoc. v. W. R. Frizzell Architects, 609 So. 2d 1349 (Fla. Dist. Ct. App. 1992), *overruled by* Moransais v. Heathman, 744 So.2d 973 (Fla.1999) 623 n.41
Sanders Co. Plumbing & Heating, Inc. v. City of Independence, 694 S.W.2d 841 (Mo. Ct. App. 1985) 661 n.108, 665 n.119
Santa Fe, Inc., VABCA No. 1983, 84-3 BCA ¶ 17,538, 480 n.69
Santa Fe, Inc., VABCA No. 2168, 87-3 B.C.A. (CCH) ¶ 20,104 (1987) 328 n.27
Santa Fe Engineers, Inc., ASBCA Nos. 24,578, et al., 94-2 B.C.A. (CCH) ¶ 26,872 (1994) 320–321, 326 n.24, 336 n.46
Santa Fe Eng'rs, Inc. v. United States, 801 F.2d 379 (Fed. Cir. 1986) 456 n.100
Saratoga Fishing Co. v. J. M. Martinac & Co., 520 U.S. 875 (1997) 645 n.69

Sasso Contracting Co. Inc. v. State of New Jersey, 173 N.J. Super. 486 (App. Div. 1980), *cert. denied*, 85 N.J. 101 (1980) 483
Saturn Constr. Co. v. Conn. Dept. of Pub. Works, 1994 WL 590604 (Conn. Super. Ct. 1994) 306 n.44
Schwartz v. Kuhn, 71 Misc 149, 358
Scott & Fetzer Co. v. Montgomery Ward & Co., 493 N.E.2d 1022 (Ill. 1986) 645–646
Scott v. Matlack, Inc., 39 P.3d 1160 (Col. 2002) 405–408, 411 n.27
Scott v. Strickland, 10 Kan. App. 2d 14 (Ct. App. 1984) 520 n.34
Scribante v. Edwards, 181 P. 75 (Cal. Ct. App. 1919) 347 n.11
Scully v. Otis Elevator Co., 275 N.E.2d 905 (Ill. App. Ct. 1971) 411 n.28
Seaman U.S.D. No. 345 v. Casson Constr. Co., 3 Kan. App. 2d 289, 594 P.2d 241 (1979) 144
Sears, Roebuck & Co. v. Enco Assocs., 43 N.Y.2d 389 (1977) 358
Secord v. Willow Ridge Stables, Inc., 690 N.Y.S.2d 375 (N.Y. App. Div. 1999, *mem. op.*) 419 n.39
Seely v. White Motor Co., 63 Cal. 2d 9, 45 Cal. Rptr. 17, 403 P.2d 145 (1965) 620 n.24, 621, 622 n.33, 36, 633–634, 657
Segari v. Uchello, 44 So. 2d 722 (La. Ct. App. 1950) 115 n.28
Semida v. Rice, 863 F.2d 1156 (4th Cir. 1988) 620 n.21
Sensenbrenner v. Rust, Orling & Neal Architects, Inc., 374 S.E.2d 55 (Va. 1988) 650 n.84
Servidone Constr. Corp. v. United States, 19 Cl. Ct. 346 (1990) 474 n.44, 476 n.56
Servidone Constr. Corp. v. United States, 931 F.2d 860 (Fed. Cir. 1991) 335 n.44
Servidone Constr. Corp. v. U.S., 931 F.2d 860, 37 Cont. Cas. Fed. (CCH) ¶ 76082 (Fed. Cir. 1991) 691 n.52
Severin v. United States, 99 Ct. Cl. 435 (1943) 272 n.82, 273–275
Seward Housing Corp. v. Conroy Bros. Co., 573 N.W.2d 364 (1998) 265 n.62
Sharp v. Coopers & Lybrand, 83 F.R.D. 343 (E.D. Pa. 1974) 620 n.21
Shea Co. v. United States, 4 Cl. Ct. 46 (1983) 198 n.40
Shea v. Keuffel & Esser, 668 F. Supp. 41 (D. Mass. 1986), *aff'd* 823 F.2d 543 (1st Cir. 1987) 607 n.44
Sheetz, Aiken & Aiken, Inc. v. Spann, Hall, Ritchie, Inc., 512 So. 2d 99 (Ala. 1987) 360 n.36
Shepard v. City of Palatka, 414 So. 2d 1077 (Fla 5th DCA 1981) 357
Sherman R. Smoot Corp., ASBCA No. 52261, 03-1 B.C.A (CCH) ¶ 32,197 (2003) 339–340
Shively v. Belleville Township High Sch. Dist. No. 201, 329 Ill. App. 3d 1156, 769 N.E.2d 1062 (2002) 61 n.7
Shoffner Indus., Inc. v. W. B. Lloyd Constr. Co., 257 S.E.2d 50 (N.C. App. 1979) 659 n.103
S & H Riggers & Erectors, Inc. v. Occupational Safety & Health Comm'n, 659 F.2d 1273 (5th Cir. 1981) 408 n.17
Sierra v. Garcia, 106 N.M. 573, 746 P.2d 1105 (1987) 698, 699
Simpson v. United States, 172 U.S. 372 (1899) 44, 462
Sims v. Lewis, 374 So. 2d 298 (Ala. 1979) 588 n.15
Sipco Serv. Marine, Inc. v. Wyatt Field Serv. Co., 857 S.W.2d 602 (Tex. App. 1993) 254 n.28, 255 n.31

S. J. Amoroso Constr. Co. v. United States, 981 F.2d 1073 (9th Cir. 1992) 201 n.58, 202 n.61
Skidmore, Owings & Merrill, 5 BNA OSHC 1762 (OSHRC 1977) 404 n.13
Skidmore, Owings & Merrill v. Intrawest I L.P., 87 Wash. App. 1054 (Div. 1 1997) 78 n.23
Skyland Developers, Inc. v. Sky Harbor Assocs., 586 S.W.2d 564 (Tex. Civ. App.–Corpus Christi 1979, no writ) 118 n.55
S & M Constructors, Inc. v. Columbus, 70 Ohio St. 2d 69, 434 N.E.2d 1349 (1982) 46
Smith; United States v., 256 U.S. 11 (1921) 464 n.13, 16
Smith v. Goff, 1958 OK 100, 325 P.2d 1061 (Okla. 1958) 151 n.30
Smith v. Seaboard Coast Line R.R. Co., 639 F.2d 1235 (5th Cir. 1981) 265 n.60
Smullian & Blumenthal, P.A. v. Futz, 762 A.2d 582 (Md. 2000) 620 n.21
Soletanche Rodio Nicholson (JV) ENG BCA Nos. 5796 and 5891, 94-1 BCA ¶ 26,472, 477 n.57
Southland Corp. v. Keating, 465 U.S. 1 (1984) 573 n.5
Southland Enterprises, Inc. v. United States, 24 Ct. Cl. 596 (1991) 456 n.99
Southwestern Bell Tel. v. PUC, 467 F.3d 418 (5th Cir. 2006) 114 n.21
Southwest Marine, Inc., ASBCA No. 36854, 95-1 BCA ¶ 27,601 (1995) 335–336
Souza & McCue Constr. Co., Inc. v. Superior Court of San Benito County, 370 P.2d 338 (Cal. 1962) 464 n.14, 16
Sowders v. M.W. Kellogg Co., 663 S.W.2d 644 (Tex. Ct. App. 1983) 611 n.55
Spearin; United States v., 248 U.S. 132, 39 S. Ct. 59 (1918) 43–49, 147–148, 461 n.2, 463–464
Spring Motors Distrib. v. Ford Motor Co., 489 A.2d 660 (N.J. 1985) 639 n.62
St. Joseph Hosp. v. Corbetta Constr. Co., Inc., 316 N.E.2d 51 (Ill. App. 1974) 146 n.25, 591–594, 695 n.69
St. Paul Fire & Marine Ins. Co. v. City of Green River 93 F. Supp. 2d 1170 (D. Wyo. 2000) 545–551
St. Paul Fire & Marine Ins. Co. v. Pearson Constr. Co., 547 N.E.2d 853 (Ind. 4th Dist. 1989) 48 n.41
St. Paul Fire & Marine Ins. Co. v. United States, 370 F.2d 870 (5th Cir. 1967) 552 n.8
Stabler Constr., Inc. v. Comm. of Pennsylvania, 692 A.2d 1150 (Pa. 1997) 46
Standard Constructors, Inc. v. Chevron Chem. Co., 101 S.W.3d 619 (Tex. App.–Houston [1st Dist.] 2003) 117–118
State Farm & Cas. Co. v. Rodriguez, 88 S.W.3d 313 (Tex. App.–San Antonio 2002, *pet. denied*) 538
State Mechanical Contractors, Inc. v. Village of Pleasant Hill, 132 Ill. App. 3d 1027, 87 Ill. Dec. 532, 477 N.E.2d 509 (Ill. App. 1985) 206 n.98
State v. *See Name of opposing party*
Stearman v. Centex Homes, 78 Cal. App. 4th 611 (2000) 621 n.26, 645 n.69
Stevens Constr. Corp. v. Carolina Corp., 217 N.W.2d 291 (Wis. 1974) 390 n.87, 90
Stewart v. C & C Excavating & Constr. Co., 877 F.2d 711 (8th Cir. 1989) 381 n.49
Stitt v. Royal Park Fashions, Inc., 546 S.W.2d 924 (Tex. Civ. App.–Dallas 1977, *writ ref'd n.r.e.*) 118 n.55
Stock & Grove, Inc. v. United States, 204 Ct. Cl. 103 (1974) 472 n.36, 473 n.42
Stoeckert v. United States, 391 F.2d 639 (Ct. Cl. 1968) 449 n.68

Stone's Throw Condominium Ass'n, Inc. v. Sand Cove Apartments, Inc., 749 So. 2d 520 (Fla App. 1999) 659 n.103
Stoney v. Franklin, 44 UCC Rep. Serv. 2d 1211 (Va. Cir. Ct. 2001) 639 n.62
Stout, Hall & Bangs v. United States, 27 C. Cls. 385 (1892) 274
Stovall v. Reliance Ins. Co., 107 P.3d 1219 (KS 2005) 694 n.64, 65
Strickland Tower Maintenance, Inc. v. AT&T Communications, Inc., 128 F.3d 1422 (10th Cir. 1997) 456 n.98
Struebing Constr. Co. v. Golub-Lake Shore Place Corp., 281 Ill. App. 2d 689, 666 N.E.2d 846 (1st Dist. 1996) 515 n.11
Sturdza v. United Arab Emirates, 281 F.3d 1287 (D.C. Cir. 2002) 172 n.37
Stuyvesant Dredging Co. v. United States, 834 F.2d 1576 (Fed. Cir. 1987) 604 n.40
Subsurfco, Inc. v. B-Y Water Dist., 337 N.W.2d 448 (S.D. 1983) 304 n.34
Suit-Kote Corp. v. City of Binghamton Bd. of Contract & Supply, 628 N.Y.S.2d 861, 216 A.D.2d 831 (N.Y. App. Div. 1995) 203 n.75
Sulem v. B.T.R. East Greenbush Inc., 589 N.Y.S.2d 969 (N.Y. App. Div. 1992) 411 n.26
Sullivan Indus., Inc. v. Double Seal Glass Co., Inc., 480 N.W.2d 623 (Mich. App. 1991) 639 n.63
Summit Custom Homes, Inc. v. Great Am. Lloyds Ins. Co. No. 05-05-00851-CV (Tex. Civ. App., July 18, 2006) 536–541
Sundstrom v. New York, 213 N.Y. 68, 44
Sunnyslope Grading, Inc. v. Bradford, Miller, & Risberg, Inc., 437 N.W.2d 213 (Wis. 1989) 625 n.49
Sun Valley Water Beds v. Herm Hughes & Son, Inc., 782 P.2d 188 (Utah 1989) 607 n.45
Sweet Co. v. White County Bridge Comm'n, 714 N.E.2d 219 (Ind. Ct. App. 1999) 445 n.48

Tabasso Homes Inc.; State v., 42 Del. 110 (1942) 516 n.16
Tandy Corp. v. Malone & Hyde, Inc., 769 F.2d 362 (6th Cir. 1985) 612 n.62
Taylor & Polk Constr., Inc., United States *ex rel* v. Mill Valley Constr., Inc., 29 F.3d 154, 161 (4th Cir. 1994) 296 n.20
TDC Mgmt. Corp.; United States v., 24 F.3d 292, 298 (D.C. Cir. 1994) 396
Temple-Eastex, Inc. v. Addison Bank, 672 S.W.2d 793 (Tex. 1984) 119, 121–122
Tharp v. Keeter/Schaefer Investments, L.P., 943 S.W.2d 811 (Mo. Ct. App. 1997) 383 n.59
Theis v. Heuer, 280 N.E.2d 300 (Ind. 1972) 588 n.15
ThermoCor, Inc. v. United States, 35 Fed. Cl. 480 (1992) 483 n.81
Thomas Shine v. David M. Childs and Skidmore Owings & Merrill, LLP, 382 F. Supp. 602 (S.D.N.Y. 2005) 172–174
Thompson v. Waters, 526 S.E.2d 650 (N.C. 2000) 620 n.18
Three Affiliated Tribes of the Fort Berthold Reservation v. Wold Eng'g, P.E., 419 N.W.2d 920 (N.D. 1988) 140 n.20
Tishman Constr. Corp. v. City of New York, 228 A.D.2d 292, 643 N.Y.S.2d 589 (1st Dep't 1996) 499–500
Titan Pac. Constr. Corp. v. United States, 17 Cl. Ct. 630, 1989 WL 78828 (1989), *aff'd* 899 F.2d 1227 (Fed. Cir. 1990) 297 n.23, 336 n.47

Tolan and Son, Inc. v. KLLM Architects, Inc., 719 N.E.2d 288 (Ill. App. 1999) 663 n.115
Tommy L. Griffin Plumbing & Heating Co. v. Jordan, Jones & Goulding, Inc., 463 S.E.2d 85 (S.C. 1995) 652, 659 n.103
Topco, Inc. v. State of Montana, 912 P.2d 805 (Mont. 1996) 271–272
Tourist Village Motel, Inc. v. Massachusetts Engineering Co., 801 F. Supp. 903 (D. N.H. 1992) 630 n.52
Trafalgar House Constr., Inc. v. United States, 73 Fed Cl. 675 (2006) 686 n.50
Tretina Printing, Inc. v. Fitzpatrick & Assocs. Inc., 640 A.2d 788 (N.J. 1994) 305 n.38
Triad Elec. & Controls, Inc. v. Power Sys. Eng'g., Inc., 117 F.3d 180 (5th Cir. 1997) 118 n.52
Tri-County Materials, Inc., *In re,* 114 BR 160 (C.D. Ill. 1990) 524 n.57
Tufano Contracting Corp. v. State, 269 N.Y.S.2d 564 (N.Y. App. Div. 1966) 452 n.81
Turnkey Enterprises Inc. v. United States, 597 F.2d 750 (Ct. Cl. 1979) 473 n.38
Tuttle/White Constructors, Inc. v. Montgomery Elevator Co., 385 So. 2d 98 (Fla. Dist. Ct. App. 1980) 289, 392 n.104
2-J Corp. v. Tice, 126 F.3d 539 (3d Cir. 1997) 645 n.70
2314 Lincoln Park West Condominium Ass'n v. Mann, Gin, Ebel & Frazier, Ltd. 555 N.E.2d 346 (Ill. 1990) 655–656, 663 n.115
Tyger Constr. Co. v. United States, 31 Fed. Cl. 177 (1994) 47 n.39

Ullmann v. May, 147 Ohio St. 468, 72 N.E. 63 (1947) 46
Umpqua River Navigation Co. v. Crescent City Harbor Dist., 618 F.2d 588 (9th Cir. 1980) 661 n.108
United Air Lines, Inc. v. CEI Indus. of Illinois, Inc., 499 N.E.2d 558 (Ill. App. 1986) 645 n.69
United Contractors v. United States, 177 Ct. Cl. 151 (1966) 483 n.81
United Pac. Ins. Co., ASBCA No.53051, 2003-2 B.C.A. (CCH) ¶ 32,267, 315 n.8
United States v. *See Name of opposing party*
U.S. Indus., Inc. v. Blake Constr. Co., 671 F.2d 539 (D.C. Cir. 1982) 307 n.51
U.S. Steel Corp. v. Missouri Pac. R.R. Co., 668 F.2d 435 (8th Cir. 1982) 308 n.54
Utah, Nevada and California Stage Co.; Utah v., 199 U.S. 414, 26 S. Ct. 69, 50 L. Ed. 251 (1905) 44, 197 n.37
Utley-James, Inc. GSBCA No. 5370, 85-1 B.C.A. (CCH) ¶ 17,816 (1985) 321–323, 325 n.20

Valhal Corp. v. Sullivan Associates, Inc. 44 F.3d 195 (3rd Cir.1995) 180–184, 698, 699
Value Group, Inc. v. Mendham Lake Estates, L.P., 800 F. Supp. 1228 (D.N.J. 1992) 172
Vann v. United States, 190 Ct. Cl. 546 (1970) 478 n.63
Vermont Marble Co. v. Baltimore Contractors, Inc., 520 F. Supp. 922 (D.D.C. 1981) 523 n.51, 52, 53
Victory Constr. Co. & Paul Krummel (JV) v. United States, 206 Ct. Cl. 274 (Ct. Cl. 1975) 481 n.74

Village of Cross Keys, Inc. v. U.S. Gypsum Co., 556 A.2d 1126 (Md. App. 1989) 665 n.123
Vincent, Missouri *ex rel,* v. Schneider, 194 S.W.3d 853 (Mo. 2006) 116 n.40
Vintage Constr., Inc. v. State Dept. of Transp., 713 P.2d 1213 (Alaska 1986) 207 n.108
Virginia Sch. of Arts, Inc. v. Eichelbaum, 493 S.E.2d 510 (Va. 1997) 254 n.25
Vulcan Painters, Inc. v. MCI Contractors, Inc., 41 F.3d 1457 (11th Cir. 1995) 456 n.98
V&V Cement Contractors, Inc. v. LaSalle Nat'l Bank, 119 Ill. App. 3d 154, 456 N.E.2d 655 (1st Dist. 1983) 517 n.24

W. Oil Fields, Inc. v. Pennzoil United, Inc., 421 F.2d 387 (5th Cir. 1970) 122 n.82
Waggoner v. W & W Steel Co. 657 P.2d 147 (Okla. 1982) 350–353
Walker v. Chancey, 117 So. 705 (Fla. 1928) 394 n.109
Ward v. Ernst & Young, 435 S.E.2d 628 (Va. 1993) 658 n.98
Wargo Builders, Inc. v. Douglas L. Cox Plumbing & Heating, Inc., 268 N.E.2d 597 (Ohio App. 1971) 254 n.26
Washington Constr. Co. v. Spinella, 84 A.2d 617 (N.J. 1951) 114 n.22
Watson, Watson, Rutland/Architects, Inc. v. Montgomery County Bd. of Educ., 559 So. 2d 168 (Ala. 1990) 598 n.24
Watson Electrical Constr. Co. v. Summit Companies, LLC, 160 N.C. App. 647 S.E.2d 87, 92 (N.C. App. 2003) 676 n.18
Wausau Tile, Inc. v. County Concrete Corp. 593 N.W.2d 445 (Wis. 1999) 626–630, 650 n.82
W. B. Constr. v. Mountains Community Hosp. Dist., 2005 Cal. App. unpub. LEXIS 5124, 12005 WL 1385571, Cal. Ct. App. 4th App. Dist., June 13, 2005 334 n.39
Weaver-Bailey Contractors, Inc. v. United States, 19 Cl. Ct. 474 (1990) 317 n.13, 326 n.24
Webb Electric Co., ASBCA No. 54293, 2007-02 BCA (CCH) 33,717 (2007) 336 n.46
Weber v. Pascarella Mason Street, LLC, 930 A.2d 779 (Conn. App. Ct. 2007) 115 n.29, 116 n.37, 38
Weekley v. Weekley, 27 S.E.2d 591 (W.Va. 1943) 114 n.19
Weeks Dredging & Contracting v. United States, 13 Cl. Ct. 193 (1987), *aff'd* 861 F.2d 728 (Fed. Cir. 1988) 472 n.34, 475 n.47, 476, 478 n.64
Wegroup PC/Architects and Planners v. State, 885 P.2d 709 (Or. 1994) 441 n.41
Wells Bros. Co. of New York v. United States, 254 U.S. 83 (1920) 307 n.50
Westchester v. United States, 28 Fed. Cl. 172 (1993) 455 n.95
WestFair Elec. Contractors v. Aetna Cas. & Surety Co., 661 N.E.2d 967 (N.Y. 1995) 268 n.69
Westinghouse Elec. Corp. v. New York City Transit Auth., 623 N.E.2d 531 (N.Y. 1993) 364 n.53
W. F. Mangann v. Diamond Mfg. Co., Inc., 580 F. Supp. 1299 (D.C.S.C. 1984), *rev'd in part on other grounds* 198 n.38
Whitesell-Green, Inc., ASBCA Nos. 53938, et al., 2006-2 B.C.A. (CCH) ¶ 33,323 (2006) 326 n.24, 332 n.33
Widett v. U.S. Fidelity and Guar. Co., 815 F.2d 885 (2d Cir. 1987) 658 n.95
Wilko v. Swan, 346 U.S. 427, 74 S. Ct. 182 (1953) 577 n.17

Willamette Crushing Co. v. Arizona Dept. of Transp., 932 P.2d 1350 (Ariz. Ct. App. 1997) 284 n.13
William Mushero, Inc. v. Hull, 667 A.2d 853 (Me. 1995) 677 n.21
William R. Clarke Corp. v. Safeco Ins. Co. of Am., 938 P.2d 372 (Cal. 1997) 268 n.69
Williams and Sons Erectors, Inc. v. South Carolina Steel Corp., 983 F.2d 1176 (2d Cir. 1993) 456 n.99
Williams & Assocs. v. Ramsey Prod. Corp., 19 N.C. App. 1, 198 S.E.2d 67, *cert. denied,* 284 N.C. 125, 199 S.E.2d 664 (1973) 163
Williams Enters, Inc. v. Strait Mfg. & Welding, Inc., 728 F.Supp. 12 (D.C. 1990), *aff'd,* 938 F.2d 230 (D.C. Cir. 1991) 315 n.8
Williams v. United States, 127 F. Supp. 617 (Ct. Cl. 1978) 440 n.31
Willms Trucking Co. v. J. W. Constr. Co., 442 S.E.2d 197 (S.C. Ct. App. 1994) 456 n.98
Wilson v. Dryvit Sys., Inc., 206 F. Supp. 2d 749 (E.D.N.C. 2002) 623 n.38, 648 n.76
Wisch & Vaughan Constr. Co. v. Melrose Properties, 21 S.W. 3d 36 (Mo. App. S.D. 2000) 432 n.2
Womack v. United States, 182 Ct. Cl. 399 (1968) 482 n.78
Woodson v. Martin, 663 So. 2d 1327 (Fla.2d DCA 1995) 624 n.43
World Trade Company v. Westinghouse Elec. Corp., 682 N.Y.S.2d 385 (1998) 621 n.32
WRB Corp. v. United States, 193 Ct. Cl. 409 (1968) 439 n.24
Wunderlich Contracting Co. v. United States, 351 F.2d 956 (Ct. Cl. 1965) 451 n.76
Wunderlich v. California, 423 P.2d 545 (Cal. 1967) 489
W. V. Nelson Constr. Co. v. City of Lindstrom, 565 N.W.2d 434 (Minn. Ct. App. 1997) 192 n.14
Wygant v. Jackson Bd. of Educ., 476 U.S. 267, 106 S. Ct. 1842 (1986) 209 n.119

Yankee Candle Co. v. New England Candle Co., 14 F. Supp. 2d 154 (D. Mass. 1998) 170–171
York Fed. Sav. & Loan Ass'n v. William A. Hazel, Inc., 256 Va. 598 (Va. 1998) 527 n.75
Youngdale & Sons Constr. Co. v. United States, 27 Fed. Cl. 516 (1993) 335 n.42, 467 n.25, 474 n.44, 476 n.56
Young v. Oak Leaf Builders, Inc., 626 S.E.2d 240 (Ga. Ct. App. 2006) 588 n.14

Zenda Grain & Supply Co. v. Farmland Indus., Inc., 894 P.2d 881 (Kan. App. 1995) 264 n.55
Zurn Constructors, Inc. v. Castaic Lake Water Agency, 2003 Cal. App. Unpub. LEXIS 11267 (2d App. Dist, Div. 3, 2003) 335 n.43

Index

AAA (American Arbitration Association), 571, 573–575
Academic scholarship, 14–15
Acceptance
  of payment, 391–392
  of work by owner, 390–391
Accessibility, under ADA, 138
Accidents
  prevention programs, 402–409
  risk allocation and transfer, 423–429
  tort liability, 411–423
  workers' compensation laws, 410
Accord and satisfaction doctrine, 391, 392, 455–456
Accountants, 620, 656
Active negligence, 411
Activity logic, 315–316
Acts of God, 291
Actual authority, 438, 439
ADA (Americans with Disabilities Act), 138
Additional services, of architect, 159–160
Additional work, defined, 432. *See also* Changes, to contract
ADM (arrow diagramming method), 317–318
Administration of contracts. *See* Contract administration
Advertisement for bids, 194–195
Advisory opinions, 570–571
Aesthetics, 355
Affirmative action, 208–211
AGC (Associated General Contractors of America), 108, 267, 469
Agency Construction Management, 71–74
Agency law, 346–347
Agreements. *See* Contracts; Design agreements
AIA (American Institute of Architects). *See* American Institute of Architects (AIA)
Air rights, 36
Alabama, statute of limitations for tort claims, 610 n.49
Alaska, archaeological and historic sites, 40
Alliances, in integrated project delivery, 88–89
Allowances, 241
Alternative dispute resolution, 13–14
Alternative payees, 384
Ambiguity, in contracts, 113–123
American Arbitration Association (AAA), 571, 573–575
American Bar Association Forum on the Construction Industry, 15
American College of Construction Lawyers, 15
American Institute of Architects (AIA)
  acceptance of payment as waiver of claims, 391
  change orders, 434–437, 440, 443, 449
  contract administration, 344
  contract document summary, 104–108

American Institute of Architects (AIA), *continued*
differing site conditions clause, 469, 470, 480
extensions for delays, 291
flow-down provisions, 257–258
founding, 4
General Conditions of the Contract for Construction, 39, 42, 65 n.12, 105
insurance requirements, 532–535
integrated project delivery, 88 n.35
mediation, 570
Standard Form of Agreement Between Contractor and Subcontractor, 257, 261, 263–264
Standard Form of Agreement Between Owner and Architect, 43, 51, 165
Standard Form of Agreement Between Owner and Contractor, 569
termination of contract, 495–498, 500
timing of payments, 379
Uniform Contract (1888), 5, 8
American Subcontractors Association, 269
Americans with Disabilities Act (ADA), 138
Ancient history, 1–3
Anti-indemnity statutes, 264–265
Apparent authority, 438, 439
Applications for payments, 373–376
Appraisers, 620
Approval procedures, 349
Arbitration
architect's role in, 366–367
attacks on right of, 575–576
binding, 571
confirmation or vacation of award by court, 576–577
contractual provision, 573–575
history of, 8–9
modification of award by arbitrators, 576
nonbinding informal summary, 571
place of, 574
principles of, 572–578
statutory basis, 572–573
use of, 14
waiver of, 575–576
Archaeological sites, 40–41
Architects and architecture. *See also* Contract administration; Design agreements
certificates of payment, 374, 375–376
as clerk of the works, 345
compensation, 159–160, 175–176
contract references, 124–125
delays by, 295
in design-bid-construct system, 64–69
in design-build system, 80, 81, 83–84
liability of, 140–151, 386, 419–423
as profession, 4
programming, 50–51
quasi-judicial role immunity, 366–367
role of, 20–21, 616–617
services, 156–164
subcontractors, 21–22, 134–135
Architectural Works Copyright Protection Act (AWCPA), 172
Arizona, lien waivers in, 528
Arkansas
retainage, 377 n.31
statute of repose, 612 n.58, 59
Army Corps of Engineers, 296, 561, 565
Arrow diagramming method (ADM), 317–318
Asbestos, 42, 522
Associated General Contractors of America (AGC), 108, 267, 469
Associated Soil and Foundation Engineers, 184
Assumption-of-risk doctrine, 428
At-risk construction management, 59 n.3, 74–78, 212–213
Attorneys, 620, 656
Audit rights, 243–244
Augustus, 2
Australia, alliance approach, 89
Authority, to issue changes, 437–440

Baby Brooks Acts, 193
Bankruptcy, 295, 378–379, 529, 553–556
Bar charts, 312–314, 333–334
Baseline schedules, 327–330
Basic services, of architect, 159–160

Beneficiaries of contract, 123–124
Benefit of the bargain principle, 669, 670–671, 672, 674
Best value awards, 213–215
Betterment, 672, 695
Bids and bidding. *See also* Competitive bidding
  contingencies for differing site conditions, 467–471
  escrow of documents, 567
  by subcontractors, 252–256
Bid shopping, 255–256
Bonds. *See also* Surety bonds
  to indemnify owner against liens, 389
  lien, 543
  in lieu of retainage, 377
  sale of, 26
Bonuses, for subcontractors, 564
Bonus/penalty provisions, 306–307
Borrow pits, 473, 482
BOT (build-operate-transfer) method, 86–87
Breach of contract
  cure, 506–510
  damages, 669–677
  for defects, 585–588, 590–594, 597–598
  by design professionals, 166
  for failure to perform changed work, 449–450
  as grounds for termination of contract, 500–505
  for late completion, 304–305
  for negligent performance, 631
  for owner's failure to make payment, 379, 387
  statute of limitations, 607
  and substantial completion, 298
  for wrongful termination, 494, 510
Breach of fiduciary duty, 167
Bridging consultants, 83–84
Brooks Act, 192–193
Budgets
  designing to, 161–170
  owner's program projection, 51
  purpose articulated in contract, 243
Builder rebates/credits, 243
Builders' risk insurance, 541
Building codes
  design professional's failure to comply with, 146–147
  farm buildings, 138
  "green" codes, 139
  historic preservation, 139
  history of, 7, 136–137
  model codes, 137
  socioeconomic, 138–139
  types, 29
Building inspectors, negligence of, 620
Building sites, 35–42. *See also* Differing site conditions; Inspections; Safety
  access to, 36–37
  archaeological and historic sites, 40–41
  environmental issues, 41–42
  owners' right to use property for construction, 37
  title requirements, 37–38
  utilities, 39
  zoning, 38–39
Build-operate-transfer (BOT) method, 86–87
Burial sites, 40
Business risk doctrine, 539
Business Rountable, 560–561, 563
"But for" analysis, 334–335
Buy American Act, 251

California
  differing site conditions, 488–490
  mechanic's liens, 526
  public procurement, 193 n.22, 194 n.30, 212 n.136–137
Cancellation, of contract, 505. *See also* Termination, of contract
Cardinal changes, 450–452
Care, standard of for design professionals, 140–151
Cause of action, 668
CERCLA (Comprehensive Environmental Response Compensation and Liability Act) (1980), 41
Certificate of occupancy, 301–302
Certificates, payment-related, 374–376, 388, 389, 393

CGL (commercial general liability) insurance, 424, 532, 536–541, 542
Changed conditions. *See* Differing site conditions
Change orders
  architect's role in, 362
  completion date, 446
  disputes over breadth, 454–456
  notice requirements for claims, 440–442
  pricing, 443–446
  priority of, 112
  process, 434–437
Changes, to contract, 431–458
  administration, 452–454
  authority to issue, 437–440
  cardinal, 450–452
  constructive, 432–433, 446–448
  contract clauses, 240, 432–433, 437–445, 449
  disputes over, 387, 437–438, 440–442, 449–450, 454–456
  documents, 111
  duty to perform, 449–450
  inevitability of, 431–432
  pricing, 221, 235, 443–446, 567
  reasons for, 387
Changes, to design agreement, 160
Civil litigation, OSHA in, 406–408
Civil projects, 84–85, 135, 222
Claims. *See also* Defective or deficient construction; Dispute resolution
  architect's role in, 363–366
  for changes, 440–442, 449–452
  CPM proof of delay, 332–341
  in design-bid-construct arrangements, 68–69
  false, 395–396
  and final completion, 304
  and multiple prime method, 71
  by subcontractors, 269–276
  by third-party beneficiaries, 123–124
  waivers of, 390–392, 455, 456
Claims industry, 693
Clean Water Act, 42
Clerk of the works, 345
Closed bidding projects, 190
CMAA (Construction Management Association of America), contract forms, 109
Codes. *See* Building codes
Collapsed as-built analysis, 334–335
Collateral assignment, 124
Colorado
  fraudulent liens, 521 n.42
  mechanic's liens, 516 n.14, 527 n.73
Commencement, time of, 283–284
Commercial general liability (CGL) insurance, 424, 532, 536–541, 542
Commercial impracticability, 297
Common law
  assumption-of-risk doctrine, 428
  breach of contract, 500
  current status, 618–619
  differing site conditions, 460, 461–467
  discovery, 574
  English source, 4 n.7
  indemnity, 262, 429
  liens, 514
  risk allocation, 307, 460
  safety of workers, 177, 401, 415
  standard of care, 140–147
  time for performance, 279–280
  unforeseen changes in costs, 238
Communication, 391, 565–566
Comparative negligence, 423, 424, 428, 429
Compensation. *See* Payments
Compensatory damages
  alternative methods of proving, 686–692
  direct methods of proving, 678–686
Competitive bidding
  advertisement for bids, 194–195
  architect's role, 158
  bid errors, 207–208
  bid package, 195–198
  bid preparation techniques, 199
  bid protests, 204–207
  bid responsiveness, 200–201
  bid security, 207
  duty to award to lowest bidder, 199–200

hiring preferences, 208–211
history of, 7
owner's right to rebid, 200
private projects, 190
responsible bidders, 202–204
unbalanced bidding, 234–235
waiver of bidding irregularities, 201–202
Completion. *See also* Time for completion
certificates of, 362–363
final, 302–304, 362–363, 389–393
late, 304–307
substantial, 297–302, 362–363, 376, 387–389
Component manufacturers, 616–617, 625–630
Comprehensive Environmental Response Compensation and Liability Act (CERCLA) (1980), 41
Concealment, 465–466
Conflict Prevention & Resolution (CPR) Rules, 571
Conflicts, within construction contract, 112–113
ConsensusDOCS, 108, 469
Consequential damages, 179, 619, 692, 700
Consideration, 505
Construction documents phase, 158
Construction industry
complexity, 615–618
standards and studies, 684–685
statistics, 9
Construction Industry Institute, 60 n.6, 560–561, 563
Construction law, generally
complexity of, 10–11, 617–618
defined, 12 n.31
history, 1–11
nature of, 11–12
scholarship, 14–15
vs. tort law, 615
Construction lenders. *See* Lenders
Construction Management Association of America (CMAA), contract forms, 109
Construction management method, 60, 71–78, 212–213
Construction managers, 620, 660–665, 672
Construction phase, 158–159
Constructive changes, 432–433, 446–448
Consultants, 21–22, 83–84, 133, 134–135, 359
Consumer protection statutes, 589
Contextual contract principles, 6–7
Contingencies, 241–243
Contingent payment provisions, 267–268, 393–394
Contract administration, 343–368
agency law principles, 346–347
applications for payment, 373–376
architect's role generally, 344–346
change orders, 362
claims process, 363–366
coordination, 359–360
improper administration claims, 635–638, 655–660
introduction, 343–344
and OSHA requirements, 404
progress payments, 360–361
quality control, 349–359
reject or stop work authority, 361
requests for information, interpretations, and clarifications, 348–349
substantial/final completion certification, 362–363
Contractors. *See also* General contractors; Subcontractors
acceptance of payment, 391–392
coordination of subcontractors, 260–261
damages, 673–674
design process, 151–152
duty to read/understand project documents, 48
insurance, 532
liability of, 381, 386–387, 412, 415–419
negligent misrepresentation, 660–665
role of, 22–24, 133
schedule, 51–52
termination of, 364–366

termination of contract for cause, 494–496
waivers of claims, 391–392

Contracts, 97–129
ambiguity in, 113–123
at-risk construction management, 78
breach of (*See* Breach of contract)
changes (*See* Changes, to contract)
commencement of services or work prior, 99–100
defined, 98
design-bid-construct approach, 66
design-build approach, 83
design document ownership, 174–175
dispute resolution system, 571–572
documents, 110–111, 112
enforceability, 99
and estimates, 165–170
flow down clause, 112
force majeure clause, 238 n.22
forms for, 102–109
hold harmless clause, 696
implied, 100–101
indemnification clause, 424–428, 695–699
integration clause, 111
internal conflicts, 112–113
interpretation, 113–123
lenders' input, 53
limitations on liability clauses, 696–699
mechanic's lien requirements, 517–518, 520
merger/entire agreement clause, 116
modifications/changes, 111, 112
no-damage-for-delay clause, 307–308, 700–701
oral, 99
parties to, 123–125
privity of, 271, 607, 618, 625–638
purpose, 97–98
quasi, 101
scope of, 125–128
sharing-of-savings clause, 240
standard of care clause, 147
statute of limitations modification, 610–611
termination (*See* Termination, of contract)
variations in estimated quantities (VEQ) clause, 232–234, 481–483

Control of the work, 412, 413, 416–419
Cooperation, 563–566
Coordination, 259–262, 359–360
Copyright Act (1976), 170–171
Copyright protection, design documents, 170–174
Cost of work, 224–231, 241, 443–445, 452–454
Cost overuns, 166, 199, 231, 241–243
Cost-plus contracts, 224–231, 240, 243, 244, 371
Cost-plus-fee contracts, 91
Cost-plus with guaranteed maximum price contracts, 91, 231–232, 371
Cost(s)
of changed work, 443–445, 452–454
and cost-plus contracts, 224–231
defined in contracts, 371
estimations, 164–170, 199
project delivery system issues, 61–62
proof/estimates when proving damages, 678–683
Course of performance, 120–121
CPM (critical path method) schedules, 280–282, 315–323, 328, 332–341
CPR (Conflict Prevention & Resolution) Rules, 571
Critical path method (CPM) schedules, 280–282, 315–323, 328, 332–341
Cure, 506–510, 589
Damages, 667–702. *See also* Economic loss rule
alternative methods of proving, 686–692
for breach of contract, 669–677
consequential, 179, 619, 692, 700
direct methods of proving, 678–686
for false claims, 395
introduction, 667–668
for late completion, 304–307
limitations, 620, 695–701
mitigation, 693–695
punitive, 511, 692–693

for tort claims, 677–678
for wrongful termination of contract, 511

Damage to other property test, 643–644, 650, 654
DART (Dispute Avoidance and Resolution Task Force), 561–562
Daubert test, 338–340
DBEs (disadvantaged business enterprises), 211 n.132, 251
DBIA (Design-Build Institute of America), contract forms, 109
Deceptive trade practices, 589
Default pricing, 237
Defaults, contract, 493–494. *See also* Termination, of contract
Defective construction products or components, 625–630, 639–649
Defective design professional services, 635–638, 655–660
Defective or deficient construction, 581–614
claims against contractors, 585–590
defenses to liability, 607–612
defined, 582
design professional's liability, 355–359, 590–599
economic loss rule, 630–635, 649–655
latent defects, 390
manifestations vs. defects, 583–584
and owner's final payment, 390–391
patent defects, 390
responsibility for, 584–585
Spearin doctrine, 43–49, 147–148, 599–607
by subcontractor, 395
types of, 582–583
Delaware, pay-if-paid clauses, 268 n.73
Delays
contract specifications, 326–327
CPM proof of delay claims, 332–341
damages for, 304–307, 388, 392–393, 678–686, 700–701
grounds for, 290–297
no-damage-for-delay clauses, 307–308, 700–701
Deletion of work, 433
Demolition, 522
Derivative suits, 528
Design agreements. *See also* Contracts
AIA Standard Form of Agreement Between Owner and Architect, 43, 51, 165
architect's compensation, 175–176
architect's services scope, 156–161
and budgets, 161–170
components/scope, 109–110, 125
design document ownership, 170–175
importance of, 155–156
insurance, 176
liability limitations, 176–185
termination, 185–186
Design-Bid-Construct method, 64–71. *See also* Competitive bidding
design professional's role, 590
drawings and specifications, 586–587
history, 7, 58
overview, 60, 64–65
payments, 370
risk factors, 67–69
selection factors, 65–67
variations, 69–71
Design-Build Institute of America (DBIA), contract forms, 109
Design-Build method, 60, 78–85, 211–212
Design development phase, 157–158
Design documents, ownership of, 170–175. *See also* Design drawings; Design specifications
Design drawings
ambiguities in, 446–447
contractor's failure to comply with, 586–587
contractor's requests for information, 348–349
copyright issues, 170–175
defects, 591–594, 600–607
mechanic's liens, 522
owner's implied warranty of, 43–49, 600–603, 605
pricing considerations, 239–240
responsibility for inaccuracies or omissions, 126

Design process, 131–153. *See also* Design professionals
contractor's implementation, 151–152
contractual relationships, 132–134
owner's role in, 133–134
phases of, 157–159
Design professionals. *See also* Architects and architecture; Engineers and engineering
breach of contract by, 166
and building codes, 136–139
damages for, 672
defective or deficient construction claims, 590–599
defective services claims, 635–638, 655–660
indemnity, 265
insurance, 176, 532, 542
liability of, 140–151, 176–185, 347, 360, 361, 366–367
negligent misrepresentation, 660–665
public projects, 192–194
regulation/licensure, 136
role and responsibilities, 132–133, 590
Design quality, 62
Design specifications
ambiguities in, 446–447
contractor's failure to comply, 586–587
copyright issues, 170–175
defects, 591–594, 600–607
owner's implied warranty of, 43–49, 600–603, 605
vs. performance specifications, 152
pricing considerations, 239–240
responsibility for inaccuracies or omissions, 126
Differing site conditions, 459–492
common law tradition, 460, 461–467
contract clauses for addressing, 467–471
defined, 459–460
Geotechnical Design Summary Report, 566–567
limitations on claims, 477–483
state survey, 483–490
Type I/II claims, 471–477
Direct costs, 679
Disabled persons, accessibility under ADA, 138
Disadvantaged business enterprises (DBEs), 211 n.132, 251
Disclaimers
differing site conditions, 479, 490
liability limitations, 176–179
owner's implied warranty to supply accurate information, 49
Discovery, 574–575, 608
Dispute resolution, 559–579. *See also* Arbitration
architect's role in, 363–364, 366–367
binding resolution stage, 563
construction industry's unique needs, 559
default pricing through, 237
dispute control stage, 562, 566–569
facilitated nonbinding resolution stage, 563, 569–571
history of, 8–9, 560–561
prevention and cooperation stage, 562, 563–566
pricing of changes, 444–445
step chart, 561–562
system for, 571–572
Dispute review or resolution boards (DRBs), 567–568, 569
Doctrine of Conditions, 373
Doctrine of Sanctity of Contract, 5
Drafters, of contracts, 119–120
Drawings. *See* Design drawings
Draw requests, 373–376
DRBs (dispute review or resolution boards), 567–568, 569

Early start letters, 100
Easements, 36–37, 525
Economic Loss Rule, 615–666
application of, 620–622
exceptions to, 598
introduction, 615–620
negligent misrepresentation application, 660–665
parties in privity of contract, 625–638
defective construction products/ components, 625–630

defective construction services, 630–635
defective professional/testing services, 635–638
parties not in privity of contract, 638–660
defective construction products/ components, 639–649
defective construction services, 649–655
defective professional/testing services, 655–660
scope and purpose of, 622–625
and subcontractor's recovery against owners, 271
Economic waste, 694
Eichleay formula, 681–682
Eight corners rule, 537
EJCDC (Engineers' Joint Contract Documents Committee), 108–109, 469
Employer status, under OSHA, 403–404
Engineer-Procure-Construct (EPC) projects, 84–85
Engineers and engineering
in civil projects, 135–136
consultants, 135, 359
contract references, 124–125
design development, 158
geotechnical, 133–134, 184
liability of, 419–423
as profession, 4
role of, 21–22
Engineers' Joint Contract Documents Committee (EJCDC), 108–109, 469
English common law, 4
Entire agreement clause, 116
Enumeration of documents, 112
Environmental consultants, 133
Environmental issues, 41–42, 139
Environmental Protection Agency (EPA), 42
EPC (engineer-procure-construct) projects, 84–85
Equal Protection Clause, 209
Equipment, 225–226, 445, 680
Equitable adjustments, 445–446, 453–454
Equitable allocation of payments, 524
Equitable liens, 397–398
Equitable subrogation, 378–379, 553
Errors and omissions (E&O), 542
Escrow, 377–376, 567
Estimates, 164–170, 199, 230–231, 252. *See also* Pricing
Estoppel, 253–254, 441–442, 674–675
Exculpatory provisions, 176–179, 479, 696 n.71
Executive Order No. 10925 (1961), 208
Expectation interest, 669–674
Expert witnesses, 12–13, 143–146, 338–341, 693
Express changes, 432–433
Express warranties, 49, 248 n.3, 374–375, 587, 605
Extra work, 432, 449–454. *See also* Changes, to contract

Fact-finders, 100, 114
False claims, 395–396
Farm buildings, 138
FARs (Federal Acquisition Regulations). *See* Federal Acquisition Regulations (FARs)
Fast track construction, 61
Federal Acquisition Regulations (FARs)
change orders, 434–437, 440, 445, 449, 453–454
competitive bidding, 251
differing site conditions, 468–469
fixed-price contracts, 461
flow-down provisions, 259
ratification principle, 440
variation in estimated quantity (VEQ), 233, 481
Federal Antiquities Act, 40
Federal Arbitration Act (1925), 8, 573, 574, 577
Federal contracts. *See also* Public procurement
bid contingencies, 467–468
changes, 437, 438–439, 441, 445–446
defaults (1893), 5–6
delays, grounds for, 291–292
OSHA requirements, 409

Federal contracts, *continued*
  ratification, 440
  variations in estimated quantities (VEQ) clauses, 232–234, 481–483
Federal False Claims Act (1863), 395–396
Federal Prompt Act, 394–395
Federal Rules of Evidence, 338–339
Fees. *See* Payments
Fee-simple interests, 524–525
FIDIC (International Federation of Consulting Engineers), 470–471
Fiduciary relationships, 371
Fill areas, 473
Final completion, 302–304, 362–363, 389–393
Financial difficulties, 295–296
Financing, 25–27, 52–55
Fire codes, 7, 137
Fixed fees, 229
Fixed-price contracts, 67–68, 370, 461
Fixtures, 523–524
Float time, 316–317, 326
Florida
  bid protests, 205
  conditional payments, 393 n.107
  fraudulent liens, 521 n.37, 41
  implied liens, 528 n.80
  improvement of property, 518 n.26
  lien waivers, 383 n.61
  public procurement, 193 n.22, 25
  statute of repose, 612 n.58, 59
  value-based contracting, 214 n.147
Flow-down obligations and rights, 112, 256–259
Force majeure clauses, 238 n.22
Foreseeability, and negligent misrepresentation, 660–665
Forms, contracts, 102–109
Fraud, 395–396
Fraudulent liens, 521
Front loading, 372–373

GAI (general agreement of indemnity), 386–387
Gantt charts, 312–314
General agreement of indemnity (GAI), 386–387
General contractors
  contractual obligations, 22–23
  in Design-Bid-Construct system, 64–69
  estimates for subcontractor work, 252
  liability of, 412, 415–419
  negligent misrepresentation, 660–665
  role in project, 22–24
  safety program responsibility, 409
General liability insurance, 424, 532, 536–541, 542
Geotechnical engineers, 133–134, 184
Geotechnical information, 43, 566–567
Government contracts. *See* Public contracts and entities
Government regulation, 29, 136, 532–533
Greenfield projects, 39
Ground leases, 38
Guaranteed maximum price
  at-risk construction, 77–78
  cost-plus contracts with, 91, 231–232, 371
  and fast track construction, 61
  negotiated design-build contracts, 83

Hammurabi's Code, 2
Hazardous waste, 42
Heard Act (1894), 5–6
Historic sites, 40–41, 139
History, of construction law, 1–11
Hold harmless clauses, 696
Holmes, Oliver Wendell, 6 n.14, 11
Home office overhead, 680–682
Human remains, 40

Idaho, statute of repose, 612 n.58
Idle equipment damages, 680
IDMs (initial decision makers), 569, 570
Illinois
  archaeological sites, 40
  bid protests, 205
  differing site conditions, 487–488
  Entry on Adjoining Land to Accomplish Repair Act, 36 n.4
  liens, 513
  lien waivers, 528 n.79

mechanic's liens, 515–516, 522, 526, 527
no-lien contracts, 529 n.85
pay-if-paid clauses, 268 n.74
professional registration laws, 4
public project set-asides for minorities, 211
Structural Work Act, 420 n.41
Impacted as-planned CPM approach, 336
Implied authority, 439
Implied contracts, 100–101
Implied warranties
breach of, 587–588
in construction generally, 147–148
of design professionals, 148–151
of drawings and specifications, 43–49, 463–464, 600–603, 605
of habitability, 588
of owner to supply accurate information, 43–49, 600–603, 605
under UCC, 248 n.3
Improper furnishing of materials, 630–635, 649–655
Improvement, 518–519
Incidental damages, 700
Incorporation by reference, 112, 257, 586
Indemnification clauses, 424–428, 695–699
Indemnity
agreements, 552
bonds, 389
subcontractor obligations, 262–265
surety's claims for, 386
Index pricing, 236
Indiana
construction bonds/security pledges in lieu of retainage, 377 n.32
mechanic's liens, 522 n.44, 526 n.70
Indications, under differing site conditions clause, 472–473, 488
Industrial projects, 84–85
Industry customs, 93–94, 121–122
Initial decision makers (IDMs), 569, 570
Inspections
by architect, 355–359, 389, 595
constructive changes arising out of overly stringent, 447–448
contractor's duty, 477–478
by lender, 396–397
by owner, 386
Virginia public contracts for highway work, 486
Inspectors, negligence of, 620
Insurance
builders' risk, 541
contractual requirements, 532–535
definitions, 531
general liability, 424, 532, 536–541, 542
policy types, 27
professional liability, 176, 542
role of, 531
state regulation, 532–533
subguard, 557
title, 28–29
waivers of subrogation, 541–542
workers' compensation, 410, 428–429, 535
Insurance companies, role in project, 27
Integrated product analysis, 644–645, 654
Integrated project delivery, 88–89
Integration clause, 111
Intent, in contract interpretation, 116–118
Interest, accrual for late payments, 387
Interested party, defined, 204
Interests
expectation, 669–674
fee-simple, 524–525
leasehold, 525
reliance, 674–675
restitution, 675–676
International Federation of Consulting Engineers (FIDIC), 470–471
International projects, 470–471
Internet-based bidding, 215–216
Investigations, contractor's duty, 477–478
Investment properties, 19

Jesus of Nazareth, 2 n.4
Johnson, Samuel, 6 n.16
Joint checks, 383–384
Joint payees, 384

Joint ventures, 80
Jury verdict method, 691–692

Kansas
  pay-if-paid clauses, 268 n.74
  public procurement, 193 n.25, 194
Knowledge, superior, 465–466

Labor costs, 228, 679–680
Laches, 612
Latent ambiguities, 447
Latent defects, 390
Leadership in Energy and Environmental Design (LEED), 159
Lean project delivery, 89 n.38
Leasehold interests, 525
LEED (Leadership in Energy and Environmental Design), 159
Legal relationships, 12
Legislation. *See* Statutes
Lenders
  collateral assignment, 124
  contract references, 124–125
  financing process, 52–53
  inspections, 396–397
  liability of, 396–398
  progress payments, 360
  role in project, 25–27
Lessees, 38
Letters of credit, 28, 329
Letters of intent, 100
Liability
  for accidents, 411–423, 429
  of architects, 140–151, 386, 419–423
  of contractors, 381, 386–387, 412, 415–419
  of design professionals, 140–151, 176–185, 347, 360, 361, 366–367
  of engineers, 419–423
  of lenders, 396–398
  limitations of, 176–185, 696–699
  of owners, 347, 385–386, 412–415
Liability insurance, 424, 532, 536–541, 542
Licensing, 136
Lien bonds, 543
Liens. *See also* Mechanic's liens
  defined, 513
  on funds by subcontractors/suppliers, 382
  against undisbursed loan proceeds, 397–398
Lien waivers, 53, 363, 382–383, 391, 528–529
Limitations on liability clauses, 696–699
Liquidated delay damages, 305–307, 388, 392–393, 700–701
Liquidating agreements, 275–276
Loans, 25–26, 52–53, 396–398
Loss of use, 305
Lost profits, 682–683
Louisiana
  public procurement, 193 n.24, 194 n.30, 212 n.138
  statute of limitations for tort claims, 610 n.48
  statute of repose, 612 n.57, 58, 60
  value-based contracting, 214 n.148
Lump sum pricing, 91, 443

Magnuson-Moss Act, 618 n.8
Maine, statute of repose, 612 n.57
Manifestation rule, 537–538
Manifestations, vs. defects, 583–584
Manufacturers. *See also* Suppliers
  approval by architect, 349
  designer's role in selection, 616–617
Maryland
  bid protests, 205
  mechanic's liens, 514
  public procurement, 193 n.22, 194
Masons, 649 n.81
Massachusetts
  performance and payment bonds, 543 n.3–4
  statute of repose, 612 n.60
Master builders, 58
Material breach
  contractor's refusal to perform, 449
  as grounds for termination of contract, 500–505
  owner's failure to make payment, 379, 387
  and substantial completion, 298
  wrongful termination, 494, 510

Material costs, 679–680
MBEs (minority business enterprises), 208–209, 251
MCAA (Mechanical Contractors Association of America), 684–685
McCarran Ferguson Act, 532 n.1
Measured mile approach, 683–684
Mechanical Contractors Association of America (MCAA), 684–685
Mechanic's liens, 513–530
  amount of, 520–521
  and bankruptcy, 529
  categories, 516–517
  history, 513–514
  materials and labor subject to, 522–524
  overview, 381
  parties entitled to, 521–522
  perfection, 526
  priorities, 526–527
  property interests subject to, 524–525
  purposes, 514–515
  requirements, 517–522
  statutory framework, 5, 515–516
  waiver and release, 528–529
Mediation, 14, 569–570
Merger clause, 116
Michigan, mechanic's liens, 526 n.67
Miller Act (1935), 6
Minitrials, 570
Minority and women business enterprises (MWBEs), 209–211, 251
Minority business enterprises (MBEs), 208–209, 251
Misrepresentation, 167, 464–465, 619–620
Mississippi, notice provisions for claims for changes, 442
Mistakes
  in bidding process, 207–208
  and recovery of costs in differing site conditions, 466–467
Mitigation of damages, 693–695
Model Procurement Code for State and Local Governments, 204–205, 205
Modifications. *See* Changes, to contract
Modified total cost method, 690–691
Mortgage liens, 527, 529
Multiple primes, 69–70, 72
Municipal projects, 525. *See also* Public contracts and entities
Mutual mistakes, 466–467
MWBEs (minority and women business enterprises), 209–211, 251

National Association of Builders, 5, 8
National Historic Preservation Act (NHPA), 41
National Register of Historic Places, 41
Nebraska, bid protests, 205
Negligence
  active, 411
  comparative, 423, 424, 428, 429
  damages, 619–620, 677–678
  design professional's liability, 140–151, 166, 360
  partial negligence statutes, 265
  sole negligence laws, 265
Negligent misrepresentation, 619–620, 660–665
Negotiation, 158, 566
Network analysis, 280–282, 315–323, 328, 332–341
New Jersey
  bid protests, 205
  differing site conditions, 483–485
  multiple primes, 69 n.19
New Mexico, indemnification clauses, 696 n.73
Newspapers, advertisement for bids, 194–195
New York
  Access to Adjoining Property to Make Improvements or Repairs Act, 36 n.4
  claims for changes, 442, 450
  indemnity prohibition, 428
  multiple primes, 69 n.19
No-damage-for-delay clauses, 307–308, 700–701
No-lien contracts, 528–529
North Carolina
  interest applied to late payments, 387 n.73, 74
  lien waivers, 383 n.61

North Carolina, *continued*
pay-if-paid/paid-when-paid provisions, 394 n.110
prompt payment, 395 n.115–118
public procurement, 193 n.21, 23, 24, 194 n.28, 30, 200 n.48, 213
retainage, 378 n.34, 379
stop notice statutes, 382 n.55
Notices
changes claims, 440–442
contract provisions to promote open communication, 565–566
contract termination/cure, 505–510
defective construction, 589–590
differing site condition claims, 479–480
Notice to proceed, 283–284

Obligee, defined, 542–543
Observation, of contractor by design professional, 595–598
Occupancy certificate, 301–302
Occupational Safety and Health Act (OSHA), 402–403
Occupational Safety and Health Administration (OSHA), 42, 402–408, 409
Occurrence rule, 538
OCIP (owner-controlled insurance program), 532
Ohio
bid protests, 205–206
multiple primes, 69 n.19
Oklahoma
public procurement, 193 n.23, 194
statute of repose, 612 n.59
Online auctions, 215–216
Open bidding projects, 190, 199–200
Operating manuals, 363
Oral contracts, 99
Oregon, construction lien statute, 525
OSHA (Occupational Safety and Health Administration), 42, 402–408, 409
Overhead, 229–230, 680–682
Overpayments, 385–387
Owner-architect agreements. *See* Design agreements
Owner-contractor agreements, 110, 125–128. *See also* Contracts
Owner-controlled insurance program (OCIP), 532
Owner-furnished items, 128
Owners, 35–55
acceptance of work, 390–391
breach of contract, 379, 387
building site, 35–42
coordination duty, 260
damages, 669–672
design process, 132, 133–134
financing, 52–55
geotechnical information, 43
implied warranty to supply accurate information, 43–49, 600–603, 605
inspections by, 386
insurance, 532
liability of, 347, 385–386, 412–415
negligent misrepresentation, 660–665
overpayment risks, 385–386
overview, 18–20, 35, 160–161
payment obligation, 379–387
program creation, 50–52
surveys, 42–43
termination of contract for cause, 496–499
waivers of claims, 390–391

Paper industry, 93–94
Parol evidence rule, 114–116
Partnering, 59 n.5, 564–565
Pass-through system, 272–275
Patent ambiguity rule, 126, 447
Patent defects, 390
Patterson, Edward, 15 n.41
Payees, 384
Pay-if-paid/pay-when-paid provisions, 267–268, 393–394
Payment bonds, 28, 53, 394, 543, 551–552
Payments, 369–400
acceptance, 391–392
applications for, 373–376
to architects, 159–160, 175–176
basis for, 370–373
certificates of, 374–376, 388, 389, 393
and contract changes, 387
introduction, 369–370
joint checks, 383–384

lien waivers, 53, 363, 382–383
methods, 91–92
overpayments, 385–387
owner's failure to make, 266–268, 387
owner's obligation, 379–387
pay-if-paid/pay-when-paid provisions, 267–268, 393–394
percentage-of-completion payments, 223–224
percentage-of-cost, 229–230
progress, 266, 360–361, 372, 380–381, 382, 391
prompt payment acts, 394–395
retainage, 376–379, 383, 388–389
to subcontractors/suppliers, 265–269
and substantial completion, 387–389
and termination of contract, 497–498
timing, 379–381
withholding of, 380–381
PDM (precedence diagramming method), 318
Penalty/bonus provisions, 306–307
Pennsylvania
mechanic's liens, 514
multiple primes, 69 n.19
public procurement, 193 n.22, 24, 194
value-based contracting, 214 n.147
Percentage-of-completion payments, 223–224
Percentage-of-cost compensation, 229–230
Perfection, of mechanic's liens, 526
Performance bonds
owner's risks under, 385–386
requirements, 543
and retainage, 378
role of, 28, 53
surety's obligations under, 511, 544–551
surety's risks under, 385
Performance specifications, 152, 604–605
Periodic payments. *See* Progress payments
Physical conditions, defined, 473
Plutarch, 1 n.1
Poor workmanship, 630–635, 649–655
P3 (public-private partnerships), 85–89, 191
Practical impossibility, 297
Precedence diagramming method (PDM), 318
Preferential hiring, 208–211
Pricing, 219–245
allowances, 241
at-risk construction management, 77–78
and builder rebates/credits, 243
and change orders, 240
changes, 443–446, 567
contract stipulations, 370–371
cost-of-work clause, 241
cost-plus contracts, 224–231, 240, 243, 244
cost-plus with guaranteed maximum price contracts, 231–232
dispute resolution, 567
interim payment clauses, 238–239
introduction, 219–220
material price escalation clauses, 238
mechanisms, 235–237
and scope of work, 239–240
self-performed work, 227–228, 241
sharing-of-savings clause, 240
stipulated-sum contracts, 91, 220–224, 232, 238, 240
unit-price contracts, 232–235
Prime contractors. *See* General contractors
Principal, defined, 542
Prior dealings, 120–121
Private procurement, 190–191
Privity of contract, 271, 607, 618, 625–638
Procurement. *See also* Public procurement
methods, 91
overview, 189
private projects, 190–191
Product liability, 620–622
Professional liability insurance, 176, 542
Professionals, design. *See* Design professionals
Professional specialization, 4
Profit
cost-plus contract stipulations, 229–230
recovery of lost, 682–683

Program creation, owners' role in, 50–52
Programmable logic controller (PLC) circuits, 152
Program management, 59 n.5
Progress payments
  acceptance as waiver of claims, 391
  approval of, 360–361
  lien waivers in conjunction with, 382
  owner's withholding of, 380–381
  schedule of values, 223–224, 266, 372–373
  to subcontractors, 266
  value of, 372
Project delivery systems, 57–95
  approaches, 58–60
  construction management, 60, 71–78, 212–213
  design-bid-construct (*See* Design-bid-construct method)
  design-build, 60, 78–85, 211–212
  importance of, 61–64
  innovative approaches, 85–89
  overview, 57–58
  selection of, 61–64, 90–94
Project manuals, 195–198
Project participants
  contractual relationships, 29–32
  overview, 17–18
  types of, 18–29
Promissory estoppel, 253–254, 674–675
Prompt payment acts, 394–395
Protests, bid, 204–207
Prudent contractor test, 473–474
Public contracts and entities. *See also* Federal contracts; Public procurement
  cardinal changes, 451–452
  contractor's refusal to perform work, grounds, 450
  design-build use, 79, 84
  dispute resolution authority, 364
  false claims, 396
  mechanic's liens, 525, 527–528
  multiple prime approach, 69–70
  project delivery, 18–19
  surety bonds, 28
Public-private partnerships (P3), 85–89, 191
Public procurement, 191–216. *See also* Competitive bidding
  best value awards, 213–215
  construction management, 212–213
  design-build contracts, 211–212
  design professional selection, 192–194
  exemptions, 191
  reverse online auctions, 215–216
  subcontractor choice, 251
Public projects. *See* Public contracts and entities
Public works projects, 85–89, 234–235
Punchlists, 303, 388, 510
Punitive damages, 511, 692–693
Pure Construction Management, 71–74

Quality control, 349–359
Quantum meruit, 101, 432, 668, 676–677, 752
Quasi-contracts, 101

Ratification, 439–440
Reasonable person standard, 118
Reasonably inferred standard, 127
Rebates, 243
Recordkeeping, 693
Regulation, 29, 136, 532–533
Releases, in change orders, 455–456
Reliance interest, 674–675
Repairs, 519, 694
Requests for information (RFIs), 348–349
Rescission, 505
Reservation of rights clause, 456
Resource Conservation and Recovery Act (RCRA) (1976), 42
Resource loading, 324–325
Restatement of Contracts, 118
Restatement (Second) of Contracts, 253–254, 380
Restatement (Second) of Torts, 37, 408 n.2, 412, 620, 660–665
Restatement (Third) of Security and Guaranty, 544
Restatement (Third) of Torts, 644

Restitution interest, 675–676
Retainage, 376–379, 383, 388–389
Reverse online auctions, 215–216
RFIs (requests for information), 348–349
Risk
  agency construction management, 74
  allocation and transfer, 423–429, 461–463, 563–564
  at-risk construction management, 76–78
  design-bid-construct approach, 67–69
  design-build approach, 81–83
  of payment provisions, 267–268
Rome, ancient, 2–3

Safety, 401–430
  accident prevention programs, 402–409
  architect's/engineer's liability for violations, 361, 419–423
  general contractor's liability for violations, 411–412, 415–419
  introduction, 401–402
  owner's liability for violations, 412–415
  recovery of compensation for accidents, 410–423
  risk allocation and transfer, 423–429
SBA (Small Business Administration), 251
SBEs (small business enterprises), 211 n.132
Schedule of values, 223–224, 266, 372–373
Schedules, 311–342
  approval, 325, 328–329
  bar charts, 312–314
  baseline, 327–330
  contractors, 51–52
  contract specifications, 323–327
  critical path method, 280–282, 315–323, 328, 332–341
  float time, 316–317, 326
  importance of, 62
  introduction, 311–312
  legal significance, 282–283
  owner's program projection, 51
  prohibitions on manipulations, 327
  revisions/time extensions, 326–327
  subcontractor involvement, 325–326, 327
  updating, 325, 330–332
Schematic design, 20 n.6, 157
Scholarship, 14–15
Sealed bid method, 7
Second-tier subcontractors, 24–25
Secured creditors, 529
Security pledges, 377
Self-performed work, 227–228, 241
Set-aside programs, 211
Settlement doctrine, 391
Sharing-of-savings clause, 240
Shop drawings, 350–355
Should-cost estimates, 685
Sites. *See* Building sites
Slack, 316–317, 326
Small Business Administration (SBA), 251
Small business enterprises (SBEs), 211 n.132
Small tool allowances, 226–227
Soil conditions, 43
Soil contamination, 41
Soil mechanics, 7–8
South Carolina
  bid protests, 205
  building codes, 137
  mechanic's liens, 527 n.74
Sovereign immunity doctrine, 456
Spearin doctrine, 43–49, 147–148, 599–607
Special damages, 179, 619, 692
Specifications. *See* Design specifications
Standard of care, design professionals, 140–151
States
  differing site conditions, 483–490
  insurance regulation, 532–533
  projects (*See* Public contracts and entities)
  safety legislation, 402
Statute of limitations, 248, 607–611
Statute of repose, 611–612

Statutes
  arbitration, 572–573
  defective construction, 589
  mechanic's liens, 5, 514–516
  trust fund, 269, 516
Step negotiations, 562, 566
Stepped claims settlement practices, 14
Stipulated-sum contracts, 220–224
  an incomplete design drawings, 240
  budgets, 243
  contingencies, 242
  contract provision, 220–221
  material price escalation clause, 238
  percentage-of-completion payments, 223–224
  scope limitations within contract price, 221–223
  vs. unit-price contracts, 232
Stop notice statutes, 382
Strict liability, 140, 620–621
Strict scrutiny, 209–210
Subconsulting agreements, 21–22
Subcontractors. *See also* Mechanic's liens
  applications for payment, 375
  approval by architect, 349
  architectural consultants, 134–135
  bids/bidding process, 252–256
  bonuses to encourage cooperation, 564
  claims, 269–276
  conditional assignment of obligation to owner, 259
  contract references, 124–125
  coordination, 259–262
  cost-effectiveness and cost-plus contracts, 225
  defective or deficient construction, 395
  in design-bid-construct system, 65
  flow-down obligations and rights, 256–259
  indemnity obligations, 262–265
  joint checks, 383–384
  liens on funds, 382
  mechanic's liens, 381
  owner's input into selection of, 249–252
  pay-if-paid/pay-when-paid provisions, 267–268, 393–394
  payments to, 265–269
  prompt payment acts, 394–395
  public projects, 199
  release of waiver, 389
  retainage by contractors, 378
  role in project, 24–25
  safety, 409, 416
  schedule involvement, 325–326, 327
  selection, 249–256
  stop notice statutes, 382
  vs. suppliers, 247–249
  and termination of contract by owner for cause, 498
  unit-price contracts, 232
  wrongful termination, 511
Subguard insurance, 557
Subrogation rights, 378–379, 541–542, 553
Substantial completion, 297–302, 362–363, 376, 387–389
Substitutions, 354
Sub-subcontractors, 24–25
Subsurface conditions, 459–461. *See also* Differing site conditions
Superior knowledge, 465–466
Suppliers
  approval by architect, 349
  contract references, 124–125
  joint checks, 384
  mechanic's liens, 381
  owner's input into selection of, 249–252
  payments to, 265–269
  release of waiver, 389
  selection, 249–256
  stop notice statutes, 382
  vs. subcontractors, 247–249
Sureties
  legal principles, 544
  overpayment, 385–386
  pay-if-paid condition, 394
  rights and remedies, 552–556
  role of, 28
  subrogation rights, 378–379
  and termination of contract, 511
Surety bonds. *See also* Performance bonds
  history, 5–6

payment, 28, 53, 394, 543, 551–552
purpose, 542–544
Surety Information Office, 69 n.18
Surveys, 37, 42–43
Sweet, Justin, 15 n.40, 365, 498, 506

Tennessee
mechanic's liens, 522 n.45
retainage, 378 n.33
Termination, of contract
by agreement, 505
for cause, 494–499
contractual provisions, 494–500
for convenience, 499–500
introduction, 493–494
notice and opportunity to cure, 505–510
and surety's role, 511
upon material breach, 500–505
wrongful, 510–511, 520
Termination, of contractor, 364–366
Terzaghi, Carl, 7
Testing services, defective, 635–638
Texas
fraudulent liens, 521 n.43
indemnity prohibition, 428
lien waivers, 383 n.61
owner's implied warranty of drawings and specifications, 601 n.30
safety systems, 402 n.5
stop notice statutes, 382 n.55
workers' compensation, 410 n.24
Third parties
claims to retainage, 378–379, 389
default pricing through, 237
Third-party beneficiaries, of contract, 123–124
Time. *See also* Schedules; Time for completion
of commencement, 283–284
of payment, 379–381
for performance, 279–280
Time and materials pricing, 91, 236–237, 444
Time and materials tickets, 453
Time and motion studies, 685–686
Time for completion
change orders, 446
contract specification, 283–284
damages for late, 304–307
extensions, 290–297, 347
no-damage-for-delay clauses, 307–308
penalty/bonus provisions, 306–307
waiver of, 284–290
Time impact analysis, 336–337
Title insurance, 28–29
Title issues, 37–38
Title reports, 37
Tool allowances, 226–227
Torts. *See also* Economic loss rule; Negligence
current industry status, 619–620
damages, 677–678
liability for accidents, 411–423, 429
owner's claims against builder, 588
owner's claims against design professional, 591
statute of limitations, 607–608
Total cost method, 686–690
Total time analysis, 335–336
Trade custom and usage, 93–94, 121–122
Trier of fact, 100, 114
Trust-fund statutes, 269, 516
Turnkey projects, 85

UCC (Uniform Commercial Code), 120–121, 248, 615, 639 n.62, 650 n.83
Unbalanced bidding, 234–235
Undeveloped areas, 39
Unified Soil Classification System, 8
Uniform Arbitration Act, 8, 573, 575, 578
Uniform Commercial Code (UCC), 120–121, 248, 615, 639 n.62, 650 n.83
Uniform Construction Lien Act, 525 n.65
Uniform Contract, 5, 8
Unit pricing, 232–235, 370, 373, 443
Unjust enrichment, 101, 515, 520, 668, 675–676
Unsecured creditors, 529
U.S. Green Building Council, 139
Utilities, 39

Value-based contracting, 213–215
Value engineering, 151
Values, schedule of, 223–224, 266, 372–373
Variations in estimated quantities (VEQ) clauses, 232–234, 481–483
Virginia
  differing site conditions, 485–486
  mechanic's liens, 527
Vitruvius, 2–3 n.5, 136–137

Waivers
  of arbitration, 575–576
  of claims, 390–392, 455, 456
  of consequential damages, 179, 700
  of liens, 53, 363, 382–383, 391, 528–529
  release of, 389
  of right to cease work, 381
  strict compliance with change notice provisions, 441–442
  of subrogation, 541–542
  time for completion, 284–290
Warranties
  breach of, 587–588
  conflicts between contractor's and owner's, 605
  by contractors, 374–375
  implied (*See* Implied warranties)
  under UCC, 248 n.3
Warranty clauses, 119, 374–375
Warranty documentation, 363
Washington
  mechanic's liens, 527
  no-damages-for-delay clauses, 308 n.55
Washington, D.C., statute of limitations for tort claims, 610 n.49
Weather delays, 296–297, 684–685
Wetlands, 42
Window analysis, 337–338
Wisconsin
  mechanic's liens, 516 n.15, 527
  multiple primes, 69 n.19
  pay-if-paid clauses, 268 n.73
Withholding of payments, 380–381
Work
  acceptance of, 390–391
  additional, 432
  control of, 412, 413, 416–419
  coordination of, 259–262, 359–360
  deletion of, 433
  description in contract, 125–128, 239–240
  extra, 432, 449–454
  self-performed, 227–228, 241
Workers' compensation, 410, 428–429, 535
Workmanlike manner, 587–588
Workplace safety. *See* Safety
Work sites. *See* Building sites
World Bank, 471
Wrongful termination, of contract, 510–511, 520

Zoning, 29, 38–39